COMPUTATIONAL DYNAMICS

COMPUTATIONAL DYNAMICS
THIRD EDITION

AHMED A. SHABANA
Richard and Loan Hill Professor of Engineering
University of Illinois at Chicago

WILEY
A John Wiley and Sons, Ltd., Publication

This edition first published 2010

Registered office
John Wiley & Sons Ltd, The Atrium, Southern Gate, Chichester, West Sussex, PO19 8SQ, United Kingdom

For details of our global editorial offices, for customer services and for information about how to apply for permission to reuse the copyright material in this book please see our website at www.wiley.com.

Library of Congress Cataloging-in-Publication Data:

Shabana, Ahmed A., 1951-
Computational dynamics / Ahmed A. Shabana. – 3rd ed.
p. cm.
Includes bibliographical references and index.
ISBN 978-0-470-68615-7 (cloth)
1. Dynamics. I. Title.
QA845.S44 2010
531′.11 – dc22

2009031442

A catalogue record for this book is available from the British Library.

ISBN 978-0-470-68615-7 (Hbk)

Typeset in 10/12pt Times by Laserwords Private Limited, Chennai, India.
Printed in Singapore by Markono Print Media Pte Ltd.

To My Students

CONTENTS

PREFACE

Computational dynamics has been the subject of extensive research over the last three decades. This subject has grown rapidly as a result of the advent of high-speed digital computers and also as a response to the need for simulation and analysis computational capabilities for physics and engineering systems that consist of interconnected bodies. These systems are highly nonlinear in nature and their analysis requires the use of matrix, numerical, and computer methods. It is the objective of this book to present an introduction to the subject of *computational dynamics* at a level suitable for senior undergraduate and first-year graduate students. The book introduces students to concepts, definitions, and techniques used in the field of multibody system dynamics. To achieve this goal, classical approaches are first discussed in order to help students review some of the fundamental concepts and techniques in the general field of mechanics. The book then builds on these concepts to demonstrate the use of the classical methods as a foundation for the study of computational dynamics. Various computational methodologies that are used in the computer-aided analysis of multibody systems are presented. In the analysis presented in this book, only rigid body dynamics is considered.

CONTENTS

The third edition of the book is organized into nine chapters that cover the basic concepts and computational methods in kinematics and dynamics of multibody systems. Simple examples are used in most chapters to demonstrate the basic ideas and procedures presented. The problem sets presented at the end of each chapter are intentionally selected to be simple in order to focus on the main concepts and computational methodologies discussed in the book. In developing the materials of this book, modest demands are made on the expertise of the reader in mathematics and dynamics.

In **Chapter 1**, some basic definitions that are used repeatedly in the book are introduced. The materials presented in this chapter also serve as a brief introduction to the materials covered in subsequent chapters. The organization of the book and the notation used are discussed at the end of this chapter. The reader is encouraged to read this introductory chapter before reading subsequent chapters.

Chapter 2 is devoted to a review of some concepts and operations in matrix and vector algebra. Matrix and vector properties and identities as well as methods for solving systems of algebraic equations are among the topics discussed in Chapter 2. **QR** decomposition and singular value decomposition, which are used in the dynamic analysis of constrained multibody systems to obtain a minimum set of independent differential equations, are discussed in two sections of this chapter. The reader with a background in linear algebra will find that most of the material presented in Chapter 2 is familiar.

In **Chapter 3**, the kinematics of constrained multibody systems is discussed. The position, velocity, and acceleration equations are developed, and the use of these equations in the kinematic analysis of multibody systems that consist of interconnected bodies is demonstrated. The number of degrees of freedom of a multibody system depends on the number and types of joints that connect the system components. Several mechanical joints and the formulation of their kinematic constraint equations are presented. It is shown that the mobility of the system depends on the number of linearly independent constraint equations of its joints. The conditions of the joint connectivity between interconnected bodies are formulated using a set of nonlinear algebraic constraint equations that depend on the system coordinates. The configuration of the system is determined by solving these nonlinear algebraic equations using numerical and computer methods.

There are several computer techniques that are currently used for the dynamic analysis of multibody systems. Some of these techniques lead to a relatively large system of loosely coupled differential and algebraic equations; others lead to smaller systems of strongly coupled equations. In **Chapter 4**, different forms of the dynamic equations are presented. In this chapter, the basic multibody system equations are developed using Newtonian mechanics without the need for using analytical Lagrangian techniques such as the principle of virtual work. The use of D'Alembert's principle to derive Euler equation is demonstrated, and the concept of Lagrange multipliers, which can be used to define the generalized constraint forces, is introduced using Newtonian mechanics and simple examples. Systematic analytical procedures for developing some of the forms of the dynamic equations presented in Chapter 4 are described in detail in the following two chapters.

In **Chapter 5**, the *principle of virtual work*, which represents the cornerstone for developing many of the existing dynamic formulations, is presented. The concepts of virtual displacement and generalized forces that are necessary for using the principle of virtual work are introduced. A systematic procedure based on the principle of virtual work for eliminating the constraint forces from the static and dynamic equations is outlined. Among the topics that are discussed in Chapter 5 are *Lagrange's equation*, the *Gibbs-Appel equation*, and the *canonical form* of the equations of motion. Chapter 5 is concluded by discussing the relationship between virtual work and Gaussian elimination.

Chapter 6 covers computational methods in dynamics. Several computer methodologies for formulating the equations of motion are discussed in this chapter. In one approach, the differential equations of motion are expressed in terms of the independent variables using the *embedding techniques* or the *recursive methods*. This approach leads to a set of ordinary differential equations in which the constraint forces are eliminated automatically. Use of **QR** decomposition and singular value decomposition to obtain a minimum set of independent differential equations is also discussed. In another computer approach, the equations of motion of the multibody systems are formulated in terms of both dependent and independent coordinates. This approach leads to a large system of loosely coupled equations in which the constraint forces appear explicitly. These constraint forces can be expressed in terms of *Lagrange multipliers*, leading to a mixed system of differential and algebraic equations that can be solved using matrix and computer methods. The numerical methods and sparse matrix formulations used for solving mixed systems of differential and algebraic equations are also discussed in this chapter.

The spatial kinematics and dynamics of rigid body systems are discussed in **Chapter 7**. The general displacement of unconstrained rigid bodies in space is defined in terms of the independent translation and rotation coordinates. Methods for defining the orientation of rigid bodies in space are described. Formulation of the spatial kinematic velocity and acceleration equations is discussed and examples of constraint equations that describe spatial mechanical joints are presented. Formulation of the augmented and recursive dynamic equations that govern the constrained spatial motion of rigid body systems is also discussed in Chapter 7. In **Chapter 8**, several topics in dynamics are discussed. These topics include the gyroscopic motion, the Rodriguez formula, Euler parameters, Rodriguez parameters, quaternions, and rigid body contact. The study of the multibody system stability using the eigenvalue analysis is also discussed in Chapter 8.

Chapter 9 is devoted to the description of general purpose multibody system codes and their capabilities. As an example, the multibody system computer code **SAMS/2000** (**S**ystematic **A**nalysis of **M**ultibody **S**ystems) is used. The structure of the code and the procedure of its use are described in the cases of planar and spatial kinematic and dynamic problems. SAMS/2000 allows the user to develop systematically virtual models of multibody systems that consist of interconnected bodies. The readers of the book are encouraged to solve many of the examples and exercise problems using multibody system codes. The use of these codes along with the study of the formulations presented in the book is necessary in order to have a good understanding and appreciation of the methods and algorithms used in developing multibody system software that are widely used in industry. The *educational version* of SAMS/2000 is limited to only four rigid bodies and does not include flexible body and rail simulation capabilities. It does not also include some other simulation options that are discussed in Chapter 9.

Several sections are marked with an asterisk, both in the table of contents and in the text. Some of these sections, such as Sections 2.7 and 2.8, can be omitted during a first reading of the book. Others, such as Section 5.10, do not contribute to the development of the main ideas presented in the book, and therefore, can be omitted entirely, since the remaining chapters do not make use of the development presented in these sections. The goal of computational and multibody dynamics is to develop general algorithms that can be applied to a large number of applications. The exercise problems are designed with this goal in mind.

UNITS AND NOTATION

The *International System of Units* (**SI**) is used in the examples and problems throughout the book. In this system, the three basic units are the meter (distance), the kilogram (mass), and the second (time). The unit of force is called the *newton* and is derived from the three basic units.

In this text, boldface letters are used to indicate vectors or matrices. Superscripts are used to indicate body numbers. For example, a^i denotes a scalar a associated with body i in the multibody system, while $\mathbf{a}^i$ denotes a vector or a matrix associated with body i. In order to distinguish between a superscript that indicates the body number and the power, we use parentheses whenever a quantity is raised to a certain power. For example, $(a^3)^2$ is a scalar a associated with body 3 and is raised to the power 2.

ACKNOWLEDGMENTS

I would like to acknowledge the help I received from many of my graduate students and research associates, who made significant contributions to the development of this text. I mention in particular, J. H. Choi, M. Gofron, K. S. Hwang, Z. Kusculuoglu, H. C. Lee, M. Omar, T. Ozaki, M. K. Sarwar, M. Shokohifard, and D. Valtora. I thank Toshikazu Nakanishi of Komatsu, Ltd. for providing the simulation results of the tracked vehicle model presented in Chapter 6; and I thank Graham Sanborn, Hiroyuki Sugiyama, and Khaled Zaazaa for their contributions to Chapter 9. I would also like to gratefully acknowledge the support received from the U.S. Army Research Office and the Federal Railroad Administration for our research in the area of computational dynamics.

Thanks are due to Ms. Denise Burt for the excellent job in typing some chapters of the original manuscript of this book. Mr. Frank Cerra, Mr. Bob Argentieri, and Mr. Eric Willner, the Senior Engineering Editors; Ms. Kimi Sugeno and Ms. Debbie Cox, the Assistant Managing Editors; Mr. Bob Hilbert, the Associate Managing Editor; Ms. Nicky Skinner, the Project Editor; and the production staff at John Wiley & Sons deserve special thanks for their cooperation and thoroughly professional work in producing the first, second, and third editions of this book. The author is also grateful to his family for their patience during the years of preparing the editions of this book.

AHMED A. SHABANA

CHAPTER 1

INTRODUCTION

Modern mechanical and aerospace systems are often very complex and consist of many components interconnected by joints and force elements such as springs, dampers, and actuators. These systems are referred to, in modern literature, as *multibody systems*. Examples of multibody systems are machines, mechanisms, robotics, vehicles, space structures, and biomechanical systems. The dynamics of such systems are often governed by complex relationships resulting from the relative motion and joint forces between the components of the system. Figure 1 shows a hydraulic excavator, which can be considered as an example of a multibody system that consists of many components. In the design of such a tracked vehicle, the engineer must deal with many interrelated questions with regard to the motion and forces of different components of the vehicle. Examples of these interrelated questions are the following: What is the relationship between the forward velocity of the vehicle and the motion of the track chains? What is the effect of the contact forces between the links of the track chains and the vehicle components on the motion of the system? What is the effect of the friction forces between the track chains and the ground on the motion and performance of the vehicle? What is the effect of the soil–track interaction on the vehicle dynamics, and how can the soil properties be characterized? How does the geometry of the track chains influence the forces and the maximum vehicle speed? These questions and many other important questions must be addressed before the design of the vehicle is completed. To provide a proper answer to many of these interrelated questions, the development of a detailed dynamic model of such a complex system becomes necessary. In this book we discuss in detail the development of the dynamic equations of complex multibody systems such as the tracked hydraulic excavator shown in Fig. 1. The methods presented in the

Computational Dynamics, Third Edition Ahmed A. Shabana

Figure 1.1 Hydraulic excavator

book will allow the reader to construct systematically the kinematic and dynamic equations of large-scale mechanical and aerospace systems that consist of interconnected bodies. The procedures for solving the resulting coupled nonlinear equations are also discussed.

1.1 COMPUTATIONAL DYNAMICS

The analysis of mechanical and aerospace systems has been carried out in the past mainly using graphical techniques. Little emphasis was given to computational methods because of the lack of powerful computing machines. The primary interest was to analyze systems that consist of relatively small numbers of bodies such that the desired solution can be obtained using graphical techniques or hand calculations. The advent of high-speed computers made it possible to analyze complex systems that consist of large numbers of bodies and joints. Classical approaches that are based on *Newtonian* or *Lagrangian mechanics* have been rediscovered and put in a form suitable for the use on high-speed digital computers.

Despite the fact that the basic theories used in developing many of the computer algorithms currently in use in the analysis of mechanical and aerospace systems are the same as those of the classical approaches, modern engineers and scientists are forced to know more about matrix and numerical methods in order to be able to utilize efficiently the computer technology available. In this book, classical and modern approaches used in the kinematic and dynamic analysis of mechanical and aerospace systems that consist of interconnected *rigid bodies* are introduced. The main focus of the presentation is on the modeling of general multibody systems and on developing the relationships that govern the dynamic motion of these systems. The objective is to develop general methodologies that can be applied to a large class of multibody system applications. Many fundamental and computational problems are discussed with the objective of addressing the merits and limitations of various procedures used in formulating and solving the equations of motion of multibody systems. This is the subject of the general area of *computational dynamics* that is concerned with the computer solution of the equations of motion of large-scale systems.

The role of computational dynamics is merely to provide tools that can be used in the dynamic simulation of multibody systems. Various tools can be used for the analysis and computer simulation of a given system. This is mainly due to the fact that the form of the kinematic and dynamic equations that govern the dynamics of a multibody system is not unique, and therefore, it is important that the analyst chooses the tool and form of the equations of motion that is most suited for his or her application. This is not always an easy task and requires familiarity of the analysts with different formulations and procedures used in the general area of computational dynamics. The forms of the equations of motion depend on the choice of the coordinates used to define the system configuration. One may choose a small or a large number of coordinates. From the computational viewpoint, there are advantages and drawbacks to each choice. The selection of a small number of coordinates always leads to a complex system of equations. Such a choice, however, has the advantage of reducing the number of equations that need to be solved. The selection of a large number of coordinates, on the other hand, has the advantage of producing simpler and less coupled equations at the expense of increasing the problem dimensionality. The main focus of this book is on the derivation and use of different forms of the equations of motion. Some formulations lead to a large system of equations expressed in terms of redundant coordinates, while others lead to a small system of equations expressed in terms of a minimum set of coordinates. The advantages and drawbacks of each of these formulations, when constrained multibody systems, are considered, are discussed in detail.

Generally speaking, multibody systems can be classified as *rigid multibody systems* or *flexible multibody systems*. Rigid multibody systems are assumed to consist only of rigid bodies. These bodies, however, may be connected by massless springs, dampers, and/or actuators. This means that when rigid multibody systems are considered, the only components that have inertia are assumed to be rigid bodies. Flexible multibody systems, on the other hand, contain rigid and deformable bodies. Deformable bodies have distributed inertia and elasticity which depend on the body deformations. As the deformable body moves, its shape changes and its inertia and elastic properties become functions of time. For this reason, the analysis of deformable bodies is more difficult than rigid body analysis. In this book, the branch of computational dynamics that deals with rigid multibody systems only is considered. The theory of flexible multibody systems is covered by the author in a more advanced text (Shabana, 2005).

1.2 MOTION AND CONSTRAINTS

Systems such as machines, mechanisms, robotics, vehicles, space structures, and biomechanical systems consist of many bodies connected by different types of joints and different types of force elements, such as springs, dampers, and actuators. The joints are often used to control the system mobility and restrict the motion of the system components in known specified directions. Using the joints and force elements, multibody systems are designed to perform certain tasks; some of these tasks are simple, whereas others can be fairly complex and may require the use of certain types of mechanical joints as well as sophisticated control algorithms. Therefore, understanding the dynamics of these systems becomes crucial at the design stage and also for performance evaluation and design improvements. To understand the dynamics of a multibody system, it is necessary to study the motion of its components. In this section, some of the basic concepts and definitions used in the motion description of rigid bodies are discussed, and examples of joints that are widely used in multibody system applications are introduced.

Unconstrained Motion A general rigid body displacement is composed of translations and rotations. The analysis of a pure *translational motion* is relatively simple and the dynamic relationships that govern this type of motion are fully understood. The problem of *finite rotation*, on the other hand, is not a trivial one since large rigid body rotations are sources of *geometric nonlinearities*. Figure 2 shows a rigid body, denoted as body i. The general displacement of this body can be conveniently described in an inertial XYZ coordinate system by introducing the body $X^iY^iZ^i$ coordinate system whose origin O^i is rigidly attached to a point on the rigid body. The general displacement of the rigid body can then be described in terms of the translation of the *reference point* O^i and also in terms of a set of coordinates that define the orientation of the body coordinate system with respect to the inertial frame of reference. For instance, the general planar motion of this body can be described using three independent coordinates that define the translation of the body along the X and Y axes as well as its rotation about the Z axis. The two translational components and the rotation are three independent coordinates since any one of them can be changed arbitrarily while keeping the other two coordinates fixed. The body may translate along the X axis while its displacement along the Y axis and its rotation about the Z axis are kept fixed.

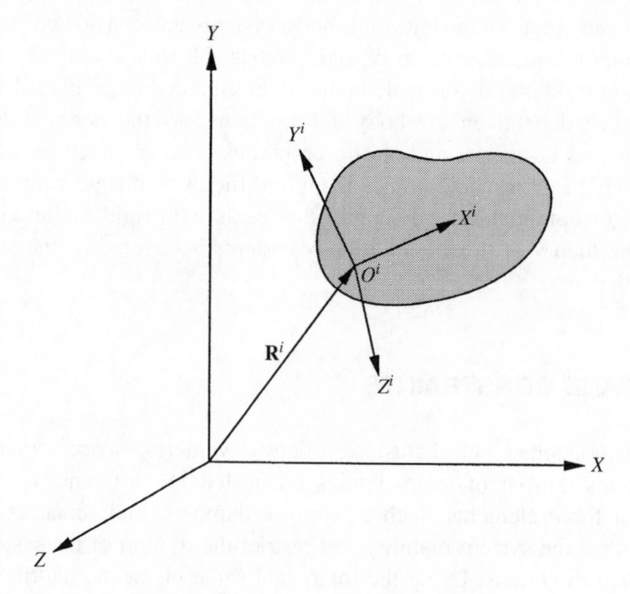

Figure 1.2 Rigid body displacement

In the spatial analysis, the configuration of an unconstrained rigid body in the three-dimensional space is identified using six coordinates. Three coordinates describe the translations of the body along the three perpendicular axes X, Y, and Z, and three coordinates describe the rotations of the body about these three axes. These again are six *independent coordinates*, since they can be varied arbitrarily.

Mechanical Joints Mechanical systems, in general, are designed for specific operations. Each of them has a topological structure that serves a certain purpose. The bodies in a mechanical system are not free to have arbitrary displacements because they are connected by *joints* or *force elements*. While force elements such as springs and dampers may significantly affect the motion of the bodies in one or more directions, such an element does not completely prevent motion in these directions. As a consequence, a force element does not reduce the number of independent coordinates required to describe the configuration of the system. On the other hand, mechanical joints as shown in Fig. 3 are used to allow motion only in certain directions. The joints reduce the number of independent coordinates of the system since they prevent motion in some directions. Figure 3a shows a *prismatic (translational) joint* that allows only relative translation between the two bodies i and j along the joint axis. The use of this joint eliminates the freedom of body i to translate relative to body j in any other direction except along the joint axis. It also eliminates the freedom of body i to rotate with respect to body j. Figure 3b shows a *revolute (pin) joint* that allows only relative rotation between bodies i and j. This joint eliminates the freedom of body i to translate with respect to body j. The *cylindrical joint* shown in Fig. 3c allows body i to translate and rotate with respect to body j along and about the joint axis. However, it eliminates the freedom of body i to translate or rotate with respect to body j along any axis other than the joint axis. Figure 3d shows the *spherical (ball) joint*, which eliminates the relative translations between bodies i and j. This joint provides body i with the freedom to rotate with respect to body j about three perpendicular axes.

Other types of joints that are often used in mechanical system applications are *cams* and *gears*. Figure 4 shows examples of cam and gear systems. In Fig. 4a, the shape of the cam is designed such that a desired motion is obtained from the *follower* when the cam rotates about its axis. Gears, on the other hand, are used to transmit a certain type of

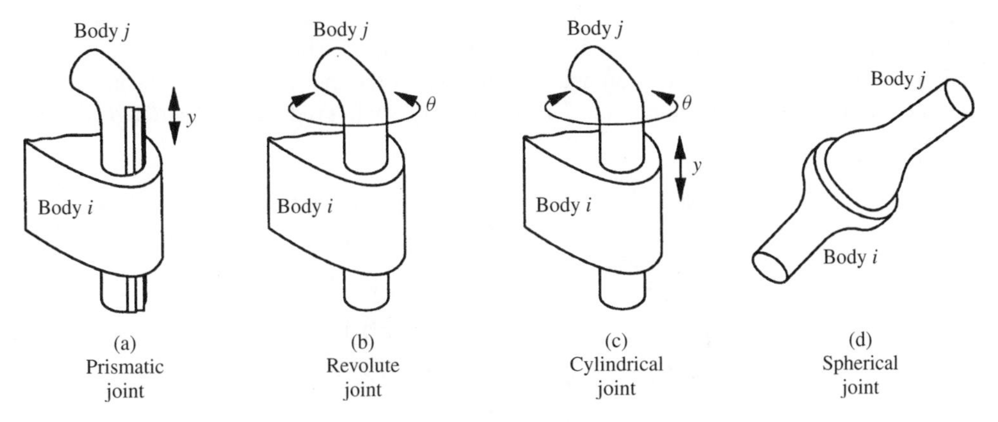

Figure 1.3 Mechanical joints

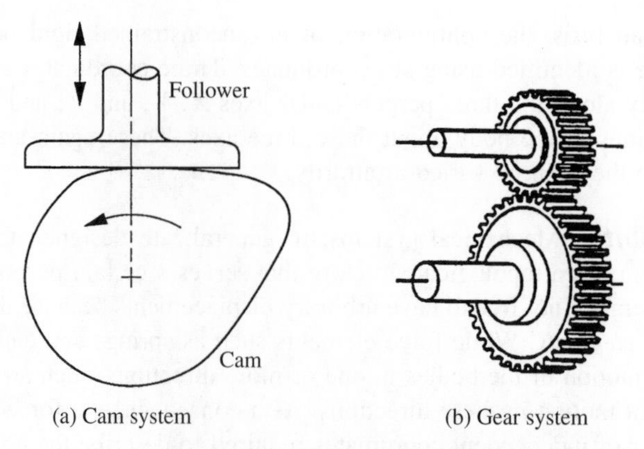

Figure 1.4 Cam and gear systems

motion (translation or rotary) from one body to another. The gears shown in Fig. 4b are used to transmit rotary motion from one shaft to another. The relationship between the rate of rotation of the driven gear to that of the driver gear depends on the diameters of the *base circles* of the two gears.

1.3 DEGREES OF FREEDOM

A mechanical system may consist of several bodies interconnected by different numbers and types of joints and force elements. The *degrees of freedom* of a system are defined to be the independent coordinates that are required to describe the configuration of the system. The number of degrees of freedom depends on the number of bodies and the number and types of joints in the system. The *slider crank mechanism* shown in Fig. 5 is used in several engineering applications, such as automobile engines and pumps. The mechanism consists of four bodies: body 1 is the cylinder frame, body 2 is the crankshaft, body 3 is the connecting rod, and body 4 is the slider block, which represents the piston. The mechanism has three revolute joints and one prismatic joint. While this mechanism has several bodies and several joints, it has only one degree of freedom; that is, the motion of all bodies in this system can be controlled and described using only one independent variable. In this case, one needs only one force input (a motor or an actuator) to control the motion of this mechanism. For instance, a specified input rotary motion to the crankshaft produces a desired rectilinear

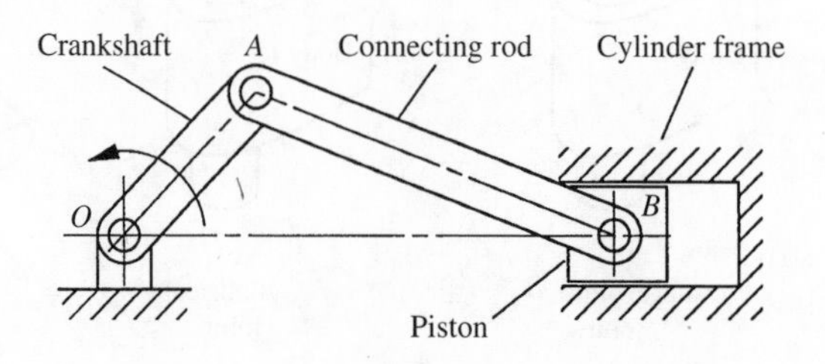

Figure 1.5 Slider crank mechanism

$n_b = 4$ $4 \times 3 = 12$

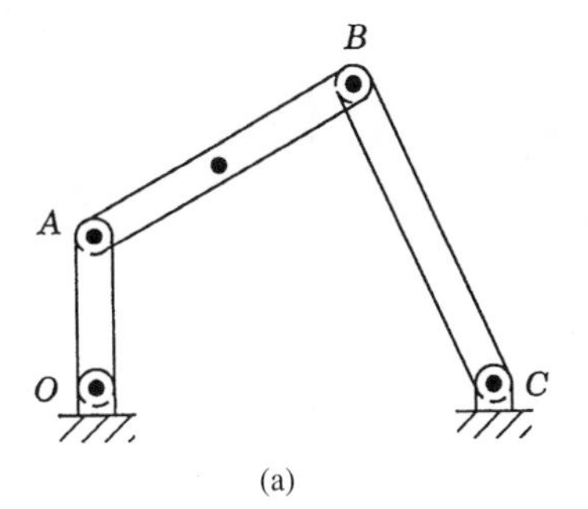

(a)

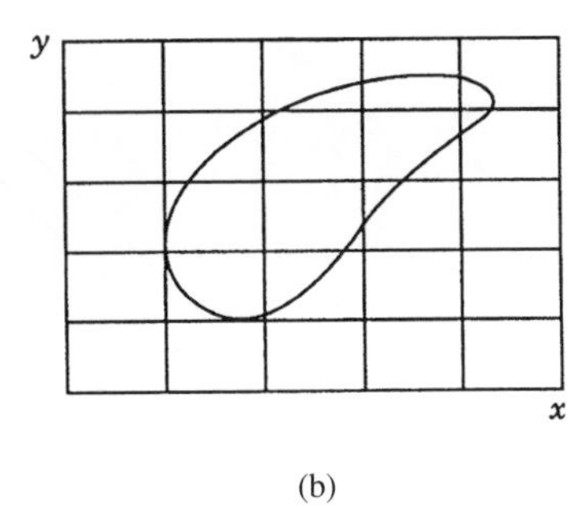

(b)

Figure 1.6 Four-bar mechanism

motion of the slider block. If the rectilinear motion of the slider block is selected to be the independent variable, the force that acts on the slider block can be chosen such that a desired output rotary motion of the crankshaft *OA* can be achieved. Similarly, two force inputs are required in order to be able to control the motion of a multibody system that has two degrees of freedom, and n force inputs are required to control the motion of an n-degree-of-freedom multibody system.

Figure 6a shows another example of a simple planar mechanism called the *four-bar mechanism*. This mechanism, which has only one degree of freedom, is used in many industrial and technological applications. The motion of the links of the four-bar mechanism can be controlled by using one force input, such as driving the crankshaft *OA* using a motor located at point O. A desired motion trajectory on the coupler link *AB* can be obtained by selecting the proper dimensions of the links of the four-bar mechanism. Figure 6b shows the motion of the center of the coupler *AB* when the crankshaft *OA* of the mechanism shown in Fig. 6a rotates one complete cycle. Different motion trajectories can be obtained by using different dimensions.

Another one-degree-of-freedom mechanism is the *Peaucellier mechanism*, shown in Fig. 7. This mechanism is designed to generate a straight-line path. The geometry of this mechanism is such that $BC = BP = EC = EP$ and $AB = AE$. Points A, C, and P should always lie on a straight line passing through A. The mechanism always satisfies the condition $AC \times AP = c$, where c is a constant called the *inversion constant*. In case $AD = CD$, point P should follow an exact straight line.

The majority of mechanism systems form single-degree-of-freedom *closed kinematic chains*, in which each member is connected to at least two other members. Robotic manipulators as shown in Fig. 8 are examples of multidegree-of-freedom open-chain systems. Robotic manipulators are designed to synthesize some aspects of human functions and are used in many applications, such as welding, painting, material transfers, and assembly tasks. Some of these applications require high precision and consequently, sophisticated sensors and control systems are used.

While the number of degrees of freedom of a system is unique and depends on the system topological structure, the set of degrees of freedom is not unique, as demonstrated previously by the slider crank mechanism. For this simple mechanism, the rotation of the crankshaft or the translation of the slider block can be considered as the system degree of freedom. Depending on the choice of the degree of freedom, a motor or an actuator can

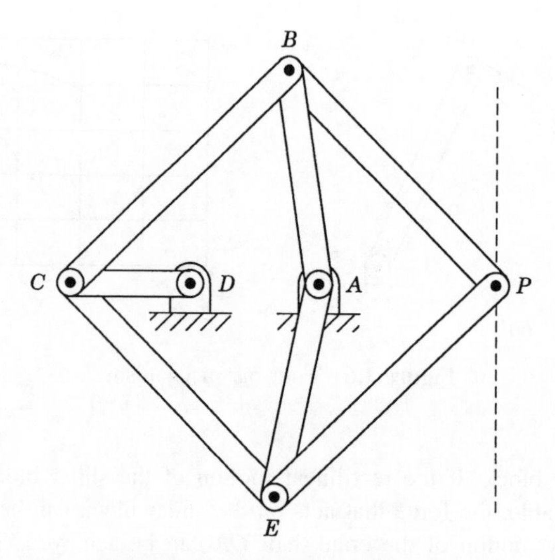

Figure 1.7 Peaucellier mechanism

be used to drive the mechanism. In the design and control of multibody systems, precise knowledge of the system degrees of freedom is crucial for motion generation and control. The number and type of degrees of freedom define the numbers and types of motors and actuators that must be used at the joints to drive and control the motion of the multibody system. In Chapter 3, simple criteria are provided for determining the number of degrees of freedom of multibody systems. These criteria depend on the number of bodies in the system as well as the number and type of the joints. When the complexity of the system increases, the identification of the system degrees of freedom using the simple criteria can be misleading. For this reason, a numerical procedure for identifying the degrees of freedom of complex multibody systems is presented in Chapter 6.

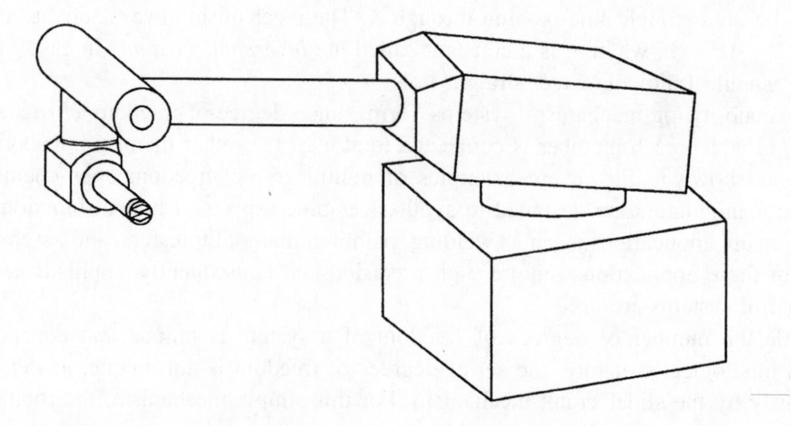

Figure 1.8 Robotic manipulators

1.4 KINEMATIC ANALYSIS

In the *kinematic analysis* we are concerned with the geometric aspects of the motion of the bodies regardless of the forces that produce this motion. In the classical approaches used in the kinematic analysis, the system degrees of freedom are first identified. Kinematic relationships are then developed and expressed in terms of the system degrees of freedom and their time derivatives. The step of determining the locations and orientations of the bodies in the mechanical system is referred to as the *position analysis*. In this first step, all the required displacement variables are determined. The second step in the kinematic analysis is the *velocity analysis*, which is used to determine the respective velocities of the bodies in the system as a function of the time rate of the degrees of freedom. This can be achieved by differentiating the kinematic relationships obtained from the position analysis. Once the displacements and velocities are determined, one can proceed to the third step in the kinematic analysis, which is referred to as the *acceleration analysis*. In the acceleration analysis, the velocity relationships are differentiated with respect to time to obtain the respective accelerations of the bodies in the system.

To demonstrate the three principal steps of the kinematic analysis, we consider the two-link manipulator shown in Fig. 9. This manipulator system has two degrees of freedom, which can be chosen as the angles θ^2 and θ^3 that define the orientation of the two links. Let l^2 and l^3 be the lengths of the two links of the manipulator. The global position of the end effector of the manipulator is defined in the coordinate system XY by the two coordinates r_x and r_y. These coordinates can be expressed in terms of the two degrees of freedom θ^2 and θ^3 as follows:

$$\left.\begin{aligned} r_x &= l^2 \cos\theta^2 + l^3 \cos\theta^3 \\ r_y &= l^2 \sin\theta^2 + l^3 \sin\theta^3 \end{aligned}\right\} \tag{1.1}$$

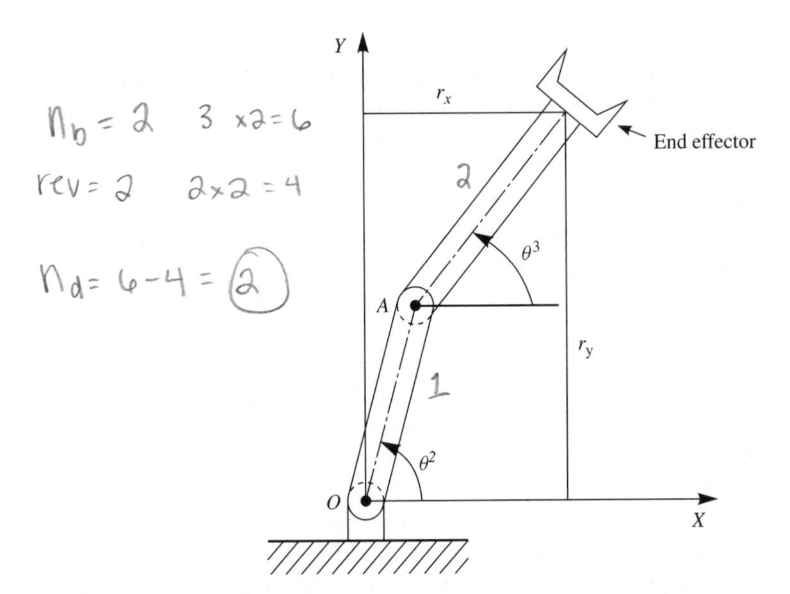

Figure 1.9 Two-degree-of-freedom robot manipulator

Note that the position of any other point on the links of the manipulator can be defined in the XY coordinate system in terms of the degrees of freedom θ^2 and θ^3. Equation 1 represents the position analysis step. Given θ^2 and θ^3, the position of the end effector or any other point on the links of the manipulator can be determined.

The velocity equations can be obtained by differentiating the position relationships of Eq. 1 with respect to time. This yields

$$\left.\begin{aligned} \dot{r}_x &= -\dot{\theta}^2 l^2 \sin\theta^2 - \dot{\theta}^3 l^3 \sin\theta^3 \\ \dot{r}_y &= \dot{\theta}^2 l^2 \cos\theta^2 + \dot{\theta}^3 l^3 \cos\theta^3 \end{aligned}\right\} \tag{1.2}$$

Given the degrees of freedom θ^2 and θ^3 and their time derivatives, the velocity of the end effector can be determined using the preceding kinematic equations. It can also be shown that the velocity of any other point on the manipulator can be determined in a similar manner.

By differentiating the velocity equations (Eq. 2), the equations that define the acceleration of the end effector can be written as follows:

$$\left.\begin{aligned} \ddot{r}_x &= -\ddot{\theta}^2 l^2 \sin\theta^2 - \ddot{\theta}^3 l^3 \sin\theta^3 - (\dot{\theta}^2)^2 l^2 \cos\theta^2 - (\dot{\theta}^3)^2 l^3 \cos\theta^3 \\ \ddot{r}_y &= \ddot{\theta}^2 l^2 \cos\theta^2 + \ddot{\theta}^3 l^3 \cos\theta^3 - (\dot{\theta}^2)^2 l^2 \sin\theta^2 - (\dot{\theta}^3)^2 l^3 \sin\theta^3 \end{aligned}\right\} \tag{1.3}$$

Therefore, given the degrees of freedom and their first and second time derivatives, the absolute acceleration of the end effector or the acceleration of any other point on the manipulator links can be determined.

Note that when the degrees of freedom and their first and second time derivatives are specified, there is no need to write force equations to determine the system configuration. The kinematic position, velocity, and acceleration equations are sufficient to define the coordinates, velocities, and accelerations of all points on the bodies of the multibody system. A system in which all the degrees of freedom are specified is called a *kinematically driven system*. If one or more of the system degrees of freedom are not known, it is necessary to develop the force equations using the laws of motion in order to determine the system configuration. Such a system will be referred to in this book as a *dynamically driven system*.

In the classical approaches, one may have to rely on intuition to select the degrees of freedom of the system. If the system has a complex topological structure or has a large number of bodies, difficulties may be encountered when classical techniques are used. While these techniques lead to simple relationships for simple mechanisms, they are not suited for the analysis of a large class of mechanical system applications. Many of the basic concepts used in the classical approaches, however, are the same as those used for modern computer techniques.

In Chapter 3, two approaches are discussed for kinematically driven multibody systems: the classical and computational approaches. In the *classical approach*, which is suited for the analysis of simple systems, it is assumed that the system degrees of freedom can easily be identified and all the kinematic variables can be expressed, in a straightforward manner, in terms of the degrees of freedom. When more complex systems are considered, the use of another computer-based method, such as the *computational approach*, becomes necessary. In the computational approach, the kinematic constraint equations that describe mechanical joints and specified motion trajectories are formulated, leading to a relatively large system of nonlinear algebraic equations that can be solved using computer and numerical methods.

This computational method can be used as the basis for developing a general-purpose computer program for the kinematic analysis of a large class of kinematically driven multibody systems, as discussed in Chapter 3.

1.5 FORCE ANALYSIS

Forces in multibody systems can be categorized as inertia, external, and joint forces. *Inertia* is the property of a body that causes it to resist any effort to change its motion. Inertia forces, in general, depend on the mass and shape of the body as well as its velocity and acceleration. If a body is at rest, its inertia forces are equal to zero. *Joint forces* are the reaction forces that arise as the result of the connectivity between different bodies in multibody systems. These forces are sometimes referred to as *internal forces* or *constraint forces*. According to *Newton's third law*, the joint reaction forces acting on two interconnected bodies are equal in magnitude and opposite in direction. In this book, *external forces* are forces that are not inertia or joint forces. Examples of external forces are spring and damper forces, motor torques, actuator forces, and gravity forces.

While in kinematics we are concerned only with motion without regard to the forces that cause it, in dynamic analysis we are interested in the motion and the forces that produce it. Unlike the case of static or kinematic analysis, where only algebraic equations are used, in dynamic analysis, the motion of a multibody system is governed by second-order differential equations. Several techniques are discussed in this book for the dynamic analysis of mechanical systems that consist of interconnected rigid bodies. Only the reader's familiarity with *Newton's second law* is assumed for understanding the developments presented in later chapters. This law states that the force that acts on a particle is equal to the rate of change of momentum of the particle. Newton's second law, with *Euler's equations* that govern the rotation of the rigid body, leads to the dynamic conditions for the rigid bodies. D'Alembert's principle, which implies that inertia forces can be treated the same as applied forces, can be used to obtain the powerful *principle of virtual work*. Lagrange used this principle as a starting point to derive his dynamic equation, which is expressed in terms of scalar work and energy quantities. D'Alembert's principle, the principle of virtual work, and Lagrange's equation are discussed in detail in Chapters 4 and 5.

1.6 DYNAMIC EQUATIONS AND THEIR DIFFERENT FORMS

Depending on the number of coordinates selected to define the configuration of a mechanical system, different equation structures can be obtained and different solution procedures can be adopted. Some of the formulations lead to equations that are expressed in terms of the constraint forces, while in other formulations, the constraint forces are eliminated automatically. For instance, the equations of motion of a simple system such as the block shown in Fig. 10 can be formulated using a minimum set of independent coordinates or using a redundant set of coordinates that are not totally independent. Since the system has one degree of freedom representing the motion in the horizontal direction, one equation suffices to define the configuration of the block. This equation can simply be written as

$$m\ddot{x} = F \tag{1.4}$$

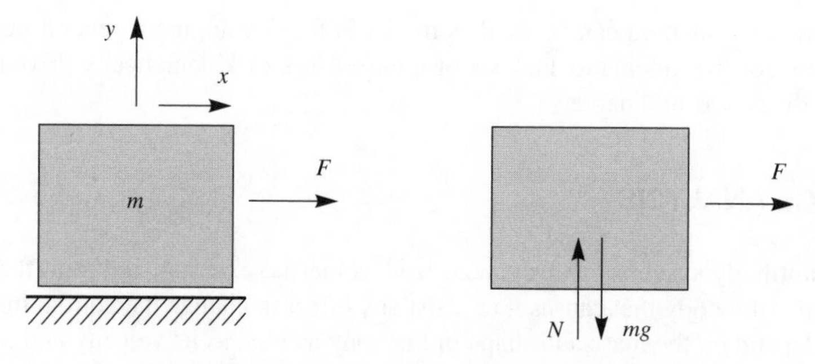

Figure 1.10 Forms of the equations of motion

where m is the mass of the block, x is the block coordinate, and F is the force acting on the block. Note that when the force is given, the preceding equation can be solved for the acceleration. We also note that the preceding equation does not include reaction forces since this equation describes motion in terms of the degree of freedom. As we will see in subsequent chapters, it is always possible to obtain a set of dynamic equations which do not include any constraint forces when the degrees of freedom are used. The *principle of virtual work in dynamics* represents a powerful tool that enables us systematically to formulate a set of dynamic equations of constrained multibody systems such that these equations do not include constraint forces. This principle is discussed in detail in Chapter 5.

Another approach that can be used to formulate the equations of motion of the simple system shown in Fig. 10 is to use redundant coordinates. For example, we may choose to describe the dynamics of the block using the following two equations:

$$\left.\begin{aligned} m\ddot{x} &= F \\ m\ddot{y} &= N - mg \end{aligned}\right\} \tag{1.5}$$

where y is the coordinate of the block in the vertical direction, N is the reaction force due to the constraint imposed on the motion of the block, and g is the gravity constant. If the force F is given, the preceding two equations have three unknowns: two acceleration components and the reaction force N. For this reason, another equation is needed to be able to solve for the three unknowns. The third equation is simply the equation of the constraint imposed on the motion of the block in the vertical direction. This equation can be written as

$$y = c \tag{1.6}$$

where c is a constant. This algebraic equation along with the two differential equations of motion (Eq. 5) form a system of *algebraic and differential equations* that can be solved for all the coordinates and forces. Here we obtained a larger system expressed in terms of a set of redundant coordinates since the y coordinate is not a degree of freedom. As we will see in this book, use of the redundant system can have computational advantages and can also increase the generality and flexibility of the formulation used. For this reason, many general-purpose multibody computer programs use formulations that employ redundant coordinates. There are, however, several general observations with regard to the use of redundant coordinates. Using our simple system, we note that the number of independent constraint (reaction) forces is equal to the number of coordinates used minus the number of the system degrees of freedom. We also note that the number of independent reaction forces is equal to the number of constraint equations. As we will see in subsequent chapters, this is always the case regardless of the complexity of the system analyzed, and the elimination of a reaction force can be equivalent to the elimination of a dependent coordinate or a constraint equation. In our example, we have one reaction force (N) and one constraint equation ($y = c$).

When the equations of motion are formulated in terms of the system degrees of freedom only, one obtains differential equations that can be solved using a simpler numerical strategy. When the equations of motion are formulated in terms of redundant coordinates, a more elaborate numerical scheme must be used to solve the resulting system of algebraic and differential equations. These algebraic and differential equations for most multibody systems are coupled and highly nonlinear. Direct numerical integration methods are used to solve for the system coordinates and velocities, and iterative numerical procedures are used to check on the violation of the constraint equations. This subject is discussed in more detail in Chapter 6, in which the Lagrangian formulation of the equations of motion is introduced. In this formulation, a symmetric structure of equations of motion expressed in terms of redundant coordinates and constraint forces is presented. To obtain this symmetric structure, the concept of *generalized constraint forces*, which are expressed in terms of multipliers known as *Lagrange multipliers*, is introduced.

The simple example of the one-degree-of-freedom block discussed in this section alludes to some of the fundamental issues in computational dynamics. However, the equations of motion of multibody systems are not likely to be as simple as the equations of the block due to the geometric nonlinearities and the kinematic constraints. As the complexity of the system topology increases, the dimensionality and nonlinearity increase. Computational methods for modeling complex and nonlinear multibody systems are discussed in Chapter 6.

1.7 FORWARD AND INVERSE DYNAMICS

In studying the dynamics of mechanical systems, there are two different types of analysis that can be performed. These are the inverse and forward dynamics. In the *inverse dynamics*, the motion trajectories of all the system degrees of freedom are specified and the objective is to determine the forces that produce this motion. This type of analysis requires only the solution of systems of algebraic equations. There is no need in this type of analysis for the use of numerical integration methods since the position coordinates, velocities, and accelerations of the system are known. In the case of the forward dynamics, however, the forces that produce the motion are given and the objective is to determine the position coordinates, velocities, and accelerations. In this type of analysis, the accelerations are first determined using the laws of motion. These accelerations must then be integrated to determine the coordinates and velocities. In most applications, a closed-form solution is difficult to obtain and, therefore, one must resort to direct numerical integration methods.

The difference between forward and inverse dynamics can be explained using a simple example. Consider a mass m which moves only in the horizontal direction with displacement x as the result of the application of a force F. The equation of motion of the mass is

$$m\ddot{x} = F \tag{1.7}$$

In the forward dynamics, the force F is given and the objective is to determine the motion of the mass as the result of the application of force. In this case, we first solve for the acceleration as

$$\ddot{x} = \frac{F}{m} \tag{1.8}$$

Knowing F and m, we integrate the acceleration to determine the velocity. Using the preceding equation, we have

$$\frac{d\dot{x}}{dt} = \frac{F}{m} \tag{1.9}$$

which yields

$$\int_{\dot{x}_0}^{\dot{x}} d\dot{x} = \int_0^t \frac{F}{m}\, dt \tag{1.10}$$

where $\dot{x}_0$ is the initial velocity of the mass. It follows that

$$\dot{x} = \dot{x}_0 + \int_0^t \frac{F}{m}\, dt \tag{1.11}$$

If the force F is known as a function of time, the preceding equation can be used to solve for the velocity of the mass. Having determined the velocity, the following equation can be used to determine the displacement:

$$\frac{dx}{dt} = \dot{x} \tag{1.12}$$

from which

$$x = x_0 + \int_0^t \dot{x}\, dt \tag{1.13}$$

where x_0 is the initial displacement of the mass. It is clear from this simple example that one needs two initial conditions: an initial displacement and an initial velocity, to be able to integrate the acceleration to determine the displacement and velocity in response to given forces. In the case of simple systems, one may be able to obtain closed-form solutions for the velocities and displacements. In more complex systems, integration of the accelerations to determine the velocities and displacements must be performed numerically as described in Chapter 6.

In the inverse dynamics, on the other hand, there is no need for performing integrations. One needs to only solve a system of algebraic equations. For instance, if the displacement of the mass is specified as a function of time, one can simply differentiate the displacement twice to obtain the acceleration and substitute the result into the equation of motion of the system to determine the force. For example, if the displacement of the mass is prescribed as

$$x = A \sin \omega t \tag{1.14}$$

where A and ω are known constants, the acceleration of the mass can be defined simply as

$$\ddot{x} = -\omega^2 A \sin \omega t \tag{1.15}$$

Using the equation of motion of the mass (Eq. 7), the force F can be determined as

$$F = m\ddot{x} = -m\omega^2 A \sin \omega t \tag{1.16}$$

This equation determines the force required to produce the prescribed displacement of the mass.

The inverse dynamics is widely used in the design and control of many industrial and technological applications, such as robot manipulators and space structures. By specifying the task to be performed by the system, the actuator forces and motor torques required to accomplish this task successfully can be predicted. Furthermore, different design alternatives and force configurations can be explored efficiently using the techniques of the inverse dynamics.

1.8 PLANAR AND SPATIAL DYNAMICS

The analysis of planar systems can be considered as a special case of spatial analysis. In *spatial analysis*, more coordinates are required to describe the configuration of an unconstrained body. As mentioned previously, six coordinates that define the location of a point on the body and the orientation of a coordinate system rigidly attached to the body are required to describe the unconstrained motion of a rigid body in space. In the *planar analysis*, only three coordinates are required, and one of these coordinates suffices to define the orientation of the body as compared to three orientation coordinates in the three-dimensional analysis. Furthermore, in the planar analysis, the order of rotation is commutative since the rotation is performed about the same axis; that is, two consecutive rotations can be added and the sequence of performing these rotations is immaterial. This is not the case, however, in three-dimensional analysis, where three independent rotations can be performed about three perpendicular axes. In this case, the order of rotation is not in general commutative, and two consecutive rotations about two different axes cannot in general be added. This can be demonstrated by using the simple block example shown in Fig. 11, which illustrates different sequences of rotations for the same block. In Fig. 11a, the block is first rotated 90° about the Y axis and then 90° about the Z axis. In Fig. 11b, the same rotations in reverse order are employed; that is, the block is first rotated 90° about the Z axis and then 90° about the Y axis. It is clear from the results presented in Figs. 11a and b that a change in the sequence of rotations leads to different final orientations. We then conclude from this simple example that the order of the finite rotations in the spatial analysis is not commutative, and for this reason, the finite rotations in the spatial analysis cannot be in general added or treated as vector quantities. The subject of the three-dimensional rotations is discussed in more detail in Chapters 7 and 8, where different sets of orientation coordinates are discussed. These sets include Euler angles, Euler parameters, direction cosines, and Rodriguez parameters. The general dynamic equations that govern the constrained and unconstrained spatial motion of rigid body systems are also developed in Chapter 7. This includes the Newton–Euler equations and recursive formulations that are often used in the computer-aided analysis of constrained multibody systems.

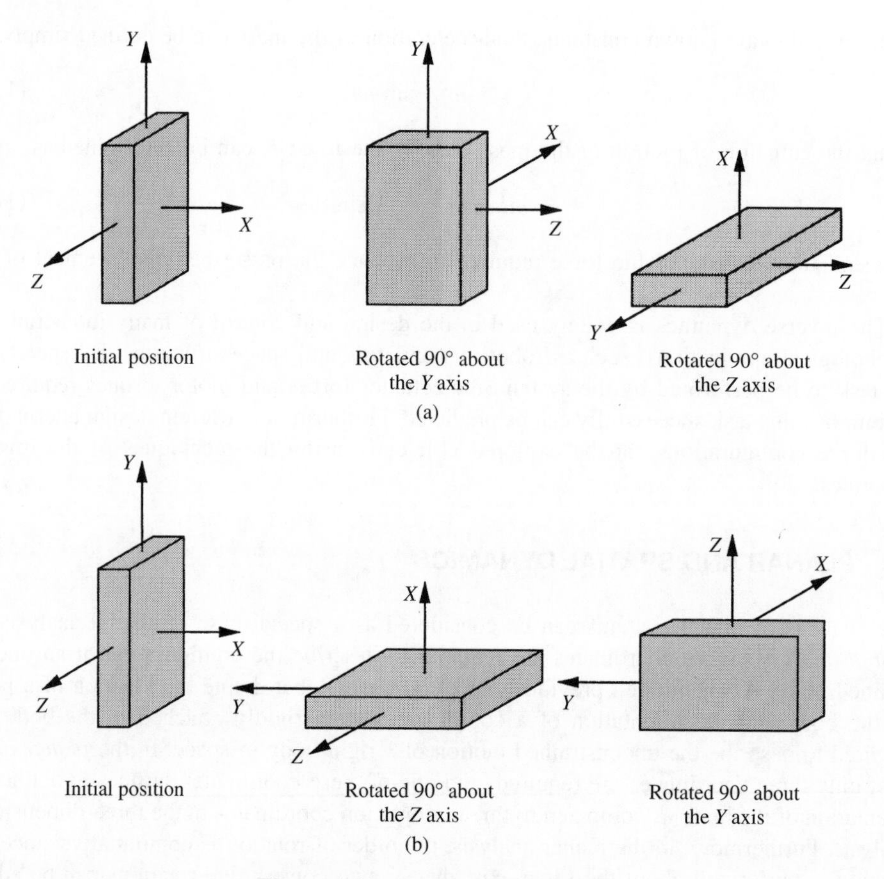

Figure 1.11 Sequence of rotations

1.9 COMPUTER AND NUMERICAL METHODS

While the analytical techniques of Newton, D'Alembert, and Lagrange were developed centuries ago, these classical approaches have proven to be suitable for implementation on high-speed digital computers when used with matrix and numerical methods. The application of these methods leads to a set of differential equations that can be expressed in a matrix form and can be solved using numerical and computer methods. Several numerical algorithms are developed based on the Newtonian or the Lagrangian approaches. These algorithms, which utilize matrix and numerical methods, are used to develop general- and special-purpose computer programs that can be used for the dynamic simulation and control of multibody systems that consist of interconnected bodies. These programs allow the user to introduce, in a systematic manner, elastic or damping elements such as springs and dampers, nonlinear general forcing functions, and/or nonlinear constraint equations. An example of these general purpose multibody system computer programs is the code SAMS/2000 (**S**ystematic **A**nalysis of **M**ultibody **S**ystems) which is described in Chapter 9. It is important that the reader becomes familiar with the capabilities of such multibody system codes in order to have an understanding of the use of the formulations and algorithms presented in this book and used in the general field of computational dynamics.

The computational efficiency of the computer programs developed for the dynamic analysis of mechanical systems depends on many factors, such as the choice of coordinates and the numerical procedure used for solving the dynamic equations. The choice of the coordinates directly influences the number and the degree of nonlinearity of the resulting dynamic equations. The use of a relatively small number of coordinates leads to a higher degree of nonlinearity and more complex dynamic equations. For this reason, in many of the computational methods developed for the dynamic analysis of mechanical systems, a larger number of displacement coordinates is used for the sake of generality.

As pointed out previously, the reader will recognize when studying this book that there are two basic dynamic formulations which are widely used in the computer simulation of multibody systems. In the first formulation, the constraint forces are eliminated from the dynamic equations by expressing these equations in terms of the system degrees of freedom. Variables that represent joint coordinates are often used as the degrees of freedom in order to be able to express the system configuration analytically in terms of these degrees of freedom. The use of the joint variables has the advantage of reducing the number of equations and the disadvantage of increasing the nonlinearity and complexity of the equations. This can be expected since all the information about the system dynamics must be included in a smaller set of equations. Formulations that use the joint variables or the degrees of freedom to obtain a minimum set of equations are referred to in this book as the *embedding techniques*. The embedding techniques are also the basis for developing the *recursive methods*, which are widely used in the analysis of robot manipulators. The recursive methods are discussed in Chapter 7.

Another dynamic formulation that is widely used in the computer simulation of multibody systems is the *augmented formulation*. In this formulation, the equations of motion are expressed in terms of redundant set of coordinates that are not totally independent. Because of this redundancy, the kinematic algebraic constraint equations that describe the relationship between these coordinates must be formulated. As a result, the constraint forces appear in the final form of the equations of motion. Clearly, one of the drawbacks of using this approach is increasing the number of coordinates and equations. Another drawback is the complexity of the numerical algorithm that must be used to solve the resulting system of differential and algebraic equations. Nonetheless, the augmented formulation has the advantage of producing simple equations that have a sparse matrix structure; therefore, these equations can be solved efficiently using sparse matrix techniques. Furthermore, the general-purpose multibody computer programs based on the augmented formulation tend to be more user-friendly since they allow the user systematic introduction of any nonlinear constraint or force function. In most general-purpose computer programs based on the augmented formulation, the motion of the bodies in the system is described using absolute Cartesian and orientation coordinates. In planar analysis, two Cartesian coordinates that define the location of the origin of the body coordinate system selected, and one orientation coordinate that defines the orientation of this coordinate system in a global inertial frame, are used. In the spatial analysis, six absolute coordinates are used to define the location and orientation of the body coordinate system. The use of similar sets of coordinates for all bodies in the system makes it easy for the user to change the model by adding or deleting bodies and joints and/or introducing nonlinear forcing and constraint functions.

Both the embedding technique and augmented formulation are discussed in detail in this book, and several examples will be used to show the structure of the equations obtained using each formulation. These two methods have been applied successfully to the analysis, design, and control of many technological and industrial applications, including vehicles, mechanisms, robot manipulators, machines, space structures, and biomechanical systems. It is hoped that by studying these two basic formulations carefully, the reader will be able to make a better choice of the method that is most suited for his or her application.

1.10 ORGANIZATION, SCOPE, AND NOTATIONS OF THE BOOK

The purpose of this book is to provide an introduction to the subject of computational dynamics. The goal is to introduce the reader to various dynamic formulations that can be implemented on the digital computer. The computer implementation is necessary to be able to study the dynamic motion of large-scale systems. The general formulations presented in this book can also be used to develop general-purpose computer codes that can be used in the analysis of a large class of multibody system applications. The book is organized in nine chapters, including this introductory chapter.

As the dimensionality and complexity of multibody systems increase, a knowledge of matrix and numerical methods becomes necessary for understanding the theory behind general- and special-purpose multibody computer programs. For this reason, Chapter 2 is devoted to a brief introduction to the subject of linear algebra. Matrix and vector operations and identities as well as methods for the numerical solution of systems of algebraic equations are discussed. The **QR** and singular value decompositions, which can be used in multibody system dynamics to determine velocity transformation matrices that relate the system velocities to the time derivatives of the degrees of freedom, are also introduced in this chapter.

The kinematics of multibody systems are discussed in Chapter 3. In this chapter, kinematically driven systems in which all the degrees of freedom are specified are investigated. For these systems, to define the system configuration, one need only formulate a set of algebraic equations. There is no need to use the laws of motion since the degrees of freedom and their time derivatives are known. Two basic approaches are discussed, the classical approach and the computational approach. The classical approach is suited for the analysis of systems that consist of small number of bodies and joints, and in which the degrees of freedom can be identified easily and intuitively. The computational approach, on the other hand, is suited for the analysis of complex systems and can be used to develop a general-purpose computer program for the kinematic analysis of varieties of multibody system applications. Based on a systematic and general description of the system topology, a general-purpose computer program can be developed and used to construct nonlinear kinematic relationships between the variables. This program can also be used to solve these relationships numerically, in order to determine the system configuration.

Various forms of the dynamic equations are presented in Chapter 4. A simple Newtonian mechanics approach is used in this chapter to derive these different forms and demonstrate the basic differences between them. It is shown in this chapter that when the equations of motion are derived in terms of a set of redundant coordinates, the constraint forces appear explicitly in the equations. This leads to the augmented form of the equations of motion. It is shown in Chapter 4 that the constraint forces can be eliminated from the system equations of motion if these equations are expressed in terms of the degrees of freedom. This procedure is referred to in this book as the embedding technique.

Although as demonstrated in Chapter 4, the embedding technique can be applied in the framework of Newtonian mechanics, the result of this technique can be obtained more elegantly by using the principle of virtual work. This principle can be used to eliminate the constraint forces systematically and obtain a minimum set of dynamic equations expressed in terms of the system degrees of freedom. The concepts of the virtual work and generalized forces that are necessary for the application of the virtual work principle, Lagrange's equation, and the Hamiltonian formulation are among the topics discussed in Chapter 5. The chapter concludes by examining the relationship between the principle of virtual work and the Gaussian elimination used in the solution of algebraic systems of equations.

The analytical methods presented in Chapter 5 are used as the foundation for the computational approaches discussed in Chapter 6. A computer-based embedding technique and an augmented formulation suitable for the analysis of large-scale constrained multibody systems are introduced. The important concepts of the generalized constraint forces and Lagrange multipliers are discussed. Numerical algorithms for solving the differential and algebraic equations of multibody systems are also presented in Chapter 6. It is important to point out that the basic methods presented in Chapter 6 are not different from the methods presented in Chapters 4 and 5 except for using a certain set of coordinates that serves our computational goals.

In Chapters 3 through 6, planar examples are used to focus on the main concepts and the development of the basic methods without delving into the details of the three-dimensional motion. The analysis of the spatial motion is presented in Chapter 7. In this chapter, methods for describing the three-dimensional rotations are developed, and the concept of angular velocity in spatial analysis is introduced. The three-dimensional form of the equations of motion is presented in terms of the generalized coordinates and used to obtain the known Newton–Euler equations. Formulations of the algebraic constraint equations of several spatial joints, such as the revolute, prismatic, cylindrical, and universal joints, are discussed. The use of Newton–Euler equations to develop a recursive formulation for multibody systems is also demonstrated in Chapter 7.

In Chapter 8, special topics are discussed. These topics include gyroscopic motion and various sets of parameters that can be used to define the orientation of the rigid body in space. These parameters include Euler parameters, Rodriguez parameters, and the quaternions. The eigenvalue analysis is widely used to study the stability of nonlinear multibody systems. The nonlinear equations of motion can be linearized at different points in time during the simulation in order to define an eigenvalue problem that can be solved to obtain information on the system stability. This important topic is also discussed in Chapter 8 of this book.

Chapter 9 describes the general purpose multibody system computer code SAMS/2000. Many of the formulations and algorithms discussed in various chapters of this book are implemented in this general purpose multibody system computer program. This code and other multibody system codes allow the user to build computer models for planar and spatial multibody systems. SAMS/2000 has also advanced multibody system capabilities based on formulations and techniques that are beyond what is covered in this introductory text.

It is important that readers become familiar with the multibody notations used in this book as described in the preface in order to follow the developments presented in different chapters. Boldface letters are used to indicate vectors or matrices. Superscripts are used to indicate body numbers. To distinguish between a superscript that indicates the body number and the power, parentheses are used whenever a quantity is raised to a certain power. For example, $(l^5)^3$ is a scalar l associated with body 5 raised to the power of 3.

CHAPTER 2

LINEAR ALGEBRA

Vector and matrix concepts have proved indispensable in the development of the subject of dynamics. The formulation of the equations of motion using the Newtonian or Lagrangian approach leads to a set of second-order simultaneous differential equations. For convenience, these equations are often expressed in vector and matrix forms. Vector and matrix identities can be utilized to provide much less cumbersome proofs of many of the kinematic and dynamic relationships. In this chapter, the mathematical tools required to understand the development presented in this book are discussed briefly. Matrices and matrix operations are discussed in the first two sections. Differentiation of vector functions and the important concept of *linear independence* are discussed in Section 3. In Section 4, important topics related to three-dimensional vectors are presented. These topics include the *cross product*, *skew-symmetric matrix representations*, *Cartesian coordinate systems*, and *conditions of parallelism*. The conditions of parallelism are used in this book to define the kinematic constraint equations of many joints in the three-dimensional analysis. Computer methods for solving algebraic systems of equations are presented in Sections 5 and 6. Among the topics discussed in these two sections are the *Gaussian elimination*, *pivoting and scaling*, *triangular factorization*, and *Cholesky decomposition*. The last two sections of this chapter deal with the **QR** *decomposition* and the *singular value decomposition*. These two types of decompositions have been used in computational dynamics to identify the independent degrees of freedom of multibody systems. The last two sections, however, can be omitted during a first reading of the book.

Computational Dynamics, Third Edition Ahmed A. Shabana

2.1 MATRICES

An $m \times n$ matrix $\mathbf{A}$ is an ordered rectangular array that has $m \times n$ elements. The matrix $\mathbf{A}$ can be written in the form

$$\mathbf{A} = (a_{ij}) = \begin{bmatrix} a_{11} & a_{12} & \cdots & a_{1n} \\ a_{21} & a_{22} & \cdots & a_{2n} \\ \vdots & \vdots & \ddots & \vdots \\ a_{m1} & a_{m2} & \cdots & a_{mn} \end{bmatrix} \tag{2.1}$$

The matrix $\mathbf{A}$ is called an $m \times n$ matrix since it has m rows and n columns. The scalar element a_{ij} lies in the ith row and jth column of the matrix $\mathbf{A}$. Therefore, the index i, which takes the values $1, 2, \ldots, m$ denotes the row number, while the index j, which takes the values $1, 2, \ldots, n$ denotes the column number.

A matrix $\mathbf{A}$ is said to be square if $m = n$. An example of a square matrix is

$$\mathbf{A} = \begin{bmatrix} 3.0 & -2.0 & 0.95 \\ 6.3 & 0.0 & 12.0 \\ 9.0 & 3.5 & 1.25 \end{bmatrix}$$

In this example, $m = n = 3$, and $\mathbf{A}$ is a 3×3 matrix.

The *transpose* of an $m \times n$ matrix $\mathbf{A}$ is an $n \times m$ matrix denoted as $\mathbf{A}^{\mathrm{T}}$ and defined as

$$\mathbf{A}^{\mathrm{T}} = \begin{bmatrix} a_{11} & a_{21} & \cdots & a_{m1} \\ a_{12} & a_{22} & \cdots & a_{m2} \\ \vdots & \vdots & \ddots & \vdots \\ a_{1n} & a_{2n} & \cdots & a_{mn} \end{bmatrix} \tag{2.2}$$

For example, let $\mathbf{A}$ be the matrix

$$\mathbf{A} = \begin{bmatrix} 2.0 & -4.0 & -7.5 & 23.5 \\ 0.0 & 8.5 & 10.0 & 0.0 \end{bmatrix}$$

The transpose of $\mathbf{A}$ is

$$\mathbf{A}^{\mathrm{T}} = \begin{bmatrix} 2.0 & 0.0 \\ -4.0 & 8.5 \\ -7.5 & 10.0 \\ 23.5 & 0.0 \end{bmatrix}$$

That is, the transpose of the matrix $\mathbf{A}$ is obtained by interchanging the rows and columns.

A square matrix $\mathbf{A}$ is said to be *symmetric* if $a_{ij} = a_{ji}$. The elements on the upper-right half of a symmetric matrix can be obtained by flipping the matrix about the diagonal. For example,

$$\mathbf{A} = \begin{bmatrix} 3.0 & -2.0 & 1.5 \\ -2.0 & 0.0 & 2.3 \\ 1.5 & 2.3 & 1.5 \end{bmatrix}$$

is a symmetric matrix. Note that if $\mathbf{A}$ is symmetric, then $\mathbf{A}$ is the same as its transpose; that is, $\mathbf{A} = \mathbf{A}^{\mathrm{T}}$.

A square matrix is said to be an *upper-triangular matrix* if $a_{ij} = 0$ for $i > j$. That is, every element below each diagonal element of an upper-triangular matrix is zero. An example of an upper-triangular matrix is

$$\mathbf{A} = \begin{bmatrix} 6.0 & 2.5 & 10.2 & -11.0 \\ 0 & 8.0 & 5.5 & 6.0 \\ 0 & 0 & 3.2 & -4.0 \\ 0 & 0 & 0 & -2.2 \end{bmatrix}$$

A square matrix is said to be a *lower-triangular matrix* if $a_{ij} = 0$ for $j > i$. That is, every element above the diagonal elements of a lower-triangular matrix is zero. An example of a lower-triangular matrix is

$$\mathbf{A} = \begin{bmatrix} 6.0 & 0 & 0 & 0 \\ 2.5 & 8.0 & 0 & 0 \\ 10.2 & 5.5 & 3.2 & 0 \\ -11.0 & 6.0 & -4.0 & -2.2 \end{bmatrix}$$

The *diagonal matrix* is a square matrix such that $a_{ij} = 0$ if $i \neq j$, which implies that a diagonal matrix has element a_{ii} along the diagonal with all other elements equal to zero. For example,

$$\mathbf{A} = \begin{bmatrix} 5.0 & 0 & 0 \\ 0 & 1.0 & 0 \\ 0 & 0 & 7.0 \end{bmatrix}$$

is a diagonal matrix.

The *null matrix* or *zero matrix* is defined to be a matrix in which all the elements are equal to zero. The *unit matrix* or identity matrix is a diagonal matrix whose diagonal elements are nonzero and equal to 1.

A *skew-symmetric matrix* is a matrix such that $a_{ij} = -a_{ji}$. Note that since $a_{ij} = -a_{ji}$ for all i and j values, the diagonal elements should be equal to zero. An example of a skew-symmetric matrix $\tilde{\mathbf{A}}$ is

$$\tilde{\mathbf{A}} = \begin{bmatrix} 0 & -3.0 & -5.0 \\ 3.0 & 0 & 2.5 \\ 5.0 & -2.5 & 0 \end{bmatrix}$$

It is clear that for a skew-symmetric matrix, $\tilde{\mathbf{A}}^{\mathrm{T}} = -\tilde{\mathbf{A}}$.

The *trace* of a square matrix is the sum of its diagonal elements. The *trace* of an $n \times n$ identity matrix is n, while the trace of a skew-symmetric matrix is zero.

2.2 MATRIX OPERATIONS

In this section we discuss some of the basic matrix operations that are used throughout the book.

Matrix Addition The sum of two matrices **A** and **B**, denoted by $\mathbf{A} + \mathbf{B}$, is given by

$$\mathbf{A} + \mathbf{B} = (a_{ij} + b_{ij}) \tag{2.3}$$

where b_{ij} are the elements of **B**. To add two matrices **A** and **B**, it is necessary that **A** and **B** have the same dimension; that is, the same number of rows and the same number of columns. It is clear from Eq. 3 that matrix addition is *commutative*, that is,

$$\mathbf{A} + \mathbf{B} = \mathbf{B} + \mathbf{A} \tag{2.4}$$

Matrix addition is also *associative*, because

$$\mathbf{A} + (\mathbf{B} + \mathbf{C}) = (\mathbf{A} + \mathbf{B}) + \mathbf{C} \tag{2.5}$$

Example 2.1

The two matrices **A** and **B** are defined as

$$\mathbf{A} = \begin{bmatrix} 3.0 & 1.0 & -5.0 \\ 2.0 & 0.0 & 2.0 \end{bmatrix}, \qquad \mathbf{B} = \begin{bmatrix} 2.0 & 3.0 & 6.0 \\ -3.0 & 0.0 & -5.0 \end{bmatrix}$$

The sum $\mathbf{A} + \mathbf{B}$ is

$$\mathbf{A} + \mathbf{B} = \begin{bmatrix} 3.0 & 1.0 & -5.0 \\ 2.0 & 0.0 & 2.0 \end{bmatrix} + \begin{bmatrix} 2.0 & 3.0 & 6.0 \\ -3.0 & 0.0 & -5.0 \end{bmatrix}$$

$$= \begin{bmatrix} 5.0 & 4.0 & 1.0 \\ -1.0 & 0.0 & -3.0 \end{bmatrix}$$

while $\mathbf{A} - \mathbf{B}$ is

$$\mathbf{A} - \mathbf{B} = \begin{bmatrix} 3.0 & 1.0 & -5.0 \\ 2.0 & 0.0 & 2.0 \end{bmatrix} - \begin{bmatrix} 2.0 & 3.0 & 6.0 \\ -3.0 & 0.0 & -5.0 \end{bmatrix}$$

$$= \begin{bmatrix} 1.0 & -2.0 & -11.0 \\ 5.0 & 0.0 & 7.0 \end{bmatrix}$$

Matrix Multiplication The product of two matrices $\mathbf{A}$ and $\mathbf{B}$ is another matrix $\mathbf{C}$, defined as

$$\mathbf{C} = \mathbf{AB} \tag{2.6}$$

The element c_{ij} of the matrix $\mathbf{C}$ is defined by multiplying the elements of the ith row in $\mathbf{A}$ by the elements of the jth column in $\mathbf{B}$ according to the rule

$$c_{ij} = a_{i1}b_{1j} + a_{i2}b_{2j} + \cdots + a_{in}b_{nj} = \sum_k a_{ik}b_{kj} \tag{2.7}$$

Therefore, the number of columns in $\mathbf{A}$ must be equal to the number of rows in $\mathbf{B}$. If $\mathbf{A}$ is an $m \times n$ matrix and $\mathbf{B}$ is an $n \times p$ matrix, then $\mathbf{C}$ is an $m \times p$ matrix. In general, $\mathbf{AB} \neq \mathbf{BA}$. That is, matrix multiplication is not commutative. Matrix multiplication, however, is distributive; that is, if $\mathbf{A}$ and $\mathbf{B}$ are $m \times p$ matrices and $\mathbf{C}$ is a $p \times n$ matrix, then

$$(\mathbf{A} + \mathbf{B})\mathbf{C} = \mathbf{AC} + \mathbf{BC} \tag{2.8}$$

Example 2.2

Let

$$\mathbf{A} = \begin{bmatrix} 0 & 4 & 1 \\ 2 & 1 & 1 \\ 3 & 2 & 1 \end{bmatrix}, \qquad \mathbf{B} = \begin{bmatrix} 0 & 1 \\ 0 & 0 \\ 5 & 2 \end{bmatrix}$$

Then

$$\mathbf{AB} = \begin{bmatrix} 0 & 4 & 1 \\ 2 & 1 & 1 \\ 3 & 2 & 1 \end{bmatrix} \begin{bmatrix} 0 & 1 \\ 0 & 0 \\ 5 & 2 \end{bmatrix} = \begin{bmatrix} 5 & 2 \\ 5 & 4 \\ 5 & 5 \end{bmatrix}$$

The product $\mathbf{BA}$ is not defined in this example since the number of columns in $\mathbf{B}$ is not equal to the number of rows in $\mathbf{A}$.

The *associative law* is valid for matrix multiplications. If $\mathbf{A}$ is an $m \times p$ matrix, $\mathbf{B}$ is a $p \times q$ matrix, and $\mathbf{C}$ is a $q \times n$ matrix, then

$$(\mathbf{AB})\mathbf{C} = \mathbf{A}(\mathbf{BC}) = \mathbf{ABC} \tag{2.9}$$

Matrix Partitioning Matrix partitioning is a useful technique that is frequently used in manipulations with matrices. In this technique, a matrix is assumed to consist of submatrices or blocks that have smaller dimensions. A matrix is divided into blocks or parts by means of horizontal and vertical lines. For example, let $\mathbf{A}$ be a 4×4 matrix. The matrix $\mathbf{A}$ can be partitioned by using horizontal and vertical lines as follows:

$$\mathbf{A} = \left[\begin{array}{ccc:c} a_{11} & a_{12} & a_{13} & a_{14} \\ a_{21} & a_{22} & a_{23} & a_{24} \\ a_{31} & a_{32} & a_{33} & a_{34} \\ \hdashline a_{41} & a_{42} & a_{43} & a_{44} \end{array}\right] \tag{2.10}$$

In this example, the matrix $\mathbf{A}$ has been partitioned into four submatrices; therefore, we can write $\mathbf{A}$ compactly in terms of these four submatrices as

$$\mathbf{A} = \begin{bmatrix} \mathbf{A}_{11} & \mathbf{A}_{12} \\ \mathbf{A}_{21} & \mathbf{A}_{22} \end{bmatrix} \tag{2.11}$$

where

$$\left.\begin{aligned} \mathbf{A}_{11} &= \begin{bmatrix} a_{11} & a_{12} & a_{13} \\ a_{21} & a_{22} & a_{23} \\ a_{31} & a_{32} & a_{33} \end{bmatrix}, \qquad \mathbf{A}_{12} = \begin{bmatrix} a_{14} \\ a_{24} \\ a_{34} \end{bmatrix}, \\ \mathbf{A}_{21} &= [a_{41} \quad a_{42} \quad a_{43}], \qquad \mathbf{A}_{22} = a_{44} \end{aligned}\right\} \tag{2.12}$$

Apparently, there are many ways by which the matrix $\mathbf{A}$ can be partitioned. As we will see in this book, the way the matrices are partitioned depends on many factors, including the applications and the selection of coordinates.

Partitioned matrices can be multiplied by treating the submatrices like the elements of the matrix. To demonstrate this, we consider another matrix $\mathbf{B}$ such that $\mathbf{AB}$ is defined. We also assume that $\mathbf{B}$ is partitioned as follows:

$$\mathbf{B} = \begin{bmatrix} \mathbf{B}_{11} & \mathbf{B}_{12} & \mathbf{B}_{13} & \mathbf{B}_{14} \\ \mathbf{B}_{21} & \mathbf{B}_{22} & \mathbf{B}_{23} & \mathbf{B}_{24} \end{bmatrix} \tag{2.13}$$

The product **AB** is then defined as follows:

$$\mathbf{AB} = \begin{bmatrix} \mathbf{A}_{11} & \mathbf{A}_{12} \\ \mathbf{A}_{21} & \mathbf{A}_{22} \end{bmatrix} \begin{bmatrix} \mathbf{B}_{11} & \mathbf{B}_{12} & \mathbf{B}_{13} & \mathbf{B}_{14} \\ \mathbf{B}_{21} & \mathbf{B}_{22} & \mathbf{B}_{23} & \mathbf{B}_{24} \end{bmatrix}$$

$$= \begin{bmatrix} \mathbf{A}_{11}\mathbf{B}_{11} + \mathbf{A}_{12}\mathbf{B}_{21} & \mathbf{A}_{11}\mathbf{B}_{12} + \mathbf{A}_{12}\mathbf{B}_{22} & \mathbf{A}_{11}\mathbf{B}_{13} + \mathbf{A}_{12}\mathbf{B}_{23} & \mathbf{A}_{11}\mathbf{B}_{14} + \mathbf{A}_{12}\mathbf{B}_{24} \\ \mathbf{A}_{21}\mathbf{B}_{11} + \mathbf{A}_{22}\mathbf{B}_{21} & \mathbf{A}_{21}\mathbf{B}_{12} + \mathbf{A}_{22}\mathbf{B}_{22} & \mathbf{A}_{21}\mathbf{B}_{13} + \mathbf{A}_{22}\mathbf{B}_{23} & \mathbf{A}_{21}\mathbf{B}_{14} + \mathbf{A}_{22}\mathbf{B}_{24} \end{bmatrix} \tag{2.14}$$

When two partitioned matrices are multiplied we must make sure that additions and products of the submatrices are defined. For example, $\mathbf{A}_{11}\mathbf{B}_{12}$ must have the same dimension as $\mathbf{A}_{12}\mathbf{B}_{22}$. Furthermore, the number of columns of the submatrix $\mathbf{A}_{ij}$ must be equal to the number of rows in the matrix $\mathbf{B}_{jk}$. It is, therefore, clear that when multiplying two partitioned matrices **A** and **B**, we must have for each vertical partitioning line in **A** a similarly placed horizontal partitioning line in **B**.

Determinant The determinant of an $n \times n$ square matrix **A**, denoted as $|\mathbf{A}|$, is a scalar defined as

$$|\mathbf{A}| = \begin{vmatrix} a_{11} & a_{12} & \cdots & a_{1n} \\ a_{21} & a_{22} & \cdots & a_{2n} \\ \vdots & \vdots & \ddots & \vdots \\ a_{n1} & a_{n2} & \cdots & a_{nn} \end{vmatrix} \tag{2.15}$$

To be able to evaluate the unique value of the determinant of **A**, some basic definitions have to be introduced. The *minor* M_{ij} corresponding to the element a_{ij} is the determinant formed by deleting the ith row and jth column from the original determinant $|\mathbf{A}|$. The *cofactor* C_{ij} of the element a_{ij} is defined as

$$C_{ij} = (-1)^{i+j} M_{ij} \tag{2.16}$$

Using this definition, the value of the determinant in Eq. 15 can be obtained in terms of the cofactors of the elements of an arbitrary row i as follows:

$$|\mathbf{A}| = \sum_{j=1}^{n} a_{ij} C_{ij} \tag{2.17}$$

Clearly, the cofactors C_{ij} are determinants of order $n - 1$. If **A** is a 2×2 matrix defined as

$$\mathbf{A} = \begin{bmatrix} a_{11} & a_{12} \\ a_{21} & a_{22} \end{bmatrix} \tag{2.18}$$

the cofactors C_{ij} associated with the elements of the first row are

$$C_{11} = (-1)^2 a_{22} = a_{22}, \qquad C_{12} = (-1)^3 a_{21} = -a_{21} \tag{2.19}$$

According to the definition of Eq. 17, the determinant of the 2×2 matrix $\mathbf{A}$ using the cofactors of the elements of the first row is

$$|\mathbf{A}| = a_{11}C_{11} + a_{12}C_{12} = a_{11}a_{22} - a_{12}a_{21} \tag{2.20}$$

If $\mathbf{A}$ is 3×3 matrix defined as

$$\mathbf{A} = \begin{bmatrix} a_{11} & a_{12} & a_{13} \\ a_{21} & a_{22} & a_{23} \\ a_{31} & a_{32} & a_{33} \end{bmatrix}, \tag{2.21}$$

the determinant of $\mathbf{A}$ in terms of the cofactors of the first row is given by

$$|\mathbf{A}| = \sum_{j=1}^{3} a_{1j}C_{1j} = a_{11}C_{11} + a_{12}C_{12} + a_{13}C_{13} \tag{2.22}$$

where

$$C_{11} = \begin{vmatrix} a_{22} & a_{23} \\ a_{32} & a_{33} \end{vmatrix}, \qquad C_{12} = -\begin{vmatrix} a_{21} & a_{23} \\ a_{31} & a_{33} \end{vmatrix}, \qquad C_{13} = \begin{vmatrix} a_{21} & a_{22} \\ a_{31} & a_{32} \end{vmatrix} \tag{2.23}$$

That is, the determinant of $\mathbf{A}$ is

$$\begin{aligned} |\mathbf{A}| &= a_{11}\begin{vmatrix} a_{22} & a_{23} \\ a_{32} & a_{33} \end{vmatrix} - a_{12}\begin{vmatrix} a_{21} & a_{23} \\ a_{31} & a_{33} \end{vmatrix} + a_{13}\begin{vmatrix} a_{21} & a_{22} \\ a_{31} & a_{32} \end{vmatrix} \\ &= a_{11}(a_{22}a_{33} - a_{23}a_{32}) - a_{12}(a_{21}a_{33} - a_{23}a_{31}) + a_{13}(a_{21}a_{32} - a_{22}a_{31}) \end{aligned} \tag{2.24}$$

One can show that the determinant of a matrix is equal to the determinant of its transpose, that is,

$$|\mathbf{A}| = |\mathbf{A}^{\mathrm{T}}| \tag{2.25}$$

and the determinant of a diagonal matrix is equal to the product of the diagonal elements. Furthermore, the interchange of any two columns or rows only changes the sign of the determinant. If a matrix has two identical rows or two identical columns, the determinant of this matrix is equal to zero. This can be demonstrated by the example of Eq. 24. For instance, if the second and third rows are identical, $a_{21} = a_{31}$, $a_{22} = a_{32}$, and $a_{23} = a_{33}$. Using these equalities in Eq. 24, one can show that the determinant of the matrix $\mathbf{A}$ is equal to zero. More generally, a square matrix in which one or more rows (columns) are linear combinations of other rows (columns) has a zero determinant. For example,

$$\mathbf{A} = \begin{bmatrix} 1 & 0 & -3 \\ 0 & 2 & 5 \\ 1 & 2 & 2 \end{bmatrix} \quad \text{and} \quad \mathbf{B} = \begin{bmatrix} 1 & 0 & 1 \\ 0 & 2 & 2 \\ -3 & 5 & 2 \end{bmatrix} \tag{2.26}$$

have zero determinants since in **A** the last row is the sum of the first two rows and in **B** the last column is the sum of the first two columns.

A matrix whose determinant is equal to zero is said to be a *singular* matrix. For an arbitrary square matrix, singular or nonsingular, it can be shown that the value of the determinant does not change if any row or column is added to or subtracted from another.

Inverse of a Matrix A square matrix $\mathbf{A}^{-1}$ that satisfies the relationship

$$\mathbf{A}^{-1}\mathbf{A} = \mathbf{A}\mathbf{A}^{-1} = \mathbf{I} \tag{2.27}$$

where **I** is the identity matrix, is called the *inverse* of the matrix **A**. The inverse of the matrix **A** is defined as

$$\mathbf{A}^{-1} = \frac{\mathbf{C}_t}{|\mathbf{A}|} \tag{2.28}$$

where $\mathbf{C}_t$ is the adjoint of the matrix **A**. The adjoint matrix $\mathbf{C}_t$ is the transpose of the matrix of the cofactors C_{ij} of the matrix **A**.

Example 2.3

Determine the inverse of the matrix

$$\mathbf{A} = \begin{bmatrix} 1 & 1 & 1 \\ 0 & 1 & 1 \\ 0 & 0 & 1 \end{bmatrix}$$

Solution. The determinant of the matrix **A** is equal to 1, that is,

$$|\mathbf{A}| = 1$$

The cofactors of the elements of the matrix **A** are

$$\begin{array}{lll} C_{11} = 1, & C_{12} = 0, & C_{13} = 0, \\ C_{21} = -1, & C_{22} = 1, & C_{23} = 0, \\ C_{31} = 0, & C_{32} = -1, & C_{33} = 1 \end{array}$$

The adjoint matrix, which is the transpose of the matrix of the cofactors, is given by

$$\mathbf{C}_t = \begin{bmatrix} C_{11} & C_{21} & C_{31} \\ C_{12} & C_{22} & C_{32} \\ C_{13} & C_{23} & C_{33} \end{bmatrix} = \begin{bmatrix} 1 & -1 & 0 \\ 0 & 1 & -1 \\ 0 & 0 & 1 \end{bmatrix}$$

Therefore,

$$\mathbf{A}^{-1} = \frac{\mathbf{C}_t}{|\mathbf{A}|} = \begin{bmatrix} 1 & -1 & 0 \\ 0 & 1 & -1 \\ 0 & 0 & 1 \end{bmatrix}$$

Matrix multiplications show that

$$\mathbf{A}^{-1}\mathbf{A} = \begin{bmatrix} 1 & -1 & 0 \\ 0 & 1 & -1 \\ 0 & 0 & 1 \end{bmatrix} \begin{bmatrix} 1 & 1 & 1 \\ 0 & 1 & 1 \\ 0 & 0 & 1 \end{bmatrix} = \begin{bmatrix} 1 & 0 & 0 \\ 0 & 1 & 0 \\ 0 & 0 & 1 \end{bmatrix}$$

$$= \mathbf{A}\mathbf{A}^{-1}$$

If $\mathbf{A}$ is the 2×2 matrix

$$\mathbf{A} = \begin{bmatrix} a_{11} & a_{12} \\ a_{21} & a_{22} \end{bmatrix} \tag{2.29}$$

the inverse of $\mathbf{A}$ can be written simply as

$$\mathbf{A}^{-1} = \frac{1}{|\mathbf{A}|} \begin{bmatrix} a_{22} & -a_{12} \\ -a_{21} & a_{11} \end{bmatrix} \tag{2.30}$$

where $|\mathbf{A}| = a_{11}a_{22} - a_{12}a_{21}$.

If the determinant of $\mathbf{A}$ is equal to zero, the inverse of $\mathbf{A}$ does not exist. This is the case of a singular matrix. It can be verified that

$$(\mathbf{A}^{-1})^{\mathrm{T}} = (\mathbf{A}^{\mathrm{T}})^{-1} \tag{2.31}$$

which implies that the transpose of the inverse of a matrix is equal to the inverse of its transpose.

If $\mathbf{A}$ and $\mathbf{B}$ are nonsingular square matrices, then

$$(\mathbf{AB})^{-1} = \mathbf{B}^{-1}\mathbf{A}^{-1} \tag{2.32}$$

In general, the inverse of the product of square nonsingular matrices $\mathbf{A}_1, \mathbf{A}_2, \ldots, \mathbf{A}_{n-1}, \mathbf{A}_n$ is

$$(\mathbf{A}_1\mathbf{A}_2 \cdots \mathbf{A}_{n-1}\mathbf{A}_n)^{-1} = \mathbf{A}_n^{-1}\mathbf{A}_{n-1}^{-1} \cdots \mathbf{A}_2^{-1}\mathbf{A}_1^{-1} \tag{2.33}$$

This equation can be used to define the inverse of matrices that arise naturally in mechanics. One of these matrices that appears in the formulations of the *recursive equations* of mechanical systems is

$$\mathbf{D} = \begin{bmatrix} \mathbf{I} & \mathbf{0} & \mathbf{0} & \mathbf{0} & \cdots & \mathbf{0} & \mathbf{0} \\ -\mathbf{D}_2 & \mathbf{I} & \mathbf{0} & \mathbf{0} & \cdots & \mathbf{0} & \mathbf{0} \\ \mathbf{0} & -\mathbf{D}_3 & \mathbf{I} & \mathbf{0} & \cdots & \mathbf{0} & \mathbf{0} \\ \vdots & \vdots & \vdots & \vdots & & \ddots & \vdots \\ \mathbf{0} & \mathbf{0} & \mathbf{0} & \mathbf{0} & \cdots & -\mathbf{D}_n & \mathbf{I} \end{bmatrix} \tag{2.34}$$

The matrix $\mathbf{D}$ can be written as the product of $n - 1$ matrices as follows:

$$\mathbf{D} = \begin{bmatrix} \mathbf{I} &&&&& \\ -\mathbf{D}_2 & \mathbf{I} &&&& \\ & \mathbf{0} & \mathbf{I} &&& \\ && \mathbf{0} & \ddots && \\ &&& \ddots && \\ &&&& \mathbf{0} & \mathbf{I} \end{bmatrix} \begin{bmatrix} \mathbf{I} &&&&& \\ \mathbf{0} & \mathbf{I} &&&& \\ & -\mathbf{D}_3 & \mathbf{I} &&& \\ && \mathbf{0} & \ddots && \\ &&& \ddots && \\ &&&& \mathbf{0} & \mathbf{I} \end{bmatrix}$$

$$\cdots \begin{bmatrix} \mathbf{I} &&&&& \\ \mathbf{0} & \mathbf{I} &&&& \\ & \mathbf{0} & \mathbf{I} &&& \\ && \mathbf{0} & \ddots && \\ &&& \ddots && \\ &&&& -\mathbf{D}_n & \mathbf{I} \end{bmatrix} \tag{2.35}$$

from which

$$\mathbf{D}^{-1} = \begin{bmatrix} \mathbf{I} &&&&& \\ \mathbf{0} & \mathbf{I} &&&& \\ & \mathbf{0} & \mathbf{I} &&& \\ && \mathbf{0} & \ddots && \\ &&& \ddots && \\ &&&& \mathbf{D}_n & \mathbf{I} \end{bmatrix} \begin{bmatrix} \mathbf{I} &&&&& \\ \mathbf{0} & \mathbf{I} &&&& \\ & \mathbf{0} & \ddots &&& \\ && \ddots &&& \\ &&& \mathbf{D}_{n-1} & \mathbf{I} & \\ &&&& \mathbf{0} & \mathbf{I} \end{bmatrix}$$

$$\cdots \begin{bmatrix} \mathbf{I} &&&&& \\ \mathbf{D}_2 & \mathbf{I} &&&& \\ & \mathbf{0} & \mathbf{I} &&& \\ && \mathbf{0} & \ddots && \\ &&& \ddots && \\ &&&& \mathbf{0} & \mathbf{I} \end{bmatrix} \tag{2.36}$$

Therefore, the inverse of the matrix $\mathbf{D}$ can be written as

$$\mathbf{D}^{-1} = \begin{bmatrix} \mathbf{I} & \mathbf{0} & \mathbf{0} & \cdots & \mathbf{0} \\ \mathbf{D}_{21} & \mathbf{I} & \mathbf{0} & \cdots & \mathbf{0} \\ \mathbf{D}_{32} & \mathbf{D}_{31} & \mathbf{I} & \cdots & \mathbf{0} \\ \mathbf{D}_{43} & \mathbf{D}_{42} & \mathbf{D}_{41} & \cdots & \mathbf{0} \\ \vdots & \vdots & \vdots & \ddots & \vdots \\ \mathbf{D}_{n(n-1)} & \mathbf{D}_{n(n-2)} & \mathbf{D}_{n(n-3)} & \cdots & \mathbf{I} \end{bmatrix} \tag{2.37}$$

where

$$\mathbf{D}_{kr} = \mathbf{D}_k \mathbf{D}_{k-1} \cdots \mathbf{D}_{k-r+1} \tag{2.38}$$

Orthogonal Matrices A square matrix $\mathbf{A}$ is said to be *orthogonal* if

$$\mathbf{A}^{\mathrm{T}}\mathbf{A} = \mathbf{A}\mathbf{A}^{\mathrm{T}} = \mathbf{I} \tag{2.39}$$

In this case $\mathbf{A}^{\mathrm{T}} = \mathbf{A}^{-1}$.

That is, the inverse of an orthogonal matrix is equal to its transpose. An example of orthogonal matrices is

$$\mathbf{A} = \mathbf{I} + \tilde{\mathbf{v}} \sin\theta + (1 - \cos\theta)(\tilde{\mathbf{v}})^2 \tag{2.40}$$

where θ is an arbitrary angle, $\tilde{\mathbf{v}}$ is the skew-symmetric matrix

$$\tilde{\mathbf{v}} = \begin{bmatrix} 0 & -v_3 & v_2 \\ v_3 & 0 & -v_1 \\ -v_2 & v_1 & 0 \end{bmatrix} \tag{2.41}$$

and v_1, v_2, and v_3 are the components of an arbitrary unit vector $\mathbf{v}$, that is, $\mathbf{v} = [v_1\ v_2\ v_3]^{\mathrm{T}}$. While $\tilde{\mathbf{v}}$ is a skew-symmetric matrix, $(\tilde{\mathbf{v}})^2$ is a symmetric matrix. The transpose of the matrix $\mathbf{A}$ of Eq. 40 can then be written as

$$\mathbf{A}^{\mathrm{T}} = \mathbf{I} - \tilde{\mathbf{v}} \sin\theta + (1 - \cos\theta)(\tilde{\mathbf{v}})^2 \tag{2.42}$$

It can be shown that

$$(\tilde{\mathbf{v}})^3 = -\tilde{\mathbf{v}}, \qquad (\tilde{\mathbf{v}})^4 = -(\tilde{\mathbf{v}})^2 \tag{2.43}$$

Using these identities, one can verify that the matrix $\mathbf{A}$ of Eq. 40 is an orthogonal matrix. In addition to the orthogonality, it can be shown that the matrix $\mathbf{A}$ and the unit vector $\mathbf{v}$ satisfy the following relationships:

$$\mathbf{A}\mathbf{v} = \mathbf{A}^{\mathrm{T}}\mathbf{v} = \mathbf{A}^{-1}\mathbf{v} = \mathbf{v} \tag{2.44}$$

In computational dynamics, the elements of a matrix can be implicit or explicit functions of time. At a given instant of time, the values of the elements of such a matrix determine whether or not a matrix is singular. For example, consider the following two matrices, which depend on the three variables ϕ, θ, ψ:

$$\mathbf{G} = \begin{bmatrix} 0 & \cos\phi & \sin\theta\sin\phi \\ 0 & \sin\phi & -\sin\theta\cos\phi \\ 1 & 0 & \cos\theta \end{bmatrix} \tag{2.45}$$

and

$$\overline{\mathbf{G}} = \begin{bmatrix} \sin\theta\sin\psi & \cos\psi & 0 \\ \sin\theta\cos\psi & -\sin\psi & 0 \\ \cos\theta & 0 & 1 \end{bmatrix} \tag{2.46}$$

The inverses of these two matrices are given as

$$\mathbf{G}^{-1} = \frac{1}{\sin\theta}\begin{bmatrix} -\sin\phi\cos\theta & \cos\phi\cos\theta & \sin\theta \\ \sin\theta\cos\phi & \sin\theta\sin\phi & 0 \\ \sin\phi & -\cos\phi & 0 \end{bmatrix} \tag{2.47}$$

and

$$\overline{\mathbf{G}}^{-1} = \frac{1}{\sin\theta}\begin{bmatrix} \sin\psi & \cos\psi & 0 \\ \sin\theta\cos\psi & -\sin\theta\sin\psi & 0 \\ -\cos\theta\sin\psi & -\cos\theta\cos\psi & \sin\theta \end{bmatrix} \tag{2.48}$$

It is clear that these two inverses do not exist if $\sin\theta = 0$. The reader, however, can show that the matrix $\mathbf{A}$, defined as $\mathbf{A} = \mathbf{G}\overline{\mathbf{G}}^{-1}$ is an orthogonal matrix and its inverse does exist regardless of the value of θ.

2.3 VECTORS

An n-dimensional vector $\mathbf{a}$ is an ordered set

$$\mathbf{a} = (a_1, a_2, \ldots, a_n) \tag{2.49}$$

of n scalars. The scalar a_i, $i = 1, 2, \ldots, n$ is called the ith component of $\mathbf{a}$. An n-dimensional vector can be considered as an $n \times 1$ matrix that consists of only one column. Therefore, the vector $\mathbf{a}$ can be written in the following column form:

$$\mathbf{a} = \begin{bmatrix} a_1 \\ a_2 \\ \vdots \\ a_n \end{bmatrix} \tag{2.50}$$

The transpose of this column vector defines the n-dimensional row vector

$$\mathbf{a}^{\mathrm{T}} = [a_1 \quad a_2 \cdots a_n] \tag{2.51}$$

The vector $\mathbf{a}$ of Eq. 50 can also be written as

$$\mathbf{a} = [a_1 \quad a_2 \cdots a_n]^{\mathrm{T}} \tag{2.52}$$

By considering the vector as special case of a matrix with only one column or one row, the rules of matrix addition and multiplication apply also to vectors. For example, if $\mathbf{a}$ and $\mathbf{b}$ are two n-dimensional vectors defined as

$$\mathbf{a} = [a_1 \quad a_2 \cdots a_n]^{\mathrm{T}}, \quad \mathbf{b} = [b_1 \quad b_2 \cdots b_n]^{\mathrm{T}} \tag{2.53}$$

then $\mathbf{a} + \mathbf{b}$ is defined as

$$\mathbf{a} + \mathbf{b} = [a_1 + b_1 \quad a_2 + b_2 \cdots a_n + b_n]^{\mathrm{T}} \tag{2.54}$$

Two vectors $\mathbf{a}$ and $\mathbf{b}$ are equal if and only if $a_i = b_i$ for $i = 1, 2, \ldots, n$.

The *product* of a vector $\mathbf{a}$ and scalar α is the vector

$$\alpha\mathbf{a} = [\alpha a_1 \quad \alpha a_2 \cdots \alpha a_n]^{\mathrm{T}} \tag{2.55}$$

The *dot*, *inner*, or *scalar* product of two vectors $\mathbf{a} = [a_1 a_2 \cdots a_n]^{\mathrm{T}}$ and $\mathbf{b} = [b_1 \; b_2 \cdots b_n]^{\mathrm{T}}$ is defined by the following scalar quantity:

$$\begin{aligned} \mathbf{a} \cdot \mathbf{b} = \mathbf{a}^{\mathrm{T}}\mathbf{b} &= [a_1 \quad a_2 \cdots a_n] \begin{bmatrix} b_1 \\ b_2 \\ \vdots \\ b_n \end{bmatrix} \\ &= a_1 b_1 + a_2 b_2 + \cdots + a_n b_n = \sum_{i=1}^{n} a_i b_i \end{aligned} \tag{2.56}$$

It follows that $\mathbf{a} \cdot \mathbf{b} = \mathbf{b} \cdot \mathbf{a}$.

Two vectors $\mathbf{a}$ and $\mathbf{b}$ are said to be *orthogonal* if their dot product is equal to zero, that is, $\mathbf{a} \cdot \mathbf{b} = \mathbf{a}^{\mathrm{T}}\mathbf{b} = 0$. The *length* of a vector $\mathbf{a}$ denoted as $|\mathbf{a}|$ is defined as the square root of the dot product of $\mathbf{a}$ with itself, that is,

$$|\mathbf{a}| = \sqrt{\mathbf{a}^{\mathrm{T}}\mathbf{a}} = [(a_1)^2 + (a_2)^2 + \cdots + (a_n)^2]^{1/2} \tag{2.57}$$

The terms *modulus*, *magnitude*, *norm*, and *absolute value* of a vector are also used to denote the length of a vector. A *unit vector* is defined to be a vector that has length equal to 1. If $\hat{\mathbf{a}}$ is a unit vector, one must have

$$|\hat{\mathbf{a}}| = [(\hat{a}_1)^2 + (\hat{a}_2)^2 + \cdots + (\hat{a}_n)^2]^{1/2} = 1 \tag{2.58}$$

If $\mathbf{a} = [a_1\ a_2\ \cdots\ a_n]^{\mathrm{T}}$ is an arbitrary vector, a unit vector $\hat{\mathbf{a}}$ collinear with the vector $\mathbf{a}$ is defined by

$$\hat{\mathbf{a}} = \frac{\mathbf{a}}{|\mathbf{a}|} = \frac{1}{|\mathbf{a}|}[a_1 \quad a_2 \ \cdots \ a_n]^{\mathrm{T}} \tag{2.59}$$

Example 2.4

Let $\mathbf{a}$ and $\mathbf{b}$ be the two vectors

$$\mathbf{a} = [0 \quad 1 \quad 3 \quad 2]^{\mathrm{T}}, \qquad \mathbf{b} = [-1 \quad 0 \quad 2 \quad 3]^{\mathrm{T}}$$

Then

$$\begin{aligned} \mathbf{a} + \mathbf{b} &= [0 \quad 1 \quad 3 \quad 2]^{\mathrm{T}} + [-1 \quad 0 \quad 2 \quad 3]^{\mathrm{T}} \\ &= [-1 \quad 1 \quad 5 \quad 5]^{\mathrm{T}} \end{aligned}$$

The dot product of $\mathbf{a}$ and $\mathbf{b}$ is

$$\begin{aligned} \mathbf{a} \cdot \mathbf{b} = \mathbf{a}^{\mathrm{T}}\mathbf{b} &= [0 \quad 1 \quad 3 \quad 2] \begin{bmatrix} -1 \\ 0 \\ 2 \\ 3 \end{bmatrix} \\ &= 0 + 0 + 6 + 6 = 12 \end{aligned}$$

Unit vectors along $\mathbf{a}$ and $\mathbf{b}$ are

$$\begin{aligned} \hat{\mathbf{a}} &= \frac{\mathbf{a}}{|\mathbf{a}|} = \frac{1}{\sqrt{14}}[0 \quad 1 \quad 3 \quad 2]^{\mathrm{T}} \\ \hat{\mathbf{b}} &= \frac{\mathbf{b}}{|\mathbf{b}|} = \frac{1}{\sqrt{14}}[-1 \quad 0 \quad 2 \quad 3]^{\mathrm{T}} \end{aligned}$$

It can be easily verified that $|\hat{\mathbf{a}}| = |\hat{\mathbf{b}}| = 1$.

Differentiation In many applications in mechanics, scalar and vector functions that depend on one or more variables are encountered. An example of a scalar function that depends on the system velocities and possibly on the system coordinates is the kinetic energy. Examples of vector functions are the coordinates, velocities, and accelerations that depend on time. Let us first consider a scalar function f that depends on several variables $q_1, q_2, \ldots,$ and q_n and the parameter t, such that

$$f = f(q_1, q_2, \ \ldots \ q_n, t) \tag{2.60}$$

where $q_1, q_2, \ldots, q_n$ are functions of t, that is, $q_i = q_i(t)$.

The total derivative of f with respect to the parameter t is

$$\frac{df}{dt} = \frac{\partial f}{\partial q_1}\frac{dq_1}{dt} + \frac{\partial f}{\partial q_2}\frac{dq_2}{dt} + \cdots + \frac{\partial f}{\partial q_n}\frac{dq_n}{dt} + \frac{\partial f}{\partial t} \tag{2.61}$$

which can be written using vector notation as

$$\frac{df}{dt} = \begin{bmatrix} \frac{\partial f}{\partial q_1} & \frac{\partial f}{\partial q_2} & \cdots & \frac{\partial f}{\partial q_n} \end{bmatrix} \begin{bmatrix} \frac{dq_1}{dt} \\ \frac{dq_2}{dt} \\ \vdots \\ \frac{dq_n}{dt} \end{bmatrix} + \frac{\partial f}{\partial t} \tag{2.62}$$

This equation can be written as

$$\frac{df}{dt} = \frac{\partial f}{\partial \mathbf{q}}\frac{d\mathbf{q}}{dt} + \frac{\partial f}{\partial t} \tag{2.63}$$

in which $\partial f/\partial t$ is the partial derivative of f with respect to t, and

$$\begin{aligned} \mathbf{q} &= [q_1 \quad q_2 \cdots q_n]^{\mathrm{T}} \\ \frac{\partial f}{\partial \mathbf{q}} &= f_{\mathbf{q}} = \begin{bmatrix} \frac{\partial f}{\partial q_1} & \frac{\partial f}{\partial q_2} & \cdots & \frac{\partial f}{\partial q_n} \end{bmatrix} \end{aligned} \tag{2.64}$$

That is, the partial derivative of a scalar function with respect to a vector is a row vector. If f is not an explicit function of t, $\partial f/\partial t = 0$.

Example 2.5

Consider the function

$$f(q_1, q_2, t) = (q_1)^2 + 3(q_2)^3 - (t)^2$$

where q_1 and q_2 are functions of the parameter t. The total derivative of f with respect to the parameter t is

$$\frac{df}{dt} = \frac{\partial f}{\partial q_1}\frac{dq_1}{dt} + \frac{\partial f}{\partial q_2}\frac{dq_2}{dt} + \frac{\partial f}{\partial t}$$

where

$$\frac{\partial f}{\partial q_1} = 2q_1, \qquad \frac{\partial f}{\partial q_2} = 9(q_2)^2, \qquad \frac{\partial f}{\partial t} = -2t$$

Hence

$$\frac{df}{dt} = 2q_1 \frac{dq_1}{dt} + 9(q_2)^2 \frac{dq_2}{dt} - 2t$$

$$= [2q_1 \quad 9(q_2)^2] \begin{bmatrix} \dfrac{dq_1}{dt} \\ \dfrac{dq_2}{dt} \end{bmatrix} - 2t$$

where $\partial f/\partial \mathbf{q}$ can be recognized as the row vector

$$\frac{\partial f}{\partial \mathbf{q}} = f_{\mathbf{q}} = [2q_1 \quad 9(q_2)^2]$$

Consider the case of several functions that depend on several variables. These functions can be written as

$$\left. \begin{aligned} f_1 &= f_1(q_1, q_2, \ldots, q_n, t) \\ f_2 &= f_2(q_1, q_2, \ldots, q_n, t) \\ &\vdots \\ f_m &= f_m(q_1, q_2, \ldots, q_n, t) \end{aligned} \right\} \tag{2.65}$$

where $q_i = q_i(t)$, $i = 1, 2, \ldots, n$. Using the procedure previously outlined in this section, the total derivative of an arbitrary function f_j can be written as

$$\frac{df_j}{dt} = \frac{\partial f_j}{\partial \mathbf{q}} \frac{d\mathbf{q}}{dt} + \frac{\partial f_j}{\partial t} \qquad j = 1, 2, \ldots, m \tag{2.66}$$

in which $\partial f_j/\partial \mathbf{q}$ is the row vector

$$\frac{\partial f_j}{\partial \mathbf{q}} = \left[\frac{\partial f_j}{\partial q_1} \; \frac{\partial f_j}{\partial q_2} \cdots \frac{\partial f_j}{\partial q_n} \right] \tag{2.67}$$

It follows that

$$\frac{d\mathbf{f}}{dt} = \begin{bmatrix} \dfrac{df_1}{dt} \\ \dfrac{df_2}{dt} \\ \vdots \\ \dfrac{df_m}{dt} \end{bmatrix} = \begin{bmatrix} \dfrac{\partial f_1}{\partial q_1} & \dfrac{\partial f_1}{\partial q_2} & \cdots & \dfrac{\partial f_1}{\partial q_n} \\ \dfrac{\partial f_2}{\partial q_1} & \dfrac{\partial f_2}{\partial q_2} & \cdots & \dfrac{\partial f_2}{\partial q_n} \\ \vdots & \vdots & \ddots & \vdots \\ \dfrac{\partial f_m}{\partial q_1} & \dfrac{\partial f_m}{\partial q_2} & \cdots & \dfrac{\partial f_m}{\partial q_n} \end{bmatrix} \begin{bmatrix} \dfrac{dq_1}{dt} \\ \dfrac{dq_2}{dt} \\ \vdots \\ \dfrac{dq_n}{dt} \end{bmatrix} + \begin{bmatrix} \dfrac{\partial f_1}{\partial t} \\ \dfrac{\partial f_2}{\partial t} \\ \vdots \\ \dfrac{\partial f_m}{\partial t} \end{bmatrix} \tag{2.68}$$

where

$$\mathbf{f} = [f_1 \quad f_2 \cdots f_m]^{\mathrm{T}} \tag{2.69}$$

Equation 68 can also be written as

$$\frac{d\mathbf{f}}{dt} = \frac{\partial \mathbf{f}}{\partial \mathbf{q}} \frac{d\mathbf{q}}{dt} + \frac{\partial \mathbf{f}}{\partial t} \tag{2.70}$$

where the $m \times n$ matrix $\partial \mathbf{f}/\partial \mathbf{q}$, the n-dimensional vector $d\mathbf{q}/dt$, and the m-dimensional vector $\partial \mathbf{f}/\partial t$ can be recognized as

$$\frac{\partial \mathbf{f}}{\partial \mathbf{q}} = \mathbf{f}_{\mathbf{q}} = \begin{bmatrix} \frac{\partial f_1}{\partial q_1} & \frac{\partial f_1}{\partial q_2} & \cdots & \frac{\partial f_1}{\partial q_n} \\ \frac{\partial f_2}{\partial q_1} & \frac{\partial f_2}{\partial q_2} & \cdots & \frac{\partial f_2}{\partial q_n} \\ \vdots & \vdots & \ddots & \vdots \\ \frac{\partial f_m}{\partial q_1} & \frac{\partial f_m}{\partial q_2} & \cdots & \frac{\partial f_m}{\partial q_n} \end{bmatrix} \tag{2.71}$$

$$\frac{d\mathbf{q}}{dt} = \begin{bmatrix} \frac{dq_1}{dt} & \frac{dq_2}{dt} & \cdots & \frac{dq_n}{dt} \end{bmatrix}^{\mathrm{T}} \tag{2.72}$$

$$\frac{\partial \mathbf{f}}{\partial t} = \mathbf{f}_t = \begin{bmatrix} \frac{\partial f_1}{\partial t} & \frac{\partial f_2}{\partial t} & \cdots & \frac{\partial f_m}{\partial t} \end{bmatrix}^{\mathrm{T}} \tag{2.73}$$

If the function f_j is not an explicit function of the parameter t, then $\partial f_j/\partial t$ is equal to zero. Note also that the partial derivative of an m-dimensional vector function $\mathbf{f}$ with respect to an n-dimensional vector $\mathbf{q}$ is the $m \times n$ matrix $\mathbf{f}_{\mathbf{q}}$ defined by Eq. 71.

Example 2.6

Consider the vector function $\mathbf{f}$ defined as

$$\mathbf{f} = \begin{bmatrix} f_1 \\ f_2 \\ f_3 \end{bmatrix} = \begin{bmatrix} (q_1)^2 + 3(q_2)^3 - (t)^2 \\ 8(q_1)^2 - 3t \\ 2(q_1)^2 - 6q_1q_2 + (q_2)^2 \end{bmatrix}$$

The total derivative of the vector function $\mathbf{f}$ is

$$\frac{d\mathbf{f}}{dt} = \begin{bmatrix} \frac{df_1}{dt} \\ \frac{df_2}{dt} \\ \frac{df_3}{dt} \end{bmatrix} = \begin{bmatrix} 2q_1 & 9(q_2)^2 \\ 16q_1 & 0 \\ (4q_1 - 6q_2) & (2q_2 - 6q_1) \end{bmatrix} \begin{bmatrix} \frac{dq_1}{dt} \\ \frac{dq_2}{dt} \end{bmatrix} + \begin{bmatrix} -2t \\ -3 \\ 0 \end{bmatrix}$$

where the matrix $\mathbf{f_q}$ can be recognized as

$$\mathbf{f_q} = \begin{bmatrix} 2q_1 & 9(q_2)^2 \\ 16q_1 & 0 \\ (4q_1 - 6q_2) & (2q_2 - 6q_1) \end{bmatrix}$$

and the vector $\mathbf{f}_t$ is

$$\frac{\partial \mathbf{f}}{\partial t} = \mathbf{f}_t = [-2t \quad -3 \quad 0]^{\mathrm{T}}$$

In the analysis of mechanical systems, we may also encounter scalar functions in the form

$$Q = \mathbf{q}^{\mathrm{T}}\mathbf{A}\mathbf{q} \tag{2.74}$$

Following a similar procedure to the one outlined previously in this section, one can show that

$$\frac{\partial Q}{\partial \mathbf{q}} = \mathbf{q}^{\mathrm{T}}(\mathbf{A} + \mathbf{A}^{\mathrm{T}}) \tag{2.75}$$

If $\mathbf{A}$ is a symmetric matrix, that is $\mathbf{A} = \mathbf{A}^{\mathrm{T}}$, one has

$$\frac{\partial Q}{\partial \mathbf{q}} = 2\mathbf{q}^{\mathrm{T}}\mathbf{A} \tag{2.76}$$

Linear Independence The vectors $\mathbf{a}_1, \mathbf{a}_2, \ldots, \mathbf{a}_n$ are said to be *linearly dependent* if there exist scalars $e_1, e_2, \ldots, e_n$, which are not all zeros, such that

$$e_1\mathbf{a}_1 + e_2\mathbf{a}_2 + \cdots + e_n\mathbf{a}_n = \mathbf{0} \tag{2.77}$$

Otherwise, the vectors $\mathbf{a}_1, \mathbf{a}_2, \ldots, \mathbf{a}_n$ are said to be *linearly independent*. Observe that in the case of linearly independent vectors, not one of these vectors can be expressed in terms of the others. On the other hand, if Eq. 77 holds, and not all the scalars $e_1, e_2, \ldots, e_n$ are equal to zeros, one or more of the vectors $\mathbf{a}_1, \mathbf{a}_2, \ldots, \mathbf{a}_n$ can be expressed in terms of the other vectors.

Equation 77 can be written in a matrix form as

$$[\mathbf{a}_1 \quad \mathbf{a}_2 \quad \cdots \quad \mathbf{a}_n] \begin{bmatrix} e_1 \\ e_2 \\ \vdots \\ e_n \end{bmatrix} = \mathbf{0} \tag{2.78}$$

which can also be written as

$$\mathbf{A}\mathbf{e} = \mathbf{0} \tag{2.79}$$

in which

$$\mathbf{A} = [\mathbf{a}_1 \quad \mathbf{a}_2 \quad \cdots \quad \mathbf{a}_n] \tag{2.80}$$

If the vectors $\mathbf{a}_1, \mathbf{a}_2, \ldots, \mathbf{a}_n$ are linearly dependent, the system of homogeneous algebraic equations defined by Eq. 79 has a nontrivial solution. On the other hand, if the vectors $\mathbf{a}_1, \mathbf{a}_2, \ldots, \mathbf{a}_n$ are linearly independent vectors, then $\mathbf{A}$ must be a nonsingular matrix since the system of homogeneous algebraic equations defined by Eq. 79 has only the trivial solution $\mathbf{e} = \mathbf{A}^{-1}\mathbf{0} = \mathbf{0}$. In the case where the vectors $\mathbf{a}_1, \mathbf{a}_2, \ldots, \mathbf{a}_n$ are linearly dependent, the square matrix $\mathbf{A}$ must be singular. The number of linearly independent columns in a matrix is called the *column rank* of the matrix. Similarly, the number of independent rows is called the *row rank* of the matrix. It can be shown that for any matrix, the row rank is equal to the column rank is equal to the *rank* of the matrix. Therefore, a square matrix that has a *full rank* is a matrix that has linearly independent rows and linearly independent columns. Thus, we conclude that a matrix that has a full rank is a nonsingular matrix.

If $\mathbf{a}_1, \mathbf{a}_2, \ldots, \mathbf{a}_n$ are n-dimensional linearly independent vectors, any other n-dimensional vector can be expressed as a linear combination of these vectors. For instance, let $\mathbf{b}$ be another n-dimensional vector. We show that this vector has a unique representation in terms of the linearly independent vectors $\mathbf{a}_1, \mathbf{a}_2, \ldots, \mathbf{a}_n$. To this end, we write $\mathbf{b}$ as

$$\mathbf{b} = x_1\mathbf{a}_1 + x_2\mathbf{a}_2 + \cdots + x_n\mathbf{a}_n \tag{2.81}$$

where $x_1, x_2, \ldots,$ and x_n are scalars. In order to show that $x_1, x_2, \ldots,$ and x_n are unique, Eq. 81 can be written as

$$\mathbf{b} = [\mathbf{a}_1 \quad \mathbf{a}_2 \quad \cdots \quad \mathbf{a}_n] \begin{bmatrix} x_1 \\ x_2 \\ \vdots \\ x_n \end{bmatrix} \tag{2.82}$$

which can be written as

$$\mathbf{b} = \mathbf{A}\mathbf{x} \tag{2.83}$$

where $\mathbf{A}$ is a square matrix defined by Eq. 80 and $\mathbf{x}$ is the vector

$$\mathbf{x} = [x_1 \quad x_2 \quad \cdots \quad x_n]^{\mathrm{T}} \tag{2.84}$$

Since the vectors $\mathbf{a}_1, \mathbf{a}_2, \ldots, \mathbf{a}_n$ are assumed to be linearly independent, the coefficient matrix $\mathbf{A}$ in Eq. 83 has a full row rank, and thus it is nonsingular. This system of algebraic equations has a unique solution $\mathbf{x}$, which can be written as $\mathbf{x} = \mathbf{A}^{-1}\mathbf{b}$. That is, an arbitrary n-dimensional vector $\mathbf{b}$ has a unique representation in terms of the linearly independent vectors $\mathbf{a}_1, \mathbf{a}_2, \ldots, \mathbf{a}_n$.

A familiar and important special case is the case of three-dimensional vectors. One can show that the three vectors

$$\mathbf{a}_1 = \begin{bmatrix} 1 \\ 0 \\ 0 \end{bmatrix}, \quad \mathbf{a}_2 = \begin{bmatrix} 0 \\ 1 \\ 0 \end{bmatrix}, \quad \mathbf{a}_3 = \begin{bmatrix} 0 \\ 0 \\ 1 \end{bmatrix} \tag{2.85}$$

are linearly independent. Any other three-dimensional vector $\mathbf{b} = [b_1 \ b_2 \ b_3]^{\mathrm{T}}$ can be written in terms of the linearly independent vectors $\mathbf{a}_1$, $\mathbf{a}_2$, and $\mathbf{a}_3$ as

$$\mathbf{b} = b_1\mathbf{a}_1 + b_2\mathbf{a}_2 + b_3\mathbf{a}_3 \tag{2.86}$$

where the coefficients x_1, x_2, and x_3 can be recognized in this special case as

$$x_1 = b_1, \qquad x_2 = b_2, \qquad x_3 = b_3 \tag{2.87}$$

The coefficients x_1, x_2, and x_3 are called the *coordinates* of the vector $\mathbf{b}$ in the *basis* defined by the vectors $\mathbf{a}_1$, $\mathbf{a}_2$, and $\mathbf{a}_3$.

Example 2.7

Show that the vectors

$$\mathbf{a}_1 = \begin{bmatrix} 1 \\ 0 \\ 0 \end{bmatrix}, \quad \mathbf{a}_2 = \begin{bmatrix} 1 \\ 1 \\ 0 \end{bmatrix}, \quad \mathbf{a}_3 = \begin{bmatrix} 1 \\ 1 \\ 1 \end{bmatrix}$$

are linearly independent. Find also the representation of the vector $\mathbf{b} = [-1 \ 3 \ 0]^{\mathrm{T}}$ in terms of the vectors $\mathbf{a}_1$, $\mathbf{a}_2$, and $\mathbf{a}_3$.

Solution. In order to show that the vectors $\mathbf{a}_1$, $\mathbf{a}_2$, and $\mathbf{a}_3$ are linearly independent, we must show that the relationship

$$e_1\mathbf{a}_1 + e_2\mathbf{a}_2 + e_3\mathbf{a}_3 = \mathbf{0}$$

holds only when $e_1 = e_2 = e_3 = 0$. To show this, we write

$$e_1 \begin{bmatrix} 1 \\ 0 \\ 0 \end{bmatrix} + e_2 \begin{bmatrix} 1 \\ 1 \\ 0 \end{bmatrix} + e_3 \begin{bmatrix} 1 \\ 1 \\ 1 \end{bmatrix} = \mathbf{0}$$

which leads to

$$\begin{aligned} e_1 + e_2 + e_3 &= 0 \\ e_2 + e_3 &= 0 \\ e_3 &= 0 \end{aligned}$$

Back substitution shows that

$$e_3 = e_2 = e_1 = 0$$

which implies that the vectors $\mathbf{a}_1$, $\mathbf{a}_2$, and $\mathbf{a}_3$ are linearly independent.

To find the unique representation of the vector $\mathbf{b}$ in terms of these linearly independent vectors, we write

$$\mathbf{b} = x_1\mathbf{a}_1 + x_2\mathbf{a}_2 + x_3\mathbf{a}_3$$

which can be written in matrix form as

$$\mathbf{b} = \mathbf{A}\mathbf{x}$$

where

$$\mathbf{A} = \begin{bmatrix} 1 & 1 & 1 \\ 0 & 1 & 1 \\ 0 & 0 & 1 \end{bmatrix}, \qquad \mathbf{b} = \begin{bmatrix} -1 \\ 3 \\ 0 \end{bmatrix}$$

Hence, the coordinate vector $\mathbf{x}$ can be obtained as

$$\mathbf{x} = \begin{bmatrix} x_1 \\ x_2 \\ x_3 \end{bmatrix} = \mathbf{A}^{-1}\mathbf{b} = \begin{bmatrix} 1 & -1 & 0 \\ 0 & 1 & -1 \\ 0 & 0 & 1 \end{bmatrix} \begin{bmatrix} -1 \\ 3 \\ 0 \end{bmatrix} = \begin{bmatrix} -4 \\ 3 \\ 0 \end{bmatrix}$$

2.4 THREE-DIMENSIONAL VECTORS

A special case of n-dimensional vectors is the three-dimensional vector. Three-dimensional vectors are important in mechanics because the position, velocity, and acceleration of a particle or an arbitrary point on a rigid or deformable body can be described in space using three-dimensional vectors. Since these vectors are a special case of the more general n-dimensional vectors, the rules of vector additions, dot products, scalar multiplications, and differentiations of these vectors are the same as discussed in the preceding section.

Cross Product Consider the three-dimensional vectors $\mathbf{a} = [a_1\ a_2\ a_3]^{\mathrm{T}}$, and $\mathbf{b} = [b_1\ b_2\ b_3]^{\mathrm{T}}$. These vectors can be defined by their components in the three-dimensional space XYZ. Therefore, the vectors $\mathbf{a}$ and $\mathbf{b}$ can be written in terms of their components along the X, Y, and Z axes as

$$\left.\begin{aligned} \mathbf{a} &= a_1\mathbf{i} + a_2\mathbf{j} + a_3\mathbf{k} \\ \mathbf{b} &= b_1\mathbf{i} + b_2\mathbf{j} + b_3\mathbf{k} \end{aligned}\right\} \tag{2.88}$$

where $\mathbf{i}$, $\mathbf{j}$, and $\mathbf{k}$ are unit vectors defined along the X, Y, and Z axes, respectively.

The *cross* or *vector product* of the vectors $\mathbf{a}$ and $\mathbf{b}$ is another vector $\mathbf{c}$ orthogonal to both $\mathbf{a}$ and $\mathbf{b}$ and is defined as

$$\begin{aligned} \mathbf{c} = \mathbf{a} \times \mathbf{b} &= \begin{vmatrix} \mathbf{i} & \mathbf{j} & \mathbf{k} \\ a_1 & a_2 & a_3 \\ b_1 & b_2 & b_3 \end{vmatrix} \\ &= (a_2b_3 - a_3b_2)\mathbf{i} + (a_3b_1 - a_1b_3)\mathbf{j} + (a_1b_2 - a_2b_1)\mathbf{k} \end{aligned} \tag{2.89}$$

which can also be written as

$$\mathbf{c} = \begin{bmatrix} c_1 \\ c_2 \\ c_3 \end{bmatrix} = \mathbf{a} \times \mathbf{b} = \begin{bmatrix} a_2b_3 - a_3b_2 \\ a_3b_1 - a_1b_3 \\ a_1b_2 - a_2b_1 \end{bmatrix} \tag{2.90}$$

This vector satisfies the following orthogonality relationships:

$$\mathbf{a} \cdot \mathbf{c} = \mathbf{a}^{\mathrm{T}}\mathbf{c} = 0, \quad \mathbf{b} \cdot \mathbf{c} = \mathbf{b}^{\mathrm{T}}\mathbf{c} = 0 \tag{2.91}$$

It can also be shown that

$$\mathbf{c} = \mathbf{a} \times \mathbf{b} = -\mathbf{b} \times \mathbf{a} \tag{2.92}$$

If $\mathbf{a}$ and $\mathbf{b}$ are parallel vectors, it can be shown that $\mathbf{c} = \mathbf{a} \times \mathbf{b} = \mathbf{0}$. It follows that $\mathbf{a} \times \mathbf{a} = \mathbf{0}$. If $\mathbf{a}$ and $\mathbf{b}$ are two orthogonal vectors, that is, $\mathbf{a}^{\mathrm{T}}\mathbf{b} = 0$, it can be shown that $|\mathbf{c}| = |\mathbf{a}||\mathbf{b}|$. The following useful identities can also be verified:

$$\left.\begin{aligned} \mathbf{a} \cdot (\mathbf{b} \times \mathbf{c}) &= (\mathbf{a} \times \mathbf{b}) \cdot \mathbf{c} \\ \mathbf{a} \times (\mathbf{b} \times \mathbf{c}) &= (\mathbf{a}^{\mathrm{T}}\mathbf{c})\mathbf{b} - (\mathbf{a}^{\mathrm{T}}\mathbf{b})\mathbf{c} \end{aligned}\right\} \tag{2.93}$$

Example 2.8

Let $\mathbf{a}$ and $\mathbf{b}$ be the three-dimensional vectors

$$\mathbf{a} = [0 \quad -5 \quad 1]^{\mathrm{T}}$$
$$\mathbf{b} = [1 \quad -2 \quad 3]^{\mathrm{T}}$$

The cross product of $\mathbf{a}$ and $\mathbf{b}$ is

$$\mathbf{c} = \mathbf{a} \times \mathbf{b} = \begin{vmatrix} \mathbf{i} & \mathbf{j} & \mathbf{k} \\ a_1 & a_2 & a_3 \\ b_1 & b_2 & b_3 \end{vmatrix} = \begin{vmatrix} \mathbf{i} & \mathbf{j} & \mathbf{k} \\ 0 & -5 & 1 \\ 1 & -2 & 3 \end{vmatrix}$$
$$= -13\mathbf{i} + \mathbf{j} + 5\mathbf{k}$$

The vector $\mathbf{c}$ can then be defined as

$$\mathbf{c} = [-13 \quad 1 \quad 5]^{\mathrm{T}}$$

It is clear that

$$\mathbf{c}^{\mathrm{T}}\mathbf{a} = \mathbf{c}^{\mathrm{T}}\mathbf{b} = 0, \quad \mathbf{a} \times \mathbf{b} = -\mathbf{b} \times \mathbf{a}$$

Skew-Symmetric Matrix Representation The vector cross product as defined by Eq. 90 can be represented using matrix notation. By using Eq. 90, one can write $\mathbf{a} \times \mathbf{b}$ as

$$\mathbf{a} \times \mathbf{b} = \begin{bmatrix} a_2b_3 - a_3b_2 \\ a_3b_1 - a_1b_3 \\ a_1b_2 - a_2b_1 \end{bmatrix} = \begin{bmatrix} 0 & -a_3 & a_2 \\ a_3 & 0 & -a_1 \\ -a_2 & a_1 & 0 \end{bmatrix} \begin{bmatrix} b_1 \\ b_2 \\ b_3 \end{bmatrix} \tag{2.94}$$

which can be written as

$$\mathbf{a} \times \mathbf{b} = \tilde{\mathbf{a}}\mathbf{b} \tag{2.95}$$

where $\tilde{\mathbf{a}}$ is the skew-symmetric matrix associated with the vector $\mathbf{a}$ and defined as

$$\tilde{\mathbf{a}} = \begin{bmatrix} 0 & -a_3 & a_2 \\ a_3 & 0 & -a_1 \\ -a_2 & a_1 & 0 \end{bmatrix} \tag{2.96}$$

Similarly, the cross product $\mathbf{b} \times \mathbf{a}$ can be written in a matrix form as

$$\mathbf{b} \times \mathbf{a} = -\mathbf{a} \times \mathbf{b} = \tilde{\mathbf{b}}\mathbf{a} \tag{2.97}$$

where $\tilde{\mathbf{b}}$ is the skew-symmetric matrix associated with the vector $\mathbf{b}$ and is defined as

$$\tilde{\mathbf{b}} = \begin{bmatrix} 0 & -b_3 & b_2 \\ b_3 & 0 & -b_1 \\ -b_2 & b_1 & 0 \end{bmatrix} \tag{2.98}$$

If $\hat{\mathbf{a}}$ is a unit vector along the vector $\mathbf{a}$, it is clear that $\hat{\mathbf{a}} \times \mathbf{a} = -\mathbf{a} \times \hat{\mathbf{a}} = \mathbf{0}$. It follows that $-\tilde{\mathbf{a}}\hat{\mathbf{a}} = \tilde{\mathbf{a}}^{\mathrm{T}}\hat{\mathbf{a}} = \mathbf{0}$.

In some of the developments presented in this book, the constraints that represent mechanical joints in the system can be expressed using a set of algebraic equations. Quite often, one encounters a system of equations that can be written in the following form:

$$\mathbf{a} \times \mathbf{x} = \mathbf{0} \tag{2.99}$$

where $\mathbf{a} = [a_1 \; a_2 \; a_3]^{\mathrm{T}}$ and $\mathbf{x} = [x_1 \; x_2 \; x_3]^{\mathrm{T}}$. Using the notation of the skew symmetric matrices, Eq. 99 can be written as

$$\tilde{\mathbf{a}}\mathbf{x} = \mathbf{0}, \tag{2.100}$$

where $\tilde{\mathbf{a}}$ is defined by Eq. 96. Equation 100 leads to the following three algebraic equations:

$$\left.\begin{aligned} a_2x_3 - a_3x_2 &= 0 \\ a_3x_1 - a_1x_3 &= 0 \\ a_1x_2 - a_2x_1 &= 0 \end{aligned}\right\} \tag{2.101}$$

These three equations are not independent because, for instance, adding a_1/a_3 times the first equation to a_2/a_3 times the second equation leads to the third equation. That is, the system of equations given by Eq. 99, or equivalently Eq. 100, has at most two independent equations. This is due primarily to the fact that the skew-symmetric matrix $\tilde{\mathbf{a}}$ of Eq. 96 is singular and its rank is at most two.

Example 2.9

Let **a** and **b** be the three-dimensional vectors

$$\mathbf{a} = [-1 \quad 7 \quad 1]^{\mathrm{T}}, \quad \mathbf{b} = [0 \quad -3 \quad 8]^{\mathrm{T}}$$

Determine the skew-symmetric matrices $\tilde{\mathbf{a}}$ and $\tilde{\mathbf{b}}$ associated, respectively, with the vectors **a** and **b** and evaluate the cross product $\mathbf{a} \times \mathbf{b}$.

Solution. The skew-symmetric matrices $\tilde{\mathbf{a}}$ and $\tilde{\mathbf{b}}$ are

$$\tilde{\mathbf{a}} = \begin{bmatrix} 0 & -1 & 7 \\ 1 & 0 & 1 \\ -7 & -1 & 0 \end{bmatrix}, \quad \tilde{\mathbf{b}} = \begin{bmatrix} 0 & -8 & -3 \\ 8 & 0 & 0 \\ 3 & 0 & 0 \end{bmatrix}$$

The cross product $\mathbf{a} \times \mathbf{b}$ can be written as

$$\mathbf{a} \times \mathbf{b} = \tilde{\mathbf{a}}\mathbf{b} = \begin{bmatrix} 0 & -1 & 7 \\ 1 & 0 & 1 \\ -7 & -1 & 0 \end{bmatrix} \begin{bmatrix} 0 \\ -3 \\ 8 \end{bmatrix} = \begin{bmatrix} 59 \\ 8 \\ 3 \end{bmatrix}$$

Example 2.10

Solve the system of equations $\tilde{\mathbf{a}}\mathbf{x} = \mathbf{0}$, where **a** is the vector $\mathbf{a} = [-1 \quad 7 \quad 1]^{\mathrm{T}}$.

Solution. As pointed out in this section, the system of equations $\tilde{\mathbf{a}}\mathbf{x} = \mathbf{0}$ has only two independent equations since the rank of the skew-symmetric matrix $\tilde{\mathbf{a}}$ is at most two. Consequently, this system of equations has a nontrivial solution that can be determined to within an arbitrary constant. The equation $\tilde{\mathbf{a}}\mathbf{x} = \mathbf{0}$ can be written explicitly as

$$\begin{aligned} a_2x_3 - a_3x_2 &= 0 \\ a_3x_1 - a_1x_3 &= 0 \\ a_1x_2 - a_2x_1 &= 0 \end{aligned}$$

Since this system has only two independent equations, we can determine x_2 and x_3 in terms of x_1. This leads to

$$x_2 = \frac{a_2}{a_1}x_1, \quad x_3 = \frac{a_3}{a_1}x_1$$

This solution satisfies the three algebraic equations, and for a given value of x_1, the other two variables x_2 and x_3 can be determined. Using the components of the vector **a**, we have

$$x_2 = \frac{a_2}{a_1} x_1 = -7x_1, \quad x_3 = \frac{a_3}{a_1} = -x_1$$

Therefore, the solution vector $\mathbf{x}$ is

$$\mathbf{x} = \begin{bmatrix} 1 \\ -7 \\ -1 \end{bmatrix} x_1$$

Cartesian Coordinate System In spatial dynamics, several sets of orientation coordinates can be used to describe the three-dimensional rotations. Some of these orientation coordinates, as will be demonstrated in Chapter 7, lack any clear physical meaning, making it difficult in many applications to define the initial configuration of the bodies using these coordinates. One method which is used in computational dynamics to define a Cartesian coordinate system is to introduce three points on the rigid body and use the vector cross product to define the location and orientation of the body coordinate system in the three-dimensional space. To illustrate the procedure for using the vector cross product to achieve this goal, we consider body i which has a coordinate system $X^iY^iZ^i$ with its origin at point O^i as shown in Fig. 1. Two other points P^i and Q^i are defined such that point P^i lies on the X^i axis and point Q^i lies in the X^iY^i plane. If the position vectors of the three points

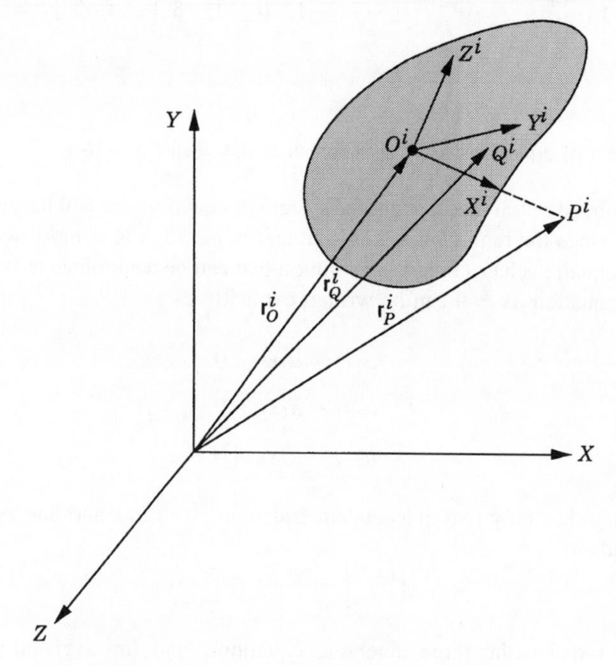

Figure 2.1 Cartesian coordinate system

O^i, P^i, and Q^i are known and defined in the XYZ coordinate system by the vectors $\mathbf{r}_O^i$, $\mathbf{r}_P^i$, and $\mathbf{r}_Q^i$, one can first define the unit vectors $\mathbf{i}^i$ and $\mathbf{i}_t^i$ as

$$\mathbf{i}^i = \frac{\mathbf{r}_P^i - \mathbf{r}_O^i}{|\mathbf{r}_P^i - \mathbf{r}_O^i|}, \quad \mathbf{i}_t^i = \frac{\mathbf{r}_Q^i - \mathbf{r}_O^i}{|\mathbf{r}_Q^i - \mathbf{r}_O^i|} \tag{2.102}$$

where $\mathbf{i}^i$ defines a unit vector along the X^i axis. It is clear that a unit vector $\mathbf{k}^i$ along the Z^i axis is defined as

$$\mathbf{k}^i = (\mathbf{i}^i \times \mathbf{i}_t^i)/|\mathbf{i}^i \times \mathbf{i}_t^i| \tag{2.103}$$

A unit vector along the Y^i axis can then be defined as

$$\mathbf{j}^i = \mathbf{k}^i \times \mathbf{i}^i \tag{2.104}$$

The vector $\mathbf{r}_O^i$ defines the position vector of the reference point O^i in the XYZ coordinate system, while the 3×3 matrix

$$\mathbf{A}^i = [\mathbf{i}^i \quad \mathbf{j}^i \quad \mathbf{k}^i] \tag{2.105}$$

as will be shown in Chapter 7, completely defines the orientation of the body coordinate system $X^iY^iZ^i$ with respect to the coordinate system XYZ. This matrix is called the *direction cosine transformation matrix*.

Conditions of Parallelism In formulating the kinematic constraint equations that describe a mechanical joint between two bodies in a multibody system, the cross product can be used to indicate the *parallelism* of two vectors on the two bodies. For instance, if $\mathbf{a}^i$ and $\mathbf{a}^j$ are two vectors defined on bodies i and j in a multibody system, the condition that these two vectors remain parallel is given by

$$\mathbf{a}^i \times \mathbf{a}^j = \mathbf{0} \tag{2.106}$$

As pointed out previously, this equation contains three scalar equations that are not independent. An alternative approach to formulate the *parallelism condition* of the two vectors $\mathbf{a}^i$ and $\mathbf{a}^j$ is to use two independent dot product equations. To demonstrate this, we form the orthogonal triad $\mathbf{a}^i$, $\mathbf{a}_1^i$, and $\mathbf{a}_2^i$, defined on body i as shown in Fig. 2. It is clear that if the vectors $\mathbf{a}^i$ and $\mathbf{a}^j$ are to remain parallel, one must have

$$\mathbf{a}_1^{i\mathrm{T}}\mathbf{a}^j = 0, \quad \mathbf{a}_2^{i\mathrm{T}}\mathbf{a}^j = 0 \tag{2.107}$$

These are two independent scalar equations that can be used instead of using the three dependent scalar equations of the cross product.

For a given nonzero vector $\mathbf{a}^i$, a simple computer procedure can be used to determine the vectors $\mathbf{a}_1^i$ and $\mathbf{a}_2^i$ such that $\mathbf{a}^i$, $\mathbf{a}_1^i$, and $\mathbf{a}_2^i$ form an orthogonal triad. In this procedure, we first define a nonzero vector $\mathbf{a}_d$ that is not parallel to the vector $\mathbf{a}^i$. The vector $\mathbf{a}_d$ can simply be defined as the three-dimensional vector that has one zero element and all other

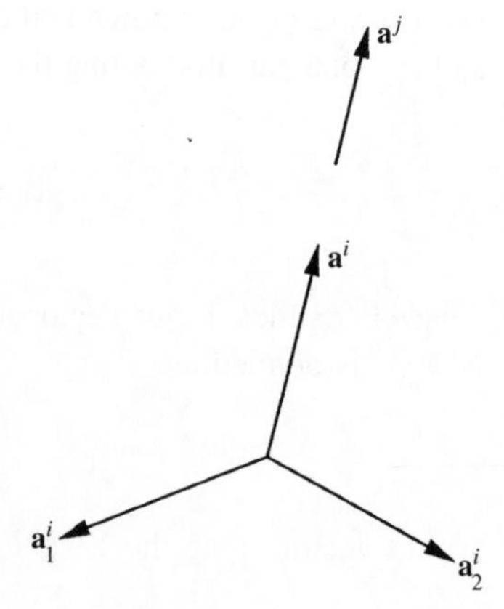

Figure 2.2 Parallelism of two vectors

elements equal to one. The location of the zero element is chosen to be the same as the location of the element of $\mathbf{a}^i$ that has the largest absolute value. The vector $\mathbf{a}_1^i$ can then be defined as

$$\mathbf{a}_1^i = \mathbf{a}^i \times \mathbf{a}_d \tag{2.108}$$

Clearly, $\mathbf{a}_1^i$ is perpendicular to $\mathbf{a}^i$. One can then define $\mathbf{a}_2^i$ that completes the orthogonal triad $\mathbf{a}^i$, $\mathbf{a}_1^i$, and $\mathbf{a}_2^i$ as

$$\mathbf{a}_2^i = \mathbf{a}^i \times \mathbf{a}_1^i \tag{2.109}$$

To demonstrate this simple procedure, consider the vector $\mathbf{a}^i$ defined as $\mathbf{a}^i = [1 \quad 0 \quad -3]^{\mathrm{T}}$. The element of $\mathbf{a}^i$ that has the largest absolute value is the third element. Therefore, the vector $\mathbf{a}_d$ is defined as $\mathbf{a}_d = [1 \quad 1 \quad 0]^{\mathrm{T}}$. The vector $\mathbf{a}_1^i$ is then defined as

$$\mathbf{a}_1^i = \mathbf{a}^i \times \mathbf{a}_d = \begin{bmatrix} 0 & 3 & 0 \\ -3 & 0 & -1 \\ 0 & 1 & 0 \end{bmatrix} \begin{bmatrix} 1 \\ 1 \\ 0 \end{bmatrix} = \begin{bmatrix} 3 \\ -3 \\ 1 \end{bmatrix}$$

and the vector $\mathbf{a}_2^i$ is

$$\mathbf{a}_2^i = \mathbf{a}^i \times \mathbf{a}_1^i = \begin{bmatrix} 0 & 3 & 0 \\ -3 & 0 & -1 \\ 0 & 1 & 0 \end{bmatrix} \begin{bmatrix} 3 \\ -3 \\ 1 \end{bmatrix} = \begin{bmatrix} -9 \\ -10 \\ -3 \end{bmatrix}$$

2.5 SOLUTION OF ALGEBRAIC EQUATIONS

The method of finding the inverse can be utilized in solving a system of *n algebraic equations* in *n* unknowns. Consider the following system of equations:

$$\left.\begin{aligned} a_{11}x_1 + a_{12}x_2 + \cdots + a_{1n}x_n &= b_1 \\ a_{21}x_1 + a_{22}x_2 + \cdots + a_{2n}x_n &= b_2 \\ &\vdots \\ a_{n1}x_1 + a_{n2}x_2 + \cdots + a_{nn}x_n &= b_n \end{aligned}\right\} \tag{2.110}$$

where a_{ij}, $i, j = 1, 2, \ldots, n$ are known coefficients, b_1, $b_2, \ldots,$ and b_n are given constants, and x_1, $x_2, \ldots,$ and x_n are unknowns. The preceding system of equations can be written in a matrix form as

$$\begin{bmatrix} a_{11} & a_{12} & \cdots & a_{1n} \\ a_{21} & a_{22} & \cdots & a_{2n} \\ \vdots & \vdots & \ddots & \vdots \\ a_{n1} & a_{n2} & \cdots & a_{nn} \end{bmatrix} \begin{bmatrix} x_1 \\ x_2 \\ \vdots \\ x_n \end{bmatrix} = \begin{bmatrix} b_1 \\ b_2 \\ \vdots \\ b_n \end{bmatrix} \tag{2.111}$$

This system of algebraic equations can be written as

$$\mathbf{Ax} = \mathbf{b} \tag{2.112}$$

where the coefficient matrix $\mathbf{A}$ is

$$\mathbf{A} = \begin{bmatrix} a_{11} & a_{12} & \cdots & a_{1n} \\ a_{21} & a_{22} & \cdots & a_{2n} \\ \vdots & \vdots & \ddots & \vdots \\ a_{n1} & a_{n2} & \cdots & a_{nn} \end{bmatrix}, \tag{2.113}$$

and the vectors $\mathbf{x}$ and $\mathbf{b}$ are given by

$$\mathbf{x} = [x_1 \quad x_2 \quad \cdots \quad x_n]^{\mathrm{T}}, \quad \mathbf{b} = [b_1 \quad b_2 \quad \cdots \quad b_n]^{\mathrm{T}} \tag{2.114}$$

If the coefficient matrix $\mathbf{A}$ in Eq. 112 has a full rank, the inverse of this matrix does exist. Multiplying Eq. 112 by the inverse of $\mathbf{A}$, one obtains

$$\mathbf{A}^{-1}\mathbf{Ax} = \mathbf{A}^{-1}\mathbf{b} \tag{2.115}$$

Since $\mathbf{A}^{-1}\mathbf{A} = \mathbf{I}$, where $\mathbf{I}$ is the identity matrix, the solution of the system of algebraic equations can be defined as

$$\mathbf{x} = \mathbf{A}^{-1}\mathbf{b} \tag{2.116}$$

It is clear from this equation that if $\mathbf{A}$ is a nonsingular matrix, the homogeneous system of algebraic equations $\mathbf{Ax} = \mathbf{0}$ has only the trivial solution $\mathbf{x} = \mathbf{0}$.

Example 2.11

Find the solution of the system of algebraic equations

$$2x_1 - x_2 = 2$$
$$-x_1 + 3x_2 - 2x_3 = -1$$
$$-2x_2 + 2x_3 = 0$$

Solution. This system of algebraic equations can be written in a matrix form as

$$\begin{bmatrix} 2 & -1 & 0 \\ -1 & 3 & -2 \\ 0 & -2 & 2 \end{bmatrix} \begin{bmatrix} x_1 \\ x_2 \\ x_3 \end{bmatrix} = \begin{bmatrix} 2 \\ -1 \\ 0 \end{bmatrix}$$

which can also be written as

$$\mathbf{Ax} = \mathbf{b}$$

where

$$\mathbf{A} = \begin{bmatrix} 2 & -1 & 0 \\ -1 & 3 & -2 \\ 0 & -2 & 2 \end{bmatrix}$$
$$\mathbf{x} = [x_1 \quad x_2 \quad x_3]^{\mathrm{T}}$$
$$\mathbf{b} = [2 \quad -1 \quad 0]^{\mathrm{T}}$$

It can be verified that the inverse of the matrix $\mathbf{A}$ is

$$\mathbf{A}^{-1} = \begin{bmatrix} 1 & 1 & 1 \\ 1 & 2 & 2 \\ 1 & 2 & 2.5 \end{bmatrix}$$

Using this inverse, the solution of the system of equations can be written as

$$\mathbf{x} = \mathbf{A}^{-1}\mathbf{b} = \begin{bmatrix} 1 & 1 & 1 \\ 1 & 2 & 2 \\ 1 & 2 & 2.5 \end{bmatrix} \begin{bmatrix} 2 \\ -1 \\ 0 \end{bmatrix} = \begin{bmatrix} 1 \\ 0 \\ 0 \end{bmatrix}$$

Since the matrix $\mathbf{A}$ is nonsingular, this solution is unique.

Gaussian Elimination The method of finding the inverse is rarely used in practice to solve a system of algebraic equations. This is mainly because the explicit construction of the inverse of a matrix by using the adjoint matrix approach, which requires the evaluation of the determinant and the cofactors, is computationally expensive and often leads to numerical errors.

The *Gaussian elimination method* is an alternative approach for solving a system of algebraic equations. This approach, which is based on the idea of eliminating variables one at a time, requires a much smaller number of arithmetic operations as compared with the method of finding the inverse. The Gaussian elimination consists of two main steps: the *forward elimination* and the *back substitution*. In the forward elimination step, the coefficient matrix is converted to an upper-triangular matrix by using *elementary row operations*. In the back substitution step, the unknown variables are determined. In order to demonstrate the use of the Gaussian elimination method, consider the following system of equations:

$$\begin{bmatrix} 2 & 1 & 1 \\ -1 & 2 & -1 \\ 4 & -3 & 1 \end{bmatrix} \begin{bmatrix} x_1 \\ x_2 \\ x_3 \end{bmatrix} = \begin{bmatrix} 6 \\ 0 \\ 2 \end{bmatrix}$$

To solve for the unknowns x_1, x_2, and x_3 using the Gaussian elimination method, we first perform the forward elimination in order to obtain an upper-triangular matrix. With this goal in mind, we multiply the first equation by $\frac{1}{2}$. This leads to

$$\begin{bmatrix} 1 & \frac{1}{2} & \frac{1}{2} \\ -1 & 2 & -1 \\ 4 & -3 & 1 \end{bmatrix} \begin{bmatrix} x_1 \\ x_2 \\ x_3 \end{bmatrix} = \begin{bmatrix} 3 \\ 0 \\ 2 \end{bmatrix}$$

By adding the first equation to the second equation, and -4 times the first equation to the third equation, one obtains

$$\begin{bmatrix} 1 & \frac{1}{2} & \frac{1}{2} \\ 0 & \frac{5}{2} & -\frac{1}{2} \\ 0 & -5 & -1 \end{bmatrix} \begin{bmatrix} x_1 \\ x_2 \\ x_3 \end{bmatrix} = \begin{bmatrix} 3 \\ 3 \\ -10 \end{bmatrix}$$

Now we multiply the second equation by $\frac{2}{5}$ to obtain

$$\begin{bmatrix} 1 & \frac{1}{2} & \frac{1}{2} \\ 0 & 1 & -\frac{1}{5} \\ 0 & -5 & -1 \end{bmatrix} \begin{bmatrix} x_1 \\ x_2 \\ x_3 \end{bmatrix} = \begin{bmatrix} 3 \\ \frac{6}{5} \\ -10 \end{bmatrix}$$

By adding 5 times the second equation to the third equation, one obtains

$$\begin{bmatrix} 1 & \frac{1}{2} & \frac{1}{2} \\ 0 & 1 & -\frac{1}{5} \\ 0 & 0 & -2 \end{bmatrix} \begin{bmatrix} x_1 \\ x_2 \\ x_3 \end{bmatrix} = \begin{bmatrix} 3 \\ \frac{6}{5} \\ -4 \end{bmatrix} \tag{2.117}$$

The coefficient matrix in this equation is an upper-triangular matrix and hence, the back substitution step can be used to solve for the variables. Using the third equation, one has

$$x_3 = 2$$

The second equation yields

$$x_2 = \tfrac{6}{5} + \tfrac{1}{5}x_3 = \tfrac{8}{5}$$

The first equation can then be used to solve for x_1 as

$$x_1 = 3 - \tfrac{1}{2}x_2 - \tfrac{1}{2}x_3 = \tfrac{6}{5}$$

Therefore, the solution of the original system of equations is

$$\mathbf{x} = [\tfrac{6}{5} \quad \tfrac{8}{5} \quad 2]^{\mathrm{T}}$$

It is clear that the Gaussian elimination solution procedure reduces the system $\mathbf{Ax} = \mathbf{b}$ to an equivalent system $\mathbf{Ux} = \mathbf{g}$, where $\mathbf{U}$ is an upper-triangular matrix. This new equivalent system is easily solved by the process of back substitution.

Gauss–Jordan Method The *Gauss–Jordan reduction method* combines the forward elimination and back substitution steps. In this case, the coefficient matrix is converted to a diagonal identity matrix, and consequently, the solution is defined by the right-hand-side vector of the resulting system of algebraic equations. To demonstrate this procedure, we consider Eq. 117. Dividing the third equation by -2, one obtains

$$\begin{bmatrix} 1 & \frac{1}{2} & \frac{1}{2} \\ 0 & 1 & -\frac{1}{5} \\ 0 & 0 & 1 \end{bmatrix} \begin{bmatrix} x_1 \\ x_2 \\ x_3 \end{bmatrix} = \begin{bmatrix} 3 \\ \frac{6}{5} \\ 2 \end{bmatrix}$$

By adding $\frac{1}{5}$ times the third equation to the second equation and $-\frac{1}{2}$ times the third equation to the first equation, one obtains

$$\begin{bmatrix} 1 & \frac{1}{2} & 0 \\ 0 & 1 & 0 \\ 0 & 0 & 1 \end{bmatrix} \begin{bmatrix} x_1 \\ x_2 \\ x_3 \end{bmatrix} = \begin{bmatrix} 2 \\ \frac{8}{5} \\ 2 \end{bmatrix}$$

Adding $-\frac{1}{2}$ times the second equation to the first equation yields

$$\begin{bmatrix} 1 & 0 & 0 \\ 0 & 1 & 0 \\ 0 & 0 & 1 \end{bmatrix} \begin{bmatrix} x_1 \\ x_2 \\ x_3 \end{bmatrix} = \begin{bmatrix} \frac{6}{5} \\ \frac{8}{5} \\ 2 \end{bmatrix}$$

The coefficient matrix in this system is the identity matrix and the right-hand side is the solution vector previously obtained using the Gaussian elimination procedure.

It is important, however, to point out that the use of Gauss–Jordan method requires 50 percent more additions and multiplications as compared to the Gaussian elimination procedure. For an $n \times n$ coefficient matrix, the Gaussian elimination method requires approximately $(n)^3/3$ multiplications and additions, while the Gauss–Jordan method requires approximately $(n)^3/2$ multiplications and additions. For this reason, use of the Gauss–Jordan procedure to solve linear systems of algebraic equations is not recommended. Nonetheless, by taking advantage of the special structure of the right-hand side of the system $\mathbf{Ax} = \mathbf{I}$, the Gauss–Jordan method can be used to produce a matrix inversion program that requires a minimum storage.

Pivoting and Scaling It is clear that the Gaussian elimination and Gauss–Jordan reduction procedures require division by the diagonal element a_{ii}. This element is called the *pivot*. The forward elimination procedure at the ith step fails if the pivot a_{ii} is equal to zero. Furthermore, if the pivot is small, the elimination procedure becomes prone to numerical errors. To avoid these problems, the equations may be reordered in order to avoid zero or small pivot elements. There are two types of pivoting strategies that are used in solving systems of algebraic equations. These are the *partial pivoting* and *full pivoting*. In the case of partial pivoting, during the ith elimination step, the equations are reordered such that the equation with the largest coefficient (magnitude) of x_i is chosen for pivoting. In the case of full or complete pivoting, the equations and the unknown variables are reordered in order to choose a pivot element that has the largest absolute value.

It has been observed that if the elements of the coefficient matrix $\mathbf{A}$ vary greatly in size, the numerical solution of the system $\mathbf{Ax} = \mathbf{b}$ can be in error. In order to avoid this problem, the coefficient matrix $\mathbf{A}$ must be *scaled* such that all the elements of the matrix have comparable magnitudes. Scaling of the matrix can be achieved by multiplying the rows and the columns of the matrix by suitable constants. That is, scaling is equivalent to performing simple row and column operations. While the row operations cause the rows of the matrix to be approximately equal in magnitude, the column operations cause the elements of the vector of the unknowns to be of approximately equal size. Let $\mathbf{C}$ be the matrix that results from scaling the matrix $\mathbf{A}$. This matrix can be written as

$$\mathbf{C} = \mathbf{B}_1\mathbf{A}\mathbf{B}_2 \tag{2.118}$$

where $\mathbf{B}_1$ and $\mathbf{B}_2$ are diagonal matrices whose diagonal elements are the scaling constants. Hence, one is interested in solving the following new system of algebraic equations:

$$\mathbf{Cy} = \mathbf{z} \tag{2.119}$$

where

$$\mathbf{y} = \mathbf{B}_2^{-1}\mathbf{x}, \quad \mathbf{z} = \mathbf{B}_1\mathbf{b} \tag{2.120}$$

The solution of the system $\mathbf{Cy} = \mathbf{z}$ defines the vector $\mathbf{y}$. This solution vector can be used to define the original vector of unknowns $\mathbf{x}$ as

$$\mathbf{x} = \mathbf{B}_2\mathbf{y} \tag{2.121}$$

The Gaussian elimination method, in addition to being widely used for solving systems of algebraic equations, can also be used to determine the rank of nonsquare matrices and also to determine the independent variables in a given system of algebraic equations. This is demonstrated by the following example.

Example 2.12

Consider the following system of algebraic equations:

$$\begin{bmatrix} 1 & 1 & 0 & 1 & 4 \\ 2 & 2 & 2 & 0 & 1 \\ 3 & 3 & 2 & 1 & 5 \end{bmatrix} \begin{bmatrix} x_1 \\ x_2 \\ x_3 \\ x_4 \\ x_5 \end{bmatrix} = \begin{bmatrix} 0 \\ 0 \\ 0 \end{bmatrix}$$

A forward elimination in the first column yields

$$\begin{bmatrix} 1 & 1 & 0 & 1 & 4 \\ 0 & 0 & 2 & -2 & -7 \\ 0 & 0 & 2 & -2 & -7 \end{bmatrix} \begin{bmatrix} x_1 \\ x_2 \\ x_3 \\ x_4 \\ x_5 \end{bmatrix} = \begin{bmatrix} 0 \\ 0 \\ 0 \end{bmatrix}$$

Since the coefficients of x_2 in the second and third equations are equal to zero, these coefficients cannot be used as pivots in the Gaussian elimination. By using an elementary column operation, the second and third columns can be interchanged leading to a reordering of the variables. Such an elementary operation yields

$$\begin{bmatrix} 1 & 0 & 1 & 1 & 4 \\ 0 & 2 & 0 & -2 & -7 \\ 0 & 2 & 0 & -2 & -7 \end{bmatrix} \begin{bmatrix} x_1 \\ x_3 \\ x_2 \\ x_4 \\ x_5 \end{bmatrix} = \begin{bmatrix} 0 \\ 0 \\ 0 \end{bmatrix}$$

Dividing the second row by 2 and performing forward elimination in the second column yields

$$\begin{bmatrix} 1 & 0 & 1 & 1 & 4 \\ 0 & 1 & 0 & -1 & -\frac{7}{2} \\ 0 & 0 & 0 & 0 & 0 \end{bmatrix} \begin{bmatrix} x_1 \\ x_3 \\ x_2 \\ x_4 \\ x_5 \end{bmatrix} = \begin{bmatrix} 0 \\ 0 \\ 0 \end{bmatrix}$$

The coefficient matrix in this equation has two independent rows and, consequently, its rank is equal to two. This is an indication that there are only two independent equations. One can then disregard the third equation and use the following system of equations:

$$\begin{bmatrix} 1 & 0 & 1 & 1 & 4 \\ 0 & 1 & 0 & -1 & -\frac{7}{2} \end{bmatrix} \begin{bmatrix} x_1 \\ x_3 \\ x_2 \\ x_4 \\ x_5 \end{bmatrix} = \begin{bmatrix} 0 \\ 0 \end{bmatrix}$$

which can also be written as

$$\begin{bmatrix} 1 & 0 \\ 0 & 1 \end{bmatrix} \begin{bmatrix} x_1 \\ x_3 \end{bmatrix} = -\begin{bmatrix} 1 & 1 & 4 \\ 0 & -1 & -\frac{7}{2} \end{bmatrix} \begin{bmatrix} x_2 \\ x_4 \\ x_5 \end{bmatrix}$$

It is clear from this equation that x_1 and x_3 can be determined if the values of x_2, x_4, and x_5 are given. Therefore, x_2, x_4, and x_5 are called the *independent variables*, while x_1 and x_3 are called the *dependent variables*.

2.6 TRIANGULAR FACTORIZATION

In the Gaussian elimination procedure used to solve the system $\mathbf{Ax} = \mathbf{b}$, the $n \times n$ coefficient matrix reduces to an upper-triangular matrix after $n - 1$ steps. The new system resulting from the forward elimination can be written as

$$\begin{bmatrix} (a_{11})_1 & (a_{12})_1 & \cdots & (a_{1n})_1 \\ 0 & (a_{22})_2 & \cdots & (a_{2n})_2 \\ \vdots & \vdots & \ddots & \vdots \\ 0 & 0 & \cdots & (a_{nn})_n \end{bmatrix} \begin{bmatrix} x_1 \\ x_2 \\ \vdots \\ x_n \end{bmatrix} = \begin{bmatrix} (b_1)_1 \\ (b_2)_2 \\ \vdots \\ (b_n)_n \end{bmatrix} \tag{2.122}$$

where $(\ )_k$ refers to step k in the forward elimination process and

$$\left.\begin{aligned} (a_{ij})_{k+1} &= (a_{ij})_k - m_{ik}(a_{kj})_k \\ (b_i)_{k+1} &= (b_i)_k - m_{ik}(b_k)_k \end{aligned}\right\} \quad i, j = k+1, \ldots, n \tag{2.123}$$

and

$$m_{ik} = (a_{ik})_k / (a_{kk})_k \qquad i = k+1, \ldots, n \tag{2.124}$$

Let $\mathbf{U}$ denote the upper-triangular coefficient matrix in Eq. 122 and define the lower-triangular matrix $\mathbf{L}$ as

$$\mathbf{L} = \begin{bmatrix} 1 & 0 & 0 & \cdots & 0 \\ m_{21} & 1 & 0 & \cdots & 0 \\ \vdots & \vdots & \vdots & \ddots & \vdots \\ m_{n1} & m_{n2} & m_{n3} & \cdots & 1 \end{bmatrix} \tag{2.125}$$

where the coefficients m_{ij} are defined by Eq. 124. Using Eqs. 123 and 124, direct matrix multiplication shows that the matrix $\mathbf{A}$ can be written as

$$\mathbf{A} = \mathbf{LU} \tag{2.126}$$

which implies that the matrix $\mathbf{A}$ can be written as the product of a lower-triangular matrix $\mathbf{L}$ and an upper-triangular matrix $\mathbf{U}$.

Example 2.13

The lower-triangular matrix **L** can also be defined using the *elementary operations* of Gaussian elimination. In order to demonstrate this, we consider the system

$$\begin{bmatrix} 2 & 1 & 1 \\ -1 & 2 & -1 \\ 4 & -3 & 1 \end{bmatrix} \begin{bmatrix} x_1 \\ x_2 \\ x_3 \end{bmatrix} = \begin{bmatrix} 6 \\ 0 \\ 2 \end{bmatrix}$$

whose solution was obtained in the preceding section using Gaussian elimination. In order to solve this system, the following three elimination steps are used.

1. Add $\frac{1}{2}$ times the first equation to the second equation.
2. Subtract 2 times the first equation from the third equation.
3. Add 2 times the second equation to the third equation.

The result of these three elementary operations is an equivalent but simpler system given by

$$\begin{bmatrix} 2 & 1 & 1 \\ 0 & \frac{5}{2} & -\frac{1}{2} \\ 0 & 0 & -2 \end{bmatrix} \begin{bmatrix} x_1 \\ x_2 \\ x_3 \end{bmatrix} = \begin{bmatrix} 6 \\ 3 \\ -4 \end{bmatrix}$$

in which the upper-triangular matrix **U** can be recognized as

$$\mathbf{U} = \begin{bmatrix} 2 & 1 & 1 \\ 0 & \frac{5}{2} & -\frac{1}{2} \\ 0 & 0 & -2 \end{bmatrix}$$

Elementary operations can also be performed using *elementary matrices*. An elementary matrix is obtained by performing the elementary operation on an identity matrix. Premultiplying the coefficient matrix **A** by an elementary matrix produces the same elementary operation for **A**. For instance, if $\frac{1}{2}$ times the first row of a 3×3 identity matrix is added to the second row, one obtains the elementary matrix

$$\mathbf{E}_1 = \begin{bmatrix} 1 & 0 & 0 \\ \frac{1}{2} & 1 & 0 \\ 0 & 0 & 1 \end{bmatrix}$$

Also, if 2 times the first row of a 3×3 identity matrix is substracted from the third row, one obtains the elementary matrix

$$\mathbf{E}_2 = \begin{bmatrix} 1 & 0 & 0 \\ 0 & 1 & 0 \\ -2 & 0 & 1 \end{bmatrix}$$

Similarly, if 2 times the second row is added to the third row, one obtains the elementary matrix

$$\mathbf{E}_3 = \begin{bmatrix} 1 & 0 & 0 \\ 0 & 1 & 0 \\ 0 & 2 & 1 \end{bmatrix}$$

The product of the three elementary matrices $\mathbf{E}_3$, $\mathbf{E}_2$, and $\mathbf{E}_1$ is

$$\mathbf{E} = \mathbf{E}_3\mathbf{E}_2\mathbf{E}_1 = \begin{bmatrix} 1 & 0 & 0 \\ 0 & 1 & 0 \\ 0 & 2 & 1 \end{bmatrix} \begin{bmatrix} 1 & 0 & 0 \\ 0 & 1 & 0 \\ -2 & 0 & 1 \end{bmatrix} \begin{bmatrix} 1 & 0 & 0 \\ \frac{1}{2} & 1 & 0 \\ 0 & 0 & 1 \end{bmatrix} = \begin{bmatrix} 1 & 0 & 0 \\ \frac{1}{2} & 1 & 0 \\ -1 & 2 & 1 \end{bmatrix}$$

Note that $\mathbf{E}$ is a lower-triangular matrix with all the diagonal elements equal to 1. Note also that premultiplying the coefficient matrix $\mathbf{A}$ by $\mathbf{E}$ leads to

$$\mathbf{EA} = \begin{bmatrix} 1 & 0 & 0 \\ \frac{1}{2} & 1 & 0 \\ -1 & 2 & 1 \end{bmatrix} \begin{bmatrix} 2 & 1 & 1 \\ -1 & 2 & -1 \\ 4 & -3 & 1 \end{bmatrix} = \begin{bmatrix} 2 & 1 & 1 \\ 0 & \frac{5}{2} & -\frac{1}{2} \\ 0 & 0 & -2 \end{bmatrix}$$

which is the same upper-triangular matrix $\mathbf{U}$ obtained previously by the elementary operations of Gaussian elimination. The diagonal elements of the matrix $\mathbf{U}$ are the pivots. Therefore, one has

$$\mathbf{EA} = \mathbf{U}$$

or

$$\mathbf{A} = \mathbf{E}^{-1}\mathbf{U}$$

where $\mathbf{E}^{-1} = (\mathbf{E}_3\mathbf{E}_2\mathbf{E}_1)^{-1} = \mathbf{E}_1^{-1}\mathbf{E}_2^{-1}\mathbf{E}_3^{-1}$ is the matrix $\mathbf{L}$ that defines the $\mathbf{LU}$ factorization of the matrix $\mathbf{A}$. The inverse of an elementary matrix is also an elementary matrix. The inverses of the matrices $\mathbf{E}_1$, $\mathbf{E}_2$, and $\mathbf{E}_3$ are defined as

$$\mathbf{E}_1^{-1} = \begin{bmatrix} 1 & 0 & 0 \\ -\frac{1}{2} & 1 & 0 \\ 0 & 0 & 1 \end{bmatrix}, \quad \mathbf{E}_2^{-1} = \begin{bmatrix} 1 & 0 & 0 \\ 0 & 1 & 0 \\ 2 & 0 & 1 \end{bmatrix}, \quad \mathbf{E}_3^{-1} = \begin{bmatrix} 1 & 0 & 0 \\ 0 & 1 & 0 \\ 0 & -2 & 1 \end{bmatrix}$$

It follows that

$$\mathbf{L} = \mathbf{E}_1^{-1}\mathbf{E}_2^{-1}\mathbf{E}_3^{-1} = \begin{bmatrix} 1 & 0 & 0 \\ -\frac{1}{2} & 1 & 0 \\ 0 & 0 & 1 \end{bmatrix} \begin{bmatrix} 1 & 0 & 0 \\ 0 & 1 & 0 \\ 2 & 0 & 1 \end{bmatrix} \begin{bmatrix} 1 & 0 & 0 \\ 0 & 1 & 0 \\ 0 & -2 & 1 \end{bmatrix}$$

$$= \begin{bmatrix} 1 & 0 & 0 \\ -\frac{1}{2} & 1 & 0 \\ 2 & -2 & 1 \end{bmatrix}$$

While Eqs. 122 through 126 present the **LU** factorization resulting from the use of Gaussian elimination, we should point out that in general such a decomposition is not unique. The triangular matrix **L** obtained by using the steps of Gaussian elimination has diagonal elements that are all equal to 1. The method that gives explicit formulas for the elements of **L** and **U** in this special case is known as *Doolittle's method*. If the upper-triangular matrix **U** is defined such that all its diagonal elements are equal to 1, we have *Crout's method*. Obviously, there is only a multiplying diagonal matrix that distinguishes between Crout's and Doolittle's methods. To demonstrate this, let us assume that **A** has the following two different decompositions:

$$\mathbf{A} = \mathbf{L}_1\mathbf{U}_1 = \mathbf{L}_2\mathbf{U}_2 \tag{2.127}$$

It is then clear that

$$\mathbf{U}_2\mathbf{U}_1^{-1} = \mathbf{L}_2^{-1}\mathbf{L}_1 \tag{2.128}$$

Since the inverse and product of lower (upper)-triangular matrices are again lower (upper) triangular, the left and right sides of Eq. 128 must be equal to a diagonal matrix **D**, that is,

$$\mathbf{U}_2\mathbf{U}_1^{-1} = \mathbf{D}, \quad \mathbf{L}_2^{-1}\mathbf{L}_1 = \mathbf{D} \tag{2.129}$$

It follows that

$$\mathbf{U}_2 = \mathbf{D}\mathbf{U}_1, \quad \mathbf{L}_2 = \mathbf{L}_1\mathbf{D}^{-1} \tag{2.130}$$

which demonstrate that there is only a multiplying diagonal matrix that distinguishes between two different methods of decomposition.

Cholesky's Method A more efficient decomposition can be found if the matrix **A** is *symmetric* and *positive definite*. The matrix **A** is said to be positive definite if $\mathbf{x}^{\mathrm{T}}\mathbf{A}\mathbf{x} > 0$ for any n-dimensional nonzero vector **x**. In the case of symmetric positive definite matrices, *Cholesky's method* can be used to obtain a simpler factorization for the matrix **A**. In this case, there exists a lower-triangular matrix **L** such that

$$\mathbf{A} = \mathbf{L}\mathbf{L}^{\mathrm{T}} \tag{2.131}$$

where the elements l_{ij} of the lower-triangular matrix **L** can be defined by equating the elements of the products of the matrices on the right side of Eq. 131 to the elements of the matrix **A**. This leads to the following general formulas for the elements of the lower-triangular matrix **L**:

$$\left.\begin{aligned} l_{ij} &= \frac{a_{ij} - \sum_{k=1}^{j-1} l_{ik} l_{jk}}{l_{jj}} \qquad j = 1, \ldots, i-1 \\ l_{ii} &= \left[a_{ii} - \sum_{k=1}^{i-1} (l_{ik})^2 \right]^{1/2} \end{aligned}\right\} \tag{2.132}$$

Cholesky's method requires only $\frac{1}{2}n(n+1)$ storage locations for the lower-triangular matrix **L** as compared to $(n)^2$ locations required by other **LU** factorization methods. Furthermore, the number of multiplications and additions required by Cholesky's method is approximately $\frac{1}{6}(n)^3$ rather than $\frac{1}{3}(n)^3$ required by other decomposition methods.

Numerical Solution Once the decomposition of **A** into its **LU** factors is defined, by whatever method, the system of algebraic equations

$$\mathbf{Ax} = \mathbf{LUx} = \mathbf{b} \tag{2.133}$$

can be solved by first solving

$$\mathbf{Ly} = \mathbf{b} \tag{2.134}$$

and then solve for **x** using the equation

$$\mathbf{Ux} = \mathbf{y} \tag{2.135}$$

The coefficient matrices in Eqs. 134 and 135 are both triangular and, consequently, the solutions of both equations can be easily obtained by back substitution.

We should point out that the accuracy of the solution obtained using the *direct methods* such as Gaussian elimination and other **LU** factorization techniques depends on the numerical properties of the coefficient matrix **A**. A linear system of algebraic equations $\mathbf{Ax} = \mathbf{b}$ is called *ill-conditioned* if the solution **x** is unstable with respect to small changes in the right-side **b**. It is important to check the effectiveness of the computer programs used to solve systems of linear algebraic equations when ill-conditioned problems are considered. An example of an ill-conditioned matrix that can be used to evaluate the performance of the computer programs is the *Hilbert matrix*. A Hilbert matrix of order n is defined by

$$\mathbf{H}_n = \begin{bmatrix} 1 & \frac{1}{2} & \frac{1}{3} & \cdots & \frac{1}{n} \\ \frac{1}{2} & \frac{1}{3} & \frac{1}{4} & \cdots & \frac{1}{n+1} \\ \vdots & \vdots & \vdots & \ddots & \vdots \\ \frac{1}{n} & \frac{1}{n+1} & \frac{1}{n+2} & \cdots & \frac{1}{2n-1} \end{bmatrix} \tag{2.136}$$

The inverse of this matrix is known explicitly. Let c_{ij} be the ijth element in the inverse of $\mathbf{H}_n$. These elements are defined as

$$c_{ij}^{(n)} = \frac{(-1)^{i+j}(n+i-1)!(n+j-1)!}{(i+j-1)[(i-1)!(j-1)!]^2(n-i)!(n-j)!} \qquad i \geq 1, j \leq n \tag{2.137}$$

The Hilbert matrix becomes more ill-conditioned as the dimension n increases.

2.7 QR DECOMPOSITION

Another important matrix factorization that is used in the computational dynamics of mechanical systems is the **QR** decomposition. In this decomposition, an arbitrary matrix **A** can be written as

$$\mathbf{A} = \mathbf{QR} \tag{2.138}$$

where the columns of **Q** are orthogonal and **R** is an upper-triangular matrix. Before examining the factorization of Eq. 138, some background material will first be presented.

In Section 3, the orthogonality of n-dimensional vectors was defined. Two vectors **a** and **b** are said to be *orthogonal* if $\mathbf{a}^{\mathrm{T}}\mathbf{b} = 0$. Orthogonal vectors are linearly independent, for if $\mathbf{a}_1, \mathbf{a}_2, \ldots, \mathbf{a}_n$ is a set of nonzero orthogonal vectors, and

$$\alpha_1\mathbf{a}_1 + \alpha_2\mathbf{a}_2 + \cdots + \alpha_n\mathbf{a}_n = \mathbf{0} \tag{2.139}$$

one can multiply this equation by $\mathbf{a}_i^{\mathrm{T}}$ and use the orthogonality condition to obtain $\alpha_i\mathbf{a}_i^{\mathrm{T}}\mathbf{a}_i = 0$. This equation implies that $\alpha_i = 0$ for any i. This proves that the orthogonal vectors $\mathbf{a}_1, \mathbf{a}_2, \ldots, \mathbf{a}_n$ are linearly independent. In fact, one can use any set of linearly independent vectors to define a set of orthogonal vectors by applying the *Gram–Schmidt orthogonalization process*.

Gram–Schmidt Orthogonalization Let $\mathbf{a}_1, \mathbf{a}_2, \ldots, \mathbf{a}_m$ be a set of n-dimensional linearly independent vectors where $m \leq n$. To use this set of vectors to define another set of orthogonal vectors $\mathbf{b}_1, \mathbf{b}_2, \ldots, \mathbf{b}_m$, we first define the unit vector

$$\mathbf{b}_1 = \mathbf{a}_1/|\mathbf{a}_1| \tag{2.140}$$

Recall that the component of the vector $\mathbf{a}_2$ in the direction of the unit vector $\mathbf{b}_1$ is defined by the dot product $\mathbf{a}_2^{\mathrm{T}}\mathbf{b}_1$. For this reason, we define $\mathbf{b}_2'$ as

$$\mathbf{b}_2' = \mathbf{a}_2 - (\mathbf{a}_2^{\mathrm{T}}\mathbf{b}_1)\mathbf{b}_1 \tag{2.141}$$

Clearly, $\mathbf{b}_2'$ has no component in the direction of $\mathbf{b}_1$ and, consequently, $\mathbf{b}_1$ and $\mathbf{b}_2'$ are orthogonal vectors. This can simply be proved by using the dot product $\mathbf{b}_1^{\mathrm{T}}\mathbf{b}_2'$ and utilizing the fact that $\mathbf{b}_1$ is a unit vector. Now the unit vector $\mathbf{b}_2$ is defined as

$$\mathbf{b}_2 = \mathbf{b}_2'/|\mathbf{b}_2'| \tag{2.142}$$

Similarly, in defining $\mathbf{b}_3$ we first eliminate the dependence of this vector on $\mathbf{b}_1$ and $\mathbf{b}_2$. This can be achieved by defining

$$\mathbf{b}_3' = \mathbf{a}_3 - (\mathbf{a}_3^{\mathrm{T}}\mathbf{b}_1)\mathbf{b}_1 - (\mathbf{a}_3^{\mathrm{T}}\mathbf{b}_2)\mathbf{b}_2 \tag{2.143}$$

The vector $\mathbf{b}_3$ can then be defined as

$$\mathbf{b}_3 = \mathbf{b}_3'/|\mathbf{b}_3'| \tag{2.144}$$

Continuing in this manner, one has

$$\begin{aligned}\mathbf{b}'_i &= \mathbf{a}_i - (\mathbf{a}_i^{\mathrm{T}}\mathbf{b}_1)\mathbf{b}_1 - (\mathbf{a}_i^{\mathrm{T}}\mathbf{b}_2)\mathbf{b}_2 \cdots - (\mathbf{a}_i^{\mathrm{T}}\mathbf{b}_{i-1})\mathbf{b}_{i-1} \\ &= \mathbf{a}_i - \sum_{j=1}^{i-1}(\mathbf{a}_i^{\mathrm{T}}\mathbf{b}_j)\mathbf{b}_j \end{aligned} \tag{2.145}$$

and

$$\mathbf{b}_i = \mathbf{b}'_i/|\mathbf{b}'_i| \tag{2.146}$$

As the result of the application of the Gram–Schmidt orthogonalization process one obtains a set of *orthonormal vectors* that satisfy

$$\mathbf{b}_i^{\mathrm{T}}\mathbf{b}_j = \begin{cases} 0 & \text{if } i \neq j \\ 1 & \text{if } i = j \end{cases} \tag{2.147}$$

The Gram–Schmidt orthogonalization process cannot be completed if the vectors are not linearly independent. In this case, it is impossible to obtain a set that consists of only nonzero orthogonal vectors.

Example 2.14

Consider the linearly independent vectors defined by the columns of the rectangular matrix

$$\mathbf{A} = \begin{bmatrix} 2 & 1 & 1 \\ -1 & 2 & -1 \\ 4 & -3 & 1 \\ 1 & 0 & 2 \end{bmatrix}$$

Let

$$\mathbf{a}_1 = \begin{bmatrix} 2 \\ -1 \\ 4 \\ 1 \end{bmatrix}, \quad \mathbf{a}_2 = \begin{bmatrix} 1 \\ 2 \\ -3 \\ 0 \end{bmatrix}, \quad \mathbf{a}_3 = \begin{bmatrix} 1 \\ -1 \\ 1 \\ 2 \end{bmatrix}$$

In order to define a set of orthogonal vectors, we first define

$$\mathbf{b}_1 = \frac{\mathbf{a}_1}{|\mathbf{a}_1|} = \frac{1}{\sqrt{22}}\begin{bmatrix} 2 \\ -1 \\ 4 \\ 1 \end{bmatrix} = \begin{bmatrix} 0.4264 \\ -0.2132 \\ 0.8528 \\ 0.2132 \end{bmatrix}$$

The vector $\mathbf{b}_2'$ is

$$\mathbf{b}_2' = \mathbf{a}_2 - (\mathbf{a}_2^T\mathbf{b}_1)\mathbf{b}_1 = \begin{bmatrix} 1 \\ 2 \\ -3 \\ 0 \end{bmatrix} - (-2.5584)\begin{bmatrix} 0.4264 \\ -0.2132 \\ 0.8528 \\ 0.2132 \end{bmatrix}$$

$$= \begin{bmatrix} 1 \\ 2 \\ -3 \\ 0 \end{bmatrix} - \begin{bmatrix} -1.0909 \\ 0.5455 \\ -2.1818 \\ -0.5455 \end{bmatrix} = \begin{bmatrix} 2.0909 \\ 1.4545 \\ -0.8182 \\ 0.5455 \end{bmatrix}$$

The vector $\mathbf{b}_2$ can then be defined as

$$\mathbf{b}_2 = \frac{\mathbf{b}_2'}{|\mathbf{b}_2'|} = \begin{bmatrix} 0.7658 \\ 0.5327 \\ -0.2997 \\ 0.1998 \end{bmatrix}$$

The vector $\mathbf{b}_3'$ is defined as

$$\mathbf{b}_3' = \mathbf{a}_3 - (\mathbf{a}_3^T\mathbf{b}_1)\mathbf{b}_1 - (\mathbf{a}_3^T\mathbf{b}_2)\mathbf{b}_2$$

$$= \begin{bmatrix} 1 \\ -1 \\ 1 \\ 2 \end{bmatrix} - (1.9188)\begin{bmatrix} 0.4264 \\ -0.2132 \\ 0.8528 \\ 0.2132 \end{bmatrix} - (0.3330)\begin{bmatrix} 0.7658 \\ 0.5327 \\ -0.2997 \\ 0.1998 \end{bmatrix}$$

$$= \begin{bmatrix} -0.0732 \\ -0.7683 \\ -0.5366 \\ 1.5244 \end{bmatrix}$$

Therefore, the vector $\mathbf{b}_3$ is

$$\mathbf{b}_3 = \frac{\mathbf{b}_3'}{|\mathbf{b}_3'|} = \begin{bmatrix} -0.0409 \\ -0.4290 \\ -0.2996 \\ 0.8512 \end{bmatrix}$$

The three orthonormal vectors are

$$\mathbf{b}_1 = \begin{bmatrix} 0.4264 \\ -0.2132 \\ 0.8528 \\ 0.2132 \end{bmatrix}, \quad \mathbf{b}_2 = \begin{bmatrix} 0.7658 \\ 0.5327 \\ -0.2997 \\ 0.1998 \end{bmatrix}, \quad \mathbf{b}_3 = \begin{bmatrix} -0.0409 \\ -0.4290 \\ -0.2996 \\ 0.8512 \end{bmatrix}$$

Q and R Matrices The Gram–Schmidt orthogonalization process can be used to demonstrate that an arbitrary rectangular matrix $\mathbf{A}$ with linearly independent columns can be expressed in the following factored form $\mathbf{A} = \mathbf{QR}$ of Eq. 138, where the columns of $\mathbf{Q}$ are orthogonal or orthonormal vectors and $\mathbf{R}$ is an upper-triangular matrix. To prove Eq. 138, we consider the $n \times m$ rectangular matrix $\mathbf{A}$. If $\mathbf{a}_1, \mathbf{a}_2, \ldots, \mathbf{a}_m$ are the columns of $\mathbf{A}$, the matrix $\mathbf{A}$ can be written as

$$\mathbf{A} = [\mathbf{a}_1 \quad \mathbf{a}_2 \quad \cdots \quad \mathbf{a}_m] \tag{2.148}$$

If the n-dimensional vectors $\mathbf{a}_1, \ldots,$ and $\mathbf{a}_m$ are linearly independent, the Gram–Schmidt orthogonalization procedure can be used to define a set of m orthogonal vectors $\mathbf{b}_1, \mathbf{b}_2, \ldots,$ and $\mathbf{b}_m$ as previously described in this section. Note that in Eq. 138, the columns of $\mathbf{A}$ are a linear combination of the columns of $\mathbf{Q}$. Thus, to obtain the factorization of Eq. 138, we attempt to write $\mathbf{a}_1, \mathbf{a}_2, \ldots, \mathbf{a}_m$ as a combination of the orthogonal vectors $\mathbf{b}_1, \mathbf{b}_2, \ldots, \mathbf{b}_m$. From Eqs. 145 and 146, one has

$$\mathbf{a}_i = \sum_{j=1}^{i-1} (\mathbf{a}_i^{\mathrm{T}} \mathbf{b}_j)\mathbf{b}_j + |\mathbf{b}_i'|\mathbf{b}_i \tag{2.149}$$

Since $\mathbf{b}_1, \mathbf{b}_2, \ldots,$ and $\mathbf{b}_m$ are orthonormal vectors, the use of Eq. 149 leads to

$$\mathbf{a}_i^{\mathrm{T}} \mathbf{b}_i = \mathbf{b}_i^{\mathrm{T}} \mathbf{a}_i = |\mathbf{b}_i'| \tag{2.150}$$

Substituting Eq. 150 into Eq. 149, one gets

$$\mathbf{a}_i = \sum_{j=1}^{i-1} (\mathbf{a}_i^{\mathrm{T}} \mathbf{b}_j)\mathbf{b}_j + (\mathbf{a}_i^{\mathrm{T}} \mathbf{b}_i)\mathbf{b}_i = \sum_{j=1}^{i} (\mathbf{a}_i^{\mathrm{T}} \mathbf{b}_j)\mathbf{b}_j = \sum_{j=1}^{i} (\mathbf{b}_j^{\mathrm{T}} \mathbf{a}_i)\mathbf{b}_j \tag{2.151}$$

Using this equation, the matrix $\mathbf{A}$, which has linearly independent columns, can be written as

$$\begin{aligned} \mathbf{A} &= [\mathbf{a}_1 \quad \mathbf{a}_2 \quad \cdots \quad \mathbf{a}_m] \\ &= [\mathbf{b}_1 \quad \mathbf{b}_2 \quad \cdots \quad \mathbf{b}_m] \begin{bmatrix} \mathbf{b}_1^{\mathrm{T}}\mathbf{a}_1 & \mathbf{b}_1^{\mathrm{T}}\mathbf{a}_2 & \cdots & \mathbf{b}_1^{\mathrm{T}}\mathbf{a}_m \\ 0 & \mathbf{b}_2^{\mathrm{T}}\mathbf{a}_2 & \cdots & \mathbf{b}_2^{\mathrm{T}}\mathbf{a}_m \\ \vdots & \vdots & \ddots & \vdots \\ 0 & 0 & \cdots & \mathbf{b}_m^{\mathrm{T}}\mathbf{a}_m \end{bmatrix} \end{aligned} \tag{2.152}$$

which can also be written as $\mathbf{A} = \mathbf{QR}$, where

$$\mathbf{Q} = [\mathbf{b}_1 \quad \mathbf{b}_2 \quad \cdots \quad \mathbf{b}_m] \tag{2.153}$$

and

$$\mathbf{R} = \begin{bmatrix} \mathbf{b}_1^{\mathrm{T}}\mathbf{a}_1 & \mathbf{b}_1^{\mathrm{T}}\mathbf{a}_2 & \cdots & \mathbf{b}_1^{\mathrm{T}}\mathbf{a}_m \\ 0 & \mathbf{b}_2^{\mathrm{T}}\mathbf{a}_2 & \cdots & \mathbf{b}_2^{\mathrm{T}}\mathbf{a}_m \\ \vdots & \vdots & \ddots & \vdots \\ 0 & 0 & \cdots & \mathbf{b}_m^{\mathrm{T}}\mathbf{a}_m \end{bmatrix} \tag{2.154}$$

The matrix $\mathbf{Q}$ is an $n \times m$ matrix that has orthonormal columns. The $m \times m$ matrix $\mathbf{R}$ is an upper triangular and is invertible.

If $\mathbf{A}$ is a square matrix, the matrix $\mathbf{Q}$ is square and orthogonal. If the $\mathbf{Q}$ and $\mathbf{R}$ factors are found for a square matrix, the solution of the system of equations $\mathbf{Ax} = \mathbf{b}$ can be determined efficiently, since in this case we have $\mathbf{QRx} = \mathbf{b}$ or $\mathbf{Rx} = \mathbf{Q}^{\mathrm{T}}\mathbf{b}$. The solution of this system can be obtained by back-substitution since $\mathbf{R}$ is an upper-triangular matrix.

Example 2.15

Consider the 4×3 matrix

$$\mathbf{A} = \begin{bmatrix} 2 & 1 & 1 \\ -1 & 2 & -1 \\ 4 & -3 & 1 \\ 1 & 0 & 2 \end{bmatrix}$$

which has the linearly independent columns

$$\mathbf{a}_1 = \begin{bmatrix} 2 \\ -1 \\ 4 \\ 1 \end{bmatrix}, \quad \mathbf{a}_2 = \begin{bmatrix} 1 \\ 2 \\ -3 \\ 0 \end{bmatrix}, \quad \mathbf{a}_3 = \begin{bmatrix} 1 \\ -1 \\ 1 \\ 2 \end{bmatrix}$$

It was shown in the preceding example that the application of the Gram–Schmidt orthogonalization process leads to the following orthonormal vectors:

$$\mathbf{b}_1 = \begin{bmatrix} 0.4264 \\ -0.2132 \\ 0.8528 \\ 0.2132 \end{bmatrix}, \quad \mathbf{b}_2 = \begin{bmatrix} 0.7658 \\ 0.5327 \\ -0.2997 \\ 0.1998 \end{bmatrix}, \quad \mathbf{b}_3 = \begin{bmatrix} -0.0409 \\ -0.4290 \\ -0.2996 \\ 0.8512 \end{bmatrix}$$

The **Q** and **R** factors of the matrix **A** are

$$\mathbf{Q} = [\mathbf{b}_1 \quad \mathbf{b}_2 \quad \mathbf{b}_3] = \begin{bmatrix} 0.4264 & 0.7658 & -0.0409 \\ -0.2132 & 0.5327 & -0.4290 \\ 0.8528 & -0.2997 & -0.2996 \\ 0.2132 & 0.1998 & 0.8512 \end{bmatrix}$$

$$\mathbf{R} = \begin{bmatrix} \mathbf{b}_1^{\mathrm{T}}\mathbf{a}_1 & \mathbf{b}_1^{\mathrm{T}}\mathbf{a}_2 & \mathbf{b}_1^{\mathrm{T}}\mathbf{a}_3 \\ 0 & \mathbf{b}_2^{\mathrm{T}}\mathbf{a}_2 & \mathbf{b}_2^{\mathrm{T}}\mathbf{a}_3 \\ 0 & 0 & \mathbf{b}_3^{\mathrm{T}}\mathbf{a}_3 \end{bmatrix} = \begin{bmatrix} 4.6904 & -2.5584 & 1.9188 \\ 0 & 2.7303 & 0.333 \\ 0 & 0 & 1.7909 \end{bmatrix}$$

Householder Transformation The application of the Gram–Schmidt orthogonalization process leads to a **QR** factorization in which the matrix **Q** has orthogonal column vectors, while the matrix **R** is a square upper-triangular matrix. In what follows, we discuss a procedure based on the Householder transformation. This procedure can be used to obtain a **QR** factorization in which the matrix **Q** is a square orthogonal matrix.

A *Householder transformation* or an *elementary reflector* associated with a unit vector $\hat{\mathbf{v}}$ is defined as

$$\mathbf{H} = \mathbf{I} - 2\hat{\mathbf{v}}\hat{\mathbf{v}}^{\mathrm{T}} \tag{2.155}$$

where **I** is an identity matrix. The matrix **H** is symmetric and also orthogonal since

$$\begin{aligned} \mathbf{H}^{\mathrm{T}}\mathbf{H} &= (\mathbf{I} - 2\hat{\mathbf{v}}\hat{\mathbf{v}}^{\mathrm{T}})(\mathbf{I} - 2\hat{\mathbf{v}}\hat{\mathbf{v}}^{\mathrm{T}}) \\ &= \mathbf{I} - 4\hat{\mathbf{v}}\hat{\mathbf{v}}^{\mathrm{T}} + 4\hat{\mathbf{v}}\hat{\mathbf{v}}^{\mathrm{T}} = \mathbf{I} \end{aligned} \tag{2.156}$$

It follows that $\mathbf{H} = \mathbf{H}^{\mathrm{T}} = \mathbf{H}^{-1}$. It is also clear that if $\mathbf{v} = |\mathbf{v}|\hat{\mathbf{v}}$, then

$$\mathbf{H}\mathbf{v} = -\mathbf{v} \tag{2.157}$$

Furthermore, if **u** is the column vector

$$\mathbf{u} = [1 \quad 0 \quad 0 \quad \cdots \quad 0]^{\mathrm{T}} \tag{2.158}$$

and

$$\mathbf{v} = \mathbf{a} + \beta\mathbf{u} \tag{2.159}$$

where

$$\beta = |\mathbf{a}| = \sqrt{\mathbf{a}^{\mathrm{T}}\mathbf{a}} \tag{2.160}$$

then

$$\mathbf{Ha} = \left[\mathbf{I} - \frac{2\mathbf{v}\mathbf{v}^{\mathrm{T}}}{(|\mathbf{v}|)^2}\right]\mathbf{a} = \mathbf{a} - \frac{2\mathbf{v}\mathbf{v}^{\mathrm{T}}}{(|\mathbf{v}|)^2}\,\mathbf{a} \tag{2.161}$$

Using Eq. 159, one obtains

$$\mathbf{Ha} = \mathbf{a} - (\mathbf{a} + \beta\mathbf{u})\,\frac{2(\mathbf{a} + \beta\mathbf{u})^{\mathrm{T}}\mathbf{a}}{(\mathbf{a} + \beta\mathbf{u})^{\mathrm{T}}(\mathbf{a} + \beta\mathbf{u})} \tag{2.162}$$

Since

$$(\beta)^2\mathbf{u}^{\mathrm{T}}\mathbf{u} = (\beta)^2 = \mathbf{a}^{\mathrm{T}}\mathbf{a}, \tag{2.163}$$

the denominator in Eq. 162 can be written as $(\mathbf{a} + \beta\mathbf{u})^{\mathrm{T}}(\mathbf{a} + \beta\mathbf{u}) = 2(\mathbf{a} + \beta\mathbf{u})^{\mathrm{T}}\mathbf{a}$. Substituting this equation into Eq. 162 yields

$$\mathbf{Ha} = -\beta\mathbf{u} = \begin{bmatrix} -\beta \\ 0 \\ 0 \\ \vdots \\ 0 \end{bmatrix} \tag{2.164}$$

This equation implies that when the Householder transformation constructed using the vector $\mathbf{v}$ of Eq. 159 is multiplied by the vector $\mathbf{a}$, the result is a vector whose only nonzero element is the first element. Using this fact, a matrix can be transformed to an upper-triangular form by the successive application of a series of Householder transformations. In order to demonstrate this process, consider the rectangular $n \times m$ matrix

$$\mathbf{A} = \begin{bmatrix} a_{11} & a_{12} & a_{13} & \cdots & a_{1m} \\ a_{21} & a_{22} & a_{23} & \cdots & a_{2m} \\ \vdots & \vdots & \vdots & \ddots & \vdots \\ a_{n1} & a_{n2} & a_{n3} & \cdots & a_{nm} \end{bmatrix} \tag{2.165}$$

where $n \geq m$. First we construct the Householder transformation associated with the first column $\mathbf{a}_1 = [a_{11}\ a_{21}\ \cdots\ a_{n1}]^{\mathrm{T}}$. This transformation matrix can be written as

$$\mathbf{H}_1 = \mathbf{I} - 2\hat{\mathbf{v}}_1\hat{\mathbf{v}}_1^{\mathrm{T}} \tag{2.166}$$

where the vector $\mathbf{v}_1$ is defined as

$$\mathbf{v}_1 = \mathbf{a}_1 + \beta_1\mathbf{u}_1 \tag{2.167}$$

in which β_1 is the norm of $\mathbf{a}_1$, and the vector $\mathbf{u}_1$ has the same dimension as $\mathbf{a}_1$ and is defined by Eq. 158. Using Eq. 164, it is clear that

$$\mathbf{H}_1\mathbf{a}_1 = -\beta_1\mathbf{u}_1 = \begin{bmatrix} -\beta_1 \\ 0 \\ 0 \\ \vdots \\ 0 \end{bmatrix} \tag{2.168}$$

It follows that

$$\mathbf{A}_1 = \mathbf{H}_1\mathbf{A} = \begin{bmatrix} -\beta_1 & (\mathbf{a}_{12})_1 & (\mathbf{a}_{13})_1 & \cdots & (\mathbf{a}_{1m})_1 \\ 0 & (\mathbf{a}_{22})_1 & (\mathbf{a}_{23})_1 & \cdots & (\mathbf{a}_{2m})_1 \\ 0 & (\mathbf{a}_{32})_1 & (\mathbf{a}_{33})_1 & \cdots & (\mathbf{a}_{3m})_1 \\ \vdots & \vdots & \vdots & \ddots & \vdots \\ 0 & (\mathbf{a}_{n2})_1 & (\mathbf{a}_{n3})_1 & \cdots & (\mathbf{a}_{nm})_1 \end{bmatrix} \tag{2.169}$$

Now we consider the second column of the matrix $\mathbf{A}_1 = \mathbf{H}_1\mathbf{A}$. We use the last $n-1$ elements of this column vector to form

$$\bar{\mathbf{a}}_2 = [(\mathbf{a}_{22})_1 \quad (\mathbf{a}_{32})_1 \quad \cdots \quad (\mathbf{a}_{n2})_1]^{\mathrm{T}} \tag{2.170}$$

A Householder transformation matrix $\overline{\mathbf{H}}_2$ can be constructed such that

$$\overline{\mathbf{H}}_2\bar{\mathbf{a}}_2 = -\beta_2\mathbf{u}_2 = \begin{bmatrix} -\beta_2 \\ 0 \\ 0 \\ \vdots \\ 0 \end{bmatrix} \tag{2.171}$$

where β_2 is the norm of the vector $\bar{\mathbf{a}}_2$, and $\mathbf{u}_2$ is an $(n-1)$-dimensional vector defined by Eq. 158. Observe that at this point, the Householder transformation $\overline{\mathbf{H}}_2$ is only of order $n-1$. This transformation can be imbedded into the lower-right corner of an $n \times n$ matrix $\mathbf{H}_2$, where

$$\mathbf{H}_2 = \begin{bmatrix} 1 & \mathbf{0} \\ \mathbf{0} & \overline{\mathbf{H}}_2 \end{bmatrix} \tag{2.172}$$

Because only the first element in the first row and first column of this matrix is nonzero and equal to 1, when this transformation is applied to an arbitrary matrix it does not change the first row or the first column of that matrix. Furthermore, $\mathbf{H}_2$ is an orthogonal symmetric matrix, since $\overline{\mathbf{H}}_2$ is both orthogonal and symmetric. By applying the matrix $\mathbf{H}_2$ to $\mathbf{A}_1$, one obtains

$$\mathbf{A}_2 = \mathbf{H}_2\mathbf{A}_1 = \mathbf{H}_2\mathbf{H}_1\mathbf{A} = \begin{bmatrix} -\beta_1 & (\mathbf{a}_{12})_1 & (\mathbf{a}_{13})_1 & \cdots & (\mathbf{a}_{1m})_1 \\ 0 & -\beta_2 & (\mathbf{a}_{23})_2 & \cdots & (\mathbf{a}_{2m})_2 \\ 0 & 0 & (\mathbf{a}_{33})_2 & \cdots & (\mathbf{a}_{3m})_2 \\ \vdots & \vdots & \vdots & \ddots & \vdots \\ 0 & 0 & (\mathbf{a}_{n3})_2 & \cdots & (\mathbf{a}_{nm})_2 \end{bmatrix} \tag{2.173}$$

We consider the third column of the matrix $\mathbf{A}_2$, and use the last $n - 2$ elements to form the vector

$$\overline{\mathbf{a}}_3 = [(\mathbf{a}_{33})_2 \quad (\mathbf{a}_{43})_2 \quad \cdots \quad (\mathbf{a}_{n3})_2]^{\mathrm{T}} \tag{2.174}$$

The Householder transformation $\overline{\mathbf{H}}_3$ associated with this $(n - 2)$-dimensional vector can be constructed. This matrix can then be imbedded into the lower-right corner of the $n \times n$ matrix

$$\mathbf{H}_3 = \begin{bmatrix} 1 & 0 & \mathbf{0} \\ 0 & 1 & \mathbf{0} \\ \mathbf{0} & \mathbf{0} & \overline{\mathbf{H}}_3 \end{bmatrix} \tag{2.175}$$

Using this matrix, one has

$$\mathbf{H}_3\mathbf{A}_2 = \mathbf{H}_3\mathbf{H}_2\mathbf{H}_1\mathbf{A}_1 = \begin{bmatrix} -\beta_1 & (\mathbf{a}_{12})_1 & (\mathbf{a}_{13})_1 & \cdots & (\mathbf{a}_{1m})_1 \\ 0 & -\beta_2 & (\mathbf{a}_{23})_2 & \cdots & (\mathbf{a}_{2m})_2 \\ 0 & 0 & -\beta_3 & \cdots & (\mathbf{a}_{3m})_3 \\ \vdots & \vdots & \vdots & \ddots & \vdots \\ 0 & 0 & 0 & \cdots & (\mathbf{a}_{nm})_3 \end{bmatrix} \tag{2.176}$$

where β_3 is the norm of the vector $\overline{\mathbf{a}}_3$. It is clear that by continuing this process, all the elements below the diagonal of the matrix $\mathbf{A}$ can be made equal to zero. If $\mathbf{A}$ is a square nonsingular matrix, the result of the Householder transformations is an upper-triangular matrix. If $\mathbf{A}$, on the other hand, is a rectangular matrix, the result of m-Householder transformations is

$$\mathbf{A}_m = \mathbf{H}_m\mathbf{H}_{m-1} \quad \cdots \quad \mathbf{H}_1\mathbf{A} = \begin{bmatrix} \mathbf{R}_1 \\ \mathbf{0} \end{bmatrix} \tag{2.177}$$

where $\mathbf{R}_1$ is an $m \times m$ upper-triangular matrix.

Example 2.16

Consider the matrix

$$\mathbf{A} = \begin{bmatrix} 2 & 1 & 1 \\ -1 & 2 & -1 \\ 4 & -3 & 1 \\ 1 & 0 & 2 \end{bmatrix}$$

First we consider the first column of this matrix:

$$\mathbf{a}_1 = [2 \quad -1 \quad 4 \quad 1]^{\mathrm{T}}$$

The norm of this vector is

$$\beta_1 = |\mathbf{a}_1| = 4.6904$$

The vector $\mathbf{v}_1$ is defined as

$$\mathbf{v}_1 = \mathbf{a}_1 + \beta_1 \mathbf{u} = \begin{bmatrix} 2 \\ -1 \\ 4 \\ 1 \end{bmatrix} + 4.6904 \begin{bmatrix} 1 \\ 0 \\ 0 \\ 0 \end{bmatrix} = \begin{bmatrix} 6.6904 \\ -1 \\ 4 \\ 1 \end{bmatrix}$$

The unit vector $\hat{\mathbf{v}}_1$ is

$$\hat{\mathbf{v}}_1 = [0.8445 \quad -0.1262 \quad 0.5049 \quad 0.1262]^{\mathrm{T}}$$

The Householder transformation $\mathbf{H}_1$ is

$$\mathbf{H}_1 = \mathbf{I} - 2\hat{\mathbf{v}}\hat{\mathbf{v}}^{\mathrm{T}} = \begin{bmatrix} -0.4264 & 0.2132 & -0.8528 & -0.2132 \\ 0.2132 & 0.9681 & 0.1274 & 0.0319 \\ -0.8528 & 0.1274 & 0.4902 & -0.1274 \\ -0.2132 & 0.0319 & -0.1274 & 0.9681 \end{bmatrix}$$

The matrix $\mathbf{A}_1$ is

$$\mathbf{A}_1 = \mathbf{H}_1\mathbf{A} = \begin{bmatrix} -4.6904 & 2.5584 & -1.9188 \\ 0 & 1.7672 & -0.5637 \\ 0 & -2.0686 & -0.7448 \\ 0 & 0.2328 & 1.5637 \end{bmatrix}$$

The vector $\bar{\mathbf{a}}_2$ can be obtained from the second column of this matrix as

$$\bar{\mathbf{a}}_2 = [1.7672 \quad -2.0686 \quad 0.2328]^{\mathrm{T}}$$

The norm of this vector is

$$\beta_2 = |\bar{\mathbf{a}}_2| = 2.7306$$

The vector $\mathbf{v}_2$ is defined as

$$\mathbf{v}_2 = \bar{\mathbf{a}}_2 + \beta_2 \mathbf{u} = \begin{bmatrix} 1.7672 \\ -2.0686 \\ 0.2328 \end{bmatrix} + 2.7306 \begin{bmatrix} 1 \\ 0 \\ 0 \end{bmatrix}$$

$$= \begin{bmatrix} 4.4978 \\ -2.0686 \\ 0.2328 \end{bmatrix}$$

It follows that

$$\hat{\mathbf{v}}_2 = [0.9075 \quad -0.4174 \quad 0.0470]^{\mathrm{T}}$$

The matrix $\overline{\mathbf{H}}_2$ is

$$\overline{\mathbf{H}}_2 = \mathbf{I} - 2\hat{\mathbf{v}}_2\hat{\mathbf{v}}_2^{\mathrm{T}} = \begin{bmatrix} -0.6471 & 0.7576 & -0.0853 \\ 0.7576 & 0.6516 & 0.0392 \\ -0.0853 & 0.0392 & 0.9956 \end{bmatrix}$$

The matrix $\mathbf{H}_2$ can be written as

$$\mathbf{H}_2 = \begin{bmatrix} 1 & \mathbf{0} \\ \mathbf{0} & \overline{\mathbf{H}}_2 \end{bmatrix} = \begin{bmatrix} 1 & 0 & 0 & 0 \\ 0 & -0.6471 & 0.7576 & -0.0853 \\ 0 & 0.7576 & 0.6516 & 0.0392 \\ 0 & -0.0853 & 0.0392 & 0.9956 \end{bmatrix}$$

and

$$\mathbf{A}_2 = \mathbf{H}_2\mathbf{A}_1 = \mathbf{H}_2\mathbf{H}_1\mathbf{A} = \begin{bmatrix} -4.6904 & 2.5584 & -1.9188 \\ 0 & -2.7306 & -0.3329 \\ 0 & 0 & -0.8511 \\ 0 & 0 & 1.5757 \end{bmatrix}$$

Using the last two elements of the third column of this matrix, one defines

$$\bar{\mathbf{a}}_3 = [-0.8511 \quad 1.5757]^{\mathrm{T}}$$

The norm of this vector is

$$\beta_3 = |\bar{\mathbf{a}}_3| = 1.7909$$

The vector $\mathbf{v}_3$ is defined as

$$\mathbf{v}_3 = \bar{\mathbf{a}}_3 + \beta_3 \mathbf{u} = \begin{bmatrix} -0.8511 \\ 1.5757 \end{bmatrix} + 1.7909 \begin{bmatrix} 1 \\ 0 \end{bmatrix} = \begin{bmatrix} 0.9398 \\ 1.5757 \end{bmatrix}$$

A unit vector in the direction of $\mathbf{v}_3$ is

$$\hat{\mathbf{v}}_3 = [0.5122 \quad 0.8588]^{\mathrm{T}}$$

The Householder transformation associated with this vector is

$$\overline{\mathbf{H}}_3 = \mathbf{I} - 2\hat{\mathbf{v}}_3\hat{\mathbf{v}}_3^{\mathrm{T}} = \begin{bmatrix} 0.4753 & -0.8798 \\ -0.8798 & -0.4751 \end{bmatrix}$$

Using this matrix, the transformation $\mathbf{H}_3$ can be defined as

$$\mathbf{H}_3 = \begin{bmatrix} 1 & 0 & \mathbf{0} \\ 0 & 1 & \mathbf{0} \\ \mathbf{0} & \mathbf{0} & \overline{\mathbf{H}}_3 \end{bmatrix} = \begin{bmatrix} 1 & 0 & 0 & 0 \\ 0 & 1 & 0 & 0 \\ 0 & 0 & 0.4753 & -0.8798 \\ 0 & 0 & -0.8798 & -0.4751 \end{bmatrix}$$

Using this matrix, one obtains

$$\mathbf{A}_3 = \mathbf{H}_3\mathbf{A}_2 = \mathbf{H}_3\mathbf{H}_2\mathbf{H}_1\mathbf{A} = \begin{bmatrix} -4.6904 & 2.5584 & -1.9188 \\ 0 & -2.7306 & -0.3329 \\ 0 & 0 & -1.7909 \\ 0 & 0 & 0 \end{bmatrix}$$

The matrix $\mathbf{A}_3$ can be written as

$$\mathbf{A}_3 = \begin{bmatrix} \mathbf{R}_1 \\ \mathbf{0} \end{bmatrix}$$

where $\mathbf{R}_1$ is the upper-triangular matrix

$$\mathbf{R}_1 = \begin{bmatrix} -4.6904 & 2.5584 & -1.9188 \\ 0 & -2.7306 & -0.3329 \\ 0 & 0 & -1.7909 \end{bmatrix}$$

Note the relationship between the matrix $\mathbf{R}_1$ obtained in this example by the successive application of Householder transformations and the matrix $\mathbf{R}$ obtained in Example 15 as the result of the application of the Gram–Schmidt orthogonalization process. The similarity between these two matrices is not surprising because the uniqueness of the **QR** factorization can easily be demonstrated.

The application of a sequence of Householder transformations to an $n \times m$ rectangular matrix $\mathbf{A}$ with linearly independent columns leads to

$$\mathbf{H}_m\mathbf{H}_{m-1} \quad \cdots \quad \mathbf{H}_1\mathbf{A} = \mathbf{R} \tag{2.178}$$

where $\mathbf{H}_i$ is the ith orthogonal Householder transformation and $\mathbf{R}$ is an $n \times m$ rectangular matrix that can be written as

$$\mathbf{R} = \begin{bmatrix} \mathbf{R}_1 \\ \mathbf{0} \end{bmatrix} \tag{2.179}$$

where $\mathbf{R}_1$ is an $m \times m$ upper-triangular matrix. If $\mathbf{A}$ is a square matrix, $\mathbf{R} = \mathbf{R}_1$. Since the Householder transformations are symmetric and orthogonal, one has

$$\mathbf{H}_i^{\mathrm{T}} = \mathbf{H}_i^{-1} = \mathbf{H}_i \tag{2.180}$$

Using this identity, Eq. 178 leads to

$$\mathbf{A} = \mathbf{H}_1\mathbf{H}_2 \quad \cdots \quad \mathbf{H}_m\mathbf{R} \tag{2.181}$$

Since the product of orthogonal matrices defines an orthogonal matrix, Eq. 181 can be written in the form of Eq. 138 as $\mathbf{A} = \mathbf{QR}$, where $\mathbf{Q}$ is an orthogonal square matrix defined as

$$\mathbf{Q} = \mathbf{H}_1\mathbf{H}_2 \quad \cdots \quad \mathbf{H}_m \tag{2.182}$$

Example 2.17

In the preceding example it was shown that the Householder transformations that reduce the matrix

$$\mathbf{A} = \begin{bmatrix} 2 & 1 & 1 \\ -1 & 2 & -1 \\ 4 & -3 & 1 \\ 1 & 0 & 2 \end{bmatrix}$$

to the matrix

$$\mathbf{R} = \begin{bmatrix} -4.6904 & 2.5584 & -1.9188 \\ 0 & -2.7306 & -0.3329 \\ 0 & 0 & -1.7909 \\ 0 & 0 & 0 \end{bmatrix}$$

are

$$\mathbf{H}_1 = \begin{bmatrix} -0.4264 & 0.2132 & -0.8528 & -0.2132 \\ 0.2132 & 0.9681 & 0.1274 & 0.0319 \\ -0.8528 & 0.1274 & 0.4902 & -0.1274 \\ -0.2132 & 0.0319 & -0.1274 & 0.9681 \end{bmatrix}$$

$$\mathbf{H}_2 = \begin{bmatrix} 1 & 0 & 0 & 0 \\ 0 & -0.6471 & 0.7576 & -0.0853 \\ 0 & 0.7576 & 0.6516 & 0.0392 \\ 0 & -0.0853 & 0.0392 & 0.9956 \end{bmatrix}$$

$$\mathbf{H}_3 = \begin{bmatrix} 1 & 0 & 0 & 0 \\ 0 & 1 & 0 & 0 \\ 0 & 0 & 0.4753 & -0.8798 \\ 0 & 0 & -0.8798 & -0.4751 \end{bmatrix}$$

In this case, the matrix $\mathbf{Q}$ of Eq. 182 is

$$\mathbf{Q} = \mathbf{H}_1\mathbf{H}_2\mathbf{H}_3$$

$$\mathbf{Q} = \begin{bmatrix} -0.4264 & -0.7659 & 0.0409 & 0.4795 \\ 0.2132 & -0.5327 & 0.4290 & -0.6977 \\ -0.8528 & 0.2997 & 0.2996 & -0.3052 \\ -0.2132 & -0.1998 & -0.8512 & -0.4359 \end{bmatrix}$$

The similarity between the first three columns of this matrix and the matrix $\mathbf{Q}$ obtained in Example 15 using the Gram–Schmidt orthogonalization process is clear.

If the fourth column of the preceding matrix is denoted as $\mathbf{Q}_2$, that is,

$$\mathbf{Q}_2 = [0.4795 \quad -0.6977 \quad -0.3052 \quad -0.4359]^{\mathrm{T}}$$

it is easy to verify that $\mathbf{A}^{\mathrm{T}}\mathbf{Q}_2 = \mathbf{0}$.

Important Identities for the QR Factors If $\mathbf{A}$ is an $n \times m$ rectangular matrix that has the $\mathbf{QR}$ decomposition given by Eq. 138, the matrix $\mathbf{R}$ takes the form given by Eq. 179. In this case, one can use matrix partitioning to write

$$\mathbf{A} = [\mathbf{Q}_1 \quad \mathbf{Q}_2]\begin{bmatrix} \mathbf{R}_1 \\ \mathbf{0} \end{bmatrix} \tag{2.183}$$

where $\mathbf{Q}_1$ and $\mathbf{Q}_2$ are partitions of the matrix $\mathbf{Q}$, that is,

$$\mathbf{Q} = [\mathbf{Q}_1 \quad \mathbf{Q}_2] \tag{2.184}$$

The matrix $\mathbf{Q}_1$ is an $n \times m$ matrix, while $\mathbf{Q}_2$ is an $n \times (n-m)$ matrix. Both $\mathbf{Q}_1$ and $\mathbf{Q}_2$ have columns that are orthogonal vectors. Furthermore,

$$\mathbf{Q}_1^{\mathrm{T}}\mathbf{Q}_2 = \mathbf{0} \tag{2.185}$$

It follows from Eq. 183 that

$$\mathbf{A} = \mathbf{Q}_1\mathbf{R}_1 \tag{2.186}$$

Consequently,

$$\mathbf{A}^{\mathrm{T}}\mathbf{A} = \mathbf{R}_1^{\mathrm{T}}\mathbf{Q}_1^{\mathrm{T}}\mathbf{Q}_1\mathbf{R}_1 = \mathbf{R}_1^{\mathrm{T}}\mathbf{R}_1 \tag{2.187}$$

If $\mathbf{A}$ has linearly independent columns, Eq. 187 represents the Cholesky factorization of the positive definitive symmetric matrix $\mathbf{A}^{\mathrm{T}}\mathbf{A}$. Therefore, $\mathbf{R}_1$ is unique. The uniqueness of $\mathbf{Q}_1$ follows immediately from Eq. 186, since

$$\mathbf{Q}_1 = \mathbf{A}\mathbf{R}_1^{-1} \tag{2.188}$$

While $\mathbf{Q}_1$ and $\mathbf{R}_1$ are unique, $\mathbf{Q}_2$ is not unique. We note, however, that

$$\mathbf{Q}_2^{\mathrm{T}}\mathbf{A} = \mathbf{Q}_2^{\mathrm{T}}\mathbf{Q}_1\mathbf{R}_1 = \mathbf{0} \tag{2.189}$$

or

$$\mathbf{A}^{\mathrm{T}}\mathbf{Q}_2 = \mathbf{0} \tag{2.190}$$

which implies that the orthogonal column vectors of the matrix $\mathbf{Q}_2$ form the basis of the *null space* of the matrix $\mathbf{A}^{\mathrm{T}}$. This fact was demonstrated by the results presented in Example 17.

2.8 SINGULAR VALUE DECOMPOSITION

Another factorization that is used in the dynamic analysis of mechanical systems is the *singular value decomposition* (SVD). The singular value decomposition of the matrix $\mathbf{A}$ can be written as

$$\mathbf{A} = \mathbf{Q}_1\mathbf{B}\mathbf{Q}_2 \tag{2.191}$$

where $\mathbf{Q}_1$ and $\mathbf{Q}_2$ are two orthogonal matrices and $\mathbf{B}$ is a diagonal matrix which has the same dimension as $\mathbf{A}$. Before we prove Eq. 191, we first discuss briefly the *eigenvalue problem*.

Eigenvalue Problem In mechanics, we frequently encounter a system of equations in the form

$$\mathbf{A}\mathbf{x} = \lambda\mathbf{x} \tag{2.192}$$

where $\mathbf{A}$ is a square matrix, $\mathbf{x}$ is an unknown vector, and λ is an unknown scalar. Equation 192 can be written as

$$(\mathbf{A} - \lambda\mathbf{I})\mathbf{x} = \mathbf{0} \tag{2.193}$$

This system of equations has a nontrivial solution if and only if the determinant of the coefficient matrix is equal to zero, that is,

$$|\mathbf{A} - \lambda \mathbf{I}| = 0 \tag{2.194}$$

This is the *characteristic equation* for the matrix $\mathbf{A}$. If $\mathbf{A}$ is an $n \times n$ matrix, Eq. 194 is a polynomial of order n in λ. This polynomial can be written in the following general form:

$$a_n \lambda^n + a_{n-1} \lambda^{n-1} + \cdots + a_0 = 0 \tag{2.195}$$

where a_k are the coefficients of the *characteristic polynomial*. The solution of Eq. 195 defines the n roots $\lambda_1, \lambda_2, \ldots, \lambda_n$. The roots λ_i, $i = 1, \ldots, n$ are called the *characteristic values* or the *eigenvalues* of the matrix $\mathbf{A}$. Corresponding to each of these eigenvalues, there is an associated *eigenvector* $\mathbf{x}_i$, which can be determined by solving the system of homogeneous equations

$$[\mathbf{A} - \lambda_i \mathbf{I}]\mathbf{x}_i = \mathbf{0} \tag{2.196}$$

If $\mathbf{A}$ is a real symmetric matrix, one can show that the eigenvectors associated with distinctive eigenvalues are orthogonal. To prove this fact, we use Eq. 196 to write

$$\mathbf{A}\mathbf{x}_i = \lambda_i \mathbf{x}_i, \quad \mathbf{A}\mathbf{x}_j = \lambda_j \mathbf{x}_j \tag{2.197}$$

Premultiplying the first equation in Eq. 197 by $\mathbf{x}_j^{\mathrm{T}}$ and postmultiplying the transpose of the second equation by $\mathbf{x}_i$, one obtains

$$\mathbf{x}_j^{\mathrm{T}} \mathbf{A}\mathbf{x}_i = \lambda_i \mathbf{x}_j^{\mathrm{T}} \mathbf{x}_i, \quad \mathbf{x}_j^{\mathrm{T}} \mathbf{A}\mathbf{x}_i = \lambda_j \mathbf{x}_j^{\mathrm{T}} \mathbf{x}_i \tag{2.198}$$

Subtracting yields

$$(\lambda_i - \lambda_j)\mathbf{x}_j^{\mathrm{T}} \mathbf{x}_i = 0 \tag{2.199}$$

which implies that

$$\left.\begin{aligned} \mathbf{x}_i^{\mathrm{T}} \mathbf{x}_j &= 0 \qquad \text{if } i \neq j \\ &\neq 0 \qquad \text{if } i = j \end{aligned}\right\} \tag{2.200}$$

This orthogonality condition guarantees that the eigenvectors associated with distinctive eigenvalues are linearly independent.

Example 2.18

Find the eigenvalues and eigenvectors of the matrix

$$\mathbf{A} = \begin{bmatrix} 4 & 1 & 2 \\ 1 & 0 & 0 \\ 2 & 0 & 0 \end{bmatrix}$$

Solution. The characteristic equation of this matrix is

$$|\mathbf{A} - \lambda\mathbf{I}| = \begin{bmatrix} 4-\lambda & 1 & 2 \\ 1 & -\lambda & 0 \\ 2 & 0 & -\lambda \end{bmatrix} = (4-\lambda)(\lambda)^2 + \lambda + 4\lambda$$
$$= -\lambda(\lambda - 5)(\lambda + 1) = 0$$

Therefore, the eigenvalues are

$$\lambda_1 = -1, \qquad \lambda_2 = 5, \qquad \lambda_3 = 0$$

To evaluate the ith eigenvector, one may use the following equation:

$$\mathbf{A}\mathbf{x}_i = \lambda_i \mathbf{x}_i$$

where $\mathbf{x}_i$ is the ith eigenvector. The preceding equation can be written as

$$(\mathbf{A} - \lambda_i \mathbf{I})\mathbf{x}_i = \mathbf{0}$$

which yields the following eigenvectors:

$$\mathbf{x}_1 = \begin{bmatrix} 1 \\ -1 \\ -2 \end{bmatrix}, \qquad \mathbf{x}_2 = \begin{bmatrix} 5 \\ 1 \\ 2 \end{bmatrix}, \qquad \mathbf{x}_3 = \begin{bmatrix} 0 \\ 2 \\ -1 \end{bmatrix}$$

These eigenvectors are orthogonal because the matrix **A** is real and symmetric. We also observe that the resulting eigenvalues are all real. It can be shown that if **A** is a real symmetric matrix, then all its eigenvalues and eigenvectors are real.

Equation 192 indicates that the eigenvectors are not unique since this equation remains valid if it is multiplied by an arbitrary nonzero scalar. In the case of a real symmetric matrix, if each of the eigenvectors is divided by its length, one obtains an orthonormal set of vectors denoted as $\hat{\mathbf{x}}_1, \hat{\mathbf{x}}_2, \ldots, \hat{\mathbf{x}}_n$. These orthonormal eigenvectors satisfy the following equation:

$$\hat{\mathbf{x}}_i^{\mathrm{T}} \mathbf{A} \hat{\mathbf{x}}_j = \begin{cases} \lambda_i & \text{if } i = j \\ 0 & \text{if } i \neq j \end{cases} \tag{2.201}$$

An orthogonal matrix whose columns are the orthonormal eigenvectors $\hat{\mathbf{x}}_1, \hat{\mathbf{x}}_2, \ldots, \hat{\mathbf{x}}_n$ can be written as

$$\mathbf{U} = [\hat{\mathbf{x}}_1 \quad \hat{\mathbf{x}}_2 \quad \cdots \quad \hat{\mathbf{x}}_n] \tag{2.202}$$

From Eq. 201, it follows that

$$\mathbf{U}^{\mathrm{T}}\mathbf{A}\mathbf{U} = \begin{bmatrix} \lambda_1 & & & \\ & \lambda_2 & \mathbf{0} & \\ & \mathbf{0} & \ddots & \\ & & & \lambda_n \end{bmatrix} \tag{2.203}$$

Example 2.19

It was shown in the preceding example that the eigenvalues of the real symmetric matrix

$$\mathbf{A} = \begin{bmatrix} 4 & 1 & 2 \\ 1 & 0 & 0 \\ 2 & 0 & 0 \end{bmatrix}$$

are $\lambda_1 = -1, \lambda_2 = 5$, and $\lambda_3 = 0$. The eigenvectors associated with these eigenvalues were found to be

$$\mathbf{x}_1 = \begin{bmatrix} 1 \\ -1 \\ -2 \end{bmatrix}, \quad \mathbf{x}_2 = \begin{bmatrix} 5 \\ 1 \\ 2 \end{bmatrix}, \quad \mathbf{x}_3 = \begin{bmatrix} 0 \\ 2 \\ -1 \end{bmatrix}$$

An orthonormal set of eigenvectors can be obtained by dividing each eigenvector by its length. This leads to

$$\hat{\mathbf{x}}_1 = \frac{1}{\sqrt{6}} \begin{bmatrix} 1 \\ -1 \\ -2 \end{bmatrix}, \quad \hat{\mathbf{x}}_2 = \frac{1}{\sqrt{30}} \begin{bmatrix} 5 \\ 1 \\ 2 \end{bmatrix}, \quad \hat{\mathbf{x}}_3 = \frac{1}{\sqrt{5}} \begin{bmatrix} 0 \\ 2 \\ -1 \end{bmatrix}$$

The orthonormal matrix $\mathbf{U}$ of Eq. 202 can then be defined as

$$\mathbf{U} = \begin{bmatrix} \frac{1}{\sqrt{6}} & \frac{5}{\sqrt{30}} & 0 \\ \frac{-1}{\sqrt{6}} & \frac{1}{\sqrt{30}} & \frac{2}{\sqrt{5}} \\ \frac{-2}{\sqrt{6}} & \frac{2}{\sqrt{30}} & \frac{-1}{\sqrt{5}} \end{bmatrix}$$

Matrix multiplications show that

$$\mathbf{U}^{\mathrm{T}}\mathbf{A}\mathbf{U} = \begin{bmatrix} -1 & 0 & 0 \\ 0 & 5 & 0 \\ 0 & 0 & 0 \end{bmatrix}$$

The zero eigenvalue that appears as the last element of the diagonal of the matrix $\mathbf{U}^{\mathrm{T}}\mathbf{A}\mathbf{U}$ indicates a rank deficiency of the matrix $\mathbf{A}$ since the second and third rows of the matrix $\mathbf{A}$ are not linearly independent. The rank of this matrix is 2.

Singular Value Decomposition In the singular value decomposition of Eq. 191, $\mathbf{A}$ can be an arbitrary $n \times m$ rectangular matrix. In this case $\mathbf{Q}_1$ is an $n \times n$ matrix, $\mathbf{B}$ is an $n \times m$ matrix, and $\mathbf{Q}_2$ is an $m \times m$ matrix. To prove the decomposition of Eq. 191, we consider the real symmetric square matrix $\mathbf{A}^{\mathrm{T}}\mathbf{A}$. The eigenvalues of $\mathbf{A}^{\mathrm{T}}\mathbf{A}$ are real and nonnegative. To demonstrate this, we assume that

$$\mathbf{A}^{\mathrm{T}}\mathbf{A}\mathbf{x} = \lambda\mathbf{x} \tag{2.204}$$

It follows that

$$\mathbf{x}^{\mathrm{T}}\mathbf{A}^{\mathrm{T}}\mathbf{A}\mathbf{x} = \lambda\mathbf{x}^{\mathrm{T}}\mathbf{x} \tag{2.205}$$

and also

$$\mathbf{x}^{\mathrm{T}}\mathbf{A}^{\mathrm{T}}\mathbf{A}\mathbf{x} = (\mathbf{A}\mathbf{x})^{\mathrm{T}}(\mathbf{A}\mathbf{x}) \tag{2.206}$$

It is clear from the preceding two equations that

$$\lambda = \frac{(\mathbf{A}\mathbf{x})^{\mathrm{T}}(\mathbf{A}\mathbf{x})}{\mathbf{x}^{\mathrm{T}}\mathbf{x}} \tag{2.207}$$

which demonstrates that λ is indeed nonnegative and $\mathbf{A}^{\mathrm{T}}\mathbf{A}\mathbf{x} = \mathbf{0}$ if and only if $\mathbf{A}\mathbf{x} = \mathbf{0}$.

It was demonstrated previously in this section that if $\hat{\mathbf{x}}_1, \hat{\mathbf{x}}_2, \ldots, \hat{\mathbf{x}}_m$ are the eigenvectors of $\mathbf{A}^{\mathrm{T}}\mathbf{A}$, there exists an orthogonal matrix $\mathbf{U} = [\hat{\mathbf{x}}_1 \quad \hat{\mathbf{x}}_2 \quad \cdots \quad \hat{\mathbf{x}}_m]$ such that

$$\mathbf{U}^{\mathrm{T}}\mathbf{A}^{\mathrm{T}}\mathbf{A}\mathbf{U} = \begin{bmatrix} \lambda_1 & 0 & \cdots & 0 \\ 0 & \lambda_2 & \cdots & 0 \\ \vdots & \vdots & \ddots & \vdots \\ 0 & 0 & \cdots & \lambda_m \end{bmatrix} \tag{2.208}$$

where $\lambda_1, \lambda_2, \ldots, \lambda_m$ are the eigenvalues of the matrix $\mathbf{A}^{\mathrm{T}}\mathbf{A}$. Equation 208 implies that the columns $\mathbf{b}_1, \mathbf{b}_2, \ldots, \mathbf{b}_m$ of the matrix $\mathbf{AU}$ satisfy

$$\mathbf{b}_i^{\mathrm{T}}\mathbf{b}_j = \begin{cases} \lambda_i & \text{if } i = j \\ 0 & \text{if } i \neq j \end{cases} \tag{2.209}$$

For the r nonzero eigenvalues, where $r \leq m$, we define

$$\hat{\mathbf{b}}_i = \frac{1}{\sqrt{\lambda_i}}\mathbf{b}_i \quad i = 1, 2, \ldots, r \tag{2.210}$$

This is an orthonormal set of vectors. If $r < m$ and $n \geq m$, we choose $\hat{\mathbf{b}}_{r+1}, \ldots, \hat{\mathbf{b}}_n$ such that

$$\mathbf{Q}_1 = [\hat{\mathbf{b}}_1 \quad \hat{\mathbf{b}}_2 \quad \cdots \quad \hat{\mathbf{b}}_n] \tag{2.211}$$

is an $n \times n$ orthogonal matrix. Recall that $\mathbf{AU} = [\mathbf{b}_1\ \mathbf{b}_2\ \cdots\ \mathbf{b}_m]$. Using this equation and Eq. 210, it can be verified that

$$\mathbf{AU} = [\sqrt{\lambda_1}\hat{\mathbf{b}}_1 \quad \sqrt{\lambda_2}\hat{\mathbf{b}}_2 \quad \cdots \quad \sqrt{\lambda_m}\hat{\mathbf{b}}_m] = \mathbf{Q}_1 \begin{bmatrix} \sqrt{\lambda_1} & 0 & \cdots & 0 \\ 0 & \sqrt{\lambda_2} & \cdots & 0 \\ \vdots & \vdots & \ddots & \vdots \\ 0 & 0 & \cdots & \sqrt{\lambda_m} \\ 0 & 0 & \cdots & 0 \\ \vdots & \vdots & \ddots & \vdots \\ 0 & 0 & \cdots & 0 \end{bmatrix} \tag{2.212}$$

or $\mathbf{A} = \mathbf{Q}_1\mathbf{B}\mathbf{Q}_2$, where $\mathbf{Q}_2 = \mathbf{U}^{\mathrm{T}}$, and $\mathbf{B}$ is the $n \times m$ matrix

$$\mathbf{B} = \begin{bmatrix} \sqrt{\lambda_1} & 0 & \cdots & 0 \\ 0 & \sqrt{\lambda_2} & \cdots & 0 \\ \vdots & \vdots & \ddots & \vdots \\ 0 & 0 & \cdots & \sqrt{\lambda_m} \\ 0 & 0 & \cdots & 0 \\ \vdots & \vdots & \ddots & \vdots \\ 0 & 0 & \cdots & 0 \end{bmatrix} \tag{2.213}$$

This completes the proof of Eq. 191.

Example 2.20

Find the singular value decomposition of the matrix

$$\mathbf{A} = \begin{bmatrix} 2 & 1 \\ 1 & 0 \\ 1 & 0 \end{bmatrix}$$

Solution. The matrix $\mathbf{A}^{\mathrm{T}}\mathbf{A}$ is a 2×2 matrix that can be calculated as

$$\mathbf{A}^{\mathrm{T}}\mathbf{A} = \begin{bmatrix} 2 & 1 & 1 \\ 1 & 0 & 0 \end{bmatrix} \begin{bmatrix} 2 & 1 \\ 1 & 0 \\ 1 & 0 \end{bmatrix} = \begin{bmatrix} 6 & 2 \\ 2 & 1 \end{bmatrix}$$

The characteristic polynomial of this matrix is

$$\begin{vmatrix} 6-\lambda & 2 \\ 2 & 1-\lambda \end{vmatrix} = (6-\lambda)(1-\lambda) - 4 = 0$$

from which the eigenvalues can be determined as

$$\lambda_1 = 0.2984, \qquad \lambda_2 = 6.7016$$

The eigenvectors are

$$\mathbf{x}_1 = \begin{bmatrix} 1 \\ -2.8508 \end{bmatrix}, \qquad \mathbf{x}_2 = \begin{bmatrix} 1 \\ 0.3508 \end{bmatrix}$$

The orthonormal eigenvectors are

$$\hat{\mathbf{x}}_1 = \begin{bmatrix} 0.3310 \\ -0.9436 \end{bmatrix}, \qquad \hat{\mathbf{x}}_2 = \begin{bmatrix} 0.9436 \\ 0.3310 \end{bmatrix}$$

The matrix $\mathbf{U}$ is then defined as

$$\mathbf{U} = \begin{bmatrix} 0.3310 & 0.9436 \\ -0.9436 & 0.3310 \end{bmatrix}$$

It follows that

$$\mathbf{AU} = \begin{bmatrix} 2 & 1 \\ 1 & 0 \\ 1 & 0 \end{bmatrix} \begin{bmatrix} 0.3310 & 0.9436 \\ -0.9436 & 0.3310 \end{bmatrix}$$
$$= \begin{bmatrix} -0.2816 & 2.2182 \\ 0.3310 & 0.9436 \\ 0.3310 & 0.9436 \end{bmatrix} = [\mathbf{b}_1 \quad \mathbf{b}_2]$$

The orthonormal vectors $\hat{\mathbf{b}}_1$ and $\hat{\mathbf{b}}_2$ are

$$\hat{\mathbf{b}}_1 = \begin{bmatrix} -0.5155 \\ 0.6059 \\ 0.6059 \end{bmatrix}, \qquad \hat{\mathbf{b}}_2 = \begin{bmatrix} 0.8569 \\ 0.3645 \\ 0.3645 \end{bmatrix}$$

The matrix $\mathbf{Q}_1$ can then be defined as

$$\mathbf{Q}_1 = [\hat{\mathbf{b}}_1 \quad \hat{\mathbf{b}}_2 \quad \hat{\mathbf{b}}_3] = \begin{bmatrix} -0.5515 & 0.8569 & 0 \\ 0.6059 & 0.3645 & 0.7071 \\ 0.6059 & 0.3645 & -0.7071 \end{bmatrix}$$

and the matrix $\mathbf{Q}_2$ is

$$\mathbf{Q}_2 = \mathbf{U}^{\mathrm{T}} = \begin{bmatrix} 0.3310 & -0.9436 \\ 0.9436 & 0.3310 \end{bmatrix}$$

The matrix $\mathbf{B}$ that contains the singular values is

$$\mathbf{B} = \begin{bmatrix} 0.5463 & 0 \\ 0 & 2.5887 \\ 0 & 0 \end{bmatrix}$$

Therefore, the matrix $\mathbf{A}$ can be written as $\mathbf{A} = \mathbf{Q}_1\mathbf{B}\mathbf{Q}_2$.

Important Results from the SVD Using the factorization of Eq. 191, one has

$$\mathbf{A}\mathbf{A}^{\mathrm{T}} = (\mathbf{Q}_1\mathbf{B}\mathbf{Q}_2)(\mathbf{Q}_1\mathbf{B}\mathbf{Q}_2)^{\mathrm{T}} = \mathbf{Q}_1\mathbf{B}\mathbf{B}^{\mathrm{T}}\mathbf{Q}_1^{\mathrm{T}} \tag{2.214}$$

which implies that the columns of the matrix $\mathbf{Q}_1$ are eigenvectors of the matrix $\mathbf{A}\mathbf{A}^{\mathrm{T}}$ and the square of the diagonal elements of the matrix $\mathbf{B}$ are the eigenvalues of the matrix $\mathbf{A}\mathbf{A}^{\mathrm{T}}$. That is, the eigenvalues of $\mathbf{A}^{\mathrm{T}}\mathbf{A}$ are the same as those of $\mathbf{A}\mathbf{A}^{\mathrm{T}}$. The eigenvectors, however, are different since

$$\mathbf{A}^{\mathrm{T}}\mathbf{A} = \mathbf{Q}_2^{\mathrm{T}}\mathbf{B}^{\mathrm{T}}\mathbf{B}\mathbf{Q}_2 \tag{2.215}$$

which implies that the rows of $\mathbf{Q}_2$ are eigenvectors of $\mathbf{A}^{\mathrm{T}}\mathbf{A}$. These conclusions can be verified using the results obtained in Example 20.

The matrix $\mathbf{B}$, whose diagonal elements are the square roots of the eigenvalues of the matrix $\mathbf{A}^{\mathrm{T}}\mathbf{A}$, takes the form of Eq. 213 if the number of rows of the matrix $\mathbf{A}$ is greater than the number of columns $(n > m)$. If $n < m$, the matrix $\mathbf{B}$ takes the form

$$\mathbf{B} = \begin{bmatrix} \sqrt{\lambda_1} & 0 & \cdots & 0 & 0 & \cdots & 0 \\ 0 & \sqrt{\lambda_2} & \cdots & 0 & 0 & \cdots & 0 \\ \vdots & \vdots & \ddots & \vdots & \vdots & \ddots & \vdots \\ 0 & 0 & \cdots & \sqrt{\lambda_n} & 0 & \cdots & 0 \end{bmatrix} \tag{2.216}$$

where the number of zero columns in this matrix is $m - n$. This fact is also clear from the definition of the transpose of the rectangular matrix $\mathbf{A}$, for if the singular value decomposition of a rectangular matrix $\mathbf{A}$ is given by Eq. 191, the singular value decomposition of its transpose is given by

$$\mathbf{A}^{\mathrm{T}} = \mathbf{Q}_2^{\mathrm{T}}\mathbf{B}^{\mathrm{T}}\mathbf{Q}_1^{\mathrm{T}} \tag{2.217}$$

Now let us consider the singular value decomposition of the $n \times m$ rectangular matrix where $n > m$. The matrix $\mathbf{B}$ of Eq. 213 can be written as

$$\mathbf{B} = \begin{bmatrix} \mathbf{B}_1 \\ \mathbf{0} \end{bmatrix} \tag{2.218}$$

where $\mathbf{B}_1$ is a diagonal matrix whose elements, the singular values, are the square roots of the eigenvalues of the matrix $\mathbf{A}^{\mathrm{T}}\mathbf{A}$. Using the partitioning of Eq. 217, Eq. 191 can be written in the following partitioned form:

$$\mathbf{A} = [\mathbf{Q}_{1d} \quad \mathbf{Q}_{1i}] \begin{bmatrix} \mathbf{B}_1 \\ \mathbf{0} \end{bmatrix} \mathbf{Q}_2 \tag{2.219}$$

where $\mathbf{Q}_{1d}$ and $\mathbf{Q}_{1i}$ are partitions of the matrix $\mathbf{Q}_1$. The columns of the matrices $\mathbf{Q}_{1d}$ and $\mathbf{Q}_{1i}$ are orthogonal vectors, and

$$\mathbf{Q}_{1d}^{\mathrm{T}}\mathbf{Q}_{1i} = \mathbf{0}, \qquad \mathbf{Q}_{1i}^{\mathrm{T}}\mathbf{Q}_{1d} = \mathbf{0} \tag{2.220}$$

It follows from Eq. 219 that

$$\mathbf{A} = \mathbf{Q}_{1d}\mathbf{B}_1\mathbf{Q}_2 \tag{2.221}$$

Multiplying this equation by $\mathbf{Q}_{1i}^{\mathrm{T}}$ and using the results of Eq. 220, one obtains

$$\mathbf{Q}_{1i}^{\mathrm{T}}\mathbf{A} = \mathbf{0} \tag{2.222}$$

which implies that the orthogonal columns of the matrix $\mathbf{Q}_{1i}$ span the null space of the matrix $\mathbf{A}^{\mathrm{T}}$. In the dynamic analysis of multibody systems, Eq. 222 can be used to obtain a minimum set of independent differential equations that govern the motion of the interconnected bodies in the system.

PROBLEMS

1. Find the sum of the following two matrices:

$$\mathbf{A} = \begin{bmatrix} -3.0 & 8.0 & -20.5 \\ 5.0 & 11.0 & 13.0 \\ 7.0 & 20.0 & 0 \end{bmatrix}, \qquad \mathbf{B} = \begin{bmatrix} 0 & 3.2 & 0 \\ -17.5 & 5.7 & 0 \\ 12.0 & 6.8 & -10.0 \end{bmatrix}$$

Evaluate also the determinant and the trace of $\mathbf{A}$ and $\mathbf{B}$.

2. Find the product $\mathbf{AB}$ and $\mathbf{BA}$, where $\mathbf{A}$ and $\mathbf{B}$ are given in problem 1.

3. Find the inverse of the following matrices.

$$\mathbf{A} = \begin{bmatrix} -1 & 2 & -1 \\ 2 & -1 & 0 \\ 0 & -1 & 1 \end{bmatrix}, \qquad \mathbf{B} = \begin{bmatrix} 0 & -3 & 5 \\ -2 & 2 & -3 \\ 6 & -2 & 0 \end{bmatrix}$$

4. Show that an arbitrary square matrix $\mathbf{A}$ can be written as

$$\mathbf{A} = \mathbf{A}_1 + \mathbf{A}_2$$

where $\mathbf{A}_1$ is a symmetric matrix and $\mathbf{A}_2$ is a skew-symmetric matrix.

5. Show that the interchange of any two rows or columns of a square matrix changes only the sign of the determinant.

6. Show that if a matrix has two identical rows or two identical columns, the determinant of this matrix is equal to zero.

7. Let

$$\mathbf{A} = \begin{bmatrix} \mathbf{A}_{11} & \mathbf{A}_{12} \\ \mathbf{A}_{21} & \mathbf{A}_{22} \end{bmatrix}$$

be a nonsingular matrix. If $\mathbf{A}_{11}$ is square and nonsingular show by direct matrix multiplications that

$$\mathbf{A}^{-1} = \begin{bmatrix} (\mathbf{A}_{11}^{-1} + \mathbf{B}_1\mathbf{H}^{-1}\mathbf{B}_2) & -\mathbf{B}_1\mathbf{H}^{-1} \\ -\mathbf{H}^{-1}\mathbf{B}_2 & \mathbf{H}^{-1} \end{bmatrix}$$

where

$$\begin{aligned} \mathbf{B}_1 &= \mathbf{A}_{11}^{-1}\mathbf{A}_{12}, \quad \mathbf{B}_2 = \mathbf{A}_{21}\mathbf{A}_{11}^{-1} \\ \mathbf{H} &= \mathbf{A}_{22} - \mathbf{B}_2\mathbf{A}_{12} = \mathbf{A}_{22} - \mathbf{A}_{21}\mathbf{B}_1 \\ &= \mathbf{A}_{22} - \mathbf{A}_{21}\mathbf{A}_{11}^{-1}\mathbf{A}_{12} \end{aligned}$$

8. Using the identity given in problem 7, find the inverse of the matrices $\mathbf{A}$ and $\mathbf{B}$ given in problem 3.

9. Let $\mathbf{a}$ and $\mathbf{b}$ be the two vectors

$$\begin{aligned} \mathbf{a} &= [1 \quad 0 \quad 3 \quad 2 \quad -5]^{\mathrm{T}} \\ \mathbf{b} &= [0 \quad -1 \quad 2 \quad 3 \quad -8]^{\mathrm{T}} \end{aligned}$$

Find $\mathbf{a} + \mathbf{b}$, $\mathbf{a} \cdot \mathbf{b}$, $|\mathbf{a}|$, and $|\mathbf{b}|$.

10. Find the total derivative of the function

$$f(q_1, q_2, q_3, t) = q_1 q_3 - 3(q_2)^2 + 5(t)^5$$

with respect to the parameter t. Define also the partial derivative of the function f with respect to the vector $\mathbf{q}(t)$ where

$$\mathbf{q}(t) = [q_1(t) \quad q_2(t) \quad q_3(t)]^{\mathrm{T}}$$

11. Find the total derivative of the vector function

$$f = \begin{bmatrix} f_1 \\ f_2 \\ f_3 \end{bmatrix} = \begin{bmatrix} (q_1)^2 + 3(q_2)^2 - 5(q_4)^3 + (t)^3 \\ (q_2)^2 - (q_3)^2 \\ q_1 q_4 + q_2 q_3 + t \end{bmatrix}$$

with respect to the parameter t. Define also the partial derivative of the function f with respect to the vector

$$\mathbf{q} = [q_1 \quad q_2 \quad q_3 \quad q_4]^{\mathrm{T}}$$

12. Let $Q = \mathbf{q}^{\mathrm{T}}\mathbf{A}\mathbf{q}$, where $\mathbf{A}$ is an $n \times n$ square matrix and $\mathbf{q}$ is an n-dimensional vector. Show that

$$\frac{\partial Q}{\partial \mathbf{q}} = \mathbf{q}^{\mathrm{T}}(\mathbf{A} + \mathbf{A}^{\mathrm{T}})$$

13. Show that the vectors

$$\mathbf{a}_1 = \begin{bmatrix} 0 \\ 0 \\ 1 \end{bmatrix}, \quad \mathbf{a}_2 = \begin{bmatrix} 0 \\ 1 \\ 1 \end{bmatrix}, \quad \mathbf{a}_3 = \begin{bmatrix} 1 \\ 1 \\ 1 \end{bmatrix}$$

are linearly independent. Determine also the coordinates of the vector $\mathbf{b} = [1 \quad -5 \quad 3]^{\mathrm{T}}$ in the basis $\mathbf{a}_1$, $\mathbf{a}_2$, and $\mathbf{a}_3$.

14. Find the rank of the following matrices:

$$\mathbf{A} = \begin{bmatrix} 2 & 5 & 1 \\ 6 & 9 & 3 \\ 4 & 0 & 2 \end{bmatrix}, \quad \mathbf{B} = \begin{bmatrix} 3 & 5 & 1 & 0 \\ 2 & 0 & -1 & 3 \\ 7 & 1 & 2 & 9 \end{bmatrix}$$

15. Find the cross product of the vectors $\mathbf{a} = [1 \quad 0 \quad 3]^{\mathrm{T}}$ and $\mathbf{b} = [9 \quad -3 \quad 1]^{\mathrm{T}}$. If $\mathbf{c} = \mathbf{a} \times \mathbf{b}$, verify that $\mathbf{c}^{\mathrm{T}}\mathbf{a} = \mathbf{c}^{\mathrm{T}}\mathbf{b} = 0$.

16. Show that if $\mathbf{a}$ and $\mathbf{b}$ are two parallel vectors, then $\mathbf{a} \times \mathbf{b} = \mathbf{0}$.

17. Show that if $\mathbf{a}$ and $\mathbf{b}$ are two orthogonal vectors and $\mathbf{c} = \mathbf{a} \times \mathbf{b}$, then $|\mathbf{c}| = |\mathbf{a}|\ |\mathbf{b}|$.

18. Find the skew-symmetric matrices associated with the vectors

$$\mathbf{a} = [-2 \quad 5 \quad -9]^{\mathrm{T}}, \quad \mathbf{b} = [18 \quad -3 \quad 10]^{\mathrm{T}}$$

Using these skew-symmetric matrices, find the cross product $\mathbf{a} \times \mathbf{b}$ and $\mathbf{b} \times \mathbf{a}$.

19. Find the solution of the homogeneous system of equations $\mathbf{a} \times \mathbf{x} = \mathbf{0}$, where $\mathbf{a}$ is the vector $\mathbf{a} = [-2 \quad 5 \quad -9]^{\mathrm{T}}$.

20. Find the solution of the system of homogeneous equations $\mathbf{a} \times \mathbf{x} = \mathbf{0}$, where $\mathbf{a}$ is the vector $\mathbf{a} = [-11 \quad -3 \quad 4]^{\mathrm{T}}$.

21. If $\mathbf{a} = [a_1 \quad a_2 \quad a_3]^{\mathrm{T}}$ and $\mathbf{b} = [b_1 \quad b_2 \quad b_3]^{\mathrm{T}}$ are given vectors, show using direct matrix multiplication that $\tilde{\mathbf{c}} = \tilde{\mathbf{a}}\tilde{\mathbf{b}} - \tilde{\mathbf{b}}\tilde{\mathbf{a}}$, where $\mathbf{c} = \mathbf{a} \times \mathbf{b}$.

22. Find the solution of the following system of algebraic equations:

$$
\begin{aligned}
-x_1 + 2x_2 - x_3 &= 2 \\
2x_1 - x_2 &= 1.5 \\
-x_2 + x_3 &= 5
\end{aligned}
$$

23. Find the solution of the following system of equations:

$$
\begin{aligned}
-3x_2 + 5x_3 &= 0 \\
-2x_1 + 2x_2 - 3x_3 &= 0 \\
6x_1 - 2x_2 &= 5.5
\end{aligned}
$$

24. Use the Gram–Schmidt orthogonalization process to determine the **QR** decomposition of the matrices

$$
\mathbf{A} = \begin{bmatrix} -1 & 2 \\ 2 & -1 \\ 0 & -1 \end{bmatrix}, \qquad \mathbf{B} = \begin{bmatrix} 0 & -3 & 5 \\ -2 & 2 & -3 \\ 6 & -2 & 0 \end{bmatrix}
$$

25. Use the Householder transformations to solve problem 24.

26. Determine the singular value decomposition of the matrices **A** and **B** of problem 24.

27. Prove that the determinant of the matrix

$$
\mathbf{G} = \frac{2}{1 + (\gamma)^2} \begin{bmatrix} 1 & -\gamma_3 & \gamma_2 \\ \gamma_3 & 1 & -\gamma_1 \\ -\gamma_2 & \gamma_1 & 1 \end{bmatrix}
$$

is 2, where $(\gamma)^2 = (\gamma_1)^2 + (\gamma_2)^2 + (\gamma_3)^2$. Prove also that the inverse of the matrix **G** is

$$
\mathbf{G}^{-1} = \frac{1}{2} \begin{bmatrix} 1 + (\gamma_1)^2 & \gamma_1\gamma_2 + \gamma_3 & \gamma_1\gamma_3 - \gamma_2 \\ \gamma_1\gamma_2 - \gamma_3 & 1 + (\gamma_2)^2 & \gamma_2\gamma_3 + \gamma_1 \\ \gamma_1\gamma_3 + \gamma_2 & \gamma_2\gamma_3 - \gamma_1 & 1 + (\gamma_3)^2 \end{bmatrix}
$$

Show that the matrix $\mathbf{A} = \mathbf{G}(\mathbf{G}^{-1})^{\mathrm{T}}$ is an orthogonal matrix. Show also that the vector

$$
\gamma = [\gamma_1 \quad \gamma_2 \quad \gamma_3]^{\mathrm{T}}
$$

is an eigenvector for the matrix **A** and determine the corresponding eigenvalue.

28. Prove the *polar decomposition theorem*, which states that a nonsingular square matrix **A** can be uniquely decomposed into $\mathbf{A} = \mathbf{Q}\mathbf{V}_1$ or $\mathbf{A} = \mathbf{V}_2\mathbf{Q}$, where **Q** is an orthogonal matrix and $\mathbf{V}_1$ and $\mathbf{V}_2$ are positive-definite symmetric matrices.

CHAPTER 3

KINEMATICS

In the kinematic analysis, we are concerned with studying motion without considering the forces that produce the motion. Unlike the case of dynamic analysis, where the motion of the system due to known forces is determined, the objective of the kinematic analysis is to determine the positions, velocities, and accelerations as the result of known prescribed input motions. Recall that the *degrees of freedom* of a mechanical system, by definition, are the smallest set of independent coordinates that are required to define the system configuration. If the degrees of freedom and their time derivatives are known, other coordinates and their time derivatives that represent the displacements, velocities, and accelerations of the bodies of the system can be expressed in terms of the system degrees of freedom and their time derivatives. This leads to the *displacement*, *velocity*, and *acceleration kinematic relationships* that can be solved for the state of the system regardless of the forces that produce the motion.

There are three stages that must be followed for the complete kinematic analysis of a mechanical system: *position*, *velocity*, and *acceleration* analyses. In the position analysis, the displacement kinematic relationships are solved assuming that the selected degrees of freedom of the system are specified. These relationships are, in general, nonlinear functions in the system coordinates and their solution may require the use of an iterative numerical procedure such as Newton–Raphson methods. The velocity and acceleration kinematic equations can be obtained by differentiating the displacement equations, once and twice, respectively. This procedure leads to a system of linear algebraic equations in the velocities and accelerations.

In this chapter, the constrained motion of systems that consist of interconnected bodies is examined. Two different, yet equivalent, approaches are presented in this chapter for the kinematic analysis of multibody systems whose degrees of freedom are specified. The first is the classical approach, which is suited for studying the kinematics of systems that consist of small numbers of bodies and joints. The use of this approach is demonstrated by several examples presented in this chapter. No detailed discussion of this approach is provided, since the focus of the book is on computational methods that can be used in the analysis of more complex systems. The second approach, on the other hand, can be used for solving large-scale applications, which consist of large numbers of bodies and joints. In this approach, the algebraic kinematic constraint relationships between coordinates are formulated and used to develop a number of equations equal to the number of unknown coordinates. These constraint equations, which are, in general, nonlinear functions of the coordinates, can be solved using iterative numerical and computer methods to determine the positions of the bodies in the system. By differentiating the constraint equations once, and twice with respect to time, linear systems of equations in the velocities and accelerations can be obtained. These linear equations can be solved in a straightforward manner to determine the first and second time derivatives of the coordinates. Computer implementation of this general computational procedure is discussed and several examples are presented in order to demonstrate its use.

3.1 KINEMATICS OF RIGID BODIES

Throughout the analysis presented in this chapter and the following chapters, it is assumed that the multibody systems consist of *rigid bodies*, such that the effect of the body deformation can be neglected. In the rigid body analysis, it is assumed that the distance between two arbitrary points on the body remains unchanged. The assumption of rigidity is justified when the components of the multibody system are made of bulky solids that experience only small deformations such that the effect of this deformation on the overall motion is negligible. If the interest, however, is to determine the stresses, or if the deformations of the bodies are large such that their effect cannot be neglected, the rigid body assumption is no longer adequate and a deformable body modeling approach must be adopted. In this section, the kinematic equations that describe the general planar motion of a rigid body are developed in terms of the body coordinates. The transformation matrix that defines the orientation of the body is derived and is used to develop the position equations that describe an arbitrary displacement of the rigid body.

Coordinate Transformation We consider the problem of a simple finite rotation about a fixed axis and develop the relationships between the axes of different coordinate systems as the result of the finite rotation. These relationships define the coordinate transformation between moving coordinate systems. Figure 1 shows two coordinate systems XY and X^iY^i. The axis X^i is assumed to make an angle θ^i with respect to the X axis of the coordinate system XY. We assume for the moment that the origins of both coordinate systems coincide. Let $\mathbf{i}$ and $\mathbf{j}$ be unit vectors along the X and Y axes, respectively, and let $\mathbf{i}^i$ and $\mathbf{j}^i$ be, respectively, unit vectors along the X^i and Y^i axes. Using Fig. 2a, the components of the

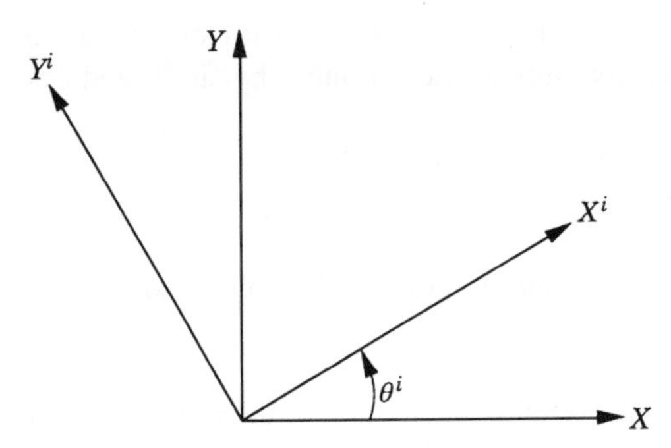

Figure 3.1 Rigid body rotation

unit vector $\mathbf{i}^i$ can be expressed in the XY coordinate system as

$$\mathbf{i}^i = \cos\theta^i\ \mathbf{i} + \sin\theta^i\ \mathbf{j} \tag{3.1}$$

From Fig. 2b, one can also show that the components of the unit vector $\mathbf{j}^i$ can be expressed in the coordinate system XY as

$$\mathbf{j}^i = -\sin\theta^i\ \mathbf{i} + \cos\theta^i\ \mathbf{j} \tag{3.2}$$

Equations 1 and 2 define the unit vectors along the axes of the coordinate system X^iY^i in terms of unit vectors along the axes of the coordinate system XY. To obtain the inverse relationship, we multiply Eqs. 1 and 2 by $\cos\theta^i$ and $\sin\theta^i$, respectively, and subtract the resulting equations. This leads to

$$\cos\theta^i\ \mathbf{i}^i - \sin\theta^i\ \mathbf{j}^i = [(\cos\theta^i)^2 + (\sin\theta^i)^2]\mathbf{i} \tag{3.3}$$

Using the trigonometric identity $(\cos\theta^i)^2 + (\sin\theta^i)^2 = 1$, the equation above leads to

$$\mathbf{i} = \cos\theta^i\ \mathbf{i}^i - \sin\theta^i\ \mathbf{j}^i \tag{3.4}$$

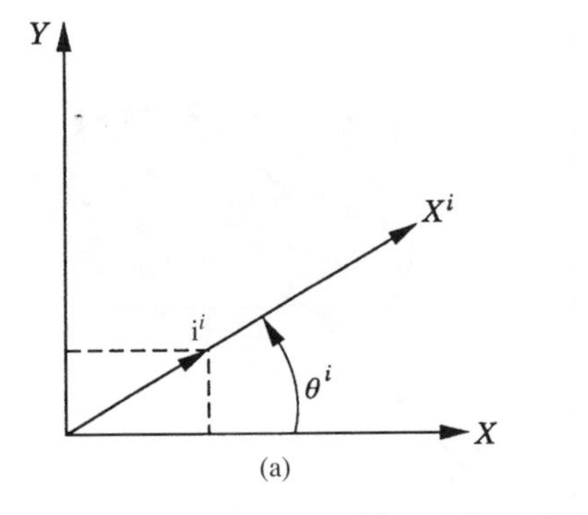

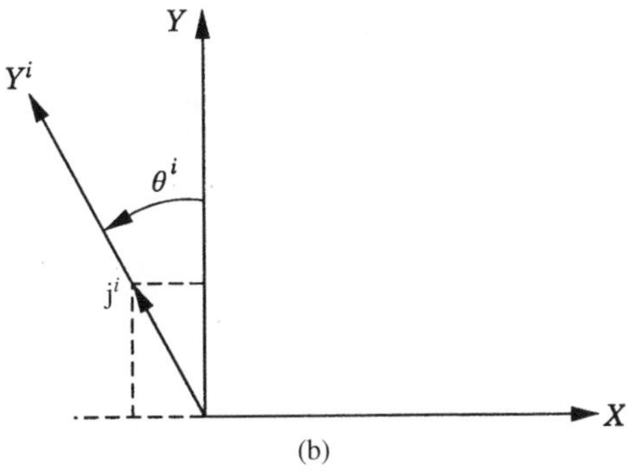

Figure 3.2 Coordinate transformation

This equation defines a unit vector along the X axis in terms of the unit vectors along the axes of the X^iY^i coordinate system. Similarly, multiplying Eqs. 1 and 2 by sin θ^i and cos θ^i, respectively, and adding leads to

$$\mathbf{j} = \sin\theta^i\ \mathbf{i}^i + \cos\theta^i\ \mathbf{j}^i \tag{3.5}$$

in which the unit vector $\mathbf{j}$ along the Y axis is expressed in terms of the unit vectors $\mathbf{i}^i$ and $\mathbf{j}^i$ of the coordinate system X^iY^i.

Position Equations For the convenience of describing the motion of the rigid bodies in the multibody system, we assign a coordinate system for each body. The origin of this body coordinate system is rigidly attached to a point on the body, and therefore, the coordinate system experiences the same rigid body motion as the body. Let X^iY^i, as shown in Fig. 3, be the body coordinate system and XY be a selected global inertial frame of reference that is fixed in time. Let P^i be an arbitrary material point on the body. The coordinates of point P^i in the body coordinate system are fixed and can be defined by the vector

$$\bar{\mathbf{u}}_P^i = [\bar{x}_P^i \quad \bar{y}_P^i]^{\mathrm{T}} \tag{3.6}$$

which can also be written in terms of unit vectors along the axes of the coordinate system X^iY^i as

$$\bar{\mathbf{u}}_P^i = \bar{x}_P^i\ \mathbf{i}^i + \bar{y}_P^i\ \mathbf{j}^i \tag{3.7}$$

where $\mathbf{i}^i$ and $\mathbf{j}^i$ are, respectively, unit vectors along the X^i and Y^i axes of the body coordinate system. Substituting Eqs. 1 and 2 into Eq. 7 yields the coordinates of the vector $\bar{\mathbf{u}}_P^i$ in the global coordinate system as

$$\mathbf{u}_P^i = \bar{x}_P^i(\cos\theta^i\ \mathbf{i} + \sin\theta^i\ \mathbf{j}) + \bar{y}_P^i(-\sin\theta^i\ \mathbf{i} + \cos\theta^i\ \mathbf{j}) \tag{3.8}$$

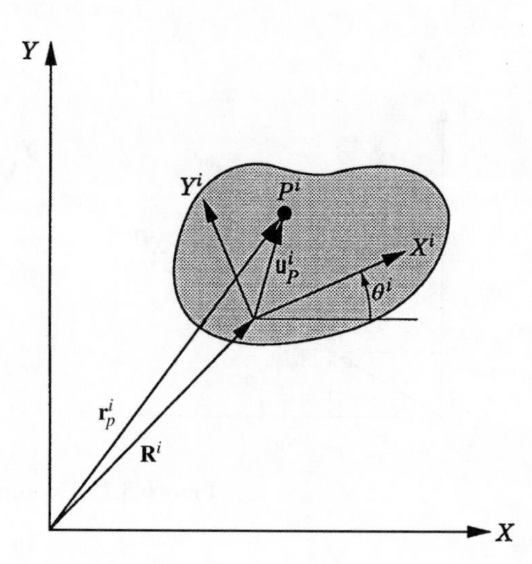

Figure 3.3 Rigid body displacement

where $\mathbf{u}_P^i$ is the global representation of the vector $\bar{\mathbf{u}}_P^i$. Equation 8 can be written as

$$\mathbf{u}_P^i = (\bar{x}_P^i \cos\theta^i - \bar{y}_P^i \sin\theta^i)\mathbf{i} + (\bar{x}_P^i \sin\theta^i + \bar{y}_P^i \cos\theta^i)\mathbf{j} \tag{3.9}$$

This equation can also be written in the following form:

$$\mathbf{u}_P^i = \begin{bmatrix} \bar{x}_P^i \cos\theta^i - \bar{y}_P^i \sin\theta^i \\ \bar{x}_P^i \sin\theta^i + \bar{y}_P^i \cos\theta^i \end{bmatrix} \tag{3.10}$$

or in the following matrix form:

$$\mathbf{u}_P^i = \begin{bmatrix} \cos\theta^i & -\sin\theta^i \\ \sin\theta^i & \cos\theta^i \end{bmatrix} \begin{bmatrix} \bar{x}_P^i \\ \bar{y}_P^i \end{bmatrix} \tag{3.11}$$

Using Eq. 6, Eq. 11 can be written simply as

$$\mathbf{u}_P^i = \mathbf{A}^i \bar{\mathbf{u}}_P^i \tag{3.12}$$

where $\bar{\mathbf{u}}_P^i$ is the local position vector of the arbitrary point P^i as defined by Eq. 6, and $\mathbf{A}^i$ is the planar transformation matrix defined as

$$\mathbf{A}^i = \begin{bmatrix} \cos\theta^i & -\sin\theta^i \\ \sin\theta^i & \cos\theta^i \end{bmatrix} \tag{3.13}$$

The matrix $\mathbf{A}^i$ is an orthogonal matrix because

$$\mathbf{A}^i \mathbf{A}^{i\mathrm{T}} = \mathbf{A}^{i\mathrm{T}} \mathbf{A}^i = \mathbf{I} \tag{3.14}$$

where $\mathbf{I}$ is the 2×2 identity matrix.

The global position vector of the arbitrary point P^i in the fixed XY coordinate system can be written as shown in Fig. 3 as the sum of the two vectors $\mathbf{R}^i$ and $\mathbf{u}_P^i$, where $\mathbf{R}^i$ is the global position vector of the origin O^i of the coordinate system X^iY^i. One can then write the following equation:

$$\mathbf{r}_P^i = \mathbf{R}^i + \mathbf{u}_P^i \tag{3.15}$$

which upon the use of Eq. 12 yields

$$\mathbf{r}_P^i = \mathbf{R}^i + \mathbf{A}^i \bar{\mathbf{u}}_P^i \tag{3.16}$$

It is clear from Eq. 16 that the global position vector of an arbitrary point on the rigid body i can be written in terms of the rotational coordinate of the body θ^i, as well as the translation of the origin of the body reference $\mathbf{R}^i$. That is, the most general rigid body displacement can be described by a translation of a reference point plus a rotation about an axis passing through this point.

3.2 VELOCITY EQUATIONS

The second step in the kinematic analysis is to determine the velocities of the bodies in the system. In the velocity analysis, it is assumed that the positions and orientations of the bodies are already known from the position analysis. The absolute velocity of a point on a rigid body that undergoes planar motion can be obtained by differentiating Eq. 16 with respect to time. This yields

$$\dot{\mathbf{r}}_P^i = \dot{\mathbf{R}}^i + \dot{\mathbf{A}}^i \bar{\mathbf{u}}_P^i \tag{3.17}$$

By using Eq. 13, the time derivative of the transformation matrix can be written as

$$\dot{\mathbf{A}}^i = \dot{\theta}^i \mathbf{A}_\theta^i \tag{3.18}$$

where $\mathbf{A}_\theta^i$ is the partial derivative of the rotation matrix with respect to the rotational coordinate θ^i and is given by

$$\mathbf{A}_\theta^i = \begin{bmatrix} -\sin\theta^i & -\cos\theta^i \\ \cos\theta^i & -\sin\theta^i \end{bmatrix} \tag{3.19}$$

The velocity vector of the arbitrary point P^i can then be expressed as

$$\dot{\mathbf{r}}_P^i = \dot{\mathbf{R}}^i + \dot{\theta}^i \mathbf{A}_\theta^i \bar{\mathbf{u}}_P^i \tag{3.20}$$

This equation defines the absolute velocity vector of point P^i in terms of the derivatives of the coordinates $\mathbf{R}^i$ and θ^i of the rigid body.

Angular Velocity Vector The second term on the right-hand side of Eq. 20 can be written explicitly as

$$\begin{aligned} \dot{\theta}^i \mathbf{A}_\theta^i \bar{\mathbf{u}}_P^i &= \dot{\theta}^i \begin{bmatrix} -\sin\theta^i & -\cos\theta^i \\ \cos\theta^i & -\sin\theta^i \end{bmatrix} \begin{bmatrix} \bar{x}_P^i \\ \bar{y}_P^i \end{bmatrix} \\ &= \dot{\theta}^i \begin{bmatrix} -\bar{x}_P^i \sin\theta^i - \bar{y}_P^i \cos\theta^i \\ \bar{x}_P^i \cos\theta^i - \bar{y}_P^i \sin\theta^i \end{bmatrix} \end{aligned} \tag{3.21}$$

Equation 21 can be written in a simple form if we define the angular velocity vector $\boldsymbol{\omega}^i$ of body i as

$$\boldsymbol{\omega}^i = \dot{\theta}^i \mathbf{k} \tag{3.22}$$

where $\mathbf{k}$ is a unit vector along the axis of rotation that is perpendicular to the plane of the motion. Equation 22 can also be written in an alternative form as

$$\boldsymbol{\omega}^i = [0 \quad 0 \quad \dot{\theta}^i]^{\mathrm{T}} \tag{3.23}$$

The velocity vector of an arbitrary point on the rigid body can be expressed in terms of the angular velocity vector. To demonstrate this, we evaluate the vector $\boldsymbol{\omega}^i \times \mathbf{u}_P^i$, where the vector $\mathbf{u}_P^i$ is given by Eq. 12 as

$$\mathbf{u}_P^i = \mathbf{A}^i \bar{\mathbf{u}}_P^i = \begin{bmatrix} \cos\theta^i & -\sin\theta^i \\ \sin\theta^i & \cos\theta^i \end{bmatrix} \begin{bmatrix} \bar{x}_P^i \\ \bar{y}_P^i \end{bmatrix}$$

$$= \begin{bmatrix} \bar{x}_P^i \cos\theta^i - \bar{y}_P^i \sin\theta^i \\ \bar{x}_P^i \sin\theta^i + \bar{y}_P^i \cos\theta^i \end{bmatrix} = \begin{bmatrix} u_x^i \\ u_y^i \end{bmatrix} \tag{3.24}$$

where u_x^i and u_y^i are the components of the vector $\mathbf{u}_P^i$ given by

$$\left. \begin{aligned} u_x^i &= \bar{x}_P^i \cos\theta^i - \bar{y}_P^i \sin\theta^i \\ u_y^i &= \bar{x}_P^i \sin\theta^i + \bar{y}_P^i \cos\theta^i \end{aligned} \right\} \tag{3.25}$$

It follows that

$$\boldsymbol{\omega}^i \times \mathbf{u}_P^i = \begin{vmatrix} \mathbf{i} & \mathbf{j} & \mathbf{k} \\ 0 & 0 & \dot{\theta}^i \\ u_x^i & u_y^i & 0 \end{vmatrix} = \begin{bmatrix} -\dot{\theta}^i u_y^i \\ \dot{\theta}^i u_x^i \\ 0 \end{bmatrix} \tag{3.26}$$

which upon using the definition of the components u_x^i and u_y^i of Eq. 25 leads to

$$\boldsymbol{\omega}^i \times \mathbf{u}_P^i = \dot{\theta}^i \begin{bmatrix} -\bar{x}_P^i \sin\theta^i - \bar{y}_P^i \cos\theta^i \\ \bar{x}_P^i \cos\theta^i - \bar{y}_P^i \sin\theta^i \end{bmatrix} \tag{3.27}$$

Comparing Eqs. 27 and 21 and using Eqs. 18 and 24, we obtain the following identity:

$$\dot{\mathbf{A}}\bar{\mathbf{u}}_P^i = \dot{\theta}^i \mathbf{A}_\theta^i \bar{\mathbf{u}}_P^i = \boldsymbol{\omega}^i \times \mathbf{u}_P^i = \boldsymbol{\omega}^i \times (\mathbf{A}^i \bar{\mathbf{u}}_P^i) \tag{3.28}$$

Using this identity, the absolute velocity vector of an arbitrary point on the rigid body i can be written in terms of the angular velocity vector as

$$\dot{\mathbf{r}}_P^i = \dot{\mathbf{R}}^i + \boldsymbol{\omega}^i \times \mathbf{u}_P^i \tag{3.29}$$

which indicates that the velocity of any point P^i on the rigid body can be written in terms of the velocity of a reference point O^i plus the relative velocity between the two points. That is

$$\mathbf{v}_P^i = \mathbf{v}_O^i + \mathbf{v}_{PO}^i \tag{3.30}$$

where $\mathbf{v}_P^i$ and $\mathbf{v}_O^i$ are, respectively, the absolute velocities of points P^i and O^i, and $\mathbf{v}_{PO}^i$ is the relative velocity of point P^i with respect to point O^i and is given by

$$\mathbf{v}_{PO}^i = \boldsymbol{\omega}^i \times \mathbf{u}_P^i \tag{3.31}$$

It will be shown in Chapter 7 that equations similar to Eqs. 29–31 can still be used in the case of the spatial motion of rigid bodies. In the more general case of spatial motion, the transformation matrix $\mathbf{A}^i$ and the angular velocity vector $\boldsymbol{\omega}^i$ take forms different from those used in the planar analysis.

3.3 ACCELERATION EQUATIONS

The absolute acceleration of a point fixed on a rigid body can be obtained by differentiating the velocity equation with respect to time. If Eq. 20 is differentiated with respect to time, one obtains the acceleration of an arbitrary point P^i on the rigid body i as

$$\ddot{\mathbf{r}}_P^i = \ddot{\mathbf{R}}^i + \dot{\theta}^i \dot{\mathbf{A}}_\theta^i \bar{\mathbf{u}}_P^i + \ddot{\theta}^i \mathbf{A}_\theta^i \bar{\mathbf{u}}_P^i \tag{3.32}$$

where $\ddot{\mathbf{R}}^i$ is the absolute acceleration of the reference point. In the case of planar motion, the following identity can be verified:

$$\dot{\mathbf{A}}_\theta^i = -\mathbf{A}^i \dot{\theta}^i \tag{3.33}$$

which upon substitution into Eq. 32 leads to

$$\ddot{\mathbf{r}}_P^i = \ddot{\mathbf{R}}^i - (\dot{\theta}^i)^2 \mathbf{A}^i \bar{\mathbf{u}}_P^i + \ddot{\theta}^i \mathbf{A}_\theta^i \bar{\mathbf{u}}_P^i \tag{3.34}$$

It can be shown that

$$\left.\begin{aligned} -(\dot{\theta}^i)^2 \mathbf{A}^i \bar{\mathbf{u}}_P^i &= \boldsymbol{\omega}^i \times (\boldsymbol{\omega}^i \times \mathbf{u}_P^i) \\ \ddot{\theta}^i \mathbf{A}_\theta^i \bar{\mathbf{u}}_P^i &= \boldsymbol{\alpha}^i \times \mathbf{u}_P^i \end{aligned}\right\} \tag{3.35}$$

where $\boldsymbol{\omega}^i$ is the *angular velocity* vector, and $\boldsymbol{\alpha}^i$ is the *angular acceleration* vector of body i defined as

$$\boldsymbol{\alpha}^i = \ddot{\theta}^i \mathbf{k} \tag{3.36}$$

By substituting Eq. 35 into Eq. 34, one obtains

$$\ddot{\mathbf{r}}_P^i = \ddot{\mathbf{R}}^i + \boldsymbol{\omega}^i \times (\boldsymbol{\omega}^i \times \mathbf{u}_P^i) + \boldsymbol{\alpha}^i \times \mathbf{u}_P^i \tag{3.37}$$

which can be rewritten as

$$\mathbf{a}_P^i = \mathbf{a}_O^i + (\mathbf{a}_{PO}^i)_n + (\mathbf{a}_{PO}^i)_t \tag{3.38}$$

where $\mathbf{a}_P^i = \ddot{\mathbf{r}}_P^i$ is the absolute acceleration vector of the arbitrary point P^i and $\mathbf{a}_O^i = \ddot{\mathbf{R}}^i$ is the absolute acceleration vector of the reference point. The vectors $(\mathbf{a}_{PO}^i)_n$ and $(\mathbf{a}_{PO}^i)_t$ are called, respectively, the *normal* and *tangential components* of the acceleration of point P^i

with respect to the reference point O^i. They are defined as

$$\left.\begin{aligned}(\mathbf{a}^i_{PO})_n &= \boldsymbol{\omega}^i \times (\boldsymbol{\omega}^i \times \mathbf{u}^i_P)\\ (\mathbf{a}^i_{PO})_t &= \boldsymbol{\alpha}^i \times \mathbf{u}^i_P\end{aligned}\right\} \tag{3.39}$$

The normal component has a magnitude $(\dot{\theta}^i)^2 l^i_P$ where l^i_P is the distance between point P^i and the reference point O^i. The direction of the normal component, however, is always along a line connecting P^i and O^i and is directed from P^i to O^i. The tangential component on the other hand has a magnitude $\ddot{\theta}^i l^i_P$ and its direction is along a line perpendicular to the line connecting points P^i and O^i.

Equation 38 can also be written as

$$\mathbf{a}^i_P = \mathbf{a}^i_O + \mathbf{a}^i_{PO} \tag{3.40}$$

where $\mathbf{a}^i_{PO}$ is the *relative acceleration* of point P^i with respect to point O^i and is defined as

$$\mathbf{a}^i_{PO} = (\mathbf{a}^i_{PO})_n + (\mathbf{a}^i_{PO})_t \tag{3.41}$$

The forms of the acceleration equations, given by Eqs. 37–41, can still be used in the case of the spatial motion of rigid bodies as will be demonstrated in Chapter 7.

3.4 KINEMATICS OF A POINT MOVING ON A RIGID BODY

In the preceding sections, the kinematic equations that define the position, velocity, and acceleration of an arbitrary point fixed on a rigid body were developed. In this section, we present the kinematic equations of a point moving on a rigid body. Figure 4 shows a particle

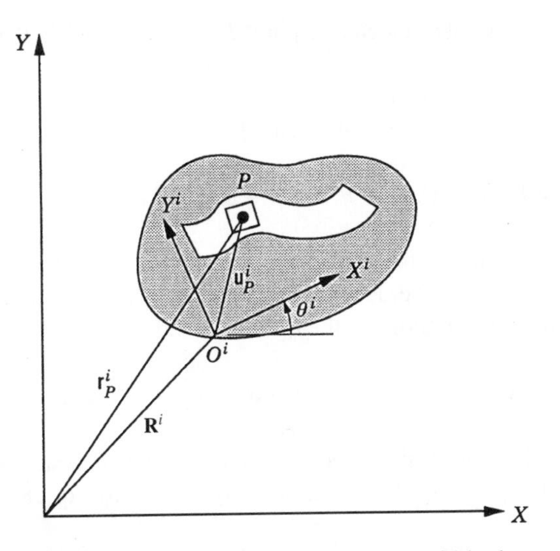

Figure 3.4 Motion of a point on a rigid body

point P moving on a rigid body i. The position vector of point P with respect to the body coordinate system X^iY^i is defined by the vector $\bar{\mathbf{u}}_P^i$. The global position vector of point P can be written as

$$\mathbf{r}_P^i = \mathbf{R}^i + \mathbf{A}^i\bar{\mathbf{u}}_P^i \tag{3.42}$$

where $\mathbf{R}^i$ is the global position vector of the reference point, and $\mathbf{A}^i$ is the transformation matrix from the body coordinate system to the global coordinate system. The vector $\bar{\mathbf{u}}_P^i$ in Eq. 42 is not a constant vector since point P moves with respect to the coordinate system of body i.

The absolute velocity of point P can be obtained by differentiating Eq. 42 with respect to time. This leads to

$$\begin{aligned}\dot{\mathbf{r}}_P^i &= \dot{\mathbf{R}}^i + \dot{\mathbf{A}}^i\bar{\mathbf{u}}_P^i + \mathbf{A}^i\dot{\bar{\mathbf{u}}}_P^i \\ &= \dot{\mathbf{R}}^i + \dot{\theta}^i\mathbf{A}_\theta^i\bar{\mathbf{u}}_P^i + \mathbf{A}^i\dot{\bar{\mathbf{u}}}_P^i\end{aligned} \tag{3.43}$$

By using the identity of Eq. 28, Eq. 43 takes the form

$$\dot{\mathbf{r}}_P^i = \dot{\mathbf{R}}^i + \boldsymbol{\omega}^i \times \mathbf{u}_P^i + (\mathbf{v}_P^i)_r \tag{3.44}$$

where

$$(\mathbf{v}_P^i)_r = \mathbf{A}^i\dot{\bar{\mathbf{u}}}_P^i \tag{3.45}$$

Equation 44 can also be written as

$$\mathbf{v}_P^i = \mathbf{v}_O^i + \boldsymbol{\omega}^i \times \mathbf{u}_P^i + (\mathbf{v}_P^i)_r \tag{3.46}$$

where $\mathbf{v}_P^i = \dot{\mathbf{r}}_P^i$ is the absolute velocity of point P, and $\mathbf{v}_O^i = \dot{\mathbf{R}}^i$ is the absolute velocity of the reference point.

The absolute acceleration of point P can be obtained by differentiating Eq. 46 or, equivalently, Eq. 43 with respect to time. This leads to

$$\ddot{\mathbf{r}}_P^i = \ddot{\mathbf{R}}^i + \dot{\theta}^i\dot{\mathbf{A}}_\theta^i\bar{\mathbf{u}}_P^i + \ddot{\theta}^i\mathbf{A}_\theta^i\bar{\mathbf{u}}_P^i + \dot{\theta}^i\mathbf{A}_\theta^i\dot{\bar{\mathbf{u}}}_P^i + \dot{\theta}^i\mathbf{A}_\theta^i\dot{\bar{\mathbf{u}}}_P^i + \mathbf{A}^i\ddot{\bar{\mathbf{u}}}_P^i \tag{3.47}$$

The second and third terms on the right-hand side of Eq. 47 are, respectively, the normal and tangential components of the acceleration defined in the preceding section by Eq. 35. Combining the fourth and fifth terms in the right-hand side of Eq. 47, this equation reduces to

$$\ddot{\mathbf{r}}_P^i = \ddot{\mathbf{R}}^i + \boldsymbol{\omega}^i \times (\boldsymbol{\omega}^i \times \mathbf{u}_P^i) + \boldsymbol{\alpha}^i \times \mathbf{u}_P^i + 2\dot{\theta}^i\mathbf{A}_\theta^i\dot{\bar{\mathbf{u}}}_P^i + \mathbf{A}^i\ddot{\bar{\mathbf{u}}}_P^i \tag{3.48}$$

where $\boldsymbol{\alpha}^i$ is the angular acceleration vector of the coordinate system of body i and $\mathbf{u}_P^i$ is as defined by Eq. 24. Using Eq. 45 and an identity similar to Eq. 28, one can show that

$$\dot{\theta}^i\mathbf{A}_\theta^i\dot{\bar{\mathbf{u}}}_P^i = \boldsymbol{\omega}^i \times (\mathbf{A}^i\dot{\bar{\mathbf{u}}}_P^i) = \boldsymbol{\omega}^i \times (\mathbf{v}_P^i)_r \tag{3.49}$$

Substituting Eq. 49 into Eq. 48, one obtains

$$\mathbf{a}_P^i = \mathbf{a}_O^i + \boldsymbol{\omega}^i \times (\boldsymbol{\omega}^i \times \mathbf{u}_P^i) + \boldsymbol{\alpha}^i \times \mathbf{u}_P^i + 2\boldsymbol{\omega}^i \times (\mathbf{v}_P^i)_r + (\mathbf{a}_P^i)_r \tag{3.50}$$

where

$$\mathbf{a}_P^i = \ddot{\mathbf{r}}_P^i, \quad \mathbf{a}_O^i = \ddot{\mathbf{R}}^i, \quad (\mathbf{a}_P^i)_r = \mathbf{A}^i \ddot{\bar{\mathbf{u}}}_P^i \tag{3.51}$$

in which $(\mathbf{a}_P^i)_r$ is the relative acceleration of point P with respect to the coordinate system of body i.

Coriolis Acceleration The fourth term in the right-hand side of Eq. 50 given by

$$(\mathbf{a}_P^i)_c = 2\boldsymbol{\omega}^i \times (\mathbf{v}_P^i)_r \tag{3.52}$$

is called the *Coriolis component* of the acceleration. This component of the acceleration has a direction along a line perpendicular to both $\boldsymbol{\omega}^i$ and $(\mathbf{v}_P^i)_r$. In the special case where point P is fixed, the vectors $(\mathbf{a}_P^i)_r$ and $(\mathbf{a}_P^i)_c$ are identically the zero vectors and Eq. 50 reduces, in this special case, to the equation that defines the acceleration of a point fixed on the rigid body.

3.5 CONSTRAINED KINEMATICS

Mechanical systems are assemblages of bodies connected by joints. The purpose of the joints is to transmit the motion from one body to another in a certain fashion. In this section, we briefly discuss the formulation of some of the joint constraints and introduce the mobility criterion that can be used to determine the number of degrees of freedom of a multibody system. A more detailed formulation of the planar joints in terms of the system coordinates is presented in Section 9 of this chapter, while a more detailed analysis of the spatial joints is presented in Chapter 7.

Planar Kinematics The kinematic relationships that describe the joint constraints can be formulated using a set of algebraic equations. As will be seen in the remainder of this book, the form of these equations depends on the *parameters* or *coordinates* used to describe the motion of the system. Figure 5a shows two bodies, i and j, in planar motion, which are connected by a *revolute joint*. The joint definition point is defined by point P. The corresponding point on body i is denoted as P^i while the corresponding point on body j is denoted as P^j. The conditions for the revolute joint require that point P^i on body i remains in contact with point P^j on body j throughout the motion. This condition can be expressed mathematically as

$$\mathbf{r}_P^i = \mathbf{r}_P^j \tag{3.53}$$

where $\mathbf{r}_P^i$ is the global position vector of point P^i, while $\mathbf{r}_P^j$ is the global position vector of point P^j. The conditions given by Eq. 53 eliminate the possibility of the relative translation

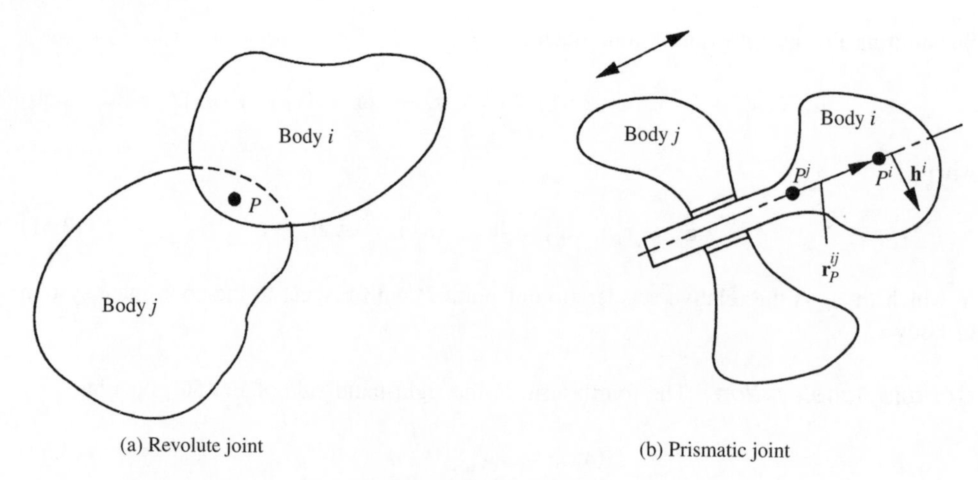

Figure 3.5 Planar joints

between the two bodies. The two bodies, however, have the freedom to rotate with respect to each other. This is the only relative motion between the two bodies that can occur as the result of their connectivity using the revolute joint. Therefore, the revolute joint in the planar analysis has one degree of freedom since it eliminates two degrees of freedom of the relative translation between the two bodies along two perpendicular axes.

Another one-degree-of-freedom joint in the planar kinematics is the *translational (prismatic) joint* shown in Fig. 5b. In this case, the only relative motion between the two bodies i and j is the relative translation along the joint axis. In the case of the prismatic joint, there are two kinematic constraint conditions that restrict two possible relative displacements. First, there should be no relative rotation between the two bodies. Second, there is no relative translation between the two bodies along an axis perpendicular to the axis of the prismatic joint. These two conditions can be stated mathematically as

$$\theta^i - \theta^j = c, \quad \mathbf{h}^{i^{\mathrm{T}}} \mathbf{r}_P^{ij} = 0 \tag{3.54}$$

where θ^i and θ^j are, respectively, the angular orientations of bodies i and j, c is a constant, $\mathbf{r}_P^{ij}$ is a vector that connects the two points P^i and P^j defined, respectively, on bodies i and j on the joint axis, and $\mathbf{h}^i$ is a vector defined on body i perpendicular to the joint axis.

Spatial Kinematics While more detailed analysis of the spatial joints will be presented in Chapter 7, as previously mentioned; in this section, a brief discussion on the formulation of the spatial joint constraints is presented in order to better understand the formulation of the mobility criteria used in this chapter. In the *spatial kinematics*, the unconstrained motion of a rigid body is described using six independent coordinates or degrees of freedom. Three of these degrees of freedom represent the translations of the body along three perpendicular axes and three degrees of freedom represent three independent rotational displacements. Figure 6 shows examples of mechanical joints in spatial kinematics. The *spherical joint* shown in Fig. 6a allows only three relative rotational motions between the two bodies i and j connected by this joint. In this case, there is no relative translation between the two bodies,

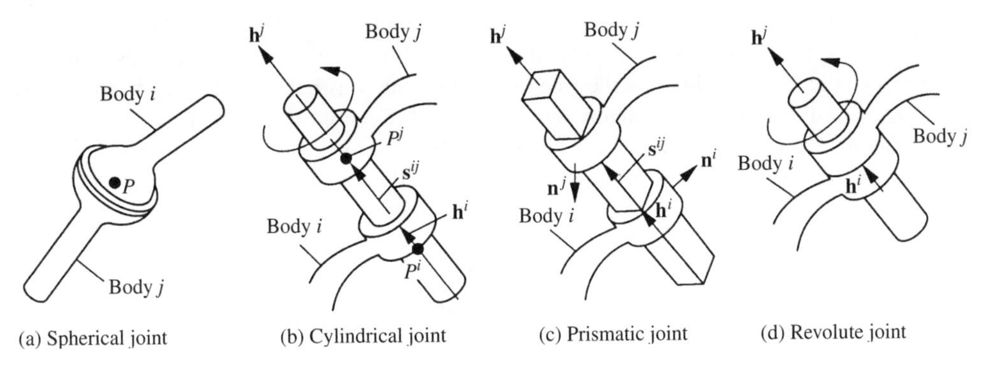

(a) Spherical joint (b) Cylindrical joint (c) Prismatic joint (d) Revolute joint

Figure 3.6 Spatial joints

and hence, one needs three kinematic constraint conditions that eliminate the freedom of the two bodies to translate with respect to each other. Let point P be the joint definition point, P^i be the corresponding point on body i, and P^j be the corresponding point on body j. The three kinematic conditions for the spherical joint require that points P^i and P^j remain in contact throughout the motion. These kinematic conditions can be stated in a vector form as

$$\mathbf{r}_P^i = \mathbf{r}_P^j \tag{3.55}$$

where $\mathbf{r}_P^i$ and $\mathbf{r}_P^j$ are three-dimensional vectors that represent, respectively, the global position vectors of points P^i and P^j. The spherical joint is considered as a three-degree-of-freedom joint, because the three kinematic constraints of Eq. 55 do not impose any restriction on the relative rotations between the two bodies.

Figure 6b shows the two-degree-of-freedom *cylindrical joint* that allows relative translational and rotational displacements between bodies i and j along the joint axis. Two components of the relative translational displacements and two components of the relative rotational displacements along two axes perpendicular to the joint axis are not allowed. In order to eliminate four degrees of freedom, four kinematic constraint conditions are imposed in the case of a cylindrical joint. Let $\mathbf{h}^i$ be a vector drawn on body i along the joint axis, and $\mathbf{h}^j$ be a vector drawn on body j along the joint axis. Also, let $\mathbf{s}^{ij}$ be a vector of variable magnitude that connects points P^i and P^j on bodies i and j, respectively. The vector $\mathbf{s}^{ij}$ is defined on the axis of the cylindrical joint as shown in Fig. 6b. Throughout the motion of the bodies i and j that are connected by the cylindrical joint, the vector $\mathbf{h}^i$ must remain collinear to the vectors $\mathbf{h}^j$ and $\mathbf{s}^{ij}$. The kinematic constraint conditions of the cylindrical joint can then be written as

$$\mathbf{h}^i \times \mathbf{h}^j = \mathbf{0}, \quad \mathbf{h}^i \times \mathbf{s}^{ij} = \mathbf{0} \tag{3.56}$$

As explained in Chapter 2, each vector equation in Eq. 56 contains only two independent equations, that is, the number of the independent kinematic constraint equations is four, leaving two degrees of freedom for the cylindrical joint.

The case of the *prismatic joint* in the spatial kinematics can be obtained as a special case from the case of the cylindrical joint in which the relative rotation between the two bodies

i and j is not allowed. In order to mathematically define this condition, two orthogonal vectors $\mathbf{n}^i$ and $\mathbf{n}^j$ are drawn perpendicular to the joint axis on bodies i and j, respectively, as shown in Fig. 6c. To eliminate the freedom of the rotations between the two bodies i and j, the vectors $\mathbf{n}^i$ and $\mathbf{n}^j$ must remain perpendicular throughout the motion. By considering the prismatic joint as a special case of the cylindrical joint, one needs to add one condition in Eq. 56, leading to the following kinematic constraint equations for the prismatic joint:

$$\left.\begin{aligned} \mathbf{h}^i \times \mathbf{h}^j &= \mathbf{0} \\ \mathbf{h}^i \times \mathbf{s}^{ij} &= \mathbf{0} \\ \mathbf{n}^{i^{\mathrm{T}}} \mathbf{n}^j &= 0 \end{aligned}\right\} \tag{3.57}$$

where $\mathbf{h}^i$, $\mathbf{h}^j$, and $\mathbf{s}^{ij}$ are as defined in Eq. 56. Equation 57 contains five independent constraint equations that define the kinematic conditions for the single-degree-of-freedom prismatic joint in the spatial analysis.

Similarly, the *revolute joint* shown in Fig. 6d can be considered a special case of the cylindrical joint where the relative translation between the two bodies is not allowed. The revolute joint in the spatial kinematics is a one-degree-of-freedom joint. In addition to the constraint equations of the cylindrical joint, one needs another condition that guarantees that the distance between the two points P^i on body i and P^j on body j, defined on the joint axis, remains constant throughout the motion. If $\mathbf{s}^{ij}$ (Fig. 6b) is the vector that connects points P^i and P^j, the kinematic conditions for the revolute joint obtained as a special case of the cylindrical joint are given by

$$\left.\begin{aligned} \mathbf{h}^i \times \mathbf{h}^j &= \mathbf{0} \\ \mathbf{h}^i \times \mathbf{s}^{ij} &= \mathbf{0} \\ \mathbf{s}^{ij^{\mathrm{T}}} \mathbf{s}^{ij} &= c \end{aligned}\right\} \tag{3.58}$$

where the vectors $\mathbf{h}^i$, $\mathbf{h}^j$, and $\mathbf{s}^{ij}$ are the same as in the case of the cylindrical joint and c is a constant. The last equation of Eq. 58 guarantees that the length of the vector $\mathbf{s}^{ij}$ remains constant throughout the motion.

The *universal (Hooke) joint* shown in Fig. 7a is a two-degree-of-freedom joint since it allows relative rotation between the bodies connected by this joint about two perpendicular axes. The constraint equations for this joint can be obtained as a special case of the spherical joint. The four conditions for the universal joint can be written as

$$\mathbf{r}_P^i = \mathbf{r}_P^j, \quad \mathbf{h}^{i^{\mathrm{T}}} \mathbf{h}^j = 0 \tag{3.59}$$

where $\mathbf{r}_P^i$ and $\mathbf{r}_P^j$ are the global position vectors of point P^i on body i and point P^j on body j that coincide with point P at the intersection of the two bars of the cross, and the vectors $\mathbf{h}^i$ and $\mathbf{h}^j$ are two vectors defined on body i and body j, respectively, along the bars of the cross as shown in Fig. 7a.

The *screw joint* shown in Fig. 7b can be considered a special case of the cylindrical joint in which the translation and rotation along the joint axis are not independent. They are related by the *pitch* of the screw. By considering the screw joint as special case of the cylindrical joint, the constraint equations of this joint can be defined using the

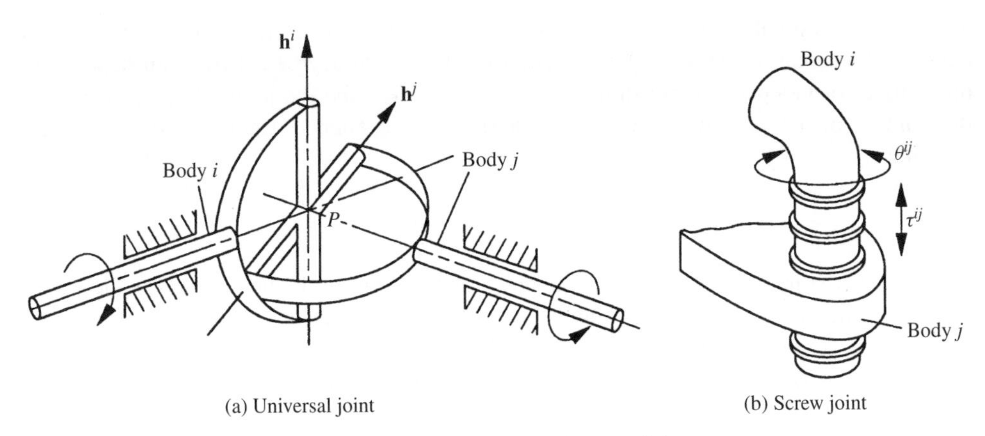

Figure 3.7 Universal and screw joints

relationships

$$\left.\begin{aligned} \mathbf{h}^i \times \mathbf{h}^j &= \mathbf{0} \\ \mathbf{h}^i \times \mathbf{s}^{ij} &= \mathbf{0} \\ \tau^{ij} - \alpha\theta^{ij} &= c \end{aligned}\right\} \tag{3.60}$$

where $\mathbf{h}^i$, $\mathbf{h}^j$, and $\mathbf{s}^{ij}$ are as defined in the case of the cylindrical joint, τ^{ij} is the relative translation, θ^{ij} is the relative rotation, α is the *pitch rate* of the screw joint, and c is a constant that accounts for the initial relative displacements between the two bodies.

Mobility Criteria It was shown in this section that the number of independent kinematic conditions of a joint is equal to the number of degrees of freedom eliminated as the result of using this joint. One of the basic steps in the kinematic and dynamic analysis of mechanical systems is to determine the number of the system *degrees of freedom* or the independent coordinates required to determine the configuration of the system. There are different types of multibody systems that consist of different numbers of bodies interconnected by different numbers and types of joints. The degrees of freedom of the system define the minimum number of independent inputs required to drive or control the system. A multibody system with zero degrees of freedom is a *structure*. The components of such a system are not permitted to undergo relative rigid body motion regardless of the forces acting on the system. Most mechanisms that are in use in industrial and technological applications are designed as single-degree-of-freedom systems. Their motion is controlled by a single input that is transmitted to a single output. Robotic manipulators, on the other hand, are multidegree-of-freedom systems. They require several inputs in order to drive the manipulator and control the position of its end effector. In this section, a simple criterion is presented for determining the number of degrees of freedom of multibody systems.

As pointed out previously, the configuration of a rigid body that undergoes unconstrained planar motion can be identified using three independent coordinates or degrees of freedom. These coordinates describe the translational motion of the body along two perpendicular axes as well as the rotation of the body. A planar system that consists of n_b unconstrained bodies has $3 \times n_b$ coordinates. If these bodies are connected by joints, the number of the system degrees of freedom decreases. The reduction in the system degrees of freedom depends on the number of independent constraint equations that describe the joints. In *planar motion*, the number of the system degrees of freedom can be evaluated according to the *mobility* criterion

$$n_d = 3 \times n_b - n_c \tag{3.61}$$

where n_d is the number of the system degrees of freedom, n_b is the number of the bodies in the system, and n_c is the total number of linearly independent constraint equations that describe the joints in the system. Each revolute or prismatic joint in the planar analysis introduces two kinematic constraints that reduce the number of degrees of freedom by two.

Example 3.1

The slider crank mechanism shown in Fig. 8 consists of four bodies, the *ground* (fixed link) denoted as body 1, the *crankshaft* denoted as body 2, the *connecting rod* denoted as body 3, and the *slider block* denoted as body 4. The system has three revolute joints at O, A, and B, each introduces two kinematic constraint equations that make the total number of kinematic constraints of the revolute joints six. The system has one prismatic joint between the slider block and the fixed link. This joint introduces two kinematic relationships. The *fixed link constraints (ground constraints)* are three since in planar motion two conditions are required to eliminate the freedom of the body to translate and one condition is required to eliminate the freedom of the body to rotate. The total number of constraints n_c is

$$n_c = 6(\text{revolute}) + 2(\text{prismatic}) + 3(\text{fixed link}) = 11$$

Thus, the use of Eq. 61 leads to

$$n_d = 3n_b - n_c = 3 \times 4 - 11 = 1$$

That is, the mechanism has only one degree of freedom.

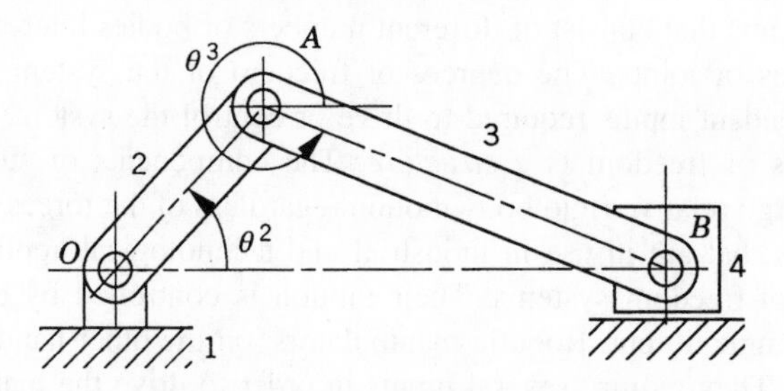

Figure 3.8 Slider crank mechanism

In spatial kinematics, the configuration of a rigid body in space is identified using six coordinates. If the mechanical system consists of n_b bodies, the mobility criterion in the spatial analysis can be written as

$$n_d = 6 \times n_b - n_c \tag{3.62}$$

The freedom of the relative translation between two bodies is eliminated if they are connected by a spherical joint. It follows that a spherical joint reduces the number of the system degrees of freedom by three. A cylindrical or a universal joint introduces four independent kinematic constraint equations that reduce the number of degrees of freedom by four. A revolute, prismatic, or a screw joint, on the other hand, introduces five kinematic constraint conditions that reduce the number of degrees of freedom by five.

Example 3.2

The spatial RSSR (revolute, spherical, spherical, revolute) mechanism shown in Fig. 9 consists of four bodies. Body 2 is connected to body 1 at O by a revolute joint, body 3 is connected to body 2 at A by a spherical joint, body 4 is connected to body 3 at B by a spherical joint, and body 4 is connected to body 1 at C by a revolute joint. Since each spherical joint introduces three kinematic constraints, the spherical joints at A and B introduce six kinematic constraint conditions. The two revolute joints at O and C introduce 10 constraints. In the spatial analysis, six conditions are required to eliminate the freedom of the body to translate or rotate. Thus, the number of *fixed link constraints (ground constraints)* for body 1 is six. The total number of constraint equations for the RSSR mechanism is

$$n_c = 6(\text{spherical}) + 10(\text{revolute}) + 6(\text{fixed link}) = 22$$

Using the mobility criterion of Eq. 62, the number of system degrees of freedom can be determined as

$$n_d = 6n_b - n_c = 6 \times 4 - 22 = 2$$

which indicates that the system has two degrees of freedom. One of these degrees of freedom is the freedom of the coupler link (body 3) to rotate about its own axis.

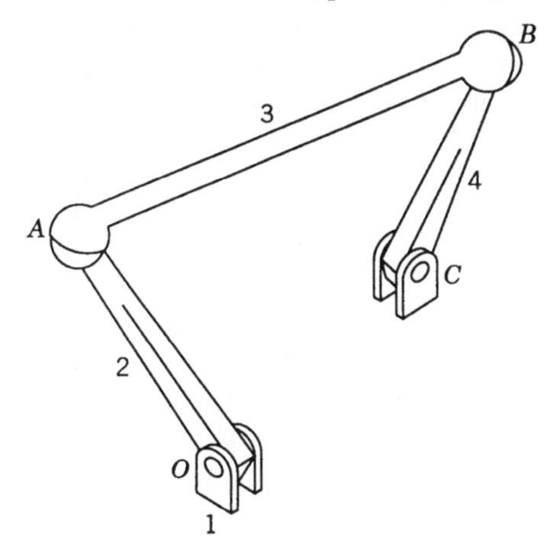

Figure 3.9 RSSR mechanism

3.6 CLASSICAL KINEMATIC APPROACH

In this section, the classical approach used for the kinematic analysis is briefly discussed. This intuitive approach, which can be used in the analysis of systems that consist of a small number of interconnected bodies, is not suited for the study of the kinematics of complex systems. In the classical approach, *analytical* or *graphical methods* can be used. If the degrees of freedom of the system are specified, one can obtain small number of equations that can be solved numerically or using graphical techniques. The use of the analytical kinematic equations will be demonstrated in this section using several examples.

In order to demonstrate the use of the graphical techniques first, we consider the slider crank mechanism shown in Fig. 10. If the dimensions of the mechanism links are given and if the rotation of the crankshaft is specified, one can use nonlinear trigonometric functions to determine the positions and orientations of all other bodies in the mechanism. This step of the kinematic analysis constitutes the step of the *position analysis*, and it is considered as the most difficult step in the classical approach since one has to deal with nonlinear algebraic equations that involve trigonometric functions. For example, if the crank angle θ^2 is given, one can write the following position equations for the slider crank mechanism shown in Fig. 10:

$$\left.\begin{aligned} l^2\cos\theta^2 + l^3\cos\theta^3 + x_B^4 &= 0 \\ l^2\sin\theta^2 + l^3\sin\theta^3 - h &= 0 \end{aligned}\right\} \tag{3.63}$$

In these equations, l^2 and l^3 are the lengths of the crankshaft and connecting rod, respectively; θ^3 and x_B^4 are the angular orientation of the connecting rod and the position of the slider block, respectively; and h is the mechanism offset. Because the mechanism has one degree of freedom, which is considered in this example to be the angular position of the crankshaft, given the crankshaft angle θ^2, the preceding two equations can be solved for the connecting rod angle θ^3 and the slider block position x_B^4. For instance, the second equation of Eq. 63 can be solved for θ^3. This value of θ^3 can be substituted into the first equation to determine x_B^4. Alternatively, since θ^2, l^2, and l^3 are known, one can draw a line diagram that represents the mechanism and from this diagram measure directly θ^3 and x_B^4.

In general, the velocity and acceleration kinematic analyses are more straightforward as compared to the position analysis since the velocity and acceleration kinematic analyses

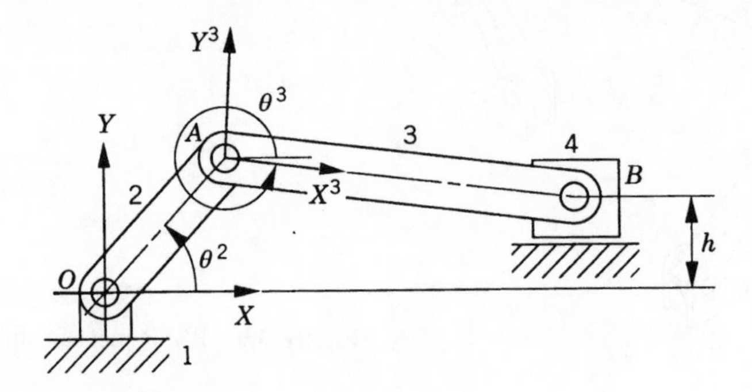

Figure 3.10 Offset slider crank mechanism

involve linear equations only, while the kinematic position equations are, in general, nonlinear. For example, the absolute velocity of point B on the connecting rod can be written as shown in Eq. 64 as

$$\mathbf{v}_B^3 = \mathbf{v}_A^3 + \mathbf{v}_{BA}^3 \tag{3.64}$$

If the angular velocity $\dot{\theta}^2$ of the connecting rod is given (say in counterclockwise direction) and knowing the position coordinates from the position kinematic analysis step, the magnitude and direction of the velocity $\mathbf{v}_A^3 = \mathbf{v}_A^2 = \boldsymbol{\omega}^2 \times \mathbf{u}_{AO}^2$ is known, where $\mathbf{u}_{AO}^2$ is the position of point A with respect to point O. The velocity component $\mathbf{v}_A^3 = \mathbf{v}_A^2$ has a magnitude $\dot{\theta}^2 l^2$ and has direction perpendicular to the line connecting points O and A. Therefore, one can draw the vector $\mathbf{v}_A^3$, which has known magnitude and direction, as shown in Fig. 11a. Since at this stage of the velocity analysis the angular velocity $\dot{\theta}^3$ of the connecting rod is not known; the velocity vector $\mathbf{v}_{BA}^3 = \boldsymbol{\omega}^3 \times \mathbf{u}_{BA}^3$, where $\mathbf{u}_{BA}^3$ is the position vector of point B with respect to point A, has only a known direction that is perpendicular to the line connecting points A and B. Therefore, as shown in Fig. 11b, one can draw the direction of this relative velocity. Because the slider block moves in the horizontal direction, the velocity $\mathbf{v}_B^3$ has a known direction that can be drawn as shown in Fig. 11c. The lines defining the directions of the velocities $\mathbf{v}_{BA}^3$ and $\mathbf{v}_B^3$ intersect at a point that defines the magnitudes of these two vectors as shown in Fig. 11c. The angular velocity of the connecting rod $\dot{\theta}^3$ can be obtained by measuring the length of $\mathbf{v}_{BA}^3$ and dividing this length by the length of the connecting rod l^3, that is, $\dot{\theta}^3 = |\mathbf{v}_{BA}^3|/l^3$. Figure 11c is called the *velocity diagram* of the mechanism. A similar graphical procedure can be used to construct the velocity diagram if the velocity of the slider block, instead of the crankshaft angular velocity, is specified.

Given the angular acceleration of the crankshaft, and knowing the mechanism coordinates from the position analysis and the mechanism velocities from the velocity analysis; one can use a procedure similar to the one used for the velocity diagram to draw the *acceleration diagram* that can be used to determine the angular acceleration $\ddot{\theta}^3$ of the connecting rod and the acceleration $\ddot{x}_B^4$ of the slider block. To this end, the acceleration equation $\mathbf{a}_B^3 = \mathbf{a}_A^3 + \mathbf{a}_{BA}^3$ can be used. In this case, the magnitude and direction of $\mathbf{a}_A^3 = \mathbf{a}_A^2 = \boldsymbol{\alpha}^2 \times \mathbf{u}_{AO}^2 + \boldsymbol{\omega}^2 \times (\boldsymbol{\omega}^2 \times \mathbf{u}_{AO}^2)$ are known since the crankshaft angular velocity and acceleration $\boldsymbol{\omega}^2$ and $\boldsymbol{\alpha}^2$ are assumed to be known. Furthermore, the magnitude and direction of the normal component of $\mathbf{a}_{BA}^3$, defined as $\boldsymbol{\omega}^3 \times (\boldsymbol{\omega}^3 \times \mathbf{u}_{BA}^3)$, is also known. Knowing the directions of the tangential component $\boldsymbol{\alpha}^3 \times \mathbf{u}_{BA}^3$ of $\mathbf{a}_{BA}^3$ and the direction of the absolute acceleration $\mathbf{a}_B^3$, one can construct the acceleration diagram in a straightforward manner.

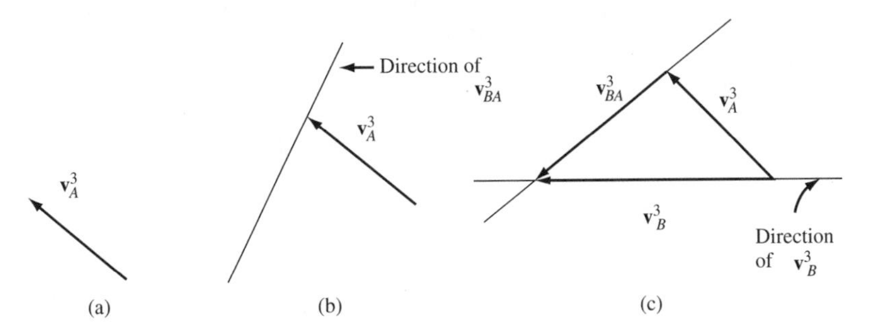

Figure 3.11 Graphical position analysis

It is clear from the analysis presented in this section that the velocity diagram cannot be constructed before the step of the position analysis is completed. It is also clear that the acceleration diagram cannot be constructed before completing the position and velocity analysis steps since the acceleration analysis requires knowing the coordinates and velocities. This order of analysis must be followed regardless of the approach (graphical, analytical, or computational) used. It is also clear that once the nonlinear position equations are solved, the velocity and acceleration diagrams can be constructed in a straightforward manner since the velocity and acceleration equations are linear, as previously mentioned.

Another alternative to the use of the graphical solution in the classical kinematic approach is to obtain a small number of analytical equations for the mechanism. These analytical equations can be solved for the same unknown variables that can be determined using the graphical technique. The use of the analytical methods of the classical kinematic approach is demonstrated in this section using several examples.

Example 3.3

Figure 10 shows an *offset slider crank mechanism* that consists of four bodies. Body 1 is the fixed link or the ground, body 2 is the crankshaft, body 3 is the connecting rod, and body 4 is the slider block at B. The system has one degree of freedom, which is selected to be the angular orientation of the crankshaft θ^2. Express the angular orientation of the connecting rod and the location of the slider block in terms of the degree of freedom. Also determine the angular velocity and acceleration of the connecting rod and the velocity and acceleration of the slider block in terms of the angular velocity $\dot{\theta}^2$ and angular acceleration $\ddot{\theta}^2$ of the crankshaft.

Solution. Consider point A to be the reference point of the connecting rod, the position vector of point B on the connecting rod can be written as

$$\mathbf{r}_B^3 = \mathbf{R}^3 + \mathbf{A}^3\bar{\mathbf{u}}_B^3$$

where $\mathbf{R}^3$ is the global position vector of the reference point A, $\mathbf{A}^3$ is the transformation matrix from the connecting rod coordinate system to the global coordinate system, and $\bar{\mathbf{u}}_B^3$ is the local position vector of point B. The vectors $\mathbf{R}^3$ and $\bar{\mathbf{u}}_B^3$ and the transformation matrix $\mathbf{A}^3$ are defined as

$$\mathbf{R}^3 = \begin{bmatrix} l^2 \cos\theta^2 \\ l^2 \sin\theta^2 \end{bmatrix}, \qquad \bar{\mathbf{u}}_B^3 = \begin{bmatrix} l^3 \\ 0 \end{bmatrix}$$

and

$$\mathbf{A}^3 = \begin{bmatrix} \cos\theta^3 & -\sin\theta^3 \\ \sin\theta^3 & \cos\theta^3 \end{bmatrix}$$

where θ^2 and θ^3 are, respectively, the angular orientations of the crankshaft and the connecting rod, and l^2 and l^3 are, respectively, the lengths of the crankshaft and the connecting rod. One can write the global position vector of point B as

$$\mathbf{r}_B^3 = \begin{bmatrix} l^2 \cos\theta^2 \\ l^2 \sin\theta^2 \end{bmatrix} + \begin{bmatrix} \cos\theta^3 & -\sin\theta^3 \\ \sin\theta^3 & \cos\theta^3 \end{bmatrix} \begin{bmatrix} l^3 \\ 0 \end{bmatrix}$$

$$= \begin{bmatrix} l^2 \cos\theta^2 + l^3 \cos\theta^3 \\ l^2 \sin\theta^2 + l^3 \sin\theta^3 \end{bmatrix}$$

From the geometry of the slider crank mechanism, it is clear that

$$\mathbf{r}_B^3 = \begin{bmatrix} x_B^4 \\ h \end{bmatrix}$$

where x_B^4 is the coordinate of the slider block in the horizontal direction and h is the magnitude of the offset. The preceding two equations lead to the following two scalar equations:

$$x_B^4 = l^2 \cos\theta^2 + l^3 \cos\theta^3$$
$$h = l^2 \sin\theta^2 + l^3 \sin\theta^3$$

which imply that

$$x_B^4 = l^2 \cos\theta^2 \pm \sqrt{(l^3)^2 - (h - l^2 \sin\theta^2)^2}$$

and

$$\theta^3 = \sin^{-1} \frac{h - l^2 \sin\theta^2}{l^3}$$

By using Eq. 30, the velocity of point A on the crankshaft can be written using point O as the reference point as

$$\mathbf{v}_A^2 = \mathbf{v}_O^2 + \mathbf{v}_{AO}^2$$

Since O is a fixed point, $\mathbf{v}_O^2 = \mathbf{0}$. Using Eqs. 27 and 31, the global velocity vector of point A is

$$\mathbf{v}_A^2 = \boldsymbol{\omega}^2 \times \mathbf{u}_A^2 = \dot{\theta}^2 l^2 \begin{bmatrix} -\sin\theta^2 \\ \cos\theta^2 \end{bmatrix}$$

The velocity of point B on the connecting rod can also be written as

$$\mathbf{v}_B^3 = \mathbf{v}_A^3 + \mathbf{v}_{BA}^3 = \mathbf{v}_A^3 + \boldsymbol{\omega}^3 \times \mathbf{u}_B^3$$

Clearly, $\mathbf{v}_A^2 = \mathbf{v}_A^3$ since both represent the global velocity vector of the same point A. Using this fact and Eq. 27, the velocity of point B is given by

$$\mathbf{v}_B^3 = \dot{\theta}^2 l^2 \begin{bmatrix} -\sin\theta^2 \\ \cos\theta^2 \end{bmatrix} + \dot{\theta}^3 l^3 \begin{bmatrix} -\sin\theta^3 \\ \cos\theta^3 \end{bmatrix}$$

The slider block at B moves only in the horizontal direction, and as a consequence

$$\mathbf{v}_B^3 = [\dot{x}_B^4 \quad 0]^{\mathrm{T}}$$

The last two vector equations lead to

$$\begin{bmatrix} \dot{x}_B^4 \\ 0 \end{bmatrix} = \dot{\theta}^2 l^2 \begin{bmatrix} -\sin\theta^2 \\ \cos\theta^2 \end{bmatrix} + \dot{\theta}^3 l^3 \begin{bmatrix} -\sin\theta^3 \\ \cos\theta^3 \end{bmatrix}$$

which can be rearranged and written as

$$\begin{bmatrix} l^3\sin\theta^3 & 1 \\ -l^3\cos\theta^3 & 0 \end{bmatrix} \begin{bmatrix} \dot{\theta}^3 \\ \dot{x}_B^4 \end{bmatrix} = \dot{\theta}^2 l^2 \begin{bmatrix} -\sin\theta^2 \\ \cos\theta^2 \end{bmatrix}$$

or

$$\begin{bmatrix} \dot{\theta}^3 \\ \dot{x}_B^4 \end{bmatrix} = \frac{\dot{\theta}^2 l^2}{l^3\cos\theta^3} \begin{bmatrix} 0 & -1 \\ l^3\cos\theta^3 & l^3\sin\theta^3 \end{bmatrix} \begin{bmatrix} -\sin\theta^2 \\ \cos\theta^2 \end{bmatrix}$$

$$= \frac{\dot{\theta}^2 l^2}{l^3\cos\theta^3} \begin{bmatrix} -\cos\theta^2 \\ l^3\sin(\theta^3 - \theta^2) \end{bmatrix}$$

The acceleration equations can be obtained by differentiating the preceding equation or by using the general expression of Eq. 40. Using Eq. 40, the acceleration of point A on the crankshaft is

$$\mathbf{a}_A^2 = \mathbf{a}_O^2 + \mathbf{a}_{AO}^2$$

Since O is a fixed point, one has $\mathbf{a}_O^2 = \mathbf{0}$, and accordingly,

$$\mathbf{a}_A^2 = \mathbf{a}_{AO}^2 = (\mathbf{a}_{AO}^2)_n + (\mathbf{a}_{AO}^2)_t$$

where

$$(\mathbf{a}_{AO}^2)_n = \boldsymbol{\omega}^2 \times (\boldsymbol{\omega}^2 \times \mathbf{u}_A^2) = -l^2(\dot{\theta}^2)^2 \begin{bmatrix} \cos\theta^2 \\ \sin\theta^2 \end{bmatrix}$$

$$(\mathbf{a}_{AO}^2)_t = \boldsymbol{\alpha}^2 \times \mathbf{u}_A^2 = l^2\ddot{\theta}^2 \begin{bmatrix} -\sin\theta^2 \\ \cos\theta^2 \end{bmatrix}$$

Thus

$$\mathbf{a}_A^2 = \begin{bmatrix} -l^2(\dot{\theta}^2)^2 \cos\theta^2 - l^2\ddot{\theta}^2 \sin\theta^2 \\ -l^2(\dot{\theta}^2)^2 \sin\theta^2 + l^2\ddot{\theta}^2 \cos\theta^2 \end{bmatrix}$$

The acceleration of point B on the connecting rod is

$$\mathbf{a}_B^3 = \mathbf{a}_A^3 + \mathbf{a}_{BA}^3$$

Since $\mathbf{a}_A^3 = \mathbf{a}_A^2$ because A represents the same point on the crankshaft and the connecting rod, one has

$$\mathbf{a}_B^3 = \mathbf{a}_A^2 + \mathbf{a}_{BA}^3 = \mathbf{a}_A^2 + (\mathbf{a}_{BA}^3)_n + (\mathbf{a}_{BA}^3)_t$$

where

$$(\mathbf{a}_{BA}^3)_n = \boldsymbol{\omega}^3 \times (\boldsymbol{\omega}^3 \times \mathbf{u}_B^3) = -l^3(\dot{\theta}^3)^2 \begin{bmatrix} \cos\theta^3 \\ \sin\theta^3 \end{bmatrix}$$

$$(\mathbf{a}_{BA}^3)_t = \boldsymbol{\alpha}^3 \times \mathbf{u}_B^3 = l^3\ddot{\theta}^3 \begin{bmatrix} -\sin\theta^3 \\ \cos\theta^3 \end{bmatrix}$$

Using the expression for $\mathbf{a}_A^2$, one has

$$\mathbf{a}_B^3 = \begin{bmatrix} -l^2(\dot{\theta}^2)^2 \cos\theta^2 - l^2\ddot{\theta}^2 \sin\theta^2 \\ -l^2(\dot{\theta}^2)^2 \sin\theta^2 + l^2\ddot{\theta}^2 \cos\theta^2 \end{bmatrix} + \begin{bmatrix} -l^3(\dot{\theta}^3)^2 \cos\theta^3 - l^3\ddot{\theta}^3 \sin\theta^3 \\ -l^3(\dot{\theta}^3)^2 \sin\theta^3 + l^3\ddot{\theta}^3 \cos\theta^3 \end{bmatrix}$$

Since the slider block moves only in the horizontal direction, one has

$$\mathbf{a}_B^3 = [\ddot{x}_B^4 \quad 0]^{\mathrm{T}}$$

The preceding two vector equations lead to the following two scalar equations:

$$\begin{aligned} \ddot{x}_B^4 &= -l^2(\dot{\theta}^2)^2 \cos\theta^2 - l^2\ddot{\theta}^2 \sin\theta^2 - l^3(\dot{\theta}^3)^2 \cos\theta^3 - l^3\ddot{\theta}^3 \sin\theta^3 \\ 0 &= -l^2(\dot{\theta}^2)^2 \sin\theta^2 + l^2\ddot{\theta}^2 \cos\theta^2 - l^3(\dot{\theta}^3)^2 \sin\theta^3 + l^3\ddot{\theta}^3 \cos\theta^3 \end{aligned}$$

Since θ^3 and $\dot{\theta}^3$ are assumed to be known from the *position* and *velocity analyses*, and θ^2, $\dot{\theta}^2$, and $\ddot{\theta}^2$ are assumed to be given, the preceding two equations are functions of the only two unknowns $\ddot{\theta}^3$ and $\ddot{x}_B^4$. These equations can be rearranged and written as

$$\begin{bmatrix} l^3 \sin\theta^3 & 1 \\ -l^3 \cos\theta^3 & 0 \end{bmatrix} \begin{bmatrix} \ddot{\theta}^3 \\ \ddot{x}_B^4 \end{bmatrix} = \begin{bmatrix} -l^2(\dot{\theta}^2)^2 \cos\theta^2 - l^2\ddot{\theta}^2 \sin\theta^2 - l^3(\dot{\theta}^3)^2 \cos\theta^3 \\ -l^2(\dot{\theta}^2)^2 \sin\theta^2 + l^2\ddot{\theta}^2 \cos\theta^2 - l^3(\dot{\theta}^3)^2 \sin\theta^3 \end{bmatrix}$$

which leads to

$$\begin{bmatrix} \ddot{\theta}^3 \\ \ddot{x}_B^4 \end{bmatrix} = \frac{1}{l^3 \cos \theta^3} \begin{bmatrix} 0 & -1 \\ l^3 \cos \theta^3 & l^3 \sin \theta^3 \end{bmatrix} \begin{bmatrix} c_1 \\ c_2 \end{bmatrix}$$

$$= \frac{1}{l^3 \cos \theta^3} \begin{bmatrix} -c_2 \\ c_1 l^3 \cos \theta^3 + c_2 l^3 \sin \theta^3 \end{bmatrix}$$

where

$$c_1 = -l^2(\dot{\theta}^2)^2 \cos \theta^2 - l^2\ddot{\theta}^2 \sin \theta^2 - l^3(\dot{\theta}^3)^2 \cos \theta^3$$

$$c_2 = -l^2(\dot{\theta}^2)^2 \sin \theta^2 + l^2\ddot{\theta}^2 \cos \theta^2 - l^3(\dot{\theta}^3)^2 \sin \theta^3$$

The kinematic equations obtained in this example can be programmed on a digital computer to obtain the values of the coordinates, velocities, and accelerations of the mechanism links for different values of θ^2, $\dot{\theta}^2$, and $\ddot{\theta}^2$. Consider, for example, the case of a slider crank mechanism which has the following data: $h = 0$, $l^2 = 0.2\,\text{m}$, and $l^3 = 0.4\,\text{m}$. The angular velocity of the crankshaft $\dot{\theta}^2$ is assumed to be constant and is equal to 50 rad/s. Table 1 shows θ^3, x_B^4, $\dot{\theta}^3$, $\dot{x}_B^4$, $\ddot{\theta}^3$, and $\ddot{x}_B^4$ for different values of the crank angle θ^2. In Table 1, angles are measured in radians and distances are in meters.

Singular Configurations In some applications, the motion simulation of the single- and multidegree-of-freedom systems does not proceed smoothly with time. A *lockup configuration* may be encountered or more than one possible motion at certain mechanism configurations can occur. These cases are called *singular configurations*. The singularity of motion may depend on the nature of the driving input. For example, consider the slider crank mechanism shown in Fig. 12. First assume that the mechanism is driven by rotating the crankshaft with a given angular velocity $\omega^2 = \dot{\theta}^2$. It was shown in the preceding example that the angular velocity of the connecting rod $\dot{\theta}^3$ and the velocity of the slider block $\dot{x}_B^4$ can be expressed in terms of the angular velocity of the crankshaft $\dot{\theta}^2$ as

$$\begin{bmatrix} l^3 \sin \theta^3 & 1 \\ -l^3 \cos \theta^3 & 0 \end{bmatrix} \begin{bmatrix} \dot{\theta}^3 \\ \dot{x}_B^4 \end{bmatrix} = \dot{\theta}^2 l^2 \begin{bmatrix} -\sin \theta^2 \\ \cos \theta^2 \end{bmatrix} \tag{3.65}$$

where l^2 and l^3 are, respectively, the lengths of the crankshaft and the connecting rod and θ^2 and θ^3 are, respectively, the angular orientations of the crankshaft and the connecting rod. Equation 65 can be used to define $\dot{\theta}^3$ and $\dot{x}_B^4$ as

$$\begin{bmatrix} \dot{\theta}^3 \\ \dot{x}_B^4 \end{bmatrix} = \frac{\dot{\theta}^2 l^2}{l^3 \cos \theta^3} \begin{bmatrix} -\cos \theta^2 \\ l^3 \sin(\theta^3 - \theta^2) \end{bmatrix} \tag{3.66}$$

Consider now the special case where $l^2 = l^3$ and $\theta^2 = \pi/2$; in this special case, one has $l^2 \sin \theta^2 + l^3 \sin \theta^3 = 0$, or $\sin \theta^2 + \sin \theta^3 = 0$. It follows that $\theta^3 = 3\pi/2$. In this

TABLE 3.1 Slider Crank Mechanism

θ^2	θ^3	X_B^4	$\dot{\theta}^3$	$\dot{X}_B^4$	$\ddot{\theta}^3$	$\ddot{X}_B^4$
0.000E + 00	0.000E + 00	0.600E + 00	−0.250E + 02	−0.000E + 00	0.000E + 00	−0.750E + 03
0.200E + 00	−0.995E − 01	0.594E + 00	−0.246E + 02	−0.297E + 01	0.189E + 03	−0.724E + 03
0.400E + 00	−0.196E + 00	0.577E + 00	−0.235E + 02	−0.572E + 01	0.387E + 03	−0.647E + 03
0.600E + 00	−0.286E + 00	0.549E + 00	−0.215E + 02	−0.808E + 01	0.600E + 03	−0.522E + 03
0.800E + 00	−0.367E + 00	0.513E + 00	−0.187E + 02	−0.985E + 01	0.827E + 03	−0.360E + 03
0.100E + 01	−0.434E + 00	0.471E + 00	−0.149E + 02	−0.109E + 02	0.106E + 04	−0.173E + 03
0.120E + 01	−0.485E + 00	0.426E + 00	−0.102E + 02	−0.112E + 02	0.126E + 04	0.169E + 03
0.140E + 01	−0.515E + 00	0.382E + 00	−0.488E + 01	−0.108E + 02	0.140E + 04	0.183E + 03
0.160E + 01	−0.523E + 00	0.341E + 00	0.843E + 00	−0.983E + 01	0.144E + 04	0.303E + 03
0.180E + 01	−0.509E + 00	0.304E + 00	0.650E + 01	−0.847E + 01	0.137E + 04	0.366E + 03
0.200E + 01	−0.472E + 00	0.273E + 00	0.117E + 02	−0.697E + 01	0.121E + 04	0.379E + 03
0.220E + 01	−0.416E + 00	0.248E + 00	0.161E + 02	−0.548E + 01	0.991E + 03	0.360E + 03
0.240E + 01	−0.345E + 00	0.229E + 00	0.196E + 02	−0.411E + 01	0.759E + 03	0.327E + 03
0.260E + 01	−0.261E + 00	0.215E + 00	0.222E + 02	−0.287E + 01	0.536E + 03	0.294E + 03
0.280E + 01	−0.168E + 00	0.206E + 00	0.239E + 02	−0.175E + 01	0.328E + 03	0.268E + 03
0.300E + 01	−0.706E − 01	0.201E + 00	0.248E + 02	−0.711E + 00	0.133E + 03	0.253E + 03
0.320E + 01	0.292E − 01	0.200E + 00	0.250E + 02	0.292E + 00	−0.548E + 02	0.251E + 03
0.340E + 01	0.128E + 00	0.203E + 00	0.244E + 02	0.131E + 01	−0.246E + 03	0.260E + 03
0.360E + 01	0.223E + 00	0.211E + 00	0.230E + 02	0.239E + 01	−0.447E + 03	0.282E + 03
0.380E + 01	0.311E + 00	0.223E + 00	0.208E + 02	0.358E + 01	−0.665E + 03	0.313E + 03
0.400E + 01	0.388E + 00	0.240E + 00	0.177E + 02	0.490E + 01	−0.895E + 03	0.347E + 03
0.420E + 01	0.451E + 00	0.262E + 00	0.136E + 02	0.634E + 01	−0.112E + 04	0.374E + 03
0.440E + 01	0.496E + 00	0.290E + 00	0.874E + 01	0.785E + 01	−0.131E + 04	0.376E + 03
0.460E + 01	0.520E + 00	0.325E + 00	0.323E + 01	0.929E + 01	−0.143E + 04	0.336E + 03
0.480E + 01	0.521E + 00	0.364E + 00	−0.252E + 01	0.105E + 02	−0.143E + 04	0.239E + 03
0.500E + 01	0.500E + 00	0.408E + 00	−0.808E + 01	0.111E + 02	−0.133E + 04	0.904E + 02
0.520E + 01	0.458E + 00	0.453E + 00	−0.131E + 02	0.111E + 02	−0.115E + 04	−0.927E + 02
0.540E + 01	0.397E + 00	0.496E + 00	−0.172E + 02	0.104E + 02	−0.923E + 03	−0.284E + 03
0.560E + 01	0.321E + 00	0.535E + 00	−0.204E + 02	0.889E + 01	−0.693E + 03	−0.459E + 03
0.580E + 01	0.234E + 00	0.566E + 00	−0.228E + 02	0.676E + 01	−0.473E + 03	−0.600E + 03
0.600E + 01	0.140E + 00	0.588E + 00	−0.242E + 02	0.415E + 01	−0.270E + 03	−0.698E + 03
0.620E + 01	0.416E − 01	0.599E + 00	−0.249E + 02	0.125E + 01	−0.781E + 02	−0.745E + 03

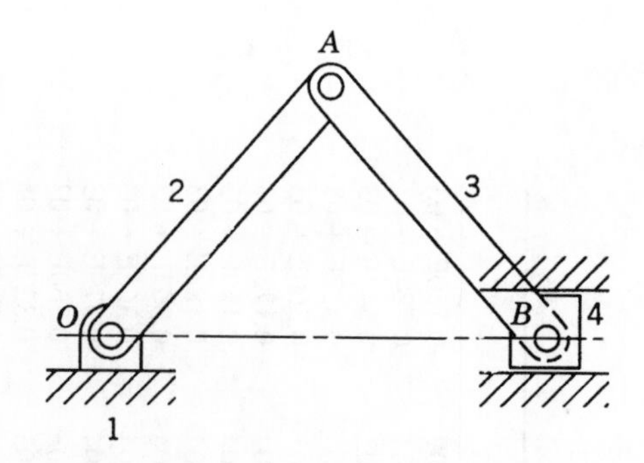

Figure 3.12 Slider crank mechanism

configuration, $\cos\theta^3 = 0$ and the coefficient matrix in Eq. 65 is singular. This implies that at this configuration, more than one possible motion can occur. One possible motion is that the crankshaft and the connecting rod are locked together and rotate as a single pendulum, as shown in Fig. 13a. Another possible motion is that the slider block moves to the right or to the left in the horizontal direction, as shown in Fig. 13b.

The configurations shown in Fig. 13 are not the only singular configurations encountered in the analysis of the slider crank mechanism. To demonstrate this, consider the case where the mechanism is driven by moving the slider block with a specified velocity $\dot{x}_B^4$. In this case, Eq. 65 can be rearranged and written as

$$\begin{bmatrix} l^2 \sin\theta^2 & l^3 \sin\theta^3 \\ -l^2 \cos\theta^2 & -l^3 \cos\theta^3 \end{bmatrix} \begin{bmatrix} \dot{\theta}^2 \\ \dot{\theta}^3 \end{bmatrix} = \begin{bmatrix} -\dot{x}_B^4 \\ 0 \end{bmatrix} \tag{3.67}$$

Now consider the configuration shown in Fig. 14, where $\theta^2 = \theta^3 = 0$. At this configuration, the coefficient matrix of Eq. 67 is singular, which indicates that it is impossible for the motion to continue by moving the slider block. The mechanism at this configuration, however, can be driven by rotating the crankshaft, since at this configuration the coefficient matrix in Eq. 65 is not singular.

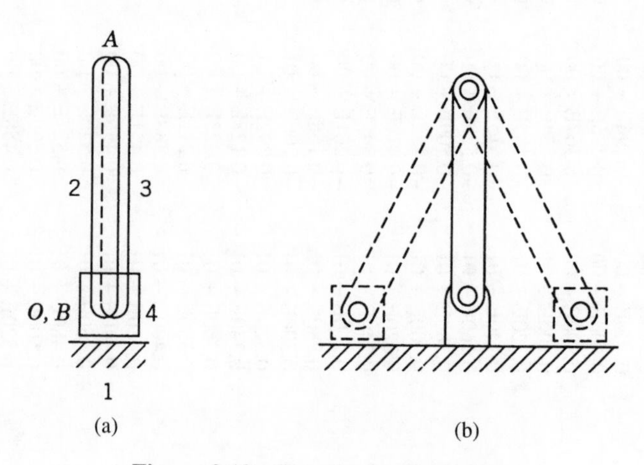

Figure 3.13 Singular configurations

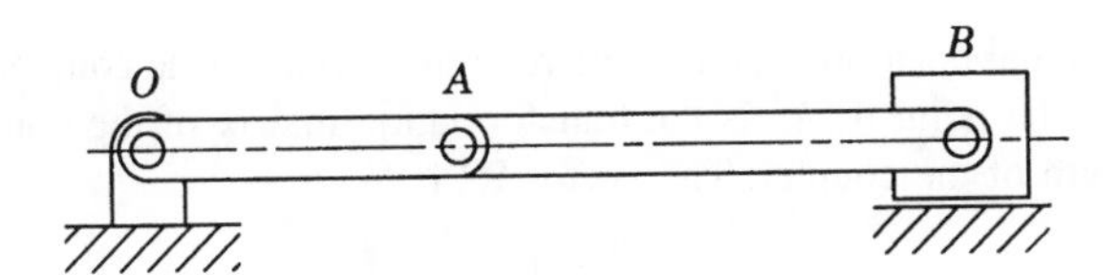

Figure 3.14 Another singular configuration

Example 3.4

Figure 15 shows a four-bar linkage. Body 1 is the fixed link or the ground, body 2 is the crankshaft OA, body 3 is the coupler AB, and body 4 is the rocker BC. Obtain an expression for the angular orientation, velocities, and accelerations of the coupler and the rocker in terms of the angular orientation, angular velocity, and angular acceleration of the crankshaft. Assuming that the angular velocity of the crankshaft is constant and is equal to 50 rad/s, determine the values of the angular coordinates, velocities and accelerations of the coupler and the rocker for different values of the angles of the crankshaft. Assume that the lengths of the crankshaft, coupler and the rocker are 0.2, 0.4, and 0.5 m, respectively, and the distance OC is 0.4 m.

Solution. The position vector of point C can be expressed in terms of the Cartesian coordinates of the rocker as

$$\mathbf{r}_C^4 = \mathbf{R}^4 + \mathbf{A}^4\bar{\mathbf{u}}_C^4$$

where $\mathbf{R}^4$ is the global position vector of the reference point of the rocker, which we select in this example to be point B, $\mathbf{A}^4$ is the transformation matrix of the rocker, $\bar{\mathbf{u}}_C^4 = [l^4 \quad 0]^{\mathrm{T}}$ is the local position vector of point C, and l^4 is the length of the rocker. The global position vector of the reference point of the rocker can be written as

$$\mathbf{R}^4 = \mathbf{r}_B^3 = \mathbf{R}^3 + \mathbf{A}^3\bar{\mathbf{u}}_B^3$$

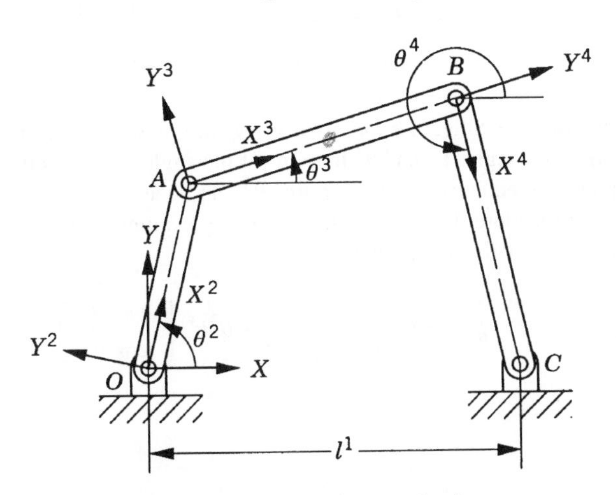

Figure 3.15 Four-bar mechanism

where $\mathbf{R}^3$ is the global position vector of the reference point of the coupler, which is selected in this example to be point A, $\mathbf{A}^3$ is the transformation matrix of the coupler, $\bar{\mathbf{u}}_B^3 = [l^3 \ 0]^{\mathrm{T}}$, and l^3 is the length of the coupler. The vector $\mathbf{R}^3$ is

$$\mathbf{R}^3 = \begin{bmatrix} l^2 \cos\theta^2 \\ l^2 \sin\theta^2 \end{bmatrix}$$

where l^2 is the length of the crankshaft. The vector $\mathbf{R}^4$ can then be written as

$$\mathbf{R}^4 = \begin{bmatrix} l^2 \cos\theta^2 \\ l^2 \sin\theta^2 \end{bmatrix} + \begin{bmatrix} \cos\theta^3 & -\sin\theta^3 \\ \sin\theta^3 & \cos\theta^3 \end{bmatrix} \begin{bmatrix} l^3 \\ 0 \end{bmatrix}$$

$$= \begin{bmatrix} l^2 \cos\theta^2 + l^3 \cos\theta^3 \\ l^2 \sin\theta^2 + l^3 \sin\theta^3 \end{bmatrix}$$

Using this equation, the global position vector of point C can be written as

$$\mathbf{r}_C^4 = \mathbf{R}^4 + \mathbf{A}^4\bar{\mathbf{u}}_C^4 = \begin{bmatrix} l^2 \cos\theta^2 + l^3 \cos\theta^3 \\ l^2 \sin\theta^2 + l^3 \sin\theta^3 \end{bmatrix} + \begin{bmatrix} \cos\theta^4 & -\sin\theta^4 \\ \sin\theta^4 & \cos\theta^4 \end{bmatrix} \begin{bmatrix} l^4 \\ 0 \end{bmatrix}$$

$$= \begin{bmatrix} l^2 \cos\theta^2 + l^3 \cos\theta^3 + l^4 \cos\theta^4 \\ l^2 \sin\theta^2 + l^3 \sin\theta^3 + l^4 \sin\theta^4 \end{bmatrix}$$

From Fig. 15 it is clear that

$$\mathbf{r}_C^4 = \begin{bmatrix} l^1 \\ 0 \end{bmatrix}$$

where l^1 is the distance OC. The preceding two equations lead to the following two scalar equations:

$$l^2 \cos\theta^2 + l^3 \cos\theta^3 + l^4 \cos\theta^4 = l^1$$
$$l^2 \sin\theta^2 + l^3 \sin\theta^3 + l^4 \sin\theta^4 = 0$$

These two equations are called the *loop closure equations* of the four-bar linkage. They can be used to express the angles θ^3 and θ^4 in terms of the angle θ^2. It is left to the reader to try to solve the loop closure equations and determine θ^3 and θ^4 as a function of the crank angle θ^2.

Following the procedure described in the preceding example, one can show that the global velocity vector of point B on the coupler is

$$\mathbf{v}_B^3 = \dot{\theta}^2 l^2 \begin{bmatrix} -\sin\theta^2 \\ \cos\theta^2 \end{bmatrix} + \dot{\theta}^3 l^3 \begin{bmatrix} -\sin\theta^3 \\ \cos\theta^3 \end{bmatrix}$$

The velocity of point C, which, in this example, is equal to zero can be expressed in terms of the velocity of point B as

$$\mathbf{v}_C^4 = \mathbf{v}_B^4 + \mathbf{v}_{CB}^4$$

Using Eqs. 27 and 31 and the fact that $\mathbf{v}_B^4 = \mathbf{v}_B^3$ and $\mathbf{v}_C^4 = \mathbf{0}$, one obtains

$$\begin{bmatrix} 0 \\ 0 \end{bmatrix} = \dot{\theta}^2 l^2 \begin{bmatrix} -\sin\theta^2 \\ \cos\theta^2 \end{bmatrix} + \dot{\theta}^3 l^3 \begin{bmatrix} -\sin\theta^3 \\ \cos\theta^3 \end{bmatrix} + \dot{\theta}^4 l^4 \begin{bmatrix} -\sin\theta^4 \\ \cos\theta^4 \end{bmatrix}$$

which can be rearranged and written as

$$\begin{bmatrix} -l^3\sin\theta^3 & -l^4\sin\theta^4 \\ l^3\cos\theta^3 & l^4\cos\theta^4 \end{bmatrix} \begin{bmatrix} \dot{\theta}^3 \\ \dot{\theta}^4 \end{bmatrix} = \dot{\theta}^2 l^2 \begin{bmatrix} \sin\theta^2 \\ -\cos\theta^2 \end{bmatrix}$$

or

$$\begin{bmatrix} \dot{\theta}^3 \\ \dot{\theta}^4 \end{bmatrix} = \frac{l^2\dot{\theta}^2}{l^3 l^4 \sin(\theta^4 - \theta^3)} \begin{bmatrix} l^4\cos\theta^4 & l^4\sin\theta^4 \\ -l^3\cos\theta^3 & -l^3\sin\theta^3 \end{bmatrix} \begin{bmatrix} \sin\theta^2 \\ -\cos\theta^2 \end{bmatrix}$$

$$= \frac{l^2\dot{\theta}^2}{l^3 l^4 \sin(\theta^4 - \theta^3)} \begin{bmatrix} l^4\sin(\theta^2 - \theta^4) \\ l^3\sin(\theta^3 - \theta^2) \end{bmatrix}$$

Following a procedure similar to the one described in the preceding example, one can show that the acceleration of point B is

$$\mathbf{a}_B^3 = \begin{bmatrix} -l^2(\dot{\theta}^2)^2\cos\theta^2 - l^2\ddot{\theta}^2\sin\theta^2 - l^3(\dot{\theta}^3)^2\cos\theta^3 - l^3\ddot{\theta}^3\sin\theta^3 \\ -l^2(\dot{\theta}^2)^2\sin\theta^2 + l^2\ddot{\theta}^2\cos\theta^2 - l^3(\dot{\theta}^3)^2\sin\theta^3 + l^3\ddot{\theta}^3\cos\theta^3 \end{bmatrix} \tag{3.68}$$

The acceleration of point C on the rocker is

$$\mathbf{a}_C^4 = \mathbf{a}_B^4 + \mathbf{a}_{CB}^4 = \mathbf{a}_B^3 + (\mathbf{a}_{CB}^4)_n + (\mathbf{a}_{CB}^4)_t$$

Since C is a fixed point, one has $\mathbf{a}_C^4 = \mathbf{0}$, and accordingly,

$$\mathbf{0} = \mathbf{a}_B^3 + (\mathbf{a}_{CB}^4)_n + (\mathbf{a}_{CB}^4)_t$$

which leads to

$$\mathbf{a}_B^3 = -\boldsymbol{\omega}^4 \times (\boldsymbol{\omega}^4 \times \mathbf{u}_C^4) - \boldsymbol{\alpha}^4 \times \mathbf{u}_C^4$$

$$= \begin{bmatrix} l^4(\dot{\theta}^4)^2\cos\theta^4 + l^4\ddot{\theta}^4\sin\theta^4 \\ l^4(\dot{\theta}^4)^2\sin\theta^4 - l^4\ddot{\theta}^4\cos\theta^4 \end{bmatrix} \tag{3.69}$$

Substituting Eq. 69 into Eq. 68, one obtains the following scalar equations:

$$l^4(\dot{\theta}^4)^2 \cos\theta^4 + l^4\ddot{\theta}^4 \sin\theta^4$$
$$= -l^2(\dot{\theta}^2)^2 \cos\theta^2 - l^2\ddot{\theta}^2 \sin\theta^2 - l^3(\dot{\theta}^3)^2 \cos\theta^3 - l^3\ddot{\theta}^3 \sin\theta^3$$
$$l^4(\dot{\theta}^4)^2 \sin\theta^4 - l^4\ddot{\theta}^4 \cos\theta^4$$
$$= -l^2(\dot{\theta}^2)^2 \sin\theta^2 + l^2\ddot{\theta}^2 \cos\theta^2 - l^3(\dot{\theta}^3)^2 \sin\theta^3 + l^3\ddot{\theta}^3 \cos\theta^3$$

Assuming that θ^3, θ^4, $\dot{\theta}^3$, and $\dot{\theta}^4$ are known from the position and velocity analyses, and θ^2, $\dot{\theta}^2$, and $\ddot{\theta}^2$ are given, there are only two unknowns $\ddot{\theta}^3$ and $\ddot{\theta}^4$ in the preceding two equations. These equations can be rearranged and rewritten as

$$\begin{bmatrix} -l^3 \sin\theta^3 & -l^4 \sin\theta^4 \\ l^3 \cos\theta^3 & l^4 \cos\theta^4 \end{bmatrix} \begin{bmatrix} \ddot{\theta}^3 \\ \ddot{\theta}^4 \end{bmatrix} = \begin{bmatrix} c_1 \\ c_2 \end{bmatrix} \tag{3.70}$$

where c_1 and c_2 are

$$c_1 = l^4(\dot{\theta}^4)^2 \cos\theta^4 + l^2(\dot{\theta}^2)^2 \cos\theta^2 + l^2\ddot{\theta}^2 \sin\theta^2 + l^3(\dot{\theta}^3)^2 \cos\theta^3$$
$$c_2 = l^4(\dot{\theta}^4)^2 \sin\theta^4 + l^2(\dot{\theta}^2)^2 \sin\theta^2 - l^2\ddot{\theta}^2 \cos\theta^2 + l^3(\dot{\theta}^3)^2 \sin\theta^3$$

Equation 70 can be solved for the angular accelerations $\ddot{\theta}^3$ and $\ddot{\theta}^4$ as follows:

$$\begin{bmatrix} \ddot{\theta}^3 \\ \ddot{\theta}^4 \end{bmatrix} = \frac{1}{l^3 l^4 \sin(\theta^4 - \theta^3)} \begin{bmatrix} l^4 \cos\theta^4 & l^4 \sin\theta^4 \\ -l^3 \cos\theta^3 & -l^3 \sin\theta^3 \end{bmatrix} \begin{bmatrix} c_1 \\ c_2 \end{bmatrix}$$
$$= \frac{1}{l^3 l^4 \sin(\theta^4 - \theta^3)} \begin{bmatrix} l^4(c_1 \cos\theta^4 + c_2 \sin\theta^4) \\ -l^3(c_1 \cos\theta^3 + c_2 \sin\theta^3) \end{bmatrix}$$

Using the dimensions of the mechanism and the kinematic equations presented in this example, the angular coordinates, velocities and accelerations of the links can be determined as functions of the crank angle as shown in Table 2. The angles presented in this table are in radians, the angular velocities are in rad/s, and the angular accelerations are in rad/s^2.

Mechanism Kinematics The position kinematic equations obtained in the preceding example can be used to express the orientations of the coupler and the rocker in terms of the crank angle θ^2. One of the important considerations in the design of many of the four-bar linkages is to ensure that the crankshaft can rotate a complete revolution. In order to determine whether the input crank of the four-bar mechanism can make a complete revolution, *Grashof's law* can be used. This law states that, for a planar four-bar linkage, if the sum of the lengths of the shortest and longest links is less than the sum of the lengths of the other two links, then a continuous relative motion between two links can be achieved. Let s and l be, respectively, the lengths of the shortest and longest links, and p and q be

TABLE 3.2 Four-Bar Mechanism

θ^2	θ^3	θ^4	$\dot{\theta}^3$	$\dot{\theta}^4$	$\ddot{\theta}^3$	$\ddot{\theta}^4$
0.157E + 01	0.795E + 00	0.495E + 01	−0.702E + 01	0.165E + 02	0.106E + 04	0.717E + 03
0.177E + 01	0.774E + 00	0.503E + 01	−0.314E + 01	0.188E + 02	0.901E + 03	0.441E + 03
0.197E + 01	0.769E + 00	0.510E + 01	0.255E + 00	0.201E + 02	0.803E + 03	0.228E + 03
0.217E + 01	0.776E + 00	0.518E + 01	0.334E + 01	0.206E + 02	0.745E + 03	0.593E + 02
0.237E + 01	0.795E + 00	0.527E + 01	0.625E + 01	0.206E + 02	0.712E + 03	−0.768E + 02
0.257E + 01	0.826E + 00	0.535E + 01	0.905E + 01	0.201E + 02	0.693E + 03	−0.187E + 03
0.277E + 01	0.868E + 00	0.543E + 01	0.118E + 02	0.191E + 02	0.677E + 03	−0.274E + 03
0.297E + 01	0.920E + 00	0.550E + 01	0.145E + 02	0.179E + 02	0.657E + 03	−0.339E + 03
0.317E + 01	0.983E + 00	0.557E + 01	0.170E + 02	0.164E + 02	0.624E + 03	−0.383E + 03
0.337E + 01	0.106E + 01	0.563E + 01	0.194E + 02	0.149E + 02	0.577E + 03	−0.408E + 03
0.357E + 01	0.114E + 01	0.569E + 01	0.216E + 02	0.132E + 02	0.515E + 03	−0.418E + 03
0.377E + 01	0.123E + 01	0.574E + 01	0.235E + 02	0.115E + 02	0.438E + 03	−0.417E + 03
0.397E + 01	0.133E + 01	0.578E + 01	0.251E + 02	0.986E + 01	0.347E + 03	−0.413E + 03
0.417E + 01	0.143E + 01	0.582E + 01	0.263E + 02	0.822E + 01	0.243E + 03	−0.411E + 03
0.437E + 01	0.154E + 01	0.585E + 01	0.270E + 02	0.656E + 01	0.123E + 03	−0.418E + 03
0.457E + 01	0.165E + 01	0.587E + 01	0.273E + 02	0.484E + 01	−0.193E + 02	−0.445E + 03
0.477E + 01	0.175E + 01	0.589E + 01	0.268E + 02	0.296E + 01	−0.200E + 03	−0.503E + 03
0.497E + 01	0.186E + 01	0.589E + 01	0.256E + 02	0.751E + 00	−0.444E + 03	−0.612E + 03
0.517E + 01	0.196E + 01	0.589E + 01	0.231E + 02	0.204E + 01	−0.799E + 03	−0.803E + 03
0.537E + 01	0.204E + 01	0.588E + 01	0.189E + 02	−0.586E + 01	−0.135E + 04	−0.113E + 04
0.557E + 01	0.211E + 01	0.584E + 01	0.121E + 02	−0.113E + 02	−0.222E + 04	−0.167E + 04
0.577E + 01	0.214E + 01	0.579E + 01	0.734E + 02	−0.195E + 02	−0.356E + 04	−0.250E + 04
0.597E + 01	0.210E + 01	0.568E + 01	−0.169E + 02	−0.314E + 02	−0.512E + 04	−0.333E + 04
0.617E + 01	0.199E + 01	0.553E + 01	−0.389E + 02	−0.447E + 02	−0.544E + 04	−0.295E + 04
0.637E + 01	0.179E + 01	0.533E + 01	−0.562E + 02	−0.516E + 02	−0.273E + 04	−0.184E + 03
0.657E + 01	0.156E + 01	0.513E + 01	−0.593E + 02	−0.457E + 02	0.100E + 04	−0.290E + 04
0.677E + 01	0.134E + 01	0.498E + 01	−0.510E + 02	−0.314E + 02	0.278E + 04	0.393E + 04
0.697E + 01	0.116E + 01	0.488E + 01	−0.393E + 02	−0.163E + 02	0.286E + 04	0.346E + 04
0.717E + 01	0.102E + 01	0.484E + 01	−0.288E + 02	−0.417E + 01	0.236E + 04	0.261E + 04
0.737E + 01	0.922E + 00	0.484E + 01	−0.205E + 02	0.471E + 01	0.184E + 04	0.186E + 04
0.757E + 01	0.853E + 00	0.488E + 01	−0.140E + 02	0.109E + 02	0.143E + 04	0.128E + 04
0.777E + 01	0.808E + 00	0.493E + 01	−0.887E + 01	0.152E + 02	0.115E + 04	0.858E + 03

the lengths of the other two links. According to Grashof's law, the shortest link will rotate continuously if

$$s + l \leq p + q \tag{3.71}$$

This inequality has to be satisfied, otherwise none of the links will make a complete revolution relative to the other links. If link 2 in the four-bar linkage (Fig. 15) can make a complete revolution while link 4 oscillates, the mechanism is called a *crank-rocker linkage*. If both link 2 and link 4 oscillate between limits, the mechanism is called a *double-rocker linkage*.

Grashof's law makes no mention of which link is fixed or of the order in which the links are connected. Several *kinematic inversions* of the four-bar mechanism, however, can be obtained by selecting which link is to be fixed and by arranging the connectivity of the links based on their lengths. When the crank is the shortest link and it is adjacent to the fixed link, the resulting mechanism is of the *crank-rocker type*. If the shortest link is the fixed link, one obtains the *double-crank mechanism*, which is also called a *drag-link mechanism*. When the link opposite to the shortest link is the fixed link, one obtains again the double-rocker mechanism. The double-rocker mechanism is also obtained if the sum of the lengths of the shortest link and longest link is larger than the sum of the other two links.

It is clear in the case of the four-bar linkage of Fig. 15 that any point on the crankshaft OA or the rocker BC moves on a circular arc that has a radius equal to the distance between this point and the fixed points O and C, respectively. During the dynamic motion of the mechanism, any point on the coupler of the four-bar linkage generates a path, called a *coupler curve*, that depends on the location of this point. Clearly, the two paths generated by points A and B are simple circles. Four-bar mechanisms can be designed such that a point on the coupler link moves in a straight line. Such mechanisms are called *straight-line mechanisms*. An example of an approximate straight-line mechanism is the four-bar *Watt's mechanism* shown in Fig. 16a. If, in this mechanism, the position of point P on the coupler is such that the ratio of the lengths of the segments AP and PB is inversely proportional to the ratio of the lengths of the links OA and BC, respectively, then the coupler curve of point P is an approximate straight line. A mechanism that generates an

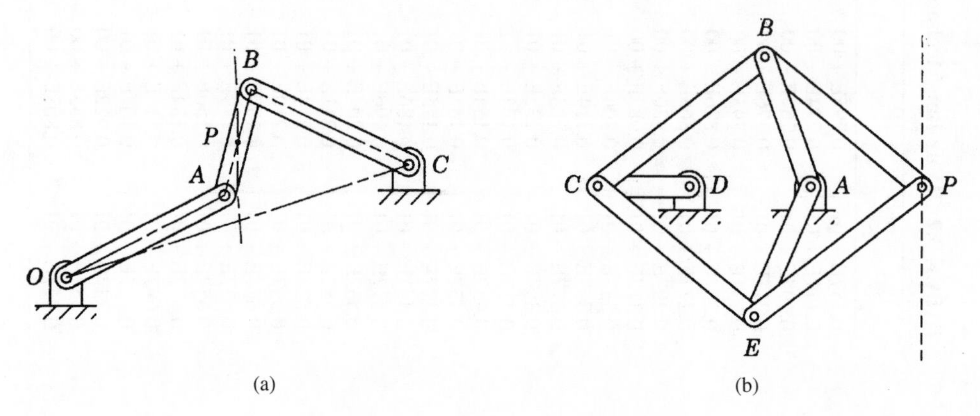

Figure 3.16 Straight-line mechanisms

exact straight line is the *Peaucellier mechanism* shown in Fig. 16b. This mechanism consists of eight links including the fixed link. As pointed out in Chapter 1, if the lengths of link AB and link AE are equal, lengths of links BC, BP, EC, and EP are equal, and the length of link AD is equal to the distance CD, point P will trace out an exact straight-line path. Straight-line mechanisms are used in many mechanical system applications such as gear switch equipment and engine indicators.

Coriolis Acceleration The classical kinematic approaches, both graphical and analytical, can also be used in the analysis of mechanisms with Coriolis acceleration. The use of the classical analytical approach in the analysis of such mechanisms is demonstrated by the following examples.

Example 3.5

Figure 17 shows a block P that slides on a slender rod i. The rod is connected to the ground by a pin joint at O and rotates with angular velocity $\dot{\theta}^i$. Determine the absolute velocity of point P and the absolute acceleration of the slider block P.

Solution. We first select the rod coordinate system to be X^iY^i with origin at O. The position vector of the block with respect to this coordinate system is

$$\bar{\mathbf{u}}_P^i = [\bar{x}_P^i \quad 0]^{\mathrm{T}}$$

Since the block is moving with respect to the rod, its velocity is described by Eq. 46 as

$$\mathbf{v}_P^i = \mathbf{v}_O^i + \boldsymbol{\omega}^i \times \mathbf{u}_P^i + (\mathbf{v}_P^i)_r$$

Since O is a fixed point, $\mathbf{v}_O^i = \mathbf{0}$, and

$$\mathbf{v}_P^i = \boldsymbol{\omega}^i \times \mathbf{u}_P^i + (\mathbf{v}_P^i)_r$$

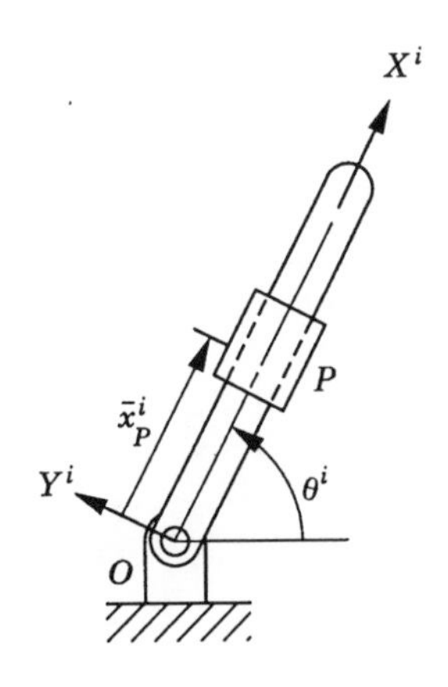

Figure 3.17 Pendulum with a sliding block

in which

$$\boldsymbol{\omega}^i \times \mathbf{u}_P^i = \dot{\theta}^i \begin{bmatrix} -\overline{x}_P^i \sin \theta^i \\ \overline{x}_P^i \cos \theta^i \end{bmatrix}$$

and

$$(\mathbf{v}_P^i)_r = \mathbf{A}^i \dot{\overline{\mathbf{u}}}_P^i = \begin{bmatrix} \cos \theta^i & -\sin \theta^i \\ \sin \theta^i & \cos \theta^i \end{bmatrix} \begin{bmatrix} \dot{\overline{x}}_P^i \\ 0 \end{bmatrix} = \begin{bmatrix} \dot{\overline{x}}_P^i \cos \theta^i \\ \dot{\overline{x}}_P^i \sin \theta^i \end{bmatrix}$$

The absolute velocity of the block P can then be written as

$$\mathbf{v}_P^i = \begin{bmatrix} -\overline{x}_P^i \dot{\theta}^i \sin \theta^i + \dot{\overline{x}}_P^i \cos \theta^i \\ \overline{x}_P^i \dot{\theta}^i \cos \theta^i + \dot{\overline{x}}_P^i \sin \theta^i \end{bmatrix}$$

The absolute acceleration vector of the block P can be obtained by differentiating the absolute velocity vector $\mathbf{v}_P^i$ or by using the general expression of Eq. 50. Both methods yield the same results. In this example, the general expression of Eq. 50 is used. Since point O is fixed, the absolute acceleration of point O is equal to zero, that is,

$$\mathbf{a}_O^i = \mathbf{0}$$

In this case, the acceleration of point P takes the form

$$\mathbf{a}_P^i = \boldsymbol{\omega}^i \times (\boldsymbol{\omega}^i \times \mathbf{u}_P^i) + \boldsymbol{\alpha}^i \times \mathbf{u}_P^i + 2\boldsymbol{\omega}^i \times (\mathbf{v}_P^i)_r + (\mathbf{a}_P^i)_r$$

in which

$$\boldsymbol{\omega}^i \times (\boldsymbol{\omega}^i \times \mathbf{u}_P^i) = -\overline{x}_P^i (\dot{\theta}^i)^2 \begin{bmatrix} \cos \theta^i \\ \sin \theta^i \end{bmatrix}, \quad \boldsymbol{\alpha}^i \times \mathbf{u}_P^i = \overline{x}_P^i \ddot{\theta}^i \begin{bmatrix} -\sin \theta^i \\ \cos \theta^i \end{bmatrix}$$

$$2\boldsymbol{\omega}^i \times (\mathbf{v}_P^i)_r = 2\dot{\overline{x}}_P^i \dot{\theta}^i \begin{bmatrix} -\sin \theta^i \\ \cos \theta^i \end{bmatrix}, \quad (\mathbf{a}_P^i)_r = \mathbf{A}^i \ddot{\overline{\mathbf{u}}}_P^i = \begin{bmatrix} \ddot{\overline{x}}_P^i \cos \theta^i \\ \ddot{\overline{x}}_P^i \sin \theta^i \end{bmatrix}$$

Substituting these equations into the expression for the acceleration of point P, one obtains

$$\mathbf{a}_P^i = -\begin{bmatrix} \cos \theta^i \\ \sin \theta^i \end{bmatrix} \overline{x}_P^i (\dot{\theta}^i)^2 + \begin{bmatrix} -\sin \theta^i \\ \cos \theta^i \end{bmatrix} \overline{x}_P^i \ddot{\theta}^i$$

$$+ 2 \begin{bmatrix} -\sin \theta^i \\ \cos \theta^i \end{bmatrix} \dot{\overline{x}}_P^i \dot{\theta}^i + \begin{bmatrix} \cos \theta^i \\ \sin \theta^i \end{bmatrix} \ddot{\overline{x}}_P^i$$

$$= \begin{bmatrix} [\ddot{\overline{x}}_P^i - \overline{x}_P^i (\dot{\theta}^i)^2] \cos \theta^i - (\overline{x}_P^i \ddot{\theta}^i + 2\dot{\overline{x}}_P^i \dot{\theta}^i) \sin \theta^i \\ [\ddot{\overline{x}}_P^i - \overline{x}_P^i (\dot{\theta}^i)^2] \sin \theta^i + (\overline{x}_P^i \ddot{\theta}^i + 2\dot{\overline{x}}_P^i \dot{\theta}^i) \cos \theta^i \end{bmatrix}$$

Example 3.6

Figure 18 shows two rotating rods that are connected by the slider block P. Given the angular velocity and angular acceleration of rod 2, determine the angular velocity and angular acceleration of rod 3 and the relative velocity and acceleration of the slider block P with respect to rod 3.

Solution. First we perform the velocity analysis of the mechanism. We first consider rod 2 as shown in Fig. 19a. The absolute velocity of point P on rod 2 is

$$\mathbf{v}_P^2 = \mathbf{v}_O^2 + \mathbf{v}_{PO}^2$$

where $\mathbf{v}_O^2 = \mathbf{0}$, and

$$\mathbf{v}_{PO}^2 = \boldsymbol{\omega}^2 \times \mathbf{u}_P^2 = \dot{\theta}^2 \begin{bmatrix} -l^2 \sin \theta^2 \\ l^2 \cos \theta^2 \end{bmatrix}$$

Here, l^2 is the length of link 2. The absolute velocity of point P can then be written as

$$\mathbf{v}_P^2 = \dot{\theta}^2 \begin{bmatrix} -l^2 \sin \theta^2 \\ l^2 \cos \theta^2 \end{bmatrix}$$

Due to the fact that the slider block P slides on link 3, the absolute velocity of point P can also be evaluated by analyzing the motion of link 3 shown in Fig. 19b. In this case, one has

$$\mathbf{v}_P^3 = \mathbf{v}_O^3 + \mathbf{v}_{PO}^3 + (\mathbf{v}_P^3)_r$$

Keeping in mind that point O^3 is a fixed point, one has $\mathbf{v}_O^3 = \mathbf{0}$. One also has

$$\mathbf{v}_{PO}^3 = \boldsymbol{\omega}^3 \times \mathbf{u}_P^3 = \dot{\theta}^3 \begin{bmatrix} -\bar{x}_P^3 \sin \theta^3 \\ \bar{x}_P^3 \cos \theta^3 \end{bmatrix}$$

$$(\mathbf{v}_P^3)_r = \mathbf{A}^3 \dot{\bar{\mathbf{u}}}_P^3 = \begin{bmatrix} \cos \theta^3 & -\sin \theta^3 \\ \sin \theta^3 & \cos \theta^3 \end{bmatrix} \begin{bmatrix} \dot{\bar{x}}_P^3 \\ 0 \end{bmatrix} = \dot{\bar{x}}_P^3 \begin{bmatrix} \cos \theta^3 \\ \sin \theta^3 \end{bmatrix}$$

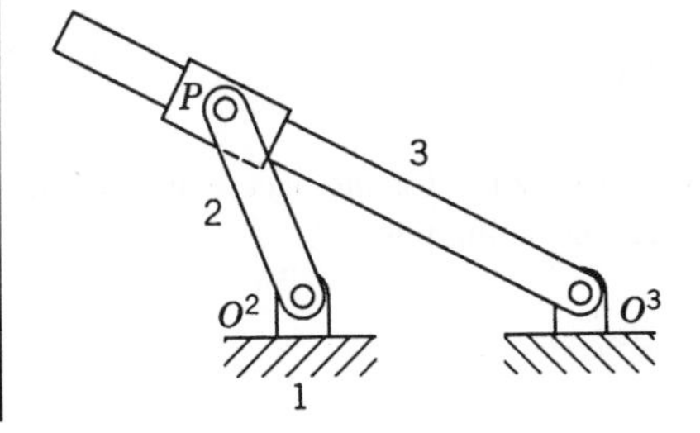

Figure 3.18 Coriolis acceleration

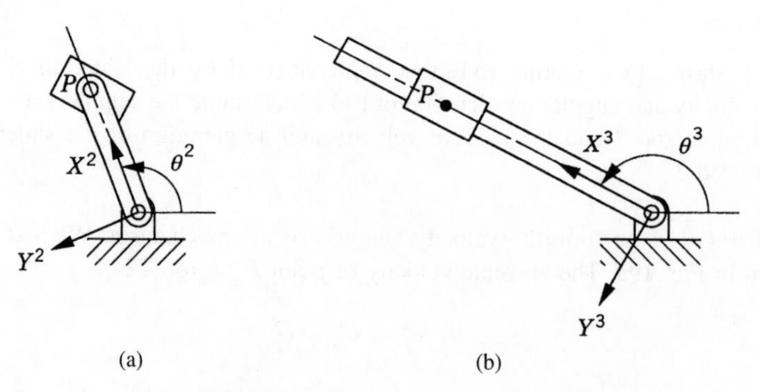

Figure 3.19 Motion of block P

where $\bar{x}_P^3$ is the distance between point P and point O^3. The absolute velocity of point P can then be written as

$$\mathbf{v}_P^3 = \dot{\theta}^3 \begin{bmatrix} -\bar{x}_P^3 \sin\theta^3 \\ \bar{x}_P^3 \cos\theta^3 \end{bmatrix} + \dot{\bar{x}}_P^3 \begin{bmatrix} \cos\theta^3 \\ \sin\theta^3 \end{bmatrix}$$

Since $\mathbf{v}_P^2 = \mathbf{v}_P^3$, one has

$$\dot{\theta}^2 \begin{bmatrix} -l^2 \sin\theta^2 \\ l^2 \cos\theta^2 \end{bmatrix} = \dot{\theta}^3 \begin{bmatrix} -\bar{x}_P^3 \sin\theta^3 \\ \bar{x}_P^3 \cos\theta^3 \end{bmatrix} + \dot{\bar{x}}_P^3 \begin{bmatrix} \cos\theta^3 \\ \sin\theta^3 \end{bmatrix}$$

This equation can be written in a matrix form as

$$\begin{bmatrix} -\bar{x}_P^3 \sin\theta^3 & \cos\theta^3 \\ \bar{x}_P^3 \cos\theta^3 & \sin\theta^3 \end{bmatrix} \begin{bmatrix} \dot{\theta}^3 \\ \dot{\bar{x}}_P^3 \end{bmatrix} = \dot{\theta}^2 \begin{bmatrix} -l^2 \sin\theta^2 \\ l^2 \cos\theta^2 \end{bmatrix}$$

This matrix equation contains two scalar algebraic equations that can be solved for the two unknowns $\dot{\theta}^3$ and $\dot{\bar{x}}_P^3$ as

$$\begin{bmatrix} \dot{\theta}^3 \\ \dot{\bar{x}}_P^3 \end{bmatrix} = \frac{\dot{\theta}^2}{\bar{x}_P^3} \begin{bmatrix} l^2 \cos(\theta^3 - \theta^2) \\ \bar{x}_P^3 l^2 \sin(\theta^3 - \theta^2) \end{bmatrix}$$

Having determined $\dot{\theta}^3$ and $\dot{\bar{x}}_P^3$, one can now proceed to solve for the accelerations. Considering rod 2, the absolute acceleration of point P can be written as

$$\mathbf{a}_P^2 = \mathbf{a}_O^2 + (\mathbf{a}_{PO}^2)_t + (\mathbf{a}_{PO}^2)_n$$

where

$$\mathbf{a}_O^2 = \mathbf{0}, \quad (\mathbf{a}_{PO}^2)_t = \boldsymbol{\alpha}^2 \times \mathbf{u}_P^2 = \ddot{\theta}^2 \begin{bmatrix} -l^2 \sin \theta^2 \\ l^2 \cos \theta^2 \end{bmatrix}$$

$$(\mathbf{a}_{PO}^2)_n = \boldsymbol{\omega}^2 \times (\boldsymbol{\omega}^2 \times \mathbf{u}_P^2) = -(\dot{\theta}^2)^2 \begin{bmatrix} l^2 \cos \theta^2 \\ l^2 \sin \theta^2 \end{bmatrix}$$

The absolute acceleration of point P can then be written as

$$\mathbf{a}_P^2 = \ddot{\theta}^2 \begin{bmatrix} -l^2 \sin \theta^2 \\ l^2 \cos \theta^2 \end{bmatrix} - (\dot{\theta}^2)^2 \begin{bmatrix} l^2 \cos \theta^2 \\ l^2 \sin \theta^2 \end{bmatrix}$$

As the slider block P moves with respect to rod 3, the absolute acceleration of point P takes the form

$$\mathbf{a}_P^3 = \mathbf{a}_O^3 + \boldsymbol{\omega}^3 \times (\boldsymbol{\omega}^3 \times \mathbf{u}_P^3) + \boldsymbol{\alpha}^3 \times \mathbf{u}_P^3 + 2\boldsymbol{\omega}^3 \times (\mathbf{v}_P^3)_r + (\mathbf{a}_P^3)_r$$

where

$$\mathbf{a}_O^3 = \mathbf{0}, \quad \boldsymbol{\omega}^3 \times (\boldsymbol{\omega}^3 \times \mathbf{u}_P^3) = -(\dot{\theta}^3)^2 \overline{x}_P^3 \begin{bmatrix} \cos \theta^3 \\ \sin \theta^3 \end{bmatrix}$$

$$\boldsymbol{\alpha}^3 \times \mathbf{u}_P^3 = \ddot{\theta}^3 \overline{x}_P^3 \begin{bmatrix} -\sin \theta^3 \\ \cos \theta^3 \end{bmatrix}, \quad 2\boldsymbol{\omega}^3 \times (\mathbf{v}_P^3)_r = 2\dot{\theta}^3 \dot{\overline{x}}_P^3 \begin{bmatrix} -\sin \theta^3 \\ \cos \theta^3 \end{bmatrix}$$

$$(\mathbf{a}_P^3)_r = \mathbf{A}^3 \ddot{\overline{\mathbf{u}}}_P^3 = \begin{bmatrix} \cos \theta^3 & -\sin \theta^3 \\ \sin \theta^3 & \cos \theta^3 \end{bmatrix} \begin{bmatrix} \ddot{\overline{x}}_P^3 \\ 0 \end{bmatrix} = \ddot{\overline{x}}_P^3 \begin{bmatrix} \cos \theta^3 \\ \sin \theta^3 \end{bmatrix}$$

The absolute acceleration of point P is

$$\mathbf{a}_P^3 = -(\dot{\theta}^3)^2 \overline{x}_P^3 \begin{bmatrix} \cos \theta^3 \\ \sin \theta^3 \end{bmatrix} + \ddot{\theta}^3 \overline{x}_P^3 \begin{bmatrix} -\sin \theta^3 \\ \cos \theta^3 \end{bmatrix}$$

$$+ 2\dot{\theta}^3 \dot{\overline{x}}_P^3 \begin{bmatrix} -\sin \theta^3 \\ \cos \theta^3 \end{bmatrix} + \ddot{\overline{x}}_P^3 \begin{bmatrix} \cos \theta^3 \\ \sin \theta^3 \end{bmatrix}$$

Using the fact that $\mathbf{a}_P^2 = \mathbf{a}_P^3$, one obtains

$$\ddot{\theta}^2 \begin{bmatrix} -l^2 \sin\theta^2 \\ l^2 \cos\theta^2 \end{bmatrix} - (\dot{\theta}^2)^2 \begin{bmatrix} l^2 \cos\theta^2 \\ l^2 \sin\theta^2 \end{bmatrix}$$

$$= -(\dot{\theta}^3)^2 \overline{x}_P^3 \begin{bmatrix} \cos\theta^3 \\ \sin\theta^3 \end{bmatrix} + \ddot{\theta}^3 \overline{x}_P^3 \begin{bmatrix} -\sin\theta^3 \\ \cos\theta^3 \end{bmatrix}$$

$$+ 2\dot{\theta}^3 \dot{\overline{x}}_P^3 \begin{bmatrix} -\sin\theta^3 \\ \cos\theta^3 \end{bmatrix} + \ddot{\overline{x}}_P^3 \begin{bmatrix} \cos\theta^3 \\ \sin\theta^3 \end{bmatrix}$$

The terms in this equation can be rearranged and written in a matrix form as

$$\begin{bmatrix} -\overline{x}_P^3 \sin\theta^3 & \cos\theta^3 \\ \overline{x}_P^3 \cos\theta^3 & \sin\theta^3 \end{bmatrix} \begin{bmatrix} \ddot{\theta}^3 \\ \ddot{\overline{x}}_P^3 \end{bmatrix}$$

$$= \begin{bmatrix} -\ddot{\theta}^2 l^2 \sin\theta^2 - (\dot{\theta}^2)^2 l^2 \cos\theta^2 + (\dot{\theta}^3)^2 \overline{x}_P^3 \cos\theta^3 + 2\dot{\theta}^3 \dot{\overline{x}}_P^3 \sin\theta^3 \\ \ddot{\theta}^2 l^2 \cos\theta^2 - (\dot{\theta}^2)^2 l^2 \sin\theta^2 + (\dot{\theta}^3)^2 \overline{x}_P^3 \sin\theta^3 - 2\dot{\theta}^3 \dot{\overline{x}}_P^3 \cos\theta^3 \end{bmatrix}$$

This matrix equation can be solved for the two unknowns $\ddot{\theta}^3$ and $\ddot{\overline{x}}_P^3$ since all the variables on the right-hand side of this equation are either given or can be determined from the position and velocity analyses.

3.7 COMPUTATIONAL KINEMATIC APPROACH

A careful examination of the classical solution procedures used for the position, velocity, and acceleration analysis of the mechanisms discussed in the examples presented thus far in this chapter reveals that, in general, there is an explicit or implicit use of a set of algebraic kinematic equations that describe the joint connectivity between the bodies of the system as well as specified motion trajectories. For instance, for the slider crank mechanism of Example 3, we explicitly or implicitly used the following algebraic equations and their derivatives

$$\left.\begin{aligned} &\dot{\mathbf{R}}^1 = \mathbf{0}, \quad \dot{\theta}^1 = 0, \quad \mathbf{v}_O^2 = \mathbf{0}, \quad \mathbf{v}_A^2 = \mathbf{v}_A^3, \\ &\mathbf{v}_B^3 = \mathbf{v}_B^4, \quad \dot{R}_y^4 = 0, \quad \dot{\theta}^4 = 0, \quad \dot{\theta}^2 = \omega^2 \end{aligned}\right\} \tag{3.72}$$

where $\mathbf{R}^1$ and θ^1 are the coordinates of the ground (fixed link) R_y^4 and θ^4 are the vertical displacement and the angular orientation of the slider block, and ω^2 is a known function of time. The last equation describes the constraint condition used to drive the crankshaft of the mechanism. Each one of the preceding equations was manipulated separately so as to yield a procedure which is tailored only for the analysis of the slider crank mechanism. Note that the number of algebraic equations given in Eq. 72 is 12 which is equal to the number of coordinates of the mechanism. The planar slider crank mechanism has four bodies ($n_b = 4$);

the ground (Body 1), the crankshaft (Body 2), the connecting rod (Body 3), and the slider block (Body 4). The total number of coordinates of the mechanism is $n = 3 \times n_b = 12$. That is, the total number of kinematic equations is equal to the total number of coordinates, and this number is equal to the number of constraint equations given in Eq. 72.

Another alternative, yet equivalent approach is to combine the preceding equations and solve them simultaneously using matrix and computer methods. While this alternative approach is not different in principle from the methods used in the preceding examples, its use, as demonstrated in the remainder of this chapter, allows us to develop a systematic computer procedure that can be used in the kinematic analysis of varieties of multibody system applications. The kinematic relationships that describe the joint connectivity between bodies as well as specified motion trajectories will be formulated so as to obtain a number of algebraic equations equal to the number of the system coordinates. The resulting system of loosely coupled equations can be solved efficiently using numerical techniques.

Absolute Coordinates As pointed out previously, the planar motion of an unconstrained rigid body can be described using three independent coordinates. Two coordinates define the translation of the body as represented by the displacement of the origin of a selected body reference and one coordinate defines the orientation of the body. The translational motion of the rigid body i can be defined by the vector $\mathbf{R}^i$ that describes the position of the origin of the body reference with respect to the global coordinate system, while the orientation of the body can be described using the angle θ^i. Using the three coordinates $\mathbf{R}^i = [R_x^i \quad R_y^i]^{\mathrm{T}}$ and θ^i, the position vector of an arbitrary point P^i on the rigid body can be written as (see Eq. 16)

$$\mathbf{r}_P^i = \mathbf{R}^i + \mathbf{A}^i \bar{\mathbf{u}}_P^i \tag{3.73}$$

where

$$\bar{\mathbf{u}}_P^i = [\bar{x}_P^i \quad \bar{y}_P^i]^{\mathrm{T}} \tag{3.74}$$

is the position vector of the arbitrary point defined in the body coordinate system, and $\mathbf{A}^i$ is the transformation matrix from the body coordinate system to the global coordinate system defined in terms of the angle of rotation θ^i as

$$\mathbf{A}^i = \begin{bmatrix} \cos\theta^i & -\sin\theta^i \\ \sin\theta^i & \cos\theta^i \end{bmatrix} \tag{3.75}$$

In this chapter and in the following chapters, the coordinates $\mathbf{R}^i$ and θ^i are referred to as the *absolute Cartesian generalized coordinates* of the rigid body i.

A multibody system consisting of n_b unconstrained rigid bodies has $3 \times n_b$ independent generalized coordinates. The vector $\mathbf{q}$ of the generalized coordinates of the multibody system is then defined as

$$\begin{aligned} \mathbf{q} &= [R_x^1 \quad R_y^1 \quad \theta^1 \quad R_x^2 \quad R_y^2 \quad \theta^2 \cdots R_x^i \quad R_y^i \quad \theta^i \cdots R_x^{n_b} \quad R_y^{n_b} \quad \theta^{n_b}]^{\mathrm{T}} \\ &= [\mathbf{R}^{1^{\mathrm{T}}} \quad \theta^1 \quad \mathbf{R}^{2^{\mathrm{T}}} \quad \theta^2 \cdots \mathbf{R}^{i^{\mathrm{T}}} \quad \theta^i \cdots \mathbf{R}^{n_b^{\mathrm{T}}} \quad \theta^{n_b}]^{\mathrm{T}} \end{aligned} \tag{3.76}$$

which can also be written in the following form:

$$\mathbf{q} = [\mathbf{q}^{1^{\mathrm{T}}} \quad \mathbf{q}^{2^{\mathrm{T}}} \quad \cdots \quad \mathbf{q}^{i^{\mathrm{T}}} \quad \cdots \quad \mathbf{q}^{n_b^{\mathrm{T}}}]^{\mathrm{T}} \tag{3.77}$$

where

$$\mathbf{q}^i = \begin{bmatrix} \mathbf{R}^i \\ \theta^i \end{bmatrix} \tag{3.78}$$

is the vector of generalized coordinates of body i.

Computational Approach As previously mentioned, in the computational kinematic approach that will be used in this text, each body in the system is assigned three absolute Cartesian coordinates in the planar analysis and six absolute Cartesian coordinates in the spatial analysis. These coordinates define the translation and the orientation of the bodies in space. Therefore, in the planar analysis, the number of system coordinates is $n = 3 \times n_b$; whereas in the spatial analysis, the number of system coordinates is $n = 6 \times n_b$, where n_b is the total number of bodies in the system. In order to solve for the system coordinates using the kinematic analysis, one must have a number of algebraic equations equal to the number of coordinates. These algebraic equations are the constraint equations imposed on the motion of the multibody system. By formulating the algebraic constraint equations in terms of the coordinates of the body, one can obtain a number of equations n_c that is equal to the number of coordinates n. These *position level* equations can be solved for the coordinates using numerical methods as will be discussed in later sections of this chapter. By differentiating the constraint equations with respect to time, one obtains the constraint equations at the *velocity level*. This differentiation leads to a system of linear algebraic equations that can be solved for the velocities. A second differentiation defines the constraint equations at the *acceleration level*. These equations, which are linear in the accelerations, can be solved in a straightforward manner to determine the second derivatives of the system coordinates. It is, therefore, important to be able to formulate the algebraic equations of the constraints in terms of the system generalized coordinates.

Kinematic constraints that impose restrictions on the relative motion between bodies in the multibody systems are classified as *joint constraints* and *driving constraints*. Joint constraints, which are the result of restrictions imposed by the mechanical joints such as revolute, prismatic, cylindrical, and spherical joints, describe the connectivity between the multibody system components, and therefore, define the system topological structure. The formulation of the driving and joint constraint equations will be discussed in more detail in the following two sections.

3.8 FORMULATION OF THE DRIVING CONSTRAINTS

While joint constraints are assumed to depend only on the system coordinates, *driving constraints* describe the specified motion trajectories, and therefore, may depend on the system generalized coordinates as well as time. An example of driving constraints is the specified motion of the crankshaft of a slider crank mechanism, as the one shown in

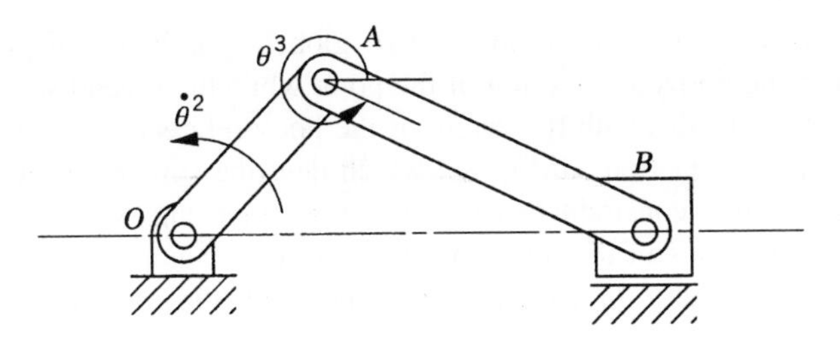

Figure 3.20 Slider crank mechanism

Fig. 20. If the crankshaft is denoted as body 2, and it is assumed to rotate with a constant angular velocity, one has

$$\dot{\theta}^2 = \omega^2 \tag{3.79}$$

where ω^2 is a constant. The preceding equation is a differential equation that can be integrated to define a kinematic constraint equation that depends on the coordinate θ^2 as well as time and can be written as

$$\theta^2 = \omega^2 t + \theta_o^2 \tag{3.80}$$

where t is time and θ_o^2 is the initial angular position of the crankshaft. Equation 80 is an example of a *simple constraint* that can be imposed on the absolute coordinates of a body in the system. For body i in the system, one may encounter situations in which one or more of the following *simple driving constraints* must be imposed:

$$R_x^i = f_1(t), \quad R_y^i = f_2(t), \quad \theta^i = f_3(t) \tag{3.81}$$

where $f_1(t), f_2(t)$, and $f_3(t)$ are time-dependent functions and R_x^i, R_y^i, and θ^i are the absolute coordinates of the rigid body i.

More complex driving constraints arise in multibody system applications when the motion of an arbitrary point on a rigid body is prescribed. Specified trajectories in the analysis of robotic manipulators, numerically controlled machine tools, and railroad vehicles are examples of such driving constraints. For example, if the coordinates of a point P^i on the rigid body i are prescribed such that this point follows a given trajectory defined by the function $\mathbf{f}(t) = [f_1(t) \quad f_2(t)]^{\mathrm{T}}$, the use of Eq. 73 leads to

$$\mathbf{r}_P^i = \mathbf{R}^i + \mathbf{A}^i \bar{\mathbf{u}}_P^i = \mathbf{f}(t) \tag{3.82}$$

This equation, in the planar analysis, leads to two scalar equations that can be written in terms of the absolute coordinates of body i as

$$\left.\begin{aligned} R_x^i + \bar{x}_P^i \cos\theta^i - \bar{y}_P^i \sin\theta^i &= f_1(t) \\ R_y^i + \bar{x}_P^i \sin\theta^i + \bar{y}_P^i \cos\theta^i &= f_2(t) \end{aligned}\right\} \tag{3.83}$$

In these two equations which constrain the two global coordinates of point P^i, the first constraint specifies the horizontal motion of the point while the second specifies the vertical motion. If point P^i coincides with the origin of the body reference, that is, $\bar{\mathbf{u}}_P^i = \mathbf{0}$, Eq. 83 reduces to the first two equations of Eq. 81, which describe simple constraints.

Other types of driving constraints may result from imposing conditions on the relative motion between two bodies in the multibody system. For example, if the relative rotation between two bodies i and j in the system is specified, the constraint equation can be written as

$$\theta^i - \theta^j = f(t) \tag{3.84}$$

where θ^i and θ^j are the angular orientation of bodies i and j, respectively, and $f(t)$ is a known function of time. Similarly, if the relative displacement between points P^i and P^j on bodies i and j is specified, the resulting kinematic constraints can be classified as driving constraints and can be written as

$$\mathbf{r}_P^i - \mathbf{r}_P^j = \mathbf{f}(t) \tag{3.85}$$

where $\mathbf{r}_P^i$ and $\mathbf{r}_P^j$ are, respectively, the global position vectors of points P^i and P^j, and $\mathbf{f}(t) = [f_1(t) \quad f_2(t)]^{\mathrm{T}}$ is a time dependent vector function. By using Eq. 73, Eq. 85 can be written as

$$\mathbf{R}^i + \mathbf{A}^i\bar{\mathbf{u}}_P^i - \mathbf{R}^j - \mathbf{A}^j\bar{\mathbf{u}}_P^j = \mathbf{f}(t) \tag{3.86}$$

This vector equation has two scalar equations that describe the constraints between the coordinates of point P^i and point P^j.

In many multibody system applications, both joint and driving constraints exist. A simple example is the slider crank mechanism shown in Fig. 20. The mechanism has four joints; three revolute and one prismatic. A driving constraint similar to the one given by Eq. 80 can still be imposed on the motion of the crankshaft. It is important, however, to point out that the maximum number of driving constraints that can be imposed on the motion of a given system must not exceed the number of the system degrees of freedom. Driving constraints depend on the applications, and they may depend explicitly on time as demonstrated in this section. For this reason, general purpose multibody system computer programs provide *user subroutines* that allow the user to introduce these nonstandard constraints. In the kinematic analysis, the motion of the system is produced by the driving constraints. Associated with these driving constraints, there are driving constraint forces that represent actuator forces or motor torques. These driving constraint forces and torques can be systematically determined as will be explained when the subject of constrained dynamics is covered in later chapters of this book.

3.9 FORMULATION OF THE JOINT CONSTRAINTS

In this section, the formulation of joint constraint equations in terms of the absolute coordinates that describe the location and orientation of the rigid bodies with respect to a fixed global coordinate system is discussed. Only planar motion constraints are considered in this section. The formulation of the joint constraints in the spatial analysis is presented in Chapter 7.

Ground Constraints A body that has zero degrees of freedom is called a *ground* or *fixed link*. The ground constraints imply that the body has no translational or rotational degrees of freedom. If body i is assumed to be a ground or a fixed link, the algebraic kinematic constraints are given by

$$R_x^i - c_1 = 0, \quad R_y^i - c_2 = 0, \quad \theta^i - c_3 = 0 \tag{3.87}$$

where c_1, c_2, and c_3 are constants. The conditions of Eq. 87 eliminate the translational and rotational degrees of freedom of the body. These conditions can be combined into one vector equation as

$$\mathbf{q}^i - \mathbf{c} = \mathbf{0} \tag{3.88}$$

where $\mathbf{q}^i = [R_x^i \quad R_y^i \quad \theta^i]^{\mathrm{T}}$ is the vector of absolute coordinates of body i and $\mathbf{c} = [c_1 \quad c_2 \quad c_3]^{\mathrm{T}}$ is a constant vector.

Revolute Joint When two bodies are connected by a revolute joint, only relative rotation between the two bodies is allowed. Figure 21 depicts two rigid bodies i and j that are connected by a *revolute joint* at point P which is called the *joint definition point*. It is clear from the figure that the position vector of this point as defined using the absolute coordinates of body i must be equal to the position vector of the same point as defined in terms of the absolute coordinates of body j. The kinematic constraint conditions of the revolute joint can then be stated mathematically as

$$\mathbf{r}_P^i = \mathbf{r}_P^j \tag{3.89}$$

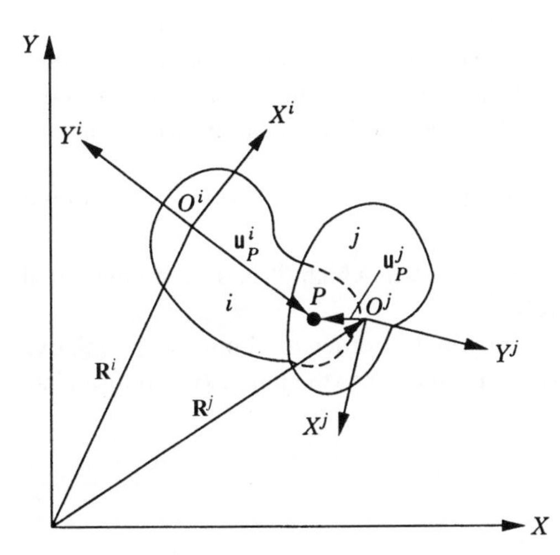

Figure 3.21 Revolute joint

or equivalently,

$$\mathbf{R}^i + \mathbf{A}^i\bar{\mathbf{u}}_P^i - \mathbf{R}^j - \mathbf{A}^j\bar{\mathbf{u}}_P^j = \mathbf{0} \tag{3.90}$$

where $\bar{\mathbf{u}}_P^i = [\bar{x}_P^i \quad \bar{y}_P^i]^{\mathrm{T}}$ and $\bar{\mathbf{u}}_P^j = [\bar{x}_P^j \quad \bar{y}_P^j]^{\mathrm{T}}$ are the local position vectors of point P defined with respect to the coordinate systems of body i and body j, respectively. Equation 90 can also be written in a more explicit form as

$$\begin{bmatrix} R_x^i \\ R_y^i \end{bmatrix} + \begin{bmatrix} \cos\theta^i & -\sin\theta^i \\ \sin\theta^i & \cos\theta^i \end{bmatrix} \begin{bmatrix} \bar{x}_P^i \\ \bar{y}_P^i \end{bmatrix} - \begin{bmatrix} R_x^j \\ R_y^j \end{bmatrix}$$

$$- \begin{bmatrix} \cos\theta^j & -\sin\theta^j \\ \sin\theta^j & \cos\theta^j \end{bmatrix} \begin{bmatrix} \bar{x}_P^j \\ \bar{y}_P^j \end{bmatrix} = \begin{bmatrix} 0 \\ 0 \end{bmatrix} \tag{3.91}$$

which yields the two scalar equations

$$\left.\begin{aligned} R_x^i + \bar{x}_P^i \cos\theta^i - \bar{y}_P^i \sin\theta^i - R_x^j - \bar{x}_P^j \cos\theta^j + \bar{y}_P^j \sin\theta^j = 0 \\ R_y^i + \bar{x}_P^i \sin\theta^i + \bar{y}_P^i \cos\theta^i - R_y^j - \bar{x}_P^j \sin\theta^j - \bar{y}_P^j \cos\theta^j = 0 \end{aligned}\right\} \tag{3.92}$$

These are the two constraint equations that eliminate the freedom of the relative translation between the two bodies.

If a rigid body i is connected to the ground by a revolute joint, Eq. 90 reduces in this special case to

$$\mathbf{R}^i + \mathbf{A}^i\bar{\mathbf{u}}_P^i - \mathbf{c} = \mathbf{0} \tag{3.93}$$

where $\mathbf{c}$ is a constant vector that defines the absolute Cartesian coordinates of point P. The kinematic conditions of Eq. 93, which are sometimes called *point constraints*, imply that point P on the rigid body i is a fixed point.

Prismatic Joint The *prismatic joint*, which is also called the *translational joint*, allows only relative translation between the two bodies along the joint axis. The constraint equations for the prismatic joint reduce the number of degrees of freedom of the system by two. Figure 22 depicts two bodies i and j that are connected by a prismatic joint. A constraint equation that eliminates the relative rotation between the two bodies can be written as

$$\theta^i - \theta^j - c = 0 \tag{3.94}$$

where c is a constant defined by the equation $c = \theta_o^i - \theta_o^j$ in which θ_o^i and θ_o^j are the initial orientation angles of bodies i and j, respectively.

A second condition for the prismatic joint is required in order to eliminate the relative translation between the two bodies along an axis perpendicular to the joint axis. To formulate

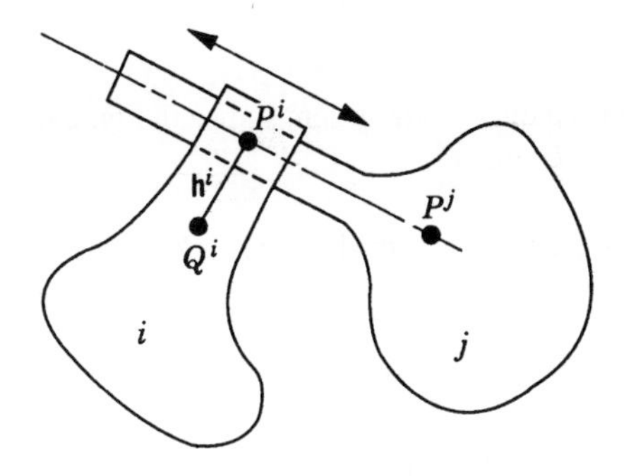

Figure 3.22 Prismatic joint

this condition, the two perpendicular vectors $\mathbf{r}_P^{ij}$ and $\mathbf{h}^i$ are defined. The vector $\mathbf{r}_P^{ij}$ connects two arbitrary points P^i and P^j that lie on the axis of the prismatic joint as shown in the figure. Point P^i is defined on body i, and therefore, its coordinates are fixed with respect to the coordinate system of body i, while point P^j is defined on body j, and accordingly, its coordinates are fixed in the coordinate system of body j. The vector $\mathbf{h}^i$, which is assumed to be perpendicular to the joint axis, may be defined on body i and can be selected to be the vector joining points P^i and Q^i, as shown in the figure. The vectors $\mathbf{r}_P^{ij}$ and $\mathbf{h}^i$ can then be defined in terms of the coordinates of body i and body j as

$$\left.\begin{aligned} \mathbf{r}_P^{ij} &= \mathbf{r}_P^i - \mathbf{r}_P^j = \mathbf{R}^i + \mathbf{A}^i\bar{\mathbf{u}}_P^i - \mathbf{R}^j - \mathbf{A}^j\bar{\mathbf{u}}_P^j \\ \mathbf{h}^i &= \mathbf{A}^i(\bar{\mathbf{u}}_P^i - \bar{\mathbf{u}}_Q^i) \end{aligned}\right\} \tag{3.95}$$

where $\bar{\mathbf{u}}_P^i$, $\bar{\mathbf{u}}_P^j$, and $\bar{\mathbf{u}}_Q^i$ are the local position vectors of points P^i, P^j, and Q^i, respectively. If there is no relative translation between the two bodies along an axis perpendicular to the joint axis, the vectors $\mathbf{r}_P^{ij}$ and $\mathbf{h}^i$ must remain perpendicular, a condition that can be written as

$$\mathbf{h}^{i\mathrm{T}}\mathbf{r}_P^{ij} = 0 \tag{3.96}$$

This is a scalar equation that can be written in a more explicit form using Eq. 95.

One can combine the two constraint equations of the prismatic joint given by Eqs. 94 and 96 in one vector equation as

$$\begin{bmatrix} \theta^i - \theta^j - c \\ \mathbf{h}^{i\mathrm{T}}\mathbf{r}_P^{ij} \end{bmatrix} = \begin{bmatrix} 0 \\ 0 \end{bmatrix} \tag{3.97}$$

While the first equation in Eq. 97 is a linear function of the rotational coordinates of body i and body j, the second equation is a nonlinear equation in the absolute coordinates of the two bodies.

Example 3.7

Derive the algebraic kinematic constraint equations of the three-body system shown in Fig. 23, and determine the number of the system degrees of freedom.

Solution. The absolute coordinates of body i in the system are assumed to be R_x^i, R_y^i, and θ^i, $i = 1, 2, 3$. The ground constraints are

$$R_x^1 - c_1 = 0, \quad R_y^1 - c_2 = 0, \quad \theta^1 - c_3 = 0$$

where c_1, c_2, and c_3 are constants. If the axes of the coordinate system of body 1 are assumed to coincide with the axes of the global coordinate system, the constants c_1, c_2, and c_3 are identically zeros. The two kinematic constraint equations for the pin (revolute) joint at O can be written in a vector form as

$$\mathbf{R}^2 + \mathbf{A}^2\bar{\mathbf{u}}_O^2 = \mathbf{0}$$

where $\bar{\mathbf{u}}_O^2$ is the position vector of point O with respect to the origin of the coordinate system of body 2. If body 2 is assumed to be a uniform rod and the origin of this body coordinate system is assumed to be at its center as shown in the figure, one has

$$\bar{\mathbf{u}}_O^2 = \left[-\frac{l^2}{2} \quad 0\right]^{\mathrm{T}}$$

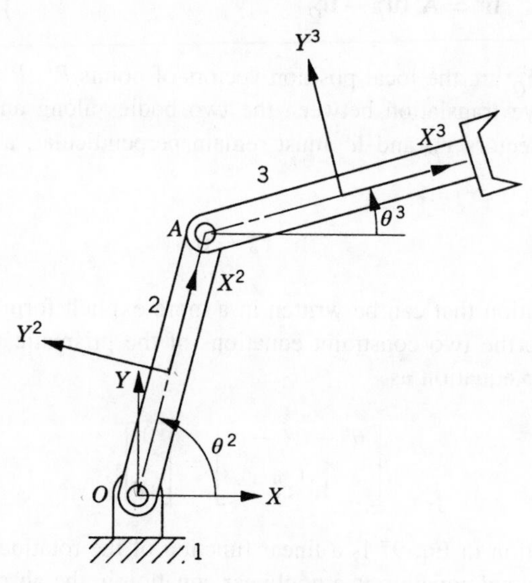

Figure 3.23 Two-degree-of-freedom system

where l^2 is the length of the rod 2. The constraint equations for the revolute joint at O lead to

$$\begin{bmatrix} R_x^2 \\ R_y^2 \end{bmatrix} + \begin{bmatrix} \cos\theta^2 & -\sin\theta^2 \\ \sin\theta^2 & \cos\theta^2 \end{bmatrix} \begin{bmatrix} -\dfrac{l^2}{2} \\ 0 \end{bmatrix} = \mathbf{0}$$

or

$$R_x^2 - \frac{l^2}{2}\cos\theta^2 = 0, \quad R_y^2 - \frac{l^2}{2}\sin\theta^2 = 0$$

Similarly, the constraint equations for the revolute joint at A are given by

$$\mathbf{R}^2 + \mathbf{A}^2\bar{\mathbf{u}}_A^2 - \mathbf{R}^3 - \mathbf{A}^3\bar{\mathbf{u}}_A^3 = \mathbf{0}$$

Assuming that body 3 is a uniform rod of length l^3 with the origin of its body coordinate system attached to its center, one has

$$\bar{\mathbf{u}}_A^2 = \begin{bmatrix} \dfrac{l^2}{2} & 0 \end{bmatrix}^{\mathrm{T}}, \quad \bar{\mathbf{u}}_A^3 = \begin{bmatrix} -\dfrac{l^3}{2} & 0 \end{bmatrix}^{\mathrm{T}}$$

which can be used to define the kinematic constraints of the revolute joint at A as

$$\begin{bmatrix} R_x^2 \\ R_y^2 \end{bmatrix} + \begin{bmatrix} \cos\theta^2 & -\sin\theta^2 \\ \sin\theta^2 & \cos\theta^2 \end{bmatrix} \begin{bmatrix} \dfrac{l^2}{2} \\ 0 \end{bmatrix} - \begin{bmatrix} R_x^3 \\ R_y^3 \end{bmatrix}$$

$$- \begin{bmatrix} \cos\theta^3 & -\sin\theta^3 \\ \sin\theta^3 & \cos\theta^3 \end{bmatrix} \begin{bmatrix} -\dfrac{l^3}{2} \\ 0 \end{bmatrix} = \begin{bmatrix} 0 \\ 0 \end{bmatrix}$$

or

$$R_x^2 + \frac{l^2}{2}\cos\theta^2 - R_x^3 + \frac{l^3}{2}\cos\theta^3 = 0$$

$$R_y^2 + \frac{l^2}{2}\sin\theta^2 - R_y^3 + \frac{l^3}{2}\sin\theta^3 = 0$$

The kinematic constraint equations of the system can be written in a vector form as

$$\mathbf{C}(\mathbf{q}^1, \mathbf{q}^2, \mathbf{q}^3) = \begin{bmatrix} R_x^1 - c_1 \\ R_y^1 - c_2 \\ \theta^1 - c_3 \\ R_x^2 - \dfrac{l^2}{2}\cos\theta^2 \\ R_y^2 - \dfrac{l^2}{2}\sin\theta^2 \\ R_x^2 + \dfrac{l^2}{2}\cos\theta^2 - R_x^3 + \dfrac{l^3}{2}\cos\theta^3 \\ R_y^2 + \dfrac{l^2}{2}\sin\theta^2 - R_y^3 + \dfrac{l^3}{2}\sin\theta^3 \end{bmatrix} = \begin{bmatrix} 0 \\ 0 \\ 0 \\ 0 \\ 0 \\ 0 \\ 0 \end{bmatrix}$$

where $\mathbf{C} = [C_1\ C_2\ C_3 \cdots C_7]^{\mathrm{T}}$ is the vector of algebraic constraints. There are seven joint constraint equations, and since the system has nine absolute coordinates (R_x^i, R_y^i, θ^i, $i = 1, 2, 3$), the number of degrees of freedom of the system is equal to two.

Cams A *cam* is a mechanical component that is used to drive another component called a *follower*. The cams are convenient and versatile mechanical devices for motion generation because of their various geometrical shapes and the various existing combinations between the cam and its follower. The cam and the follower mechanism constitute an important part in many mechanism systems such as instruments, internal combustion engines, and machine tools. Cams may be classified according to their basic shapes or according to the basic shapes of the followers. Figure 24 shows two different types of cam systems classified according to the shape of the cam, while Fig. 25 shows different types of cam systems classified according to the shape of the follower element. As the result of the rotation of the camshaft, the output motion of the follower can be translating or rotating motion. In most cam applications, as shown in Figs. 24 and 25, the shape of the follower in the contact region with the cam is chosen to be of simple geometry, while the desired motion is achieved by the proper design of the cam shape. The cam and follower, however, must remain in contact at all times. This can be achieved by using a suitable spring, by utilizing the effect of gravity, or by using any mechanical constraints.

In order to demonstrate the formulation of the algebraic kinematic constraints in the case of cam systems, as an example, the *offset reciprocating knife-edge follower* shown in Fig. 26 is first considered. The cam and the follower denoted, respectively, as bodies i and j may be connected with other bodies in a system by different types of joints. The point of contact between the cam and the follower is assumed to be point P, where point P is a fixed point in the follower coordinate system. The coordinates of this point, however, in the cam coordinate system depend on the cam shape. The global position of point P as

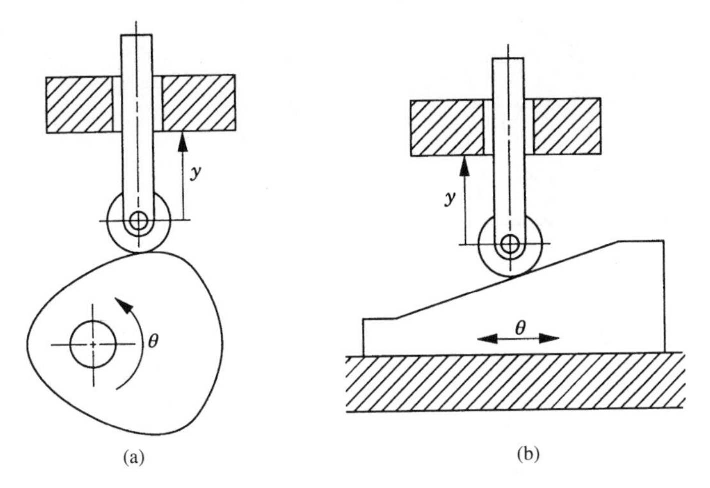

Figure 3.24 Cams

defined using the absolute coordinates of the cam (body i) and the follower (body j) can be written as

$$\left.\begin{aligned} \mathbf{r}_P^i &= \mathbf{R}^i + \mathbf{A}^i \bar{\mathbf{u}}_P^i \\ \mathbf{r}_P^j &= \mathbf{R}^j + \mathbf{A}^j \bar{\mathbf{u}}_P^j \end{aligned}\right\} \tag{3.98}$$

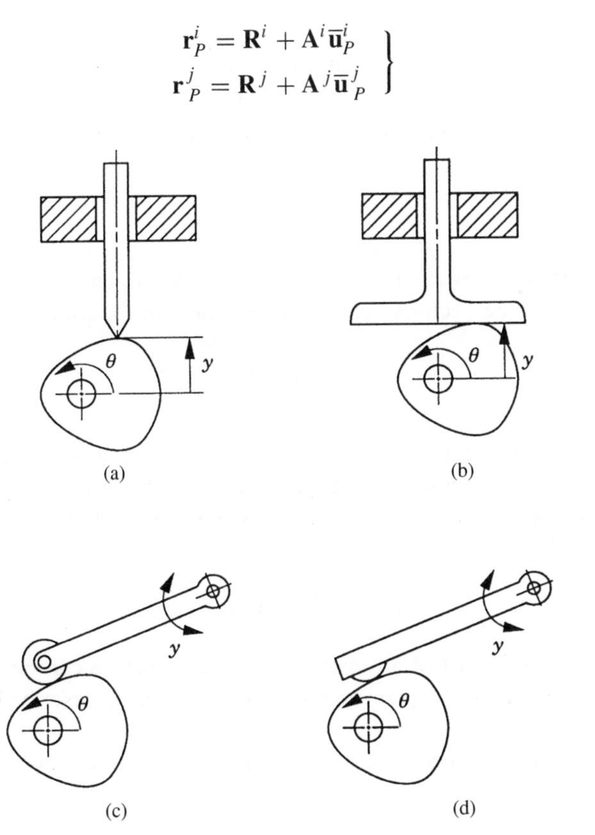

Figure 3.25 Follower motion

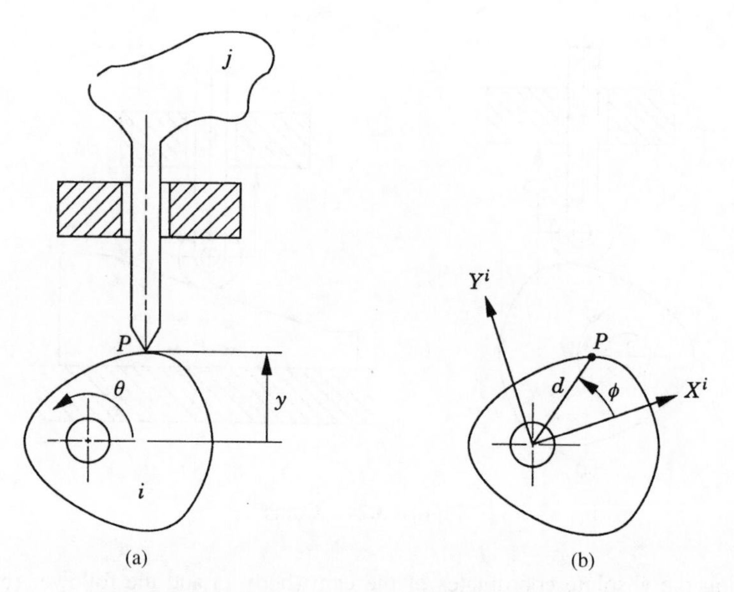

Figure 3.26 Offset reciprocating knife-edge follower

where $\bar{\mathbf{u}}_P^i$ and $\bar{\mathbf{u}}_P^j$ are the local coordinates of point P as defined in the coordinate systems of the cam and the follower, respectively. While $\bar{\mathbf{u}}_P^j$ is a constant vector, $\bar{\mathbf{u}}_P^i$ depends on the shape of the cam and, therefore, it is not a constant vector. It is clear from Fig. 26b that $\bar{\mathbf{u}}_P^i$ depends on the parameter ϕ, and can be written as

$$\bar{\mathbf{u}}_P^i = \bar{\mathbf{u}}_P^i(\phi) \tag{3.99}$$

This equation defines the exact nature of the shape of the cam. By using the functional relationship of Eq. 99, the vector $\bar{\mathbf{u}}_P^i$ can be represented in the following parametric forms

$$\bar{\mathbf{u}}_P^i = \begin{bmatrix} \bar{x}_P^i \\ \bar{y}_P^i \end{bmatrix} = \begin{bmatrix} d \cos \phi \\ d \sin \phi \end{bmatrix} \tag{3.100}$$

which implies that any point on the surface of the cam corresponds to a unique set of the two parameters d and ϕ, which define the shape of the cam that produces the desired follower motion. The shape of the cam can be defined by expressing d analytically or numerically in terms of the angle ϕ as

$$d = d(\phi) \tag{3.101}$$

Consequently, the coordinates of any point on the cam surface can be defined in terms of the angle ϕ. In the computer implementation, Eq. 101 can be described numerically using the *cubic spline functions*.

By using Eq. 98, the kinematic constraint equations for the cam system shown in Fig. 26 can be written as

$$\mathbf{r}_P^i - \mathbf{r}_P^j = \mathbf{0} \tag{3.102}$$

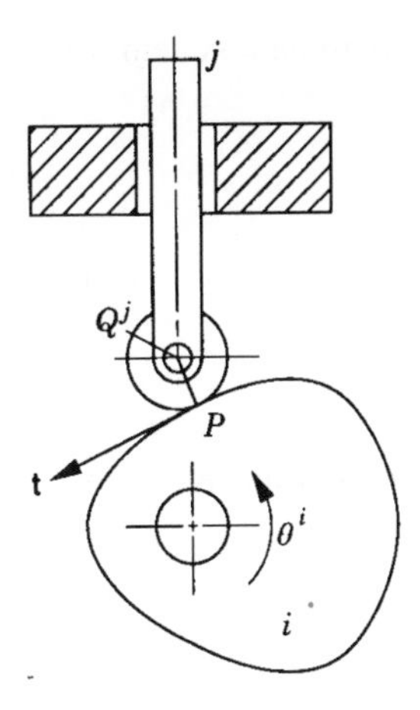

Figure 3.27 Roller follower

Note that in addition to the system generalized coordinates, the geometric nongeneralized surface parameter ϕ that defines the cam surface was introduced in order to be able to formulate the constraints of Eq. 102 (Shabana and Sany, 2001). Therefore, the number of unknown variables increases as a result of including the geometric parameter ϕ, and the two constraint equations of Eq. 102 eliminate only one degree of freedom from the original n coordinates of the system. That is, one of the equations in Eq. 102 can be used to eliminate the parameter ϕ from the other equation, leading to only one constraint equation that is a function of the original generalized coordinates of the system. By so doing, it becomes clear that the system of Eq. 102 is equivalent to one equation that is function of the generalized coordinates of the cam and the follower, and therefore, the resulting cam/follower constraint equations eliminate only one degree of freedom of the system.

Another type of cam systems is the *roller follower cam*, shown in Fig. 27. The contact point between the cam and the roller is point P, while the center of the roller is defined by point Q^j on the follower, which is denoted as body j. The vector $\mathbf{n}$ connecting the two points P and Q^j is perpendicular to the vector $\mathbf{t}$, which is tangent to the roller at the contact point. The two kinematic constraint equations for this type of cam system can be written as

$$\mathbf{n}^{\mathrm{T}}\mathbf{n} - (r)^2 = 0, \quad \mathbf{t}^{\mathrm{T}}\mathbf{n} = 0 \tag{3.103}$$

where r is the radius of the roller and the vector $\mathbf{n}$ is defined in terms of the absolute coordinates of the cam and the follower as

$$\mathbf{n} = \mathbf{R}^i + \mathbf{A}^i\bar{\mathbf{u}}_P^i - \mathbf{R}^j - \mathbf{A}^j\bar{\mathbf{u}}_Q^j \tag{3.104}$$

The vectors $\bar{\mathbf{u}}_P^i$ and $\bar{\mathbf{u}}_Q^j$ are the local position vectors of points P and Q^j defined, respectively, in the cam and follower coordinate systems. The vector $\bar{\mathbf{u}}_Q^j$ has fixed components in the follower coordinate system, while $\bar{\mathbf{u}}_P^i$ has components that depend on the shape of the cam. The first condition given by Eq. 103 ensures no separation or penetration between the cam and the follower.

The tangent vector $\mathbf{t}$ in Eq. 103 can also be defined in the cam coordinate system using Eq. 100 as

$$\bar{\mathbf{t}}^i = \frac{\partial \bar{\mathbf{u}}_P^i}{\partial \phi} = \begin{bmatrix} -d \sin\phi + \dfrac{\partial d}{\partial \phi} \cos\phi \\ d \cos\phi + \dfrac{\partial d}{\partial \phi} \sin\phi \end{bmatrix} \tag{3.105}$$

Using this equation, the tangent vector can be defined in the global coordinate system as $\mathbf{t} = \mathbf{A}^i \bar{\mathbf{t}}^i$.

Example 3.8

Derive the kinematic constraint equations of the *offset flat-faced follower* shown in Fig. 28.

Solution. Let point P^i and P^j denote the contact point on the cam and the flat-faced follower, respectively. Two perpendicular vectors $\mathbf{n}$ and $\mathbf{t}$ are defined at the contact point. The vector $\mathbf{r}_P^{ij}$ that connects point P^i with point P^j can be written as

$$\mathbf{r}_P^{ij} = \mathbf{r}_P^i - \mathbf{r}_P^j = \mathbf{R}^i + \mathbf{A}^i \bar{\mathbf{u}}_P^i - \mathbf{R}^j - \mathbf{A}^j \bar{\mathbf{u}}_P^j$$

This vector must remain perpendicular to $\mathbf{n}$, that is, the first constraint equation is given by

$$\mathbf{r}_P^{ij\mathrm{T}} \mathbf{n} = 0$$

Another vector $\mathbf{t}$ that is tangent to the cam at point P^i may be defined. This vector is also perpendicular to the vector $\mathbf{n}$, leading to the second condition

$$\mathbf{t}^{\mathrm{T}} \mathbf{n} = 0$$

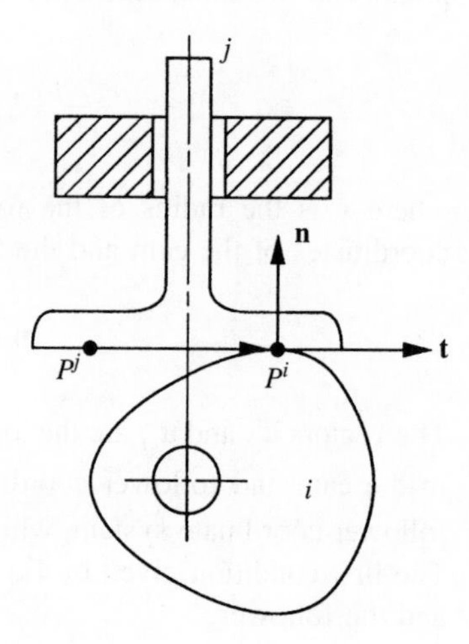

Figure 3.28 Offset flat-faced follower

Using the preceding two equations, the constraint equations for this type of cam can be written in a vector form as

$$\begin{bmatrix} \mathbf{r}_P^{ij\mathrm{T}}\mathbf{n} \\ \mathbf{t}^{\mathrm{T}}\mathbf{n} \end{bmatrix} = \begin{bmatrix} 0 \\ 0 \end{bmatrix}$$

While the second condition guarantees that the flat-faced follower remains parallel to the tangent to the cam surface at the contact point, the first condition guarantees that there is no separation or penetration between the cam and the follower. The vector $\mathbf{r}_P^{ij}$ depends on the absolute coordinates of the cam and the follower as well as the shape of the cam, while the vector $\mathbf{t}$ depends on the absolute coordinates and the shape of the cam only.

Gears *Gears* are widely used in machines for the purpose of transmission of rotary or rectilinear motion from one component to another (Litvin, 1994). Gears are used in a variety of industrial and technological applications such as automobiles, tractors, electric drills, helicopter rotor systems, machine tools, kitchen appliances, aircrafts, alarming clocks, and others. The theory of gearing is based on the fact that power can be transmitted from one body to another if the bodies have rolling contact. A rotary motion, for instance, can be transmitted from one body to another by *friction* if the two bodies are pressed against each other. If the friction force is high enough such that the two bodies roll without slipping, the velocities of the two bodies at the point of contact are equal. In this case, there is a definite relationship between the input and the output motions. The friction between the two bodies can be increased by increasing the roughness of the two surfaces in contact. A more reliable approach is to cut teeth on the surfaces of the two bodies. In this case, motion is transmitted by successive engagement of the teeth. *Spur gears* as shown in Fig. 29a are formed if the teeth are cut in a direction parallel to the axis of rotation. Gears can also be formed by cutting the teeth along a helix generated around the axis of the gear. In this case, the gear is called a *helical gear* and is shown in Fig. 29b. Both spur and helical gears are used to transmit power between two parallel axes. If the diameter of one of the spur gears goes to infinity, one obtains the rack and pinion system. Another type of gear that is widely used are *bevel gears* which, as shown in Fig. 29c, are cut from cones and are used in the case

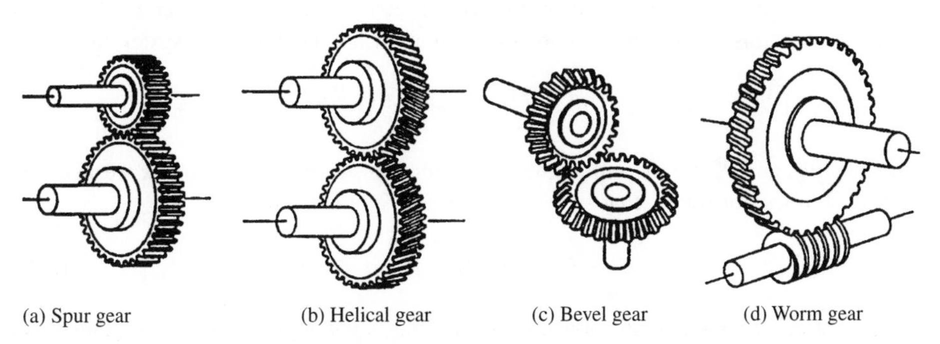

Figure 3.29 Gears

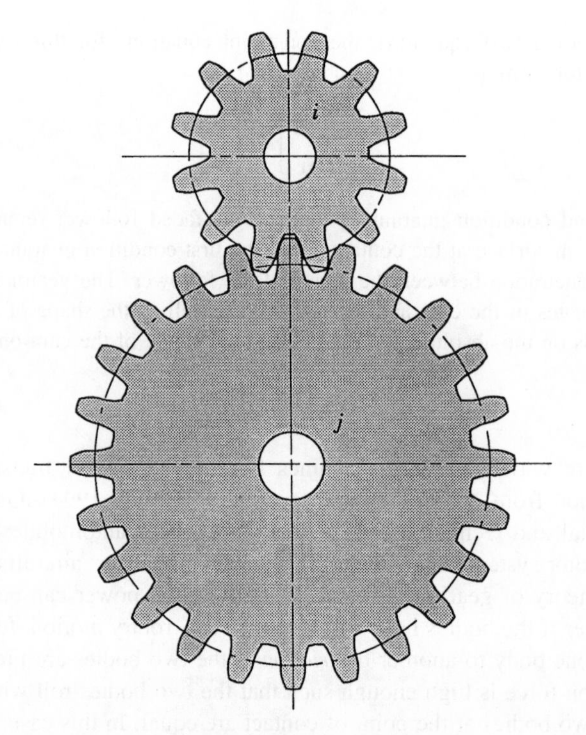

Figure 3.30 Spur gears

of intersecting shafts. In the case of nonintersecting and nonparallel shafts, the *skew gears*, *hypoid gears*, and *worm gears* (Fig. 29d) are used for the purpose of power transmission.

The simple case of spur gears is considered as an example in this section to demonstrate the formulation of the kinematic constraints in gear systems. Figure 30 shows a pair of spur gears which are assumed to be attached to a third body k. The condition that no sliding occurs between the two gears i and j at the contact point can be written as

$$(\dot{\theta}^i - \dot{\theta}^k)a^i = -(\dot{\theta}^j - \dot{\theta}^k)a^j \tag{3.106}$$

where a^i and a^j are, respectively, the radius of the pitch circles of gears i and j. Integrating Eq. 106, one obtains the kinematic constraint equation for the spur gear system as

$$[(\theta^i - \theta^i_o) - (\theta^k - \theta^k_o)]a^i + [(\theta^j - \theta^j_o) - (\theta^k - \theta^k_o)]a^j = 0 \tag{3.107}$$

in which θ^i_o, θ^j_o, and θ^k_o are the initial angular orientations of bodies i, j, and k, respectively.

If body k does not rotate, that is, $\theta^k = \theta^k_o$, Eq. 107 reduces to

$$(\theta^i - \theta^i_o)a^i + (\theta^j - \theta^j_o)a^j = 0 \tag{3.108}$$

The formulation of the kinematic constraints for other types of gears can be developed using a similar procedure as demonstrated by the following example.

Example 3.9

Figure 31 shows a rack-and-pinion mechanism in which the pinion is denoted as body i while the rack is denoted as body j. The pinion is assumed to have only rotational motion about its own axis, while the rack is assumed to have only translational motion along the horizontal direction. The condition at the point of contact that ensures no sliding between the two bodies is given by

$$\dot{\theta}^i a^i = \dot{x}^j_P$$

where a^i is the radius of the pitch circle of the pinion, $\dot{\theta}^i$ is the angular velocity of the pinion, and $\dot{x}^j_P$ is the velocity of the contact point on the rack. By integrating the preceding equation, one obtains the kinematic constraint equation for this simple mechanism as

$$(\theta^i - \theta^i_o)a^i - (x^j_P - x^j_{P_o}) = 0$$

where θ^i_o and $x^j_{P_o}$ are the constants of integration that represent the initial conditions.

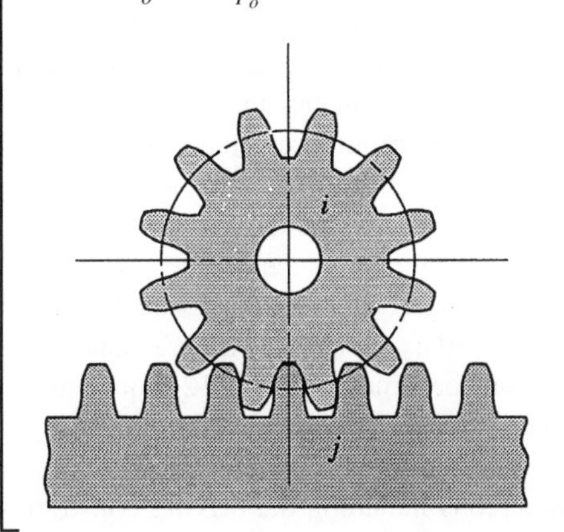

Figure 3.31 Rack and pinion

3.10 COMPUTATIONAL METHODS IN KINEMATICS

The formulations of the algebraic constraint equations presented in the preceding sections are used in this section to develop computer methods for the position, velocity, and acceleration analyses of multibody systems consisting of interconnected rigid bodies. In the analysis presented in this section, the configuration of the system is described by n coordinates, which can be written in a vector form as

$$\mathbf{q} = [q_1 \quad q_2 \quad q_3 \quad \cdots \quad q_n]^{\mathrm{T}} \tag{3.109}$$

In the planar analysis, if the system consists of n_b bodies, and the *absolute coordinates* are selected to describe the system configuration, one has $n = 3 \times n_b$ and the coordinates are defined as

$$\begin{aligned}\mathbf{q} &= [q_1 \quad q_2 \quad q_3 \quad q_4 \quad q_5 \quad q_6 \quad \cdots \quad q_{n-2} \quad q_{n-1} \quad q_n]^{\mathrm{T}} \\ &= [R_x^1 \quad R_y^1 \quad \theta^1 \quad R_x^2 \quad R_y^2 \quad \theta^2 \quad \cdots \quad R_x^{nb} \quad R_y^{nb} \quad \theta^{nb}]^{\mathrm{T}}\end{aligned} \tag{3.110}$$

In multibody system applications, these coordinates are not independent as the result of the kinematic constraint equations that describe system joints as well as specified motion trajectories. Examples of these constraints are the driving constraints, prismatic joints, revolute joints, and cam and gear constraints discussed in the preceding sections. These constraint equations can be written in a vector form as

$$\mathbf{C}(\mathbf{q}, t) = [C_1(\mathbf{q}, t) \quad C_2(\mathbf{q}, t) \quad \cdots \quad C_{n_c}(\mathbf{q}, t)]^{\mathrm{T}} = \mathbf{0} \tag{3.111}$$

where n_c is the total number of constraint equations and t is time.

Kinematically Driven Systems There are two cases that are encountered in the dynamic analysis of multibody systems. In the first case, the number of linearly independent constraint equations is equal to the number of the system coordinates, that is, $n_c = n$. This situation arises when all the degrees of freedom of the system are prescribed using driving constraints. For example, a slider crank mechanism that consists of four bodies (including the ground) has 12 absolute coordinates. The three ground constraints, the three revolute joints, and the prismatic joint in the mechanism make the number of the degrees of freedom of the mechanism equal to one. One may select this degree of freedom to be the crank angle. If a driving constraint is used to specify the angular velocity of the crankshaft, the number of joint constraints plus the driving constraint becomes equal to 12, equal to the number of absolute coordinates of the system. When the number of constraint equations is equal to the number of system coordinates, the system is said to be *kinematically driven*.

In the second case, the number of constraint equations including the driving constraints is less than the number of the system coordinates, that is, $n_c < n$. This situation arises when some of the degrees of freedom of the system are *dynamically driven* using force inputs. In this case, some of the degrees of freedom are not specified, and the system in this case is said to be *dynamically driven*. In the case of a dynamically driven system, a force analysis is required in order to obtain the position, velocity, and acceleration of the system components.

In this chapter, only kinematically driven systems are discussed. Dynamically driven systems are discussed in later chapters when the force analysis of multibody systems is considered. In the case of kinematically driven systems, the vector of the constraint equations, defined by Eq. 111, includes the joint and driving constraints as demonstrated by the following simple example.

Example 3.10

For the three-body system of Example 7, bodies 2 and 3 are assumed to rotate with constant angular velocities $\dot{\theta}^2 = \omega^2$ and $\dot{\theta}^3 = \omega^3$. Determine the vector of the constraint functions of the system.

Solution. If the absolute coordinates are used, the number of coordinates n is equal to 9 and the number of the joint constraints is equal to 7. By imposing constraints on the angular velocities of bodies 2 and 3, the system becomes kinematically driven. The two driving constraints

$$\dot{\theta}^2 - \omega^2 = 0, \quad \dot{\theta}^3 - \omega^3 = 0$$

can be integrated, yielding

$$\theta^2 - \theta_o^2 - \omega^2 t = 0, \quad \theta^3 - \theta_o^3 - \omega^3 t = 0$$

where θ_o^2 and θ_o^3 are the initial angular orientations of bodies 2 and 3, respectively. These two driving constraint equations can be combined with the joint constraints obtained in Example 7, leading to

$$\mathbf{C}(\mathbf{q}, t) = \begin{bmatrix} C_1(\mathbf{q}, t) \\ C_2(\mathbf{q}, t) \\ C_3(\mathbf{q}, t) \\ C_4(\mathbf{q}, t) \\ C_5(\mathbf{q}, t) \\ C_6(\mathbf{q}, t) \\ C_7(\mathbf{q}, t) \\ C_8(\mathbf{q}, t) \\ C_9(\mathbf{q}, t) \end{bmatrix} = \begin{bmatrix} R_x^1 - c_1 \\ R_y^1 - c_2 \\ \theta^1 - c_3 \\ R_x^2 - \dfrac{l^2}{2} \cos \theta^2 \\ R_y^2 - \dfrac{l^2}{2} \sin \theta^2 \\ R_x^2 + \dfrac{l^2}{2} \cos \theta^2 - R_x^3 + \dfrac{l^3}{2} \cos \theta^3 \\ R_y^2 + \dfrac{l^2}{2} \sin \theta^2 - R_y^3 + \dfrac{l^3}{2} \sin \theta^3 \\ \theta^2 - \theta_o^2 - \omega^2 t \\ \theta^3 - \theta_o^3 - \omega^3 t \end{bmatrix} = \begin{bmatrix} 0 \\ 0 \\ 0 \\ 0 \\ 0 \\ 0 \\ 0 \\ 0 \\ 0 \end{bmatrix}$$

Given $\omega^2, \omega^3, \theta_0^2$, and θ_0^3, the vector of constraints of the preceding equation defines nine algebraic equations that can be solved for the nine coordinates of the system R_x^i, R_y^i, and θ^i, $i = 1, 2, 3$.

Position Analysis For kinematically driven systems, the total number of constraint equations n_c is equal to the number of system coordinates n. Consequently, the vector of the constraint equations defined by Eq. 111 contains n algebraic equations that describe joint constraints as well as driving constraints. This vector of the constraint equations represents n scalar equations that can be solved for the n unknown coordinates of Eq. 109. The equations, however, can be nonlinear functions of the system coordinates and time, a fact which was demonstrated in the preceding example where nonlinear trigonometric functions of the system coordinates appear in the algebraic kinematic constraint equations. Because a closed form solution cannot be obtained, in general, in the case of nonlinear systems; numerical methods are often used.

The numerical procedure often used for solving a system of nonlinear algebraic equations is the *Newton–Raphson algorithm*. This iterative procedure which can be employed to solve the nonlinear kinematic constraint equations starts by making an estimate of the desired solution vector. If this estimate at certain point in time t is denoted as $\mathbf{q}_i$, the exact solution can be written as $\mathbf{q}_i + \Delta\mathbf{q}_i$. By using Taylor's theorem, the vector of constraint equations defined by Eq. 111 can be written as

$$\mathbf{C}(\mathbf{q}_i + \Delta\mathbf{q}_i, t) = \mathbf{C}(\mathbf{q}_i, t) + \mathbf{C}_{\mathbf{q}_i}\ \Delta\mathbf{q}_i + \tfrac{1}{2}(\mathbf{C}_{\mathbf{q}_i}\ \Delta\mathbf{q}_i)_{\mathbf{q}_i}\ \Delta\mathbf{q}_i + \cdots \tag{3.112}$$

where $\Delta\mathbf{q} = [\Delta q_1 \quad \Delta q_2 \cdots \Delta q_n]^{\mathrm{T}}$ is called the vector of *Newton differences*, and $\mathbf{C}_{\mathbf{q}_i}$ is the *constraint Jacobian matrix*, defined as

$$\mathbf{C}_{\mathbf{q}_i} = \begin{bmatrix} \dfrac{\partial C_1}{\partial q_1} & \dfrac{\partial C_1}{\partial q_2} & \dfrac{\partial C_1}{\partial q_3} & \cdots & \dfrac{\partial C_1}{\partial q_n} \\ \dfrac{\partial C_2}{\partial q_1} & \dfrac{\partial C_2}{\partial q_2} & \dfrac{\partial C_2}{\partial q_3} & \cdots & \dfrac{\partial C_2}{\partial q_n} \\ \vdots & \vdots & \vdots & \ddots & \vdots \\ \dfrac{\partial C_{n_c}}{\partial q_1} & \dfrac{\partial C_{n_c}}{\partial q_2} & \dfrac{\partial C_{n_c}}{\partial q_3} & \cdots & \dfrac{\partial C_{n_c}}{\partial q_n} \end{bmatrix} \tag{3.113}$$

For a kinematically driven system, the Jacobian matrix is a square matrix since $n_c = n$, and additionally, if the constraint equations are assumed to be linearly independent, $\mathbf{C}_{\mathbf{q}}$ is a nonsingular matrix. If the vector $\mathbf{q}_i + \Delta\mathbf{q}_i$ is assumed to be the desired exact solution, $\mathbf{C}(\mathbf{q}_i + \Delta\mathbf{q}_i, t) = \mathbf{0}$, and Eq. 112 reduces to

$$\mathbf{C}(\mathbf{q}_i, t) + \mathbf{C}_{\mathbf{q}_i}\ \Delta\mathbf{q}_i + \tfrac{1}{2}(\mathbf{C}_{\mathbf{q}_i}\ \Delta\mathbf{q}_i)_{\mathbf{q}_i}\ \Delta\mathbf{q}_i + \cdots = \mathbf{0} \tag{3.114}$$

If the assumed solution is close to the exact solution, the norm of the vector $\Delta\mathbf{q}_i$ becomes small and higher-order terms in the vector $\Delta\mathbf{q}$ in Eq. 114 can be neglected. This assumption defines the first-order approximation of Eq. 114 as

$$\mathbf{C}(\mathbf{q}_i, t) + \mathbf{C}_{\mathbf{q}_i}\ \Delta\mathbf{q}_i \approx \mathbf{0} \tag{3.115}$$

which yields

$$\mathbf{C}_{\mathbf{q}_i}\ \Delta\mathbf{q}_i = -\mathbf{C}(\mathbf{q}_i, t) \tag{3.116}$$

Since the constraint Jacobian matrix $\mathbf{C}_{\mathbf{q}_i}$ is assumed to be nonsingular, Eq. 116 can be solved for the vector of Newton differences $\Delta\mathbf{q}_i$. This vector can be used to iteratively update the vector of the system coordinates as

$$\mathbf{q}_{i+1} = \mathbf{q}_i + \Delta\mathbf{q}_i \tag{3.117}$$

where i is the iteration number. The updated vector $\mathbf{q}_{i+1}$ can then be used to reconstruct Eq. 116 and solve this system of equations for the new vector of Newton differences $\Delta\mathbf{q}_{i+1}$, which can be used again to update the vector of the system coordinates, thus defining the vector $\mathbf{q}_{i+2}$. This process continues until the norm of the vector of Newton differences or the norm of the vector of constraint equations becomes less than a specified tolerance, that is,

$$|\Delta\mathbf{q}_k| < \epsilon_1 \quad \text{or} \quad |\mathbf{C}(\mathbf{q}_k, t)| < \epsilon_2 \tag{3.118}$$

where ϵ_1 and ϵ_2 are specified tolerances and k is the iteration number.

It is important to mention at this point that due to the fact that the Newton–Raphson method does not always converge, one must specify an upper limit on the number of iteractions used in this numerical algorithm. Failure to achieve convergence may be due to several factors, such as the initial estimate of the desired solution is not close enough to the exact solution, an error is made in the definition of the system constraints, and/or the multibody system is close to a singular configuration.

Example 3.11

For the three-body system of Example 10, it was shown that the vector of constraint equations is

$$\mathbf{C}(\mathbf{q},t) = \begin{bmatrix} C_1(\mathbf{q},t) \\ C_2(\mathbf{q},t) \\ C_3(\mathbf{q},t) \\ C_4(\mathbf{q},t) \\ C_5(\mathbf{q},t) \\ C_6(\mathbf{q},t) \\ C_7(\mathbf{q},t) \\ C_8(\mathbf{q},t) \\ C_9(\mathbf{q},t) \end{bmatrix} = \begin{bmatrix} R_x^1 - c_1 \\ R_y^1 - c_2 \\ \theta^1 - c_3 \\ R_x^2 - \dfrac{l^2}{2}\cos\theta^2 \\ R_y^2 - \dfrac{l^2}{2}\sin\theta^2 \\ R_x^2 + \dfrac{l^2}{2}\cos\theta^2 - R_x^3 + \dfrac{l^3}{2}\cos\theta^3 \\ R_y^2 + \dfrac{l^2}{2}\sin\theta^2 - R_y^3 + \dfrac{l^3}{2}\sin\theta^3 \\ \theta^2 - \theta_o^2 - \omega^2 t \\ \theta^3 - \theta_o^3 - \omega^3 t \end{bmatrix} = \begin{bmatrix} 0 \\ 0 \\ 0 \\ 0 \\ 0 \\ 0 \\ 0 \\ 0 \\ 0 \end{bmatrix}$$

where the vector $\mathbf{q}$ is selected to be the vector of absolute coordinates given by

$$\mathbf{q} = [R_x^1 \quad R_y^1 \quad \theta^1 \quad R_x^2 \quad R_y^2 \quad \theta^2 \quad R_x^3 \quad R_y^3 \quad \theta^3]^{\mathrm{T}}$$

The Jacobian matrix as defined by Eq. 113 can be developed for this system as

$$\mathbf{C_q} = \begin{bmatrix} 1 & 0 & 0 & 0 & 0 & 0 & 0 & 0 & 0 \\ 0 & 1 & 0 & 0 & 0 & 0 & 0 & 0 & 0 \\ 0 & 0 & 1 & 0 & 0 & 0 & 0 & 0 & 0 \\ 0 & 0 & 0 & 1 & 0 & \frac{l^2}{2}\sin\theta^2 & 0 & 0 & 0 \\ 0 & 0 & 0 & 0 & 1 & -\frac{l^2}{2}\cos\theta^2 & 0 & 0 & 0 \\ 0 & 0 & 0 & 1 & 0 & -\frac{l^2}{2}\sin\theta^2 & -1 & 0 & -\frac{l^3}{2}\sin\theta^3 \\ 0 & 0 & 0 & 0 & 1 & \frac{l^2}{2}\cos\theta^2 & 0 & -1 & \frac{l^3}{2}\cos\theta^3 \\ 0 & 0 & 0 & 0 & 0 & 1 & 0 & 0 & 0 \\ 0 & 0 & 0 & 0 & 0 & 0 & 0 & 0 & 1 \end{bmatrix}$$

The matrix $\mathbf{C_q}$ is a square matrix, since the number of the constraint equations n_c is equal to the number of coordinates n.

Velocity Analysis Differentiating the vector of constraint equations defined by Eq. 111 with respect to time, and using the *chain rule of differentiation* leads to

$$\mathbf{C_q}\dot{\mathbf{q}} + \mathbf{C}_t = \mathbf{0} \tag{3.119}$$

where $\mathbf{C_q}$ is the constraint Jacobian matrix defined by Eq. 113 and $\mathbf{C}_t$ is the vector of partial derivative of the constraint equations with respect to time. This vector is defined as

$$\mathbf{C}_t = \begin{bmatrix} \frac{\partial C_1}{\partial t} & \frac{\partial C_2}{\partial t} & \cdots & \frac{\partial C_n}{\partial t} \end{bmatrix}^{\mathrm{T}} \tag{3.120}$$

If the constraint equations are not explicit functions of time, the vector $\mathbf{C}_t$ is identically the zero vector.

Since the coordinates of the system components are assumed to be known from the position analysis, the Jacobian matrix $\mathbf{C_q}$ and the vector $\mathbf{C}_t$ which can be functions of the coordinates and time only can be evaluated. Equation 119, which can be considered as a linear system of algebraic equations in the velocity vector $\dot{\mathbf{q}}$, can be written as

$$\mathbf{C_q}\dot{\mathbf{q}} = -\mathbf{C}_t \tag{3.121}$$

Because $\mathbf{C_q}$ is assumed to be a square and nonsingular matrix in the case of kinematically driven systems, Eq. 121 can be solved for the velocity vector $\dot{\mathbf{q}}$.

Acceleration Analysis The acceleration equations can be obtained by differentiating Eq. 119 with respect to time, leading to

$$\frac{d}{dt}(\mathbf{C_q}\dot{\mathbf{q}} + \mathbf{C}_t) = \mathbf{0} \tag{3.122}$$

This equation, by using the chain rule of differentiation, yields

$$\mathbf{C_q}\ddot{\mathbf{q}} + (\mathbf{C_q}\dot{\mathbf{q}})_{\mathbf{q}}\dot{\mathbf{q}} + 2\mathbf{C}_{\mathbf{q}_t}\dot{\mathbf{q}} + \mathbf{C}_{tt} = \mathbf{0} \tag{3.123}$$

This is a linear system of algebraic equations in the acceleration vector $\ddot{\mathbf{q}}$, which can be written in the following form:

$$\mathbf{C_q}\ddot{\mathbf{q}} = \mathbf{Q}_d \tag{3.124}$$

where the vector $\mathbf{Q}_d$ absorbs terms that are quadratic in the velocities and is defined as

$$\mathbf{Q}_d = -(\mathbf{C_q}\dot{\mathbf{q}})_{\mathbf{q}}\dot{\mathbf{q}} - 2\mathbf{C}_{\mathbf{q}_t}\dot{\mathbf{q}} - \mathbf{C}_{tt} \tag{3.125}$$

Having determined the coordinate and velocity vectors $\mathbf{q}$ and $\dot{\mathbf{q}}$ using the position and velocity analysis methods discussed previously, the coefficient matrix $\mathbf{C_q}$ and the vector $\mathbf{Q}_d$ in Eq. 124 can be evaluated. Assuming $\mathbf{C_q}$ to be nonsingular for a kinematically driven system, Eq. 124 can be solved for the acceleration vector $\ddot{\mathbf{q}}$.

Example 3.12

For the three-body system of Example 11, one can verify that the vector $\mathbf{C}_t$ of Eq. 120 is given by

$$\mathbf{C}_t = \begin{bmatrix} \frac{\partial C_1}{\partial t} & \frac{\partial C_2}{\partial t} & \cdots & \frac{\partial C_n}{\partial t} \end{bmatrix}^{\mathrm{T}}$$
$$= [0 \quad 0 \quad 0 \quad 0 \quad 0 \quad 0 \quad 0 \quad -\omega^2 \quad -\omega^3]^{\mathrm{T}}$$

Using Eq. 121 and the constraint Jacobian matrix obtained in Example 11, it can be shown that the velocity vector $\dot{\mathbf{q}}$ is given at any point in time for this example by

$$\dot{\mathbf{q}} = [\dot{R}_x^1 \quad \dot{R}_y^1 \quad \dot{\theta}^1 \quad \dot{R}_x^2 \quad \dot{R}_y^2 \quad \dot{\theta}^2 \quad \dot{R}_x^3 \quad \dot{R}_y^3 \quad \dot{\theta}^3]^{\mathrm{T}}$$
$$= [0 \quad 0 \quad 0 \quad h_1 \quad h_2 \quad \omega^2 \quad h_3 \quad h_4 \quad \omega^3]^{\mathrm{T}}$$

where

$$h_1 = -\frac{\omega^2 l^2}{2}\sin\theta^2, \quad h_2 = \frac{\omega^2 l^2}{2}\cos\theta^2,$$
$$h_3 = -\omega^2 l^2 \sin\theta^2 - \frac{\omega^3 l^3}{2}\sin\theta^3,$$
$$h_4 = \omega^2 l^2 \cos\theta^2 + \frac{\omega^3 l^3}{2}\cos\theta^3$$

For the acceleration analysis, one has to evaluate the vector $\mathbf{Q}_d$ of Eq. 125. For this system, if the angular velocities are assumed to be constant, the vector $\mathbf{C}_t$ is not an explicit function of the system coordinates or time. It follows that

$$\mathbf{C}_{\mathbf{q}}t = \mathbf{0}, \quad \mathbf{C}_{tt} = \mathbf{0}$$

The vector $\mathbf{C}_{\mathbf{q}}\dot{\mathbf{q}}$ is

$$\mathbf{C}_{\mathbf{q}}\dot{\mathbf{q}} = \begin{bmatrix} \dot{R}_x^1 \\ \dot{R}_y^1 \\ \dot{\theta}^1 \\ \dot{R}_x^2 + \dfrac{\dot{\theta}^2 l^2}{2} \sin\theta^2 \\ \dot{R}_y^2 - \dfrac{\dot{\theta}^2 l^2}{2} \cos\theta^2 \\ \dot{R}_x^2 - \dfrac{\dot{\theta}^2 l^2}{2} \sin\theta^2 - \dot{R}_x^3 - \dfrac{\dot{\theta}^3 l^3}{2} \sin\theta^3 \\ \dot{R}_y^2 + \dfrac{\dot{\theta}^2 l^2}{2} \cos\theta^2 - \dot{R}_y^3 + \dfrac{\dot{\theta}^3 l^3}{2} \cos\theta^3 \\ \dot{\theta}^2 \\ \dot{\theta}^3 \end{bmatrix}$$

which can be used to define the matrix $(\mathbf{C}_{\mathbf{q}}\dot{\mathbf{q}})_{\mathbf{q}}$ as

$$(\mathbf{C}_{\mathbf{q}}\dot{\mathbf{q}})_{\mathbf{q}} = \begin{bmatrix} 0 & 0 & 0 & 0 & 0 & 0 & 0 & 0 & 0 \\ 0 & 0 & 0 & 0 & 0 & 0 & 0 & 0 & 0 \\ 0 & 0 & 0 & 0 & 0 & 0 & 0 & 0 & 0 \\ 0 & 0 & 0 & 0 & 0 & \dfrac{\dot{\theta}^2 l^2}{2} \cos\theta^2 & 0 & 0 & 0 \\ 0 & 0 & 0 & 0 & 0 & \dfrac{\dot{\theta}^2 l^2}{2} \sin\theta^2 & 0 & 0 & 0 \\ 0 & 0 & 0 & 0 & 0 & -\dfrac{\dot{\theta}^2 l^2}{2} \cos\theta^2 & 0 & 0 & -\dfrac{\dot{\theta}^3 l^3}{2} \cos\theta^3 \\ 0 & 0 & 0 & 0 & 0 & -\dfrac{\dot{\theta}^2 l^2}{2} \sin\theta^2 & 0 & 0 & -\dfrac{\dot{\theta}^3 l^3}{2} \sin\theta^3 \\ 0 & 0 & 0 & 0 & 0 & 0 & 0 & 0 & 0 \\ 0 & 0 & 0 & 0 & 0 & 0 & 0 & 0 & 0 \end{bmatrix}$$

The vector $\mathbf{Q}_d = -(\mathbf{C}_\mathbf{q}\dot{\mathbf{q}})_\mathbf{q}\dot{\mathbf{q}}$ is given by

$$\mathbf{Q}_d = -(\mathbf{C}_\mathbf{q}\dot{\mathbf{q}})_\mathbf{q}\dot{\mathbf{q}}$$

$$= -\left[0 \quad 0 \quad 0 \quad \frac{(\dot{\theta}^2)^2 l^2}{2}\cos\theta^2 \quad \frac{(\dot{\theta}^2)^2 l^2}{2}\sin\theta^2 \right.$$

$$-\left(\frac{(\dot{\theta}^2)^2 l^2}{2}\cos\theta^2 + \frac{(\dot{\theta}^3)^2 l^3}{2}\cos\theta^3\right)$$

$$\left. -\left(\frac{(\dot{\theta}^2)^2 l^2}{2}\sin\theta^2 + \frac{(\dot{\theta}^3)^2 l^3}{2}\sin\theta^3\right) \quad 0 \quad 0\right]^{\mathrm{T}}$$

in which θ^2, θ^3, $\dot{\theta}^2$, and $\dot{\theta}^3$ are assumed to be known from the position and velocity analyses. Substituting the vector $\mathbf{Q}_d$ into Eq. 124, one obtains the acceleration vector $\ddot{\mathbf{q}}$ as

$$\ddot{R}_x^1 = \ddot{R}_y^1 = \ddot{\theta}^1 = \ddot{\theta}^2 = \ddot{\theta}^3 = 0$$

$$\ddot{R}_x^2 = -(\dot{\theta}^2)^2 \frac{l^2}{2}\cos\theta^2$$

$$\ddot{R}_y^2 = -(\dot{\theta}^2)^2 \frac{l^2}{2}\sin\theta^2$$

$$\ddot{R}_x^3 = -(\dot{\theta}^2)^2 l^2 \cos\theta^2 - (\dot{\theta}^3)^2 \frac{l^3}{2}\cos\theta^3$$

$$\ddot{R}_y^3 = -(\dot{\theta}^2)^2 l^2 \sin\theta^2 - (\dot{\theta}^3)^2 \frac{l^3}{2}\sin\theta^3$$

The results obtained in this simple example, using the general procedure outlined in this section, can also be obtained by simply differentiating the following kinematic relationships:

$$\begin{bmatrix} R_x^2 \\ R_y^2 \\ R_x^3 \\ R_y^3 \end{bmatrix} = \begin{bmatrix} \frac{l^2}{2}\cos\theta^2 \\ \frac{l^2}{2}\sin\theta^2 \\ l^2\cos\theta^2 + \frac{l^3}{2}\cos\theta^3 \\ l^2\sin\theta^2 + \frac{l^3}{2}\sin\theta^3 \end{bmatrix}$$

3.11 COMPUTER IMPLEMENTATION

The formulation presented in the preceding section for kinematically driven systems leads to a number of nonlinear algebraic constraint equations which is equal to the number of coordinates. These constraint equations can be solved for the coordinates using a Newton–Raphson algorithm. Differentiation of the kinematic constraints once and twice leads to a linear system of algebraic equations in the velocities and accelerations which can be solved in a straightforward manner to determine the vector of system velocities and accelerations. Clearly, the number of resulting algebraic constraint equations depends on the choice of coordinates. Different sets of coordinates lead to different sets of algebraic equations. The number of the system degrees of freedom, however, remains the same, regardless of the type of coordinates used. As a consequence, the use of a larger number of coordinates requires the use of a larger number of constraint equations. Consider, for example, the four-bar mechanism shown in Fig. 32. One may select the system coordinates as

$$\mathbf{q} = [\theta^2 \quad \theta^3 \quad \theta^4]^{\mathrm{T}} \tag{3.126}$$

which is a subset of the absolute Cartesian coordinates of the mechanism. In this case, fewer constraint equations are required in order to describe the kinematic relationships between the angles θ^2, θ^3, and θ^4. The constraint equations can be expressed in terms of these coordinates as

$$\left.\begin{aligned} l^2 \cos\theta^2 + l^3 \cos\theta^3 + l^4 \cos\theta^4 &= l^1 \\ l^2 \sin\theta^2 + l^3 \sin\theta^3 + l^4 \sin\theta^4 &= 0 \end{aligned}\right\} \tag{3.127}$$

where l^1 is the length OC. If the system is kinematically driven, a driving constraint must be introduced in order to control the degree of freedom of the system, thereby making the number of constraint equations equal to the number of coordinates. In this case, the general procedure outlined in the preceding section can still be used to solve a smaller system of equations as demonstrated by the following example.

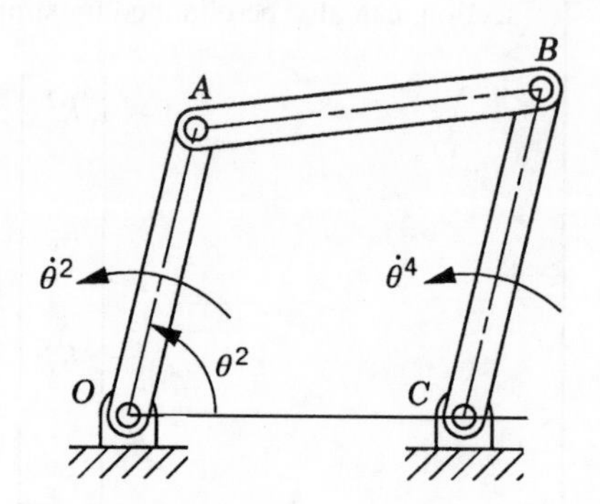

Figure 3.32 Four-bar mechanism

Example 3.13

For the four-bar linkage shown in Fig. 32, let $l^1 = OC = 0.35\,\text{m}$, $l^2 = 0.2\,\text{m}$, $l^3 = 0.35\,\text{m}$, and $l^4 = 0.25\,\text{m}$. The crankshaft of the mechanism is assumed to rotate with a constant angular velocity $\dot{\theta}^2 = 5$ rad/s. Determine the angular orientations, velocities, and accelerations of the coupler and the rocker when $t = 0.02$ s. Use θ^2, θ^3, and θ^4 as the system coordinates. Assume that the initial angular orientation of the crankshaft is 57.27°.

Solution. Since the mechanism has only one degree of freedom, the coordinates θ^2, θ^3, and θ^4 are not independent. They are related by the following loop-closure equations and the driving constraint equation

$$\mathbf{C}(\mathbf{q}, t) = \begin{bmatrix} C_1(\mathbf{q}, t) \\ C_2(\mathbf{q}, t) \\ C_3(\mathbf{q}, t) \end{bmatrix} = \begin{bmatrix} l^2 \cos\theta^2 + l^3 \cos\theta^3 + l^4 \cos\theta^4 - l^1 \\ l^2 \sin\theta^2 + l^3 \sin\theta^3 + l^4 \sin\theta^4 \\ \theta^2 - \omega^2 t - \theta^2_o \end{bmatrix} = \begin{bmatrix} 0 \\ 0 \\ 0 \end{bmatrix}$$

where $\omega^2 = \dot{\theta}^2 = 5$ rad/s, θ^2_o is the initial angular orientation of the crankshaft, and

$$\mathbf{q} = [\theta^2 \quad \theta^3 \quad \theta^4]^{\mathrm{T}}$$

The Jacobian matrix of the constraint equations is

$$\mathbf{C_q} = \begin{bmatrix} -l^2 \sin\theta^2 & -l^3 \sin\theta^3 & -l^4 \sin\theta^4 \\ l^2 \cos\theta^2 & l^3 \cos\theta^3 & l^4 \cos\theta^4 \\ 1 & 0 & 0 \end{bmatrix}$$

At $t = 0.02$ s, $\theta^2 = 63° = 1.0996$ rad. To start the numerical solution, we make an initial guess for the angles θ^2, θ^3, and θ^4 as

$$\mathbf{q}_0 = [63° \quad 10° \quad 245°]^{\mathrm{T}} = [1.0996 \quad 0.1745 \quad 4.2761]^{\mathrm{T}} \text{ rad}$$

Using the initial guess, the Jacobian matrix is

$$\mathbf{C_q}(\mathbf{q}_0) = \begin{bmatrix} -0.1782 & -0.06077 & 0.22658 \\ 0.09079 & 0.34468 & -0.105646 \\ 1.0 & 0.0 & 0.0 \end{bmatrix}$$

and the constraint equations are

$$\mathbf{C}(\mathbf{q}_0, t) = \begin{bmatrix} -0.020176 \\ 0.01239 \\ 0.0 \end{bmatrix}$$

For this mechanism, Eq. 116 can be written as

$$\begin{bmatrix} -0.1782 & -0.06077 & 0.22658 \\ 0.09079 & 0.34468 & -0.105646 \\ 1.0 & 0.0 & 0.0 \end{bmatrix} \begin{bmatrix} \Delta\theta_0^2 \\ \Delta\theta_0^3 \\ \Delta\theta_0^4 \end{bmatrix} = - \begin{bmatrix} -0.020176 \\ 0.01239 \\ 0.0 \end{bmatrix}$$

It is clear that $\Delta\theta_0^2 = 0$, and

$$\begin{bmatrix} -0.06077 & 0.22658 \\ 0.34468 & -0.105646 \end{bmatrix} \begin{bmatrix} \Delta\theta_0^3 \\ \Delta\theta_0^4 \end{bmatrix} = - \begin{bmatrix} -0.020176 \\ 0.01239 \end{bmatrix}$$

which can be solved for $\Delta\theta_0^3$, and $\Delta\theta_0^4$ as

$$\Delta\theta_0^3 = -9.4285 \times 10^{-3}, \qquad \Delta\theta_0^4 = 0.086517$$

The vector of the system coordinates can be updated according to Eq. 117 as

$$\mathbf{q}_1 = \mathbf{q}_0 + \Delta\mathbf{q}_0 = \begin{bmatrix} 1.0996 \\ 0.1745 \\ 4.2761 \end{bmatrix} + \begin{bmatrix} 0.0 \\ -9.4285 \times 10^{-3} \\ 0.086517 \end{bmatrix} = \begin{bmatrix} 1.0966 \\ 0.16507 \\ 4.362617 \end{bmatrix}$$

Using the vector $\mathbf{q}_1 = [\theta_1^2 \;\; \theta_1^3 \;\; \theta_1^4]^{\mathrm{T}}$, the Jacobian matrix and the vector of constraint equations can be evaluated as

$$\mathbf{C_q}(\mathbf{q}_1) = \begin{bmatrix} -0.1782 & -0.05751 & 0.23486 \\ 0.09079 & 0.34524 & -0.08567 \\ 1.0 & 0.0 & 0.0 \end{bmatrix}, \quad \mathbf{C}(\mathbf{q}_1, t) = \begin{bmatrix} 3.6 \times 10^{-4} \\ 8.5 \times 10^{-4} \\ 0.0 \end{bmatrix}$$

Equation 116 yields

$$\begin{bmatrix} -0.1782 & -0.05751 & 0.23486 \\ 0.09079 & 0.34524 & -0.08567 \\ 1.0 & 0.0 & 0.0 \end{bmatrix} \begin{bmatrix} \Delta\theta_1^2 \\ \Delta\theta_1^3 \\ \Delta\theta_1^4 \end{bmatrix} = - \begin{bmatrix} 3.6 \times 10^{-4} \\ 8.5 \times 10^{-4} \\ 0.0 \end{bmatrix}$$

The solution of this system of equations yields

$$\Delta\mathbf{q}_1 = [\Delta\theta_1^2 \;\; \Delta\theta_1^3 \;\; \Delta\theta_1^4]^{\mathrm{T}} = [0.0 \quad -3.026 \times 10^{-3} \quad -2.2738 \times 10^{-3}]^{\mathrm{T}}$$

We may define the norms of $\mathbf{C}$ and $\Delta\mathbf{q}$ as

$$|\mathbf{C}| = \frac{\left[\sum_{i=1}^{n_c} (C_i)^2\right]^{1/2}}{n_c} = \frac{1}{3}\sqrt{(C_1)^2 + (C_2)^2 + (C_3)^2}$$

$$|\Delta\mathbf{q}| = \frac{\left[\sum_{i=1}^{n} (\Delta q_i)^2\right]^{1/2}}{n} = \frac{1}{3}\sqrt{(\Delta\theta^2)^2 + (\Delta\theta^3)^2 + (\Delta\theta^4)^2}$$

where n_c and n are, respectively, the number of constraint equations and the number of coordinates. It follows that

$$|\mathbf{C}(\mathbf{q}_1, t)| = \frac{10^{-4}}{3}\sqrt{(3.6)^2 + (8.5)^2 + (0)^2} = 3.077 \times 10^{-4}$$

$$|\Delta\mathbf{q}_1| = \frac{10^{-3}}{3}\sqrt{(0)^2 + (3.026)^2 + (2.2738)^2} = 1.261 \times 10^{-3}$$

Since the norms of $\mathbf{C}$ and $\Delta\mathbf{q}_1$ are relatively small, one may decide to accept $\mathbf{q}_1$ as the correct answer for the position analysis and proceed to perform the velocity analysis. In this example, the vector $\mathbf{C}_t$ is

$$\mathbf{C}_t = \begin{bmatrix} 0 \\ 0 \\ -\omega^2 \end{bmatrix} = \begin{bmatrix} 0 \\ 0 \\ -5.0 \end{bmatrix}$$

Using the Jacobian matrix previously evaluated at $\mathbf{q} = \mathbf{q}_1$, the linear algebraic velocity equations can be written as

$$\begin{bmatrix} -0.1782 & -0.05751 & 0.23486 \\ 0.09079 & 0.34524 & -0.08567 \\ 1.0 & 0.0 & 0.0 \end{bmatrix} \begin{bmatrix} \dot{\theta}^2 \\ \dot{\theta}^3 \\ \dot{\theta}^4 \end{bmatrix} = -\begin{bmatrix} 0.0 \\ 0.0 \\ -5.0 \end{bmatrix}$$

This equation defines the angular velocities $\dot{\theta}^2$, $\dot{\theta}^3$, and $\dot{\theta}^4$ at $t = 0.02$ s as

$$\dot{\theta}^2 = 5 \text{ rad/s}, \qquad \dot{\theta}^3 = -0.39764 \text{ rad/s}, \qquad \dot{\theta}^4 = 3.6964 \text{ rad/s}$$

One can verify that the vector $\mathbf{Q}_d$ of Eq. 124 is given for this example by

$$\mathbf{Q}_d = \begin{bmatrix} (\dot{\theta}^2)^2 l^2 \cos\theta^2 + (\dot{\theta}^3)^2 l^3 \cos\theta^3 + (\dot{\theta}^4)^2 l^4 \cos\theta^4 \\ (\dot{\theta}^2)^2 l^2 \sin\theta^2 + (\dot{\theta}^3)^2 l^3 \sin\theta^3 + (\dot{\theta}^4)^2 l^4 \sin\theta^4 \\ 0 \end{bmatrix}$$

Using the values of the coordinates and velocities obtained previously in the position and velocity analyses at time $t = 0.02$ s, the vector $\mathbf{Q}_d$ can be evaluated as

$$\mathbf{Q}_d = \begin{bmatrix} (5)^2(0.2)\cos(1.0996) + (-0.39764)^2(0.35)\cos(0.16507) \\ + (3.6964)^2(0.25)\cos(4.362617) \\ (5)^2(0.2)\sin(1.0996) + (-0.39764)^2(0.35)\sin(0.16507) \\ + (3.6964)^2(0.25)\sin(4.362617) \\ 0 \end{bmatrix}$$

$$= \begin{bmatrix} 1.153798 \\ 1.255206 \\ 0.0 \end{bmatrix}$$

The linear algebraic equations for the angular accelerations can then be written as

$$\begin{bmatrix} -0.1782 & -0.05751 & 0.23486 \\ 0.09079 & 0.34524 & -0.08567 \\ 1.0 & 0.0 & 0.0 \end{bmatrix} \begin{bmatrix} \ddot{\theta}^2 \\ \ddot{\theta}^3 \\ \ddot{\theta}^4 \end{bmatrix} = - \begin{bmatrix} 1.153798 \\ 1.255206 \\ 0.0 \end{bmatrix}$$

The solution of this system yields

$$[\ddot{\theta}^2 \quad \ddot{\theta}^3 \quad \ddot{\theta}^4]^{\mathrm{T}} = [0 \quad 5.1689 \quad 6.1784]^{\mathrm{T}} \ \mathrm{rad/s^2}$$

Choice of the Coordinates The choice of the coordinates θ^2, θ^3, and θ^4 in the preceding example was made by examining the topological structure of the four-bar mechanism. If another mechanism is considered, this set of coordinates is no longer suitable, and different constraint equations that take different forms must be formulated. For this reason, the choice of the coordinates based on the topological structure of the system under consideration makes it difficult to develop a general-purpose computer program for the kinematic analysis of multibody systems.

As discussed before, an alternative method for choosing the coordinates of the four-bar mechanism shown in Fig. 33 is to assume that the mechanism consists of four bodies, the fixed link or ground (body 1), the crankshaft (body 2), the coupler (body 3), and the rocker (body 4). Every body is assigned an identical set of coordinates that we select to be the absolute Cartesian coordinates R_x^i, R_y^i, and θ^i ($i = 1, 2, 3, 4$). As the result of this choice, the configuration of the mechanism is described using 12 coordinates, and consequently, the number of constraint equations increases. In this case, we have 11 joint constraints which include three ground constraints and eight constraints resulting from the revolute joints at O, A, B, and C. The formulation for the ground constraints can be simply written as

$$R_x^1 = R_y^1 = \theta^1 = 0 \tag{3.128}$$

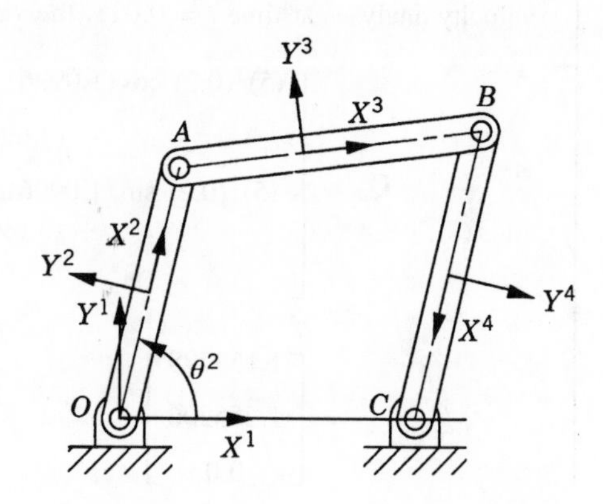

Figure 3.33 Four-bar mechanism

while the revolute joint constraints can be written as

$$\left.\begin{aligned}\mathbf{R}^1 + \mathbf{A}^1\bar{\mathbf{u}}_O^1 - \mathbf{R}^2 - \mathbf{A}^2\bar{\mathbf{u}}_O^2 &= \mathbf{0}\\ \mathbf{R}^2 + \mathbf{A}^2\bar{\mathbf{u}}_A^2 - \mathbf{R}^3 - \mathbf{A}^3\bar{\mathbf{u}}_A^3 &= \mathbf{0}\\ \mathbf{R}^3 + \mathbf{A}^3\bar{\mathbf{u}}_B^3 - \mathbf{R}^4 - \mathbf{A}^4\bar{\mathbf{u}}_B^4 &= \mathbf{0}\\ \mathbf{R}^4 + \mathbf{A}^4\bar{\mathbf{u}}_C^4 - \mathbf{R}^1 - \mathbf{A}^1\bar{\mathbf{u}}_C^1 &= \mathbf{0}\end{aligned}\right\} \tag{3.129}$$

where $\bar{\mathbf{u}}_O^i$, $\bar{\mathbf{u}}_A^i$, $\bar{\mathbf{u}}_B^i$, and $\bar{\mathbf{u}}_C^i$ are the local position vectors of points O, A, B, and C, respectively, defined with respect to the selected coordinate system of body i, $\mathbf{A}^i$ is the planar transformation matrix from the body i coordinate system to the global coordinate system, and $\mathbf{R}^i = [R_x^i \quad R_y^i]^{\mathrm{T}}$ is the global position vector of the origin of the body coordinate system. Note that the loop closure equations of Eq. 127 can be obtained by adding the joint constraints of Eq. 129.

The use of the three absolute coordinates R_x^i, R_y^i, and θ^i for each body in the system, however, makes it possible to develop a general-purpose computer program for the dynamic analysis of multibody systems. In order to formulate the constraints of Eqs. 128 and 129, one only needs to identify which body is the ground, the bodies connected by the revolute joints, and the local position vectors of the joint definition points at O, A, B, and C. The local position vectors $\bar{\mathbf{u}}_O^i$, $\bar{\mathbf{u}}_A^i$, $\bar{\mathbf{u}}_B^i$, $\bar{\mathbf{u}}_C^i$ are constant in the body coordinate systems and their values remain constant during the simulation time. Also note that the form of the constraint equations in this case does not depend on the topological structure of the mechanism. In fact, links can be added or removed from this mechanism by simply adding or deleting some of the kinematic constraints.

Standard Constraint Library If three absolute coordinates are used for each body in the mechanical system, the kinematic algebraic constraint equations of the joints can be formulated in a general form. This general formulation for the joint constraints does not depend on a specific application problem and can be used in the computer-aided kinematic analysis of varieties of multibody systems that consist of interconnected rigid bodies. Standard formulations for typical joints such as the ground constraints as well as revolute and prismatic joints can be implemented on the digital computer and made available for the analysis of a large class of multibody systems. For example, the kinematic constraints of a revolute joint between any pair of two rigid bodies i and j are given by

$$\mathbf{C}(\mathbf{q}, t) = \mathbf{R}^i + \mathbf{A}^i\bar{\mathbf{u}}_P^i - \mathbf{R}^j - \mathbf{A}^j\bar{\mathbf{u}}_P^j = \mathbf{0} \tag{3.130}$$

where P is the joint definition point. If one provides the body numbers, and the two constant vectors $\bar{\mathbf{u}}_P^i$ and $\bar{\mathbf{u}}_P^j$, a general purpose routine can be used to automatically generate Eq. 130. Note that the constraints of Eq. 129 are in the same form as Eq. 130 except for the superscripts, which indicate the body numbers.

The position, velocity, and acceleration analyses require the evaluation of the constraint Jacobian matrix. The Jacobian matrix of constraints such as the ground as well as revolute and prismatic joints can be also made as part of a standard constraint library in a general-purpose computer program. For instance, the Jacobian matrix of the revolute joint constraints of Eq. 130 can be written using vector notation as

$$\mathbf{C_q} = \frac{\partial \mathbf{C}}{\partial \mathbf{q}} = \left[\frac{\partial \mathbf{C}}{\partial \mathbf{R}^i} \; \frac{\partial \mathbf{C}}{\partial \theta^i} \; \frac{\partial \mathbf{C}}{\partial \mathbf{R}^j} \; \frac{\partial \mathbf{C}}{\partial \theta^j} \right] = [\mathbf{I} \quad \mathbf{A}^i_\theta \bar{\mathbf{u}}^i_P \quad -\mathbf{I} \quad -\mathbf{A}^j_\theta \bar{\mathbf{u}}^j_P] \tag{3.131}$$

where $\mathbf{I}$ is the 2×2 identity matrix, and $\mathbf{A}^i_\theta$ and $\mathbf{A}^j_\theta$ are the partial derivatives of $\mathbf{A}^i$ and $\mathbf{A}^j$ with respect to the rotational coordinates of bodies i and j, respectively. It is clear that the form of the Jacobian matrix of Eq. 131 remains the same for any pair of two rigid bodies connected by a revolute joint and this equation can be used for an arbitrary revolute joint in a multibody system. One only has to change the superscripts, which indicate the body numbers, and the constant vectors that define the local positions of the joint definition points.

For the velocity analysis, one has to evaluate the vector $\mathbf{C}_t$ of Eq. 121. This vector is equal to zero in the case of the revolute joint constraints of Eq. 130 because these constraints are not explicit functions of time. For the acceleration analysis, one must evaluate the vector $\mathbf{Q}_d$ of Eq. 124. In the case of revolute joint, this vector reduces to

$$\mathbf{Q}_d = -(\mathbf{C_q}\dot{\mathbf{q}})_\mathbf{q}\dot{\mathbf{q}} \tag{3.132}$$

By using the Jacobian matrix of Eq. 131, one can show that

$$\mathbf{Q}_d = \mathbf{A}^i \bar{\mathbf{u}}^i_P (\dot{\theta}^i)^2 - \mathbf{A}^j \bar{\mathbf{u}}^j_P (\dot{\theta}^j)^2 \tag{3.133}$$

This equation has the same form for any pair of rigid bodies connected by a revolute joint.

Example 3.14

The constraint equations of the *prismatic joint* were defined by Eq. 97 as

$$\mathbf{C} = \begin{bmatrix} C_1 \\ C_2 \end{bmatrix} = \begin{bmatrix} \theta^i - \theta^j - c \\ \mathbf{h}^{i^{\mathrm{T}}} \mathbf{r}^{ij}_P \end{bmatrix} = \begin{bmatrix} 0 \\ 0 \end{bmatrix}$$

where

$$\mathbf{h}^i = \mathbf{A}^i \bar{\mathbf{h}}^i$$

$$\mathbf{r}^{ij}_P = \mathbf{R}^i + \mathbf{A}^i \bar{\mathbf{u}}^i_P - \mathbf{R}^j - \mathbf{A}^j \bar{\mathbf{u}}^j_P$$

in which

$$\bar{\mathbf{h}}^i = (\bar{\mathbf{u}}^i_P - \bar{\mathbf{u}}^i_Q)$$

The nonzero elements of the Jacobian matrix are

$$C_{13} = \frac{\partial C_1}{\partial \theta^i} = 1, \quad C_{16} = \frac{\partial C_1}{\partial \theta^j} = -1$$

$$C_{21} = \frac{\partial C_2}{\partial R_x^i} = h_x^i, \quad C_{22} = \frac{\partial C_2}{\partial R_y^i} = h_y^i$$

$$C_{23} = \frac{\partial C_2}{\partial \theta^i} = \mathbf{r}_P^{ij^{\mathrm{T}}} \mathbf{A}_\theta^i \overline{\mathbf{h}}^i + \mathbf{h}^{i^{\mathrm{T}}} \mathbf{A}_\theta^i \overline{\mathbf{u}}_P^i$$

$$C_{24} = \frac{\partial C_2}{\partial R_x^j} = -h_x^i, \quad C_{25} = \frac{\partial C_2}{\partial R_y^j} = -h_y^i$$

$$C_{26} = \frac{\partial C_2}{\partial \theta^j} = -\mathbf{h}^{i^{\mathrm{T}}} \mathbf{A}_\theta^j \overline{\mathbf{u}}_P^j$$

where h_x^i and h_y^i are the components of the vector $\mathbf{h}^i$, that is,

$$\mathbf{h}^i = [h_x^i \quad h_y^i]^{\mathrm{T}}$$

The Jacobian matrix of the prismatic joint can then be defined as

$$\mathbf{C_q} = \begin{bmatrix} \frac{\partial C_1}{\partial R_x^i} & \frac{\partial C_1}{\partial R_y^i} & \frac{\partial C_1}{\partial \theta^i} & \frac{\partial C_1}{\partial R_x^j} & \frac{\partial C_1}{\partial R_y^j} & \frac{\partial C_1}{\partial \theta^j} \\ \frac{\partial C_2}{\partial R_x^i} & \frac{\partial C_2}{\partial R_y^i} & \frac{\partial C_2}{\partial \theta^i} & \frac{\partial C_2}{\partial R_x^j} & \frac{\partial C_2}{\partial R_y^j} & \frac{\partial C_2}{\partial \theta^j} \end{bmatrix}$$

$$= \begin{bmatrix} 0 & 0 & C_{13} & 0 & 0 & C_{16} \\ C_{21} & C_{22} & C_{23} & C_{24} & C_{25} & C_{26} \end{bmatrix}$$

Since the prismatic joint constraints are not explicit functions of time, one has $\mathbf{C}_t = \mathbf{0}$. It follows that

$$\mathbf{C}_{tt} = \mathbf{0}, \quad \mathbf{C}_{\mathbf{q}t} = \mathbf{0}$$

and

$$\mathbf{C_q}\dot{\mathbf{q}} = \begin{bmatrix} \dot{\theta}^i - \dot{\theta}^j \\ d_1 \end{bmatrix}$$

where

$$d_1 = \mathbf{h}^{i^{\mathrm{T}}}(\dot{\mathbf{R}}^i - \dot{\mathbf{R}}^j) + \dot{\theta}^i [\mathbf{r}_P^{ij^{\mathrm{T}}} \mathbf{A}_\theta^i \overline{\mathbf{h}}^i + \mathbf{h}^{i^{\mathrm{T}}} \mathbf{A}_\theta^i \overline{\mathbf{u}}_P^i] - \dot{\theta}^j \mathbf{h}^{i^{\mathrm{T}}} \mathbf{A}_\theta^j \overline{\mathbf{u}}_P^j$$

Hence,

$$(\mathbf{C_q}\dot{\mathbf{q}})_\mathbf{q} = \begin{bmatrix} 0 & 0 & 0 & 0 & 0 & 0 \\ \overline{C}_{21} & \overline{C}_{22} & \overline{C}_{23} & \overline{C}_{24} & \overline{C}_{25} & \overline{C}_{26} \end{bmatrix}$$

where

$$\begin{bmatrix} \overline{C}_{21} \\ \overline{C}_{22} \end{bmatrix} = \dot{\theta}^i \mathbf{A}_\theta^i \overline{\mathbf{h}}^i$$

$$\overline{C}_{23} = (\mathbf{A}_\theta^i \overline{\mathbf{h}}^i)^{\mathrm{T}} (\dot{\mathbf{R}}^i - \dot{\mathbf{R}}^j) + \dot{\theta}^i [\overline{\mathbf{u}}_P^{i^{\mathrm{T}}} \overline{\mathbf{h}}^i - \mathbf{r}_P^{ij^{\mathrm{T}}} \mathbf{h}^i] - \dot{\theta}^j (\mathbf{A}_\theta^i \overline{\mathbf{h}}^i)^{\mathrm{T}} (\mathbf{A}_\theta^j \overline{\mathbf{u}}_P^j)$$

$$\begin{bmatrix} \overline{C}_{24} \\ \overline{C}_{25} \end{bmatrix} = -\dot{\theta}^i \mathbf{A}_\theta^i \overline{\mathbf{h}}^i$$

$$\overline{C}_{26} = -\dot{\theta}^i (\mathbf{A}_\theta^j \overline{\mathbf{u}}_P^j)^{\mathrm{T}} \mathbf{A}_\theta^i \overline{\mathbf{h}}^i + \dot{\theta}^j \mathbf{h}^{i^{\mathrm{T}}} \mathbf{u}_P^j$$

in which

$$\mathbf{u}_P^i = \mathbf{A}^i \overline{\mathbf{u}}_P^i, \qquad \mathbf{u}_P^j = \mathbf{A}^j \overline{\mathbf{u}}_P^j$$

The vector $\mathbf{Q}_d$ can then be defined as

$$\mathbf{Q}_d = \begin{bmatrix} Q_{d1} \\ Q_{d2} \end{bmatrix} = -(\mathbf{C_q}\dot{\mathbf{q}})_{\mathbf{q}}\dot{\mathbf{q}}$$

$$= -\begin{bmatrix} 0 \\ \overline{C}_{21}\dot{R}_x^i + \overline{C}_{22}\dot{R}_y^i + \overline{C}_{23}\dot{\theta}^i + \overline{C}_{24}\dot{R}_x^j + \overline{C}_{25}\dot{R}_y^j + \overline{C}_{26}\dot{\theta}^j \end{bmatrix}$$

where Q_{d2} can be written explicitly as

$$Q_{d2} = -2\dot{\theta}^i (\mathbf{A}_\theta^i \overline{\mathbf{h}}^i)^{\mathrm{T}} (\dot{\mathbf{R}}^i - \dot{\mathbf{R}}^j) - (\dot{\theta}^i)^2 [\overline{\mathbf{u}}_P^{i^{\mathrm{T}}} \overline{\mathbf{h}}^i - \mathbf{r}_P^{ij^{\mathrm{T}}} \mathbf{h}^i]$$

$$+ 2\dot{\theta}^i \dot{\theta}^j (\mathbf{A}_\theta^i \overline{\mathbf{h}}^i)^{\mathrm{T}} \mathbf{A}_\theta^j \overline{\mathbf{u}}_P^j - (\dot{\theta}^j)^2 \mathbf{h}^{i^{\mathrm{T}}} \mathbf{u}_P^j$$

Computer Algorithm It is clear from the discussion presented in this section that the kinematic constraint equations of joints, such as revolute, prismatic, and other joints that are commonly used in multibody systems, can be derived systematically in a general form in terms of the absolute Cartesian coordinates. These joints can be made available as standard elements in a general-purpose computer program that can be used for the kinematic analysis of varieties of multibody system applications. The driving constraints, on the other hand, depend on the application and can be introduced to the general-purpose computer program by using *user subroutines*.

In what follows, a numerical algorithm that can be implemented in a general-purpose computer program for the position, velocity, and acceleration analyses of multibody systems is presented. The basic kinematic equations used in this numerical algorithm are first summarized.

In the position analysis, one needs to solve the system of nonlinear algebraic constraint equations

$$\mathbf{C}(\mathbf{q}, t) = \mathbf{0} \tag{3.134}$$

As pointed out in this chapter, a Newton–Raphson algorithm can be used for solving this equation. Thus, one must construct the matrix equation

$$\mathbf{C_q}\ \Delta\mathbf{q} = -\mathbf{C} \tag{3.135}$$

where $\Delta\mathbf{q}$ is the vector of Newton differences.

The basic equations used in the velocity and acceleration analyses are

$$\mathbf{C_q}\dot{\mathbf{q}} = -\mathbf{C}_t \tag{3.136}$$

$$\mathbf{C_q}\ddot{\mathbf{q}} = \mathbf{Q}_d \tag{3.137}$$

where the vector $\mathbf{Q}_d$ is defined by Eq. 125.

The computational scheme for the analysis of *kinematically driven* systems that consist of interconnected rigid bodies is shown in Fig. 34 and proceeds in the following routine.

Step 1 At a given point in time t, an estimate for the desired solution is made. This estimate must be close to the exact solution in order to avoid divergence.

Step 2 The Jacobian matrix $\mathbf{C_q}$ and the vector of constraint equations $\mathbf{C}$ of Eq. 135 can be evaluated.

Step 3 Equation 135 is solved for the vector of Newton differences $\Delta\mathbf{q}$.

Step 4 If the norm of the vector $\Delta\mathbf{q}$ or the norm of the vector of constraint equations $\mathbf{C}$ is small and less than specified tolerances (Eq. 118), proceed to step 5. Otherwise, update the vector of coordinates, that is, $\mathbf{q} = \mathbf{q} + \Delta\mathbf{q}$ and go to step 2 if the number of specified iterations is not exceeded.

Step 5 Having determined the vector of system coordinates, this vector can be used to evaluate the Jacobian matrix $\mathbf{C_q}$ and the vector $\mathbf{C}_t$ of Eq. 136.

Step 6 Equation 136 is a linear system of algebraic equations in the velocity vector. This system of equations can be solved for the vector $\dot{\mathbf{q}}$.

Step 7 Using the vectors $\mathbf{q}$ and $\dot{\mathbf{q}}$ determined from the position and velocity analyses, evaluate the Jacobian matrix $\mathbf{C_q}$ and the vector $\mathbf{Q}_d$ of Eq. 137.

Step 8 Equation 137 is a linear system of algebraic equations in the acceleration vector. This equation can be solved for the vector $\ddot{\mathbf{q}}$.

Step 9 Steps 1 through 8 are repeated until the simulation time ends.

For the most part, in a well-posed problem, there are no numerical problems encountered in the solution for the velocities and accelerations because only the solution of a linear system of equations is required. If the constraints are linearly independent, the constraint Jacobian matrix is a nonsingular square matrix, and consequently, there is a unique solution for Eqs. 136 and 137. Some numerical problems, however, may be encountered in the

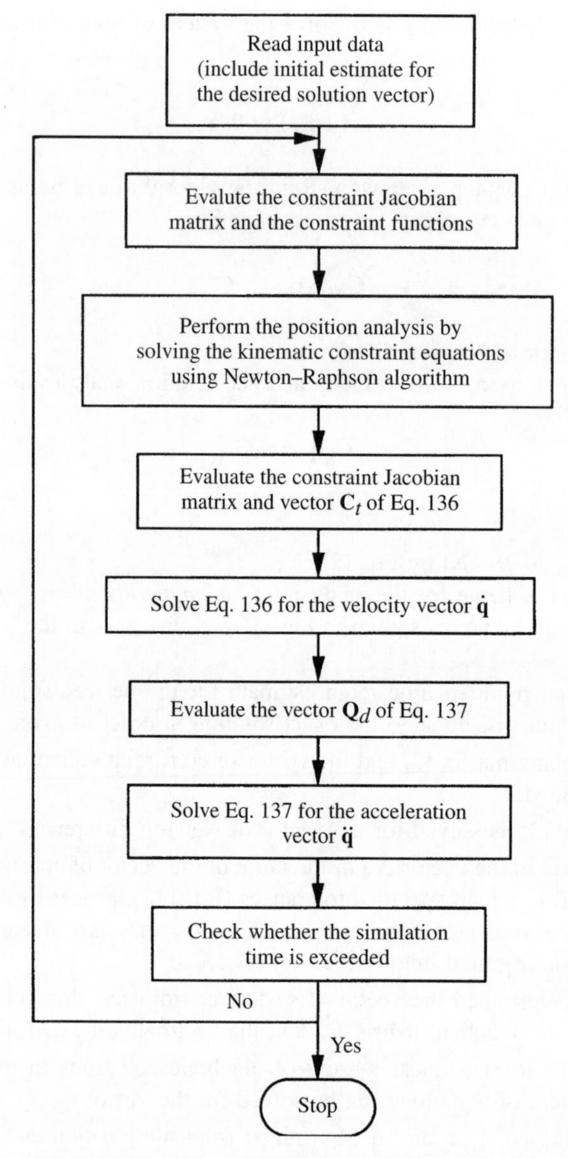

Figure 3.34 Flowchart for the kinematic analysis

position analysis since an iterative scheme is required for the solution of a nonlinear system of equations. As pointed out previously, the Newton–Raphson algorithm may diverge if the assumed initial guess at each time step is not close enough to the exact solution. The use of the solution obtained in a previous step as the initial guess for the iterative Newton–Raphson algorithm at the current step may require the use of a very small step size in order to achieve convergence. One method for obtaining an improved initial approximation for the solution at a given step is to use the velocity and acceleration vectors obtained in the previous step to predict the solution. For example, the following truncated Taylor series expressed in terms of the coordinates, velocities, and accelerations at the previous step can be used:

$$\overline{\mathbf{q}}_{i+1} = \mathbf{q}_i + \Delta t \dot{\mathbf{q}}^i + \frac{(\Delta t)^2}{2} \ddot{\mathbf{q}}^i \tag{3.138}$$

where $\mathbf{q}_i$, $\dot{\mathbf{q}}_i$, and $\ddot{\mathbf{q}}_i$ are the coordinate, velocity, and acceleration vectors determined at step i, Δt is the time step, and $\overline{\mathbf{q}}_{i+1}$ is the improved initial guess for the solution at step $i+1$. The use of Eq. 138 to obtain the initial guess for the Newton–Raphson algorithm may significantly reduce the number of iterations required to achieve convergence.

3.12 KINEMATIC MODELING AND ANALYSIS

In this section, the single-degree-of-freedom slider crank mechanism shown in Fig. 35 is used to demonstrate the use of the computer methods presented in the preceding section in the kinematic analysis of mechanism systems. The mechanism consists of four bodies; the fixed link denoted as body 1, the crankshaft *OA* denoted as body 2, the connecting rod *AB* denoted as body 3, and the slider block *B* denoted as body 4. Bodies 1 and 2 are connected by a revolute joint at O, bodies 2 and 3 are connected by a revolute joint at A, and bodies 3 and 4 are connected by a revolute joint at B. Bodies 1 and 4 are connected by a prismatic joint. The length of the crankshaft is assumed to be 0.15 m and the length of the connecting rod is assumed to be 0.35 m. As shown in Fig. 36 a body fixed coordinate system is attached to every body in the system including the ground. The vector of absolute coordinates of the slider crank mechanism is given by

$$\mathbf{q} = [R_x^1 \quad R_y^1 \quad \theta^1 \quad R_x^2 \quad R_y^2 \quad \theta^2 \quad R_x^3 \quad R_y^3 \quad \theta^3 \quad R_x^4 \quad R_y^4 \quad \theta^4]^{\mathrm{T}}$$

Initially, the crankshaft is assumed to make an angle $\theta_0^2 = 30^\circ$. The local position vectors of the revolute joint definition points are as follows:

$$\begin{aligned}
\overline{\mathbf{u}}_O^1 &= [0.0 \quad 0.0]^{\mathrm{T}}, & \overline{\mathbf{u}}_O^2 &= [-0.075 \quad 0.0]^{\mathrm{T}} \\
\overline{\mathbf{u}}_A^2 &= [0.075 \quad 0.0]^{\mathrm{T}}, & \overline{\mathbf{u}}_A^3 &= [-0.175 \quad 0.0]^{\mathrm{T}} \\
\overline{\mathbf{u}}_B^3 &= [0.175 \quad 0.0]^{\mathrm{T}}, & \overline{\mathbf{u}}_B^4 &= [0.0 \quad 0.0]^{\mathrm{T}}
\end{aligned}$$

In this example, the prismatic constraints between bodies 1 and 4 reduce to

$$R_y^4 = 0, \quad \theta^4 = 0$$

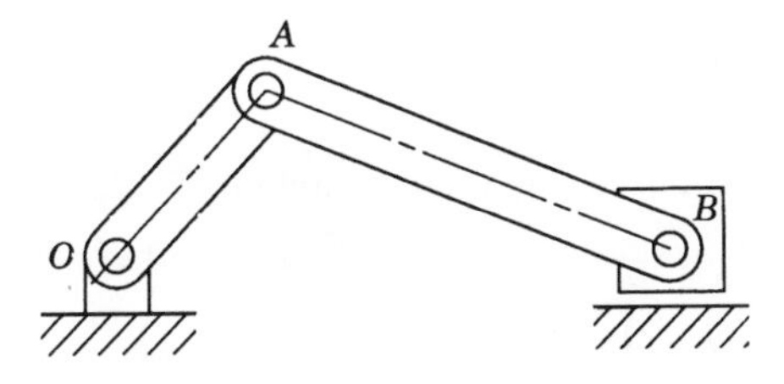

Figure 3.35 Slider crank mechanism

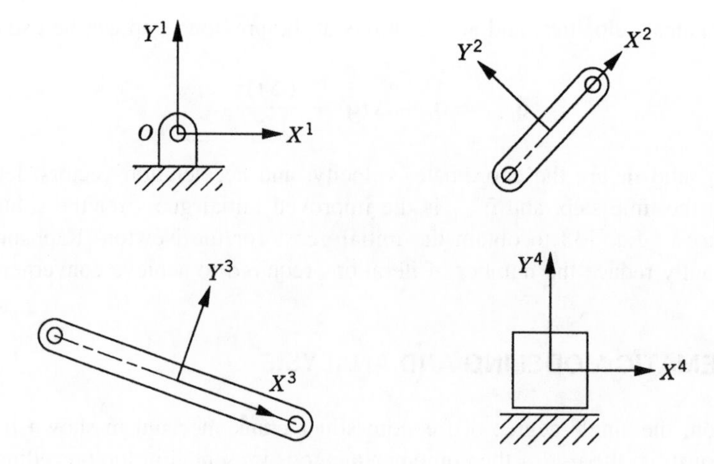

Figure 3.36 Body coordinate systems

It is clear that the total number of joint constraints is 11 (three ground constraints, six revolute joint constraints, and two prismatic joint constraints). Recall that in the kinematic analysis, the number of coordinates must be equal to the number of constraint equations so as to have a kinematically driven system. There are two alternatives for introducing a driving constraint for this mechanism. The first alternative is to drive the mechanism by rotating the crankshaft, while in the second alternative, the mechanism is driven by moving the slider block. In each case, one driving constraint can be defined. This driving constraint, when it is added to the joint constraints, makes the total number of kinematic constraint equations equal to the total number of absolute coordinates of the mechanism. The simulation results presented in this section are obtained using the general purpose multibody computer program **SAMS/2000** (**S**ystematic **A**nalysis of **M**ultibody **S**ystems) which is described in Chapter 9 of this book.

Prescribed Rotation of the Crankshaft First, we consider the case where the angular velocity of the crankshaft is prescribed as

$$\dot{\theta}^2 = f(t)$$

where $f(t)$ is a specified function of time. This equation can also be written as

$$d\theta^2 = f(t)\ dt$$

which upon integration yields

$$\theta^2 - \theta_0^2 = \int_0^t f(t)\ dt$$

where θ_0^2 is the angular orientation of the crankshaft at the initial configuration. The preceding equation can be written as

$$C_d = \theta^2 - \theta_0^2 - g(t) = 0$$

where C_d is the driving constraint and

$$g(t) = \int_0^t f(t)\, dt$$

Using the definition of the ground and revolute joint constraints presented in the preceding sections and the special form of the prismatic joint and driving constraints presented in this section, the vector of the constraint equations of the slider crank mechanism can be written as

$$\mathbf{C}(\mathbf{q}, t) = \begin{bmatrix} R_x^1 \\ R_y^1 \\ \theta^1 \\ \mathbf{R}^2 + \mathbf{A}^2\bar{\mathbf{u}}_O^2 \\ \mathbf{R}^2 + \mathbf{A}^2\bar{\mathbf{u}}_A^2 - \mathbf{R}^3 - \mathbf{A}^3\bar{\mathbf{u}}_A^3 \\ \mathbf{R}^3 + \mathbf{A}^3\bar{\mathbf{u}}_B^3 - \mathbf{R}^4 - \mathbf{A}^4\bar{\mathbf{u}}_B^4 \\ R_y^4 \\ \theta^4 \\ \theta^2 - \theta_O^2 - g(t) \end{bmatrix} = \mathbf{0}$$

in which the first three constraints are the ground constraints, the fourth to the ninth constraints represented by the three vector equations are the revolute joint constraints, the tenth and eleventh constraints are the prismatic joint constraints, and the twelfth constraint is the driving constraint.

The Newton–Raphson iterative procedure used for the position analysis requires the evaluation of the constraint Jacobian matrix, which is defined for this example as

$$\mathbf{C_q} = \begin{bmatrix} 1 & 0 & 0 & 0 & 0 & 0 & 0 & 0 & 0 & 0 & 0 & 0 \\ 0 & 1 & 0 & 0 & 0 & 0 & 0 & 0 & 0 & 0 & 0 & 0 \\ 0 & 0 & 1 & 0 & 0 & 0 & 0 & 0 & 0 & 0 & 0 & 0 \\ 0 & 0 & 0 & 1 & 0 & C_{4,6} & 0 & 0 & 0 & 0 & 0 & 0 \\ 0 & 0 & 0 & 0 & 1 & C_{5,6} & 0 & 0 & 0 & 0 & 0 & 0 \\ 0 & 0 & 0 & 1 & 0 & C_{6,6} & -1 & 0 & C_{6,9} & 0 & 0 & 0 \\ 0 & 0 & 0 & 0 & 1 & C_{7,6} & 0 & -1 & C_{7,9} & 0 & 0 & 0 \\ 0 & 0 & 0 & 0 & 0 & 0 & 1 & 0 & C_{8,9} & -1 & 0 & 0 \\ 0 & 0 & 0 & 0 & 0 & 0 & 0 & 1 & C_{9,9} & 0 & -1 & 0 \\ 0 & 0 & 0 & 0 & 0 & 0 & 0 & 0 & 0 & 0 & 1 & 0 \\ 0 & 0 & 0 & 0 & 0 & 0 & 0 & 0 & 0 & 0 & 0 & 1 \\ 0 & 0 & 0 & 0 & 0 & 1 & 0 & 0 & 0 & 0 & 0 & 0 \end{bmatrix}$$

where

$$C_{4,6} = \frac{l^2}{2} \sin \theta^2, \qquad C_{5,6} = -\frac{l^2}{2} \cos \theta^2,$$

$$C_{6,6} = -\frac{l^2}{2} \sin \theta^2, \qquad C_{7,6} = \frac{l^2}{2} \cos \theta^2$$

$$C_{6,9} = -\frac{l^3}{2} \sin \theta^3, \qquad C_{7,9} = \frac{l^3}{2} \cos \theta^3,$$

$$C_{8,9} = -\frac{l^3}{2} \sin \theta^3, \qquad C_{9,9} = \frac{l^3}{2} \cos \theta^3$$

Using the Jacobian matrix and the vector of constraint equations, an iterative Newton–Raphson procedure can be used to determine the coordinates of the bodies of the slider crank mechanism. Figures 37a and 37b show, respectively, the angular orientation of the connecting rod and the displacement of the slider block as a function of the crank angle when the crankshaft rotates with a constant angular velocity equal to 150 rad/s. In this special case, $g(t) = 150t$.

For the velocity analysis, one needs to evaluate the vector $\mathbf{C}_t$, which is given in this case by

$$\mathbf{C}_t = [0 \quad 0 \quad 0 \quad 0 \quad 0 \quad 0 \quad 0 \quad 0 \quad 0 \quad 0 \quad 0 \quad f(t)]^{\mathrm{T}}$$

where $f(t) = 150$. Figures 38a and 38b show, respectively, the angular velocity of the connecting rod and the velocity of the slider block as functions of the crank angle.

For the acceleration analysis, the Jacobian matrix and the vector $\mathbf{Q}_d$ must be evaluated. The vector $\mathbf{Q}_d$ is

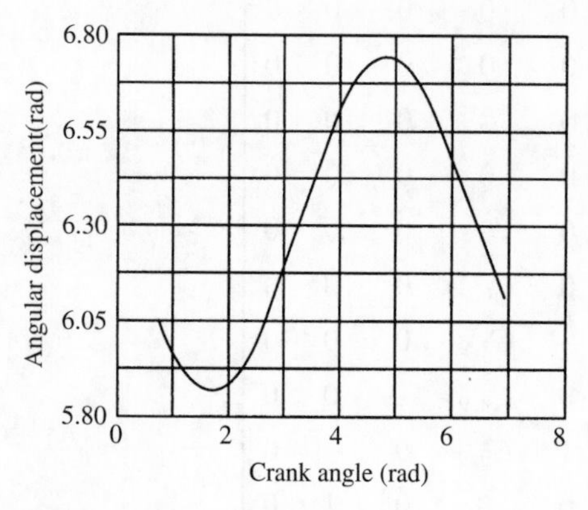

Figure 3.37a Orientation of the connecting rod

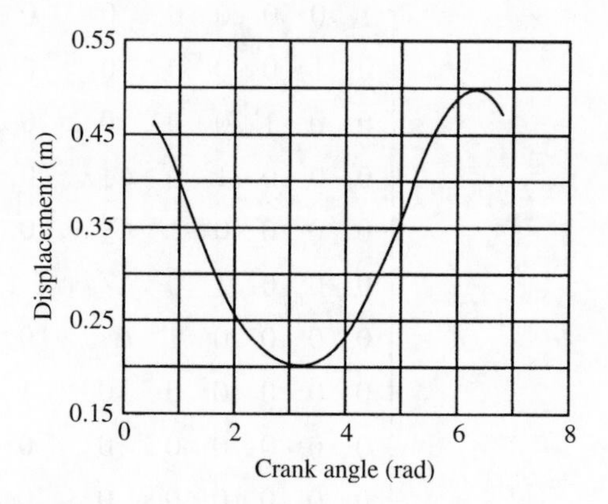

Figure 3.37b Displacement of the slider block

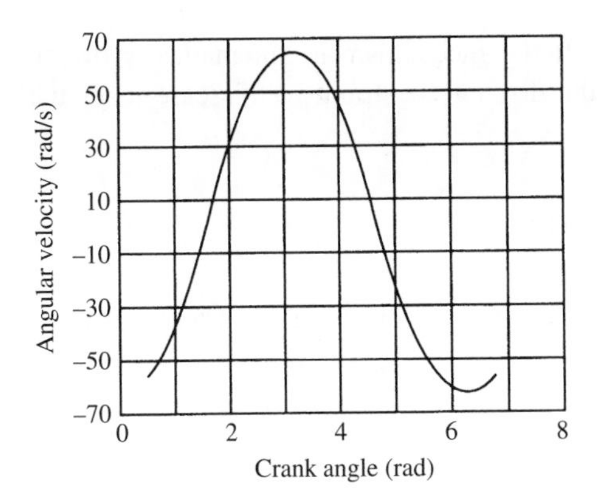

Figure 3.38a Angular velocity of the connecting rod

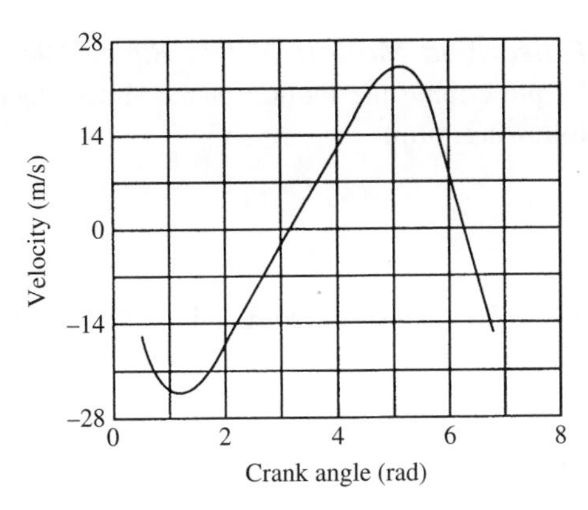

Figure 3.38b Velocity of the slider block

$$\mathbf{Q}_d = \begin{bmatrix} 0 \\ 0 \\ 0 \\ (\dot{\theta}^2)^2\mathbf{A}^2\overline{\mathbf{u}}_O^2 \\ (\dot{\theta}^2)^2\mathbf{A}^2\overline{\mathbf{u}}_A^2 - (\dot{\theta}^3)^2\mathbf{A}^3\overline{\mathbf{u}}_A^3 \\ (\dot{\theta}^3)^2\mathbf{A}^3\overline{\mathbf{u}}_B^3 - (\dot{\theta}^4)^2\mathbf{A}^4\overline{\mathbf{u}}_B^4 \\ 0 \\ 0 \\ 0 \end{bmatrix}$$

Using this vector and the Jacobian matrix, Eq. 137 can be solved for the accelerations. Figures 39a and 39b show, respectively, the angular acceleration of the connecting rod and the acceleration of the slider block as functions of the crank angle.

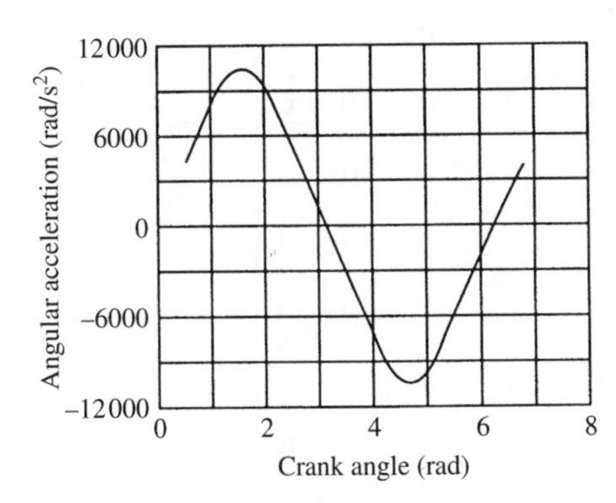

Figure 3.39a Angular acceleration of the connecting rod

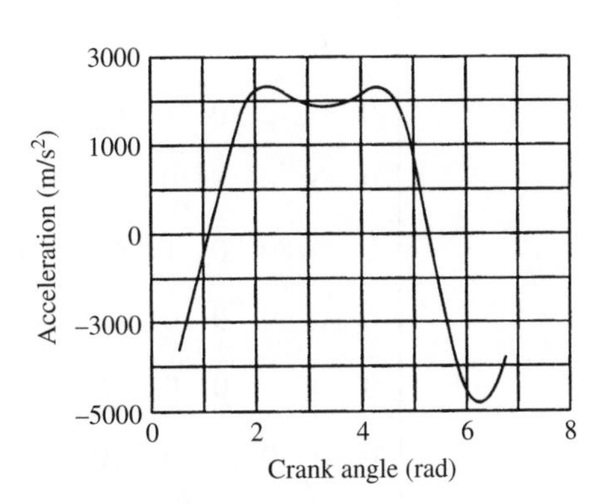

Figure 3.39b Acceleration of the slider block

Prescribed Motion of the Slider Block If the mechanism is kinematically driven by prescribing the motion of the slider block, the driving constraint in this case takes the following form:

$$C_d = R_x^4 - f(t) = 0$$

where $f(t)$ is a specified function of time. In this case, the vector of the constraint equations can be written as

$$\mathbf{C}(\mathbf{q}, t) = \begin{bmatrix} R_x^1 \\ R_y^1 \\ \theta^1 \\ \mathbf{R}^2 + \mathbf{A}^2\bar{\mathbf{u}}_O^2 \\ \mathbf{R}^2 + \mathbf{A}^2\bar{\mathbf{u}}_A^2 - \mathbf{R}^3 - \mathbf{A}^3\bar{\mathbf{u}}_A^3 \\ \mathbf{R}^3 + \mathbf{A}^3\bar{\mathbf{u}}_B^3 - \mathbf{R}^4 - \mathbf{A}^4\bar{\mathbf{u}}_B^4 \\ R_y^4 \\ \theta^4 \\ R_x^4 - f(t) \end{bmatrix} = \mathbf{0}$$

and the Jacobian matrix of the kinematic constraints is

$$\mathbf{C_q} = \begin{bmatrix} 1 & 0 & 0 & 0 & 0 & 0 & 0 & 0 & 0 & 0 & 0 & 0 \\ 0 & 1 & 0 & 0 & 0 & 0 & 0 & 0 & 0 & 0 & 0 & 0 \\ 0 & 0 & 1 & 0 & 0 & 0 & 0 & 0 & 0 & 0 & 0 & 0 \\ 0 & 0 & 0 & 1 & 0 & C_{4,6} & 0 & 0 & 0 & 0 & 0 & 0 \\ 0 & 0 & 0 & 0 & 1 & C_{5,6} & 0 & 0 & 0 & 0 & 0 & 0 \\ 0 & 0 & 0 & 1 & 0 & C_{6,6} & -1 & 0 & C_{6,9} & 0 & 0 & 0 \\ 0 & 0 & 0 & 0 & 1 & C_{7,6} & 0 & -1 & C_{7,9} & 0 & 0 & 0 \\ 0 & 0 & 0 & 0 & 0 & 0 & 1 & 0 & C_{8,9} & -1 & 0 & 0 \\ 0 & 0 & 0 & 0 & 0 & 0 & 0 & 1 & C_{9,9} & 0 & -1 & 0 \\ 0 & 0 & 0 & 0 & 0 & 0 & 0 & 0 & 0 & 0 & 1 & 0 \\ 0 & 0 & 0 & 0 & 0 & 0 & 0 & 0 & 0 & 0 & 0 & 1 \\ 0 & 0 & 0 & 0 & 0 & 0 & 0 & 0 & 0 & 1 & 0 & 0 \end{bmatrix}$$

where the coefficients that appear in this matrix are the same as those defined in the case of the prescribed rotation of the crankshaft. The vectors $\mathbf{C}_t$ and $\mathbf{Q}_d$ are

$$\mathbf{C}_t = \begin{bmatrix} 0 & 0 & 0 & 0 & 0 & 0 & 0 & 0 & 0 & 0 & 0 & \dfrac{\partial f}{\partial t} \end{bmatrix}^{\mathrm{T}}$$

$$\mathbf{Q}_d = \begin{bmatrix} 0 \\ 0 \\ 0 \\ (\dot{\theta}^2)^2\mathbf{A}^2\overline{\mathbf{u}}_O^2 \\ (\dot{\theta}^2)^2\mathbf{A}^2\overline{\mathbf{u}}_A^2 - (\dot{\theta}^3)^2\mathbf{A}^3\overline{\mathbf{u}}_A^3 \\ (\dot{\theta}^3)^2\mathbf{A}^3\overline{\mathbf{u}}_B^3 - (\dot{\theta}^4)^2\mathbf{A}^4\overline{\mathbf{u}}_B^4 \\ 0 \\ 0 \\ \dfrac{\partial^2 f}{\partial t^2} \end{bmatrix}$$

Figures 40a and 40b show, respectively, the angular orientations of the crankshaft and the connecting rod as functions of time when

$$f(t) = 0.35 - 0.8l^2 \sin 150t$$

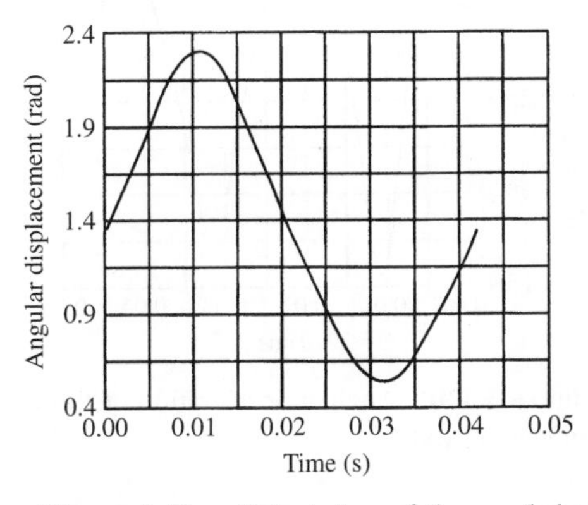

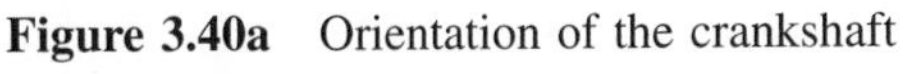

Figure 3.40a Orientation of the crankshaft

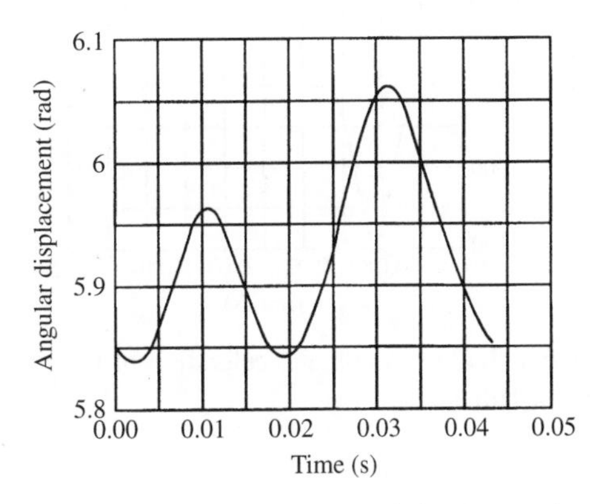

Figure 3.40b Orientation of the connecting rod

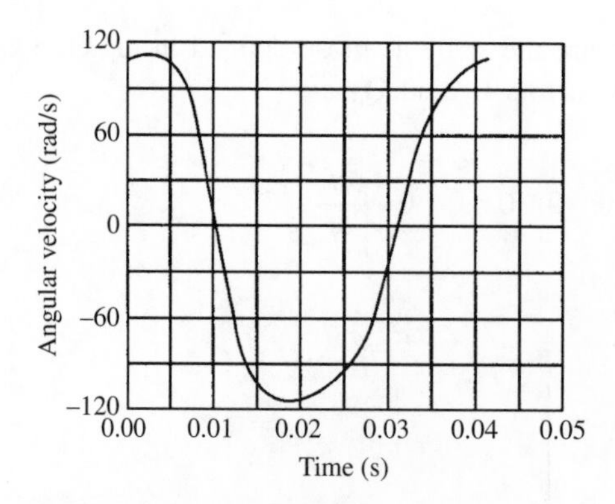

Figure 3.41a Angular velocity of the crankshaft

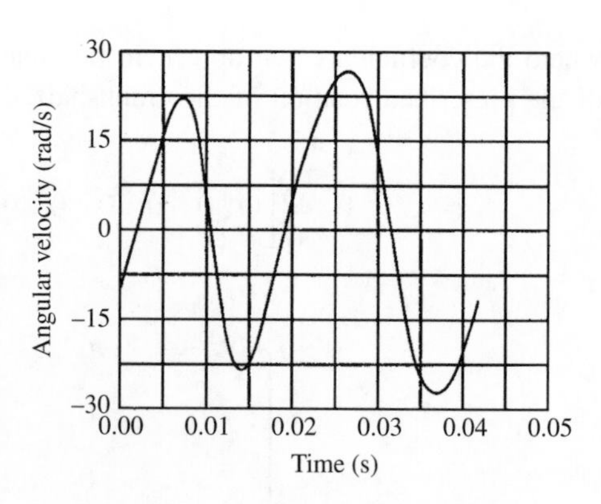

Figure 3.41b Angular velocity of the connecting rod

Figures 41a and 41b show, respectively, the angular velocities of the crankshaft and the connecting rod, while Figs. 42a and 42b show their angular accelerations as the result of the specified motion of the slider block.

At the special configuration in which $\theta^2 = \theta^3 = 0$, one has

$$C_{4,6} = C_{6,6} = C_{6,9} = C_{8,9} = 0$$

$$C_{5,6} = -C_{7,6} = -\frac{l^2}{2}, \quad C_{7,9} = C_{9,9} = \frac{l^3}{2}$$

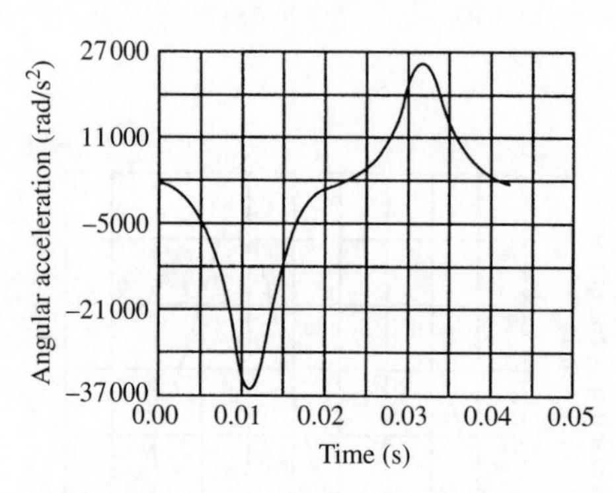

Figure 3.42a Angular acceleration of the crankshaft

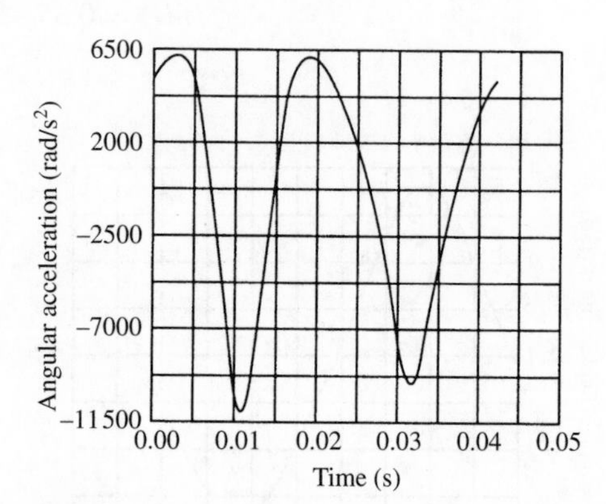

Figure 3.42b Angular acceleration of the connecting rod

and at this special configuration, the Jacobian matrix becomes

$$
\mathbf{C_q} = \begin{bmatrix}
1 & 0 & 0 & 0 & 0 & 0 & 0 & 0 & 0 & 0 & 0 & 0 \\
0 & 1 & 0 & 0 & 0 & 0 & 0 & 0 & 0 & 0 & 0 & 0 \\
0 & 0 & 1 & 0 & 0 & 0 & 0 & 0 & 0 & 0 & 0 & 0 \\
0 & 0 & 0 & 1 & 0 & 0 & 0 & 0 & 0 & 0 & 0 & 0 \\
0 & 0 & 0 & 0 & 1 & -\dfrac{l^2}{2} & 0 & 0 & 0 & 0 & 0 & 0 \\
0 & 0 & 0 & 1 & 0 & 0 & -1 & 0 & 0 & 0 & 0 & 0 \\
0 & 0 & 0 & 0 & 1 & \dfrac{l^2}{2} & 0 & -1 & \dfrac{l^3}{2} & 0 & 0 & 0 \\
0 & 0 & 0 & 0 & 0 & 0 & 1 & 0 & 0 & -1 & 0 & 0 \\
0 & 0 & 0 & 0 & 0 & 0 & 0 & 1 & \dfrac{l^3}{2} & 0 & -1 & 0 \\
0 & 0 & 0 & 0 & 0 & 0 & 0 & 0 & 0 & 0 & 1 & 0 \\
0 & 0 & 0 & 0 & 0 & 0 & 0 & 0 & 0 & 0 & 0 & 1 \\
0 & 0 & 0 & 0 & 0 & 0 & 0 & 0 & 0 & 1 & 0 & 0
\end{bmatrix}
$$

This matrix is singular since the sum of the sixth, eighth and twelfth rows is the fourth row. This is an indication that it is impossible to drive the mechanism at this *singular configuration* by specifying the motion of the slider block. This singularity, however, does not occur at this configuration when the mechanism is driven by specifying the angular orientation of the crankshaft.

3.13 CONCLUDING REMARKS

In this chapter, methods for the analysis of kinematically driven systems are presented. Two distinctive, yet equivalent, procedures can be recognized from the kinematic development presented in this chapter. The first is the classical approach, which is suited for the analysis of multibody systems that consist of small numbers of bodies and joints. The kinematic analysis using the classical approach, as demonstrated by Examples 3 through 6, starts by developing trigonometric relationships between the angles that define the orientation of the bodies in the system. These trigonometric relationships, which depend on the topological structure of the system, define a set of nonlinear algebraic equations that can be solved for the position variables. Once the position coordinates are defined, the kinematic velocity analysis starts by considering the velocities of a body in the system and their relationships to the velocities of other bodies as the result of the joint connections. By utilizing the topological structure of a given system, a set of linear algebraic constraint equations in the velocities can be determined and solved for the time derivatives of the coordinates. The acceleration analysis can be performed once the velocities are determined, and in the classical approach it heavily utilizes the particular topological structure of the system, as in the case of the position and velocity analyses.

The second approach presented in this chapter is more general and can be applied to a wide class of multibody system applications. In this approach, the nonlinear constraint equations that describe mechanical joints and specified motion trajectories of the system are formulated. In the case of kinematically driven systems, the resulting number of equations is equal to the number of the system coordinates. Therefore, these equations can be solved using numerical and computer methods to determine the system coordinates. By differentiating the constraint equations once and twice with respect to time, one obtains linear systems of algebraic equations in the velocities and accelerations, respectively. These equations can be solved in a straightforward manner to determine the first and second time derivatives of the system coordinates. This procedure for the kinematic analysis can be implemented on the digital computer and used in the kinematic analysis of varieties of multibody system applications as demonstrated in this chapter.

In the analysis of kinematically driven systems, it is assumed that all the system degrees of freedom are specified. In this case, one obtains a number of algebraic equations equal to the number of system coordinates, and therefore, a complete kinematic analysis can be performed without the need for developing the differential dynamic equations of motion, which are expressed in terms of the inertia, applied, and/or joint forces. If one or more of the degrees of freedom are not specified, the algebraic constraint equations are not sufficient to solve for the system coordinates. This is the case of *dynamically driven systems* whose analysis requires use of the *laws of motion* and formulation of the differential equations of the system as well as consideration of the force relationships in addition to the kinematic relationships. In the following chapters, different techniques for formulating the dynamic equations of constrained multibody systems are presented and the computer implementation of these techniques is discussed.

PROBLEMS

1. Figure P1 shows a rigid body i that has a body fixed coordinate system X^iY^i. The global position vector of the origin of the body coordinate system O^i is defined by the vector $\mathbf{R}^i$, and the orientation of the body i coordinate system in the global coordinate system is defined by the angle θ^i. The local position vector of point P^i on the body is defined by the vector $\bar{\mathbf{u}}_P^i$. If $\mathbf{R}^i = [5 \quad 3]^{\mathrm{T}}$ m, $\theta^i = 45^\circ$, and $\bar{\mathbf{u}}_P^i = [0.2 \quad 1.5]^{\mathrm{T}}$ m,

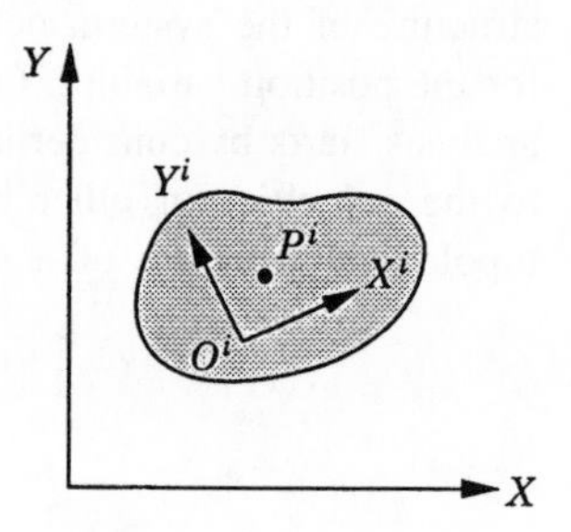

Figure P3.1

determine the global position vector of point P^i. If the body rotates with a constant angular velocity $\omega^i = 150$ rad/s, determine the absolute velocity of point P^i assuming that the absolute velocity of the reference point $\dot{\mathbf{R}}^i = [32 \; -10]^{\mathrm{T}}$ m/s.

2. In problem 1, let $\mathbf{R}^i = t[5 \quad 3]^{\mathrm{T}}$ m, $\theta^i = \omega t$, $\omega = 50$ rad/s, and $\bar{\mathbf{u}}^i_P = [0.2 \quad 1.5]^{\mathrm{T}}$ m, determine the global position vector of point P^i at $t = 0$, 0.25, and 1 s. Also determine the absolute velocity of point P^i at these points in time.

3. In the slider crank mechanism shown in Fig. P2, the lengths of the crankshaft OA and the connecting rod AB are, respectively, 0.3 and 0.5 m. The crankshaft is assumed to rotate with a constant angular velocity $\dot{\theta}^i = 100$ rad/s (counterclockwise). Assuming that the offset $h = 0$, use analytical methods to determine the orientation and angular velocity and acceleration of the connecting rod and the position, velocity, and acceleration of the slider block when the angular orientation θ^2 of the crankshaft is 45°. Also determine the absolute velocity and acceleration of the center of the connecting rod.

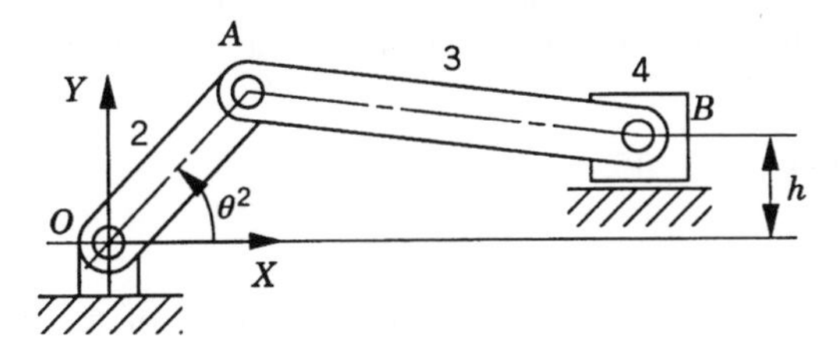

Figure P3.2

4. Repeat problem 3 assuming that the offset $h = 0.05$ m.

5. The lengths of the crankshaft OA and the connecting rod AB of the slider crank mechanism shown in Fig. P2 are 0.3 and 0.5 m, respectively, and the offset $h = 0.0$. The motion of the crankshaft is such that $\theta^2 = 150t + 3.0$ rad. Determine the orientation of the connecting rod and the position of the slider block at time $t = 0$, 0.01, and 0.03 s.

6. The lengths of the crankshaft OA, coupler AB, and rocker BC of the four-bar linkage shown in Fig. P3 are, respectively, 0.3, 0.35, and 0.4 m. The distance OC is 0.32 m. The crankshaft rotates with a constant angular velocity of 100 rad/s. Determine the orientation and angular velocity and acceleration of the coupler and the rocker when

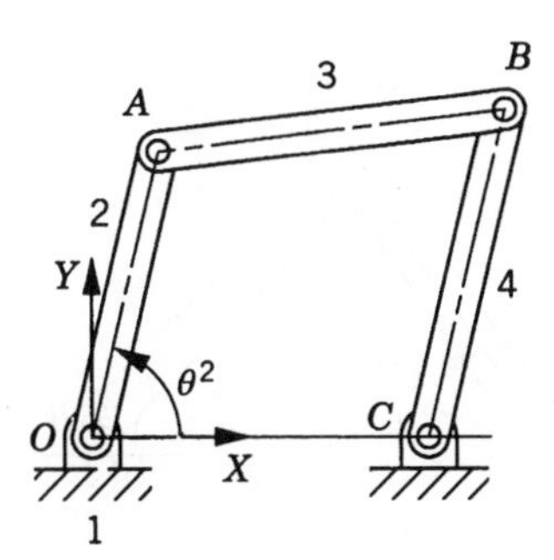

Figure P3.3

the orientation of the crankshaft $\theta^2 = 60°$. Also determine the orientation and angular velocity and angular acceleration of the crankshaft and the coupler when the orientation of the rocker $\theta^4 = 60°$.

7. Using Eq. 29, show that the velocities of two points A and B on a rigid body have equal components along the line AB.

8. Repeat problem 1 assuming that the absolute velocity of the reference point is zero. Prove in this case, by using vector algebra, that the absolute velocity of point P^i is perpendicular to the line O^iP^i, where O^i is the reference point.

9. Show that in the case of a general rigid body displacement, there exists a point on the rigid body whose velocity is instantaneously equal to zero. This point is called the *instantaneous center of rotation*.

10. In problem 3, let the velocity of the slider block be constant and equal to 5 m/s. Determine the angular velocities and angular accelerations of the crankshaft and the connecting rod. Also determine the absolute velocity and acceleration of the center of the connecting rod.

11. Prove the identities of Eq. 35.

12. Prove the identity of Eq. 49.

13. The motion of a rigid body i is such that the location of the origin of its reference is defined by the vector $\mathbf{R}^i = [t \quad 8(t)^3]^{\mathrm{T}}$ m, its angular velocity $\dot{\theta}^i = -150$ rad/s (clockwise), and its angular acceleration $\ddot{\theta}^i = 0$. Determine the position, velocity, and acceleration of a point P that moves with respect to the body such that its coordinates are defined in the body coordinate system by the vector $\bar{\mathbf{u}}_P^i = [1.5(t)^2 \quad -3(t)^3]^{\mathrm{T}}$ m. Find the solution at time $t = 0$, 1, and 1.5 s.

14. Solve problem 13 assuming that $\dot{\theta}^i = -150(t)^2$ rad/s.

15. Examine the singular configurations of the four-bar mechanism.

16. In the system shown in Fig. P4, $OA = 0.2$ m, $OB = 0.3$ m, and $BC = 0.6$ m. Link OA is assumed to rotate with an angular velocity 50 rpm counter-clockwise. Find the velocity and acceleration of point C and the angular velocity and acceleration of link BC.

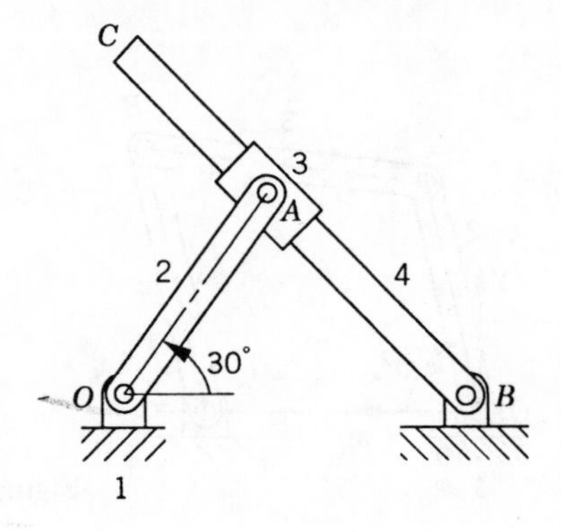

Figure P3.4

17. Figure P5 shows a gear system that consists of gears i, j, and k, which are pinned at their centers to the rod r at points O, A, and B, respectively. Gear i is fixed with $r^i = 0.3$ m, while $r^j = r^k = 0.1$ m. If the rod r rotates counterclockwise with a constant angular velocity of 15 rad/s, determine the angular velocities and angular accelerations of the gears j and k.

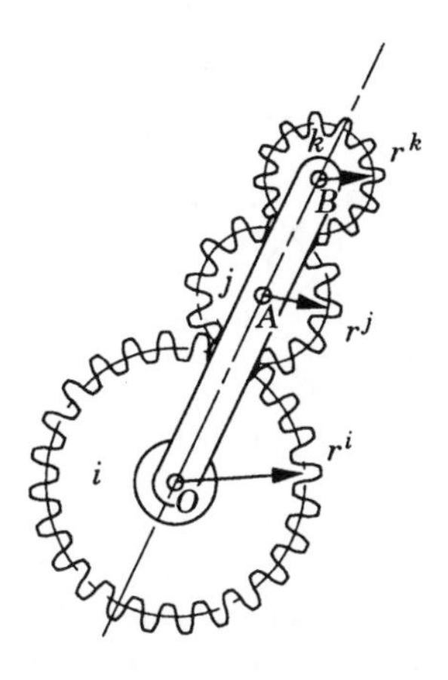

Figure P3.5

18. Solve problem 17 if the angular velocity and the angular acceleration of the rod are 15 rad/s, and 120 rad/s^2, respectively. The angular velocity and acceleration of the rod are assumed to be counterclockwise.

19. The motion of a rigid body i is such that the global coordinates of point P on the rigid body is given by $\mathbf{r}_P^i = [vt \quad 0]^{\mathrm{T}}$, where v is a constant. The angular velocity of the rigid body is assumed to be $\dot{\theta}^i = a_0 + a_1 t$. Derive an expression for the kinematic constraint equations of this system in terms of the absolute coordinates R_x^i, R_y^i, and θ^i. Assume that $\bar{\mathbf{u}}_P^i = [0.3 \quad 1.2]^{\mathrm{T}}$ m. Also determine the first and the second derivatives of the constraint equations. Use the resulting equations to determine the velocities $\dot{R}_x^i$, $\dot{R}_y^i$, and $\dot{\theta}^i$ and the accelerations $\ddot{R}_x^i$, $\ddot{R}_y^i$, and $\ddot{\theta}^i$ at $t = 0$, and 2 s. Use the data $v = 5$ m/s, $a_0 = 0$, $a_1 = 15$ rad/s^2.

20. The motion of two bodies i and j is such that $R_x^i = 5\,\text{m} = constant$, $R_y^i = 3 \sin 5t$ m, $\dot{\theta}^i - \dot{\theta}^j = 5\,\text{rad/s} = constant$, and the position vector of point P^j on body j with respect to point P^i on body i is defined by

$$\mathbf{r}_P^j - \mathbf{r}_P^i = \begin{bmatrix} 0.5 \sin 3t \\ 0.1 \cos 3t \end{bmatrix}$$

where

$$\bar{\mathbf{u}}_P^i = [0.7 \quad 1.2]^{\mathrm{T}}, \qquad \bar{\mathbf{u}}_P^j = [0.5 \quad -0.8]^{\mathrm{T}}$$

Derive the vector of the constraint equations of this system and determine the number of degrees of freedom.

21. For the three-body system shown in Fig. P6, use three absolute coordinates for each body to write the kinematic constraint equations of the revolute and prismatic joints. Also derive the constraint Jacobian matrix for the system.

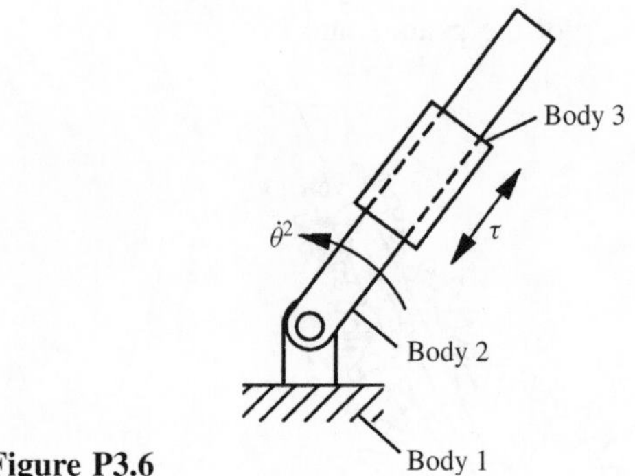

Figure P3.6

22. Figure P7 shows two bodies i and j connected by a revolute-revolute joint that keeps the distance between points P^i and P^j constant. Derive the constraint equations for this type of joint using the absolute Cartesian coordinates. Derive also the constraint Jacobian matrix for this joint.

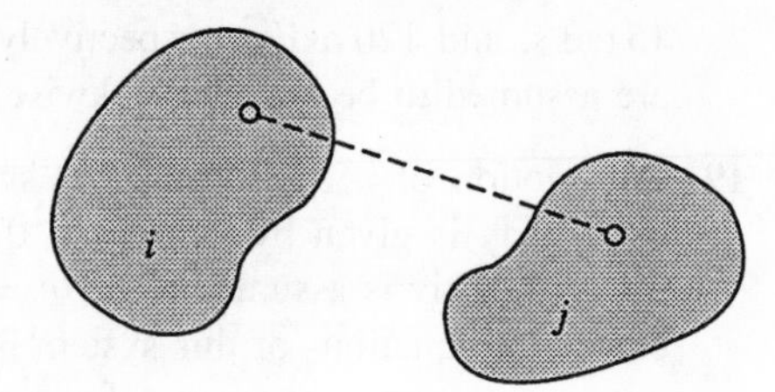

Figure P3.7

23. Derive the constraint Jacobian matrix of the reciprocating knife-edge follower cam system, and the roller follower cam system.

24. Derive the kinematic constraint Jacobian matrix of the offset flat-faced follower of Example 8.

25. Determine the vectors $\mathbf{C}_t$ and $\mathbf{Q}_d$ of Eqs. 121 and 124, respectively, for the revolute–revolute joint of problem 22.

26. Determine the vectors $\mathbf{C}_t$ and $\mathbf{Q}_d$ of Eqs. 121 and 124, respectively, in the case of the reciprocating knife-edge follower cam system and the roller follower cam system of problem 23.

27. Determine the vectors $\mathbf{C}_t$ and $\mathbf{Q}_d$ of Eqs. 121 and 124, respectively, in the case of the offset flat-faced follower of Example 8.

28. Derive the constraint equations, the Jacobian matrix, the vector $\mathbf{C}_t$, and the vector $\mathbf{Q}_d$ for the system shown in Fig. P8 using the absolute Cartesian coordinates.

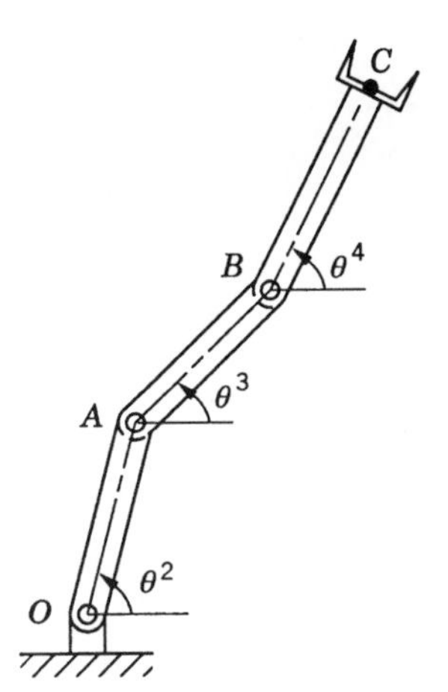

Figure P3.8

29. Figure P9 shows a slider crank mechanism. The lengths of the crankshaft and the connecting rod are 0.2 m and 0.4 m, respectively. The crankshaft is assumed to rotate with a constant angular velocity $\dot{\theta}^2 = 30$ rad/s. The initial angle θ_o^2 of the crankshaft is assumed to be 30°. Using three absolute coordinates R_x^i, R_y^i, and θ^i for each body in the system and the methods of constrained kinematics, determine the positions, velocities, and accelerations of the bodies at time $t = 0$, 1, and 2 s.

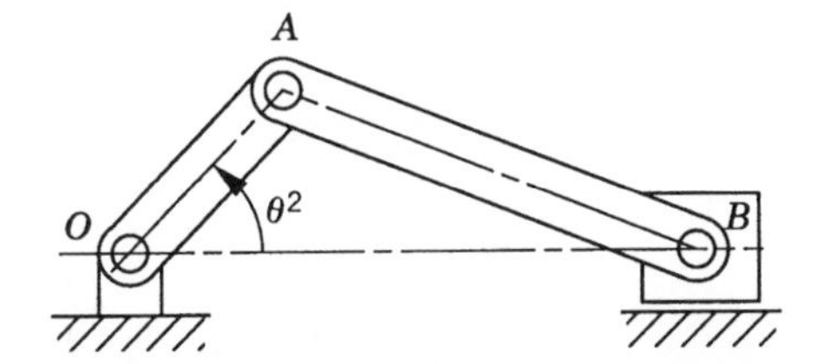

Figure P3.9

30. Figure P10 shows a four-bar mechanism. The lengths of the crankshaft, coupler, and rocker are 0.2 m, 0.4 m, and 0.3 m, respectively. The crankshaft of the mechanism is assumed to rotate with a constant angular velocity $\dot{\theta}^2 = 15$ rad/s. The initial orientation of the crankshaft is assumed to be $\theta_o^2 = 45°$. By using three absolute coordinates R_x^i, R_y^i, and θ^i for each body in the system, determine the position, velocity, and acceleration of each body at time $t = 0$, 1, and 2 s.

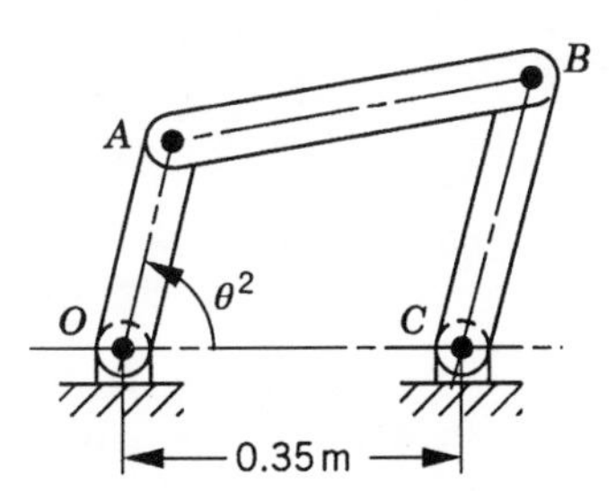

Figure P3.10

CHAPTER 4

FORMS OF THE DYNAMIC EQUATIONS

The focus of Chapter 3 was on the analysis of kinematically driven systems in which all the degrees of freedom are specified. Since the system configuration can be completely determined when the degrees of freedom are known, the analysis of kinematically driven systems leads to a system of algebraic equations that can be solved for the coordinates, velocities, and accelerations without the need for a force analysis. However, if one or more of the system degrees of freedom are not known a priori, the force analysis becomes necessary and the system equations of motion must be formulated to obtain a number of equations equal to the number of the unknown variables. In the case of unconstrained motion, the equations of motion of the system take a simple known form defined by Newton–Euler equations, and therefore, the selection of the system coordinates is not the subject of much argument. In the case of constrained multibody dynamics, on the other hand, different numbers of coordinates can be selected, leading to different forms of the dynamic equations. Some formulations that employ redundant coordinates lead to a relatively large system of equations expressed in terms of the constraint forces, while some other formulations lead to a minimum set of differential equations of motion expressed in terms of the degrees of freedom. Since the degrees of freedom, by definition, are independent and are not related by kinematic relationships, it is expected that the constraint forces are automatically eliminated when the equations of motion are formulated in terms of the degrees of freedom.

Much of the research on computational dynamics has been focused on the selection of the coordinates and on studying the advantages and drawbacks of different formulations. Despite the drawback of increasing the number of equations and the dimensionality of the problem, the use of redundant coordinates instead of using the degrees of freedom has the advantage

Computational Dynamics, Third Edition Ahmed A. Shabana

of increasing the generality of the formulation and achieving a sparse matrix structure. On the other hand, the formulations in terms of the degrees of freedom have the advantage of reducing the number of equations at the expense of increasing the complexity of the inertia and force coefficients that appear in these equations. In this chapter, a brief introduction to some forms of the dynamic equations of motion is presented. D'Alembert's principle is introduced and its use in formulating the equations of motion of mechanical systems is demonstrated using simple examples. Different matrix formulations for the equations of motion are then obtained using D'Alembert's principle. Some of these formulations lead to a large number of equations which include the constraint forces, while others lead to a smaller set of equations which do not include any constraint forces. Among the matrix formulations presented in this chapter are the *augmented formulation*, the *embedding technique*, and the *amalgamated formulation*. This chapter also includes a brief discussion on the analysis of *open-* and *closed-chain systems*. The material presented in this chapter can be considered as an introduction to some of the concepts and computational methods which are discussed in more detail in the remainder of the book. Simple procedures are used in this chapter to introduce different formulations; a more systematic and rigorous development of some of these formulations is presented in the following chapters.

4.1 D'ALEMBERT'S PRINCIPLE

As demonstrated in Chapter 3, the unconstrained planar motion of a rigid body can be described using three independent coordinates. As shown in Fig. 1, two coordinates R_x^i and R_y^i can be used to define the translation of the reference point and one coordinate θ^i defines the orientation of the rigid body coordinate system X^iY^i with respect to the global coordinate system XY. Associated with these three coordinates, there are three independent differential equations that govern the unconstrained planar motion of a rigid body. If the reference point

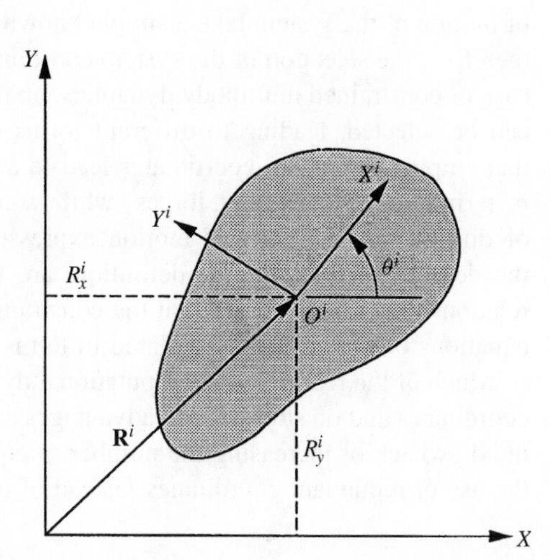

Figure 4.1 Rigid body coordinates

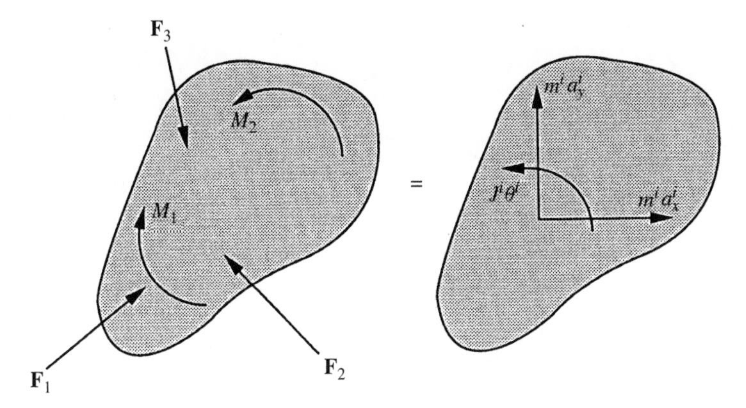

Figure 4.2 D'Alembert's principle

is selected to be the center of mass of the body, these equations can be written as

$$\left.\begin{aligned} m^i a_x^i &= F_x^i \\ m^i a_y^i &= F_y^i \\ J^i \ddot{\theta}^i &= M^i \end{aligned}\right\} \tag{4.1}$$

where m^i is the total mass of the rigid body, J^i is the mass moment of inertia defined with respect to the center of mass, a_x^i and a_y^i are the components of the absolute acceleration of the center of mass of the body, $\ddot{\theta}^i$ is the angular acceleration, F_x^i and F_y^i are the components of the resultant force acting at the center of mass, and M^i is the resultant moment. The first two equations in Eq. 1 are called *Newton's equations*, while the last equation is called *Euler's equation*. The left-hand side of the first two equations in Eq. 1 is called the *inertia* or *effective force*, and the left-hand side of the third equation is called the *inertia* or *effective moment*. D'Alembert's principle states that the effective or inertia forces and moments of a rigid body are equal to the external forces acting on the body; that is, the inertia forces and moments can be treated in the same way as the externally applied forces. This fact is demonstrated by the simple diagram shown in Fig. 2. Note that the system of forces and moments acting on body i as shown in Fig. 2 can be replaced by an equivalent (equipollent) system that consists of one force vector acting at the center of mass and a moment acting on the rigid body.

Example 4.1

Figure 3 shows a simple system that consists of two moving bodies. Body 1 is a slider block that has a specified motion defined by the function $z(t)$. Body 2 is a uniform slender rod that has mass m^2, mass moment of inertia about its center of mass J^2, and length l. The rod that is subjected to the external moment M^2 is connected to the sliding block by a pin joint at O.

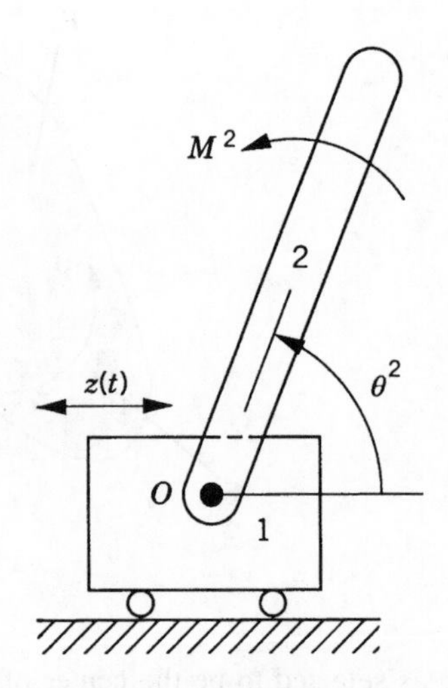

Figure 4.3 Pendulum with moving base

Figure 4 shows a free-body diagram for the rod. By applying Eq. 1, one obtains

$$m^2 a_x^2 = F_x^{12}$$

$$m^2 a_y^2 = F_y^{12} - m^2 g$$

$$J^2 \ddot{\theta}^2 = M^2 + F_x^{12} \frac{l}{2} \sin \theta^2 - F_y^{12} \frac{l}{2} \cos \theta^2$$

where F_x^{12} and F_y^{12} are the joint reaction forces at the pin joint. Since the motion of the sliding block is specified, the system has only one degree of freedom, which can be considered as the angular orientation of the rod θ^2.

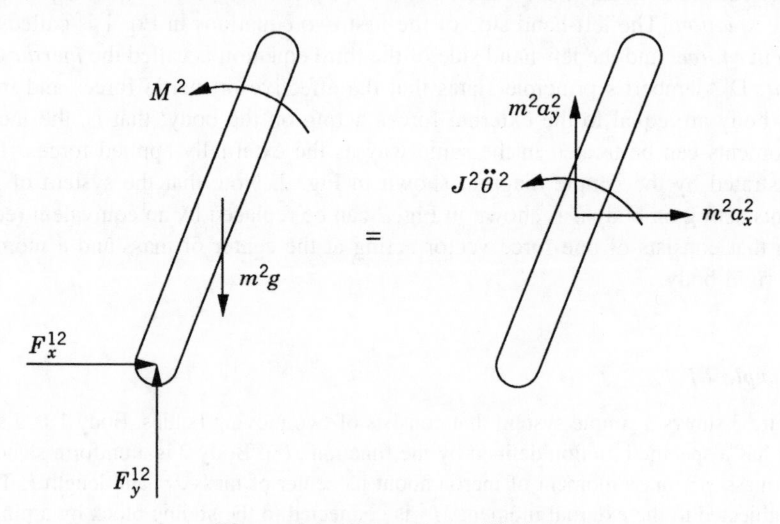

Figure 4.4 Dynamic equilibrium

If the applied forces and moments are given, as in the case of the forward dynamic analysis, the direct application of the Newton–Euler equations, in this example, leads to three differential equations of motion which are expressed in terms of five unknown acceleration components and reaction forces. These unknowns are a_x^2, a_y^2, $\ddot{\theta}^2$, F_x^{12}, and F_y^{12}. Therefore, in order to be able to solve the equations of motion for the five unknowns, two additional equations are required. These additional equations represent the constraint equations imposed on the motion of the rod. Because of the pin joint at point O, the coordinates of the center of mass of the rod must satisfy the following two conditions:

$$R_x^2 = z(t) + \frac{l}{2}\cos\theta^2, \quad R_y^2 = \frac{l}{2}\sin\theta^2$$

which upon differentiation once and twice lead to

$$\dot{R}_x^2 = \dot{z}(t) - \dot{\theta}^2\frac{l}{2}\sin\theta^2, \quad \dot{R}_y^2 = \dot{\theta}^2\frac{l}{2}\cos\theta^2$$

$$a_x^2 = \ddot{R}_x^2 = \ddot{z}(t) - \ddot{\theta}^2\frac{l}{2}\sin\theta^2 - (\dot{\theta}^2)^2\frac{l}{2}\cos\theta^2$$

$$a_y^2 = \ddot{R}_y^2 = \ddot{\theta}^2\frac{l}{2}\cos\theta^2 - (\dot{\theta}^2)^2\frac{l}{2}\sin\theta^2$$

Using the expressions for a_x^2 and a_y^2 in these equations, a *forward elimination* process can be applied to reduce the number of unknowns in the equations of motion. To this end, the expressions for a_x^2 and a_y^2 in the preceding equations are substituted into the differential equations of motion. This leads to

$$m^2\left[\ddot{z}(t) - \ddot{\theta}^2\frac{l}{2}\sin\theta^2 - (\dot{\theta}^2)^2\frac{l}{2}\cos\theta^2\right] = F_x^{12}$$

$$m^2\left[\ddot{\theta}^2\frac{l}{2}\cos\theta^2 - (\dot{\theta}^2)^2\frac{l}{2}\sin\theta^2\right] = F_y^{12} - m^2 g$$

$$J^2\ddot{\theta}^2 = M^2 + F_x^{12}\frac{l}{2}\sin\theta^2 - F_y^{12}\frac{l}{2}\cos\theta^2$$

These are three equations that can be solved for the three unknowns $\ddot{\theta}^2$, F_x^{12}, and F_y^{12}. For instance, multiplying the first equation by $-(l/2)\sin\theta^2$, the second equation by $(l/2)\cos\theta^2$, and adding the resulting two equations to the third, one obtains

$$\left[J^2 + m^2\left(\frac{l}{2}\right)^2\right]\ddot{\theta}^2 = M^2 - m^2 g\frac{l}{2}\cos\theta^2 + m^2\ddot{z}\frac{l}{2}\sin\theta^2$$

This equation does not contain the joint reaction forces and can be solved for the angular acceleration $\ddot{\theta}^2$. Once the angular acceleration $\ddot{\theta}^2$ is determined, a back substitution process can be used to determine the accelerations a_x^2 and a_y^2 as well as the reaction forces F_x^{12} and F_y^{12}.

Another alternative for obtaining the preceding equation is to apply D'Alembert's principle. According to this principle, the moment of the inertia forces about O is equal to the moment of the externally applied forces about the same point. This leads to one differential equation that does not include the joint reaction forces. The moment of the inertia (effective) forces about point O is

$$(M_e^2)_O = J^2\ddot{\theta}^2 - m^2 a_x^2 \frac{l}{2} \sin\theta^2 + m^2 a_y^2 \frac{l}{2} \cos\theta^2$$

The moment of the externally applied forces about O is

$$(M_a^2)_O = M^2 - m^2 g \frac{l}{2} \cos\theta^2$$

D'Alembert's principle implies that

$$(M_e^2)_O = (M_a^2)_O$$

or

$$J^2\ddot{\theta}^2 - m^2 a_x^2 \frac{l}{2} \sin\theta^2 + m^2 a_y^2 \frac{l}{2} \cos\theta^2 = M^2 - m^2 g \frac{l}{2} \cos\theta^2$$

Substituting for the accelerations of the center of mass of the rod using the previously obtained kinematic relationships, one obtains

$$\left[J^2 + m^2\left(\frac{l}{2}\right)^2\right]\ddot{\theta}^2 = M^2 - m^2 g \frac{l}{2} \cos\theta^2 + m^2\ddot{z} \frac{l}{2} \sin\theta^2$$

which is the same equation obtained previously by eliminating the joint reaction forces.

It is clear from the analysis presented in Example 1 that the direct application of Newton's second law or D'Alembert's principle to derive the dynamic conditions for a body in a mechanical system leads to a set of equations expressed in terms of the accelerations, the applied forces and moments, and the joint reaction forces. If the applied forces and moments are known, the resulting dynamic equations can be considered as a linear system of algebraic equations that can be solved for the accelerations and the joint reaction forces. The accelerations can be integrated forward in time in order to determine the coordinates and velocities. There are, however, several methods for formulating the acceleration equations. These methods are discussed briefly in this chapter to provide a motivation for the study of the materials covered in the following chapters. Before these different methods are presented, we first show in the following section how to use D'Alembert's principle to derive the Newton–Euler equations in the case of rigid bodies.

4.2 D'ALEMBERT'S PRINCIPLE AND NEWTON–EULER EQUATIONS

The Newton–Euler equations defined in Eq. 1 can be derived in a straightforward manner by applying D'Alembert's principle. In fact, D'Alembert's principle can be used to develop a set of equations, which do not restrict the choice of the reference point to be the center of mass of the body. In this section, the derivations of both Newton and Euler equations for rigid bodies are presented.

Newton Equations As shown in Fig. 5, a rigid body i can be considered to consist of a large number of particles, each of which has mass $\rho^i dV^i$ where ρ^i is the mass density, and dV^i is the infinitesimal volume. The inertia force of a particle whose position vector $\mathbf{r}^i$ is then equal to $(\rho^i dV^i)\ddot{\mathbf{r}}^i$. The body inertia force can be obtained by integrating the inertia forces of its particles over the volume of the body. By equating the inertia forces of the body to the applied forces, one obtains

$$\int_{V^i} \rho^i \ddot{\mathbf{r}}^i dV^i = \mathbf{F}^i \tag{4.2}$$

In this equation, $\mathbf{F}^i$ is the vector of resultant forces acting on the body. The preceding equation is general and allows using any point as the reference point. In the planar analysis, Eq. 2 has two scalar equations. Recall that the absolute acceleration of an arbitrary point on the rigid body i can be written as

$$\ddot{\mathbf{r}}^i = \ddot{\mathbf{R}}^i + \boldsymbol{\alpha}^i \times \mathbf{u}^i + \boldsymbol{\omega}^i \times (\boldsymbol{\omega}^i \times \mathbf{u}^i) \tag{4.3}$$

where $\mathbf{R}^i$ is the global position vector of the reference point on the body, $\boldsymbol{\omega}^i$ and $\boldsymbol{\alpha}^i$ are, respectively, the angular velocity and angular acceleration vectors of the body, $\mathbf{u}^i = \mathbf{A}^i \overline{\mathbf{u}}^i$ is the position vector of the arbitrary point with respect to the reference point, $\mathbf{A}^i$ is the transformation matrix that defines the orientation of the body, and $\overline{\mathbf{u}}^i$ is the position of the

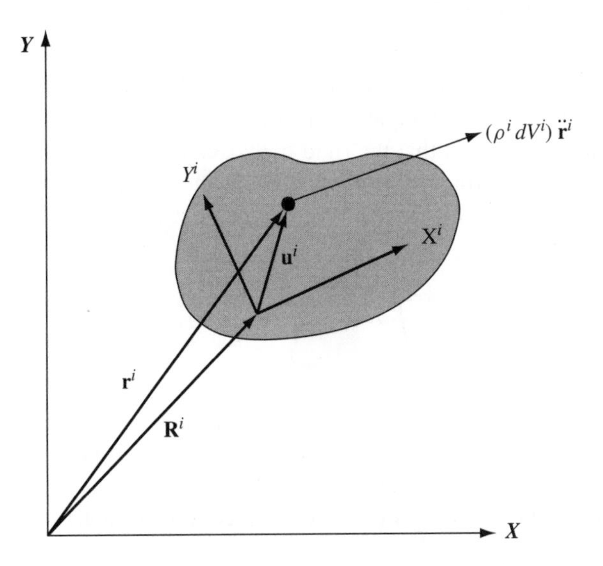

Figure 4.5 Inertia forces of the rigid body

arbitrary point with respect to the reference point. Note that if the reference point is selected to be the *center of mass* of the body, one has

$$\int_{V^i} \rho^i \mathbf{u}^i dV^i = \int_{V^i} \rho^i \mathbf{A}^i \bar{\mathbf{u}}^i dV^i = \mathbf{A}^i \int_{V^i} \rho^i \bar{\mathbf{u}}^i dV^i = \mathbf{0} \tag{4.4}$$

Substituting Eq. 3 into Eq. 2 and using the identity of Eq. 4 and the fact that $\boldsymbol{\omega}^i$ and $\boldsymbol{\alpha}^i$ do not depend on the location of the arbitrary point, one obtains

$$\int_{V^i} \rho^i \ddot{\mathbf{R}}^i dV^i = \mathbf{F}^i \tag{4.5}$$

The vector of the center of mass acceleration $\ddot{\mathbf{R}}^i = [\ddot{R}_x^i \quad \ddot{R}_y^i]^{\mathrm{T}}$ does not depend on the location of the arbitrary point, and therefore, this vector can be factored out of the integration sign. Using this fact and the definition of the body mass

$$m^i = \int_{V^i} \rho^i dV^i \tag{4.6}$$

Eq. 5 yields

$$m^i \ddot{\mathbf{R}}^i = \mathbf{F}^i \tag{4.7}$$

This is the Newton equation of motion for the rigid body i, it is the same as the first two scalar equations given by Eq. 1. It is worth mentioning that Eq. 7 is a special case of Eq. 2. In Eq. 2, the reference point can be an arbitrary point; while in Eq. 7, the reference point must be the center of mass of the body.

Euler Equation The third equation in Eq. 1 is Euler equation. This equation can be obtained from D'Alembert's principle by treating the inertia forces as the applied forces. To this end, it is assumed again that the rigid body i consists of a large number of particles. The moment of the inertia force of a particle about the reference point is defined as $\mathbf{u}^i \times \{(\rho^i dV^i)\ddot{\mathbf{r}}^i\}$. It follows from D'Alembert's principle that the moments of the inertia forces of the body are given by

$$\int_{V^i} \rho^i \mathbf{u}^i \times \ddot{\mathbf{r}}^i dV^i = \mathbf{M}_R^i \tag{4.8}$$

In this equation, $\mathbf{M}_R^i = [0 \quad 0 \quad M^i]^{\mathrm{T}}$ is the vector of the resultant moment about the reference point. This vector has one nonzero component only because the case of planar analysis is considered. Three-dimensional vectors are used in this section because of the convenience of using the cross product, which is defined for three-dimensional vectors only.

It is important to mention that Eq. 8 does not restrict the choice of the reference point to the center of mass. In this equation, any point on the body can be used as the reference point. The choice of the reference point as the center of mass, however, leads to significant simplifications in the resulting dynamic equation. Note also that in the case of planar motion, Eq. 8 reduces to only one nontrivial scalar equation that represents the moment about the Z^i axis since $\mathbf{u}^i$ and $\ddot{\mathbf{r}}^i$ are two planar vectors and their cross product defines a vector that is perpendicular to the plane containing these two vectors.

In order to obtain Euler equation, the reference point is chosen to be the body center of mass. Substituting Eq. 3 into Eq. 8, and using the identity of Eq. 4, and the fact that $\mathbf{u}^i$ and $\boldsymbol{\omega}^i \times (\boldsymbol{\omega}^i \times \mathbf{u}^i)$ are two parallel vectors, one obtains

$$\int_{V^i} \rho^i \mathbf{u}^i \times (\boldsymbol{\alpha}^i \times \mathbf{u}^i) dV^i = \mathbf{M}_R^i \tag{4.9}$$

Note that in the case of planar motion, $\boldsymbol{\alpha}^i = \ddot{\theta}^i \mathbf{k}$ and $\mathbf{u}^i = [\overline{x}^i \quad \overline{y}^i \quad 0]^{\mathrm{T}}$ is a planar vector that has zero component along the Z^i axis. One can then show that the preceding equation reduces to the following scalar equation:

$$\int_{V^i} \rho^i ((\overline{x}^i)^2 + (\overline{y}^i)^2) \ddot{\theta}^i dV^i = M^i \tag{4.10}$$

Note that the mass moment of inertia of the body J^i is defined as

$$J^i = \int_{V^i} \rho^i ((\overline{x}^i)^2 + (\overline{y}^i)^2) dV^i \tag{4.11}$$

It follows from the preceding two equations that

$$J^i \ddot{\theta}^i = M^i \tag{4.12}$$

This is Euler equation, the third equation of Eq. 1.

Remarks In deriving the Newton–Euler equations using D'Alembert's principle, it is assumed that the reference point is the body center of mass. This assumption leads to significant simplifications of the resulting equations of motion. It also leads to a formulation that does not include inertia coupling between the translation of the center of mass and the rotation of the body. If the center of mass is not considered as the reference point, Eqs. 2 and 8 represent the equations of motion of the body. One can show in this more general case that the resulting equations include inertia coupling between the translation of the reference point and the rotation of the body. In the case of rigid body dynamics, there is no clear advantage of using a reference point different from the body center of mass since the resulting equations become more complex. Similar comment applies to the case of spatial analysis, where a procedure similar to the one described in this section can be used to develop the three-dimensional Newton–Euler equations that govern the spatial motion of rigid bodies. This subject will be discussed in detail in a latter chapter of this book.

4.3 CONSTRAINED DYNAMICS

Mechanical joints and specified motion trajectories in multibody systems impose restrictions on the motion of the system components. Because of the kinematic constraints of the joints and specified motion trajectories, the selection of coordinates and the form of the equations of motion is not a trivial task and has been the subject of extensive research in the area of computational multibody system dynamics. The efficiency, generality, and numerical algorithm of a solution procedure strongly depend on the choice of the coordinates and the resulting form of the equations of motion. Joints and specified motion trajectories introduce constraint forces that may explicitly appear in the formulation or can be systematically eliminated by expressing the dynamic equations in terms of a chosen set of independent coordinates or degrees of freedom. As will be demonstrated in this section using a simple example, the number of independent constraint forces is always equal to the number of independent constraint equations, which is equal to the number of dependent coordinates. Obviously, if there are no constraints between the coordinates, there are no constraint forces and there are no dependent coordinates. This simple fact is crucial in understanding the basis of different forms of the dynamic equations of motion.

Consider the simple system shown in Fig. 6. This system consists of the ground denoted as body 1, a rod *OA* denoted as body 2, and a disk denoted as body 3. The rod is connected to the ground by a pin joint at O, while the disk is connected to the rod by a pin joint at A. The rod is assumed to be uniform and its length is l. Figures 7a and b show the free-body

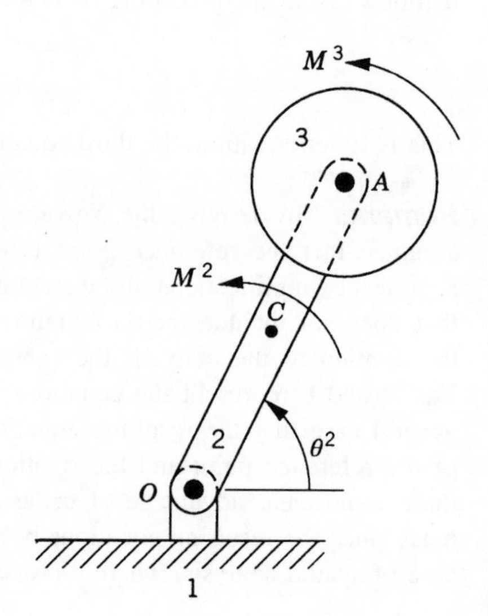

Figure 4.6 Illustrative example

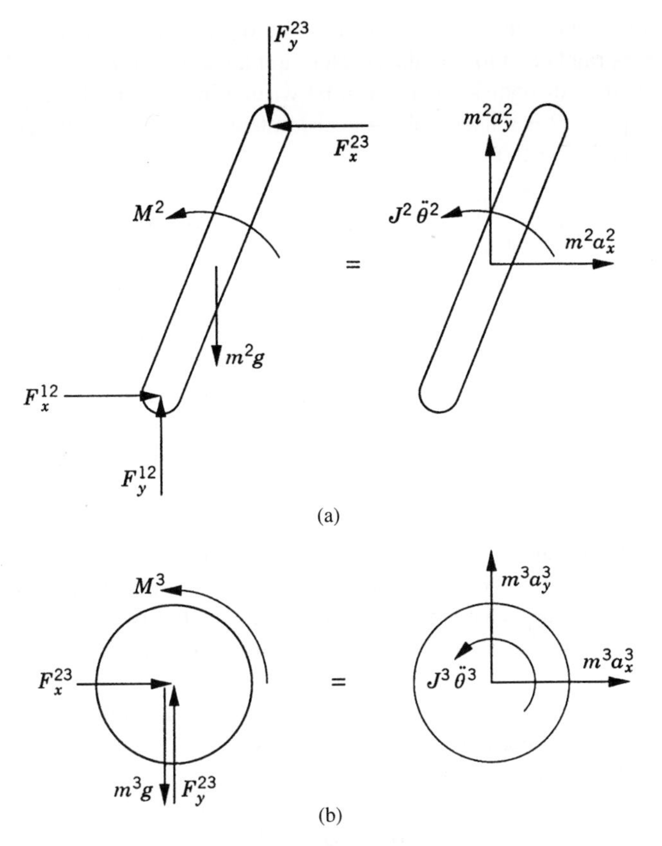

Figure 4.7 Dynamic equilibrium

diagrams of the rod and the disk. It is clear from these diagrams that

$$\left.\begin{aligned}
m^2 a_x^2 &= F_x^{12} - F_x^{23} \\
m^2 a_y^2 &= F_y^{12} - F_y^{23} - m^2 g \\
J^2 \ddot{\theta}^2 &= M^2 + F_x^{12} \frac{l}{2} \sin \theta^2 - F_y^{12} \frac{l}{2} \cos \theta^2 \\
&\quad + F_x^{23} \frac{l}{2} \sin \theta^2 - F_y^{23} \frac{l}{2} \cos \theta^2 \\
m^3 a_x^3 &= F_x^{23} \\
m^3 a_y^3 &= F_y^{23} - m^3 g \\
J^3 \ddot{\theta}^3 &= M^3
\end{aligned}\right\} \tag{4.13}$$

These are six dynamic equations in 10 unknowns; six acceleration components and four components of the reaction forces. Since the system has two degrees of freedom, the six acceleration components can be expressed in terms of two independent accelerations using the kinematic constraint equations at the acceleration level. In this example, the accelerations of the centers of mass of bodies 2 and 3 can be written in terms of the angular acceleration of body 2. Since point O is a fixed point, the absolute accelerations of the centers of mass of bodies 2 and 3 can be written as

$$\begin{aligned} \mathbf{a}^2 &= \begin{bmatrix} a_x^2 \\ a_y^2 \end{bmatrix} = \boldsymbol{\alpha}^2 \times \mathbf{u}_{CO}^2 + \boldsymbol{\omega}^2 \times (\boldsymbol{\omega}^2 \times \mathbf{u}_{CO}^2) \\ &= \ddot{\theta}^2 \frac{l}{2} \begin{bmatrix} -\sin\theta^2 \\ \cos\theta^2 \end{bmatrix} - (\dot{\theta}^2)^2 \frac{l}{2} \begin{bmatrix} \cos\theta^2 \\ \sin\theta^2 \end{bmatrix} \end{aligned} \tag{4.14}$$

$$\begin{aligned} \mathbf{a}^3 &= \begin{bmatrix} a_x^3 \\ a_y^3 \end{bmatrix} = \boldsymbol{\alpha}^2 \times \mathbf{u}_{AO}^2 + \boldsymbol{\omega}^2 \times (\boldsymbol{\omega}^2 \times \mathbf{u}_{AO}^2) \\ &= \ddot{\theta}^2 l \begin{bmatrix} -\sin\theta^2 \\ \cos\theta^2 \end{bmatrix} - (\dot{\theta}^2)^2 l \begin{bmatrix} \cos\theta^2 \\ \sin\theta^2 \end{bmatrix} \end{aligned} \tag{4.15}$$

where $\mathbf{u}_{CO}^2$ and $\mathbf{u}_{AO}^2$ are the vectors that define the locations of the centers of the rod and the disk with respect to point O. Note that the number of constraints of Eqs. 14 and 15 is equal to the number of the four independent reaction forces of the two pin joints. This number is also equal to the number of dependent coordinates used to formulate the equations of motion of the system.

In general, one can show that for any given constrained multibody system, the direct application of Newton–Euler equations leads to a system of differential equations that can be written in the following general matrix form:

$$\mathbf{M}\ddot{\mathbf{q}} = \mathbf{Q}_e + \mathbf{Q}_c \tag{4.16}$$

In this equation, $\mathbf{M}$ is the mass matrix of the system, $\mathbf{q}$ is the vector of the system coordinates, $\mathbf{Q}_e$ is the vector of applied forces, and $\mathbf{Q}_c$ is the vector of constraint forces. The number of scalar equations in the matrix equation of Eq. 16 is equal to the number of accelerations. In the case of forward dynamics, the applied forces defined by the vector $\mathbf{Q}_e$ are given. The unknowns in this case of forward dynamics are the accelerations and the constraint forces that enter into the formulation of the vector $\mathbf{Q}_c$. The number of independent constraint forces is equal to the number of the algebraic equations that represent the constraints imposed on the motion of the system. These algebraic constraint equations, as demonstrated in the preceding chapter, can be written in the following vector form:

$$\mathbf{C}(\mathbf{q}, t) = \mathbf{0} \tag{4.17}$$

The second derivatives of these constraint functions with respect to time define the constraint equations at the acceleration level, which in the example discussed in this section are equivalent to Eqs. 14 and 15.

In the special case of the example discussed in this section, one can recognize the vector of coordinates $\mathbf{q}$ as

$$\mathbf{q} = [R_x^2 \quad R_y^2 \quad \theta^2 \quad R_x^3 \quad R_y^3 \quad \theta^3]^{\mathrm{T}} \tag{4.18}$$

The mass matrix **M** can be defined using the coefficients of the accelerations in Eq. 13 as

$$\mathbf{M} = \begin{bmatrix} m^2 & 0 & 0 & 0 & 0 & 0 \\ 0 & m^2 & 0 & 0 & 0 & 0 \\ 0 & 0 & J^2 & 0 & 0 & 0 \\ 0 & 0 & 0 & m^3 & 0 & 0 \\ 0 & 0 & 0 & 0 & m^3 & 0 \\ 0 & 0 & 0 & 0 & 0 & J^3 \end{bmatrix} \tag{4.19}$$

The vector of applied forces $\mathbf{Q}_e$ and the vector of constraint forces $\mathbf{Q}_c$ can also be defined using Eq. 13 as

$$\mathbf{Q}_e = \begin{bmatrix} 0 \\ -m^2 g \\ M^2 \\ 0 \\ -m^3 g \\ M^3 \end{bmatrix}, \quad \mathbf{Q}_c = \begin{bmatrix} F_x^{12} - F_x^{23} \\ F_y^{12} - F_y^{23} \\ F_x^{12}\frac{l}{2}\sin\theta^2 - F_y^{12}\frac{l}{2}\cos\theta^2 + F_x^{23}\frac{l}{2}\sin\theta^2 - F_y^{23}\frac{l}{2}\cos\theta^2 \\ F_x^{23} \\ F_y^{23} \\ 0 \end{bmatrix} \tag{4.20}$$

For this example, the constraints of Eq. 17 are defined as

$$\mathbf{C} = \begin{bmatrix} \mathbf{R}^2 + \mathbf{A}^2\bar{\mathbf{u}}_O^2 \\ \mathbf{R}^2 + \mathbf{A}^2\bar{\mathbf{u}}_A^2 - \mathbf{R}^3 \end{bmatrix} = \mathbf{0} \tag{4.21}$$

In these constraint equations, $\mathbf{R}^i = [R_x^i \quad R_y^i]^{\mathrm{T}}, i = 2, 3$; and

$$\bar{\mathbf{u}}_O^2 = \begin{bmatrix} -l/2 \\ 0 \end{bmatrix}, \quad \bar{\mathbf{u}}_A^2 = \begin{bmatrix} l/2 \\ 0 \end{bmatrix}, \quad \mathbf{A}^2 = \begin{bmatrix} \cos\theta^2 & -\sin\theta^2 \\ \sin\theta^2 & \cos\theta^2 \end{bmatrix} \tag{4.22}$$

Using these definitions (Eq. 22), the system constraint equations at the position level, as defined by Eq. 21, can be written more explicitly as

$$\mathbf{C} = \begin{bmatrix} R_x^2 - \frac{l}{2}\cos\theta^2 \\ R_y^2 - \frac{l}{2}\sin\theta^2 \\ R_x^2 + \frac{l}{2}\cos\theta^2 - R_x^3 \\ R_y^2 + \frac{l}{2}\sin\theta^2 - R_y^3 \end{bmatrix} \tag{4.23}$$

There are several matrix methods for solving Eqs. 16 and 17 for the accelerations and the joint reaction forces. In this chapter we discuss briefly some of the matrix methods that are used to formulate the acceleration equations. Among these methods are the *augmented formulation*, the *embedding technique*, and the *amalgamated formulation*. Equations 13–15, which describe the dynamics of the system shown in Fig. 6, are used as an example to illustrate the concepts underlying these methods.

4.4 AUGMENTED FORMULATION

In the augmented formulation, the constraint forces explicitly appear in the dynamic equations, which are expressed in this case, in terms of redundant coordinates. The constraint relationships are used with the differential equations of motion to solve for the unknown accelerations and constraint forces. This approach leads to a sparse matrix structure and can be used as the basis for developing more general multibody system codes. Nonetheless, the augmented formulation has the drawback of increasing the problem dimensionality and it requires more sophisticated numerical algorithms to solve the resulting system of differential and algebraic equations, as discussed in the following chapters. In this section, the simple example discussed in the preceding section is used to introduce the augmented formulation.

In the augmented formulation, Eqs. 13–15 are combined in order to form a system of 10 scalar equations that can be solved for the 10 unknown accelerations and joint reaction forces. This leads to the following system:

$$\begin{bmatrix} m^2 & 0 & 0 & 0 & 0 & 0 & -1 & 0 & 1 & 0 \\ 0 & m^2 & 0 & 0 & 0 & 0 & 0 & -1 & 0 & 1 \\ 0 & 0 & J^2 & 0 & 0 & 0 & -\frac{l}{2}\sin\theta^2 & \frac{l}{2}\cos\theta^2 & -\frac{l}{2}\sin\theta^2 & \frac{l}{2}\cos\theta^2 \\ 0 & 0 & 0 & m^3 & 0 & 0 & 0 & 0 & -1 & 0 \\ 0 & 0 & 0 & 0 & m^3 & 0 & 0 & 0 & 0 & -1 \\ 0 & 0 & 0 & 0 & 0 & J^3 & 0 & 0 & 0 & 0 \\ -1 & 0 & -\frac{l}{2}\sin\theta^2 & 0 & 0 & 0 & 0 & 0 & 0 & 0 \\ 0 & -1 & \frac{l}{2}\cos\theta^2 & 0 & 0 & 0 & 0 & 0 & 0 & 0 \\ 0 & 0 & -l\sin\theta^2 & -1 & 0 & 0 & 0 & 0 & 0 & 0 \\ 0 & 0 & l\cos\theta^2 & 0 & -1 & 0 & 0 & 0 & 0 & 0 \end{bmatrix}$$

$$\times \begin{bmatrix} a_x^2 \\ a_y^2 \\ \ddot{\theta}^2 \\ a_x^3 \\ a_y^3 \\ \ddot{\theta}^3 \\ F_x^{12} \\ F_y^{12} \\ F_x^{23} \\ F_y^{23} \end{bmatrix} = \begin{bmatrix} 0 \\ -m^2 g \\ M^2 \\ 0 \\ -m^3 g \\ M^3 \\ (\dot{\theta}^2)^2 \frac{l}{2}\cos\theta^2 \\ (\dot{\theta}^2)^2 \frac{l}{2}\sin\theta^2 \\ (\dot{\theta}^2)^2 l\cos\theta^2 \\ (\dot{\theta}^2)^2 l\sin\theta^2 \end{bmatrix} \tag{4.24}$$

Note that in this form of the equations of motion, the constraint equations are not used to eliminate the dependent accelerations. As a result, a relatively large system of equations is obtained. It is also clear that the coefficient matrix in this equation is a sparse matrix since it has many zero elements. Sparse matrix techniques can then be used to solve the preceding form of the dynamic equations efficiently in order to determine the accelerations and the constraint forces.

4.5 LAGRANGE MULTIPLIERS

A more systematic and general procedure for developing the augmented equations of motion is based on the *Lagrangian dynamics*. In the Lagrangian approach, the technique of Lagrange multipliers is used to define generalized constraint forces and to obtain an augmented formulation in which the coefficient matrix is symmetric.

Equation 24 can be used to introduce the powerful technique of Lagrange multipliers and to demonstrate the basic differences between the Lagrangian mechanics and the Newtonian mechanics. In the Lagrangian mechanics, one does not need to make cuts at the joints and be concerned from the outset with the actual reaction forces. Instead, the equations of motion can be developed using the assembled system and the connectivity conditions (constraint equations). In order to demonstrate this approach, the example shown in Fig. 6 is used again. To this end, the Jacobian matrix of the constraints of Eq. 23 is written as

$$\mathbf{C_q} = \begin{bmatrix} 1 & 0 & \frac{l}{2}\sin\theta^2 & 0 & 0 & 0 \\ 0 & 1 & -\frac{l}{2}\cos\theta^2 & 0 & 0 & 0 \\ 1 & 0 & -\frac{l}{2}\sin\theta^2 & -1 & 0 & 0 \\ 0 & 1 & \frac{l}{2}\cos\theta^2 & 0 & -1 & 0 \end{bmatrix} \tag{4.25}$$

The columns of this constraint Jacobian matrix correspond to the vector of the system coordinates defined in Eq. 18. Differentiating the constraint functions of Eq. 17 twice with respect to time, one obtains

$$\mathbf{C_q}\ddot{\mathbf{q}} = \mathbf{Q}_d \tag{4.26}$$

Using the constraints of Eq. 23, one can show that

$$\mathbf{Q}_d = (\dot{\theta}^2)^2\frac{l}{2}[-\cos\theta^2 \quad -\sin\theta^2 \quad \cos\theta^2 \quad \sin\theta^2\,]^{\mathrm{T}} \tag{4.27}$$

After developing the expressions for the constraint equations and their second derivatives with respect to time, we return again to Eq. 24. We note that this equation can be, after multiplying some equations by a negative sign, rewritten as

$$\begin{bmatrix} m^2 & 0 & 0 & 0 & 0 & 0 & 1 & 0 & 1 & 0 \\ 0 & m^2 & 0 & 0 & 0 & 0 & 0 & 1 & 0 & 1 \\ 0 & 0 & J^2 & 0 & 0 & 0 & \frac{l}{2}\sin\theta^2 & -\frac{l}{2}\cos\theta^2 & -\frac{l}{2}\sin\theta^2 & \frac{l}{2}\cos\theta^2 \\ 0 & 0 & 0 & m^3 & 0 & 0 & 0 & 0 & -1 & 0 \\ 0 & 0 & 0 & 0 & m^3 & 0 & 0 & 0 & 0 & -1 \\ 0 & 0 & 0 & 0 & 0 & J^3 & 0 & 0 & 0 & 0 \\ 1 & 0 & \frac{l}{2}\sin\theta^2 & 0 & 0 & 0 & 0 & 0 & 0 & 0 \\ 0 & 1 & -\frac{l}{2}\cos\theta^2 & 0 & 0 & 0 & 0 & 0 & 0 & 0 \\ 1 & 0 & -\frac{l}{2}\sin\theta^2 & -1 & 0 & 0 & 0 & 0 & 0 & 0 \\ 0 & 1 & \frac{l}{2}\cos\theta^2 & 0 & -1 & 0 & 0 & 0 & 0 & 0 \end{bmatrix}$$

$$\times \begin{bmatrix} \ddot{R}_x^2 \\ \ddot{R}_y^2 \\ \ddot{\theta}^2 \\ \ddot{R}_x^3 \\ \ddot{R}_y^3 \\ \ddot{\theta}^3 \\ -F_x^{12} \\ -F_y^{12} \\ F_x^{23} \\ F_y^{23} \end{bmatrix} = \begin{bmatrix} 0 \\ -m^2 g \\ M^2 \\ 0 \\ -m^3 g \\ M^3 \\ -(\dot{\theta}^2)^2 \frac{l}{2} \cos\theta^2 \\ -(\dot{\theta}^2)^2 \frac{l}{2} \sin\theta^2 \\ (\dot{\theta}^2)^2 \frac{l}{2} \cos\theta^2 \\ (\dot{\theta}^2)^2 \frac{l}{2} \sin\theta^2 \end{bmatrix} \tag{4.28}$$

By using the expressions given in Eqs. 19, 20, 25 and 27, it is clear that Eq. 28 can be written in the following form:

$$\begin{bmatrix} \mathbf{M} & \mathbf{C}_\mathbf{q}^\mathrm{T} \\ \mathbf{C}_\mathbf{q} & \mathbf{0} \end{bmatrix} \begin{bmatrix} \ddot{\mathbf{q}} \\ \boldsymbol{\lambda} \end{bmatrix} = \begin{bmatrix} \mathbf{Q}_e \\ \mathbf{Q}_d \end{bmatrix} \tag{4.29}$$

In this equation

$$\boldsymbol{\lambda} = [\, -F_x^{12} \quad -F_y^{12} \quad F_x^{23} \quad F_y^{23} \,]^\mathrm{T} \tag{4.30}$$

While in this section, the simple example of Fig. 6 is used to derive Eq. 29; this equation is general and can be applied to any system subject to constraints. The coefficient matrix in this equation is always symmetric and positive definite for a well-posed problem. The vector $\boldsymbol{\lambda}$ is called the *vector of Lagrange multipliers*. In this simple example, Lagrange multipliers take a simple form expressed in terms of the actual reaction forces at the joints. While this is not always the case, the constraint forces associated with the system coordinates can always be written as

$$\mathbf{Q}_c = -\mathbf{C}_\mathbf{q}^\mathrm{T} \boldsymbol{\lambda} \tag{4.31}$$

The number of Lagrange multipliers is always equal to the number of constraint equations, which is equal to the number of dependent variables. Lagrange multipliers, which replace the independent reaction forces, are treated as unknowns, and therefore, one does not need to be concerned with the reaction forces from the outset. Instead, in the Lagrangian formulation, one needs to write the constraint equations (connectivity conditions) and use them to determine the constraint Jacobian matrix and the vector $\mathbf{Q}_d$ as was described in Chapter 3 of this book. The use of the constraint equations instead of the reaction forces, in addition to being one of the fundamental differences between the Lagrangian and Newtonian mechanics, allows for the systematic development of general computer algorithms that can be used in the analysis of complex and large-scale multibody systems. More detailed discussion on the augmented form of the equations of motion as given in Eq. 29 will be presented in later chapters of this book.

4.6 ELIMINATION OF THE DEPENDENT ACCELERATIONS

The approach discussed in this section is not one of the basic methods used in computational dynamics and is not widely used for solving multibody system applications. It is discussed here to serve as an intermediate step and as a brief introduction to the more widely used technique, the *embedding technique*, introduced in the following section. In this section, the constraint equations are used to eliminate the dependent accelerations leading to a system of equations that can be solved for the independent accelerations and the constraint forces. To demonstrate the use of this procedure, we consider the same example that was discussed in the preceding sections.

To solve Eqs. 13–15, the kinematic relationships of Eqs. 14 and 15 are used to eliminate the dependent components of the accelerations. To this end, we substitute Eqs. 14 and 15 into Eq. 13, and arrange the terms to obtain

$$\left.\begin{aligned}
-m^2 \tfrac{l}{2}\,\ddot{\theta}^2 \sin\theta^2 - F_x^{12} + F_x^{23} &= m^2 \tfrac{l}{2}\,(\dot{\theta}^2)^2 \cos\theta^2 \\
m^2 \tfrac{l}{2}\,\ddot{\theta}^2 \cos\theta^2 - F_y^{12} + F_y^{23} &= m^2 \tfrac{l}{2}\,(\dot{\theta}^2)^2 \sin\theta^2 - m^2 g \\
J^2\ddot{\theta}^2 - F_x^{12}\tfrac{l}{2}\sin\theta^2 + F_y^{12}\tfrac{l}{2}\cos\theta^2 - F_x^{23}\tfrac{l}{2}\sin\theta^2 + F_y^{23}\tfrac{l}{2}\cos\theta^2 &= M^2 \\
-m^3\ddot{\theta}^2 l \sin\theta^2 - F_x^{23} &= m^3 l(\dot{\theta}^2)^2 \cos\theta^2 \\
m^3\ddot{\theta}^2 l \cos\theta^2 - F_y^{23} &= m^3 l(\dot{\theta}^2)^2 \sin\theta^2 - m^3 g \\
J^3\ddot{\theta}^3 &= M^3
\end{aligned}\right\} \tag{4.32}$$

These six equations, which have two unknown angular accelerations and four unknown reaction forces, can be written in the following matrix form:

$$\begin{bmatrix} a_{11} & 0 & -1 & 0 & 1 & 0 \\ a_{21} & 0 & 0 & -1 & 0 & 1 \\ a_{31} & 0 & a_{33} & a_{34} & a_{35} & a_{36} \\ a_{41} & 0 & 0 & 0 & -1 & 0 \\ a_{51} & 0 & 0 & 0 & 0 & -1 \\ 0 & a_{62} & 0 & 0 & 0 & 0 \end{bmatrix} \begin{bmatrix} \ddot{\theta}^2 \\ \ddot{\theta}^3 \\ F_x^{12} \\ F_y^{12} \\ F_x^{23} \\ F_y^{23} \end{bmatrix} = \begin{bmatrix} b_1 \\ b_2 \\ M^2 \\ b_4 \\ b_5 \\ M^3 \end{bmatrix} \tag{4.33}$$

where

$$\left.\begin{aligned}
&a_{11} = -m^2 \frac{l}{2} \sin\theta^2, \; a_{21} = m^2 \frac{l}{2} \cos\theta^2, \; a_{31} = J^2 \\
&a_{41} = -m^3 l \sin\theta^2, \; a_{51} = m^3 l \cos\theta^2, \; a_{62} = J^3 \\
&a_{33} = a_{35} = -\frac{l}{2} \sin\theta^2, \; a_{34} = a_{36} = \frac{l}{2} \cos\theta^2
\end{aligned}\right\} \tag{4.34}$$

and

$$\left.\begin{aligned} b_1 &= m^2 \frac{l}{2} (\dot{\theta}^2)^2 \cos \theta^2 \\ b_2 &= m^2 \frac{l}{2} (\dot{\theta}^2)^2 \sin \theta^2 - m^2 g \\ b_4 &= m^3 l (\dot{\theta}^2)^2 \cos \theta^2 \\ b_5 &= m^3 l (\dot{\theta}^2)^2 \sin \theta^2 - m^3 g \end{aligned}\right\} \tag{4.35}$$

The system of matrix equations defined by Eq. 33 can be solved for the unknown independent angular accelerations and the joint reaction forces. The dependent accelerations can be determined using the constraint equations at the acceleration level (Eqs. 14 and 15).

Generalization As previously mentioned in this chapter, the vector of constraint forces associated with the system coordinates can be expressed in terms of multipliers, called *Lagrange multipliers* $\boldsymbol{\lambda}$ (see Eq. 31). The number of these multipliers is equal to the number of constraint equations and is also equal to the number of independent constraint forces. In order to develop a general procedure for the elimination of the dependent accelerations, Eq. 31 is substituted into Eq. 16. This leads to

$$\mathbf{M}\ddot{\mathbf{q}} = \mathbf{Q}_e - \mathbf{C}_{\mathbf{q}}^{\mathrm{T}} \boldsymbol{\lambda} \tag{4.36}$$

Note that this equation is the same as the equation defined in the first row in Eq. 29. In the case of forward dynamics, the unknowns in Eq. 36 are the vector of accelerations $\ddot{\mathbf{q}}$ and the vector of Lagrange multipliers $\boldsymbol{\lambda}$. Using the constraint equations at the acceleration level, one can always write the vector of system accelerations $\ddot{\mathbf{q}}$ in terms of a set of independent accelerations $\ddot{\mathbf{q}}_i$. The relationship between the system accelerations and the independent accelerations can always be written in the following form:

$$\ddot{\mathbf{q}} = \mathbf{B}_i \ddot{\mathbf{q}}_i + \boldsymbol{\gamma}_i \tag{4.37}$$

In this equation, the matrix $\mathbf{B}_i$ is called the *velocity transformation matrix*. This matrix plays a central role in the *embedding technique* discussed in the following section. The vector $\boldsymbol{\gamma}_i$ is always quadratic in the velocities. Substituting Eq. 37 into Eq. 36 and arranging the terms, one obtains

$$[\mathbf{M}\mathbf{B}_i \quad \mathbf{C}_{\mathbf{q}}^{\mathrm{T}}] \begin{bmatrix} \ddot{\mathbf{q}}_i \\ \boldsymbol{\lambda} \end{bmatrix} = \mathbf{Q}_e - \mathbf{M}\boldsymbol{\gamma}_i \tag{4.38}$$

This system of equations, which has a square coefficient matrix, can be solved for the independent accelerations $\ddot{\mathbf{q}}_i$ and the vector of Lagrange multipliers $\boldsymbol{\lambda}$.

In order to demonstrate the formulation of the velocity transformation matrix $\mathbf{B}_i$ and the vector $\boldsymbol{\gamma}_i$ used in Eq. 37, we select the vector of independent coordinates as $\mathbf{q}_i = [\theta^2 \quad \theta^3]^{\mathrm{T}}$

and use the results of Eqs. 14 and 15 to write

$$\begin{bmatrix} \ddot{R}_x^2 \\ \ddot{R}_y^2 \\ \ddot{\theta}^2 \\ \ddot{R}_x^3 \\ \ddot{R}_y^3 \\ \ddot{\theta}^3 \end{bmatrix} = \begin{bmatrix} -\frac{l}{2}\sin\theta^2 & 0 \\ \frac{l}{2}\cos\theta^2 & 0 \\ 1 & 0 \\ -l\sin\theta^2 & 0 \\ l\cos\theta^2 & 0 \\ 0 & 1 \end{bmatrix} \begin{bmatrix} \ddot{\theta}^2 \\ \ddot{\theta}^3 \end{bmatrix} + (\dot{\theta}^2)^2 l \begin{bmatrix} -\frac{1}{2}\cos\theta^2 \\ -\frac{1}{2}\sin\theta^2 \\ 0 \\ -\cos\theta^2 \\ -\sin\theta^2 \\ 0 \end{bmatrix} \tag{4.39}$$

Using this equation, the velocity transformation matrix $\mathbf{B}_i$ and the quadratic velocity vector $\boldsymbol{\gamma}_i$ are recognized as

$$\mathbf{B}_i = \begin{bmatrix} -\frac{l}{2}\sin\theta^2 & 0 \\ \frac{l}{2}\cos\theta^2 & 0 \\ 1 & 0 \\ -l\sin\theta^2 & 0 \\ l\cos\theta^2 & 0 \\ 0 & 1 \end{bmatrix}, \quad \boldsymbol{\gamma}_i = (\dot{\theta}^2)^2 l \begin{bmatrix} -\frac{1}{2}\cos\theta^2 \\ -\frac{1}{2}\sin\theta^2 \\ 0 \\ -\cos\theta^2 \\ -\sin\theta^2 \\ 0 \end{bmatrix} \tag{4.40}$$

Note that the product $\mathbf{MB}_i$, which appears in Eq. 38, is a 6×2 matrix and it is left to the reader to verify that this product is the same as the first two columns of the coefficient matrix that appears in Eq. 33. The matrix $\mathbf{C}_\mathbf{q}^\mathrm{T}$, on the other hand, is a 6×4 matrix since the number of constraint equations is 4 and the number of coordinates is 6. One can show by substituting the results of Eq. 40 into Eq. 38 and using the definition of Lagrange multipliers given for this example by Eq. 30 that the use of Eq. 38 will lead to the same equations as presented in Eq. 33.

4.7 EMBEDDING TECHNIQUE

In the formulations discussed in the preceding sections, the equations of motion are expressed in terms of the constraint forces. By using the embedding technique, the constraint forces can be eliminated systematically and a number of equations of motion equal to the number of the system degrees of freedom can be obtained. To obtain this minimum set of differential equations, it is necessary to use the *velocity transformation matrix* defined in the preceding section. This matrix can be defined systematically when the total vector of the system accelerations is expressed in terms of the independent accelerations. In the embedding technique, Eq. 38 is premultiplied by the transpose of the velocity transformation matrix $\mathbf{B}_i$. This leads to

$$[\mathbf{B}_i^\mathrm{T}\mathbf{M}\mathbf{B}_i \quad \mathbf{B}_i^\mathrm{T}\mathbf{C}_\mathbf{q}^\mathrm{T}] \begin{bmatrix} \ddot{\mathbf{q}}_i \\ \boldsymbol{\lambda} \end{bmatrix} = \mathbf{B}_i^\mathrm{T}\mathbf{Q}_e - \mathbf{B}_i^\mathrm{T}\mathbf{M}\boldsymbol{\gamma}_i \tag{4.41}$$

The following important identity holds

$$\mathbf{B}_i^{\mathrm{T}}\mathbf{C}_{\mathbf{q}}^{\mathrm{T}} = \mathbf{0} \tag{4.42}$$

This is an expected result that will be demonstrated using the two-body example discussed in this chapter. Substituting Eq. 42 into Eq. 41, one obtains

$$(\mathbf{B}_i^{\mathrm{T}}\mathbf{M}\mathbf{B}_i)\ddot{\mathbf{q}}_i = \mathbf{B}_i^{\mathrm{T}}\mathbf{Q}_e - \mathbf{B}_i^{\mathrm{T}}\mathbf{M}\boldsymbol{\gamma}_i \tag{4.43}$$

This equation can be written as

$$\mathbf{M}_i\ddot{\mathbf{q}}_i = \mathbf{Q}_i \tag{4.44}$$

where

$$\mathbf{M}_i = \mathbf{B}_i^{\mathrm{T}}\mathbf{M}\mathbf{B}_i, \quad \mathbf{Q}_i = \mathbf{B}_i^{\mathrm{T}}\mathbf{Q}_e - \mathbf{B}_i^{\mathrm{T}}\mathbf{M}\boldsymbol{\gamma}_i \tag{4.45}$$

Note that Eq. 44 does not include any constraint forces, and the number of the scalar equations in this matrix equation is equal to the number of the system degrees of freedom. Using the procedure described in this section, one can always obtain a number of equations equal to the number of degrees of freedom, and these equations do not include any constraint forces. The matrix $\mathbf{M}_i$ is the *generalized inertia matrix* associated with the system degrees of freedom, and the vector $\mathbf{Q}_i$ is the vector of *generalized forces* associated with these degrees of freedom.

Illustrative Example In order to demonstrate the use of the embedding technique described in this section, we return to the two-body example shown in Fig. 6. The constraint Jacobian matrix and the velocity transformation matrix of this system are given, respectively, by Eqs. 25 and 40. Using these two equations, one can show that $\mathbf{B}_i^{\mathrm{T}}\mathbf{C}_{\mathbf{q}}^{\mathrm{T}} = \mathbf{0}$. Using the definition of the mass matrix of Eq. 19 and the velocity transformation matrix of Eq. 40, one can show that the generalized mass matrix $\mathbf{M}_i$ defined in Eq. 45 is given as

$$\mathbf{M}_i = \begin{bmatrix} J^2 + m^2\left(\dfrac{l}{2}\right)^2 + m^3(l)^2 & 0 \\ 0 & J^3 \end{bmatrix} \tag{4.46}$$

Using the definitions of Eqs. 20 and 40, Eq. 45 can be used to define the vector $\mathbf{Q}_i$ as

$$\mathbf{Q}_i = \mathbf{B}_i^{\mathrm{T}}\mathbf{Q}_e - \mathbf{B}_i^{\mathrm{T}}\mathbf{M}\boldsymbol{\gamma}_i = \begin{bmatrix} M^2 - m^2 g\frac{l}{2}\cos\theta^2 - m^3 gl\cos\theta^2 \\ M^3 \end{bmatrix} \tag{4.47}$$

The equations of motion of the two-body system can then be obtained using Eq. 44 as

$$\begin{bmatrix} J^2 + m^2\left(\dfrac{l}{2}\right)^2 + m^3(l)^2 & 0 \\ 0 & J^3 \end{bmatrix}\begin{bmatrix} \ddot{\theta}^2 \\ \ddot{\theta}^3 \end{bmatrix} = \begin{bmatrix} M^2 - m^2 g\frac{l}{2}\cos\theta^2 - m^3 gl\cos\theta^2 \\ M^3 \end{bmatrix} \tag{4.48}$$

These equations, which do not include constraint forces, can be solved for the independent angular accelerations. The total vector of the system accelerations can be determined using Eq. 37. Knowing all the accelerations, one can substitute into the equations of motion before eliminating the dependent accelerations in order to determine the constraint forces.

As previously mentioned, the velocity transformation matrix $\mathbf{B}_i$ plays a fundamental role in the embedding formulation. This matrix allows for the systematic elimination of the constraint forces as demonstrated in this section. By so doing, a minimum set of dynamic differential equations of motion can be defined and expressed solely in terms of the system degrees of freedom. In Chapter 5, a systematic procedure based on the concept of the *virtual displacement* is used to define the velocity transformation matrix. The *principle of virtual work* is also used in Chapter 5 to systematically develop the embedding technique, which is introduced in this section using the familiar Newtonian mechanics approach.

4.8 AMALGAMATED FORMULATION

Another method for solving for the accelerations and the joint reaction forces is to obtain a very large system of loosely coupled algebraic equations. To this end, Eq. 16 is reproduced here for convenience

$$\mathbf{M}\ddot{\mathbf{q}} = \mathbf{Q}_e + \mathbf{Q}_c \tag{4.49}$$

It was previously shown that the vector of accelerations can be expressed in terms of the independent accelerations using Eq. 37, which is repeated here

$$\ddot{\mathbf{q}} = \mathbf{B}_i\ddot{\mathbf{q}}_i + \gamma_i \tag{4.50}$$

One can verify by using Eq. 42 that

$$\mathbf{B}_i^{\mathrm{T}}\mathbf{Q}_c = \mathbf{0} \tag{4.51}$$

where $\mathbf{B}_i$ is the velocity transformation matrix.

Equations 49–51 can be combined in one matrix equation to yield

$$\begin{bmatrix} \mathbf{M} & \mathbf{I} & \mathbf{0} \\ \mathbf{I} & \mathbf{0} & -\mathbf{B}_i \\ \mathbf{0} & -\mathbf{B}_i^{\mathrm{T}} & \mathbf{0} \end{bmatrix} \begin{bmatrix} \ddot{\mathbf{q}} \\ -\mathbf{Q}_c \\ \ddot{\mathbf{q}}_i \end{bmatrix} = \begin{bmatrix} \mathbf{Q}_e \\ \gamma_i \\ \mathbf{0} \end{bmatrix} \tag{4.52}$$

This large system, which has a sparse symmetric coefficient matrix, can be solved for the accelerations and the joint forces as well as the independent accelerations.

4.9 OPEN-CHAIN SYSTEMS

As will be demonstrated in Chapter 6, one of the advantages of using the augmented formulation is that open and closed kinematic chains can be treated alike. When other methods are used, special attention must be given to closed kinematic chains, which can have singular configurations, as also discussed in Chapter 6. In this section and the following section, we discuss some of the basic differences between open and closed chains to demonstrate the difficulties encountered in the analysis of closed chains and to have an appreciation of some of the advantages of the technique of Lagrange multipliers, which is discussed in more detail in Chapter 6.

Two methods are used in this section to develop the dynamic equations of open-chain systems. In the first method, the dynamic conditions are developed for each body in the system, leading to a set of equations expressed explicitly in terms of the joint reaction forces. The resulting number of equations is equal to the number of the system degrees of freedom plus the number of the joint reaction forces. These equations can be solved for the reaction forces in addition to a number of unknowns equal to the number of degrees of freedom of the system as previously demonstrated in Section 6 of this chapter. For example, if all the external forces are specified, the resulting dynamic equations can be solved for the reaction forces and a number of unknown accelerations equal to the number of degrees of freedom of the system. In the second method discussed in this section, cuts are made at selected joints and the dynamic conditions are formulated for selected subsystems leading to minimum number of differential equations. The number of these equations, which do not contain the joint reaction forces, is equal to the number of degrees of freedom of the system. Clearly, this minimum number of equations can be obtained using the first approach by eliminating the joint reaction forces, as demonstrated in Section 7.

Equilibrium of the Separate Bodies First, we consider the formal application of the dynamic conditions to each body in the system. The two-degree-of-freedom two-arm manipulator system shown in Fig. 8 is used. Body 1 represents the ground or the fixed link. Body 2 represents the first movable link in the manipulator, and its orientation is defined by the angle θ^2. Body 3 represents the second movable link in the manipulator, and the orientation of this body is defined by the angle θ^3. Let M^2 and M^3 be the joint torques that

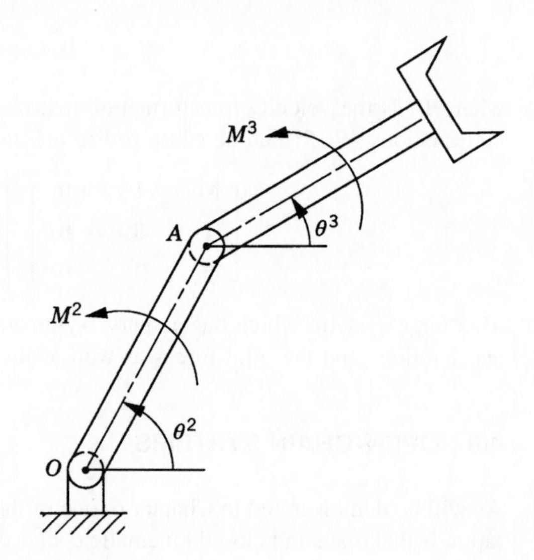

Figure 4.8 Open-chain system

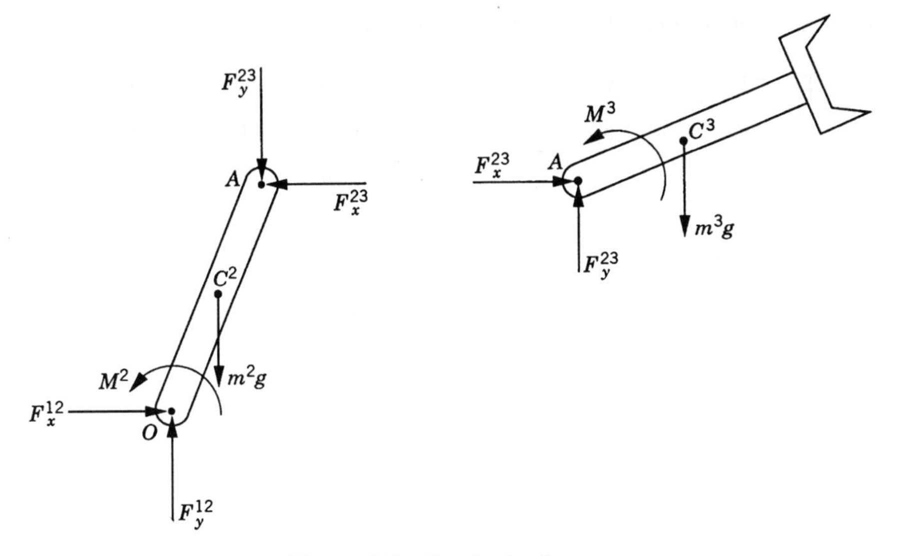

Figure 4.9 Free-body diagram

act on body 2 and body 3, respectively. Figure 9 shows the free-body diagrams of the two bodies. From this figure the dynamic conditions of body 2 are

$$\left.\begin{aligned} F_x^{12} - F_x^{23} &= m^2 a_x^2 \\ F_y^{12} - m^2 g - F_y^{23} &= m^2 a_y^2 \\ F_x^{12} l_O^2 \sin\theta^2 - F_y^{12} l_O^2 \cos\theta^2 + M^2 + F_x^{23} l_A^2 \sin\theta^2 & \\ - F_y^{23} l_A^2 \cos\theta^2 &= J^2 \ddot{\theta}^2 \end{aligned}\right\} \tag{4.53}$$

where m^2 and J^2 are, respectively, the mass and mass moment of inertia of body 2; a_x^2 and a_y^2 are the components of the acceleration of the center of mass of this body, g is the gravity constant; and F_x^{ij} and F_y^{ij} are the components of the reaction force acting on body i as the result of its connection with body j. In a similar manner, one may write the dynamic equations for body 3 as

$$\left.\begin{aligned} F_x^{23} &= m^3 a_x^3 \\ F_y^{23} - m^3 g &= m^3 a_y^3 \\ F_x^{23} l_A^3 \sin\theta^3 - F_y^{23} l_A^3 \cos\theta^3 + M^3 &= J^3 \ddot{\theta}^3 \end{aligned}\right\} \tag{4.54}$$

In this section, the case of the inverse dynamics (kinematically driven system) is considered to focus the attention on the basic differences between the open and closed kinematic chains. In this case, the motion of the system is assumed to be known and the goal is to solve for the external and joint forces. Assuming that the accelerations are known, Eqs. 53 and 54 can be arranged and combined in one matrix equation as

$$\begin{bmatrix} 1 & 0 & -1 & 0 & 0 & 0 \\ 0 & 1 & 0 & -1 & 0 & 0 \\ 0 & 0 & 1 & 0 & 0 & 0 \\ 0 & 0 & 0 & 1 & 0 & 0 \\ l_O^2 \sin\theta^2 & -l_O^2 \cos\theta^2 & l_A^2 \sin\theta^2 & -l_A^2 \cos\theta^2 & 1 & 0 \\ 0 & 0 & l_A^3 \sin\theta^3 & -l_A^3 \cos\theta^3 & 0 & 1 \end{bmatrix} \begin{bmatrix} F_x^{12} \\ F_y^{12} \\ F_x^{23} \\ F_y^{23} \\ M^2 \\ M^3 \end{bmatrix}$$

$$= \begin{bmatrix} m^2 a_x^2 \\ m^2 a_y^2 + m^2 g \\ m^3 a_x^3 \\ m^3 a_y^3 + m^3 g \\ J^2 \ddot{\theta}^2 \\ J^3 \ddot{\theta}^3 \end{bmatrix} \tag{4.55}$$

which can be written as

$$\mathbf{Ax} = \mathbf{b} \tag{4.56}$$

where the coefficient matrix $\mathbf{A}$ is defined as

$$\mathbf{A} = \begin{bmatrix} 1 & 0 & -1 & 0 & 0 & 0 \\ 0 & 1 & 0 & -1 & 0 & 0 \\ 0 & 0 & 1 & 0 & 0 & 0 \\ 0 & 0 & 0 & 1 & 0 & 0 \\ A_{51} & A_{52} & A_{53} & A_{54} & 1 & 0 \\ 0 & 0 & A_{63} & A_{64} & 0 & 1 \end{bmatrix} \tag{4.57}$$

and the vectors $\mathbf{x}$ and $\mathbf{b}$ are

$$\left.\begin{aligned} \mathbf{x} &= [F_x^{12} \quad F_y^{12} \quad F_x^{23} \quad F_y^{23} \quad M^2 \quad M^3]^{\mathrm{T}} \\ \mathbf{b} &= [m^2 a_x^2 \quad (m^2 a_y^2 + m^2 g) \quad m^3 a_x^3 \quad (m^3 a_y^3 + m^3 g) \quad J^2 \ddot{\theta}^2 \quad J^3 \ddot{\theta}^3]^{\mathrm{T}} \end{aligned}\right\} \tag{4.58}$$

In these equations

$$\left.\begin{aligned} A_{51} &= l_O^2 \sin\theta^2, & A_{52} &= -l_O^2 \cos\theta^2 \\ A_{53} &= l_A^2 \sin\theta^2, & A_{54} &= -l_A^2 \cos\theta^2 \\ A_{63} &= l_A^3 \sin\theta^3, & A_{64} &= -l_A^3 \cos\theta^3 \end{aligned}\right\} \tag{4.59}$$

and l_O^i and l_A^i are the distances of points O and A from the center of mass of link i.

The solution of Eq. 56 is given by $\mathbf{x} = \mathbf{A}^{-1}\mathbf{b}$, where the matrix $\mathbf{A}^{-1}$ is the inverse of the matrix $\mathbf{A}$ given by

$$\mathbf{A}^{-1} = \begin{bmatrix} 1 & 0 & 1 & 0 & 0 & 0 \\ 0 & 1 & 0 & 1 & 0 & 0 \\ 0 & 0 & 1 & 0 & 0 & 0 \\ 0 & 0 & 0 & 1 & 0 & 0 \\ -A_{51} & -A_{52} & -(A_{51}+A_{53}) & -(A_{52}+A_{54}) & 1 & 0 \\ 0 & 0 & -A_{63} & -A_{64} & 0 & 1 \end{bmatrix} \tag{4.60}$$

It follows that the solution $\mathbf{x}$ is given as

$$\left.\begin{aligned} F_x^{12} &= m^2 a_x^2 + m^3 a_x^3 \\ F_y^{12} &= m^2 a_y^2 + m^2 g + m^3 a_y^3 + m^3 g \\ F_x^{23} &= m^3 a_x^3 \\ F_y^{23} &= m^3 a_y^3 + m^3 g \\ M^2 &= J^2 \ddot{\theta}^2 - A_{51} m^2 a_x^2 - A_{52}(m^2 a_y^2 + m^2 g) - (A_{51} + A_{53}) m^3 a_x^3 \\ &\quad - (A_{52} + A_{54})(m^3 a_y^3 + m^3 g) \\ M^3 &= J^3 \ddot{\theta}^3 - A_{63} m^3 a_x^3 - A_{64}(m^3 a_y^3 + m^3 g) \end{aligned}\right\} \tag{4.61}$$

The last two equations in Eq. 61 do not include the reaction forces which are given by the first four equations.

Equilibrium of the Subsystems In the analysis of open-chain systems, the last two equations in Eq. 61 can be obtained directly, without considering the internal reaction forces by studying the equilibrium of selected subsystems, as shown in Fig. 10. For example, we may consider the dynamic equilibrium of link 3 in our example and take the moment about point A for both the systems of external and effective forces. The moments of the external forces and moments are

$$M_e = M^3 - m^3 g l_A^3 \cos\theta^3 \tag{4.62}$$

The moments of the inertia forces about A are

$$M_i = -m^3 a_x^3 l_A^3 \sin\theta^3 + m^3 a_y^3 l_A^3 \cos\theta^3 + J^3 \ddot{\theta}^3 \tag{4.63}$$

Applying D'Alembert's principle, which implies that the inertia or effective moment is equal to the moment of the applied forces, one obtains

$$M^3 - m^3 g l_A^3 \cos\theta^3 = -m^3 a_x^3 l_A^3 \sin\theta^3 + m^3 a_y^3 l_A^3 \cos\theta^3 + J^3 \ddot{\theta}^3 \tag{4.64}$$

which can be rearranged as

$$M^3 = J^3 \ddot{\theta}^3 - m^3 a_x^3 l_A^3 \sin\theta^3 + (m^3 a_y^3 + m^3 g) l_A^3 \cos\theta^3 \tag{4.65}$$

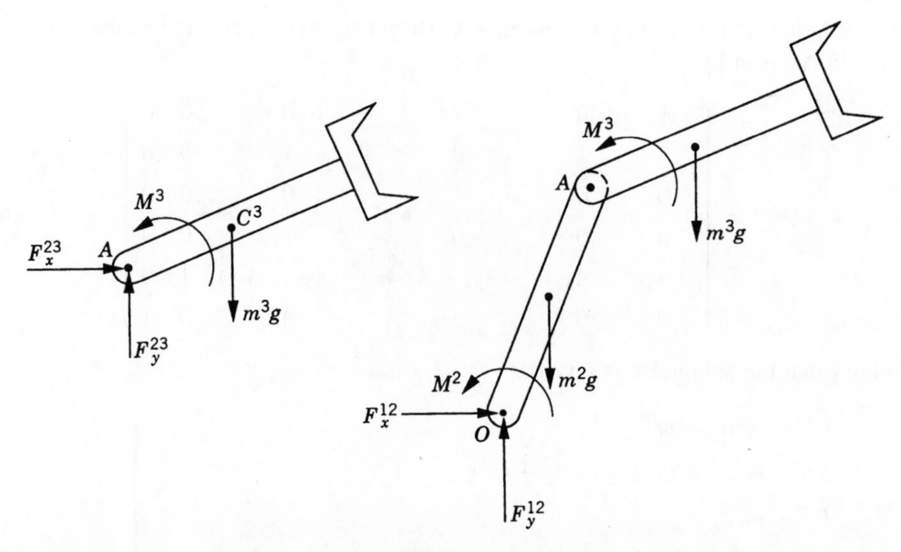

Figure 4.10 Equilibrium of the subsystems

This equation is the same as the last equation in Eq. 61. A second equation can be obtained by studying the equilibrium of bodies 2 and 3 together, as shown in Fig. 10. By taking the moments about point O, we can eliminate the internal reactions. For the external forces and moments, we have

$$M_e = M^2 + M^3 - m^2 g l_O^2 \cos\theta^2 - m^3 g(l^2 \cos\theta^2 + l_A^3 \cos\theta^3) \tag{4.66}$$

The moments of the inertia forces and moments about O yield

$$\begin{aligned} M_i &= J^2\ddot{\theta}^2 + J^3\ddot{\theta}^3 - m^2 a_x^2 l_O^2 \sin\theta^2 + m^2 a_y^2 l_O^2 \cos\theta^2 \\ &\quad - m^3 a_x^3 (l^2 \sin\theta^2 + l_A^3 \sin\theta^3) + m^3 a_y^3 (l^2 \cos\theta^2 + l_A^3 \cos\theta^3) \end{aligned} \tag{4.67}$$

Thus, the dynamic equilibrium condition for this subsystem is

$$\begin{aligned} M^2 &+ M^3 - m^2 g l_O^2 \cos\theta^2 - m^3 g(l^2 \cos\theta^2 + l_A^3 \cos\theta^3) \\ &= J^2\ddot{\theta}^2 + J^3\ddot{\theta}^3 - m^2 a_x^2 l_O^2 \sin\theta^2 + m^2 a_y^2 l_O^2 \cos\theta^2 \\ &\quad - m^3 a_x^3 (l^2 \sin\theta^2 + l_A^3 \sin\theta^3) + m^3 a_y^3 (l^2 \cos\theta^2 + l_A^3 \cos\theta^3) \end{aligned} \tag{4.68}$$

which can be rearranged and written as

$$\begin{aligned} M^2 + M^3 &= J^2\ddot{\theta}^2 + J^3\ddot{\theta}^3 - m^2 a_x^2 l_O^2 \sin\theta^2 + (m^2 a_y^2 + m^2 g) l_O^2 \cos\theta^2 \\ &\quad - m^3 a_x^3 (l^2 \sin\theta^2 + l_A^3 \sin\theta^3) + (m^3 a_y^3 + m^3 g) \\ &\quad \cdot (l^2 \cos\theta^2 + l_A^3 \cos\theta^3) \end{aligned} \tag{4.69}$$

Equations 65 and 69 represent the dynamic conditions for the two-degree-of-freedom system. They are two independent equations that can be solved for two unknowns. It is clear that upon subtracting Eq. 65 from Eq. 69, one obtains

$$\begin{aligned} M^2 = J^2\ddot{\theta}^2 - m^2 a_x^2 l_O^2 \sin\theta^2 + (m^2 a_y^2 + m^2 g) l_O^2 \cos\theta^2 \\ - m^3 a_x^3 l^2 \sin\theta^2 + (m^3 a_y^3 + m^3 g) l^2 \cos\theta^2 \end{aligned} \quad (4.70)$$

This is the same equation as the fifth equation in Eq. 61 obtained here from the application of the dynamic conditions to each body in the system separately. Therefore, the two methods discussed in this section lead to the same results. The second method, however, represents the foundation for some of the *recursive methods*, which allow elimination of the joint reaction forces in the analysis of open kinematic chains. Another systematic and straightforward approach to obtain Eqs. 65 and 70, which are the same as the last two equations of Eq. 61, is to use the principle of virtual work in dynamics, which will be discussed in the following chapter.

4.10 CLOSED-CHAIN SYSTEMS

It was demonstrated by the analysis presented in the preceding section that the joint reaction forces can be eliminated from the dynamic equations of open-chain systems by considering the equilibrium of selected subsystems. The analysis that follows will demonstrate the differences between open- and closed-chain systems, and as in the case of open-chain systems, two methods will be considered. In the first method, the dynamic equations are developed for each body in the system leading to a set of equations which are explicit functions of the joint reaction forces. In the second approach, cuts are made at selected *secondary joints* and the dynamics of the resulting subsystems is examined.

Equilibrium of the Separate Bodies As in the case of open-chain systems, the dynamic equations of a closed-chain system are first obtained by developing the dynamic equations of each body in the system separately. This leads to a number of equations equal to the number of reaction forces plus the number of degrees of freedom of the system. To demonstrate the use of this approach, we consider the closed-chain four-bar linkage shown in Fig. 11. The fixed link is denoted as body 1, the crankshaft is denoted as body 2, the coupler is denoted as body 3, and the rocker is denoted as body 4. The dynamic equations of the crankshaft, which is subjected to an external moment M^2 as shown in Fig. 12, are given by the following three equations:

$$\left.\begin{aligned} F_x^{12} - F_x^{23} = m^2 a_x^2 \\ F_y^{12} - m^2 g - F_y^{23} = m^2 a_y^2 \\ F_x^{12} l_O^2 \sin\theta^2 - F_y^{12} l_O^2 \cos\theta^2 + M^2 + F_x^{23} l_A^2 \sin\theta^2 - F_y^{23} l_A^2 \cos\theta^2 = J^2\ddot{\theta}^2 \end{aligned}\right\} \quad (4.71)$$

where the moment equation is defined with respect to the center of mass of the crankshaft.

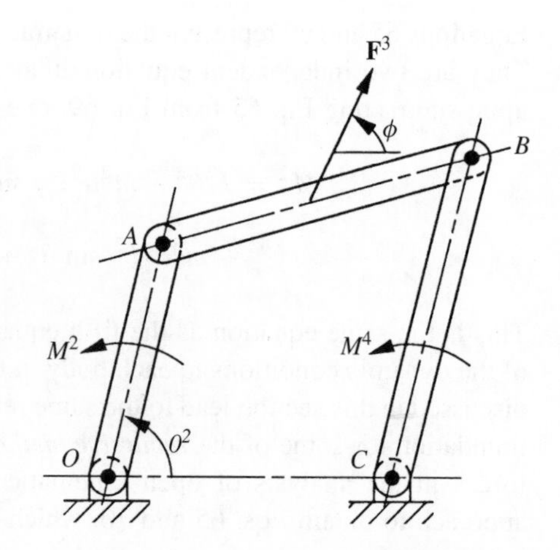

Figure 4.11 Closed-chain mechanisms

The coupler, as shown in Fig. 12, is subjected to an external force $\mathbf{F}^3$ that acts at its center of mass. The dynamic equations for the coupler can be written as

$$\left.\begin{aligned} F_x^{23} + F_x^3 - F_x^{34} &= m^3 a_x^3 \\ F_y^{23} - m^3 g + F_y^3 - F_y^{34} &= m^3 a_y^3 \\ F_x^{23} l_A^3 \sin\theta^3 - F_y^{23} l_A^3 \cos\theta^3 + F_x^{34} l_B^3 \sin\theta^3 - F_y^{34} l_B^3 \cos\theta^3 &= J^3 \ddot{\theta}^3 \end{aligned}\right\} \tag{4.72}$$

where F_x^3 and F_y^3 are the components of the external force vector $\mathbf{F}^3$.

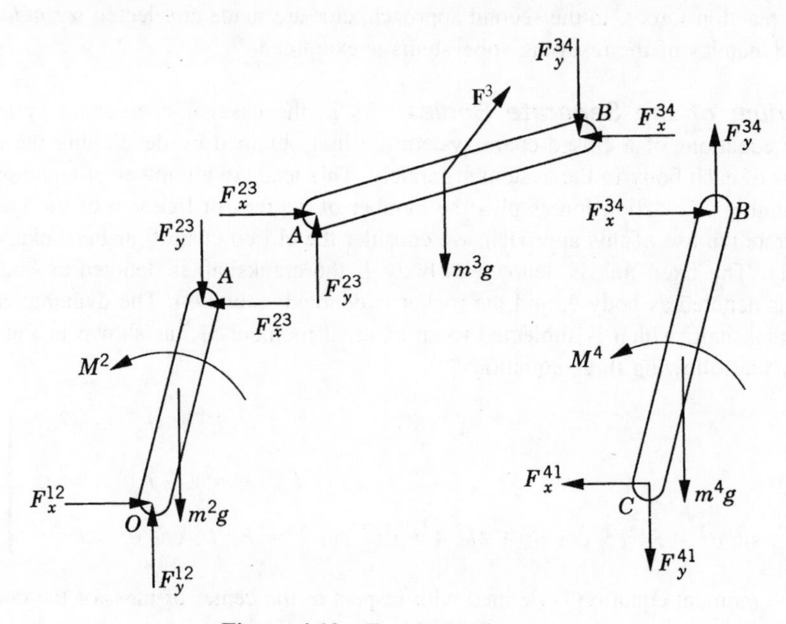

Figure 4.12 Free-body diagrams

As shown in Fig. 11, the rocker is subjected to the external moment M^4. The dynamic equations of the rocker are

$$\left.\begin{aligned} F_x^{34} - F_x^{41} &= m^4 a_x^4 \\ F_y^{34} - m^4 g - F_y^{41} &= m^4 a_y^4 \\ F_x^{34} l_B^4 \sin\theta^4 - F_y^{34} l_B^4 \cos\theta^4 + M^4 + F_x^{41} l_C^4 \sin\theta^4 - F_y^{41} l_C^4 \cos\theta^4 &= J^4 \ddot{\theta}^4 \end{aligned}\right\} \quad (4.73)$$

The dynamic conditions of the four-bar mechanism lead to nine equations that can be solved for nine unknowns; eight of them are the reaction forces at the joints. We arrange these nine equations presented in the preceding three equations and write them in the following matrix form:

$$\begin{bmatrix} 1 & 0 & -1 & 0 & 0 & 0 & 0 & 0 & 0 \\ 0 & 1 & 0 & -1 & 0 & 0 & 0 & 0 & 0 \\ 0 & 0 & 1 & 0 & -1 & 0 & 0 & 0 & 0 \\ 0 & 0 & 0 & 1 & 0 & -1 & 0 & 0 & 0 \\ 0 & 0 & 0 & 0 & 1 & 0 & -1 & 0 & 0 \\ 0 & 0 & 0 & 0 & 0 & 1 & 0 & -1 & 0 \\ A_{71} & A_{72} & A_{73} & A_{74} & 0 & 0 & 0 & 0 & 0 \\ 0 & 0 & A_{83} & A_{84} & A_{85} & A_{86} & 0 & 0 & 0 \\ 0 & 0 & 0 & 0 & A_{95} & A_{96} & A_{97} & A_{98} & 1 \end{bmatrix} \begin{bmatrix} F_x^{12} \\ F_y^{12} \\ F_x^{23} \\ F_y^{23} \\ F_x^{34} \\ F_y^{34} \\ F_x^{41} \\ F_y^{41} \\ M^4 \end{bmatrix} = \begin{bmatrix} m^2 a_x^2 \\ m^2 a_y^2 + m^2 g \\ m^3 a_x^3 - F_x^3 \\ m^3 a_y^3 + m^3 g - F_y^3 \\ m^4 a_x^4 \\ m^4 a_y^4 + m^4 g \\ J^2 \ddot{\theta}^2 - M^2 \\ J^3 \ddot{\theta}^3 \\ J^4 \ddot{\theta}^4 \end{bmatrix} \quad (4.74)$$

where

$$\left.\begin{aligned} A_{71} &= l_O^2 \sin\theta^2, & A_{72} &= -l_O^2 \cos\theta^2 \\ A_{73} &= l_A^2 \sin\theta^2, & A_{74} &= -l_A^2 \cos\theta^2 \\ A_{83} &= l_A^3 \sin\theta^3, & A_{84} &= -l_A^3 \cos\theta^3 \\ A_{85} &= l_B^3 \sin\theta^3, & A_{86} &= -l_B^3 \cos\theta^3 \\ A_{95} &= l_B^4 \sin\theta^4, & A_{96} &= -l_B^4 \cos\theta^4 \\ A_{97} &= l_C^4 \sin\theta^4, & A_{98} &= -l_C^4 \cos\theta^4 \end{aligned}\right\} \quad (4.75)$$

Equation 74 can be written as

$$\mathbf{Ax} = \mathbf{b} \tag{4.76}$$

where $\mathbf{A}$ is the coefficient matrix

$$\mathbf{A} = \begin{bmatrix} 1 & 0 & -1 & 0 & 0 & 0 & 0 & 0 & 0 \\ 0 & 1 & 0 & -1 & 0 & 0 & 0 & 0 & 0 \\ 0 & 0 & 1 & 0 & -1 & 0 & 0 & 0 & 0 \\ 0 & 0 & 0 & 1 & 0 & -1 & 0 & 0 & 0 \\ 0 & 0 & 0 & 0 & 1 & 0 & -1 & 0 & 0 \\ 0 & 0 & 0 & 0 & 0 & 1 & 0 & -1 & 0 \\ A_{71} & A_{72} & A_{73} & A_{74} & 0 & 0 & 0 & 0 & 0 \\ 0 & 0 & A_{83} & A_{84} & A_{85} & A_{86} & 0 & 0 & 0 \\ 0 & 0 & 0 & 0 & A_{95} & A_{96} & A_{97} & A_{98} & 1 \end{bmatrix} \tag{4.77}$$

and the vectors $\mathbf{x}$ and $\mathbf{b}$ are

$$\left.\begin{aligned} \mathbf{x} &= [F_x^{12} \quad F_y^{12} \quad F_x^{23} \quad F_y^{23} \quad F_x^{34} \quad F_y^{34} \quad F_x^{41} \quad F_y^{41} \quad M^4]^{\mathrm{T}} \\ \mathbf{b} &= [m^2a_x^2 \quad (m^2a_y^2 + m^2g) \quad (m^3a_x^3 - F_x^3) \quad (m^3a_y^3 + m^3g - F_y^3) \quad m^4a_x^4 \\ &\quad (m^4a_y^4 + m^4g) \quad (J^2\ddot{\theta}^2 - M^2) \quad J^3\ddot{\theta}^3 \quad J^4\ddot{\theta}^4]^{\mathrm{T}} \end{aligned}\right\} \tag{4.78}$$

The solution of Eq. 76 can be defined as in the case of the open chain system as $\mathbf{x} = \mathbf{A}^{-1}\mathbf{b}$.

Equilibrium of the Subsystems As in the case of the analysis of open-chain systems, a reduced number of equations can be obtained by studying the equilibrium of subsystems resulting from cuts at selected joints. For example, to obtain three independent equations in terms of M^4 and the reactions F_x^{12} and F_y^{12}, we make a cut at the revolute joint at O. We first study the equilibrium of the crankshaft shown in Fig. 12. By taking the moments of the forces acting on the crankshaft about point A, we obtain the following equation:

$$\begin{aligned} &F_x^{12}l^2 \sin\theta^2 - F_y^{12}l^2 \cos\theta^2 + M^2 + m^2gl_A^2 \cos\theta^2 \\ &\quad = J^2\ddot{\theta}^2 + m^2a_x^2l_A^2 \sin\theta^2 - m^2a_y^2l_A^2 \cos\theta^2 \end{aligned} \tag{4.79}$$

A second equation can be obtained by studying the equilibrium of the system shown in Fig. 13. By taking the moments about joint B, one obtains

$$\begin{aligned} &F_x^{12}(l^2 \sin\theta^2 + l^3 \sin\theta^3) - F_y^{12}(l^2 \cos\theta^2 + l^3 \cos\theta^3) + M^2 \\ &\quad + m^2g(l_A^2 \cos\theta^2 + l^3 \cos\theta^3) + F_x^3l_B^3 \sin\theta^3 \\ &\quad + (m^3g - F_y^3)l_B^3 \cos\theta^3 \end{aligned}$$

$$
\begin{aligned}
&= J^2\ddot{\theta}^2 + m^2a_x^2(l_A^2 \sin\theta^2 + l^3 \sin\theta^3) \\
&\quad - m^2a_y^2(l_A^2 \cos\theta^2 + l^3 \cos\theta^3) \\
&\quad + J^3\ddot{\theta}^3 + m^3a_x^3 l_B^3 \sin\theta^3 - m^3a_y^3 l_B^3 \cos\theta^3
\end{aligned} \tag{4.80}
$$

which, upon using Eq. 79, can be reduced to

$$
\begin{aligned}
&F_x^{12}l^3 \sin\theta^3 - F_y^{12}l^3 \cos\theta^3 + m^2gl^3 \cos\theta^3 + F_x^3 l_B^3 \sin\theta^3 \\
&\quad + (m^3g - F_y^3)l_B^3 \cos\theta^3 \\
&= m^2a_x^2 l^3 \sin\theta^3 - m^2a_y^2 l^3 \cos\theta^3 + J^3\ddot{\theta}^3 + m^3a_x^3 l_B^3 \sin\theta^3 \\
&\quad - m^3a_y^3 l_B^3 \cos\theta^3
\end{aligned} \tag{4.81}
$$

A third equation can be obtained by examining the system shown in Fig. 14. By taking the moments about point C, the following equation can be obtained:

$$
\begin{aligned}
&-F_y^{12}l^1 + M^2 + m^2g(l_A^2 \cos\theta^2 + l^3 \cos\theta^3 + l^4 \cos\theta^4) \\
&\quad + F_x^3(l_B^3 \sin\theta^3 + l^4 \sin\theta^4) + (m^3g - F_y^3)(l_B^3 \cos\theta^3 + l^4 \cos\theta^4) \\
&\quad + M^4 + m^4gl_C^4 \cos\theta^4 \\
&= J^2\ddot{\theta}^2 + m^2a_x^2(l_A^2 \sin\theta^2 + l^3 \sin\theta^3 + l^4 \sin\theta^4) \\
&\quad - m^2a_y^2(l_A^2 \cos\theta^2 + l^3 \cos\theta^3 + l^4 \cos\theta^4) + J^3\ddot{\theta}^3 \\
&\quad + m^3a_x^3(l_B^3 \sin\theta^3 + l^4 \sin\theta^4) \\
&\quad - m^3a_y^3(l_B^3 \cos\theta^3 + l^4 \cos\theta^4) + J^4\ddot{\theta}^4 + m^4a_x^4 l_C^4 \sin\theta^4 \\
&\quad - m^4a_y^4 l_C^4 \cos\theta^4
\end{aligned} \tag{4.82}
$$

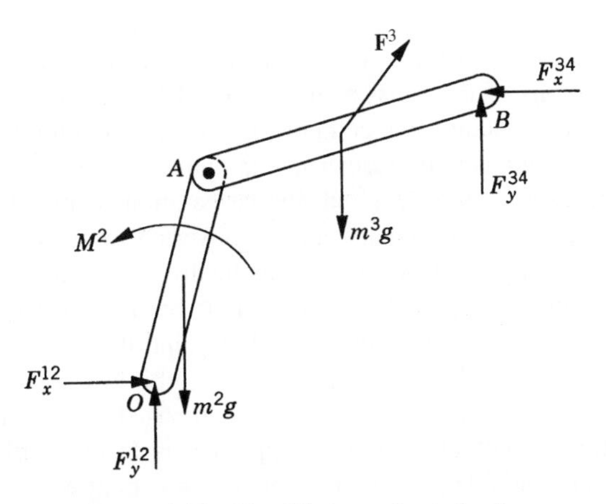

Figure 4.13 Equilibrium of two bodies

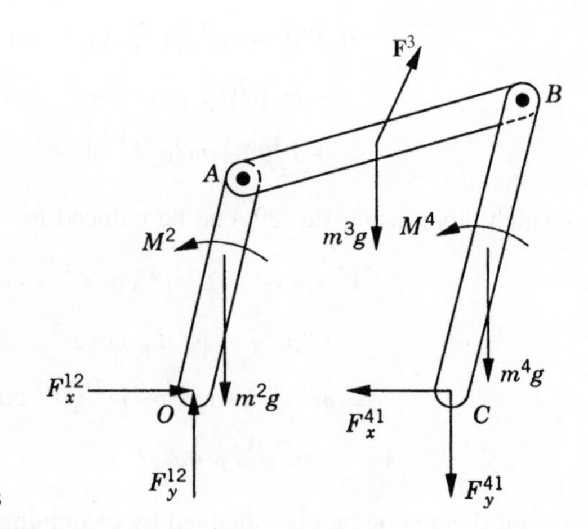

Figure 4.14 Equilibrium of three bodies

where

$$l^1 = l^2 \cos\theta^2 + l^3 \cos\theta^3 + l^4 \cos\theta^4 \tag{4.83}$$

By using the first two moment equations about A and B (Eqs. 79 and 81), the third equation (Eq. 82) reduces to

$$\begin{aligned} & F_x^{12} l^4 \sin\theta^4 + (m^2 g - F_y^{12}) l^4 \cos\theta^4 + F_x^3 l^4 \sin\theta^4 \\ & + (m^3 g - F_y^3) l^4 \cos\theta^4 + M^4 + m^4 g l_C^4 \cos\theta^4 \\ & = (m^2 a_x^2 + m^3 a_x^3) l^4 \sin\theta^4 - (m^2 a_y^2 + m^3 a_y^3) l^4 \cos\theta^4 \\ & + J^4 \ddot{\theta}^4 + m^4 a_x^4 l_C^4 \sin\theta^4 - m^4 a_y^4 l_C^4 \cos\theta^4 \end{aligned} \tag{4.84}$$

Equations 79, 81, and 84 can be solved for the three unknowns: the two reactions F_x^{12} and F_y^{12} and the external moment M^4. Using a similar procedure, another set of three equations in terms of F_x^{23}, F_y^{23}, and M^4, or in terms of F_x^{34}, F_y^{34}, and M^4, or in terms of F_x^{41}, F_y^{41}, and M^4 can be obtained. It is clear, however, that unlike the case of open kinematic chains the equilibrium conditions of the subsystems of closed kinematic chains will always lead to a set of equations that contain some components of the reaction forces. Obviously, these reaction forces can be eliminated by further manipulations of the resulting equations.

The source of the extra efforts required for the solution of the closed-chain equations can be understood because such a chain can be converted to an open chain by making a cut at a selected secondary joint. One can systematically derive the equations of motion of the resulting open chain and augment these equations by a set of algebraic equations that describe the connectivity conditions at the secondary joint, thereby defining the differential and algebraic equations of the closed chain. Lagrange multipliers and the augmented formulation, which is discussed in detail in Chapter 6, can be used to solve the chain differential and algebraic equations. Another alternative approach is to use further manipulations to eliminate the dependent variables using the algebraic constraint equations of the secondary joint. In the latter case, a procedure similar to the embedding technique can be employed.

4.11 CONCLUDING REMARKS

In this chapter, different forms of the dynamic equations of motion were presented. These different forms were developed using elementary Newtonian mechanics. Among the forms discussed in this chapter, two forms are widely used in computational dynamics: the augmented formulation and the embedding technique. The augmented formulation leads to a relatively large system of equations expressed in terms of a redundant set of coordinates. As a result of this redundancy, the coordinates are not independent and they are related by a set of kinematic constraints. As was pointed out, the number of dependent coordinates used in the augmented formulation is equal to the number of independent constraint forces. By using the equations of motion and the constraint equations, a number of equations equal to the number of unknown variables can be obtained. The augmented formulation leads to a sparse matrix structure and is used as the basis for developing many of the general-purpose multibody computer programs. Its drawbacks are the increase in problem dimensionality and the need for using a more elaborate numerical algorithm to solve the resulting system of differential and algebraic equations, as discussed in Chapter 6. A systematic construction of the equations of motion of multibody systems using the augmented formulation is also presented in detail in Chapter 6.

In the embedding technique, the vector of the system accelerations is expressed in terms of independent accelerations using the velocity transformation matrix. This kinematic relationship is used to obtain a minimum set of differential equations expressed in terms of the independent accelerations only. It was demonstrated that the use of the embedding technique leads to elimination of the constraint forces. In the following chapter we discuss the principle of virtual work, which can be used to systematically eliminate the constraint forces and obtain a minimum set of differential equations of motion.

PROBLEMS

1. Discuss the relationship between the number of dependent coordinates used to describe the dynamics of a multibody system and the number of the constraint forces that appear in the dynamic equations.
2. What are the advantages and drawbacks of the augmented formulation?
3. Discuss the sparse matrix structure of the augmented formulation and how such a structure can be utilized in the computer implementation of this formulation.
4. Can you formulate the pin joint constraints of Eqs. 14 and 15 to obtain a symmetric coefficient matrix in Eq. 24?
5. Develop the equations of motion of the two-link robotic system shown in Fig. 8 using the augmented formulation.
6. Develop the equations of motion of the four-bar mechanism shown in Fig. 11 using the augmented formulation.
7. What are the advantages and drawbacks of the embedding technique?
8. What is the role of the velocity transformation matrix in the embedding technique?

9. Discuss the basic differences between the techniques presented in Sections 6 and 7.

10. Develop the equations of motion of the two-link robotic system shown in Fig. 8 using the embedding technique.

11. Develop the equations of motion of the four-bar mechanism shown in Fig. 11 using the embedding technique.

12. What are the differences between the augmented and amalgamated formulations?

13. What are the differences between open and closed kinematic chains when the equations of motion are formulated?

CHAPTER 5

VIRTUAL WORK AND LAGRANGIAN DYNAMICS

The principle of virtual work represents a powerful tool for deriving the static and dynamic equations of multibody systems. Unlike Newtonian mechanics, the principle of virtual work does not require considering the constraint forces, and it requires only scalar work quantities to define the static and dynamic equations. This principle can be used to systematically derive a minimum set of equations of motion of the multibody systems by eliminating the constraint forces. To use the principle of virtual work, the important concepts of the *virtual displacements* and *generalized forces* are first introduced and used to formulate the generalized forces of several force elements, such as springs and dampers and friction forces. It is shown in this chapter that the principle of virtual work can be used to obtain a number of equations equal to the number of the system degrees of freedom, thereby providing a systematic procedure for obtaining the embedding form of the equations of motion of the mechanical system. Use of the principle of virtual work in statics and dynamics is demonstrated using several applications. The principle of virtual work is also used in this chapter to derive the well-known Lagrange's equation, in which the generalized inertia force is expressed in terms of the scalar kinetic energy. Several other forms of the generalized inertia forces are also presented, including the form that appears in the *Gibbs–Appel equation*, in which the generalized inertia is expressed in terms of an acceleration function. The *Hamiltonian formulation* and the relationship between the virtual work and the Gaussian elimination are discussed in the last two sections of this chapter.

Computational Dynamics, Third Edition Ahmed A. Shabana

5.1 VIRTUAL DISPLACEMENTS

An important step in the application of the principle of virtual work is the definition of the virtual displacements and generalized forces. The concept of the generalized forces is introduced in Section 3, while the concept of the virtual displacement is discussed in this and the following section. Throughout this section and the sections that follow, the term *generalized coordinates* is used to refer to any set of coordinates used to describe the system configuration.

The *virtual displacement* is defined to be an infinitesimal displacement that is consistent with the kinematic constraints imposed on the motion of the system. Virtual displacements are imaginary in the sense that they are assumed to occur while time is held fixed. Consider, for instance, the displacement of the unconstrained body shown in Fig. 1. The position vector of an arbitrary point P^i on the rigid body is given by

$$\mathbf{r}_P^i = \mathbf{R}^i + \mathbf{A}^i\overline{\mathbf{u}}_P^i \tag{5.1}$$

where $\mathbf{R}^i$ is the position vector of the reference point, $\overline{\mathbf{u}}_P^i$ is the position vector of point P^i with respect to the reference point O^i, and $\mathbf{A}^i$ is the transformation matrix given by

$$\mathbf{A}^i = \begin{bmatrix} \cos\theta^i & -\sin\theta^i \\ \sin\theta^i & \cos\theta^i \end{bmatrix} \tag{5.2}$$

where θ^i is the angle that defines the orientation of the body. A virtual change in the position vector of point P^i of Eq. 1 is denoted as $\delta\mathbf{r}_P^i$ and is given by

$$\delta\mathbf{r}_P^i = \delta\mathbf{R}^i + \delta(\mathbf{A}^i\overline{\mathbf{u}}_P^i) \tag{5.3}$$

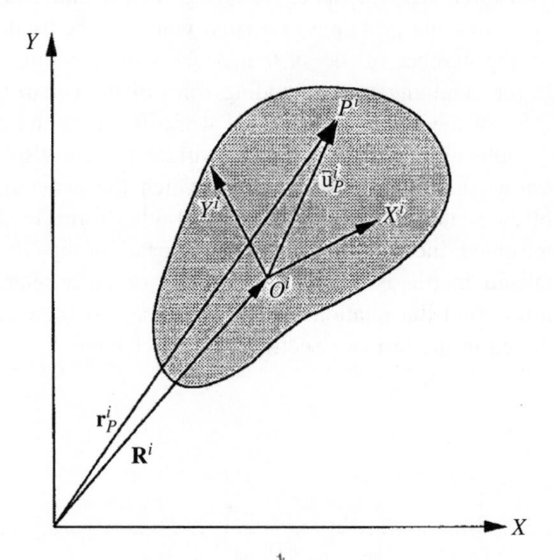

Figure 5.1 Position vector

Since the vector $\mathbf{A}^i\bar{\mathbf{u}}_P^i$ depends only on one variable, namely the angular orientation θ^i, Eq. 3 can be rewritten as

$$\delta\mathbf{r}_P^i = \delta\mathbf{R}^i + \mathbf{A}_\theta^i\bar{\mathbf{u}}_P^i\,\delta\theta^i \tag{5.4}$$

where $\mathbf{A}_\theta^i$ is the partial derivative of $\mathbf{A}^i$ with respect to the angle θ^i, that is,

$$\mathbf{A}_\theta^i = \frac{\partial\mathbf{A}^i}{\partial\theta^i} = \begin{bmatrix} -\sin\theta^i & -\cos\theta^i \\ \cos\theta^i & -\sin\theta^i \end{bmatrix} \tag{5.5}$$

In Eq. 4, the virtual change in the position vector of an arbitrary point on the body is expressed in terms of the virtual changes in the body coordinates, or in this case the body degrees of freedom. Equation 4 can also be written as

$$\delta\mathbf{r}_P^i = \mathbf{r}_{\mathbf{q}^i}^i\,\delta\mathbf{q}^i \tag{5.6}$$

where

$$\left.\begin{aligned} \mathbf{q}^i &= [\mathbf{R}^{i^{\mathrm{T}}} \quad \theta^i]^{\mathrm{T}} \\ \mathbf{r}_{\mathbf{q}^i}^i &= \frac{\partial\mathbf{r}^i}{\partial\mathbf{q}^i} = [\mathbf{I} \quad \mathbf{A}_\theta^i\bar{\mathbf{u}}_P^i] \end{aligned}\right\} \tag{5.7}$$

Clearly, if the reference point O^i is fixed, as in the case of a simple pendulum, we have $\delta\mathbf{R}^i = \mathbf{0}$ and Eq. 4 reduces to $\delta\mathbf{r}_P^i = \mathbf{A}_\theta^i\bar{\mathbf{u}}_P^i\,\delta\theta^i$.

Example 5.1

For the two-degree-of-freedom manipulator shown in Fig. 2, express the virtual change in the position of point P (end effector) in terms of the virtual changes in the system degrees of freedom.

Solution. The position vector of point P is given by

$$\mathbf{r}_P = \begin{bmatrix} l^2\cos\theta^2 + l^3\cos\theta^3 \\ l^2\sin\theta^2 + l^3\sin\theta^3 \end{bmatrix}$$

where l^2 and l^3 are, respectively, the lengths of links 2 and 3 (the fixed link is denoted as body 1), and θ^2 and θ^3 are, respectively, the angular orientations of links 2 and 3. By taking a virtual change in the position vector of point P, we obtain

$$\delta\mathbf{r}_P = \begin{bmatrix} -l^2\sin\theta^2\delta\theta^2 - l^3\sin\theta^3\delta\theta^3 \\ l^2\cos\theta^2\delta\theta^2 + l^3\cos\theta^3\delta\theta^3 \end{bmatrix}$$

which can be written as

$$\delta\mathbf{r}_P = \begin{bmatrix} -l^2\sin\theta^2 & -l^3\sin\theta^3 \\ l^2\cos\theta^2 & l^3\cos\theta^3 \end{bmatrix}\begin{bmatrix} \delta\theta^2 \\ \delta\theta^3 \end{bmatrix}$$

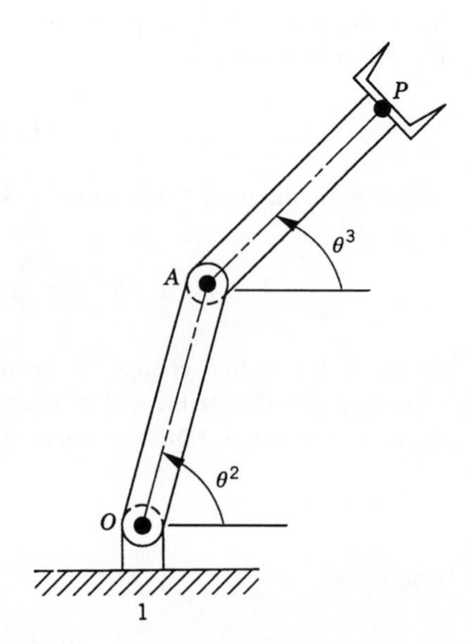

Figure 5.2 Two-degree-of-freedom manipulator

Virtual displacements can be regarded as partial differentials with time assumed to be fixed. Thus, the differential of time is taken to be zero. To explain the difference between the actual displacement and the virtual displacement, we consider the case of a position vector that is an explicit function of the generalized coordinates $\mathbf{q}$ and time t. This vector can be written as

$$\mathbf{r} = \mathbf{r}(\mathbf{q}, t) \tag{5.8}$$

Differentiating this equation with respect to time, one obtains

$$\frac{d\mathbf{r}}{dt} = \frac{\partial \mathbf{r}}{\partial \mathbf{q}} \dot{\mathbf{q}} + \frac{\partial \mathbf{r}}{\partial t} \tag{5.9}$$

Multiplying both sides of this equation by dt yields the actual differential displacement as

$$d\mathbf{r} = \frac{\partial \mathbf{r}}{\partial \mathbf{q}} d\mathbf{q} + \frac{\partial \mathbf{r}}{\partial t} dt \tag{5.10}$$

If $\mathbf{r}$ is not an explicit function of time, the virtual displacement $\delta\mathbf{r}$ and the actual differential displacement $d\mathbf{r}$ are the same provided that the partial differential $\delta\mathbf{q}$ is the same as the total differential $d\mathbf{q}$. It follows that in the case of an n-dimensional vector of generalized coordinates, one has

$$\begin{aligned} \delta\mathbf{r} &= \frac{\partial \mathbf{r}}{\partial q_1} \delta q_1 + \frac{\partial \mathbf{r}}{\partial q_2} \delta q_2 + \cdots + \frac{\partial \mathbf{r}}{\partial q_n} \delta q_n \\ &= \sum_{j=1}^{n} \frac{\partial \mathbf{r}}{\partial q_j} \delta q_j = \frac{\partial \mathbf{r}}{\partial \mathbf{q}} \delta\mathbf{q} \end{aligned} \tag{5.11}$$

where q_j is the jth generalized coordinate, and $\frac{\partial \mathbf{r}}{\partial \mathbf{q}} = \left[\frac{\partial \mathbf{r}}{\partial q_1} \; \frac{\partial \mathbf{r}}{\partial q_2} \; \cdots \; \frac{\partial \mathbf{r}}{\partial q_n} \right]$.

5.2 KINEMATIC CONSTRAINTS AND COORDINATE PARTITIONING

In constrained multibody systems, the system coordinates are related by a set of kinematic constraint equations as the result of mechanical joints or specified motion trajectories. If the system is not kinematically driven, the number of the kinematic constraint equations n_c is less than the number of the system coordinates n. In this case, the constraint kinematic relationships can be used to write a subset of the coordinates in terms of the others. Therefore, the coordinates of a multibody system can be divided into two groups: the first group is the set of *dependent coordinates* $\mathbf{q}_d$ and the second group is the set of *independent coordinates* or the *degrees of freedom* of the system $\mathbf{q}_i$. The number of dependent coordinates is equal to the number of the kinematic constraint equations n_c and the number of independent coordinates is equal to $n - n_c$. By using the kinematic relationships, the virtual changes in the dependent coordinates can be expressed in terms of the virtual changes of the independent coordinates. Consider, for example, the slider crank mechanism shown in Fig. 3. The loop-closure equations for this mechanism can be written in a vector form as

$$\mathbf{r}^2 + \mathbf{r}^3 + \mathbf{r}^1 = \mathbf{0} \tag{5.12}$$

which can be written more explicitly as

$$\left.\begin{aligned} l^2 \cos\theta^2 + l^3 \cos\theta^3 &= R_x^4 \\ l^2 \sin\theta^2 + l^3 \sin\theta^3 &= 0 \end{aligned}\right\} \tag{5.13}$$

where l^2 and l^3 are the lengths of the crankshaft and the connecting rod, and R_x^4 is the position of the slider block with respect to point O. By taking the virtual changes of the coordinates, the preceding equation leads to

$$\left.\begin{aligned} -l^2 \sin\theta^2\, \delta\theta^2 - l^3 \sin\theta^3\, \delta\theta^3 &= \delta R_x^4 \\ l^2 \cos\theta^2\, \delta\theta^2 + l^3 \cos\theta^3\, \delta\theta^3 &= 0 \end{aligned}\right\} \tag{5.14}$$

One may select the dependent coordinates to be θ^3 and R_x^4, that is,

$$\mathbf{q}_d = [\theta^3 \quad R_x^4]^{\mathrm{T}} \tag{5.15}$$

where $\mathbf{q}_d$ is the n_c-dimensional vector of dependent coordinates. Since the mechanism has only one degree of freedom, there is only one independent coordinate that can be selected

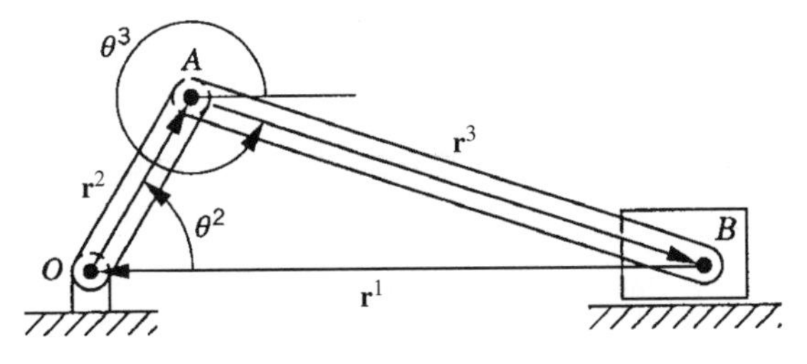

Figure 5.3 Slider crank mechanism

as θ^2, that is, $\mathbf{q}_i = \theta^2$. One may then rearrange Eq. 14 and write this equation using matrix notation as

$$\begin{bmatrix} -l^3 \sin\theta^3 & -1 \\ -l^3 \cos\theta^3 & 0 \end{bmatrix} \begin{bmatrix} \delta\theta^3 \\ \delta R_x^4 \end{bmatrix} = \begin{bmatrix} \sin\theta^2 \\ \cos\theta^2 \end{bmatrix} l^2\,\delta\theta^2 \tag{5.16}$$

which defines $\delta\theta^3$ and δR_x^4 in terms of the independent coordinate $\delta\theta^2$ as

$$\begin{bmatrix} \delta\theta^3 \\ \delta R_x^4 \end{bmatrix} = \frac{-1}{l^3\cos\theta^3} \begin{bmatrix} l^2\cos\theta^2 \\ l^2 l^3 \sin(\theta^2 - \theta^3) \end{bmatrix} \delta\theta^2 \tag{5.17}$$

It is clear from this equation that a singular configuration occurs when $\theta^3 = \pi/2$ or $3\pi/2$. At this singular configuration $\delta\theta^3$ and δR_x^4 cannot be expressed in terms of $\delta\theta^2$.

Alternatively, one may try to express $\delta\theta^2$ and $\delta\theta^3$ in terms of δR_x^4 using Eq. 14. This leads to

$$\begin{bmatrix} -l^2\sin\theta^2 & -l^3\sin\theta^3 \\ l^2\cos\theta^2 & l^3\cos\theta^3 \end{bmatrix} \begin{bmatrix} \delta\theta^2 \\ \delta\theta^3 \end{bmatrix} = \begin{bmatrix} \delta R_x^4 \\ 0 \end{bmatrix} \tag{5.18}$$

The solution of this matrix equation is

$$\begin{bmatrix} \delta\theta^2 \\ \delta\theta^3 \end{bmatrix} = \frac{1}{l^2 l^3 \sin(\theta^2 - \theta^3)} \begin{bmatrix} -l^3\cos\theta^3 \\ l^2\cos\theta^2 \end{bmatrix} \delta R_x^4 \tag{5.19}$$

In this case, singularity occurs whenever θ^2 is equal to θ^3.

Example 5.2

For the four-bar linkage shown in Fig. 4, obtain an expression for the virtual changes in the angular orientations of the coupler and the rocker in terms of the virtual change of the angular orientation of the crankshaft.

Solution. The loop-closure equations for this mechanism are

$$l^2\cos\theta^2 + l^3\cos\theta^3 + l^4\cos\theta^4 - l^1 = 0$$

$$l^2\sin\theta^2 + l^3\sin\theta^3 + l^4\sin\theta^4 = 0$$

where l^2, l^3, and l^4 are, respectively, the lengths of the crankshaft, coupler, and rocker; l^1 is the distance between points O and C; and θ^2, θ^3, and θ^4 are, respectively, the angular orientations of the crankshaft, coupler, and rocker. By taking virtual changes in the coordinates θ^2, θ^3, and θ^4, and keeping in mind that l^1 is constant, the loop-closure equations yield

$$l^2\sin\theta^2\,\delta\theta^2 + l^3\sin\theta^3\,\delta\theta^3 + l^4\sin\theta^4\,\delta\theta^4 = 0$$

$$l^2\cos\theta^2\,\delta\theta^2 + l^3\cos\theta^3\,\delta\theta^3 + l^4\cos\theta^4\,\delta\theta^4 = 0$$

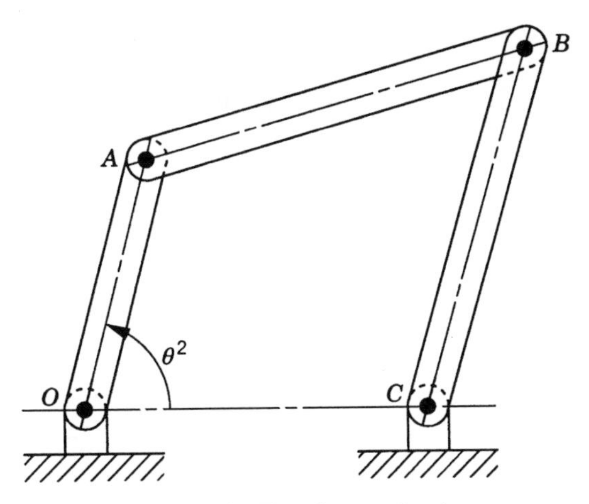

Figure 5.4 Four-bar mechanism

The coordinates θ^3 and θ^4 may be selected as dependent coordinates and θ^2 as the independent coordinate leading to the following relationship:

$$\begin{bmatrix} l^3 \sin\theta^3 & l^4 \sin\theta^4 \\ l^3 \cos\theta^3 & l^4 \cos\theta^4 \end{bmatrix} \begin{bmatrix} \delta\theta^3 \\ \delta\theta^4 \end{bmatrix} = -\begin{bmatrix} \sin\theta^2 \\ \cos\theta^2 \end{bmatrix} l^2\,\delta\theta^2$$

or

$$\begin{bmatrix} \delta\theta^3 \\ \delta\theta^4 \end{bmatrix} = \frac{1}{l^3 l^4 \sin(\theta^3 - \theta^4)} \begin{bmatrix} l^2 l^4 \sin(\theta^4 - \theta^2) \\ l^2 l^3 \sin(\theta^2 - \theta^3) \end{bmatrix} \delta\theta^2$$

A similar procedure can be used if $\delta\theta^3$ or $\delta\theta^4$ is selected to be the independent coordinate.

Constraint Jacobian Matrix One may generalize the procedure described in this section for expressing the virtual changes of the dependent coordinates in terms of the virtual changes of the independent ones. This can be demonstrated by writing the algebraic kinematic constraint equations between coordinates in the following general form:

$$\mathbf{C}(\mathbf{q}, t) = \mathbf{0} \tag{5.20}$$

where $\mathbf{q} = [q_1 \quad q_2 \cdots q_n]^{\mathrm{T}}$ is the vector of the system coordinates, t is time, and $\mathbf{C}$ is the vector of constraint functions, which can be written as

$$\mathbf{C} = [C_1(\mathbf{q}, t) \quad C_2(\mathbf{q}, t) \quad \cdots \quad C_{n_c}(\mathbf{q}, t)]^{\mathrm{T}} \tag{5.21}$$

where n_c is the total number of constraint equations that are assumed to be linearly independent. If the system is dynamically driven, the number of constraint equations n_c is less than the number of the coordinates n.

Equation 20, as the result of a virtual change in the vector of system coordinates, leads to

$$\mathbf{C_q}\,\delta\mathbf{q} = \mathbf{0} \tag{5.22}$$

where $\mathbf{C_q}$ is the *constraint Jacobian matrix* defined as

$$\mathbf{C_q} = \frac{\partial \mathbf{C}}{\partial \mathbf{q}} = \begin{bmatrix} \dfrac{\partial C_1}{\partial q_1} & \dfrac{\partial C_1}{\partial q_2} & \cdots & \dfrac{\partial C_1}{\partial q_n} \\ \dfrac{\partial C_2}{\partial q_1} & \dfrac{\partial C_2}{\partial q_2} & \cdots & \dfrac{\partial C_2}{\partial q_n} \\ \vdots & \vdots & \ddots & \vdots \\ \dfrac{\partial C_{n_c}}{\partial q_1} & \dfrac{\partial C_{n_c}}{\partial q_2} & \cdots & \dfrac{\partial C_{n_c}}{\partial q_n} \end{bmatrix} \tag{5.23}$$

The constraint Jacobian matrix has a number of rows equal to the number of constraint equations and a number of columns equal to the number of the system coordinates. The vector of coordinates $\mathbf{q}$ can be written in the following partitioned form:

$$\mathbf{q} = [\mathbf{q}_d^{\mathrm{T}} \quad \mathbf{q}_i^{\mathrm{T}}]^{\mathrm{T}} \tag{5.24}$$

where $\mathbf{q}_d$ is an n_c-dimensional vector of the dependent coordinates, and $\mathbf{q}_i$ is an $(n - n_c)$-dimensional vector of independent coordinates. According to this coordinate partitioning, Eq. 22 can be rewritten as

$$\mathbf{C}_{\mathbf{q}_d}\,\delta\mathbf{q}_d + \mathbf{C}_{\mathbf{q}_i}\,\delta\mathbf{q}_i = \mathbf{0} \tag{5.25}$$

where the dependent and independent coordinates are chosen such that the $n_c \times n_c$ matrix $\mathbf{C}_{\mathbf{q}_d}$ is nonsingular. Equation 25 can then be used to write $\delta\mathbf{q}_d$ in terms of $\delta\mathbf{q}_i$ as

$$\delta\mathbf{q}_d = -\mathbf{C}_{\mathbf{q}_d}^{-1}\mathbf{C}_{\mathbf{q}_i}\,\delta\mathbf{q}_i \tag{5.26}$$

or simply as

$$\delta\mathbf{q}_d = \mathbf{C}_{di}\,\delta\mathbf{q}_i \tag{5.27}$$

where

$$\mathbf{C}_{di} = -\mathbf{C}_{\mathbf{q}_d}^{-1}\mathbf{C}_{\mathbf{q}_i} \tag{5.28}$$

By using Eqs. 24 and 27, the virtual change in the total vector of system coordinates can be expressed in terms of the virtual change of the independent coordinates as

$$\delta\mathbf{q} = \begin{bmatrix} \delta\mathbf{q}_d \\ \delta\mathbf{q}_i \end{bmatrix} = \begin{bmatrix} \mathbf{C}_{di} \\ \mathbf{I} \end{bmatrix} \delta\mathbf{q}_i \tag{5.29}$$

This equation can be written as

$$\delta \mathbf{q} = \mathbf{B}_i \, \delta \mathbf{q}_i \tag{5.30}$$

where the matrix $\mathbf{B}_i$ is an $n \times (n - n_c)$ matrix defined as

$$\mathbf{B}_i = \begin{bmatrix} \mathbf{C}_{di} \\ \mathbf{I} \end{bmatrix} \tag{5.31}$$

Example 5.3

The use of the general procedure described in this section can be demonstrated using the four-bar linkage discussed in Example 2. In this example, the vector of coordinates $\mathbf{q}$ is selected to be

$$\mathbf{q} = [\theta^2 \quad \theta^3 \quad \theta^4]^{\mathrm{T}}$$

The kinematic constraint equations that relate these coordinates are defined by the loop-closure equations of Example 2. These constraint equations are

$$\mathbf{C}(\mathbf{q}, t) = \begin{bmatrix} C_1(\mathbf{q}, t) \\ C_2(\mathbf{q}, t) \end{bmatrix} = \begin{bmatrix} l^2 \cos\theta^2 + l^3 \cos\theta^3 + l^4 \cos\theta^4 - l^1 \\ l^2 \sin\theta^2 + l^3 \sin\theta^3 + l^4 \sin\theta^4 \end{bmatrix} = \begin{bmatrix} 0 \\ 0 \end{bmatrix}$$

By taking a virtual change in the system coordinates, one has

$$\mathbf{C}_{\mathbf{q}} \, \delta \mathbf{q} = \begin{bmatrix} -l^2 \sin\theta^2 \, \delta\theta^2 - l^3 \sin\theta^3 \, \delta\theta^3 - l^4 \sin\theta^4 \, \delta\theta^4 \\ l^2 \cos\theta^2 \, \delta\theta^2 + l^3 \cos\theta^3 \, \delta\theta^3 + l^4 \cos\theta^4 \, \delta\theta^4 \end{bmatrix} = \begin{bmatrix} 0 \\ 0 \end{bmatrix}$$

which can also be written as

$$\mathbf{C}_{\mathbf{q}} \, \delta \mathbf{q} = \begin{bmatrix} -l^2 \sin\theta^2 & -l^3 \sin\theta^3 & -l^4 \sin\theta^4 \\ l^2 \cos\theta^2 & l^3 \cos\theta^3 & l^4 \cos\theta^4 \end{bmatrix} \begin{bmatrix} \delta\theta^2 \\ \delta\theta^3 \\ \delta\theta^4 \end{bmatrix} = \begin{bmatrix} 0 \\ 0 \end{bmatrix}$$

From which the Jacobian matrix of the kinematic constraints can be identified as

$$\mathbf{C}_{\mathbf{q}} = \begin{bmatrix} -l^2 \sin\theta^2 & -l^3 \sin\theta^3 & -l^4 \sin\theta^4 \\ l^2 \cos\theta^2 & l^3 \cos\theta^3 & l^4 \cos\theta^4 \end{bmatrix}$$

If θ^2 is selected as the independent coordinate, one has

$$\mathbf{q}_i = \theta^2, \quad \mathbf{q}_d = [\theta^3 \quad \theta^4]^{\mathrm{T}}$$

It follows that

$$\mathbf{C}_{\mathbf{q}_d} \, \delta \mathbf{q}_d + \mathbf{C}_{\mathbf{q}_i} \, \delta \mathbf{q}_i = \begin{bmatrix} -l^3 \sin\theta^3 & -l^4 \sin\theta^4 \\ l^3 \cos\theta^3 & l^4 \cos\theta^4 \end{bmatrix} \begin{bmatrix} \delta\theta^3 \\ \delta\theta^4 \end{bmatrix} + \begin{bmatrix} -l^2 \sin\theta^2 \\ l^2 \cos\theta^2 \end{bmatrix} \delta\theta^2 = \mathbf{0}$$

where

$$\mathbf{C}_{\mathbf{q}_d} = \begin{bmatrix} -l^3 \sin\theta^3 & -l^4 \sin\theta^4 \\ l^3 \cos\theta^3 & l^4 \cos\theta^4 \end{bmatrix}, \qquad \mathbf{C}_{\mathbf{q}_i} = \begin{bmatrix} -l^2 \sin\theta^2 \\ l^2 \cos\theta^2 \end{bmatrix}$$

The virtual changes in the dependent coordinates can then be expressed in terms of the virtual change in the independent coordinate as

$$\delta \mathbf{q}_d = -\mathbf{C}_{\mathbf{q}_d}^{-1} \mathbf{C}_{\mathbf{q}_i}\, \delta \mathbf{q}_i$$

Since in this example

$$\mathbf{C}_{\mathbf{q}_d}^{-1} = \frac{1}{l^3 l^4 \sin(\theta^4 - \theta^3)} \begin{bmatrix} l^4 \cos\theta^4 & l^4 \sin\theta^4 \\ -l^3 \cos\theta^3 & -l^3 \sin\theta^3 \end{bmatrix}$$

the preceding equation yields

$$\begin{aligned} \delta \mathbf{q}_d &= \begin{bmatrix} \delta\theta^3 \\ \delta\theta^4 \end{bmatrix} \\ &= \frac{1}{l^3 l^4 \sin(\theta^3 - \theta^4)} \begin{bmatrix} l^4 \cos\theta^4 & l^4 \sin\theta^4 \\ -l^3 \cos\theta^3 & -l^3 \sin\theta^3 \end{bmatrix} \begin{bmatrix} -l^2 \sin\theta^2 \\ l^2 \cos\theta^2 \end{bmatrix} \delta\theta^2 \\ &= \frac{1}{l^3 l^4 \sin(\theta^3 - \theta^4)} \begin{bmatrix} l^2 l^4 \sin(\theta^4 - \theta^2) \\ l^2 l^3 \sin(\theta^2 - \theta^3) \end{bmatrix} \delta\theta^2 \end{aligned}$$

which is the same result obtained in Example 2. The matrix $\mathbf{C}_{di}$ of Eq. 28 is recognized as

$$\mathbf{C}_{di} = \frac{1}{l^3 l^4 \sin(\theta^3 - \theta^4)} \begin{bmatrix} l^2 l^4 \sin(\theta^4 - \theta^2) \\ l^2 l^3 \sin(\theta^2 - \theta^3) \end{bmatrix}$$

The virtual change in the total vector of system coordinates can be expressed in terms of the virtual change in the independent coordinate as

$$\delta \mathbf{q} = \begin{bmatrix} \delta\theta^3 \\ \delta\theta^4 \\ \delta\theta^2 \end{bmatrix} = \begin{bmatrix} l^2 \sin(\theta^4 - \theta^2)/l^3 \sin(\theta^3 - \theta^4) \\ l^2 \sin(\theta^2 - \theta^3)/l^4 \sin(\theta^3 - \theta^4) \\ 1 \end{bmatrix} \delta\theta^2$$

where the matrix $\mathbf{B}_i$ of Eq. 31 can be recognized as

$$\mathbf{B}_i = \begin{bmatrix} l^2 \sin(\theta^4 - \theta^2)/l^3 \sin(\theta^3 - \theta^4) \\ l^2 \sin(\theta^2 - \theta^3)/l^4 \sin(\theta^3 - \theta^4) \\ 1 \end{bmatrix}$$

Absolute Coordinates As pointed out in Chapter 3, in many general-purpose multibody computer algorithms the *absolute coordinates* are employed for the sake of generality. In this case, the configuration of the rigid body is identified by the global position vector of the

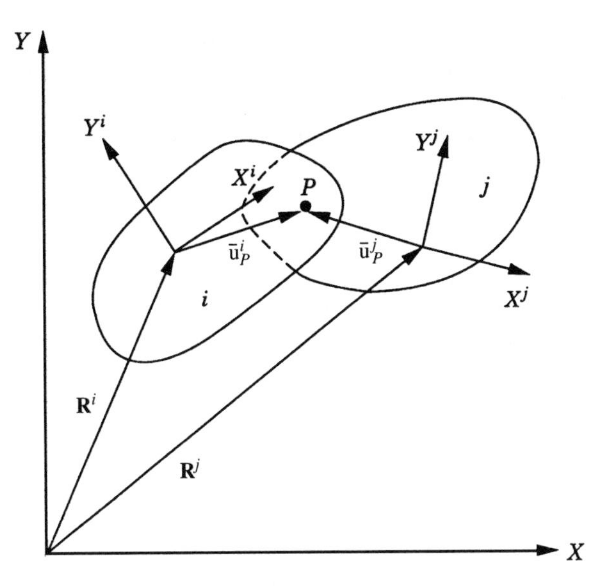

Figure 5.5 Two-body system

origin of the body reference (reference point) and by a set of orientational coordinates that define the orientation of the body in a global fixed frame of reference. Kinematic constraints that represent mechanical joints in the system can be formulated in terms of the absolute coordinates. For example, the algebraic kinematic constraint equations that describe the revolute joint between body i and body j in Fig. 5 can be expressed as

$$\mathbf{r}^i_P - \mathbf{r}^j_P = \mathbf{0} \tag{5.32}$$

where $\mathbf{r}^i_P$ is the position vector of the joint definition point P expressed in terms of the coordinates of body i, and $\mathbf{r}^j_P$ is the position vector of the same point P expressed in terms of the coordinates of body j. Equation 32 can be written in a more explicit form in terms of the absolute coordinates as

$$\mathbf{R}^i + \mathbf{A}^i\bar{\mathbf{u}}^i_P - \mathbf{R}^j - \mathbf{A}^j\bar{\mathbf{u}}^j_P = \mathbf{0} \tag{5.33}$$

where $\mathbf{R}^i$ and $\mathbf{R}^j$ are, respectively, the position vectors of the reference points of body i and body j, $\mathbf{A}^i$ and $\mathbf{A}^j$ are the transformation matrices of body i and body j, and $\bar{\mathbf{u}}^i_P$ and $\bar{\mathbf{u}}^j_P$, as shown in Fig. 5, are the local position vectors of point P defined in the coordinate systems of bodies i and j, respectively.

By taking a virtual change in the absolute coordinates of body i and body j, Eq. 33 leads to

$$\delta\mathbf{R}^i + (\mathbf{A}^i_\theta\bar{\mathbf{u}}^i_P)\,\delta\theta^i - \delta\mathbf{R}^j - (\mathbf{A}^j_\theta\bar{\mathbf{u}}^j_P)\,\delta\theta^j = \mathbf{0} \tag{5.34}$$

Since the revolute joint eliminates two degrees of freedom, one may select $\delta\mathbf{R}^j$ as the vector of dependent coordinates and write this vector in terms of the other absolute coordinates as

$$\delta\mathbf{R}^j = \delta\mathbf{R}^i + \mathbf{A}_\theta^i\bar{\mathbf{u}}_P^i\,\delta\theta^i - \mathbf{A}_\theta^j\bar{\mathbf{u}}_P^j\,\delta\theta^j \tag{5.35}$$

Example 5.4

Figure 6 shows a two-body system that consists of the ground and a rigid rod with a uniform cross-sectional area and length l^2. In this example, three absolute coordinates R_x^i, R_y^i, and θ^i are selected for each body i in the system. The reference point of the rod is assumed to be at its geometric center. If the system is assumed to be dynamically driven, there are only joint constraints that represent the ground and the revolute joint constraints. The ground constraints are

$$R_x^1 = 0, \quad R_y^1 = 0, \quad \theta^1 = 0$$

The revolute joint constraints are

$$\mathbf{R}^2 + \mathbf{A}^2\bar{\mathbf{u}}_O^2 = \mathbf{0}$$

where $\bar{\mathbf{u}}_O^2 = [-l^2/2 \quad 0]^{\mathrm{T}}$, and $\mathbf{A}^2$ is the planar transformation matrix. The revolute joint constraints can be written more explicitly as

$$\begin{bmatrix} R_x^2 - \dfrac{l^2}{2}\cos\theta^2 \\ R_y^2 - \dfrac{l^2}{2}\sin\theta^2 \end{bmatrix} = \begin{bmatrix} 0 \\ 0 \end{bmatrix}$$

The vector of the system generalized coordinates is

$$\mathbf{q} = [q_1 \quad q_2 \quad q_3 \quad \cdots \quad q_6]^{\mathrm{T}} = [R_x^1 \quad R_y^1 \quad \theta^1 \quad R_x^2 \quad R_y^2 \quad \theta^2]^{\mathrm{T}}$$

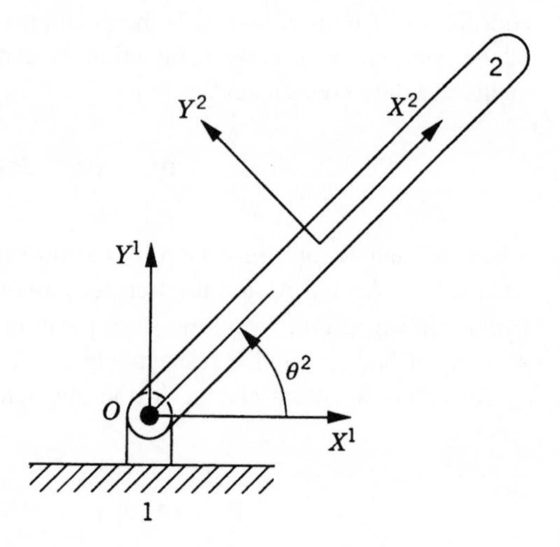

Figure 5.6 Simple pendulum

The vector of the system constraint equations is

$$\mathbf{C}(\mathbf{q}, t) = \begin{bmatrix} R_x^1 \\ R_y^1 \\ \theta^1 \\ R_x^2 - \dfrac{l^2}{2}\cos\theta^2 \\ R_y^2 - \dfrac{l^2}{2}\sin\theta^2 \end{bmatrix} = \begin{bmatrix} 0 \\ 0 \\ 0 \\ 0 \\ 0 \end{bmatrix}$$

and the constraint Jacobian matrix is

$$\mathbf{C_q} = \begin{bmatrix} 1 & 0 & 0 & 0 & 0 & 0 \\ 0 & 1 & 0 & 0 & 0 & 0 \\ 0 & 0 & 1 & 0 & 0 & 0 \\ 0 & 0 & 0 & 1 & 0 & \dfrac{l^2}{2}\sin\theta^2 \\ 0 & 0 & 0 & 0 & 1 & -\dfrac{l^2}{2}\cos\theta^2 \end{bmatrix}$$

In this example, there are six coordinates ($n = 6$) and five constraint equations ($n_c = 5$). Therefore, the system has one degree of freedom. One may select this degree of freedom to be θ^2 and write Eq. 25 as

$$\mathbf{C}_{\mathbf{q}_d}\,\delta\mathbf{q}_d + \mathbf{C}_{\mathbf{q}_i}\,\delta\mathbf{q}_i = \mathbf{0}$$

where in this case $\mathbf{q}_d = [R_x^1 \quad R_y^1 \quad \theta^1 \quad R_x^2 \quad R_y^2]^{\mathrm{T}}$ and $\mathbf{q}_i = \theta^2$. According to this generalized coordinate partitioning, the matrices $\mathbf{C}_{\mathbf{q}_d}$ and $\mathbf{C}_{\mathbf{q}_i}$ can be identified as

$$\mathbf{C}_{\mathbf{q}_d} = \begin{bmatrix} 1 & 0 & 0 & 0 & 0 \\ 0 & 1 & 0 & 0 & 0 \\ 0 & 0 & 1 & 0 & 0 \\ 0 & 0 & 0 & 1 & 0 \\ 0 & 0 & 0 & 0 & 1 \end{bmatrix}, \qquad \mathbf{C}_{\mathbf{q}_i} = \begin{bmatrix} 0 \\ 0 \\ 0 \\ \dfrac{l^2}{2}\sin\theta^2 \\ -\dfrac{l^2}{2}\cos\theta^2 \end{bmatrix}$$

It follows that the matrix $\mathbf{C}_{di}$ of Eq. 28 can be written in this case as

$$\mathbf{C}_{di} = -\mathbf{C}_{\mathbf{q}_d}^{-1}\mathbf{C}_{\mathbf{q}_i} = -\begin{bmatrix} 0 \\ 0 \\ 0 \\ \dfrac{l^2}{2}\sin\theta^2 \\ -\dfrac{l^2}{2}\cos\theta^2 \end{bmatrix}$$

and the matrix $\mathbf{B}_i$ in Eq. 30 is

$$\mathbf{B}_i = \begin{bmatrix} 0 \\ 0 \\ 0 \\ -\dfrac{l^2}{2}\sin\theta^2 \\ \dfrac{l^2}{2}\cos\theta^2 \\ 1 \end{bmatrix}$$

Using Eq. 30, the virtual change in the total vector of the system coordinates can be expressed in terms of the virtual change in the system degrees of freedom as

$$\delta\mathbf{q} = \begin{bmatrix} \delta R_x^1 \\ \delta R_y^1 \\ \delta\theta^1 \\ \delta R_x^2 \\ \delta R_y^2 \\ \delta\theta^2 \end{bmatrix} = \begin{bmatrix} 0 \\ 0 \\ 0 \\ -\dfrac{l^2}{2}\sin\theta^2 \\ \dfrac{l^2}{2}\cos\theta^2 \\ 1 \end{bmatrix}\delta\theta^2$$

Nonholonomic Constraints Joint and driving constraints that can be described by Eq. 20 are called *holonomic constraints* since they can be expressed as algebraic equations in the system coordinates and time. There are other types of constraints that cannot be expressed as functions of the coordinates and time only. These types of constraints, which are known as *nonholonomic constraints*, can be expressed in terms of differentials of the coordinates as

$$\sum_{k=1}^{n} a_{jk}\,dq_k + b_j\,dt = 0 \qquad j = 1, 2, \ldots, n_{cn} \tag{5.36}$$

where n_{cn} is the number of nonholonomic constraint equations, and a_{jk} and b_k can be functions of the system coordinates $\mathbf{q} = [q_1 \quad q_2 \cdots q_n]^{\mathrm{T}}$ as well as time. One should not be able to integrate the preceding equation and write it in terms of the coordinates and time only; otherwise we obtain the form of the holonomic constraints. Hence, one cannot use nonholonomic constraints to eliminate dependent coordinates and, consequently, in this case of a nonholonomic system, the number of independent coordinates is more than the number of independent velocities.

Recall that a differential form is integrable if it is an *exact differential*. In this case, the following conditions hold:

$$\frac{\partial a_{jk}}{\partial q_l} = \frac{\partial a_{jl}}{\partial q_k}, \qquad \frac{\partial a_{jk}}{\partial t} = \frac{\partial b_j}{\partial q_k} \tag{5.37}$$

If these conditions are not satisfied, Eq. 36 is of the nonholonomic type since this equation cannot be integrated and written in the form of Eq. 20. It follows that in the case of a nonholonomic system, none of the constraint equations of Eq. 36 can be written in the form

$$dC(\mathbf{q}, t) = \frac{\partial C}{\partial q_1} dq_1 + \frac{\partial C}{\partial q_2} dq_2 + \cdots + \frac{\partial C}{\partial q_n} dq_n + \frac{\partial C}{\partial t} dt = 0 \tag{5.38}$$

An example of nonholonomic constraints is

$$\left. \begin{array}{l} dq_1 - \sin q_3 \, dq_4 = 0 \\ dq_2 - \cos q_3 \, dq_4 = 0 \end{array} \right\} \tag{5.39}$$

These are two independent constraint equations expressed in terms of the differentials of the four coordinates q_1, q_2, q_3, and q_4. These two equations do not satisfy the conditions of exact differentials of Eq. 37, and therefore, they cannot be integrated and expressed in the form of Eq. 20.

5.3 VIRTUAL WORK

The virtual work of a force vector is defined to be the dot (scalar) product of the force vector and the vector of the virtual change of the position vector of the point of application of the force. Both vectors must be defined in the same coordinate system. The virtual work of a moment that acts on a rigid body is defined to be the product of the moment and the virtual change in the angular orientation of the body. Figure 7 shows a rigid body i that is acted upon by a moment M^i and a force vector $\mathbf{F}^i$ whose point of application is denoted as P^i. The virtual work of this system of forces is given by

$$\delta W^i = \mathbf{F}^{i^{\mathrm{T}}} \delta \mathbf{r}_P^i + M^i \, \delta\theta^i \tag{5.40}$$

where $\mathbf{r}_P^i$ is the position vector of point P^i, and θ^i is the angular orientation of the body i.

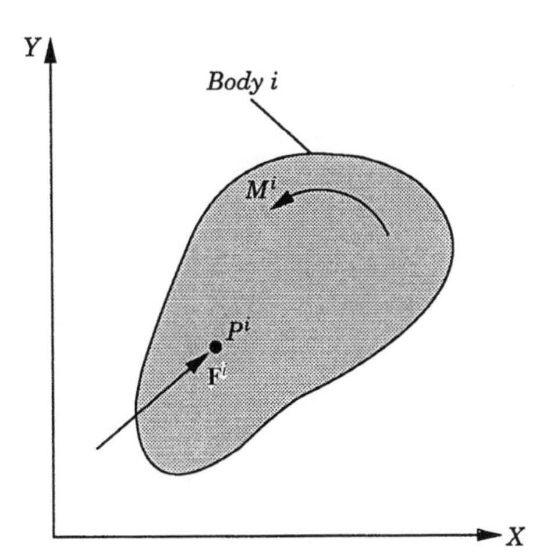

Figure 5.7 Virtual work

Generalized Forces The position vector of an arbitrary point on a rigid body can be expressed in terms of the position vector of the reference point as well as the angular orientation of the body. The coordinates of the rigid body may be defined by the vector $\mathbf{q}^i$ where

$$\mathbf{q}^i = [\mathbf{R}^{i^{\mathrm{T}}} \quad \theta^i]^{\mathrm{T}} \tag{5.41}$$

where $\mathbf{R}^i$ is the position vector of the reference point and θ^i is the angular orientation of the body. In terms of these coordinates, the position of point P^i given by the vector $\mathbf{r}_P^i$ of Eq. 40 can be written as

$$\mathbf{r}_P^i = \mathbf{R}^i + \mathbf{A}^i \bar{\mathbf{u}}_P^i \tag{5.42}$$

where $\mathbf{A}^i$ is the transformation matrix from the body coordinate system to the global coordinate system, and $\bar{\mathbf{u}}_P^i = [\bar{u}_x^i \quad \bar{u}_y^i]^{\mathrm{T}}$ is the local position vector of the point of application of the force $\mathbf{F}^i$. By taking a virtual change in the body coordinates, Eq. 42 yields

$$\delta \mathbf{r}_P^i = \delta \mathbf{R}^i + \mathbf{A}_\theta^i \bar{\mathbf{u}}_P^i \, \delta\theta^i \tag{5.43}$$

Substituting Eq. 43 into Eq. 40, the virtual work of the force $\mathbf{F}^i$ and the moment M^i can be expressed as

$$\begin{aligned} \delta W^i &= \mathbf{F}^{i^{\mathrm{T}}} (\delta \mathbf{R}^i + \mathbf{A}_\theta^i \bar{\mathbf{u}}_P^i \, \delta\theta^i) + M^i \, \delta\theta^i \\ &= \mathbf{F}^{i^{\mathrm{T}}} \delta \mathbf{R}^i + (\mathbf{F}^{i^{\mathrm{T}}} \mathbf{A}_\theta^i \bar{\mathbf{u}}_P^i + M^i) \, \delta\theta^i \end{aligned} \tag{5.44}$$

This equation can be written as

$$\delta W^i = \mathbf{Q}_R^{i^{\mathrm{T}}} \delta \mathbf{R}^i + Q_\theta^i \, \delta\theta^i \tag{5.45}$$

where

$$\mathbf{Q}_R^i = \mathbf{F}^i, \quad Q_\theta^i = M^i + \bar{\mathbf{u}}_P^{i^{\mathrm{T}}} \mathbf{A}_\theta^{i^{\mathrm{T}}} \mathbf{F}^i \tag{5.46}$$

The vector $\mathbf{Q}_R^i$ is called the vector of *generalized forces* associated with the coordinates of the reference point, and the scalar Q_θ^i is called the generalized force associated with the rotation of the body. The second term in the second equation of Eq. 46, which is the contribution of the force $\mathbf{F}^i$ to the generalized force associated with the rotation of the body, can be written as

$$\begin{aligned} \bar{\mathbf{u}}_P^{i^{\mathrm{T}}} \mathbf{A}_\theta^{i^{\mathrm{T}}} \mathbf{F}^i &= [\bar{u}_x^i \quad \bar{u}_y^i] \begin{bmatrix} -\sin\theta^i & \cos\theta^i \\ -\cos\theta^i & -\sin\theta^i \end{bmatrix} \begin{bmatrix} F_x^i \\ F_y^i \end{bmatrix} \\ &= -F_x^i(\bar{u}_x^i \sin\theta^i + \bar{u}_y^i \cos\theta^i) + F_y^i(\bar{u}_x^i \cos\theta^i - \bar{u}_y^i \sin\theta^i) \end{aligned} \tag{5.47}$$

One can verify that this equation also takes the following form:

$$\bar{\mathbf{u}}_P^{i^{\mathrm{T}}} \mathbf{A}_\theta^{i^{\mathrm{T}}} \mathbf{F}^i = (\mathbf{u}_P^i \times \mathbf{F}^i) \cdot \mathbf{k} \tag{5.48}$$

or

$$\bar{\mathbf{u}}_P^{i\mathrm{T}} \mathbf{A}_\theta^{i\mathrm{T}} \mathbf{F}^i = [\mathbf{A}^i (\bar{\mathbf{u}}_P^i \times \bar{\mathbf{F}}^i)] \cdot \mathbf{k} \tag{5.49}$$

where $\mathbf{k}$ is a unit vector along the Z axis, and $\mathbf{u}_P^i = \mathbf{A}^i \bar{\mathbf{u}}_P^i$, $\bar{\mathbf{F}}^i = \mathbf{A}^{i\mathrm{T}} \mathbf{F}^i$. It follows that the second equation of Eq. 46 can simply be written as

$$Q_\theta^i = M^i + (\mathbf{u}_P^i \times \mathbf{F}^i) \cdot \mathbf{k} \tag{5.50}$$

Equations 46 and 50 imply that a force vector $\mathbf{F}^i$ that acts at an arbitrary point P^i is equivalent (equipollent) to another system of forces that consists of the same force $\mathbf{F}^i$ acting at the reference point and a moment $(\mathbf{u}_p^i \times \mathbf{F}^i) \cdot \mathbf{k}$ associated with the rotation of the body.

Generalization The method discussed in this section for obtaining the generalized forces can be generalized to any number of forces and moments. The procedure is to express the position vectors of the points of application of the forces in terms of the system coordinates. Substituting the resulting kinematic relationships into the expression for the virtual work leads to the definition of the generalized forces associated with the system coordinates. For example, if the configuration of the multibody system is described by the n coordinates

$$\mathbf{q} = [q_1 \quad q_2 \quad \cdots \quad q_n]^{\mathrm{T}} \tag{5.51}$$

The virtual work of the forces acting on the system can be expressed in the general form

$$\delta W = Q_1 \, \delta q_1 + Q_2 \, \delta q_2 + \cdots + Q_n \, \delta q_n \tag{5.52}$$

where Q_j is the generalized force associated with the jth coordinate q_j.

Equation 52 can be written in a vector form as

$$\delta W = \mathbf{Q}^{\mathrm{T}} \delta \mathbf{q} \tag{5.53}$$

where $\mathbf{Q}$ is the vector of generalized forces and $\delta \mathbf{q}$ is the vector of the virtual changes in the coordinates. The vectors $\mathbf{Q}$ and $\delta \mathbf{q}$ are

$$\left.\begin{aligned} \mathbf{Q} &= [Q_1 \quad Q_2 \quad \cdots \quad Q_n]^{\mathrm{T}} \\ \delta \mathbf{q} &= [\delta q_1 \quad \delta q_2 \quad \cdots \quad \delta q_n]^{\mathrm{T}} \end{aligned}\right\} \tag{5.54}$$

Coordinate Transformation Equation 52 or its equivalent vector form of Eq. 53 defines the generalized forces associated with the coordinates $\mathbf{q} = [q_1 \quad q_2 \cdots q_n]^{\mathrm{T}}$. The generalized forces associated with another set of coordinates can be obtained if the transformation between the two sets of coordinates is defined. Let $\mathbf{p} = [p_1 \quad p_2 \cdots p_m]^{\mathrm{T}}$ be another set of coordinates such that the vector of virtual changes in $\mathbf{q}$ can be expressed in terms of the virtual changes in $\mathbf{p}$ as

$$\delta \mathbf{q} = \mathbf{B}_{qp} \, \delta \mathbf{p} \tag{5.55}$$

By substituting Eq. 55 into Eq. 53, one obtains

$$\delta W = \mathbf{Q}^{\mathrm{T}}\mathbf{B}_{qp}\,\delta\mathbf{p} = \mathbf{Q}_p^{\mathrm{T}}\,\delta\mathbf{p} \tag{5.56}$$

where

$$\mathbf{Q}_p = \mathbf{B}_{qp}^{\mathrm{T}}\mathbf{Q} = [Q_{p1} \quad Q_{p2} \quad \cdots \quad Q_{pm}]^{\mathrm{T}} \tag{5.57}$$

is the vector of generalized forces associated with the vector of coordinates $\mathbf{p}$.

Example 5.5

As an illustrative example, the slider crank mechanism shown in Fig. 8 is considered. The virtual work of the external forces acting on the links of this mechanism is

$$\delta W = M^2\,\delta\theta^2 + \mathbf{F}^{3^{\mathrm{T}}}\,\delta\mathbf{r}_C^3 + F^4\,\delta R_x^4$$

where $\mathbf{F}^3 = [F_x^3 \quad F_y^3]^{\mathrm{T}}$. The vector $\mathbf{r}_C^3$ is

$$\mathbf{r}_C^3 = \begin{bmatrix} l^2\cos\theta^2 + l_A^3\cos\theta^3 \\ l^2\sin\theta^2 + l_A^3\sin\theta^3 \end{bmatrix}$$

where l_A^3 is the distance between points A and C. Therefore, one has

$$\delta\mathbf{r}_C^3 = \begin{bmatrix} -l^2\sin\theta^2 & -l_A^3\sin\theta^3 \\ l^2\cos\theta^2 & l_A^3\cos\theta^3 \end{bmatrix}\begin{bmatrix} \delta\theta^2 \\ \delta\theta^3 \end{bmatrix}$$

Substituting this equation into the expression for the virtual work, one obtains

$$\delta W = M^2\delta\theta^2 + [F_x^3 \quad F_y^3]\begin{bmatrix} -l^2\sin\theta^2 & -l_A^3\sin\theta^3 \\ l^2\cos\theta^2 & l_A^3\cos\theta^3 \end{bmatrix}\begin{bmatrix} \delta\theta^2 \\ \delta\theta^3 \end{bmatrix} + F^4\,\delta R_x^4$$

or

$$\delta W = Q_{\theta^2}\,\delta\theta^2 + Q_{\theta^3}\,\delta\theta^3 + Q_{R_x^4}\,\delta R_x^4 = [Q_{\theta^2} \quad Q_{\theta^3} \quad Q_{R_x^4}]\begin{bmatrix} \delta\theta^2 \\ \delta\theta^3 \\ \delta R_x^4 \end{bmatrix}$$

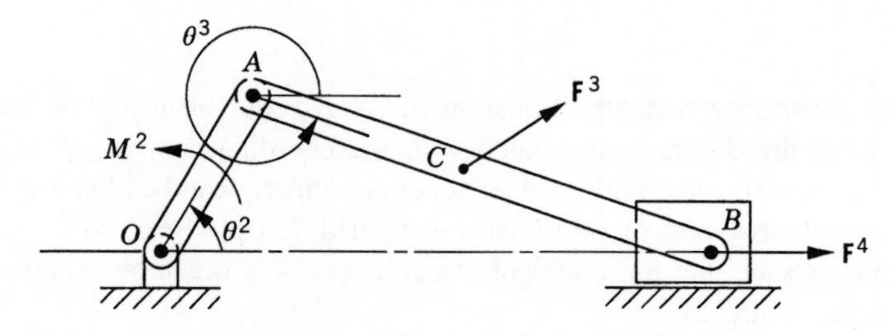

Figure 5.8 Slider crank mechanism

where

$$Q_{\theta^2} = M^2 - F_x^3 l^2 \sin\theta^2 + F_y^3 l^2 \cos\theta^2$$

$$Q_{\theta^3} = -F_x^3 l_A^3 \sin\theta^3 + F_y^3 l_A^3 \cos\theta^3, \quad Q_{R_x^4} = F^4$$

It was shown in Section 2 that $\delta\theta^3$ and δR_x^4 can be expressed in terms of $\delta\theta^2$ as

$$\begin{bmatrix} \delta\theta^3 \\ \delta R_x^4 \end{bmatrix} = \frac{1}{l^3 \cos\theta^3} \begin{bmatrix} -l^2 \cos\theta^2 \\ -l^2 l^3 \sin(\theta^2 - \theta^3) \end{bmatrix} \delta\theta^2$$

One may define the vector $\mathbf{q}$ as

$$\mathbf{q} = [\theta^2 \quad \theta^3 \quad R_x^4]^{\mathrm{T}}$$

The virtual change in this vector can be written in terms of the virtual change in θ^2 as

$$\begin{bmatrix} \delta\theta^2 \\ \delta\theta^3 \\ \delta R_x^4 \end{bmatrix} = \begin{bmatrix} 1 \\ -l^2 \cos\theta^2 / l^3 \cos\theta^3 \\ -l^2 \sin(\theta^2 - \theta^3)/\cos\theta^3 \end{bmatrix} \delta\theta^2$$

Substituting this equation into the expression of δW leads to the definition of the generalized force, of all the forces and moments acting on the slider crank mechanism, associated with the coordinate θ^2 as

$$\begin{aligned} \delta W &= [Q_{\theta^2} \quad Q_{\theta^3} \quad Q_{R_x^4}] \begin{bmatrix} 1 \\ -l^2 \cos\theta^2 / l^3 \cos\theta^3 \\ -l^2 \sin(\theta^2 - \theta^3)/\cos\theta^3 \end{bmatrix} \delta\theta^2 \\ &= \mathbf{Q}_p^{\mathrm{T}} \, \delta\mathbf{p} \end{aligned}$$

where in this case $\delta\mathbf{p}$ reduces to $\delta\mathbf{p} = \delta\theta^2$, and

$$\mathbf{Q}_p = Q_{\theta^2} - (Q_{\theta^3} l^2 \cos\theta^2 / l^3 \cos\theta^3) - Q_{R_x^4} l^2 \sin(\theta^2 - \theta^3)/\cos\theta^3$$

in which Q_{θ^2}, Q_{θ^3}, and $Q_{R_x^4}$ are defined previously in this example.

Example 5.6

Determine the generalized force associated with the rotation of the crank shaft, due to the system of forces acting on the four-bar linkage shown in Fig. 9.

Solution. The virtual work of the forces shown in Fig. 9 is given by

$$\delta W = M^2 \, \delta\theta^2 + \mathbf{F}^{3^{\mathrm{T}}} \, \delta\mathbf{r}_C^3 + M^4 \, \delta\theta^4$$

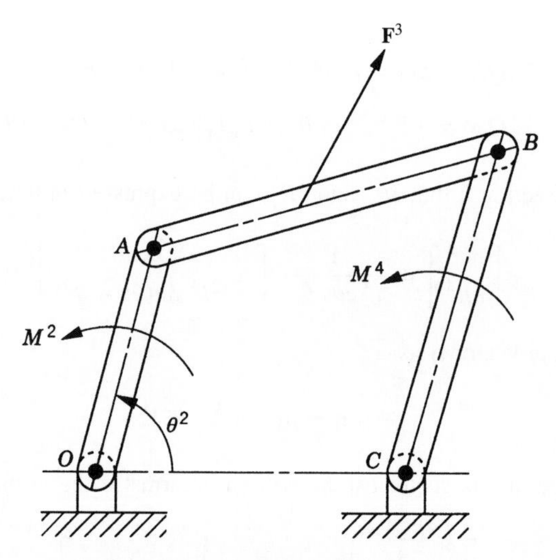

Figure 5.9 Four-bar mechanism

where $\mathbf{F}^3 = [F_x^3 \quad F_y^3]^{\mathrm{T}}$ and

$$\mathbf{r}_C^3 = \begin{bmatrix} l^2 \cos\theta^2 + l_A^3 \cos\theta^3 \\ l^2 \sin\theta^2 + l_A^3 \sin\theta^3 \end{bmatrix}$$

where l_A^3 is the distance between point A and the center of the coupler. It follows that

$$\delta\mathbf{r}_C^3 = \begin{bmatrix} -l^2 \sin\theta^2 & -l_A^3 \sin\theta^3 \\ l^2 \cos\theta^2 & l_A^3 \cos\theta^3 \end{bmatrix} \begin{bmatrix} \delta\theta^2 \\ \delta\theta^3 \end{bmatrix}$$

Substituting this into the expression for the virtual work, one obtains

$$\begin{aligned} \delta W &= (M^2 - F_x^3 l^2 \sin\theta^2 + F_y^3 l^2 \cos\theta^2)\, \delta\theta^2 \\ &\quad + (-F_x^3 l_A^3 \sin\, \theta^3 + F_y^3 l_A^3 \cos\, \theta^3)\, \delta\theta^3 + M^4\, \delta\theta^4 \end{aligned}$$

By using the results of Example 2, $\delta\theta^3$ and $\delta\theta^4$ can be expressed in terms of $\delta\theta^2$ as

$$\begin{bmatrix} \delta\theta^3 \\ \delta\theta^4 \end{bmatrix} = \frac{1}{l^3 l^4 \sin(\theta^3 - \theta^4)} \begin{bmatrix} l^2 l^4 \sin(\theta^4 - \theta^2) \\ l^2 l^3 \sin(\theta^2 - \theta^3) \end{bmatrix} \delta\theta^2$$

which upon substitution into the expression of the virtual work yields

$$\begin{aligned} \delta W &= (M^2 - F_x^3 l^2 \sin\, \theta^2 + F_y^3 l^2 \cos\, \theta^2)\, \delta\theta^2 \\ &\quad + \frac{l^2 \sin(\theta^4 - \theta^2)}{l^3 \sin(\theta^3 - \theta^4)} (-F_x^3 l_A^3 \sin\, \theta^3 + F_y^3 l_A^3 \cos\, \theta^3)\, \delta\theta^2 \\ &\quad + \frac{M^4 l^2 \sin(\theta^2 - \theta^3)}{l^4 \sin(\theta^3 - \theta^4)}\, \delta\theta^2 \end{aligned}$$

which can be written as

$$\delta W = Q_p \, \delta\theta^2$$

where

$$\begin{aligned} Q_p &= M^2 - F_x^3 l^2 \sin\theta^2 + F_y^3 l^2 \cos\theta^2 \\ &+ \frac{l^2 \sin(\theta^4 - \theta^2)}{l^3 \sin(\theta^3 - \theta^4)} (-F_x^3 l_A^3 \sin\theta^3 + F_y^3 l_A^3 \cos\theta^3) \\ &+ \frac{M^4 l^2 \sin(\theta^2 - \theta^3)}{l^4 \sin(\theta^3 - \theta^4)} \end{aligned}$$

Conservative Forces Before concluding this section it is important to point out that, in general, the virtual work is not an *exact differential*. That is, the virtual work is not, in general, the variation of a certain function. In the special case where the virtual work is an exact differential one has

$$Q_j = \frac{\partial W}{\partial q_j} \tag{5.58}$$

and consequently,

$$\frac{\partial Q_j}{\partial q_k} = \frac{\partial Q_k}{\partial q_j} \tag{5.59}$$

In this special case, the forces are said to be *conservative* since they can be obtained using a potential function. *Nonconservative forces*, however, cannot be derived from a potential function, and hence their virtual work is not equal to the variation of a certain function. Examples of conservative forces are the gravity forces and linear spring forces. Examples of nonconservative forces are the damping, friction, and actuator forces. In this book, for the sake of generality, we use the general expression of Eq. 53 to define the generalized forces regardless of whether these forces are conservative or nonconservative.

5.4 EXAMPLES OF FORCE ELEMENTS

In this section, the generalized forces associated with some of the commonly encountered forces in multibody dynamics are developed. The definitions of these generalized forces are obtained by using the virtual work expression.

Gravity The virtual work of the gravity force acting on body i is given by

$$\delta W^i = -m^i g \, \delta y^i \tag{5.60}$$

where m^i is the mass of body i, g is the gravity constant, and y^i is the vertical coordinate of the position vector of the body center of mass. If the reference point is the same as the center of mass, one has $\delta y^i = \delta R_y^i$. If, on the other hand, the reference point is different from the center of mass, y^i can be expressed in terms of the coordinates of body i as

$$y^i = R_y^i + \bar{u}_x^i \sin\theta^i + \bar{u}_y^i \cos\theta^i \tag{5.61}$$

where $\bar{\mathbf{u}}^i = [\bar{u}_x^i \quad \bar{u}_y^i]^{\mathrm{T}}$ is the local position vector of the center of mass with respect to the reference point of body i. The virtual change in y^i in terms of the virtual change in the coordinates of body i is

$$\delta y^i = \delta R_y^i + (\bar{u}_x^i \cos\theta^i - \bar{u}_y^i \sin\theta^i)\,\delta\theta^i \tag{5.62}$$

which upon substitution into the expression for the virtual work of Eq. 60 leads to

$$\begin{aligned} \delta W^i &= -m^i g\,\delta R_y^i - m^i g(\bar{u}_x^i \cos\theta^i - \bar{u}_y^i \sin\theta^i)\,\delta\theta^i \\ &= Q_y^i\,\delta R_y^i + Q_\theta^i\,\delta\theta^i \end{aligned} \tag{5.63}$$

where Q_y^i and Q_θ^i are, respectively, the generalized forces associated with the coordinates R_y^i and θ^i, and are given by

$$Q_y^i = -m^i g, \quad Q_\theta^i = -m^i g(\bar{u}_x^i \cos\theta^i - \bar{u}_y^i \sin\theta^i) \tag{5.64}$$

If the reference point is selected to be the center of mass of the body, Q_θ^i is identically zero since $\bar{u}_x^i = \bar{u}_y^i = 0$.

Spring–Damper–Actuator Element Figure 10 shows two bodies i and j connected by a force element that consists of a translational spring, damper, and actuator. The *spring stiffness* is assumed to be k, the *damping coefficient* is c, and the *actuator force* is f_a. The point of attachment of this force element on body i is assumed to be P^i, while on body j it is assumed to be P^j. The position vectors of these points with respect to their respective body coordinate systems are denoted as $\bar{\mathbf{u}}_P^i$ and $\bar{\mathbf{u}}_P^j$. The resultant force of the spring, damper, and actuator acting along a line connecting points P^i and P^j is given by

$$f_s = k(l - l_0) + c\dot{l} + f_a \tag{5.65}$$

where l is the current spring length, l_0 is the undeformed length of the spring, and $\dot{l}$ is the time derivative of l with respect to time. In Eq. 65, $k(l - l_0)$ is the spring force and $c\dot{l}$ is the damper force, which is assumed to be proportional to the relative velocity between points P^i and P^j. The spring stiffness, the damping coefficient, and the actuator force can be nonlinear functions of the system coordinates and velocities as well as time.

The virtual work due to the force defined by Eq. 65 can be written as

$$\delta W = -f_s\,\delta l \tag{5.66}$$

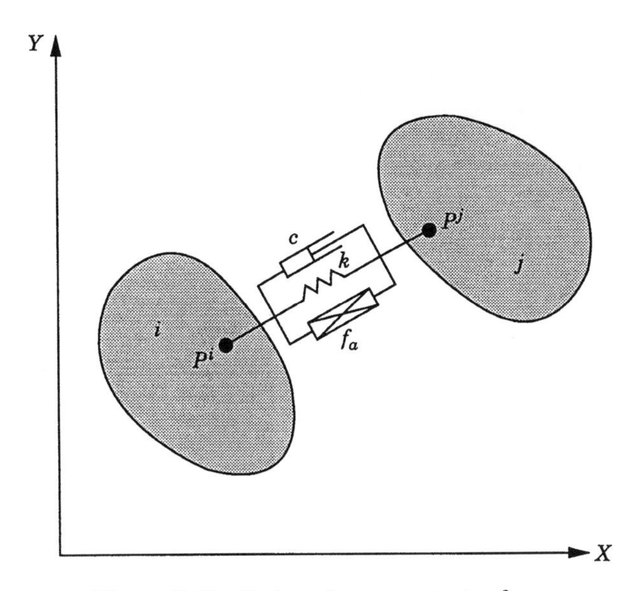

Figure 5.10 Spring–damper–actuator force

where δl is the virtual change in the distance between points P^i and P^j. In terms of the absolute coordinates of the two bodies, the position vector of point P^j with respect to point P^i can be written as

$$\mathbf{r}_P^{ij} = \mathbf{R}^i + \mathbf{A}^i \bar{\mathbf{u}}_P^i - \mathbf{R}^j - \mathbf{A}^j \bar{\mathbf{u}}_P^j \tag{5.67}$$

One can, therefore, define the current spring length as

$$l = (\mathbf{r}_P^{ij^{\mathrm{T}}} \mathbf{r}_P^{ij})^{1/2} \tag{5.68}$$

and the virtual change in this length as

$$\delta l = \frac{\partial l}{\partial \mathbf{q}} \delta \mathbf{q} = (\mathbf{r}_P^{ij^{\mathrm{T}}} \mathbf{r}_P^{ij})^{-1/2} \mathbf{r}_P^{ij^{\mathrm{T}}} \frac{\partial \mathbf{r}_P^{ij}}{\partial \mathbf{q}} \delta \mathbf{q} \tag{5.69}$$

where $\mathbf{q}$ is the vector of coordinates of body i and body j given by

$$\mathbf{q} = [\mathbf{q}^{i^{\mathrm{T}}} \quad \mathbf{q}^{j^{\mathrm{T}}}]^{\mathrm{T}} = [\mathbf{R}^{i^{\mathrm{T}}} \quad \theta^i \quad \mathbf{R}^{j^{\mathrm{T}}} \quad \theta^j]^{\mathrm{T}} \tag{5.70}$$

Equation 69, upon the use of Eq. 68, can be expressed as

$$\begin{aligned}
\delta l &= \frac{\mathbf{r}_P^{ij^{\mathrm{T}}}}{l} \frac{\partial \mathbf{r}_P^{ij}}{\partial \mathbf{q}} \delta \mathbf{q} \\
&= \hat{\mathbf{l}}^{\mathrm{T}} \left[\frac{\partial \mathbf{r}_P^{ij}}{\partial \mathbf{q}^i} \delta \mathbf{q}^i + \frac{\partial \mathbf{r}_P^{ij}}{\partial \mathbf{q}^j} \delta \mathbf{q}^j \right]
\end{aligned}$$

$$= \hat{\mathbf{l}}^{\mathrm{T}} \begin{bmatrix} \dfrac{\partial \mathbf{r}_P^{ij}}{\partial \mathbf{q}^i} & \dfrac{\partial \mathbf{r}_P^{ij}}{\partial \mathbf{q}^j} \end{bmatrix} \begin{bmatrix} \delta \mathbf{q}^i \\ \delta \mathbf{q}^j \end{bmatrix} \tag{5.71}$$

where $\hat{\mathbf{l}}$ is a unit vector in the direction of the vector $\mathbf{r}_P^{ij}$, and $\partial \mathbf{r}_P^{ij}/\partial \mathbf{q}^i$ and $\partial \mathbf{r}_P^{ij}/\partial \mathbf{q}^j$ can be obtained using Eq. 67 as

$$\frac{\partial \mathbf{r}_P^{ij}}{\partial \mathbf{q}^i} = [\mathbf{I} \quad \mathbf{A}_\theta^i \bar{\mathbf{u}}_P^i], \qquad \frac{\partial \mathbf{r}_P^{ij}}{\partial \mathbf{q}^j} = -[\mathbf{I} \quad \mathbf{A}_\theta^j \bar{\mathbf{u}}_P^j] \tag{5.72}$$

The generalized forces associated with the coordinates of body i and body j can be obtained by substituting Eq. 71 into Eq. 66, yielding

$$\delta W = -f_s \hat{\mathbf{l}}^{\mathrm{T}} \begin{bmatrix} \dfrac{\partial \mathbf{r}_P^{ij}}{\partial \mathbf{q}^i} & \dfrac{\partial \mathbf{r}_P^{ij}}{\partial \mathbf{q}^j} \end{bmatrix} \begin{bmatrix} \delta \mathbf{q}^i \\ \delta \mathbf{q}^j \end{bmatrix} = \mathbf{Q}^{i^{\mathrm{T}}} \delta \mathbf{q}^i + \mathbf{Q}^{j^{\mathrm{T}}} \delta \mathbf{q}^j \tag{5.73}$$

where $\mathbf{Q}^i$ and $\mathbf{Q}^j$ are the vectors of the generalized forces associated with the coordinates of body i and body j, respectively. Using Eq. 72, these vectors are

$$\left.\begin{aligned} \mathbf{Q}^i &= \begin{bmatrix} \mathbf{Q}_R^i \\ \mathbf{Q}_\theta^i \end{bmatrix} = -f_s \begin{bmatrix} \dfrac{\partial \mathbf{r}_P^{ij}}{\partial \mathbf{q}^i} \end{bmatrix}^{\mathrm{T}} \hat{\mathbf{l}} = -f_s \begin{bmatrix} \mathbf{I} \\ \bar{\mathbf{u}}_P^{i^{\mathrm{T}}} \mathbf{A}_\theta^{i^{\mathrm{T}}} \end{bmatrix} \hat{\mathbf{l}} \\ \mathbf{Q}^j &= \begin{bmatrix} \mathbf{Q}_R^j \\ \mathbf{Q}_\theta^j \end{bmatrix} = f_s \begin{bmatrix} \dfrac{\partial \mathbf{r}_P^{ij}}{\partial \mathbf{q}^j} \end{bmatrix}^{\mathrm{T}} \hat{\mathbf{l}} = f_s \begin{bmatrix} \mathbf{I} \\ \bar{\mathbf{u}}_P^{j^{\mathrm{T}}} \mathbf{A}_\theta^{j^{\mathrm{T}}} \end{bmatrix} \hat{\mathbf{l}} \end{aligned}\right\} \tag{5.74}$$

in which f_s is defined by Eq. 65. In the expression for the force f_s, l is defined by Eq. 68, and $\dot{l}$ can be obtained according to

$$\dot{l} = \frac{\partial l}{\partial \mathbf{q}} \dot{\mathbf{q}} = \hat{\mathbf{l}}^{\mathrm{T}} \frac{\partial \mathbf{r}_P^{ij}}{\partial \mathbf{q}} \dot{\mathbf{q}} \tag{5.75}$$

The special case, in which there is only a spring element, can be obtained from the general development presented in this section by assuming that $c = f_a = 0$. Similarly, if the force element consists of a damper or an actuator only, one has $k = f_a = 0$ or $k = c = 0$, respectively.

Example 5.7

Figure 11 shows a spring–damper force element that is connected between the crankshaft and the rocker of a four-bar linkage. The stiffness coefficient k of the spring is assumed to be 250 N/m and the damping coefficient c is assumed to be 10 N · s/m. The undeformed length of the spring is assumed to be 0.35 m. The local positions of the attachment points of the spring–damper element with respect to the crankshaft and the rocker coordinate systems are given, respectively, by $\bar{\mathbf{u}}_P^2 = [0.03 \quad 0]^{\mathrm{T}}$ and $\bar{\mathbf{u}}_P^4 = [-0.05 \quad 0]^{\mathrm{T}}$. The respective lengths of

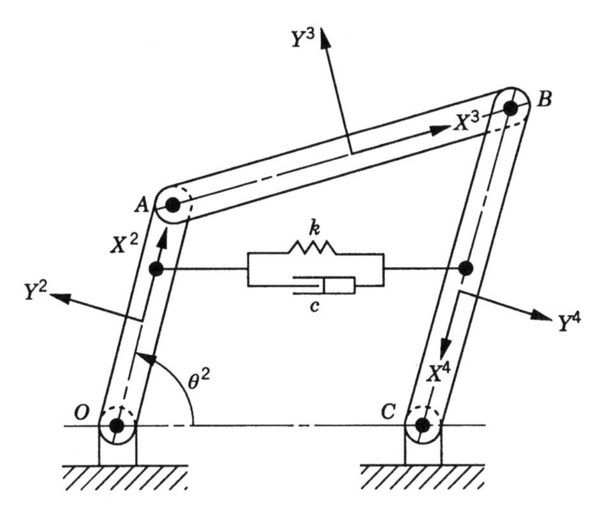

Figure 5.11 Spring–damper force

the crankshaft, rocker, and coupler are $l^2 = 0.2\,\text{m}$, $l^3 = 0.4\,\text{m}$, and $l^4 = 0.3\,\text{m}$. The distance between points O and C is assumed to be 0.35 m. At a particular configuration, the angular orientation of the crankshaft $\theta^2 = 70^\circ$ and its angular velocity $\dot{\theta}^2 = 150$ rad/s. Determine the generalized forces of the spring–damper element associated with the absolute Cartesian coordinates R_x^i, R_y^i, and θ^i. If the generalized coordinates are selected to be θ^2, θ^3, and θ^4, determine the generalized forces of the spring–damper associated with this set of the generalized coordinates.

Solution. By performing a position analysis for the four-bar mechanism, one obtains

$$\theta^2 = 70^\circ, \quad \theta^3 = 13.31^\circ, \quad \theta^4 = 248.97^\circ$$

The velocity analysis leads to

$$\dot{\theta}^2 = 150 \text{ rad/s}, \quad \dot{\theta}^3 = 1.6328 \text{ rad/s}, \quad \dot{\theta}^4 = 101.212 \text{ rad/s}$$

The spring–damper force is

$$f_s = k(l - l_0) + c\dot{l}$$

The virtual work of the spring–damper force is

$$\delta W = -f_s\,\delta l$$

Note that

$$\mathbf{r}_P^{24} = \mathbf{R}^2 + \mathbf{A}^2\bar{\mathbf{u}}_P^2 - \mathbf{R}^4 - \mathbf{A}^4\bar{\mathbf{u}}_P^4$$

$$= \begin{bmatrix} \frac{l^2}{2}\cos\theta^2 \\ \frac{l^2}{2}\sin\theta^2 \end{bmatrix} + \begin{bmatrix} \cos\theta^2 & -\sin\theta^2 \\ \sin\theta^2 & \cos\theta^2 \end{bmatrix} \begin{bmatrix} 0.03 \\ 0 \end{bmatrix}$$

$$- \begin{bmatrix} l^2\cos\theta^2 + l^3\cos\theta^3 + \frac{l^4}{2}\cos\theta^4 \\ l^2\sin\theta^2 + l^3\sin\theta^3 + \frac{l^4}{2}\sin\theta^4 \end{bmatrix}$$

$$- \begin{bmatrix} \cos\theta^4 & -\sin\theta^4 \\ \sin\theta^4 & \cos\theta^4 \end{bmatrix} \begin{bmatrix} -0.05 \\ 0 \end{bmatrix} = \begin{bmatrix} -0.3773 \\ -0.0927 \end{bmatrix}$$

It follows that

$$|\mathbf{r}_P^{24}| = l = \sqrt{(-0.3773)^2 + (-0.0927)^2} = 0.3885$$

A unit vector along a line connecting the attachment points of the spring–damper element is

$$\hat{\mathbf{l}} = \frac{\mathbf{r}_P^{24}}{l} = \begin{bmatrix} -0.9712 \\ -0.2386 \end{bmatrix}$$

The vector of the generalized coordinates of the crankshaft and the rocker can be written as

$$\mathbf{q} = [\mathbf{R}^{2^{\mathrm{T}}} \quad \theta^2 \quad \mathbf{R}^{4^{\mathrm{T}}} \quad \theta^4]^{\mathrm{T}}$$

The time derivative of the spring length is

$$\dot{l} = \hat{\mathbf{l}}^{\mathrm{T}} \frac{\partial \mathbf{r}_P^{24}}{\partial \mathbf{q}} \dot{\mathbf{q}}$$

$$\frac{\partial \mathbf{r}_P^{24}}{\partial \mathbf{q}} = [\mathbf{I} \quad \mathbf{A}_\theta^2 \bar{\mathbf{u}}_P^2 \quad -\mathbf{I} \quad -\mathbf{A}_\theta^4 \bar{\mathbf{u}}_P^4]$$

in which

$$\mathbf{A}_\theta^2 \bar{\mathbf{u}}_P^2 = \begin{bmatrix} -\sin\theta^2 & -\cos\theta^2 \\ \cos\theta^2 & -\sin\theta^2 \end{bmatrix} \begin{bmatrix} 0.03 \\ 0 \end{bmatrix} = \begin{bmatrix} -0.0282 \\ 0.0103 \end{bmatrix}$$

$$\mathbf{A}_\theta^4 \bar{\mathbf{u}}_P^4 = \begin{bmatrix} -\sin\theta^4 & -\cos\theta^4 \\ \cos\theta^4 & -\sin\theta^4 \end{bmatrix} \begin{bmatrix} -0.05 \\ 0 \end{bmatrix} = \begin{bmatrix} -0.0467 \\ 0.0179 \end{bmatrix}$$

$$\dot{\mathbf{q}} = [\dot{\mathbf{R}}^{2^{\mathrm{T}}} \quad \dot{\theta}^2 \quad \dot{\mathbf{R}}^{4^{\mathrm{T}}} \quad \dot{\theta}^4]^{\mathrm{T}}$$

$$= [-14.0954 \quad 5.1303 \quad 150 \quad -14.1706 \quad 5.4481 \quad 101.212]^{\mathrm{T}}$$

which yield

$$\dot{l} = -0.4159 \text{ m/s}$$

$$f_s = 250(0.3885 - 0.35) + 10(-0.4159) = 5.466 \text{ N}$$

The generalized forces can then be written as

$$\mathbf{Q}^2 = \begin{bmatrix} \mathbf{Q}_R^2 \\ Q_\theta^2 \end{bmatrix} = -f_s \begin{bmatrix} \mathbf{I} \\ \bar{\mathbf{u}}_P^{2^\mathrm{T}} \mathbf{A}_\theta^{2^\mathrm{T}} \end{bmatrix} \hat{\mathbf{I}} = \begin{bmatrix} 5.3086 \\ 1.3042 \\ 0.1363 \end{bmatrix}$$

$$\mathbf{Q}^4 = \begin{bmatrix} \mathbf{Q}_R^4 \\ Q_\theta^4 \end{bmatrix} = f_s \begin{bmatrix} \mathbf{I} \\ \bar{\mathbf{u}}_P^{4^\mathrm{T}} \mathbf{A}_\theta^{4^\mathrm{T}} \end{bmatrix} \hat{\mathbf{I}} = \begin{bmatrix} -5.3086 \\ -1.3042 \\ 0.2246 \end{bmatrix}$$

These are the generalized forces associated with the absolute coordinates of the crankshaft and the rocker. Note that the forces associated with the translational coordinates are equal in magnitude and opposite in direction.

To determine the generalized forces associated with the angles θ^2, θ^3, and θ^4, we first evaluate $\delta\mathbf{r}_P^{24}$ as

$$\delta\mathbf{r}_P^{24} = \frac{\partial\mathbf{r}_P^{24}}{\partial\mathbf{q}}\,\delta\mathbf{q} = \frac{\partial\mathbf{r}_P^{24}}{\partial\mathbf{q}}\,\frac{\partial\mathbf{q}}{\partial\boldsymbol{\theta}}\,\delta\boldsymbol{\theta} = \frac{\partial\mathbf{r}_P^{24}}{\partial\boldsymbol{\theta}}\,\delta\boldsymbol{\theta}$$

where

$$\boldsymbol{\theta} = [\theta^2 \quad \theta^3 \quad \theta^4]^\mathrm{T}$$

and

$$\frac{\partial\mathbf{r}_P^{24}}{\partial\boldsymbol{\theta}} = \frac{\partial\mathbf{r}_P^{24}}{\partial\mathbf{q}}\,\frac{\partial\mathbf{q}}{\partial\boldsymbol{\theta}}$$

Using this equation, it can be shown that

$$\delta\mathbf{r}_P^{24} = \begin{bmatrix} 0.0658 & 0.0921 & -0.0933 \\ -0.0342 & -0.3893 & 0.0359 \end{bmatrix} \begin{bmatrix} \delta\theta^2 \\ \delta\theta^3 \\ \delta\theta^4 \end{bmatrix}$$

The spring–damper force vector is

$$\mathbf{F}_s = f_s\hat{\mathbf{I}} = \begin{bmatrix} -5.3086 \\ -1.3042 \end{bmatrix} \text{N}$$

The virtual work of this force is

$$\delta W = -\mathbf{F}_s^{\mathrm{T}} \delta \mathbf{r}_P^{24} = -[-5.3086 \quad -1.3042] \begin{bmatrix} 0.0658 & 0.0921 & -0.0933 \\ -0.0342 & -0.3893 & 0.0359 \end{bmatrix} \cdot \begin{bmatrix} \delta\theta^2 \\ \delta\theta^3 \\ \delta\theta^4 \end{bmatrix}$$

$$= -[-0.3047 \quad 0.0188 \quad 0.4485] \begin{bmatrix} \delta\theta^2 \\ \delta\theta^3 \\ \delta\theta^4 \end{bmatrix}$$

where the generalized forces associated with the angles θ^2, θ^3, and θ^4 are recognized as

$$Q_{\theta^1} = 0.3047, \quad Q_{\theta^2} = -0.0188, \quad Q_{\theta^3} = -0.4485$$

Rotational Spring–Damper Element Figure 12 depicts two bodies i and j that are connected by rotational spring and damper. The stiffness coefficient of the spring is assumed to be k_r and the damping coefficient is assumed to be c_r. The resultant moment of the spring and damper can be expressed as

$$M_s = k_r(\theta - \theta_0) + c_r\dot{\theta} \tag{5.76}$$

where θ_0 is the angle between body i and body j before displacements, and θ is the relative angular displacement between the two bodies, that is, $\theta = \theta^i - \theta^j$. The virtual work due to the moment of Eq. 76 is

$$\delta W = -M_s\, \delta\theta = -[k_r(\theta - \theta_0) + c_r\dot{\theta}]\, \delta\theta \tag{5.77}$$

Using the preceding two equations,

$$\delta W = -[k_r(\theta - \theta_0) + c_r\dot{\theta}](\delta\theta^i - \delta\theta^j) \tag{5.78}$$

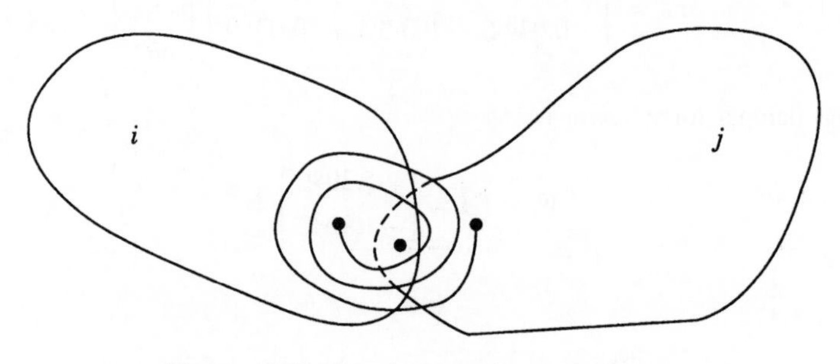

Figure 5.12 Torsional spring and damper

which can be written as

$$\delta W = Q_\theta^i \, \delta\theta^i + Q_\theta^j \, \delta\theta^j \tag{5.79}$$

where Q_θ^i and Q_θ^j are the generalized forces associated with the rotational coordinates of bodies i and j, respectively. These generalized forces are

$$\left.\begin{aligned} Q_\theta^i &= -[k_r(\theta - \theta_0) + c_r\dot{\theta}] \\ Q_\theta^j &= k_r(\theta - \theta_0) + c_r\dot{\theta} \end{aligned}\right\} \tag{5.80}$$

If the force element consists of a spring only, $c_r = 0$. On the other hand, if the force element consists of a damper only, $k_r = 0$

Coulomb Friction In the case of ideal joints, the reaction forces are assumed to be normal to the contact surfaces. Although this assumption is valid in many situations and its use leads to a relatively small error, there are many applications wherein the interaction between the contact surfaces must be described by normal and tangential components. The tangential component that opposes the relative motion between the two surfaces is called the friction force. In many types of multibody systems, such as gears and bearings, it is desirable to minimize the effect of the friction forces, while in other applications, such as brakes and clutches, one desires to maximize the friction effect.

In the case of *Coulomb* or *dry friction*, the friction force is not an explicit function of the displacement and its derivatives. Figure 13a shows two bodies i and j that are in contact. Let $\mathbf{t}$ be a unit vector along the flat contact surface, and $\mathbf{v}^i$ and $\mathbf{v}^j$ be the absolute velocities of the reference points of the two bodies. The velocity of body i with respect to body j along the vector $\mathbf{t}$ is

$$v_r = (\mathbf{v}^i - \mathbf{v}^j)^{\mathrm{T}}\mathbf{t} \tag{5.81}$$

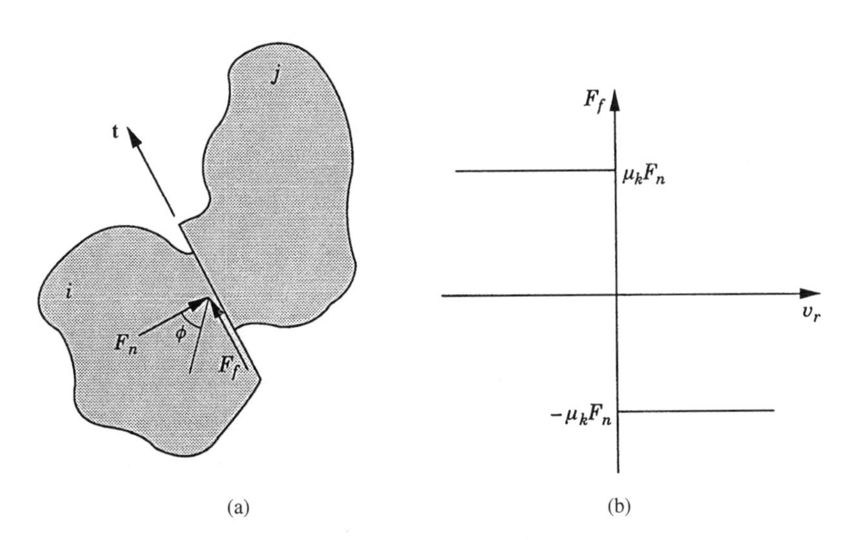

Figure 5.13 Friction force

TABLE 5.1 Coefficient of Statis Friction

Rubber on concrete	0.60–0.90
Metal on stone	0.30–0.70
Metal on wood	0.20–0.60
Metal on metal	0.15–0.60
Stone on stone	0.40–0.70

As shown in Fig. 13a, the contact force is represented by two components; the component F_n, which is normal to the flat contact surface, and the component F_f, which is parallel to that surface. The component F_f, which is developed by friction, opposes the relative motion. In the classical theory of dry friction, the friction force F_f is directly proportional to the normal force F_n. Depending on the materials of the two bodies, there is a limit to the magnitude of the force F_f. In the special case where $v_r = 0$, one has

$$F_f \leq \mu_s F_n \tag{5.82}$$

where μ_s, called the *coefficient of static friction*, depends on the properties of the materials in contact. The values of the coefficient of static friction can be found experimentally. Table 1 shows approximate values of this coefficient in several cases of dry surfaces.

If body i slides with respect to body j with a relative velocity v_r, the friction force takes on the value

$$F_f = \mu_k F_n \operatorname{sgn}(v_r) \tag{5.83}$$

where μ_k is called the *coefficient of sliding friction*. This coefficient can also be determined experimentally and its value is slightly less than μ_s for most materials. The function $\operatorname{sgn}(v_r)$ has the value ± 1 depending on the sign of its argument v_r. Figure 13b shows the friction force F_f as a function of the relative velocity v_r. It is clear from this figure that when v_r is equal to zero, the friction force F_f can have any magnitude. The actual magnitude of this force can be determined from the static equilibrium conditions. While it is often assumed in the analysis of systems involving dry friction that the maximum friction force is $\mu_k F_n$, in reality the force required to initiate the motion is slightly larger than the force required to maintain it.

It is clear from Eq. 83 that

$$\frac{F_f}{F_n} = \mu_k = \tan \phi \tag{5.84}$$

where the angle ϕ shown in Fig. 13a is called the *friction angle*.

Example 5.8

The mass–spring system shown in Fig. 14 has mass $m = 5$ kg, stiffness coefficient $k = 5 \times 10^3$ N/m, coefficient of friction $\mu_k = 0.1$, initial displacement $x_o = 0.03$ m, and zero initial velocity. Determine the number of cycles of oscillations of the mass before it comes to rest.

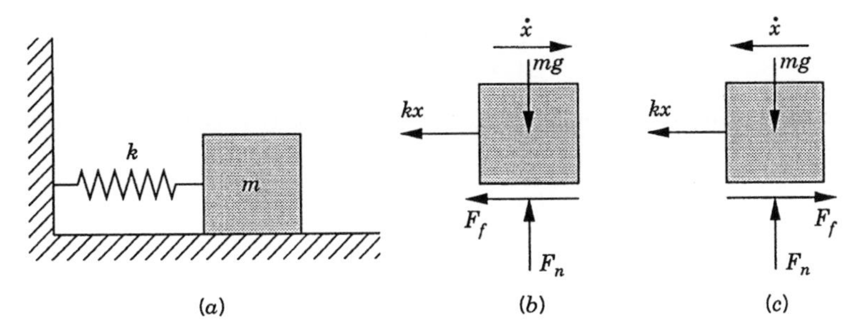

Figure 5.14 Mass–spring system

Solution. The equation of motion of the mass is

$$m\ddot{x} + kx = \mp F_f$$

where the negative sign is used when the mass moves to the right and the positive sign is used when the mass moves to the left. The solution of the differential equation of motion can be written as

$$x(t) = \begin{cases} A_1 \sin \omega t + A_2 \cos \omega t - \dfrac{F_f}{k} & \dot{x} \geq 0 \quad (5.85a) \\ B_1 \sin \omega t + B_2 \cos \omega t + \dfrac{F_f}{k} & \dot{x} \leq 0 \quad (5.85b) \end{cases}$$

where ω is the natural frequency of the system defined as

$$\omega = \sqrt{\frac{k}{m}} = \sqrt{\frac{5 \times 10^3}{5}} = 31.6228 \text{ rad/s}$$

Substituting with the initial conditions into Eq. 85b, which describes the dynamics of the system when the mass moves to the left, one obtains

$$x_o = 0.03 = B_2 + \frac{F_f}{k}, \quad \dot{x}_o = 0 = \omega B_1$$

where

$$F_f = \mu_k mg = (0.1)(5)(9.81) = 4.905 \text{ N}$$

It follows that

$$B_1 = 0, \qquad B_2 = x_o - \frac{F_f}{k} = 0.02902$$

Therefore, the displacement and velocity of the mass when it first moves to the left can be described by the equations

$$x(t) = \left(x_o - \frac{F_f}{k}\right) \cos \omega t + \frac{F_f}{k}$$

$$\dot{x}(t) = -\omega\left(x_o - \frac{F_f}{k}\right)\sin\omega t$$

The direction of the motion will change when the velocity is equal to zero, that is

$$0 = -\omega\left(x_o - \frac{F_f}{k}\right)\sin\omega t_1$$

which yields

$$t_1 = \frac{\pi}{\omega} = 0.0993 \text{ s}$$

At this time the displacement is determined from Eq. 85b, which describes the motion to the left, as

$$x(t_1) = x\left(\frac{\pi}{\omega}\right) = -x_o + \frac{2F_f}{k} = -0.028038 \text{ m}$$

This equation shows that the amplitude in the first half cycle is reduced by the amount $2F_f/k$ as the result of dry friction.

In the second half cycle, the mass moves to the right and the motion is governed by Eq. 85a, which describes the motion to the right with the initial conditions

$$x\left(\frac{\pi}{\omega}\right) = -x_o + \frac{2F_f}{k} = -0.028038 \text{ m}$$

$$\dot{x}_o\left(\frac{\pi}{\omega}\right) = 0$$

These initial conditions yield

$$A_1 = 0, \qquad A_2 = x_o - \frac{3F_f}{k} = 0.027057 \text{ m}$$

The displacement $x(t)$ in the second half cycle can then be written as

$$x(t) = \left(x_o - \frac{3F_f}{k}\right)\cos\omega t - \frac{F_f}{k}$$

and the velocity

$$\dot{x}(t) = -\omega\left(x_o - \frac{3F_f}{k}\right)\sin\omega t$$

The velocity is zero at time $t_2 = 2\pi/\omega = \tau$, where τ is the periodic time of the natural oscillations. At time t_2, the end of the first cycle, the displacement is

$$x(t_2) = x\left(\frac{2\pi}{\omega}\right) = x_o - \frac{4F_f}{k} = 0.026076 \text{ m}$$

which indicates that the amplitude decreases in the second half cycle by the amount $2F_f/k$, as shown in Fig. 15. By continuing in this manner, one can verify that there is a constant decrease in the amplitude of $2F_f/k$ every half cycle. It is not necessary that the system comes to rest at the undeformed spring position. The final position will be at an amplitude X_f, where the spring force $F_s = kX_f$ is less than or equal to the friction force. In this example, the motion will stop if

$$kX_f \leq 4.905$$

or

$$X_f \leq \frac{4.905}{k} = \frac{4.905}{5 \times 10^3} = 0.981 \times 10^{-3} \text{ m}$$

The amplitude loss per half cycle is

$$\frac{2F_f}{k} = \frac{2(4.905)}{5 \times 10^3} = 1.962 \times 10^{-3} \text{ m}$$

The number of half cycles n_r completed before the mass comes to rest can be obtained from the following equation:

$$x_o - n_f \left(\frac{2F_f}{k} \right) \leq 0.981 \times 10^{-3} \text{ m}$$

which implies that

$$0.03 - n_r(1.962 \times 10^{-3}) \leq 0.981 \times 10^{-3} \text{ m}$$

The smallest n_r that satisfies this inequality is $n_r = 15$ half cycles; that is, the number of cycles completed before the mass comes to rest is 7.5.

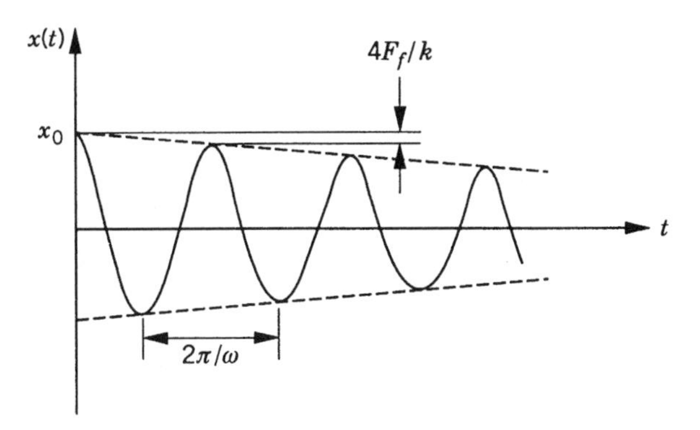

Figure 5.15 Effect of the friction force

The preceding example demonstrates the complexity of the analysis of dry friction using a simple mass spring system. In a more complex mechanical system that consists of a set of interconnected bodies, the *generalized friction forces* associated with the system generalized coordinates can be systematically determined. Recall that in rigid body dynamics, a force is a *sliding vector* that can be moved along its line of action without changing its effect. It follows that once the friction force along the flat contact surface is determined, the generalized forces associated with the generalized coordinates of two bodies i and j in contact can be simply obtained using the concept of equipollent systems of forces or using the expression of the virtual work of the friction force where the point of application of the force is assumed to be an arbitrary point on the contact surface.

Further generalization of the development presented in this section can be made if the friction force is considered as arising from uniformly or nonuniformly distributed shear stress at the contact area (Greenwood, 1988). In this case, the frictional shear stress is equal to μ_k times the normal pressure. While this approach gives the same results for the simpler case of a flat contact surface, it can also be used in the analysis of more complicated systems. In order to demonstrate the use of this approach, consider the case of a circular rotating disk of radius a being pressed against another disk with a force F_n. If the compressive stress σ_n is assumed to be uniform, one has

$$\sigma_n = \frac{\mu_k F_n}{\pi(a)^2} \tag{5.86}$$

The moment due to the friction force acting on an annular element of radius dr and area $dA = 2\pi r\ dr$ as shown in Fig. 16 is

$$dM = \sigma_n(2\pi r\ dr)r = \frac{\mu_k F_n}{\pi(a)^2} 2\pi(r)^2\ dr \tag{5.87}$$

which upon integration leads to

$$M = \frac{2\mu_k F_n}{(a)^2} \int_0^a (r)^2\ dr = \frac{2a}{3}\ \mu_k F_n \tag{5.88}$$

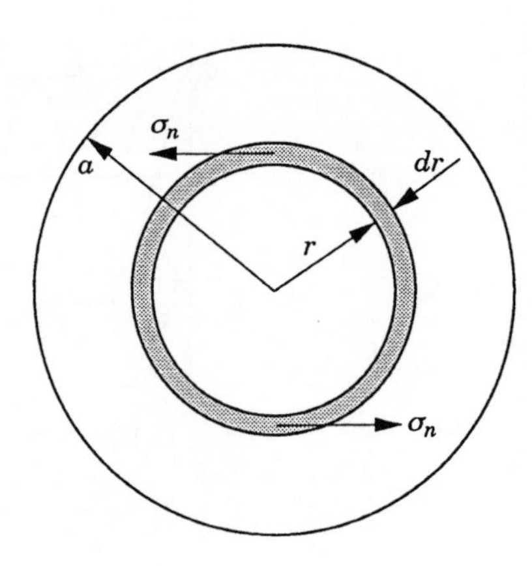

Figure 5.16 Friction stress

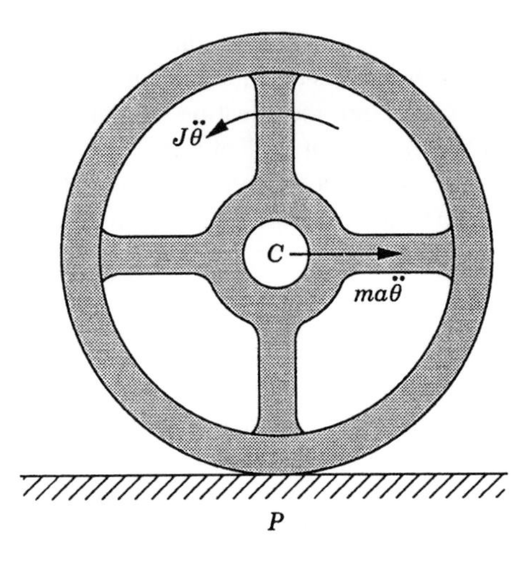

Figure 5.17 Rolling contact

Another important friction application pertains to wheeled vehicles, which depend on the friction forces for starting, moving, and stopping. A point on a moving wheel as the one shown in Fig. 17 can have an instantaneous zero velocity while its acceleration is different from zero. Generally, there are two different situations that may occur. In the first situation, the wheel rolls and slides on the ground such that the wheel motion can be described using two degrees of freedom; one describes the rolling motion, while the other describes the sliding displacement. Because of the sliding motion, the velocity of the point of contact P on the wheel is not equal to zero. In the case of *pure rolling motion*, on the other hand, no sliding occurs and the instantaneous velocity of the contact point P on the wheel is equal to zero. In this case, the instantaneous velocity of the center of the wheel C is

$$\mathbf{v}_C = \boldsymbol{\omega} \times \mathbf{u}_{CP} \tag{5.89}$$

where $\boldsymbol{\omega}$ is the angular velocity of the wheel, and $\mathbf{u}_{CP}$ is the position vector of point C with respect to point P. We must keep in mind that while in the preceding equation the velocity of point C is obtained using the instantaneous velocity of point P, $\mathbf{v}_C$ takes on the same value so long as $\mathbf{u}_{CP}$ represents the position vector of point C with respect to the contact point P. The direction of this velocity is always perpendicular to CP and its magnitude is equal to $|\mathbf{v}_C| = \dot{\theta}a$, where θ is the angle of rotation of the wheel and a is the radius of the wheel. The absolute acceleration of point C is

$$\mathbf{a}_C = \boldsymbol{\alpha} \times \mathbf{u}_{CP} \tag{5.90}$$

where $\boldsymbol{\alpha}$ is the angular acceleration of the wheel. The absolute acceleration of the contact point P can then be written as

$$\begin{aligned}
\mathbf{a}_P &= \mathbf{a}_C + \mathbf{a}_{PC} \\
&= \boldsymbol{\alpha} \times \mathbf{u}_{CP} + \boldsymbol{\alpha} \times \mathbf{u}_{PC} + \boldsymbol{\omega} \times (\boldsymbol{\omega} \times \mathbf{u}_{PC}) \\
&= \boldsymbol{\omega} \times (\boldsymbol{\omega} \times \mathbf{u}_{PC})
\end{aligned} \tag{5.91}$$

Assuming that the wheel is balanced such that its center of mass is the same as its geometric center, the absolute acceleration of the center of mass is equal to $\mathbf{a}_C$. Since in the case of pure rolling the wheel has only one degree of freedom, the equation of motion of the wheel can be obtained by taking the moments of the inertia and applied forces about the contact point P. This equation can be used to determine the unknown force or acceleration. The reaction force at the contact point can then be determined by evaluating the sum of the forces in the vertical direction.

5.5 WORKLESS CONSTRAINTS

Mechanical joints in multibody systems give rise to constraint forces that influence the motion of the system components. These forces appear in the static and dynamic equations when the equilibrium conditions are developed for each body in the system. As demonstrated in the remainder of this book, the resulting system of equations can be solved for a number of unknowns equal to the number of the constraint reaction forces plus the number of the system degrees of freedom. These equations, by eliminating the reaction forces, reduce to a number of equations equal to the number of degrees of freedom of the system, and therefore, the constraint reaction forces may be considered as auxiliary quantities that we are forced to introduce when we study the equilibrium of each body in the system separately. These forces, which can be eliminated by considering the equilibrium of the entire system of bodies, are the result of *workless* or *ideal constraints*. The internal reaction forces between the particles that form a rigid body are constraint forces which do no work. This can be demonstrated by using the fact that the distance between two particles i and j on a rigid body remains constant, that is,

$$(\mathbf{r}^i - \mathbf{r}^j)^{\mathrm{T}}(\mathbf{r}^i - \mathbf{r}^j) = c_1 \tag{5.92}$$

where $\mathbf{r}^i$ and $\mathbf{r}^j$ are, respectively, the position vectors of the particles i and j, and c_1 is a constant. By assuming a virtual change in the position vector of the two particles, Eq. 92 yields

$$(\mathbf{r}^i - \mathbf{r}^j)^{\mathrm{T}}(\delta\mathbf{r}^i - \delta\mathbf{r}^j) = 0 \tag{5.93}$$

Let $\mathbf{F}_c^{ij}$ be the constraint force acting on particle i as the result of the kinematic constraint of Eq. 93. Newton's third law states that when two particles exert forces on each other, the resulting interaction forces are equal in magnitude, opposite in direction, and directed along the straight line joining the two particles. According to this law, one may write $\mathbf{F}_c^{ji} = -\mathbf{F}_c^{ij}$ as the reaction force that acts on particle j. Furthermore,

$$\mathbf{F}_c^{ij} = c_2(\mathbf{r}^i - \mathbf{r}^j) \tag{5.94}$$

where c_2 is a constant. The virtual work of the forces $\mathbf{F}_c^{ij}$ and $\mathbf{F}_c^{ji}$ can be written as

$$\delta W = \mathbf{F}_c^{ij^{\mathrm{T}}}\,\delta\mathbf{r}^i + \mathbf{F}_c^{ji^{\mathrm{T}}}\,\delta\mathbf{r}^j = \mathbf{F}_c^{ij^{\mathrm{T}}}\,\delta\mathbf{r}^i - \mathbf{F}_c^{ij^{\mathrm{T}}}\,\delta\mathbf{r}^j = \mathbf{F}_c^{ij^{\mathrm{T}}}(\delta\mathbf{r}^i - \delta\mathbf{r}^j) \tag{5.95}$$

Substituting Eq. 94 into Eq. 95 and using Eq. 93, yields

$$\delta W = c_2(\mathbf{r}^i - \mathbf{r}^j)(\delta\mathbf{r}^i - \delta\mathbf{r}^j) = 0 \tag{5.96}$$

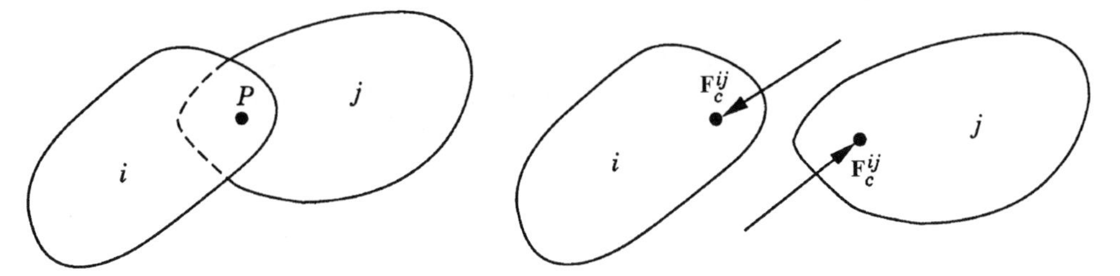

Figure 5.18 Constraint forces

which implies that the connection forces resulting from the constraints between the particles forming the rigid body do no work.

Similar comments apply to the case of other friction-free mechanical joints such as the revolute joint shown in Fig. 18. For instance, the virtual work of the reaction forces acting on body i is

$$\delta W^i = -\mathbf{F}_c^{ij^{\mathrm{T}}} \delta \mathbf{r}_P \tag{5.97}$$

where $\mathbf{r}_P$ is the global position vector of the joint definition point P shown in Fig. 18. The virtual work of the joint reaction forces acting on body j is

$$\delta W^j = \mathbf{F}_c^{ij^{\mathrm{T}}} \delta \mathbf{r}_P \tag{5.98}$$

Using the preceding two equations, one has

$$\delta W^i + \delta W^j = 0 \tag{5.99}$$

This simple fact will be utilized in developing the principle of virtual work to eliminate the reaction forces from the equilibrium conditions leading to a number of equations equal to the number of degrees of freedom of the system.

5.6 PRINCIPLE OF VIRTUAL WORK IN STATICS

The concepts and definitions presented in the preceding sections are used in this section to develop the *principle of virtual work for static equilibrium*. The principle of virtual work in dynamics is discussed in the following section.

Equipollent Systems of Forces The first step in deriving the principle of virtual work is to prove that two equipollent systems of forces produce the same virtual work. It was shown in Section 3 that a force $\mathbf{F}^i$ acting at an arbitrary point P^i on a rigid body i is equipollent to a force $\mathbf{F}^i$ that acts at the reference point and a moment M^i given by

$$M^i = \bar{\mathbf{u}}_P^{i^{\mathrm{T}}} \mathbf{A}_\theta^{i^{\mathrm{T}}} \mathbf{F}^i \tag{5.100}$$

where $\bar{\mathbf{u}}_P^i$ is the local position vector of point P^i defined with respect to the reference point, and $\mathbf{A}_\theta^i$ is the partial derivative of the transformation matrix with respect to the angle θ^i. The virtual work of the system of forces that consists of the force $\mathbf{F}^i$ and the moment M^i is

$$\delta W_r^i = \mathbf{F}^{i^{\mathrm{T}}} \delta \mathbf{R}^i + M^i \delta\theta^i = \mathbf{F}^{i^{\mathrm{T}}} \delta \mathbf{R}^i + \bar{\mathbf{u}}_P^{i^{\mathrm{T}}} \mathbf{A}_\theta^{i^{\mathrm{T}}} \mathbf{F}^i \delta\theta^i \tag{5.101}$$

The virtual work of the original system of forces that consists of the force $\mathbf{F}^i$ is

$$\delta W^i = \mathbf{F}^{i^{\mathrm{T}}} \delta \mathbf{r}_P^i \tag{5.102}$$

where $\mathbf{r}_P^i$ is the global position vector of the arbitrary point P^i, which can be expressed in terms of the coordinates of the reference point and the angular orientation of the body as

$$\mathbf{r}_P^i = \mathbf{R}^i + \mathbf{A}^i \bar{\mathbf{u}}_P^i \tag{5.103}$$

Substituting Eq. 103 into Eq. 102 yields

$$\delta W^i = \mathbf{F}^{i^{\mathrm{T}}} \delta \mathbf{R}^i + \mathbf{F}^{i^{\mathrm{T}}} \mathbf{A}_\theta^i \bar{\mathbf{u}}_P^i \delta\theta^i \tag{5.104}$$

By comparing Eqs. 101 and 104 one concludes that

$$\delta W^i = \delta W_r^i \tag{5.105}$$

which implies that two equipollent systems of forces do the same work.

Principle of Virtual Work The fact that two equipollent systems of forces do the same work can be utilized to provide a systematic development of the principle of virtual work. Consider a body i that is acted upon by the system of forces $\mathbf{F}_1^i, \mathbf{F}_2^i, \ldots, \mathbf{F}_{n_f}^i$ and the system of moments $M_1^i, M_2^i, \ldots, M_{n_m}^i$. This system of forces and moments that also includes the reaction forces and moments can be replaced by an equipollent system that consists of one force $\mathbf{F}_e^i$ and one moment M_e^i as shown in Fig. 19. The virtual work of the original system of forces shown in Fig. 19a is

$$\begin{aligned}\delta W^i &= \mathbf{F}_1^{i^{\mathrm{T}}} \delta \mathbf{r}_1^i + \mathbf{F}_2^{i^{\mathrm{T}}} \delta \mathbf{r}_2^i + \cdots + \mathbf{F}_{n_f}^{i^{\mathrm{T}}} \delta \mathbf{r}_n^i \\ &\quad + (M_1^i + M_2^i + \cdots + M_{n_m}^i)\, \delta\theta^i\end{aligned} \tag{5.106}$$

where $\mathbf{r}_j^i$ is the position vector of the point of application of the force $\mathbf{F}_j^i$ and θ^i is the angular orientation of the body i. Equation 106 can be written as

$$\delta W^i = \sum_{j=1}^{n_f} \mathbf{F}_j^{i^{\mathrm{T}}} \delta \mathbf{r}_j^i + \left(\sum_{j=1}^{n_m} M_j^i \right) \delta\theta^i \tag{5.107}$$

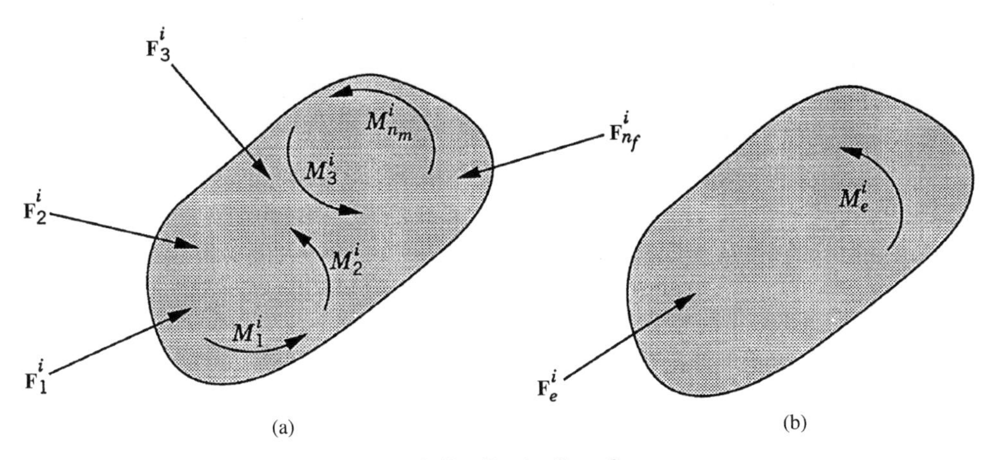

Figure 5.19 Equipollent forces

The virtual work of the system of forces shown in Fig. 19b can simply be written as

$$\delta W_r^i = \mathbf{F}_e^{i^{\mathrm{T}}} \delta \mathbf{r}_e^i + M_e^i \delta \theta^i \tag{5.108}$$

where $\mathbf{r}_e^i$ is the position vector of the point of application of the resultant force $\mathbf{F}_e^i$.

Since the two systems of forces shown in Fig. 19 are equipollent, one has $\delta W^i = \delta W_r^i$, or

$$\sum_{j=1}^{n_f} \mathbf{F}_j^{i^{\mathrm{T}}} \delta \mathbf{r}_j^i + \left(\sum_{j=1}^{n_m} M_j^i \right) \delta \theta^i = \mathbf{F}_e^{i^{\mathrm{T}}} \delta \mathbf{r}_e^i + M_e^i \delta \theta^i \tag{5.109}$$

If body i is to be in static equilibrium, the following conditions must hold:

$$\mathbf{F}_e^i = \mathbf{0}, \quad M_e^i = 0 \tag{5.110}$$

which also yield

$$\mathbf{F}_e^{i^{\mathrm{T}}} \delta \mathbf{r}_e^i = 0, \quad M_e^i \delta \theta^i = 0 \tag{5.111}$$

Substituting these two equations into Eq. 109, one obtains

$$\sum_{j=1}^{n_f} \mathbf{F}_j^{i^{\mathrm{T}}} \delta \mathbf{r}_j^i + \left(\sum_{j=1}^{n_m} M_j^i \right) \delta \theta^i = 0 \tag{5.112}$$

This is a mathematical statement of the principle of virtual work for the static equilibrium of body i. Equation 112 states that if body i is in static equilibrium, the virtual work of all the forces and moments that act on this body must be equal to zero. This equation can be written as

$$\delta W^i = 0 \tag{5.113}$$

Constraint Forces Equation 113 includes the effect of the external and constraint forces and moments. One may write Eq. 113 as

$$\delta W^i = \delta W_e^i + \delta W_c^i = 0 \tag{5.114}$$

where δW_e^i is the virtual work of the external forces and moments, and δW_c^i is the virtual work of the constraint forces and moments.

If the mechanical system consists of n_b bodies, an equation similar to Eq. 114 can be obtained for each body in the system. By summing up these equations, one obtains

$$\sum_{i=1}^{n_b} \delta W^i = \sum_{i=1}^{n_b} \delta W_e^i + \sum_{i=1}^{n_b} \delta W_c^i = 0 \tag{5.115}$$

Since joint constraint forces are equal in magnitude and opposite in direction, the virtual work of the constraint forces that act on the system must be equal to zero, that is,

$$\sum_{i=1}^{n_b} \delta W_c^i = 0 \tag{5.116}$$

Substituting Eq. 116 into Eq. 115 leads to the *principle of virtual work for the static equilibrium* of multibody systems as

$$\delta W_e = \sum_{i=1}^{n_b} \delta W_e^i = 0 \tag{5.117}$$

which implies that the multibody system that consists of interconnected bodies is in static equilibrium if the virtual work of all the external forces acting on the system is equal to zero.

Equilibrium Equations Let a multibody system that consists of n_b bodies be subjected to a system of *external forces and moments* given by

$$\mathbf{F} = [\mathbf{F}_1^{\mathrm{T}} \quad \mathbf{F}_2^{\mathrm{T}} \cdots \mathbf{F}_{n_f}^{\mathrm{T}}]^{\mathrm{T}}, \quad \mathbf{M} = [M_1 \quad M_2 \cdots M_{n_m}]^{\mathrm{T}} \tag{5.118}$$

The virtual work of this system of forces and moments is

$$\delta W_e = \sum_{j=1}^{n_f} \mathbf{F}_j^{\mathrm{T}}\, \delta \mathbf{r}_j + \sum_{j=1}^{n_m} M_j\, \delta\theta_j \tag{5.119}$$

As demonstrated previously, $\mathbf{r}_j$ and θ_j can be expressed in terms of the independent coordinates of the system, that is,

$$\mathbf{r}_j = \mathbf{r}_j(\mathbf{q}_i), \quad \theta_j = \theta_j(\mathbf{q}_i) \tag{5.120}$$

where $\mathbf{q}_i$ is the vector of system independent coordinates or degrees of freedom. Virtual changes in the system coordinates yield

$$\delta \mathbf{r}_j = \frac{\partial \mathbf{r}_j}{\partial \mathbf{q}_i} \delta \mathbf{q}_i, \quad \delta \theta_j = \frac{\partial \theta_j}{\partial \mathbf{q}_i} \delta \mathbf{q}_i \tag{5.121}$$

Substituting Eq. 121 into Eq. 119 leads to

$$\delta W_e = \left(\sum_{j=1}^{n_f} \mathbf{F}_j^{\mathrm{T}} \frac{\partial \mathbf{r}_j}{\partial \mathbf{q}_i} + \sum_{j=1}^{n_m} M_j \frac{\partial \theta_j}{\partial \mathbf{q}_i} \right) \delta \mathbf{q}_i \tag{5.122}$$

which can be written as

$$\delta W_e = \mathbf{Q}_e^{\mathrm{T}} \delta \mathbf{q}_i \tag{5.123}$$

where $\mathbf{Q}_e$ is the vector of generalized external forces defined as

$$\mathbf{Q}_e = \sum_{j=1}^{n_f} \left(\frac{\partial \mathbf{r}_j}{\partial \mathbf{q}_i} \right)^{\mathrm{T}} \mathbf{F}_j + \sum_{j=1}^{n_m} M_j \left(\frac{\partial \theta_j}{\partial \mathbf{q}_i} \right)^{\mathrm{T}} \tag{5.124}$$

If the system is in static equilibrium, Eqs. 117 and 123 yield

$$\delta W_e = \mathbf{Q}_e^{\mathrm{T}} \delta \mathbf{q}_i = 0 \tag{5.125}$$

Since the components of the vector $\mathbf{q}_i$ are assumed to be independent, one has

$$\mathbf{Q}_e = \mathbf{0} \tag{5.126}$$

This equation implies that if the system is in static equilibrium, the vector of generalized external forces associated with the system degrees of freedom must be equal to zero. Equation 126 represents a system of algebraic equilibrium equations that has a number of equations equal to the number of degrees of freedom of the system. Therefore, these equations can be solved for a number of unknowns equal to the number of degrees of freedom of the system.

It is clear that several basic steps are used in deriving the principle of virtual work for static equilibrium. In the first step, the fact that equipollent systems of forces do the same work is utilized. In the second step, the static equilibrium conditions are used to obtain the principle of virtual work as applied to each body in the system. At this intermediate step, the virtual work of the joint reaction forces must be considered because each body is treated separately. In the third step, the principle of virtual work for the multibody system that consists of a set of interconnected bodies is developed. Since in this step the equilibrium of the entire system is considered, the fact that the virtual work of the joint reaction forces acting on the system is equal to zero is utilized. This step leads to Eq. 117, which is valid regardless of the set of coordinates used. Finally, a set of independent equilibrium conditions is obtained by writing the principle of virtual work in terms of the virtual change in the system degrees of freedom. This step leads to the static equilibrium conditions of Eq. 126.

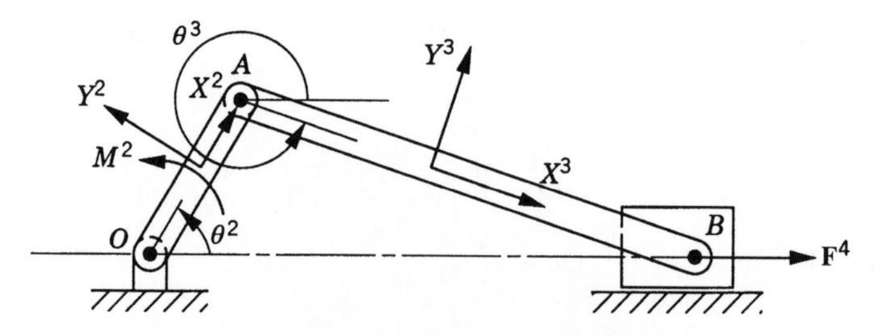

Figure 5.20 Slider crank mechanism

Illustrative Example To demonstrate the use of the principle of virtual work in the static equilibrium analysis of multibody systems, the slider crank mechanism shown in Fig. 20 is used. The crankshaft is subjected to an external moment M^2, while the slider block is acted upon by a force F^4. We assume that the origins of the body coordinate systems are attached to the body centers of mass. To illustrate the process of eliminating the reaction forces, Eq. 113 is first used for the static equilibrium analysis of each body in the mechanism. For the crankshaft, Eq. 113 is given by

$$\mathbf{F}^{12^{\mathrm{T}}}\,\delta\mathbf{r}_O^2 - \mathbf{F}^{23^{\mathrm{T}}}\,\delta\mathbf{r}_A^2 - m^2 g\,\delta R_y^2 + M^2\,\delta\theta^2 = 0 \tag{5.127}$$

where m^2 is the mass of the crankshaft, $\mathbf{r}_O^2$ and $\mathbf{r}_A^2$ are, respectively, the global position vectors of points O and A, R_y^2 is the vertical component of the position vector of the center of mass of the crankshaft, and $\mathbf{F}^{ij}$ is the vector of the joint reaction forces acting on body j as the result of its connection with body i.

Similarly, the virtual work of the forces acting on the connecting rod can be written as

$$\mathbf{F}^{23^{\mathrm{T}}}\,\delta\mathbf{r}_A^3 - \mathbf{F}^{34^{\mathrm{T}}}\,\delta\mathbf{r}_B^3 - m^3 g\,\delta R_y^3 = 0 \tag{5.128}$$

where m^3 is the mass of the connecting rod, $\mathbf{r}_B^3$ is the global position vector of point B, and R_y^3 is the vertical component of the position vector of the center of mass of the connecting rod.

The virtual work of the forces acting on the slider block is also equal to zero. This leads to

$$\mathbf{F}^{34^{\mathrm{T}}}\,\delta\mathbf{r}_B^4 + (F^{41} - m^4 g)\,\delta R_y^4 + F^4\,\delta R_x^4 = 0 \tag{5.129}$$

Observe that $\delta\mathbf{r}_O^2 = \mathbf{0}$ since point O is a fixed point, and that $\delta\mathbf{r}_A^2 = \delta\mathbf{r}_A^3$ and $\delta\mathbf{r}_B^3 = \delta\mathbf{r}_B^4$ as the result of the conditions of the revolute joints at points A and B, respectively. Also, $\delta R_y^4 = 0$, since the slider block moves only in the horizontal direction. Keeping this in mind and adding the preceding three equations leads to Eq. 117 for this mechanism as

$$-m^2 g\,\delta R_y^2 - m^3 g\,\delta R_y^3 + M^2\,\delta\theta^2 + F^4\,\delta R_x^4 = 0 \tag{5.130}$$

While the reaction forces appear in the static equilibrium equations when the principle of virtual work is applied to each link separately, these reactions are automatically eliminated by adding the resulting equilibrium equations leading to Eq. 130 which contains only the virtual work of the external forces. This equation in its current form is not very useful. In order to make use of this equation we express δR_y^2, δR_y^3, $\delta\theta^2$, and δR_x^4 in terms of the system degree of freedom which we may select as θ^2. In this case, one has

$$\left.\begin{aligned}\delta R_y^2 &= l_O^2 \cos\theta^2\,\delta\theta^2\\ \delta R_y^3 &= l^2 \cos\theta^2\,\delta\theta^2 + l_A^3 \cos\theta^3\delta\theta^3\\ \delta R_x^4 &= -l^2 \sin\theta^2\,\delta\theta^2 - l^3 \sin\theta^3\,\delta\theta^3\end{aligned}\right\} \tag{5.131}$$

where l_O^2 is the distance from point O to the center of the crankshaft and l_A^3 is the distance from point A to the center of the connecting rod. Since

$$\sin\theta^3 = -\frac{l^2}{l^3}\sin\theta^2 \tag{5.132}$$

one has

$$\delta\theta^3 = -\frac{l^2\cos\theta^2}{l^3\cos\theta^3}\,\delta\theta^2 \tag{5.133}$$

Substituting this equation into Eq. 131 yields

$$\left.\begin{aligned}\delta R_y^2 &= l_O^2 \cos\theta^2\,\delta\theta^2\\ \delta R_y^3 &= l^2\left(1-\frac{l_A^3}{l^3}\right)\cos\theta^2\,\delta\theta^2\\ \delta R_x^4 &= l^2(-\sin\theta^2 + \cos\theta^2\tan\theta^3)\,\delta\theta^2\end{aligned}\right\} \tag{5.134}$$

Substituting these equations into Eq. 130 leads to

$$\begin{aligned}&\left[-m^2 g l_O^2 \cos\theta^2 - m^3 g l^2\left(1-\frac{l_A^3}{l^3}\right)\cos\theta^2 + M^2\right.\\ &\left.\quad + F^4 l^2(-\sin\theta^2 + \cos\theta^2\tan\theta^3)\right]\delta\theta^2 = 0\end{aligned} \tag{5.135}$$

which can be written in the form of Eq. 125 as

$$Q_e\,\delta\theta^2 = 0 \tag{5.136}$$

where Q_e is the generalized force associated with the independent coordinate θ^2 and is given by

$$\begin{aligned}Q_e = &-m^2 g l_O^2 \cos\theta^2 - m^3 g l^2\left(1-\frac{l_A^3}{l^3}\right)\cos\theta^2 + M^2\\ &+ F^4 l^2(-\sin\theta^2 + \cos\theta^2\tan\theta^3)\end{aligned} \tag{5.137}$$

Since θ^2 is an independent coordinate, the scalar Q_e of Eq. 136 must be equal to zero, leading to the following algebraic equation:

$$
\begin{aligned}
&-m^2 g l_O^2 \cos\theta^2 - m^3 g l^2 \left(1 - \frac{l_A^3}{l^3}\right) \cos\theta^2 + M^2 \\
&+ F^4 l^2 (-\sin\theta^2 + \cos\theta^2 \tan\theta^3) = 0
\end{aligned} \tag{5.138}
$$

This equation does not include any reaction forces and can be solved for one unknown. For example, if the external moment M^2 is given, one can use the preceding equation to determine the force F^4, which is required in order to keep the mechanism in a given static equilibrium configuration.

Example 5.9

The four-bar linkage shown in Fig. 21 is subjected to two external moments M^2 and M^4 that act, respectively, on the crankshaft and the rocker. If the effect of gravity of the links is neglected and if M^2 is assumed to be known, determine the moment M^4 that acts on the rocker such that the mechanism is in static equilibrium.

Solution. Since the mechanism is in static equilibrium, the virtual work of all the external forces and moments that act on the mechanism must be equal to zero. This yields

$$\delta W_e = -M^2\,\delta\theta^2 + M^4\,\delta\theta^4 = 0$$

In Example 2, it was shown that

$$\delta\theta^4 = \frac{l^2 \sin(\theta^2 - \theta^3)}{l^4 \sin(\theta^3 - \theta^4)}\,\delta\theta^2$$

Substituting this equation into the expression for the virtual work, one obtains

$$\delta W_e = \left[-M^2 + \frac{l^2 \sin(\theta^2 - \theta^3)}{l^4 \sin(\theta^3 - \theta^4)}\,M^4\right]\delta\theta^2 = 0$$

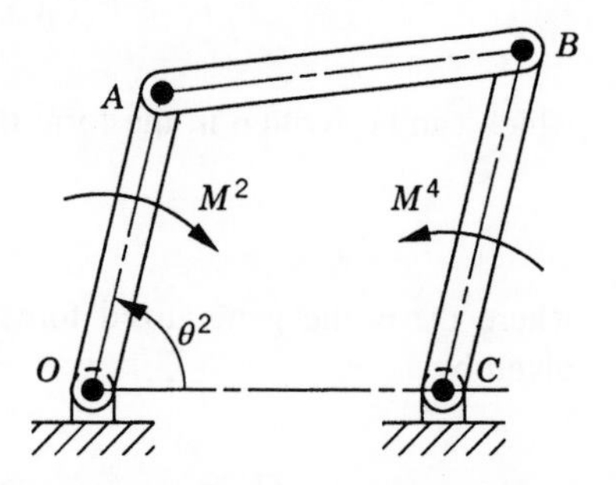

Figure 5.21 Four-bar mechanism

Since the mechanism has one degree of freedom, θ^2 can be selected as the independent coordinate. The coefficient of $\delta\theta^2$ in the preceding equation must then be equal to zero, leading to the following equilibrium condition:

$$-M^2 + \frac{l^2 \sin(\theta^2 - \theta^3)}{l^4 \sin(\theta^3 - \theta^4)} M^4 = 0$$

from which M^4 can be determined as

$$M^4 = \frac{l^4 \sin(\theta^3 - \theta^4)}{l^2 \sin(\theta^2 - \theta^3)} M^2$$

Example 5.10

The two-arm robotic manipulator shown in Fig. 22 is subjected to a torque M^2 that acts on link 2 and a force $\mathbf{F}^3$ that acts at the tip point of link 3, as shown in the figure. The force $\mathbf{F}^3$ is assumed to have a known direction defined by the angle ϕ. The mass of link 2 is assumed to be m^2, while the mass of link 3 is m^3. Considering the effect of gravity, determine M^2 and $\mathbf{F}^3$ such that the system remains in static equilibrium.

Solution. The virtual work of the forces and moments that act on the system is

$$\delta W_e = M^2\,\delta\theta^2 - m^2 g\,\delta R_y^2 - m^3 g\,\delta R_y^3 + \mathbf{F}^{3^{\mathrm{T}}}\,\delta\mathbf{r}_P^3$$

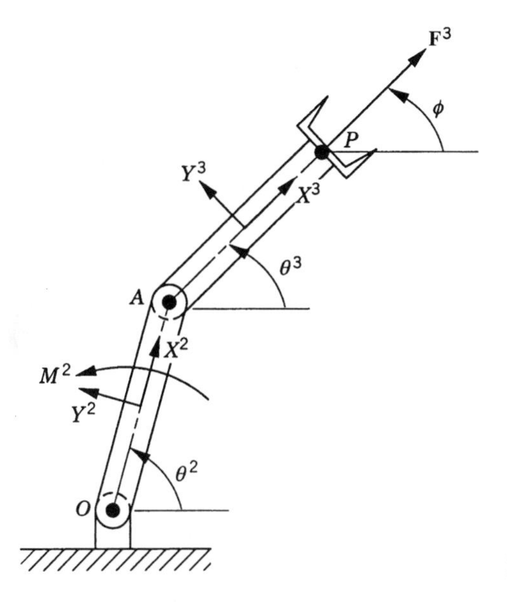

Figure 5.22 Robotic manipulator

where

$$R_y^2 = l_O^2 \sin \theta^2$$
$$R_y^3 = l^2 \sin \theta^2 + l_A^3 \sin \theta^3$$
$$\mathbf{r}_P^3 = \begin{bmatrix} l^2 \cos\theta^2 + l^3 \cos\theta^3 \\ l^2 \sin\theta^2 + l^3 \sin\theta^3 \end{bmatrix}$$

These equations yield

$$\delta R_y^2 = l_O^2 \cos\theta^2 \, \delta\theta^2$$
$$\delta R_y^3 = l^2 \cos \theta^2 \, \delta\theta^2 + l_A^3 \cos \theta^3 \, \delta\theta^3$$
$$\delta\mathbf{r}_P^3 = \begin{bmatrix} -l^2 \sin\theta^2 & -l^3 \sin\theta^3 \\ l^2 \cos\theta^2 & l^3 \cos\theta^3 \end{bmatrix} \begin{bmatrix} \delta\theta^2 \\ \delta\theta^3 \end{bmatrix}$$

Substituting these virtual changes into the expression for the virtual work, and writing $\mathbf{F}^3$ in terms of its components $F^3 \cos \phi$ and $F^3 \sin \phi$, one obtains

$$\begin{aligned} \delta W_e = {} & M^2 \, \delta\theta^2 - m^2 g l_O^2 \cos \theta^2 \, \delta\theta^2 - m^3 g (l^2 \cos \theta^2 \, \delta\theta^2 + l_A^3 \cos \theta^3 \, \delta\theta^3) \\ & + [F^3 \cos \phi \quad F^3 \sin \phi] \begin{bmatrix} -l^2 \sin\theta^2 & -l^3 \sin\theta^3 \\ l^2 \cos\theta^2 & l^3 \cos\theta^3 \end{bmatrix} \begin{bmatrix} \delta\theta^2 \\ \delta\theta^3 \end{bmatrix} \end{aligned}$$

If the system is in static equilibrium, one has $\delta W_e = 0$, and consequently

$$\begin{aligned} \delta W_e = {} & [M^2 - m^2 g l_O^2 \cos \theta^2 - m^3 g l^2 \cos \theta^2 \\ & - F^3 l^2 \sin \theta^2 \cos \phi + F^3 l^2 \cos \theta^2 \sin \phi] \, \delta\theta^2 \\ & + [-m^3 g l_A^3 \cos \theta^3 - F^3 l^3 \sin \theta^3 \cos \phi + F^3 l^3 \cos \theta^3 \sin \phi] \, \delta\theta^3 \\ = {} & 0 \end{aligned}$$

Since θ^2 and θ^3 are independent coordinates, their coefficients in the preceding equation must be equal to zero. This leads to the following algebraic equations:

$$M^2 + F^3 l^2 \sin(\phi - \theta^2) = m^2 g l_O^2 \cos \theta^2 + m^3 g l^2 \cos \theta^2$$
$$F^3 l^3 \sin(\phi - \theta^3) = m^3 g l_A^3 \cos \theta^3$$

The solution of these two equations defines F^3 and M^2 as

$$F^3 = \frac{m^3 g l_A^3 \cos \theta^3}{l^3 \sin(\phi - \theta^3)}$$

$$M^2 = m^2 g l_O^2 \cos \theta^2 + m^3 g l^2 \cos \theta^2 - m^3 g l_A^3 \cos \theta^3 \, \frac{l^2 \sin(\phi - \theta^2)}{l^3 \sin(\phi - \theta^3)}$$

Note that if the gravity of the two links are neglected, M^2 and F^3 are equal to zero. In this special case, external forces and moments are not required to keep the system in static equilibrium.

5.7 PRINCIPLE OF VIRTUAL WORK IN DYNAMICS

The principle of virtual work can be generalized to study the dynamics of multibody systems that consist of interconnected bodies. In this case, the inertia forces and D'Alembert's principle can be used to establish the *principle of virtual work in dynamics*. For the rigid body i, the equations of motion are

$$\left.\begin{aligned} \mathbf{F}^i - m^i\mathbf{a}^i &= \mathbf{0} \\ M^i - J^i\ddot{\theta}^i &= 0 \end{aligned}\right\} \tag{5.139}$$

where $\mathbf{F}^i$ is the vector of resultant forces that act on body i, M^i is the sum of the moments about the center of mass, $\mathbf{a}^i$ is the acceleration vector of the center of mass, $\ddot{\theta}^i$ is the angular acceleration, m^i is the mass of body i, and J^i is the mass moment of inertia of the body about an axis passing through the body center of mass.

Using the concept of the equipollent systems of forces discussed in the preceding sections, and without any loss of generality, the force vector $\mathbf{F}^i$ can be selected in such a manner that its point of application is the center of mass of the body. One may then multiply the first equation in Eq. 139 by $\delta\mathbf{R}^i$ and the second equation by $\delta\theta^i$, where $\mathbf{R}^i$ is the global position vector of the center of mass of the body. This yields

$$\left.\begin{aligned} (\mathbf{F}^i - m^i\mathbf{a}^i)^{\mathrm{T}}\,\delta\mathbf{R}^i &= 0 \\ (M^i - J^i\ddot{\theta}^i)\,\delta\theta^i &= 0 \end{aligned}\right\} \tag{5.140}$$

By adding these two equations, one obtains

$$(\mathbf{F}^i - m^i\mathbf{a}^i)^{\mathrm{T}}\,\delta\mathbf{R}^i + (M^i - J^i\ddot{\theta}^i)\,\delta\theta^i = 0 \tag{5.141}$$

or

$$\mathbf{F}^{i^{\mathrm{T}}}\,\delta\mathbf{R}^i + M^i\,\delta\theta^i - m^i\mathbf{a}^{i^{\mathrm{T}}}\,\delta\mathbf{R}^i - J^i\ddot{\theta}^i\,\delta\theta^i = 0 \tag{5.142}$$

This equation can be written as

$$\delta W^i - \delta W_i^i = 0 \tag{5.143}$$

where δW^i is the virtual work of the external and reaction forces and moments that act on body i, and δW_i^i is the virtual work of the inertia forces and moments of this body, that is,

$$\left.\begin{aligned} \delta W^i &= \mathbf{F}^{i^{\mathrm{T}}}\,\delta\mathbf{R}^i + M^i\,\delta\theta^i \\ \delta W_i^i &= m^i\mathbf{a}^{i^{\mathrm{T}}}\,\delta\mathbf{R}^i + J^i\ddot{\theta}^i\,\delta\theta^i \end{aligned}\right\} \tag{5.144}$$

Note that δW^i can also be written as

$$\delta W^i = \delta W_c^i + \delta W_e^i \tag{5.145}$$

where δW_c^i is the virtual work of the constraint forces and moments, and δW_e^i is the virtual work of the external forces and moments. Substituting Eq. 145 into Eq. 143, one obtains

$$\delta W_c^i + \delta W_e^i - \delta W_i^i = 0 \tag{5.146}$$

which implies that when the dynamics of a body is considered separately, the virtual work of the reaction forces acting on the body must be included. It is also important to reiterate at this point that the virtual work of Eq. 146 may be expressed in terms of the actual system of external and reaction forces acting on this body instead of the equipollent system, since both systems produce the same virtual work.

Connectivity Conditions If the mechanical system consists of n_b interconnected rigid bodies, the use of Eq. 146 leads to

$$\sum_{i=1}^{n_b} (\delta W_c^i + \delta W_e^i - \delta W_i^i) = 0 \tag{5.147}$$

Due to the fact that the joint constraint forces that act on two adjacent bodies are equal in magnitude and opposite in direction, one has

$$\sum_{i=1}^{n_b} \delta W_c^i = 0 \tag{5.148}$$

Substituting this equation into Eq. 147 yields

$$\sum_{i=1}^{n_b} \delta W_e^i - \sum_{i=1}^{n_b} \delta W_i^i = 0 \tag{5.149}$$

This is the principle of virtual work for dynamics, which states that the virtual work of the external forces and moments acting on the system is equal to the virtual work of the inertia forces and moments of the system. In Eq. 149, one does not need to consider the reaction forces of the workless constraints. Equation 149 can be written as

$$\delta W_e - \delta W_i = 0 \tag{5.150}$$

where

$$\delta W_e = \sum_{i=1}^{n_b} \delta W_e^i, \quad \delta W_i = \sum_{i=1}^{n_b} \delta W_i^i \tag{5.151}$$

The coefficients of the virtual displacements in Eq. 150 cannot be set equal to zero because these displacements are not independent. To make use of Eq. 150, the virtual changes in the system degrees of freedom are used.

Dynamic Equations As in the case of the static equilibrium, the virtual displacements can be expressed in terms of the virtual displacements of the independent coordinates. By so doing, the virtual work of the external and inertia forces and moments can be expressed as

$$\delta W_e = \mathbf{Q}_e^{\mathrm{T}}\, \delta \mathbf{q}_i, \quad \delta W_i = \mathbf{Q}_i^{\mathrm{T}}\, \delta \mathbf{q}_i \tag{5.152}$$

where $\mathbf{Q}_e$ and $\mathbf{Q}_i$ are, respectively, the vectors of generalized external and inertia forces associated with the system independent coordinates $\mathbf{q}_i$. Substituting Eq. 152 into Eq. 150, one obtains

$$(\mathbf{Q}_e^{\mathrm{T}} - \mathbf{Q}_i^{\mathrm{T}})\, \delta \mathbf{q}_i = 0 \tag{5.153}$$

Since the components of the vector $\mathbf{q}_i$ are assumed to be independent, Eq. 153 leads to

$$\mathbf{Q}_e = \mathbf{Q}_i \tag{5.154}$$

These are the dynamic equations for the multibody system, which imply that the vectors of the generalized external and inertia forces associated with the independent coordinates must be equal. The number of scalar equations in Eq. 154 is equal to the number of degrees of freedom of the system. Consequently, Eq. 154 can be used to solve for a number of unknowns equal to the number of the system degrees of freedom.

Illustrative Example The use of the principle of virtual work in dynamics can be demonstrated using the slider crank mechanism discussed in the preceding section and shown in Fig. 20. Since the mechanism has one degree of freedom, the vectors $\mathbf{Q}_e$ and $\mathbf{Q}_i$ of Eq. 154 reduce to scalars. It was shown in the preceding section that the generalized external force Q_e associated with the angular rotation of the crankshaft is given by

$$\begin{aligned} Q_e = {} & -m^2 g l_O^2 \cos\theta^2 - m^3 g l^2 \left(1 - \frac{l_A^3}{l^3}\right) \cos\theta^2 + M^2 \\ & + F^4 l^2 (-\sin\theta^2 + \cos\theta^2 \tan\theta^3) \end{aligned} \tag{5.155}$$

where all the variables in this equation are as defined in the preceding section. The virtual work of the inertia forces is given by

$$\begin{aligned} \delta W_i = {} & m^2 a_x^2\, \delta R_x^2 + m^2 a_y^2\, \delta R_y^2 + J^2 \ddot{\theta}^2\, \delta\theta^2 \\ & + m^3 a_x^3\, \delta R_x^3 + m^3 a_y^3\, \delta R_y^3 + J^3 \ddot{\theta}^3\, \delta\theta^3 \\ & + m^4 a_x^4\, \delta R_x^4 \end{aligned} \tag{5.156}$$

where a_x^i and a_y^i are the components of the acceleration of the center of mass of link i, $\ddot{\theta}^i$ is its angular acceleration, and R_x^i and R_y^i are the components of the global position vector of the center of mass of link i. The components R_x^2 and R_x^3 can be written as

$$R_x^2 = l_O^2 \cos\theta^2, \quad R_x^3 = l^2 \cos\theta^2 + l_A^3 \cos\theta^3 \tag{5.157}$$

which upon using Eq. 133 yields

$$\left.\begin{aligned}\delta R_x^2 &= -l_O^2 \sin\theta^2\, \delta\theta^2 \\ \delta R_x^3 &= -l^2\left(\sin\theta^2 - \frac{l_A^3}{l^3}\cos\theta^2\tan\theta^3\right)\delta\theta^2\end{aligned}\right\} \tag{5.158}$$

Substituting Eqs. 133, 134, and 158 into Eq. 156 yields

$$\begin{aligned}\delta W_i = \Bigg[&-m^2 a_x^2 l_O^2 \sin\theta^2 + m^2 a_y^2 l_O^2 \cos\theta^2 + J^2\ddot{\theta}^2 \\ &- m^3 a_x^3 l^2\left(\sin\theta^2 - \frac{l_A^3}{l^3}\cos\theta^2\tan\theta^3\right) \\ &+ m^3 a_y^3 l^2\left(1 - \frac{l_A^3}{l^3}\right)\cos\theta^2 - J^3\ddot{\theta}^3\,\frac{l^2\cos\theta^2}{l^3\cos\theta^3} \\ &+ m^4 a_x^4 l^2(-\sin\theta^2 + \cos\theta^2\tan\theta^3)\Bigg]\delta\theta^2\end{aligned} \tag{5.159}$$

which can be written as $\delta W_i = Q_i\, \delta\theta^2$, where

$$\begin{aligned}Q_i = &-m^2 a_x^2 l_O^2 \sin\theta^2 + m^2 a_y^2 l_O^2 \cos\theta^2 + J^2\ddot{\theta}^2 \\ &- m^3 a_x^3 l^2\left(\sin\theta^2 - \frac{l_A^3}{l^3}\cos\theta^2\tan\theta^3\right) \\ &+ m^3 a_y^3 l^2\left(1 - \frac{l_A^3}{l^3}\right)\cos\theta^2 - J^3\ddot{\theta}^3\,\frac{l^2\cos\theta^2}{l^3\cos\theta^3} \\ &+ m^4 a_x^4 l^2(-\sin\theta^2 + \cos\theta^2\tan\theta^3)\end{aligned} \tag{5.160}$$

By using Eqs. 154, 155, and 160, the dynamic condition for the slider crank mechanism can be written as

$$\begin{aligned}&-m^2 a_x^2 l_O^2 \sin\theta^2 + m^2 a_y^2 l_O^2 \cos\theta^2 + J^2\ddot{\theta}^2 \\ &\quad- m^3 a_x^3 l^2\left(\sin\theta^2 - \frac{l_A^3}{l^3}\cos\theta^2\tan\theta^3\right) \\ &\quad+ m^3 a_y^3 l^2\left(1 - \frac{l_A^3}{l^3}\right)\cos\theta^2 - J^3\ddot{\theta}^3\left(\frac{l^2\cos\theta^2}{l^3\cos\theta^3}\right) \\ &\quad+ m^4 a_x^4 l^2(-\sin\theta^2 + \cos\theta^2\tan\theta^3) \\ &\quad= -m^2 g l_O^2 \cos\theta^2 - m^3 g l^2\left(1 - \frac{l_A^3}{l^3}\right)\cos\theta^2 + M^2 \\ &\quad\quad+ F^4 l^2(-\sin\theta^2 + \cos\theta^2\tan\theta^3)\end{aligned} \tag{5.161}$$

The acceleration components that appear in this equation are not independent by virtue of the kinematic constraints. The relationships between these accelerations can be found by differentiating the algebraic constraint equations, as previously explained. All the acceleration components of the slider crank mechanism can be expressed in terms of θ^2, $\dot{\theta}^2$, and $\ddot{\theta}^2$ because the mechanism has one degree of freedom only. If all the forces in Eq. 161 are given, this equation can be solved for the angular acceleration of the crankshaft in terms of θ^2 and $\dot{\theta}^2$. Given a set of initial conditions, the angular acceleration $\ddot{\theta}^2$ can be integrated to determine the angular displacement θ^2 and the angular velocity $\dot{\theta}^2$. Having determined the degree of freedom and its time derivatives, other coordinates and their time derivatives can be determined using the kinematic equations.

Example 5.11

Obtain the dynamic equations of the two-arm robotic manipulator of Example 10.

Solution. Since the system has two degrees of freedom, the virtual work of the applied forces can be expressed in terms of the virtual changes in these two degrees of freedom. It was shown in Example 10 that the virtual work of the external forces is

$$\begin{aligned}\delta W_e = {} & [M^2 - m^2 g l_O^2 \cos\theta^2 - m^3 g l^2 \cos\theta^2 \\ & - F^3 l^2 \sin\theta^2 \cos\phi + F^3 l^2 \cos\theta^2 \sin\phi]\,\delta\theta^2 \\ & + [-m^3 g l_A^3 \cos\theta^3 - F^3 l^3 \sin\theta^3 \cos\phi + F^3 l^3 \cos\theta^3 \sin\phi]\,\delta\theta^3\end{aligned}$$

or

$$\delta W_e = \mathbf{Q}_e^{\mathrm{T}}\,\delta\mathbf{q}_i$$

where $\mathbf{q}_i = [\theta^2 \quad \theta^3]^{\mathrm{T}}$, and $\mathbf{Q}_e = [Q_1 \quad Q_2]^{\mathrm{T}}$, where

$$\begin{aligned}Q_1 &= M^2 - m^2 g l_O^2 \cos\theta^2 - m^3 g l^2 \cos\theta^2 + F^3 l^2 \sin(\phi - \theta^2) \\ Q_2 &= -m^3 g l_A^3 \cos\theta^3 + F^3 l^3 \sin(\phi - \theta^3)\end{aligned}$$

The virtual work of the inertia forces is

$$\delta W_i = m^2 a_x^2\,\delta R_x^2 + m^2 a_y^2\,\delta R_y^2 + J^2\ddot{\theta}^2\,\delta\theta^2 + m^3 a_x^3\,\delta R_x^3 + m^3 a_y^3\,\delta R_y^3 + J^3\ddot{\theta}^3\,\delta\theta^3$$

where

$$\begin{aligned}\delta R_x^2 &= -l_O^2 \sin\theta^2\,\delta\theta^2 \\ \delta R_y^2 &= l_O^2 \cos\theta^2\,\delta\theta^2 \\ \delta R_x^3 &= -l^2 \sin\theta^2\,\delta\theta^2 - l_A^3 \sin\theta^3\,\delta\theta^3 \\ \delta R_y^3 &= l^2 \cos\theta^2\,\delta\theta^2 + l_A^3 \cos\theta^3\,\delta\theta^3\end{aligned}$$

Substituting these virtual changes into the expression of the virtual work of the inertia forces, one obtains

$$\delta W_i = (-m^2 a_x^2 l_O^2 \sin\theta^2 + m^2 a_y^2 l_O^2 \cos\theta^2 + J^2\ddot{\theta}^2$$

$$- m^3a_x^3l^2 \sin \theta^2 + m^3a_y^3l^2 \cos \theta^2)\, \delta\theta^2$$
$$+ (-m^3a_x^3l_A^3 \sin \theta^3 + m^3a_y^3l_A^3 \cos \theta^3 + J^3\ddot{\theta}^3)\, \delta\theta^3$$

which can be written as

$$\delta W_i = \mathbf{Q}_i^{\mathrm{T}}\, \delta\mathbf{q}_i$$

where

$$\mathbf{Q}_i = [Q_{i1} \quad Q_{i2}]^{\mathrm{T}}$$

in which

$$Q_{i1} = -m^2a_x^2l_O^2 \sin \theta^2 + m^2a_y^2l_O^2 \cos \theta^2 + J^2\ddot{\theta}^2 - m^3a_x^3l^2 \sin \theta^2 + m^3a_y^3l^2 \cos \theta^2$$

$$Q_{i2} = -m^3a_x^3l_A^3 \sin \theta^3 + m^3a_y^3l_A^3 \cos \theta^3 + J^3\ddot{\theta}^3$$

Applying Eq. 154, the dynamic equations for this system are

$$- m^2a_x^2l_O^2 \sin \theta^2 + m^2a_y^2l_O^2 \cos \theta^2 + J^2\ddot{\theta}^2 - m^3a_x^3l^2 \sin \theta^2 + m^3a_y^3l^2 \cos \theta^2$$
$$= M^2 - m^2gl_O^2 \cos \theta^2 - m^3gl^2 \cos \theta^2 + F^3l^2 \sin(\phi - \theta^2)$$

$$- m^3a_x^3l_A^3 \sin \theta^3 + m^3a_y^3l_A^3 \cos \theta^3 + J^3\ddot{\theta}^3$$
$$= -m^3gl_A^3 \cos \theta^3 + F^3l^3 \sin(\phi - \theta^3)$$

5.8 LAGRANGE'S EQUATION

The principle of virtual work allows us to formulate the dynamic equations using any set of independent generalized coordinates. Lagrange (1736–1813) created this powerful tool, recognized its superiority in formulating the dynamic equations, and used it as the starting point to formulate *Lagrange's equation*, which we will discuss in this section.

The virtual work of the inertia forces of a rigid body i is defined as

$$\delta W_i^i = \int_{V^i} \rho^i \ddot{\mathbf{r}}^{i\mathrm{T}}\, \delta\mathbf{r}^i\, dV^i \tag{5.162}$$

where ρ^i and V^i are, respectively, the mass density and volume of the rigid body, and $\mathbf{r}^i$ is the global position vector of an arbitrary point on the rigid body. The global position vector of the arbitrary point can be written in terms of the system generalized coordinates as

$$\mathbf{r}^i = \mathbf{r}^i(\mathbf{q}, t) \tag{5.163}$$

It follows that

$$\delta \mathbf{r}^i = \frac{\partial \mathbf{r}^i}{\partial \mathbf{q}} \delta \mathbf{q} \tag{5.164}$$

Substituting Eq. 164 into Eq. 162, the virtual work of the inertia forces of the rigid body i can be written as

$$\delta W_i^i = \int_{V^i} \rho^i \ddot{\mathbf{r}}^{i\mathrm{T}} \frac{\partial \mathbf{r}^i}{\partial \mathbf{q}} \delta \mathbf{q} \, dV^i \tag{5.165}$$

or

$$\delta W_i^i = \mathbf{Q}_i^{i\mathrm{T}} \delta \mathbf{q} \tag{5.166}$$

where

$$\mathbf{Q}_i^i = \int_{V^i} \rho^i \left(\frac{\partial \mathbf{r}^i}{\partial \mathbf{q}} \right)^{\mathrm{T}} \ddot{\mathbf{r}}^i \, dV^i \tag{5.167}$$

is the vector of generalized inertia forces of body i associated with the system generalized coordinates $\mathbf{q}$.

We note also that the absolute velocity vector of the arbitrary point on the rigid body is

$$\begin{aligned} \dot{\mathbf{r}}^i &= \frac{\partial \mathbf{r}^i}{\partial q_1} \dot{q}_1 + \frac{\partial \mathbf{r}^i}{\partial q_2} \dot{q}_2 + \cdots + \frac{\partial \mathbf{r}^i}{\partial q_n} \dot{q}_n + \frac{\partial \mathbf{r}^i}{\partial t} \\ &= \sum_{j=1}^{n} \frac{\partial \mathbf{r}^i}{\partial q_j} \dot{q}_j + \frac{\partial \mathbf{r}^i}{\partial t} = \frac{\partial \mathbf{r}^i}{\partial \mathbf{q}} \dot{\mathbf{q}} + \frac{\partial \mathbf{r}^i}{\partial t} \end{aligned} \tag{5.168}$$

from which one can deduce the following identity:

$$\frac{\partial \dot{\mathbf{r}}^i}{\partial \dot{\mathbf{q}}} = \frac{\partial \mathbf{r}^i}{\partial \mathbf{q}} \tag{5.169}$$

By using a similar procedure, one can also show that

$$\frac{\partial \mathbf{r}^i}{\partial \mathbf{q}} = \frac{\partial \dot{\mathbf{r}}^i}{\partial \dot{\mathbf{q}}} = \frac{\partial \ddot{\mathbf{r}}^i}{\partial \ddot{\mathbf{q}}} \tag{5.170}$$

By using the identity of Eq. 169, the generalized inertia forces of the rigid body i given by Eq. 167 can be written as

$$\mathbf{Q}_i^i = \int_{V^i} \rho^i \left(\frac{\partial \dot{\mathbf{r}}^i}{\partial \dot{\mathbf{q}}} \right)^{\mathrm{T}} \ddot{\mathbf{r}}^i \, dV^i \tag{5.171}$$

Note that

$$\frac{d}{dt}\left\{\left(\frac{\partial \dot{\mathbf{r}}^i}{\partial \dot{\mathbf{q}}}\right)^{\mathrm{T}} \dot{\mathbf{r}}^i\right\} = \left\{\frac{d}{dt}\left(\frac{\partial \dot{\mathbf{r}}^i}{\partial \dot{\mathbf{q}}}\right)^{\mathrm{T}}\right\} \dot{\mathbf{r}}^i + \left(\frac{\partial \dot{\mathbf{r}}^i}{\partial \dot{\mathbf{q}}}\right)^{\mathrm{T}} \ddot{\mathbf{r}}^i \tag{5.172}$$

which upon utilizing the identity of Eq. 169 leads to

$$\begin{aligned}
\left(\frac{\partial \dot{\mathbf{r}}^i}{\partial \dot{\mathbf{q}}}\right)^{\mathrm{T}} \ddot{\mathbf{r}}^i &= \frac{d}{dt}\left\{\left(\frac{\partial \dot{\mathbf{r}}^i}{\partial \dot{\mathbf{q}}}\right)^{\mathrm{T}} \dot{\mathbf{r}}^i\right\} - \frac{d}{dt}\left\{\frac{\partial \mathbf{r}^i}{\partial \mathbf{q}}\right\} \dot{\mathbf{r}}^i \\
&= \frac{d}{dt}\left\{\frac{\partial}{\partial \dot{\mathbf{q}}}\left(\frac{1}{2}\, \dot{\mathbf{r}}^{i^{\mathrm{T}}} \dot{\mathbf{r}}^i\right)\right\} - \frac{\partial}{\partial \mathbf{q}}\left\{\frac{1}{2}\, \dot{\mathbf{r}}^{i^{\mathrm{T}}} \dot{\mathbf{r}}^i\right\}
\end{aligned} \tag{5.173}$$

Substituting Eq. 173 into Eq. 171 and using the definition of the kinetic energy of body i

$$T^i = \frac{1}{2} \int_{V^i} \rho^i \dot{\mathbf{r}}^{i^{\mathrm{T}}} \dot{\mathbf{r}}^i \, dV^i, \tag{5.174}$$

the generalized inertia forces of the rigid body i can be expressed in terms of the body kinetic energy as

$$\mathbf{Q}_i^i = \frac{d}{dt}\left(\frac{\partial T^i}{\partial \dot{\mathbf{q}}}\right)^{\mathrm{T}} - \left(\frac{\partial T^i}{\partial \mathbf{q}}\right)^{\mathrm{T}} \tag{5.175}$$

The inertia forces of a system of n_b rigid bodies can then be written as

$$\begin{aligned}
\mathbf{Q}_i &= \sum_{i=1}^{n_b} \mathbf{Q}_i^i = \sum_{i=1}^{n_b} \frac{d}{dt}\left(\frac{\partial T^i}{\partial \dot{\mathbf{q}}}\right)^{\mathrm{T}} - \left(\frac{\partial T^i}{\partial \mathbf{q}}\right)^{\mathrm{T}} \\
&= \frac{d}{dt}\left(\frac{\partial T}{\partial \dot{\mathbf{q}}}\right)^{\mathrm{T}} - \left(\frac{\partial T}{\partial \mathbf{q}}\right)^{\mathrm{T}}
\end{aligned} \tag{5.176}$$

where T is the system kinetic energy defined as $T = \sum_{i=1}^{n_b} T^i$.

Using the principle of virtual work in dynamics, one concludes that if the generalized coordinates are independent, the system equations of motions can be written as

$$\frac{d}{dt}\left(\frac{\partial T}{\partial \dot{q}_j}\right) - \frac{\partial T}{\partial q_j} = Q_j \qquad j = 1, 2, \ldots, n \tag{5.177}$$

where $q_j, j = 1, 2, \ldots, n$ are the independent coordinates or the system degrees of freedom and Q_j is the generalized applied force associated with the independent coordinate q_j. Equation 177 is called *Lagrange's equation of motion*.

Example 5.12

To demonstrate the use of Lagrange's equation, we consider the system shown in Fig. 23. The rod in this system is assumed to be uniform and has mass m^2, mass moment of inertia about its center of mass J^2, and length l. The block is assumed to have a specified motion $z(t)$ and, consequently, the system has one degree of freedom, which we select to be the angular orientation of the rod θ^2. The kinetic energy of the rod can be written as

$$T = \tfrac{1}{2}\, m^2[(\dot{R}_x^2)^2 + (\dot{R}_y^2)^2] + \tfrac{1}{2}\, J^2(\dot{\theta}^2)^2$$

where R_x^2 and R_y^2 are the Cartesian components of the position vector of the center of mass of the rod. These coordinates are defined as

$$R_x^2 = z(t) + \frac{l}{2}\cos\theta^2, \quad R_y^2 = \frac{l}{2}\sin\theta^2$$

and their time derivatives are

$$\dot{R}_x^2 = \dot{z}(t) - \dot{\theta}^2\,\frac{l}{2}\sin\theta^2$$

$$\dot{R}_y^2 = \dot{\theta}^2\,\frac{l}{2}\cos\theta^2$$

The kinetic energy of the rod takes the form

$$\begin{aligned} T &= \frac{1}{2}\, m^2\left[\left(\dot{z}(t) - \dot{\theta}^2\,\frac{l}{2}\sin\theta^2\right)^2 + \left(\dot{\theta}^2\,\frac{l}{2}\cos\theta^2\right)^2\right] + \frac{1}{2}\, J^2(\dot{\theta}^2)^2 \\ &= \frac{1}{2}\, m^2\left[(\dot{z})^2 - \dot{z}\dot{\theta}^2 l\sin\theta^2 + \left(\frac{\dot{\theta}^2 l}{2}\right)^2\right] + \frac{1}{2}\, J^2(\dot{\theta}^2)^2 \end{aligned}$$

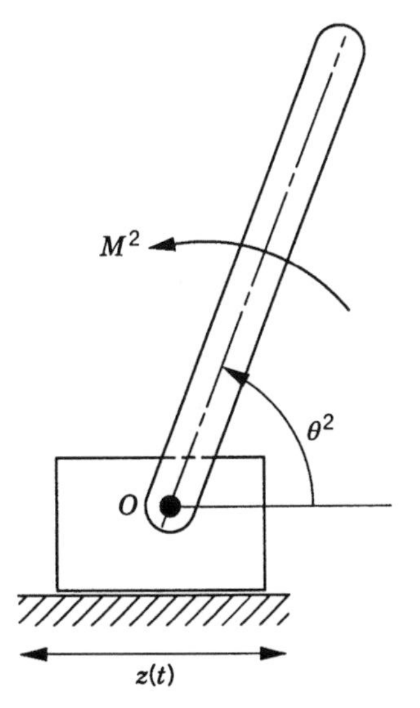

Figure 5.23 Pendulum with moving base

The virtual work of the external forces acting on the rod is

$$\delta W_e = -m^2 g\ \delta R_y^2 + M^2\ \delta\theta^2 = \left(-m^2 g\ \frac{l}{2}\ \cos\theta^2 + M^2\right)\delta\theta^2 = Q_e\ \delta\theta^2$$

where Q_e is the generalized force associated with the system degree of freedom θ^2 and is defined as

$$Q_e = -m^2 g\ \frac{l}{2}\ \cos\theta^2 + M^2$$

Lagrange's equation of motion of this system can then be written as

$$\frac{d}{dt}\left(\frac{\partial T}{\partial\dot{\theta}^2}\right) - \frac{\partial T}{\partial\theta^2} = Q_e$$

where

$$\frac{\partial T}{\partial\dot{\theta}^2} = -\frac{1}{2}\ m^2\dot{z}l\ \sin\theta^2 + \left[J^2 + m^2\ \frac{(l)^2}{4}\right]\dot{\theta}^2$$

It follows that

$$\frac{d}{dt}\left(\frac{\partial T}{\partial\dot{\theta}^2}\right) = -m^2\ddot{z}\ \frac{l}{2}\ \sin\theta^2 - m^2\dot{z}\dot{\theta}^2\ \frac{l}{2}\ \cos\theta^2 + J_O^2\ddot{\theta}^2$$

in which

$$J_O^2 = J^2 + \frac{m^2(l)^2}{4}$$

One also has

$$\frac{\partial T}{\partial\theta^2} = -m^2\dot{z}\dot{\theta}^2\ \frac{l}{2}\ \cos\theta^2$$

Substituting the preceding equation into Eq. 177, one obtains

$$J_O^2\ddot{\theta}^2 + m^2 g\ \frac{l}{2}\ \cos\theta^2 = M^2 + m^2\ddot{z}\ \frac{l}{2}\ \sin\theta^2$$

Clearly, this equation can also be derived using D'Alembert's principle by equating the moments of the applied and inertia forces about point O. The use of D'Alembert's principle to obtain the above equation was demonstrated in the preceding chapter. A third alternative for deriving this equation is to use the principle of virtual work, which is the starting point in deriving Lagrange's equation.

5.9 GIBBS–APPEL EQUATION

The identity of Eq. 170 clearly demonstrates that the virtual work of the inertia forces of the rigid body i can be evaluated using any of the following expressions:

$$\left.\begin{aligned} \mathbf{Q}_i^i &= \int_{V^i} \rho^i \left(\frac{\partial \mathbf{r}^i}{\partial \mathbf{q}}\right)^{\mathrm{T}} \ddot{\mathbf{r}}^i \, dV^i \\ \mathbf{Q}_i^i &= \int_{V^i} \rho^i \left(\frac{\partial \dot{\mathbf{r}}^i}{\partial \dot{\mathbf{q}}}\right)^{\mathrm{T}} \ddot{\mathbf{r}}^i \, dV^i \\ \mathbf{Q}_i^i &= \int_{V^i} \rho^i \left(\frac{\partial \ddot{\mathbf{r}}^i}{\partial \ddot{\mathbf{q}}}\right)^{\mathrm{T}} \ddot{\mathbf{r}}^i \, dV^i \end{aligned}\right\} \tag{5.178}$$

or by using the expression of the kinetic energy in Lagrange's equation as previously discussed. All of these forms are equivalent and lead to the same results when the same set of coordinates are used.

While in Lagrange's equation, the generalized inertia forces are expressed in terms of the kinetic energy which is quadratic in the velocities; in *Gibbs–Appel equation*, the generalized inertia forces are expressed in terms of an *acceleration function*. The Gibbs–Appel form of the inertia forces can be obtained using the third equation in Eq. 178. This equation can be written as

$$\mathbf{Q}_i^i = \int_{V^i} \rho^i \left[\frac{\partial}{\partial \ddot{\mathbf{q}}}\left(\frac{1}{2}\ddot{\mathbf{r}}^{i\mathrm{T}}\ddot{\mathbf{r}}^i\right)\right]^{\mathrm{T}} dV^i \tag{5.179}$$

which can also be written as

$$\mathbf{Q}_i^i = \frac{\partial S^i}{\partial \ddot{\mathbf{q}}} \tag{5.180}$$

where S^i is an acceleration function defined as

$$S^i = \frac{1}{2}\int_{V^i} \rho^i \ddot{\mathbf{r}}^{i\mathrm{T}}\ddot{\mathbf{r}}^i \, dV^i \tag{5.181}$$

5.10 HAMILTONIAN FORMULATION

The forces acting on a mechanical system can be classified as *conservative* and *nonconservative forces*. If $\mathbf{Q}_e$ is the vector of generalized forces acting on the multibody system, this vector can be written as

$$\mathbf{Q}_e = \mathbf{Q}_{co} + \mathbf{Q}_{nc} \tag{5.182}$$

where $\mathbf{Q}_{co}$ and $\mathbf{Q}_{nc}$ are, respectively, the vectors of conservative and nonconservative forces. As pointed out previously, conservative forces can be derived from a potential function V, that is,

$$\mathbf{Q}_{co} = -\left(\frac{\partial V}{\partial \mathbf{q}}\right)^{\mathrm{T}} \tag{5.183}$$

where $\mathbf{q} = [q_1 \quad q_2 \cdots q_n]^{\mathrm{T}}$ is the vector of generalized coordinates of the multibody system. Substituting Eq. 183 into Eq. 182, one obtains

$$\mathbf{Q}_e = -\left(\frac{\partial V}{\partial \mathbf{q}}\right)^{\mathrm{T}} + \mathbf{Q}_{nc} \tag{5.184}$$

If the generalized coordinates are independent, Lagrange's equation can be written in the following form:

$$\frac{d}{dt}\left(\frac{\partial T}{\partial \dot{\mathbf{q}}}\right)^{\mathrm{T}} - \left(\frac{\partial T}{\partial \mathbf{q}}\right)^{\mathrm{T}} = \mathbf{Q}_e \tag{5.185}$$

Using Eqs. 184 and 185 and keeping in mind that the potential function V does not depend on the generalized velocities, one gets

$$\frac{d}{dt}\left[\frac{\partial (T-V)}{\partial \dot{\mathbf{q}}}\right]^{\mathrm{T}} - \left[\frac{\partial (T-V)}{\partial \mathbf{q}}\right]^{\mathrm{T}} = \mathbf{Q}_{nc} \tag{5.186}$$

Define the *Lagrangian* L as

$$L = T - V \tag{5.187}$$

In terms of the Lagrangian, Lagrange's equation takes the form

$$\frac{d}{dt}\left(\frac{\partial L}{\partial \dot{\mathbf{q}}}\right)^{\mathrm{T}} - \left(\frac{\partial L}{\partial \mathbf{q}}\right)^{\mathrm{T}} = \mathbf{Q}_{nc} \tag{5.188}$$

Canonical Equations If the multibody system has n degrees of freedom, Lagrange's equation defines n second-order differential equations of motion expressed in terms of the degrees of freedom and their time derivatives. The Hamiltonian formulation, on the other hand, leads to a set of first-order differential equations defined in terms of the generalized coordinates and generalized momenta P_i, which are defined as

$$P_i = \frac{\partial T}{\partial \dot{q}_i} = \frac{\partial L}{\partial \dot{q}_i} \tag{5.189}$$

In obtaining Eq. 189, the fact that the potential energy is independent of the velocities is utilized. This equation can be used to define the vector of generalize momenta as

$$\mathbf{P} = \left(\frac{\partial T}{\partial \dot{\mathbf{q}}}\right)^{\mathrm{T}} = \left(\frac{\partial L}{\partial \dot{\mathbf{q}}}\right)^{\mathrm{T}} \tag{5.190}$$

The Hamiltonian H is defined as

$$H = \dot{\mathbf{q}}^{\mathrm{T}}\mathbf{P} - L \tag{5.191}$$

and

$$\delta H = \dot{\mathbf{q}}^{\mathrm{T}}\delta\mathbf{P} + \mathbf{P}^{\mathrm{T}}\delta\dot{\mathbf{q}} - \left(\frac{\partial L}{\partial \dot{\mathbf{q}}}\right)\delta\dot{\mathbf{q}} - \left(\frac{\partial L}{\partial \mathbf{q}}\right)\delta\mathbf{q} \tag{5.192}$$

It is clear from the definition of the vector of the generalized momenta of Eq. 190 that the second and third terms on the right side of the preceding equation cancel. Consequently,

$$\delta H = \dot{\mathbf{q}}^{\mathrm{T}}\,\delta\mathbf{P} - \left(\frac{\partial L}{\partial \mathbf{q}}\right)\delta\mathbf{q} \tag{5.193}$$

Using Eq. 190, the generalized velocity vector can be expressed in terms of the generalized momenta. If the results are substituted into Eq. 191, the Hamiltonian can be expressed in terms of the generalized coordinates and momenta as

$$H = H(\mathbf{P}, \mathbf{q}) \tag{5.194}$$

If follows that

$$\delta H = \frac{\partial H}{\partial \mathbf{P}}\,\delta\mathbf{P} + \frac{\partial H}{\partial \mathbf{q}}\,\delta\mathbf{q} \tag{5.195}$$

Comparing Eqs. 193 and 195, it is clear that

$$\dot{\mathbf{q}} = \left(\frac{\partial H}{\partial \mathbf{P}}\right)^{\mathrm{T}}, \quad \frac{\partial L}{\partial \mathbf{q}} = -\frac{\partial H}{\partial \mathbf{q}} \tag{5.196}$$

Using these identities and Eq. 188, one obtains the following set of $2n$ first-order differential equations:

$$\left.\begin{aligned}\dot{\mathbf{P}} &= -\left(\frac{\partial H}{\partial \mathbf{q}}\right)^{\mathrm{T}} + \mathbf{Q}_{nc}\\ \dot{\mathbf{q}} &= \left(\frac{\partial H}{\partial \mathbf{P}}\right)^{\mathrm{T}}\end{aligned}\right\} \tag{5.197}$$

These equations are called the *canonical equations of Hamilton*. The $2n$ first-order differential equations replace the n second-order differential equations that can be derived using the principle of virtual work or Lagrange's equation. In order to use the canonical equations, one first needs to define the Lagrangian L as a function of the coordinates and velocities using Eq. 187. The vector of generalized momenta can then be defined using Eq. 190. The generalized momenta and the Lagrangian can be used to define the Hamiltonian H using Eq. 191. The first-order differential equations of the system can be obtained by substituting the Hamiltonian H into Eq. 197.

In principle, the principle of virtual work, Lagrange's equation, Gibbs–Appel equation, and the canonical equations of Hamilton can be used to define the same set of differential equations. The answer to the question of which method is better for formulating the dynamic equations depends on the application. In computational dynamics wherein the interest is focused on developing general solution procedures, it is not clear what advantage each method can provide as a computational tool since all these methods can be used to obtain the same equations.

Example 5.13

It was shown in Example 12 that the kinetic energy of the system shown in Fig. 23 is given by

$$T = \frac{1}{2} m^2 \left[(\dot{z})^2 - \dot{z}\dot{\theta}^2 l \sin \theta^2 + \left(\frac{\dot{\theta}^2 l}{2} \right)^2 \right] + \frac{1}{2} J^2 (\dot{\theta}^2)^2$$

The potential energy of the system is

$$V = m^2 g R_y = m^2 g \frac{l}{2} \sin \theta^2$$

The Lagrangian L is then given by

$$\begin{aligned} L &= T - V \\ &= \frac{1}{2} m^2 [(\dot{z})^2 - \dot{z}\dot{\theta}^2 l \sin \theta^2] + \frac{1}{2} J_O^2 (\dot{\theta}^2)^2 - m^2 g \frac{l}{2} \sin \theta^2 \end{aligned}$$

where J_O^2 is the mass moment of inertia of the rod about point O. Since the system has one degree of freedom, the generalized momentum is defined as

$$P = \frac{\partial L}{\partial \dot{\theta}^2} = \frac{\partial T}{\partial \dot{\theta}^2} = -\frac{1}{2} m^2 \dot{z} l \sin \theta^2 + J_O^2 \dot{\theta}^2$$

The preceding equation can be used to express $\dot{\theta}^2$ in terms of the generalized momentum as

$$\dot{\theta}^2 = \frac{1}{J_O^2} \left[P + \frac{1}{2} m^2 \dot{z} l \sin \theta^2 \right]$$

The Hamiltonian is defined as

$$\begin{aligned} H &= \dot{\theta}^2 P - L \\ &= \dot{\theta}^2 P - \frac{1}{2} m^2 [(\dot{z})^2 - \dot{z}\dot{\theta}^2 l \sin \theta^2] - \frac{1}{2} J_O^2 (\dot{\theta}^2)^2 + m^2 g \frac{l}{2} \sin \theta^2 \end{aligned}$$

Substituting for $\dot{\theta}^2$ in terms of the generalized momentum P, one obtains

$$H = \frac{P}{J_O^2} \left[P + \frac{1}{2} m^2 \dot{z} l \sin \theta^2 \right]$$

$$- \frac{1}{2} m^2 \left\{ (\dot{z})^2 - \frac{1}{J_O^2} \left[P + \frac{1}{2} m^2 \dot{z} l \sin \theta^2 \right] \dot{z} l \sin \theta^2 \right\}$$

$$- \frac{1}{2J_O^2} \left(P + \frac{1}{2} m^2 \dot{z} l \sin \theta^2 \right)^2 + m^2 g \frac{l}{2} \sin \theta^2$$

The first-order equations of the system can then be obtained using Eq. 197 as

$$\begin{aligned} \dot{P} &= -\frac{\partial H}{\partial \theta^2} + M^2 \\ &= -\frac{1}{2J_O^2} m^2 P \dot{z} l \cos \theta^2 - \frac{1}{4J_O^2} (m^2 l \dot{z})^2 \sin \theta^2 \cos \theta^2 \\ &\quad - m^2 g \frac{l}{2} \cos \theta^2 + M^2 \\ \dot{\theta}^2 &= \frac{\partial H}{\partial P} \\ &= \frac{1}{J_O^2} \left(P + \frac{1}{2} m^2 \dot{z} l \sin \theta^2 \right) \end{aligned}$$

Conservation Theorem A generalized coordinate that does not appear in the Lagrangian L is called *cyclic* or *ignorable*. Cyclic or ignorable coordinates are also absent from the Hamiltonian H. For a cyclic coordinate q_k, one has

$$\frac{\partial L}{\partial q_k} = \frac{\partial H}{\partial q_k} = 0 \tag{5.198}$$

If there are no nonconservative forces associated with the cyclic coordinate q_k, Eq. 197 yields $\dot{P}_k = 0$, which implies that the generalized momentum associated with the cyclic coordinate is conserved, that is, P_k is a constant. We also note that if all the forces acting on the system are conservative, one has from Lagrange's equation

$$\frac{d}{dt} \left(\frac{\partial L}{\partial \dot{\mathbf{q}}} \right) = \frac{\partial L}{\partial \mathbf{q}} \tag{5.199}$$

The total time derivative of the Lagrangian L is given by

$$\frac{dL}{dt} = \left(\frac{\partial L}{\partial \dot{\mathbf{q}}} \right) \ddot{\mathbf{q}} + \left(\frac{\partial L}{\partial \mathbf{q}} \right) \dot{\mathbf{q}} \tag{5.200}$$

Substituting Eq. 199 into the preceding equation, one gets

$$\frac{dL}{dt} = \frac{\partial L}{\partial \dot{\mathbf{q}}} \ddot{\mathbf{q}} + \left\{ \frac{d}{dt} \left(\frac{\partial L}{\partial \dot{\mathbf{q}}} \right) \right\} \dot{\mathbf{q}} \tag{5.201}$$

which yields

$$\frac{dL}{dt} = \frac{d}{dt}\left\{\frac{\partial L}{\partial \dot{\mathbf{q}}}\dot{\mathbf{q}}\right\} \tag{5.202}$$

or equivalently,

$$\frac{d}{dt}\left(L - \frac{\partial L}{\partial \dot{\mathbf{q}}}\dot{\mathbf{q}}\right) = 0 \tag{5.203}$$

Since the potential energy function is independent of the velocities, one has

$$\frac{\partial L}{\partial \dot{\mathbf{q}}} = \frac{\partial T}{\partial \dot{\mathbf{q}}} = \mathbf{P}^{\mathrm{T}} \tag{5.204}$$

Using this equation, Eq. 203 can be written as

$$\frac{d}{dt}(L - \mathbf{P}^{\mathrm{T}}\dot{\mathbf{q}}) = \frac{d}{dt}(-H) = 0 \tag{5.205}$$

which implies that

$$H = \mathbf{P}^{\mathrm{T}}\dot{\mathbf{q}} - L = \text{constant} \tag{5.206}$$

That is, in the case of a conservative system, the Hamiltonian H is a constant of motion. We also note that since

$$\mathbf{P}^{\mathrm{T}}\dot{\mathbf{q}} = \frac{\partial T}{\partial \dot{\mathbf{q}}}\dot{\mathbf{q}} = 2T, \tag{5.207}$$

the Hamiltonian H takes the following form:

$$H = 2T - L = 2T - T + V = T + V \tag{5.208}$$

which implies that, for a conservative system, the Hamiltonian is the sum of the kinetic and potential energies of the system and it remains constant throughout the system motion.

Example 5.14

Figure 24 shows a homogeneous circular cylinder of radius r, mass m, and mass moment of inertia J about its center of mass, where $J = m(r)^2/2$. The cylinder rolls without slipping on a curved surface of radius R. Use the principle of conservation of energy to derive the equation of motion of the cylinder.

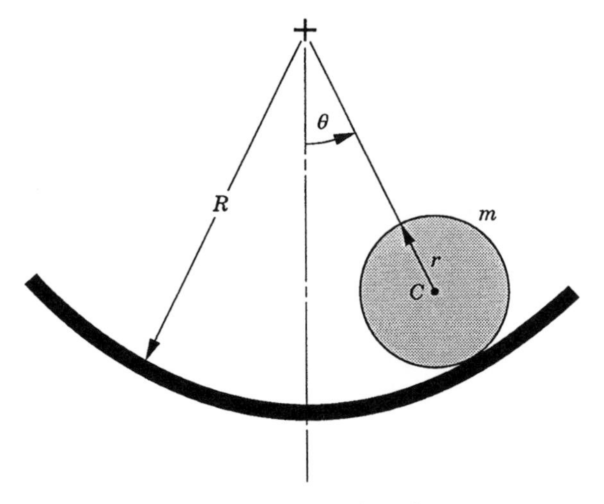

Figure 5.24 Conservation of energy

Solution. The kinetic and potential energies of the cylinder are

$$T = \tfrac{1}{2}m(v_c)^2 + \tfrac{1}{2}J(\omega)^2$$
$$V = mg(R-r)(1-\cos\theta)$$

where v_c is the absolute velocity of the center of mass of the cylinder and ω is its angular velocity, both defined as

$$v_c = (R-r)\dot{\theta}$$
$$\omega = \frac{v_c}{r} = \frac{(R-r)\dot{\theta}}{r}$$

Substituting from these two equations into the expression of the kinetic energy, we obtain

$$T = \tfrac{3}{4}\, m(R-r)^2(\dot{\theta})^2$$

The Hamiltonian H is

$$H = T + V = \tfrac{3}{4}\, m(R-r)^2(\dot{\theta})^2 + mg(R-r)(1-\cos\theta)$$

Since the Hamiltonian is constant,

$$\frac{dH}{dt} = \frac{3}{2}\, m(R-r)^2\dot{\theta}\ddot{\theta} + mg(R-r)\dot{\theta}\sin\theta = 0$$

which yields the equation of motion of the cylinder

$$\tfrac{3}{2}(R-r)\ddot{\theta} + g\sin\theta = 0$$

5.11 RELATIONSHIP BETWEEN VIRTUAL WORK AND GAUSSIAN ELIMINATION

The results obtained previously in this chapter for the slider crank mechanism shown in Fig. 20 using the principle of virtual work can also be obtained using the Gaussian elimination and the equations of the static equilibrium. Figure 25 shows the forces acting on the links of the slider crank mechanism. The equations of the static equilibrium of the crankshaft can be written as

$$\left.\begin{aligned} F_x^{12} - F_x^{23} = 0 \\ F_y^{12} - F_y^{23} - m^2 g = 0 \\ F_x^{12} l_O^2 \sin\theta^2 - F_y^{12} l_O^2 \cos\theta^2 + F_x^{23} l_A^2 \sin\theta^2 - F_y^{23} l_A^2 \cos\theta^2 + M^2 = 0 \end{aligned}\right\} \tag{5.209}$$

The equations of the static equilibrium of the connecting rod are

$$\left.\begin{aligned} F_x^{23} - F_x^{34} = 0 \\ F_y^{23} - F_y^{34} - m^3 g = 0 \\ F_x^{23} l_A^3 \sin\theta^3 - F_y^{23} l_A^3 \cos\theta^3 + F_x^{34} l_B^3 \sin\theta^3 - F_y^{34} l_B^3 \cos\theta^3 = 0 \end{aligned}\right\} \tag{5.210}$$

The equation of the static equilibrium fo the slider block is

$$F_x^{34} + F^4 = 0 \tag{5.211}$$

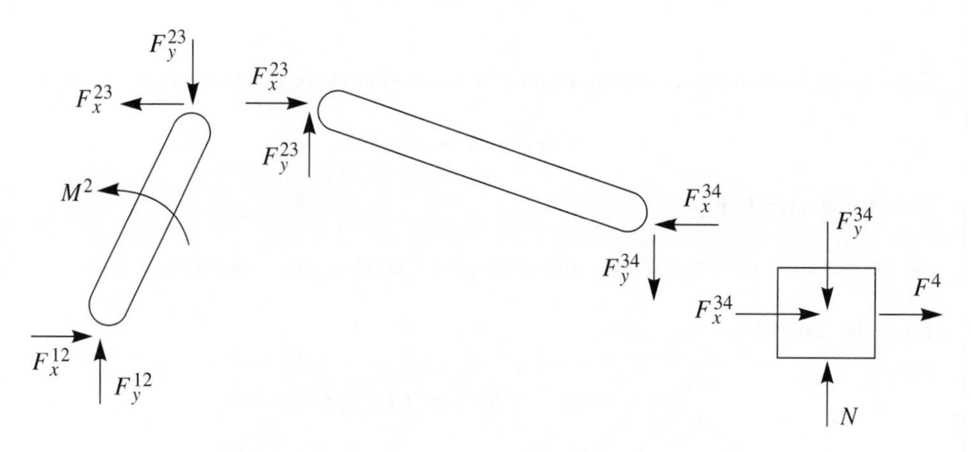

Figure 5.25 Forces of the slider crank mechanism

The preceding seven equations of the static equilibrium of the three links of the slider crank mechanism can be rearranged and written in the following matrix form:

$$\begin{bmatrix} 1 & 0 & -1 & 0 & 0 & 0 & 0 \\ 0 & 1 & 0 & -1 & 0 & 0 & 0 \\ 0 & 0 & 1 & 0 & -1 & 0 & 0 \\ 0 & 0 & 0 & 1 & 0 & -1 & 0 \\ 0 & 0 & 0 & 0 & 1 & 0 & 0 \\ l_O^2 \sin\theta^2 & -l_O^2 \cos\theta^2 & l_A^2 \sin\theta^2 & -l_A^2 \cos\theta^2 & 0 & 0 & 1 \\ 0 & 0 & l_A^3 \sin\theta^3 & -l_A^3 \cos\theta^3 & l_B^3 \sin\theta^3 & -l_B^3 \cos\theta^3 & 0 \end{bmatrix} \cdot \begin{bmatrix} F_x^{12} \\ F_y^{12} \\ F_x^{23} \\ F_y^{23} \\ F_x^{34} \\ F_y^{34} \\ M^2 \end{bmatrix} = \begin{bmatrix} 0 \\ m^2 g \\ 0 \\ m^3 g \\ -F^4 \\ 0 \\ 0 \end{bmatrix} \tag{5.212}$$

In this matrix equation it is assumed that the unknowns are the external moment M^2 and the components of the reaction forces F_x^{12}, F_y^{12}, F_x^{23}, F_y^{23}, F_x^{34}, and F_y^{34}. A standard Gaussian elimination procedure can be used to obtain an upper triangular form of the coefficient matrix in the preceding equation. This Gaussian elimination procedure leads to

$$\begin{bmatrix} 1 & 0 & -1 & 0 & 0 & 0 & 0 \\ 0 & 1 & 0 & -1 & 0 & 0 & 0 \\ 0 & 0 & 1 & 0 & -1 & 0 & 0 \\ 0 & 0 & 0 & 1 & 0 & -1 & 0 \\ 0 & 0 & 0 & 0 & 1 & 0 & 0 \\ 0 & 0 & 0 & 0 & 0 & 1 & -1/l^2 \cos\theta^2 \\ 0 & 0 & 0 & 0 & 0 & 0 & -l^3 \cos\theta^3 / l^2 \cos\theta^2 \end{bmatrix} \cdot \begin{bmatrix} F_x^{12} \\ F_y^{12} \\ F_x^{23} \\ F_y^{23} \\ F_x^{34} \\ F_y^{34} \\ M^2 \end{bmatrix} = \begin{bmatrix} 0 \\ m^2 g \\ 0 \\ m^3 g \\ -F^4 \\ -A_1 / l^2 \cos\theta^2 \\ A_2 \end{bmatrix} \tag{5.213}$$

where

$$\left.\begin{aligned} A_1 &= m^2 g l_O^2 \cos\theta^2 + m^3 g l^2 \cos\theta^2 + F^4 l^2 \sin\theta^2 \\ A_2 &= m^3 g l_A^3 \cos\theta^3 + F^4 l^3 \sin\theta^3 \end{aligned}\right\} \tag{5.214}$$

Equation 213 can be solved for M^2 as

$$M^2 = m^2 g l_O^2 \cos\theta^2 + m^3 g l^2 \left(1 - \frac{l_A^3}{l^3}\right) \cos\theta^2 - F^4 l^2(-\sin\theta^2 + \cos\theta^2 \tan\theta^3) \tag{5.215}$$

which is the same equation obtained earlier in this chapter (Section 6) using the principle of virtual work.

PROBLEMS

1. For the rod shown in Fig. P1, assume that $F = 5$ N, $M = 3$ N·m, and $\phi = 45°$. Assuming that the generalized coordinates of the rod are the Cartesian coordinates of point A and the angular orientation of the rod, determine the generalized forces associated with these generalized coordinates.

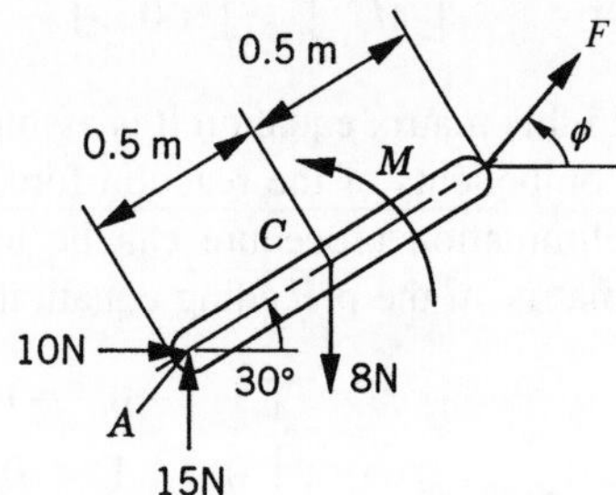

Figure P5.1

2. Repeat problem 1 assuming that the generalized coordinates of the rod are the Cartesian coordinates of point C and the angular orientation of the rod.

3. For the system shown in Fig. P2, let θ be the independent generalized coordinate of the rod. Determine the generalized forces associated with this generalized coordinate. Neglect the effect of gravity.

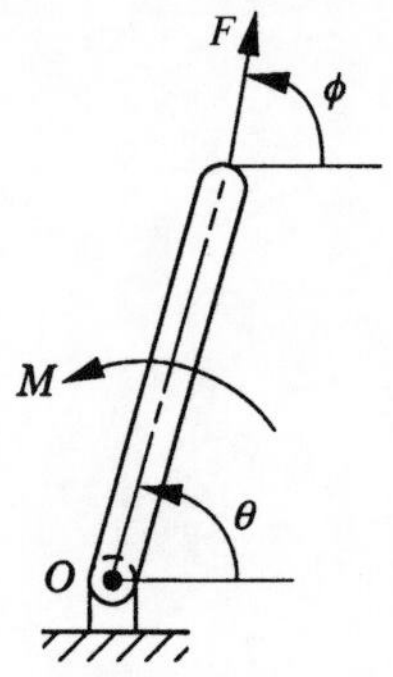

Figure P5.2

4. Repeat problem 3 taking into consideration the effect of gravity.

5. For the system shown in Fig. P3, let $F = 10\,\text{N}$, $M^2 = 3\ \text{N}\cdot\text{m}$, $M^3 = 3\ \text{N}\cdot\text{m}$, $\theta^2 = 45^\circ$, and $\theta^3 = 30^\circ$. Determine the generalized forces associated with the generalized coordinates θ^2 and θ^3. Neglect the effect of gravity.

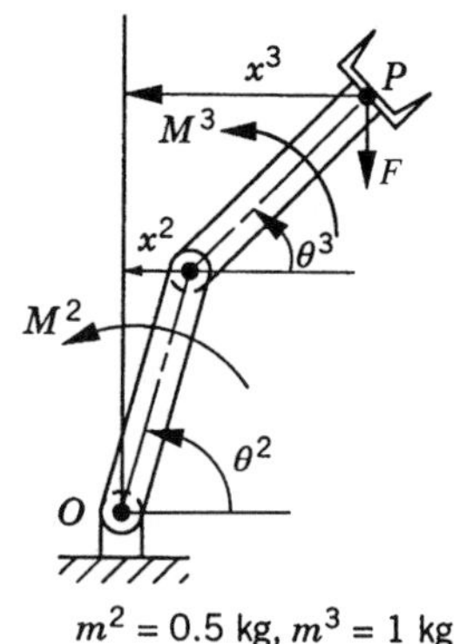

Figure P5.3

6. Repeat problem 5 taking into account the effect of gravity.

7. Repeat problem 5 assuming that the generalized coordinates are x^2 and x^3.

8. For the system shown in Fig. P4, write the virtual changes in the position vector of point C in terms of the virtual change in the independent generalized coordinates θ^2, θ^3, and θ^4.

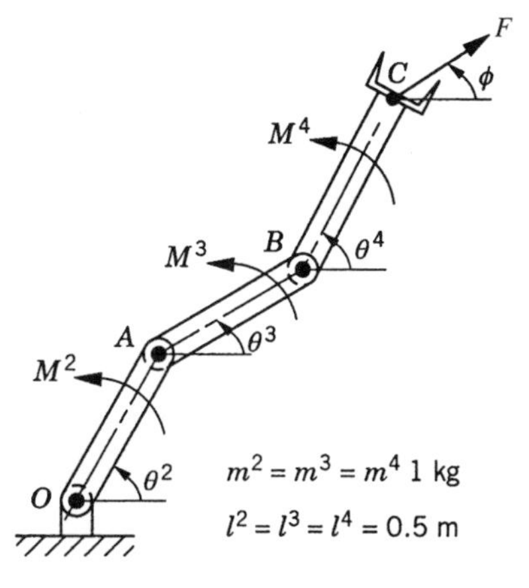

Figure P5.4

9. For the system shown in Fig. P4, obtain the generalized forces associated with the generalized coordinates θ^2, θ^3, and θ^4. Assume $\theta^2 = 45^\circ$, $\theta^3 = 30^\circ$, $\theta^4 = 45^\circ$, $\phi = 30^\circ$, $F = 10\,\text{N}$, $M^2 = 10\ \text{N}\cdot\text{m}$, $M^3 = 8\ \text{N}\cdot\text{m}$, and $M^4 = 3\ \text{N}\cdot\text{m}$. Consider the effect of gravity.

10. For the slider crank mechanism shown in Fig. P5, assume that $\theta^2 = 45^\circ$, $M^2 = 10\ \text{N}\cdot\text{m}$, and $F^4 = 10\,\text{N}$. Determine the generalized force associated with the independent generalized coordinate θ^2. Consider the effect of gravity.

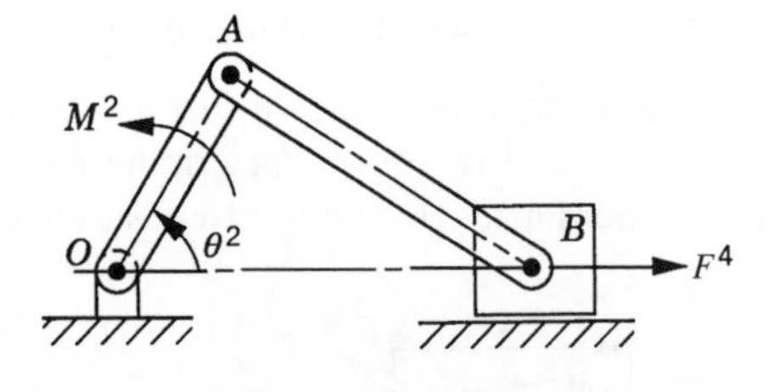

Figure P5.5

11. Repeat problem 10 assuming that the independent generalized coordinate is the location of the slider block.

12. For the four-bar linkage shown in Fig. P6, obtain the generalized force associated with the independent generalized coordinate θ^2. Assume that $\theta^2 = 60^\circ$, $\phi = 30^\circ$, $M^2 = 5$ N $\cdot$ m, $F^3 = 3$ N, and $M^4 = 5$ N $\cdot$ m. Consider the effect of gravity.

13. Repeat problem 12 assuming that the generalized coordinate is selected to be the angular orientation of the coupler θ^3.

14. Repeat problem 12 assuming that the generalized coordinate is selected to be the angular orientation of the rocker θ^4.

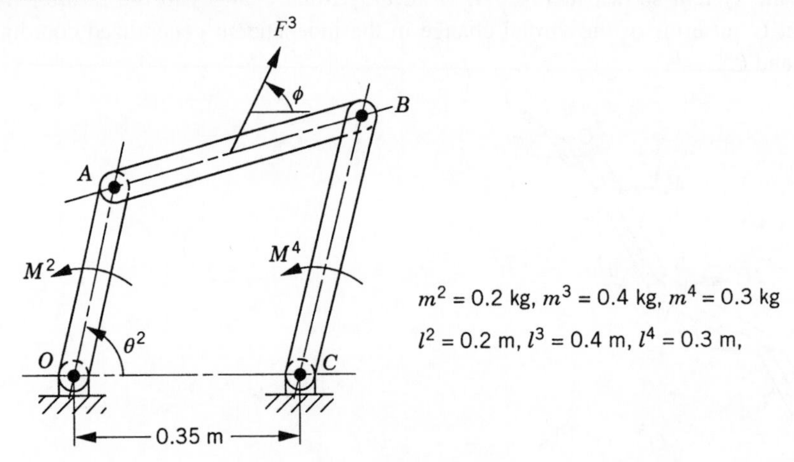

Figure P5.6

15. For the unconstrained rod shown in Fig. P1, determine F, M, and ϕ such that the rod is in static equilibrium. Use the principle of virtual work.

16. For the system shown in Fig. P2, let $\theta = 45^\circ$, $F = 5$ N, and $\phi = 60^\circ$. Assume that the mass and length of the rod are 1 kg and 1 m, respectively. Determine the moment M such that the system is in static equilibrium position. Use the principle of virtual work and consider the effect of gravity.

17. For the system shown in Fig. P2, let the length of the rod be 1 m, $\theta = 45^\circ$, $F = 5\,\text{N}$, $M = 3\ \text{N} \cdot \text{m}$, and $\phi = 60^\circ$. Using the principle of virtual work determine the weight of the rod such that the system is in static equilibrium position.

18. Use the principle of virtual work in statics to determine the joint torque M^2 and M^3 of the system shown in Fig. P3. Assume that $\theta^2 = 45^\circ$, $\theta^3 = 30^\circ$, and $F = 5\,\text{N}$. Consider the effect of gravity.

19. For the system shown in Fig. P3, let $M^2 = 3\ \text{N} \cdot \text{m}$, $M^3 = 5\ \text{N} \cdot \text{m}$, $F = 5\,\text{N}$. Does a static equilibrium configuration exist for this system? Use the principle of virtual work to find the answer and consider the effect of gravity.

20. For the system shown in Fig. P4, let $\theta^2 = 45^\circ$, $\theta^3 = 30^\circ$, $\theta^4 = 45^\circ$, $\phi = 30^\circ$, and $F = 5\,\text{N}$. Use the principle of virtual work to determine M^2, M^3, and M^4 such that the system is in static equilibrium. Consider the effect of gravity.

21. Use the principle of virtual work in statics to determine the input torque M^2 of the slider crank mechanism shown Fig. P5. Assume that $\theta^2 = 45^\circ$ and $F^4 = 4\,\text{N}$. Take into consideration the effect of the gravity. Assume that the generalized coordinate is the joint angle θ^2.

22. Repeat problem 21 assuming that the generalized coordinate is the horizontal position of the slider block.

23. Use the principle of virtual work in statics to determine the input torque M^2 of the four-bar linkage shown in Fig. P6. Consider the generalized coordinate to be the crank angle θ^2. Consider the effect of gravity and assume that $\theta^2 = 60^\circ$, $\phi = 45^\circ$, $F^3 = 5\,\text{N}$, and $M^4 = 5\ \text{N} \cdot \text{m}$.

24. Repeat problem 23 assuming that the generalized coordinate is the orientation of the coupler θ^3.

25. Repeat problem 23 assuming that the generalized coordinate is the orientation of the rocker θ^4.

26. Use the principle of virtual work in statics to determine the output torque M^4 that acts on the rocker of the four-bar linkage shown in Fig. P6. Assume that $\theta^2 = 60^\circ$, $\phi = 45^\circ$, $F^3 = 5\,\text{N}$, and $M^2 = 5\ \text{N} \cdot \text{m}$. Consider the crank angle θ^2 as the generalized coordinate. Take the effect of gravity into consideration.

27. Repeat problem 26 assuming that the generalized coordinate is the rocker angle θ^4.

28. The components of the acceleration of the center of mass of the rod shown in Fig. P1 are $a_x = 50\ \text{m/s}^2$, $a_y = 120\ \text{m/s}^2$. The angular acceleration of the rod is assumed to be $500\ \text{rad/s}^2$. The rod is assumed to be slender and uniform with mass 1 kg. Use the principle of virtual work in dynamics to determine M, F, and ϕ. Consider the effect of gravity.

29. For the system shown in Fig. P2, let $\theta = 45^\circ$, $\phi = 60^\circ$, $F = 10\,\text{N}$. The rod shown in the figure is assumed to be uniform and slender with mass 1 kg and length 1 m. The angular velocity and angular acceleration of the rod are assumed to be $\dot{\theta} = 150\ \text{rad/s}$

and $\ddot{\theta} = 0$ rad/s^2, respectively. Considering the effect of gravity, use the principle of virtual work in dynamics to determine the moment M.

30. Repeat problem 29 assuming that the angular acceleration $\ddot{\theta} = 500$ rad/s^2.

31. Use the principle of virtual work in dynamics to determine the joint torques M^2 and M^3 for the system shown in Fig. P3. Use the following data: $\theta^2 = 45°$, $\theta^3 = 30°$, $\dot{\theta}^2 = 70$ rad/s, $\dot{\theta}^3 = 40$ rad/s, $\ddot{\theta}^2 = 120$ rad/s^2, $\ddot{\theta}^3 = 180$ rad/s^2, and $F = 10$ N. Assume that the two links shown in the figure are uniform slender rods. Consider the effect of gravity.

32. The system shown in Fig. P4 consists of three uniform slender rods that are connected by revolute joints. Let $\theta^2 = 45°$, $\theta^3 = 30°$, $\theta^4 = 45°$, $\dot{\theta}^2 = 10$ rad/s, $\dot{\theta}^3 = 8$ rad/s, $\dot{\theta}^4 = 4$ rad/s, $\ddot{\theta}^2 = 200$ rad/s^2, $\ddot{\theta}^3 = 250$ rad/s^2, $\ddot{\theta}^4 = 500$ rad/s^2, $\phi = 30°$, and $F = 8$ N. Considering the effect of gravity, use the principle of virtual work in dynamics to determine the torques M^2, M^3, and M^4.

33. Determine the joint reaction forces in problem 31.

34. Determine the joint reaction forces in problem 32.

35. The crankshaft and connecting rod of the slider crank mechanism shown in Fig. P5 are assumed to be uniform slender rods. Use the principle of virtual work in dynamics to determine the input torque M^2. Use the following data: $\theta^2 = 45°$, $\dot{\theta}^2 = 150$ rad/s, $\ddot{\theta}^2 = 0$ rad/s^2, and $F^4 = 10$ N. Consider the effect of gravity.

36. Repeat problem 35 assuming that $\ddot{\theta}^2 = 500$ rad/s^2.

37. Determine the joint reaction forces in problem 35.

38. Determine the joint reaction forces in problem 36.

39. The crankshaft, coupler, and rocker of the four-bar linkage shown in Fig. P6 are assumed to be uniform slender rods. Assume that $\theta^2 = 60°$, $\dot{\theta}^2 = 150$ rad/s, $\ddot{\theta}^2 = 0$, $\phi = 30°$, $F = 10$ N, and $M^4 = 5$ N $\cdot$ m. Considering the effect of gravity, use the principle of virtual work in dynamics to determine the input torque M^2. Use θ^2 as the generalized coordinate.

40. Repeat problem 39, assuming the rocker angle θ^3 as the generalized coordinate.

41. Repeat problem 39, assuming the coupler angle θ^4 as the generalized coordinate.

42. Determine the joint reaction forces in problem 39.

43. Repeat problem 39 assuming that $\ddot{\theta}^2 = 700$ rad/s^2.

44. The system shown in Fig. P7 consists of a slider block of mass m^2 and a uniform slender rod of mass m^3, length l^3, and mass moment of inertia about its center of mass J^3. The slider block is connected to the ground by a spring that has a stiffness coefficient k. The slider block is subjected to the force $F(t)$, while the rod is subjected to the moment M. Obtain the differential equations of motion of this two-degree-of-freedom system using Lagrange's equation.

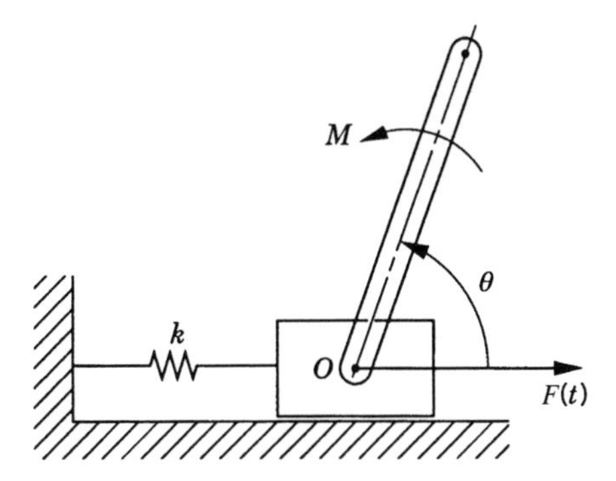

Figure P5.7

45. Determine the generalized inertia forces of the system of problem 44 using the three different equations given in Eq. 178. Compare the results obtained using these three equations with the results obtained using Lagrange's equation.

46. Derive the differential equations of motion of the system of problem 44 using the Gibbs–Appel equation.

CHAPTER 6

CONSTRAINED DYNAMICS

As shown in the preceding chapter, when the kinematic relationships are expressed in terms of the system degrees of freedom, the application of the principle of virtual work in dynamics leads to a number of dynamic differential equations equal to the number of the system degrees of freedom. In these equations the forces of the workless constraint are automatically eliminated. Constraint forces, however, appear in the dynamic equations if these equations are formulated in terms of a set of coordinates that are not totally independent, as discussed in Chapter 4. The number of independent constraint forces that appear in these equations is equal to the number of dependent coordinates used in the dynamic formulation.

In this chapter, the concepts and techniques, presented in Chapter 4 based on the Newtonian approach and D'Alembert's principle, are generalized using the technique of the virtual work presented in the preceding chapter. The principle of virtual work can be used to provide a proof for the existence of *Lagrange multipliers* when redundant coordinates are used with the *augmented formulation*. The virtual work principle, as demonstrated in the preceding chapter, can also be used to systematically define the *velocity transformation* required in the case of the *embedding technique*. In this chapter, the general augmented form of the equations of motion of multibody systems that consist of interconnected bodies is developed using the *absolute Cartesian coordinates*. The use of these coordinates allows treating the bodies as free when the unknown constraint forces are added to the other known applied forces. The generalized inertia and applied forces associated with the redundant Cartesian coordinates are first defined using the virtual work. Using the principle of virtual work, it is shown that the Newton–Euler formulation of the equations of motion can be obtained as a special case of a more general form of the equations of motion. In the

Computational Dynamics, Third Edition Ahmed A. Shabana

case of the Newton–Euler formulation, the reference point of the body is assumed to be the body center of mass. Using this assumption, the inertia coupling between the translation and rotation of the body can be eliminated. A set of algebraic equations that describe the kinematic relationships between the redundant variables is formulated and used to systematically define the generalized constraint forces. The equations of motion of the system can then be defined in terms of the redundant coordinates and the generalized constraint forces. The computational and numerical procedures used for the computer-aided dynamic analysis of multibody systems are also discussed in this chapter, including topics such as *sparse matrix techniques* which are required for the efficient solution of the constrained multibody system dynamic equations. The use of the methods developed in this chapter is demonstrated using *planar systems* in order to emphasize the concepts and procedures presented without delving into the details of the three-dimensional motion. The analysis of the spatial systems is presented in the following chapter.

6.1 GENERALIZED INERTIA

In this section, we demonstrate that the concept of *equipollent systems of forces* can also be applied to the inertia forces. This concept allows us to use the simple definition of the inertia forces defined at the center of mass of the body to obtain the inertia forces associated with the coordinates of an arbitrary point. It is important to understand how to apply the principle of virtual work to obtain different forms of the equations of motion, and to understand how the choice of the body reference point affects the forms of the inertia matrix and forces. Being able to use the virtual work principle effectively allows for the systematic formulation of the dynamic equations of motion of more complex systems as in the case when the bodies deform and the assumption of rigidity is no longer valid.

If the reference point O^i of the rigid body is selected to be its center of mass, the inertia force and the inertia moment are

$$\mathbf{F}_i^i = m^i \begin{bmatrix} \ddot{R}_x^i \\ \ddot{R}_y^i \end{bmatrix}, \quad M_i^i = J^i \ddot{\theta}^i \tag{6.1}$$

where m^i is the mass of the rigid body i, J^i is the *polar mass moment of inertia* of the body about an axis passing through its mass center, R_x^i and R_y^i are the coordinates of the reference point, and θ^i is the angular orientation of the body. The mass moment of inertia J^i is defined as

$$J^i = \int_{V^i} \rho^i \bar{\mathbf{u}}^{i\mathrm{T}} \bar{\mathbf{u}}^i \, dV^i \tag{6.2}$$

where ρ^i and V^i are, respectively, the mass density and volume of the rigid body i, and $\bar{\mathbf{u}}^i$ is the position vector of an arbitrary point on the rigid body i defined with respect to the center of mass.

The virtual work of the inertia force $\mathbf{F}_i^i$ and the inertia moment M^i is given by

$$\delta W_i^i = \mathbf{F}_i^{i\mathrm{T}} \, \delta \mathbf{R}^i + M_i^i \, \delta\theta^i \tag{6.3}$$

where $\mathbf{R}^i = [R_x^i \quad R_y^i]^\mathrm{T}$.

Equipollent Systems The inertia forces and moment acting at the center of mass may be replaced by an equipollent system of inertia forces and moments acting at another point on the body. The original and the equivalent systems must do the same work and accordingly, the selection of the reference point is a matter of preference or convenience. To demonstrate this fact, let P^i be an arbitrary point on the rigid body. As shown in Fig. 1, the position vector of point P^i in terms of the coordinates of the reference point is

$$\mathbf{r}_P^i = \mathbf{R}^i + \mathbf{A}^i \bar{\mathbf{u}}_P^i \tag{6.4}$$

where $\mathbf{A}^i$ is the transformation matrix from the body coordinate system to the global coordinate system, and $\bar{\mathbf{u}}_P^i$ is the local position vector of point P^i with respect to the reference point. By assuming a virtual change in the coordinates, Eq. 4 leads to

$$\delta \mathbf{r}_P^i = \delta \mathbf{R}^i + \mathbf{A}_\theta^i \bar{\mathbf{u}}_P^i \, \delta\theta^i \tag{6.5}$$

or equivalently

$$\delta \mathbf{R}^i = \delta \mathbf{r}_P^i - \mathbf{A}_\theta^i \bar{\mathbf{u}}_P^i \, \delta\theta^i \tag{6.6}$$

Substituting Eq. 6 into Eq. 3 leads to

$$\delta W_i^i = \mathbf{F}_i^{i\mathrm{T}} \, \delta \mathbf{r}_P^i + (M_i^i - \mathbf{F}_i^{i\mathrm{T}} \mathbf{A}_\theta^i \bar{\mathbf{u}}_P^i) \, \delta\theta^i \tag{6.7}$$

One can show that

$$\mathbf{F}_i^{i\mathrm{T}} \mathbf{A}_\theta^i \bar{\mathbf{u}}_P^i = (\mathbf{u}_P^i \times \mathbf{F}_i^i) \cdot \mathbf{k} \tag{6.8}$$

where $\mathbf{u}_P^i = \mathbf{A}^i \bar{\mathbf{u}}_P^i$. Substituting Eq. 8 into Eq. 7, one obtains

$$\delta W_i^i = \mathbf{F}_i^{i\mathrm{T}} \, \delta \mathbf{r}_P^i + [M_i^i - (\mathbf{u}_P^i \times \mathbf{F}_i^i) \cdot \mathbf{k}] \, \delta\theta^i \tag{6.9}$$

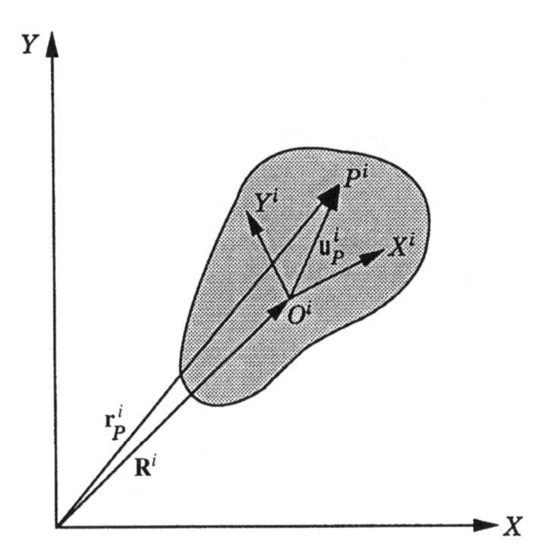

Figure 6.1 Reference coordinates

This equation states that the system of inertia forces and moments acting at the reference point (center of mass) is equipollent to another system, defined at the arbitrary point P^i, which consists of the force $\mathbf{F}_i^i$ and the moment $M_i^i - (\mathbf{u}_P^i \times \mathbf{F}_i^i) \cdot \mathbf{k}$. This fact is the familiar result obtained for the equivalence of systems of applied forces.

Parallel Axis Theorem The *parallel axis theorem* can be obtained from Eq. 9 as a special case. To demonstrate this, we consider the special case where point P^i is a fixed point. In this special case, $\delta \mathbf{r}_P^i = \mathbf{0}$, and Eq. 9 reduces to

$$\delta W_i^i = (M_i^i - \mathbf{F}_i^{i\mathrm{T}} \mathbf{A}_\theta^i \bar{\mathbf{u}}_P^i)\, \delta\theta^i = [M_i^i - (\mathbf{u}_P^i \times \mathbf{F}_i^i) \cdot \mathbf{k}]\, \delta\theta^i \tag{6.10}$$

where

$$\mathbf{F}_i^i = m^i \begin{bmatrix} \ddot{R}_x^i \\ \ddot{R}_y^i \end{bmatrix} = m^i \{-\ddot{\theta}^i \mathbf{A}_\theta^i \bar{\mathbf{u}}_P^i + (\dot{\theta}^i)^2 \mathbf{A}^i \bar{\mathbf{u}}_P^i\} \tag{6.11}$$

Using this equation, one can verify that

$$-\mathbf{F}_i^{i\mathrm{T}} \mathbf{A}_\theta^i \bar{\mathbf{u}}_P^i = m^i (l_P^i)^2 \ddot{\theta}^i \tag{6.12}$$

where $l_P^i = (\bar{\mathbf{u}}_P^{i\mathrm{T}} \bar{\mathbf{u}}_P^i)^{1/2}$. Therefore, Eq. 10 can be written as $\delta W_i^i = \{M_i^i + m^i (l_P^i)^2 \ddot{\theta}^i\}\, \delta\theta^i$ which upon the use of Eq. 1 yields

$$\delta W_i^i = \{J^i + m^i (l_P^i)^2\} \ddot{\theta}^i\, \delta\theta^i = J_P^i \ddot{\theta}^i\, \delta\theta^i \tag{6.13}$$

where J_P^i is the mass moment of inertia about an axis passing through point P^i and is defined as

$$J_P^i = J^i + m^i (l_P^i)^2 \tag{6.14}$$

This equation is the *parallel axis theorem*, which states that the mass moment of inertia defined with respect to an arbitrary point P^i on the rigid body is equal to the mass moment of inertia defined with respect to the center of mass plus the product of the mass and the square of the distance between point P^i and the center of mass. This familiar result was obtained in this section using the general expression of the virtual work of the inertia forces as defined by Eq. 9. It is important to emphasize, however, that if point P^i is not a fixed point, the general expression of Eq. 9 must be used in order to define the equivalent system of inertia forces at P^i. In this case, the resulting system consists of an inertia force vector as well as a moment, as demonstrated by the following example.

Example 6.1

Figure 2 shows a composite body i that consists of a slender rod, a rectangular prism, and a removed round disk. The length of the rod is $l^i = 1$ m, its mass is $m_1^i = 0.62$ kg, and its mass moment of inertia about its center of mass is $J_1^i = 5.167 \times 10^{-2}$ kg · m^2. The length of the rectangular prism is $b^i = 1$ m, its mass before the removal of the disk is $m_2^i = 39.45$ kg, and

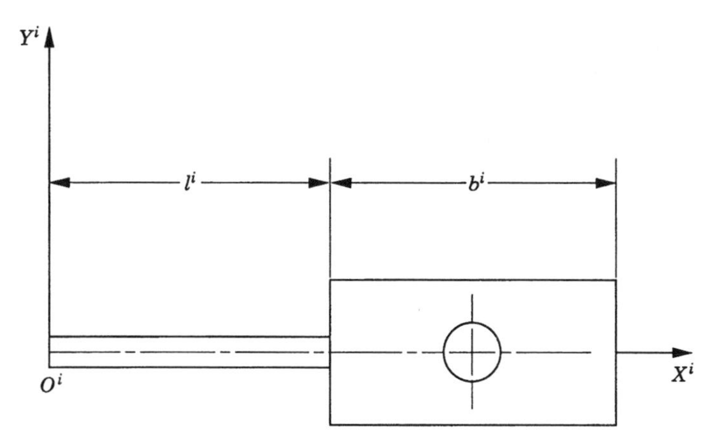

Figure 6.2 Composite body

its mass moment of inertia about its center of mass is $J_2^i = 4.109\,\text{kg}\cdot\text{m}^2$. The diameter of the removed round disk is $D^i = 0.25\,\text{m}$, its mass is $m_3^i = 3.873\,\text{kg}$, and its mass moment of inertia about its center of mass is $J_3^i = 3.026 \times 10^{-2}\,\text{kg}\cdot\text{m}^2$. Using the parallel axis theorem, determine the mass moment of inertia of the composite body about the body center of mass. If the absolute acceleration of the center of mass of the body at a given instant of time is defined by the vector $\mathbf{a}_c^i = [100 \quad -45]^{\text{T}}\,\text{m/s}^2$, and the angular acceleration of the body is $150\,\text{rad/s}^2$, determine the generalized inertia forces associated with the coordinates of the reference point O^i and the orientation of the body θ^i. Assume that $\theta^i = \pi/2$.

Solution. The location of the center of mass of the composite body in the coordinate system shown in the figure is given by

$$\begin{aligned}
\overline{x}_c^i &= \frac{m_1^i \dfrac{l^i}{2} + m_2^i\left(l^i + \dfrac{b^i}{2}\right) - m_3^i\left(l^i + \dfrac{b^i}{2}\right)}{m_1^i + m_2^i - m_3^i} \\
&= \frac{0.62\left(\frac{1}{2}\right) + (39.45)\left(1 + \frac{1}{2}\right) - 3.873\left(1 + \frac{1}{2}\right)}{0.62 + 39.45 - 3.873} \\
&= 1.483\ \text{m}
\end{aligned}$$

One can then define the location of the center of mass of the rod, the rectangular prism, and the removed round disk, from the center of mass of the composite system, respectively, as

$$\begin{aligned}
d_1^i &= \frac{l^i}{2} - \overline{x}_c^i = 0.5 - 1.483 = -0.983\ \text{m} \\
d_2^i &= \left(l^i + \frac{b^i}{2}\right) - \overline{x}_c^i = 1.5 - 1.483 = 0.017\ \text{m} \\
d_3^i &= \left(l^i + \frac{b^i}{2}\right) - \overline{x}_c^i = 1.5 - 1.483 = 0.017\ \text{m}
\end{aligned}$$

Using the parallel axis theorem, the mass moment of inertia of the composite system about its center of mass can be obtained as

$$\begin{aligned} J^i &= [J_1^i + m_1^i(d_1^i)^2] + [J_2^i + m_2^i(d_2^i)^2] - [J_3^i + m_3^i(d_3^i)^2] \\ &= [5.176 \times 10^{-2} + 0.62(-0.983)^2] + [4.109 + 39.45(0.017)^2] \\ &\quad - [3.026 \times 10^{-2} + 3.873(0.017)^2] \\ &= 0.651 + 4.120 - 0.314 \times 10^{-1} = 4.7396 \text{ kg} \cdot \text{m}^2 \end{aligned}$$

The parallel axis theorem can also be used to determine the mass moment of inertia about the reference point O^i as

$$J_O^i = J^i + m^i(\bar{x}_c^i)^2$$

where m^i is the total mass of the composite body defined as

$$m^i = m_1^i + m_2^i - m_3^i = 0.62 + 39.45 - 3.873 = 36.227 \text{ kg}$$

It follows that

$$J_O^i = 4.7396 + 36.227(1.483)^2 = 84.413 \text{ kg} \cdot \text{m}^2$$

The virtual work of the inertia forces is

$$\delta W_i^i = m^i \mathbf{a}_c^{i\mathrm{T}} \delta \mathbf{r}_c^i + J^i \ddot{\theta}^i \delta\theta^i$$

where $\mathbf{r}_c^i$ is the global position vector of the center of mass. This vector can be expressed in terms of the reference coordinates as

$$\mathbf{r}_c^i = \mathbf{R}^i + \mathbf{A}^i \bar{\mathbf{u}}_c^i$$

where $\mathbf{R}^i$ is the global position vector of the reference point, and $\bar{\mathbf{u}}_c^i$ is the position vector of the center of mass with respect to the reference point. Using the preceding equation, the virtual work of the inertia forces is

$$\delta W_i^i = m^i \mathbf{a}_c^{i\mathrm{T}} \delta \mathbf{R}^i + (J^i \ddot{\theta}^i + m^i \mathbf{a}_c^{i\mathrm{T}} \mathbf{A}_\theta^i \bar{\mathbf{u}}_c^i)\, \delta\theta^i$$

which defines the generalized inertia force associated with the translation of the body reference as

$$\mathbf{F}_i^i = m^i \mathbf{a}_c^i = 36.227 \begin{bmatrix} 100 \\ -45 \end{bmatrix} = \begin{bmatrix} 3.6227 \times 10^3 \\ -1.6302 \times 10^3 \end{bmatrix} \text{N}$$

and the generalized inertia moment as

$$M_i^i = J^i \ddot{\theta}^i + m^i \mathbf{a}_c^{i\mathrm{T}} \mathbf{A}_\theta^i \bar{\mathbf{u}}_c^i$$

where

$$\mathbf{A}_\theta^i = \begin{bmatrix} -\sin\theta^i & -\cos\theta^i \\ \cos\theta^i & -\sin\theta^i \end{bmatrix} = \begin{bmatrix} -1 & 0 \\ 0 & -1 \end{bmatrix}$$

$$\overline{\mathbf{u}}_c^i = [\overline{x}_c^i \quad 0]^{\mathrm{T}} = [1.483 \quad 0]^{\mathrm{T}}$$

Thus

$$M_i^i = (4.7396)(150) + [3.6227 \times 10^3 \quad -1.6302 \times 10^3] \cdot \begin{bmatrix} -1 & 0 \\ 0 & -1 \end{bmatrix} \begin{bmatrix} 1.483 \\ 0 \end{bmatrix} = -4.6615 \times 10^3 \text{ N} \cdot \text{m}$$

Note that $J_0^i \ddot{\theta}^i = (84.413)(150) = 1.2662 \times 10^3$ N · m, which is not the same as M_i^i.

6.2 MASS MATRIX AND CENTRIFUGAL FORCES

In this section, an expression for the kinetic energy for the rigid body i is defined and used to develop the general form of the mass matrix of a rigid body that undergoes an arbitrary large displacement. The effect of the selection of the reference point on the form of the kinetic energy and the mass matrix of the rigid body is also examined.

The kinetic energy of the rigid body i is defined as

$$T^i = \frac{1}{2} \int_{V^i} \rho^i \dot{\mathbf{r}}^{i^{\mathrm{T}}} \dot{\mathbf{r}}^i \, dV^i \tag{6.15}$$

where ρ^i and V^i are, respectively, the mass density and volume of the body, and $\mathbf{r}^i$ is the global position vector of an arbitrary point on the rigid body. The vector $\mathbf{r}^i$ can be expressed in terms of the coordinates $\mathbf{R}^i$ of the reference point and the angle of rotation θ^i of the body as

$$\mathbf{r}^i = \mathbf{R}^i + \mathbf{A}^i \overline{\mathbf{u}}^i \tag{6.16}$$

where $\mathbf{A}^i$ is the planar transformation matrix, and $\overline{\mathbf{u}}^i$ is the local position vector of the arbitrary point on the body. Differentiating Eq. 16 with respect to time yields

$$\dot{\mathbf{r}}^i = \dot{\mathbf{R}}^i + \mathbf{A}_\theta^i \overline{\mathbf{u}}^i \dot{\theta}^i \tag{6.17}$$

This equation can be written in a matrix form as

$$\dot{\mathbf{r}}^i = [\mathbf{I} \quad \mathbf{A}_\theta^i \overline{\mathbf{u}}^i] \begin{bmatrix} \dot{\mathbf{R}}^i \\ \dot{\theta}^i \end{bmatrix} \tag{6.18}$$

where $\mathbf{I}$ is the 2×2 identity matrix. Substituting Eq. 18 into Eq. 15, one obtains

$$T^i = \frac{1}{2}\int_{V^i} \rho^i [\dot{\mathbf{R}}^{i^{\mathrm{T}}} \quad \dot{\theta}^i] \begin{bmatrix} \mathbf{I} \\ \bar{\mathbf{u}}^{i^{\mathrm{T}}} \mathbf{A}_\theta^{i^{\mathrm{T}}} \end{bmatrix} [\mathbf{I} \quad \mathbf{A}_\theta^i \bar{\mathbf{u}}^i] \begin{bmatrix} \dot{\mathbf{R}}^i \\ \dot{\theta}^i \end{bmatrix} dV^i \tag{6.19}$$

which upon carrying out the matrix multiplication and utilizing the fact that $\mathbf{A}_\theta^{i^{\mathrm{T}}} \mathbf{A}_\theta^i = \mathbf{I}$, one obtains

$$T^i = \frac{1}{2}[\dot{\mathbf{R}}^{i^{\mathrm{T}}} \quad \dot{\theta}^i] \left\{ \int_{V^i} \rho^i \begin{bmatrix} \mathbf{I} & \mathbf{A}_\theta^i \bar{\mathbf{u}}^i \\ \bar{\mathbf{u}}^{i^{\mathrm{T}}} \mathbf{A}_\theta^{i^{\mathrm{T}}} & \bar{\mathbf{u}}^{i^{\mathrm{T}}} \bar{\mathbf{u}}^i \end{bmatrix} dV^i \right\} \begin{bmatrix} \dot{\mathbf{R}}^i \\ \dot{\theta}^i \end{bmatrix} \tag{6.20}$$

which can be written as

$$T^i = \tfrac{1}{2} \dot{\mathbf{q}}^{i^{\mathrm{T}}} \mathbf{M}^i \dot{\mathbf{q}}^i \tag{6.21}$$

where $\mathbf{q}^i$ and $\mathbf{M}^i$ are, respectively, the vector of coordinates and mass matrix of the rigid body i given by

$$\mathbf{q}^i = [\mathbf{R}^{i^{\mathrm{T}}} \quad \theta^i]^{\mathrm{T}}, \quad \mathbf{M}^i = \begin{bmatrix} \mathbf{m}_{RR}^i & \mathbf{m}_{R\theta}^i \\ \mathbf{m}_{\theta R}^i & m_{\theta\theta}^i \end{bmatrix} \tag{6.22}$$

in which

$$\left.\begin{aligned} \mathbf{m}_{RR}^i &= \int_{V^i} \rho^i \mathbf{I} \, dV^i = m^i \mathbf{I} \\ \mathbf{m}_{R\theta}^i &= \mathbf{m}_{\theta R}^{i^{\mathrm{T}}} = \mathbf{A}_\theta^i \int_{V^i} \rho^i \bar{\mathbf{u}}^i \, dV^i \\ m_{\theta\theta}^i &= \textstyle\int_{V^i} \rho^i \bar{\mathbf{u}}^{i^{\mathrm{T}}} \bar{\mathbf{u}}^i \, dV^i \end{aligned}\right\} \tag{6.23}$$

and m^i is the total mass of the body. Note that $m_{\theta\theta}^i$ is a scalar that defines the body *mass moment of inertia* with respect to an axis passing through the reference point of the body. The matrix $\mathbf{m}_{R\theta}^i$ and its transpose $\mathbf{m}_{\theta R}^i$ represent the inertia coupling between the translation of the reference point and the rotation of the body.

Example 6.2

In order to demonstrate the use of Eqs. 22 and 23, we consider the case of a uniform slender rod that has mass density ρ^i, cross-sectional area a^i, and length l^i. We assume that the reference point is selected to be one of the endpoints at which $\bar{x}^i = 0$, where $\bar{x}^i$ is the coordinate along the rod axis. Keeping in mind that in this case

$$dV^i = a^i \, d\bar{x}^i$$

and

$$m^i = \int_{V^i} \rho^i \, dV^i = \int_0^{l^i} \rho^i a^i \, d\bar{x}^i = \rho^i a^i l^i$$

where V^i is the volume, and m^i is the total mass of the rod. The matrix $\mathbf{m}_{RR}^i$ can be evaluated as

$$\mathbf{m}_{RR}^i = \int_{V^i} \rho^i \mathbf{I}\, dV^i = m^i \mathbf{I} = \begin{bmatrix} m^i & 0 \\ 0 & m^i \end{bmatrix}$$

The matrix $\mathbf{m}_{R\theta}^i$, which represents the inertia coupling between the translation and rotation of the rod, can be written as

$$\mathbf{m}_{R\theta}^i = \mathbf{A}_\theta^i \int_{V^i} \rho^i \overline{\mathbf{u}}^i \, dV^i$$

in which

$$\mathbf{A}_\theta^i = \begin{bmatrix} -\sin\theta^i & -\cos\theta^i \\ \cos\theta^i & -\sin\theta^i \end{bmatrix}, \qquad \overline{\mathbf{u}}^i = [\overline{x}^i \quad 0]^{\mathrm{T}}$$

It follows that

$$\mathbf{m}_{R\theta}^i = \frac{m^i l^i}{2} \begin{bmatrix} -\sin\theta^i \\ \cos\theta^i \end{bmatrix}$$

The mass moment of inertia of the rod defined with respect to an axis passing through the reference point is given by

$$m_{\theta\theta}^i = \int_{V^i} \rho^i \overline{\mathbf{u}}^{i^{\mathrm{T}}} \overline{\mathbf{u}}^i \, dV^i = \int_0^{l^i} \rho^i a^i (\overline{x}^i)^2 \, d\overline{x}^i = \frac{m^i (l^i)^2}{3}$$

Therefore, the mass matrix of the rod is given by

$$\mathbf{M}^i = \begin{bmatrix} m^i & 0 & -\frac{1}{2} m^i l^i \sin\theta^i \\ 0 & m^i & \frac{1}{2} m^i l^i \cos\theta^i \\ -\frac{1}{2} m^i l^i \sin\theta^i & \frac{1}{2} m^i l^i \cos\theta^i & \frac{1}{3} m^i (l^i)^2 \end{bmatrix}$$

Centrifugal Inertia Forces The general expression of the kinetic energy obtained in Eq. 21 can be used to define the generalized inertia forces of the rigid body i using Lagrange's equation. In this case, one has

$$\mathbf{Q}_i^i = \frac{d}{dt}\left(\frac{\partial T^i}{\partial \dot{\mathbf{q}}^i}\right)^{\mathrm{T}} - \left(\frac{\partial T^i}{\partial \mathbf{q}^i}\right)^{\mathrm{T}} \tag{6.24}$$

Another elegant way for defining the generalized inertia forces associated with the absolute coordinates is to use the virtual work. Recall that the virtual work of the inertia forces of the rigid body i is defined as

$$\delta W_i^i = \int_{V^i} \rho^i \ddot{\mathbf{r}}^{i^{\mathrm{T}}} \delta \mathbf{r}^i \, dV^i \tag{6.25}$$

where $\mathbf{r}^i$ is the global position vector of an arbitrary point on the rigid body as defined by Eq. 16. It follows that

$$\delta\mathbf{r}^i = \delta\mathbf{R}^i + \mathbf{A}_\theta^i\bar{\mathbf{u}}^i\,\delta\theta^i = [\mathbf{I} \quad \mathbf{A}_\theta^i\bar{\mathbf{u}}^i]\begin{bmatrix}\delta\mathbf{R}^i\\ \delta\theta^i\end{bmatrix} \tag{6.26}$$

This equation can be written as

$$\delta\mathbf{r}^i = \mathbf{L}^i\,\delta\mathbf{q}^i \tag{6.27}$$

where

$$\mathbf{L}^i = [\mathbf{I} \quad \mathbf{A}_\theta^i\bar{\mathbf{u}}^i], \quad \delta\mathbf{q}^i = [\delta\mathbf{R}^{i^{\mathrm{T}}} \quad \delta\theta^i]^{\mathrm{T}} \tag{6.28}$$

Using Eq. 18, the absolute velocity and acceleration vectors of the arbitrary point can be written as

$$\dot{\mathbf{r}}^i = \mathbf{L}^i\dot{\mathbf{q}}^i, \quad \ddot{\mathbf{r}}^i = \mathbf{L}^i\ddot{\mathbf{q}}^i + \dot{\mathbf{L}}^i\dot{\mathbf{q}}^i \tag{6.29}$$

in which

$$\dot{\mathbf{L}}^i = [\mathbf{0} \quad \dot{\theta}^i\mathbf{A}_{\theta\theta}^i\bar{\mathbf{u}}^i] = [\mathbf{0} \quad -\dot{\theta}^i\mathbf{A}^i\bar{\mathbf{u}}^i] \tag{6.30}$$

Substituting Eqs. 27 and 29 into Eq. 25, one obtains

$$\delta W_i^i = \int_{V^i}\rho^i\ddot{\mathbf{q}}^{i^{\mathrm{T}}}\mathbf{L}^{i^{\mathrm{T}}}\mathbf{L}^i\,\delta\mathbf{q}^i\,dV^i + \int_{V^i}\rho^i\dot{\mathbf{q}}^{i^{\mathrm{T}}}\dot{\mathbf{L}}^{i^{\mathrm{T}}}\mathbf{L}^i\,\delta\mathbf{q}^i\,dV^i \tag{6.31}$$

Note that the symmetric mass matrix of Eq. 22 is

$$\mathbf{M}^i = \int_{V^i}\rho^i\mathbf{L}^{i^{\mathrm{T}}}\mathbf{L}^i\,dV^i \tag{6.32}$$

Equation 31 can then be written as

$$\delta W_i^i = [\mathbf{M}^i\ddot{\mathbf{q}}^i - \mathbf{Q}_v^i]^{\mathrm{T}}\,\delta\mathbf{q}^i \tag{6.33}$$

where $\mathbf{Q}_v^i$ is the vector of *centrifugal inertia forces* defined as

$$\mathbf{Q}_v^i = -\int_{V^i}\rho^i\mathbf{L}^{i^{\mathrm{T}}}\dot{\mathbf{L}}^i\dot{\mathbf{q}}^i\,dV^i \tag{6.34}$$

Substituting Eqs. 28 and 30 into Eq. 34, one obtains

$$\mathbf{Q}_v^i = \begin{bmatrix}(\mathbf{Q}_v^i)_R\\ (Q_v^i)_\theta\end{bmatrix} = -\int_{V^i}\rho^i\begin{bmatrix}\mathbf{I}\\ \bar{\mathbf{u}}^{i^{\mathrm{T}}}\mathbf{A}_\theta^{i^{\mathrm{T}}}\end{bmatrix}[\mathbf{0} \quad -\dot{\theta}^i\mathbf{A}^i\bar{\mathbf{u}}^i]\begin{bmatrix}\dot{\mathbf{R}}^i\\ \dot{\theta}^i\end{bmatrix}dV^i$$

$$= -\int_{V^i}\rho^i\begin{bmatrix}-(\dot{\theta}^i)^2\mathbf{A}^i\bar{\mathbf{u}}^i\\ -(\dot{\theta}^i)^2\bar{\mathbf{u}}^{i^{\mathrm{T}}}\mathbf{A}_\theta^{i^{\mathrm{T}}}\mathbf{A}^i\bar{\mathbf{u}}^i\end{bmatrix}dV^i \tag{6.35}$$

The product $\mathbf{A}_\theta^{i^{\mathrm{T}}}\mathbf{A}^i$ is a skew-symmetric matrix defined as

$$\mathbf{A}_\theta^{i^{\mathrm{T}}}\mathbf{A}^i = \begin{bmatrix} 0 & 1 \\ -1 & 0 \end{bmatrix} \tag{6.36}$$

and as a consequence $\bar{\mathbf{u}}^{i^{\mathrm{T}}}\mathbf{A}_\theta^{i^{\mathrm{T}}}\mathbf{A}^i\bar{\mathbf{u}}^i = 0$. Therefore, the vector of centrifugal inertia forces of Eq. 35 reduces to

$$\mathbf{Q}_v^i = \begin{bmatrix} (\mathbf{Q}_v^i)_R \\ (\mathbf{Q}_v^i)_\theta \end{bmatrix} = (\dot{\theta}^i)^2\mathbf{A}^i \begin{bmatrix} \int_{V^i} \rho^i \bar{\mathbf{u}}^i \, dV^i \\ 0 \end{bmatrix} \tag{6.37}$$

For the simple rod discussed in Example 2 one can show that the vector of centrifugal inertia forces is $\mathbf{Q}_v^i = (\dot{\theta}^i)^2 \, m^i (l^i/2) \, [\cos\theta^i \quad \sin\theta^i \quad 0]^{\mathrm{T}}$

Centroidal Body Coordinate System A special case of the foregoing development is the case in which the reference point is selected to be the center of mass of the body. This is the case of a *centroidal body coordinate system*.

The integral in the second equation of Eq. 23 represents the moment of mass of the body. If the reference point is chosen to be the center of mass, this integral is identically zero; that is,

$$\int_{V^i} \rho^i \bar{\mathbf{u}}^i \, dV^i = \mathbf{0} \tag{6.38}$$

and, as a consequence, the matrices $\mathbf{m}_{R\theta}^i$ and $\mathbf{m}_{\theta R}^i$ defined by Eq. 23 are identically equal to zero. In this special case, the mass matrix $\mathbf{M}^i$ of the body reduces to

$$\mathbf{M}^i = \begin{bmatrix} \mathbf{m}_{RR}^i & \mathbf{0} \\ \mathbf{0} & m_{\theta\theta}^i \end{bmatrix} \tag{6.39}$$

and the kinetic energy of the body can be written as

$$T^i = \tfrac{1}{2} m^i \dot{\mathbf{R}}^{i^{\mathrm{T}}} \dot{\mathbf{R}}^i + \tfrac{1}{2} m_{\theta\theta}^i (\dot{\theta}^i)^2 \tag{6.40}$$

In this special case, there is no coupling between the translation and rotation of the body in the mass matrix. Furthermore, the vector of centrifugal inertia forces is identically equal to zero, that is

$$\mathbf{Q}_v^i = \mathbf{0} \tag{6.41}$$

Thus, in the case of a centroidal body coordinate system, the mass matrix of a rigid body in *planar motion* is diagonal, the vector of centrifugal inertia forces vanishes, and the kinetic energy of the rigid body consists of two terms; one is due to the translation of the center of mass and the other is due to the planar rigid body rotation.

Example 6.3

Consider the slender rod of Example 2. The rod is assumed to have mass density ρ^i, cross-sectional area a^i, and length l^i. We consider the case where the reference point is selected to be the center of mass of the rod. In this case, one has

$$\int_{V^i} \rho^i \bar{\mathbf{u}}^i \, dV^i = \int_{-l^i/2}^{l^i/2} \rho^i a^i \begin{bmatrix} \bar{x}^i \\ 0 \end{bmatrix} d\bar{x}^i = \mathbf{0}$$

which implies that

$$\mathbf{m}^i_{R\theta} = \mathbf{m}^{i\mathrm{T}}_{\theta R} = \mathbf{0}$$

The matrix $\mathbf{m}^i_{RR}$ is

$$\mathbf{m}^i_{RR} = \int_{V^i} \rho^i \mathbf{I} \, dV^i = \begin{bmatrix} m^i & 0 \\ 0 & m^i \end{bmatrix}$$

The mass moment of inertia $m^i_{\theta\theta}$ is given by

$$m^i_{\theta\theta} = \int_{V^i} \rho^i \bar{\mathbf{u}}^{i\mathrm{T}} \bar{\mathbf{u}}^i \, dV^i = \int_{-l^i/2}^{l^i/2} \rho^i a^i (\bar{x}^i)^2 \, d\bar{x}^i = \frac{m^i (l^i)^2}{12}$$

The mass matrix of the rod can then be written as

$$\mathbf{M}^i = \begin{bmatrix} m^i & 0 & 0 \\ 0 & m^i & 0 \\ 0 & 0 & \dfrac{m^i (l^i)^2}{12} \end{bmatrix}$$

Comparing the results obtained in this example and the results of Example 2, we see that the mass matrix in the case of a centroidal body coordinate system is diagonal, as compared to the nondiagonal mass matrix obtained in the preceding example. Furthermore, the mass moment of inertia obtained when the reference point is selected to be at one of the endpoints of the rod is equal to the mass moment of inertia defined with respect to the center of mass plus $m^i (l^i/2)^2$.

6.3 EQUATIONS OF MOTION

The equations of motion of the rigid body are developed in this section in terms of the absolute Cartesian coordinates that represent the translation of the reference point of the body as well as its orientation with respect to the global inertial frame of reference. This representation will prove useful in developing general-purpose computer algorithms for the dynamic analysis of interconnected sets of rigid bodies, since practically speaking, there is no limitation on the number of bodies or the types of forces and constraints that can be introduced to this formulation.

In the preceding chapter, it was shown that the conditions for the dynamic equilibrium for the rigid body i can be developed using the *principle of virtual work* as

$$\delta W_i^i = \delta W_e^i + \delta W_c^i \tag{6.42}$$

where δW_i^i is the virtual work of the inertia forces, δW_e^i is the virtual work of the externally applied forces, and δW_c^i is the virtual work of the constraint forces. The virtual work of the externally applied forces can be expressed in terms of the vector of generalized coordinates as

$$\delta W_e^i = \mathbf{Q}_e^{i\mathrm{T}} \, \delta \mathbf{q}^i \tag{6.43}$$

where $\mathbf{Q}_e^i$ is the vector of generalized forces, and $\mathbf{q}^i$ is the vector of generalized coordinates of the rigid body i. Using the Cartesian coordinates, the vector $\mathbf{q}^i$ is given by

$$\mathbf{q}^i = [\mathbf{R}^{i\mathrm{T}} \quad \theta^i]^\mathrm{T} \tag{6.44}$$

where $\mathbf{R}^i$ is the position vector of the reference point, and θ^i is the angular orientation of the body.

One can also write the virtual work of the joint constraint forces acting on the rigid body i as

$$\delta W_c^i = \mathbf{Q}_c^{i\mathrm{T}} \, \delta \mathbf{q}^i \tag{6.45}$$

where $\mathbf{Q}_c^i$ is the vector of the generalized constraint forces associated with the body generalized coordinates.

The virtual work of the inertia forces can be obtained using the development of the preceding section as

$$\delta W_i^i = [\ddot{\mathbf{q}}^{i\mathrm{T}} \mathbf{M}^i - \mathbf{Q}_v^{i\mathrm{T}}] \, \delta \mathbf{q}^i \tag{6.46}$$

where $\mathbf{M}^i$ is the symmetric mass matrix of the body defined in Eq. 22 or equivalently by Eq. 32, and $\mathbf{Q}_v^i$ is the vector of centrifugal forces defined by Eq. 37.

Substituting Eqs. 43, 45 and 46 into Eq. 42, one obtains

$$[\ddot{\mathbf{q}}^{i\mathrm{T}} \mathbf{M}^i - \mathbf{Q}_v^{i\mathrm{T}}] \, \delta \mathbf{q}^i = \mathbf{Q}_e^{i\mathrm{T}} \, \delta \mathbf{q}^i + \mathbf{Q}_c^{i\mathrm{T}} \, \delta \mathbf{q}^i \tag{6.47}$$

which, upon utilizing the fact that the mass matrix is symmetric, leads to

$$[\mathbf{M}^i \ddot{\mathbf{q}}^i - \mathbf{Q}_v^i - \mathbf{Q}_e^i - \mathbf{Q}_c^i]^\mathrm{T} \, \delta \mathbf{q}^i = 0 \tag{6.48}$$

Since the constraint forces, as result of the connection of this body with other bodies in the system, are included in this equation and represented by the vector $\mathbf{Q}_c^i$, the elements of the vector $\mathbf{q}^i$ can be treated as independent. Consequently, Eq. 48 leads to

$$\mathbf{M}^i \ddot{\mathbf{q}}^i = \mathbf{Q}_e^i + \mathbf{Q}_c^i + \mathbf{Q}_v^i \tag{6.49}$$

which can be rewritten according to the coordinate partitioning of Eq. 44 as

$$\begin{bmatrix} \mathbf{m}^i_{RR} & \mathbf{m}^i_{R\theta} \\ \mathbf{m}^i_{\theta R} & m^i_{\theta\theta} \end{bmatrix} \begin{bmatrix} \ddot{\mathbf{R}}^i \\ \ddot{\theta}^i \end{bmatrix} = \begin{bmatrix} (\mathbf{Q}^i_e)_R \\ (\mathbf{Q}^i_e)_\theta \end{bmatrix} + \begin{bmatrix} (\mathbf{Q}^i_c)_R \\ (\mathbf{Q}^i_c)_\theta \end{bmatrix} + \begin{bmatrix} (\mathbf{Q}^i_v)_R \\ 0 \end{bmatrix} \tag{6.50}$$

Equation 49, or its equivalent form of Eq. 50, represents the dynamic equations of motion of the rigid body i, developed using an arbitrary reference point.

Newton–Euler Equations If the reference point is selected to be the center of mass of the body, one has $\mathbf{m}^i_{R\theta} = \mathbf{m}^{i\mathrm{T}}_{\theta R} = \mathbf{0}$ and $\mathbf{Q}^i_v = \mathbf{0}$. In this case, Eq. 49 reduces to

$$\mathbf{M}^i\ddot{\mathbf{q}}^i = \mathbf{Q}^i_e + \mathbf{Q}^i_c \tag{6.51}$$

which can be written in a more explicit form as

$$\begin{bmatrix} m^i\mathbf{I} & \mathbf{0} \\ \mathbf{0} & J^i \end{bmatrix} \begin{bmatrix} \ddot{\mathbf{R}}^i \\ \ddot{\theta}^i \end{bmatrix} = \begin{bmatrix} (\mathbf{Q}^i_e)_R \\ (\mathbf{Q}^i_e)_\theta \end{bmatrix} + \begin{bmatrix} (\mathbf{Q}^i_c)_R \\ (\mathbf{Q}^i_c)_\theta \end{bmatrix} \tag{6.52}$$

where m^i is the mass of the body, and J^i is its mass moment of inertia about an axis passing through the center of mass. Clearly, Eq. 52 is the same as the fundamental Newton and Euler equations that govern the motion of the rigid bodies. These equations are obtained, however, as a special case of the general form represented by Eq. 50.

6.4 SYSTEM OF RIGID BODIES

It was shown in the preceding section that affixing the origin of the body coordinate system to the body center of mass leads to a significant simplification in the resulting dynamic equations. Therefore, without any loss of generality, we consider the case of a *centroidal body reference* where the origin of the body coordinate system is rigidly attached to the body center of mass. In this case, the mass matrix is diagonal and the vector of centrifugal forces is identically equal to zero. By using Eq. 51, the equations of motion of a multibody system consisting of n_b interconnected bodies are given by

$$\mathbf{M}^i\ddot{\mathbf{q}}^i = \mathbf{Q}^i_e + \mathbf{Q}^i_c \qquad i = 1, 2, \ldots, n_b \tag{6.53}$$

which can also be written as

$$\begin{bmatrix} m^i\mathbf{I} & \mathbf{0} \\ \mathbf{0} & J^i \end{bmatrix} \begin{bmatrix} \ddot{\mathbf{R}}^i \\ \ddot{\theta}^i \end{bmatrix} = \begin{bmatrix} (\mathbf{Q}^i_e)_R \\ (\mathbf{Q}^i_e)_\theta \end{bmatrix} + \begin{bmatrix} (\mathbf{Q}^i_c)_R \\ (\mathbf{Q}^i_c)_\theta \end{bmatrix} \qquad i = 1, 2, \ldots, n_b \tag{6.54}$$

where all the scalars, vectors, and matrices that appear in Eqs. 53 and 54 are the same as defined in the preceding section. The number of scalar equations given by the matrix equation of Eq. 53 or Eq. 54 is $3 \times n_b$. These equations can be combined into one matrix equation given by

$$\begin{bmatrix} \mathbf{M}^1 & & & & & \\ & \mathbf{M}^2 & & \mathbf{0} & & \\ & & \ddots & & & \\ & & & \mathbf{M}^i & & \\ & \mathbf{0} & & & \ddots & \\ & & & & & \mathbf{M}^{n_b} \end{bmatrix} \begin{bmatrix} \ddot{\mathbf{q}}^1 \\ \ddot{\mathbf{q}}^2 \\ \vdots \\ \ddot{\mathbf{q}}^i \\ \vdots \\ \ddot{\mathbf{q}}^{n_b} \end{bmatrix} = \begin{bmatrix} \mathbf{Q}_e^1 \\ \mathbf{Q}_e^2 \\ \vdots \\ \mathbf{Q}_e^i \\ \vdots \\ \mathbf{Q}_e^{n_b} \end{bmatrix} + \begin{bmatrix} \mathbf{Q}_c^1 \\ \mathbf{Q}_c^2 \\ \vdots \\ \mathbf{Q}_c^i \\ \vdots \\ \mathbf{Q}_c^{n_b} \end{bmatrix} \tag{6.55}$$

which can be written as

$$\mathbf{M}\ddot{\mathbf{q}} = \mathbf{Q}_e + \mathbf{Q}_c \tag{6.56}$$

where $\mathbf{M}$ is the system mass matrix, $\mathbf{q}$ is the total vector of system generalized coordinates, $\mathbf{Q}_e$ is the vector of system generalized external forces, and $\mathbf{Q}_c$ is the vector of the system generalized constraint forces. The mass matrix $\mathbf{M}$ and the vectors $\mathbf{q}$, $\mathbf{Q}_e$, and $\mathbf{Q}_c$ are

$$\left.\begin{aligned} \mathbf{M} &= \begin{bmatrix} \mathbf{M}^1 & & & & & \\ & \mathbf{M}^2 & & \mathbf{0} & & \\ & & \ddots & & & \\ & & & \mathbf{M}^i & & \\ & \mathbf{0} & & & \ddots & \\ & & & & & \mathbf{M}^{n_b} \end{bmatrix} \\ \mathbf{q} &= [\mathbf{q}^{1^{\mathrm{T}}} \quad \mathbf{q}^{2^{\mathrm{T}}} \quad \cdots \quad \mathbf{q}^{i^{\mathrm{T}}} \quad \cdots \quad \mathbf{q}^{n_b^{\mathrm{T}}}]^{\mathrm{T}} \\ \mathbf{Q}_e &= [\mathbf{Q}_e^{1^{\mathrm{T}}} \quad \mathbf{Q}_e^{2^{\mathrm{T}}} \quad \cdots \quad \mathbf{Q}_e^{i^{\mathrm{T}}} \quad \cdots \quad \mathbf{Q}_e^{n_b^{\mathrm{T}}}]^{\mathrm{T}} \\ \mathbf{Q}_c &= [\mathbf{Q}_c^{1^{\mathrm{T}}} \quad \mathbf{Q}_c^{2^{\mathrm{T}}} \quad \cdots \quad \mathbf{Q}_c^{i^{\mathrm{T}}} \quad \cdots \quad \mathbf{Q}_c^{n_b^{\mathrm{T}}}]^{\mathrm{T}} \end{aligned}\right\} \tag{6.57}$$

Note that the equations of motion of the multibody system given by Eq. 56 contain the generalized constraint forces, since these equations are not expressed in terms of the system degrees of freedom.

Example 6.4

Figure 3a shows a two link planar manipulator. The origins of the body coordinate systems are assumed to be rigidly attached to the centers of mass of the links. Let m^i and J^i, and m^j and J^j be, respectively, the mass and mass moment of inertia of links i and j, and M^i and M^j be, respectively, the external torque applied to links i and j. Obtain the differential equations of motion of the system in terms of the absolute coordinates. Neglect the effect of gravity.

Solution. The virtual work of the reaction forces acting on link i can be expressed as

$$\delta W_c^i = [F_x^{1i} \quad F_y^{1i}]\delta\mathbf{r}_O^i - [F_x^{ij} \quad F_y^{ij}]\,\delta\mathbf{r}_A^i$$

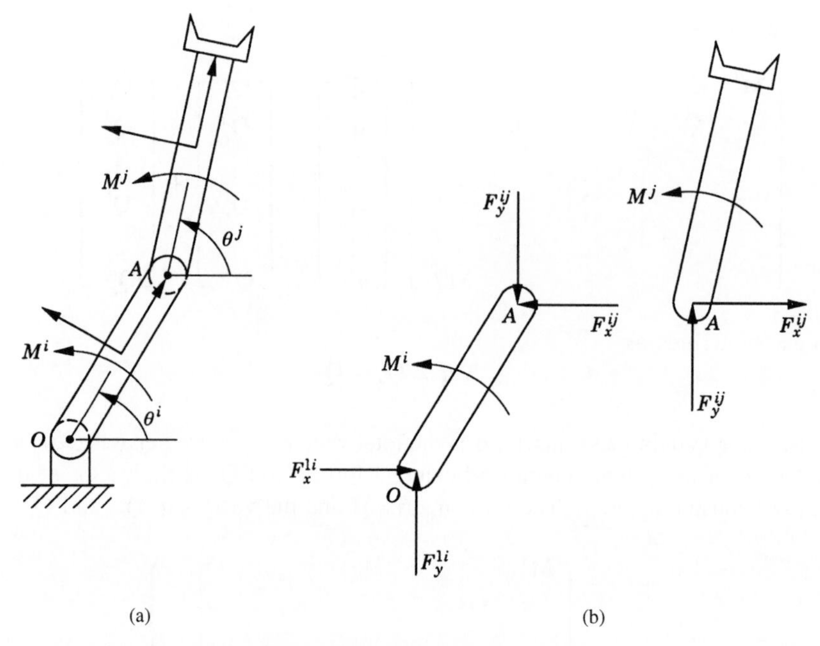

Figure 6.3 Planar manipulator

where F_x^{1i}, F_y^{1i}, F_x^{ij}, and F_y^{ij} are the reaction forces acting on link i as shown in Fig. 3b, and $\delta \mathbf{r}_O^i$ and $\delta \mathbf{r}_A^i$ can be expressed in terms of the absolute coordinates of link i as

$$\delta \mathbf{r}_O^i = \delta \mathbf{R}^i + \mathbf{A}_\theta^i \overline{\mathbf{u}}_O^i \, \delta\theta^i$$

$$\delta \mathbf{r}_A^i = \delta \mathbf{R}^i + \mathbf{A}_\theta^i \overline{\mathbf{u}}_A^i \, \delta\theta^i$$

where

$$\mathbf{R}^i = \begin{bmatrix} R_x^i \\ R_y^i \end{bmatrix}, \quad \overline{\mathbf{u}}_O^i = \begin{bmatrix} -l_O^i \\ 0 \end{bmatrix}, \quad \overline{\mathbf{u}}_A^i = \begin{bmatrix} l_A^i \\ 0 \end{bmatrix}, \quad \mathbf{A}_\theta^i = \begin{bmatrix} -\sin\theta^i & -\cos\theta^i \\ \cos\theta^i & -\sin\theta^i \end{bmatrix}$$

and l_O^i and l_A^i are, respectively, the distances of points O and A from the center of mass of link i. The virtual work of the reaction forces of link i is

$$\begin{aligned} \delta W_c^i = {} & [F_x^{1i} \quad F_y^{1i}] \begin{bmatrix} \delta R_x^i \\ \delta R_y^i \end{bmatrix} + [F_x^{1i} \quad F_y^{1i}] \begin{bmatrix} -\sin\theta^i & -\cos\theta^i \\ \cos\theta^i & -\sin\theta^i \end{bmatrix} \begin{bmatrix} -l_O^i \\ 0 \end{bmatrix} \delta\theta^i \\ & - [F_x^{ij} \quad F_y^{ij}] \begin{bmatrix} \delta R_x^i \\ \delta R_y^i \end{bmatrix} - [F_x^{ij} \quad F_y^{ij}] \begin{bmatrix} -\sin\theta^i & -\cos\theta^i \\ \cos\theta^i & -\sin\theta^i \end{bmatrix} \begin{bmatrix} l_A^i \\ 0 \end{bmatrix} \delta\theta^i \end{aligned}$$

This equation leads to

$$\begin{aligned} \delta W_c^i = {} & (F_x^{1i} - F_x^{ij}) \delta R_x^i + (F_y^{1i} - F_y^{ij}) \delta R_y^i \\ & + \{(F_x^{1i} \sin\theta^i - F_y^{1i} \cos\theta^i) l_O^i + (F_x^{ij} \sin\theta^i - F_y^{ij} \cos\theta^i) l_A^i\} \, \delta\theta^i \end{aligned}$$

from which the vector of generalized reactions $\mathbf{Q}_c^i$ associated with the absolute coordinates of link i can be expressed as

$$\mathbf{Q}_c^i = \begin{bmatrix} F_x^{1i} - F_x^{ij} \\ F_y^{1i} - F_y^{ij} \\ (F_x^{1i}\sin\theta^i - F_y^{1i}\cos\theta^i)l_O^i + (F_x^{ij}\sin\theta^i - F_y^{ij}\cos\theta^i)l_A^i \end{bmatrix}$$

Using Eq. 54, the equation of motion of link i can be written as

$$\begin{bmatrix} m^i & 0 & 0 \\ 0 & m^i & 0 \\ 0 & 0 & J^i \end{bmatrix} \begin{bmatrix} \ddot{R}_x^i \\ \ddot{R}_y^i \\ \ddot{\theta}^i \end{bmatrix} = \begin{bmatrix} 0 \\ 0 \\ M^i \end{bmatrix} + \begin{bmatrix} F_x^{1i} - F_x^{ij} \\ F_y^{1i} - F_y^{ij} \\ (F_x^{1i}\sin\theta^i - F_y^{1i}\cos\theta^i)l_O^i + (F_x^{ij}\sin\theta^i - F_y^{ij}\cos\theta^i)l_A^i \end{bmatrix}$$

Similarly, the virtual work of the generalized reactions acting on link j is

$$\delta W_c^j = [F_x^{ij} \quad F_y^{ij}]\, \delta\mathbf{r}_A^j$$

where

$$\delta\mathbf{r}_A^j = \delta\mathbf{R}^j + \mathbf{A}_\theta^j \bar{\mathbf{u}}_A^j\, \delta\theta^j$$

in which

$$\mathbf{R}^j = [R_x^j \quad R_y^j]^{\mathrm{T}}, \quad \bar{\mathbf{u}}_A^j = [-l_A^j \quad 0]^{\mathrm{T}}$$

and

$$\mathbf{A}_\theta^j = \begin{bmatrix} -\sin\theta^j & -\cos\theta^j \\ \cos\theta^j & -\sin\theta^j \end{bmatrix}$$

where l_A^j is the distance of point A from the center of mass of link j. It follows that

$$\begin{aligned} \delta W_c^j &= [F_x^{ij} \quad F_y^{ij}] \begin{bmatrix} \delta R_x^j \\ \delta R_y^j \end{bmatrix} \\ &\quad + [F_x^{ij} \quad F_y^{ij}] \begin{bmatrix} -\sin\theta^j & -\cos\theta^j \\ \cos\theta^j & -\sin\theta^j \end{bmatrix} \begin{bmatrix} -l_A^j \\ 0 \end{bmatrix} \delta\theta^j \\ &= F_x^{ij}\,\delta R_x^j + F_y^{ij}\,\delta R_y^j + (F_x^{ij}\sin\theta^j - F_y^{ij}\cos\theta^j)l_A^j\,\delta\theta^j \end{aligned}$$

from which the vector of joint reaction forces $\mathbf{Q}_c^j$ associated with the absolute coordinates of link j can be defined as

$$\mathbf{Q}_c^j = \begin{bmatrix} F_x^{ij} \\ F_y^{ij} \\ (F_x^{ij} \sin\theta^j - F_y^{ij} \cos\theta^j) l_A^j \end{bmatrix}$$

The equations of motion of link j are

$$\begin{bmatrix} m^j & 0 & 0 \\ 0 & m^j & 0 \\ 0 & 0 & J^j \end{bmatrix} \begin{bmatrix} \ddot{R}_x^j \\ \ddot{R}_y^j \\ \ddot{\theta}^j \end{bmatrix} = \begin{bmatrix} 0 \\ 0 \\ M^j \end{bmatrix} + \begin{bmatrix} F_x^{ij} \\ F_y^{ij} \\ (F_x^{ij} \sin\theta^j - F_y^{ij} \cos\theta^j) l_A^j \end{bmatrix}$$

The system equations of motion can be defined as

$$\begin{bmatrix} m^i & 0 & 0 & 0 & 0 & 0 \\ 0 & m^i & 0 & 0 & 0 & 0 \\ 0 & 0 & J^i & 0 & 0 & 0 \\ 0 & 0 & 0 & m^j & 0 & 0 \\ 0 & 0 & 0 & 0 & m^j & 0 \\ 0 & 0 & 0 & 0 & 0 & J^j \end{bmatrix} \begin{bmatrix} \ddot{R}_x^i \\ \ddot{R}_y^i \\ \ddot{\theta}^i \\ \ddot{R}_x^j \\ \ddot{R}_y^j \\ \ddot{\theta}^j \end{bmatrix}$$

$$= \begin{bmatrix} 0 \\ 0 \\ M^i \\ 0 \\ 0 \\ M^j \end{bmatrix} + \begin{bmatrix} F_x^{1i} - F_x^{ij} \\ F_y^{1i} - F_y^{ij} \\ (F_x^{1i} \sin\theta^i - F_y^{1i} \cos\theta^i) l_O^i + (F_x^{ij} \sin\theta^i - F_y^{ij} \cos\theta^i) l_A^i \\ F_x^{ij} \\ F_y^{ij} \\ (F_x^{ij} \sin\theta^j - F_y^{ij} \cos\theta^j) l_A^j \end{bmatrix}$$

These are six scalar equations in 10 unknowns $\ddot{R}_x^i$, $\ddot{R}_y^i$, $\ddot{\theta}^i$, $\ddot{R}_x^j$, $\ddot{R}_y^j$, $\ddot{\theta}^j$, F_x^{1i}, F_y^{1i}, F_x^{ij}, and F_y^{ij}. The first six unknowns, however, can be expressed in terms of two acceleration components only, since the system has two degrees of freedom. This reduces the number of unknowns to six and consequently the preceding six equations can be solved for the six independent unknowns. Note also that

$$\delta W_c^i + \delta W_c^j = 0,$$

since $\delta\mathbf{r}_O^i = \mathbf{0}$, and $\delta\mathbf{r}_A^i = \delta\mathbf{r}_A^j$.

6.5 ELIMINATION OF THE CONSTRAINT FORCES

The equations of motion obtained for the multibody system in the preceding section and given by Eq. 56 contain the generalized constraint forces due mainly to the fact that these equations are formulated using a redundant set of coordinates. It is important, however, to point out that the force of constraints can be eliminated from the dynamic formulation if the redundant coordinates are expressed in terms of the independent coordinates, as discussed in the preceding chapter. To illustrate the use of such a procedure, Eq. 42, which is a statement of the principle of virtual work for body i, is used. If the multibody system consists of n_b interconnected bodies, Eq. 42 leads to

$$\sum_{i=1}^{n_b} \delta W_i^i = \sum_{i=1}^{n_b} \delta W_e^i + \sum_{i=1}^{n_b} \delta W_c^i \tag{6.58}$$

As demonstrated in the preceding chapter, $\sum_{i=1}^{n_b} \delta W_c^i = 0$. Using this equation, Eq. 58 reduces to

$$\sum_{i=1}^{n_b} (\delta W_i^i - \delta W_e^i) = 0 \tag{6.59}$$

Substituting Eqs. 43 and 46 into this equation, and keeping in mind that the origin of the body i coordinate system is rigidly attached to the body center of mass, that is $\mathbf{Q}_v^i = \mathbf{0}$, one obtains

$$\sum_{i=1}^{n_b} [\mathbf{M}^i \ddot{\mathbf{q}}^i - \mathbf{Q}_e^i]^{\mathrm{T}} \, \delta\mathbf{q}^i = 0 \tag{6.60}$$

which can be written in a matrix form as

$$\begin{bmatrix} \mathbf{M}^1\ddot{\mathbf{q}}^1 - \mathbf{Q}_e^1 \\ \mathbf{M}^2\ddot{\mathbf{q}}^2 - \mathbf{Q}_e^2 \\ \vdots \\ \mathbf{M}^i\ddot{\mathbf{q}}^i - \mathbf{Q}_e^i \\ \vdots \\ \mathbf{M}^{n_b}\ddot{\mathbf{q}}^{n_b} - \mathbf{Q}_e^{n_b} \end{bmatrix}^{\mathrm{T}} \begin{bmatrix} \delta\mathbf{q}^1 \\ \delta\mathbf{q}^2 \\ \vdots \\ \delta\mathbf{q}^i \\ \vdots \\ \delta\mathbf{q}^{n_b} \end{bmatrix} = 0 \tag{6.61}$$

This equation can also be written as

$$\begin{bmatrix} \delta\mathbf{q}^1 \\ \delta\mathbf{q}^2 \\ \vdots \\ \delta\mathbf{q}^i \\ \vdots \\ \delta\mathbf{q}^{n_b} \end{bmatrix}^{\mathrm{T}} \left\{ \begin{bmatrix} \mathbf{M}^1 & & & & \\ & \mathbf{M}^2 & & & \mathbf{0} \\ & & \ddots & & \\ & & & \mathbf{M}^i & \\ & \mathbf{0} & & & \ddots \\ & & & & & \mathbf{M}^{n_b} \end{bmatrix} \begin{bmatrix} \ddot{\mathbf{q}}^1 \\ \ddot{\mathbf{q}}^2 \\ \vdots \\ \ddot{\mathbf{q}}^i \\ \vdots \\ \ddot{\mathbf{q}}^{n_b} \end{bmatrix} - \begin{bmatrix} \mathbf{Q}_e^1 \\ \mathbf{Q}_e^2 \\ \vdots \\ \mathbf{Q}_e^i \\ \vdots \\ \mathbf{Q}_e^{n_b} \end{bmatrix} \right\} = 0 \tag{6.62}$$

or

$$\delta\mathbf{q}^{\mathrm{T}}[\mathbf{M}\ddot{\mathbf{q}} - \mathbf{Q}_e] = 0 \tag{6.63}$$

where $\mathbf{M}$ is the system mass matrix, $\mathbf{q}$ is the vector of system absolute coordinates, and $\mathbf{Q}_e$ is the vector of system generalized forces associated with the absolute coordinates. The matrix $\mathbf{M}$ and the vectors $\mathbf{q}$ and $\mathbf{Q}_e$ are defined in the preceding section.

Coordinate Partitioning Equation 63 is a scalar equation that does not contain the constraint forces. The coefficient vector $[\mathbf{M}\ddot{\mathbf{q}} - \mathbf{Q}_e]$ of the vector $\delta\mathbf{q}$ cannot, however, be set equal to zero, since the components of the vector of coordinates $\mathbf{q}$ are not totally independent because of the kinematic constraints that represent specified motion trajectories and mechanical joints in the system. These constraints can be expressed mathematically as

$$\mathbf{C}(\mathbf{q}, t) = \mathbf{0} \tag{6.64}$$

where $\mathbf{C} = [C_1(\mathbf{q}, t)\ C_2(\mathbf{q}, t) \cdots C_{n_c}(\mathbf{q}, t)]^{\mathrm{T}}$ is the vector of linearly independent constraint equations, and n_c is the number of constraint functions.

For a virtual change in the system coordinates, Eq. 64 yields

$$\mathbf{C}_{\mathbf{q}}\,\delta\mathbf{q} = \mathbf{0} \tag{6.65}$$

where $\mathbf{C}_{\mathbf{q}}$ is the constraint Jacobian matrix defined as

$$\mathbf{C}_{\mathbf{q}} = \begin{bmatrix} \dfrac{\partial C_1}{\partial q_1} & \dfrac{\partial C_1}{\partial q_2} & \cdots & \dfrac{\partial C_1}{\partial q_n} \\ \dfrac{\partial C_2}{\partial q_1} & \dfrac{\partial C_2}{\partial q_2} & \cdots & \dfrac{\partial C_2}{\partial q_n} \\ \vdots & \vdots & \ddots & \vdots \\ \dfrac{\partial C_{n_c}}{\partial q_1} & \dfrac{\partial C_{n_c}}{\partial q_2} & \cdots & \dfrac{\partial C_{n_c}}{\partial q_n} \end{bmatrix} \tag{6.66}$$

in which $\mathbf{q} = [q_1 \quad q_2 \cdots q_n]^{\mathrm{T}}$ is the n-dimensional vector of system coordinates.

Because of the kinematic constraints of Eq. 64, the components of the vector $\mathbf{q}$ are not independent. One, therefore, may write the vector $\mathbf{q}$ in the following partitioned form:

$$\mathbf{q} = [\mathbf{q}_d^{\mathrm{T}} \quad \mathbf{q}_i^{\mathrm{T}}]^{\mathrm{T}} \tag{6.67}$$

where $\mathbf{q}_d$ is the n_c-dimensional vector of dependent coordinates and $\mathbf{q}_i$ is the vector of independent coordinates, which has the dimension $(n - n_c)$. According to the coordinate partitioning of Eq. 67, Eq. 65 can be written as

$$\mathbf{C}_{\mathbf{q}_d}\,\delta\mathbf{q}_d + \mathbf{C}_{\mathbf{q}_i}\,\delta\mathbf{q}_i = \mathbf{0} \tag{6.68}$$

where the vector $\mathbf{q}_d$ is selected such that the matrix $\mathbf{C}_{\mathbf{q}_d}$ is nonsingular. This choice of the matrix $\mathbf{C}_{\mathbf{q}_d}$ is always possible since the constraint equations are assumed to be linearly independent. Equation 68 can then be used, as described in the preceding chapter, to write the virtual changes of the dependent coordinates in terms of the virtual changes of the independent ones as

$$\delta\mathbf{q}_d = -\mathbf{C}_{\mathbf{q}_d}^{-1}\mathbf{C}_{\mathbf{q}_i}\,\delta\mathbf{q}_i \tag{6.69}$$

or

$$\delta \mathbf{q}_d = \mathbf{C}_{di} \, \delta \mathbf{q}_i \tag{6.70}$$

in which

$$\mathbf{C}_{di} = -\mathbf{C}_{\mathbf{q}_d}^{-1} \mathbf{C}_{\mathbf{q}_i} \tag{6.71}$$

The virtual changes in the total vector of system coordinates can be written in terms of the virtual changes of the independent coordinates using Eq. 70 as

$$\delta \mathbf{q} = \begin{bmatrix} \delta \mathbf{q}_d \\ \delta \mathbf{q}_i \end{bmatrix} = \begin{bmatrix} \mathbf{C}_{di} \\ \mathbf{I} \end{bmatrix} \delta \mathbf{q}_i \tag{6.72}$$

which can be written as

$$\delta \mathbf{q} = \mathbf{B}_i \, \delta \mathbf{q}_i \tag{6.73}$$

where

$$\mathbf{B}_i = \begin{bmatrix} \mathbf{C}_{di} \\ \mathbf{I} \end{bmatrix} \tag{6.74}$$

Embedding Technique Equations 63 and 73 can be used to obtain a minimum number of differential equations expressed in terms of the independent coordinates. In order to demonstrate this, Eq. 73 is substituted into Eq. 63, leading to

$$\delta \mathbf{q}_i^{\mathrm{T}} \mathbf{B}_i^{\mathrm{T}} [\mathbf{M}\ddot{\mathbf{q}} - \mathbf{Q}_e] = 0 \tag{6.75}$$

Since the components of the vector $\delta \mathbf{q}_i$ are independent, their coefficients in Eq. 75 must be equal to zero. This leads to

$$\mathbf{B}_i^{\mathrm{T}} \mathbf{M}\ddot{\mathbf{q}} - \mathbf{B}_i^{\mathrm{T}} \mathbf{Q}_e = \mathbf{0} \tag{6.76}$$

This is a system of $n - n_c$ differential equations that can be expressed in terms of the independent accelerations. For instance, by differentiating Eq. 64 once and twice with respect to time, one obtains

$$\left.\begin{aligned} \mathbf{C}_{\mathbf{q}} \dot{\mathbf{q}} &= -\mathbf{C}_t \\ \mathbf{C}_{\mathbf{q}} \ddot{\mathbf{q}} &= -[(\mathbf{C}_{\mathbf{q}} \dot{\mathbf{q}})_{\mathbf{q}} \dot{\mathbf{q}} + 2\mathbf{C}_{\mathbf{q}t} \dot{\mathbf{q}} + \mathbf{C}_{tt}] \end{aligned}\right\} \tag{6.77}$$

where subscript t indicates a partial differentiation with respect to time. By using the coordinate partitioning of Eq. 67, the second equation in Eq. 77 can be written as

$$\ddot{\mathbf{q}}_d = \mathbf{C}_{di} \ddot{\mathbf{q}}_i + \mathbf{C}_d \tag{6.78}$$

where

$$\mathbf{C}_d = -\mathbf{C}_{\mathbf{q}_d}^{-1}[(\mathbf{C}_\mathbf{q}\dot{\mathbf{q}})_\mathbf{q}\dot{\mathbf{q}} + 2\mathbf{C}_{\mathbf{q}t}\dot{\mathbf{q}} + \mathbf{C}_{tt}] \tag{6.79}$$

Equation 78 can then be used to write the total vector of system accelerations in terms of the independent ones as

$$\begin{aligned}\ddot{\mathbf{q}} = \begin{bmatrix}\ddot{\mathbf{q}}_d \\ \ddot{\mathbf{q}}_i\end{bmatrix} &= \begin{bmatrix}\mathbf{C}_{di}\ddot{\mathbf{q}}_i + \mathbf{C}_d \\ \ddot{\mathbf{q}}_i\end{bmatrix} \\ &= \begin{bmatrix}\mathbf{C}_{di} \\ \mathbf{I}\end{bmatrix}\ddot{\mathbf{q}}_i + \begin{bmatrix}\mathbf{C}_d \\ \mathbf{0}\end{bmatrix}\end{aligned} \tag{6.80}$$

which upon using Eq. 74 leads to

$$\ddot{\mathbf{q}} = \mathbf{B}_i\ddot{\mathbf{q}}_i + \boldsymbol{\gamma}_i \tag{6.81}$$

where

$$\boldsymbol{\gamma}_i = \begin{bmatrix}\mathbf{C}_d \\ \mathbf{0}\end{bmatrix} \tag{6.82}$$

Using Eqs. 76 and 81, one obtains

$$\mathbf{B}_i^{\mathrm{T}}\mathbf{M}\mathbf{B}_i\ddot{\mathbf{q}}_i + \mathbf{B}_i^{\mathrm{T}}\mathbf{M}\boldsymbol{\gamma}_i - \mathbf{B}_i^{\mathrm{T}}\mathbf{Q}_e = \mathbf{0} \tag{6.83}$$

which can be written as

$$\overline{\mathbf{M}}_i\ddot{\mathbf{q}}_i = \overline{\mathbf{Q}}_i \tag{6.84}$$

where $\overline{\mathbf{M}}_i$ is an $(n - n_c) \times (n - n_c)$ mass matrix associated with the independent coordinates, and $\overline{\mathbf{Q}}_i$ is the vector of generalized forces associated with the independent coordinates. This vector also contains terms that are quadratic in the first time derivatives of the coordinates. The matrix $\overline{\mathbf{M}}_i$ and the vector $\overline{\mathbf{Q}}_i$ are

$$\overline{\mathbf{M}}_i = \mathbf{B}_i^{\mathrm{T}}\mathbf{M}\mathbf{B}_i, \quad \overline{\mathbf{Q}}_i = \mathbf{B}_i^{\mathrm{T}}\mathbf{Q}_e - \mathbf{B}_i^{\mathrm{T}}\mathbf{M}\boldsymbol{\gamma}_i \tag{6.85}$$

For a given system of forces, Eq. 84 can be solved for the independent accelerations as follows:

$$\ddot{\mathbf{q}}_i = \overline{\mathbf{M}}_i^{-1}\overline{\mathbf{Q}}_i \tag{6.86}$$

For a well-posed problem with linearly independent constraint equations, the inverse of the matrix $\overline{\mathbf{M}}_i$ does exist. Equation 86 can then be used to define the independent accelerations that can be integrated forward in time in order to determine the independent coordinates and velocities. Dependent coordinates, velocities, and accelerations can be obtained by using Eqs. 64, 77, and 78, respectively. Each of these matrix equations, by providing the known independent variables, represents n_c algebraic scalar equations, which can be solved for the dependent variables.

It is important to mention that Eq. 84 can be obtained directly from Eq. 56 by using Eq. 81. The transformation of Eq. 81 automatically eliminates the constraint force vector $\mathbf{Q}_c$ of Eq. 56. Substituting Eq. 81 into Eq. 56, one obtains

$$\mathbf{M}(\mathbf{B}_i\ddot{\mathbf{q}}_i + \boldsymbol{\gamma}_i) = \mathbf{Q}_e + \mathbf{Q}_c \tag{6.87}$$

Premultiplying this equation by $\mathbf{B}_i^{\mathrm{T}}$, one obtains

$$\mathbf{B}_i^{\mathrm{T}}\mathbf{M}\mathbf{B}_i\ddot{\mathbf{q}}_i = \mathbf{B}_i^{\mathrm{T}}\mathbf{Q}_e + \mathbf{B}_i^{\mathrm{T}}\mathbf{Q}_c - \mathbf{B}_i^{\mathrm{T}}\mathbf{M}\boldsymbol{\gamma}_i \tag{6.88}$$

Since this equation is expressed in terms of the independent accelerations, one must have

$$\mathbf{B}_i^{\mathrm{T}}\mathbf{Q}_c = \mathbf{0} \tag{6.89}$$

which implies that the columns of the matrix $\mathbf{B}_i$ are orthogonal to the constraint force vector $\mathbf{Q}_c$. Consequently, the system differential equation reduces to

$$\mathbf{B}_i^{\mathrm{T}}\mathbf{M}\mathbf{B}_i\ddot{\mathbf{q}}_i = \mathbf{B}_i^{\mathrm{T}}\mathbf{Q}_e - \mathbf{B}_i^{\mathrm{T}}\mathbf{M}\boldsymbol{\gamma}_i \tag{6.90}$$

The matrices and vectors that appear in this equation are exactly the same as those of Eq. 83 which was used to define Eq. 84.

Example 6.5

For the two-link manipulator of Example 4, the revolute joint constraints at points O and A can be written as

$$\mathbf{r}_O^i = \mathbf{0}, \qquad \mathbf{r}_A^i - \mathbf{r}_A^j = \mathbf{0}$$

which can be expressed in terms of the absolute coordinates of the two links as

$$\mathbf{R}^i + \mathbf{A}^i\bar{\mathbf{u}}_O^i = \mathbf{0}, \qquad \mathbf{R}^i + \mathbf{A}^i\bar{\mathbf{u}}_A^i - \mathbf{R}^j - \mathbf{A}^j\bar{\mathbf{u}}_A^j = \mathbf{0}$$

These are four scalar constraint equations in which

$$\bar{\mathbf{u}}_O^i = [-l_O^i \quad 0]^{\mathrm{T}}, \quad \bar{\mathbf{u}}_A^i = [l_A^i \quad 0]^{\mathrm{T}}, \quad \bar{\mathbf{u}}_A^j = [-l_A^j \quad 0]^{\mathrm{T}}$$

The vector of constraints $\mathbf{C}(\mathbf{q}, t)$ can then be written as

$$\mathbf{C}(\mathbf{q}, t) = \begin{bmatrix} \mathbf{R}^i + \mathbf{A}^i\bar{\mathbf{u}}_O^i \\ \mathbf{R}^i + \mathbf{A}^i\bar{\mathbf{u}}_A^i - \mathbf{R}^j - \mathbf{A}^j\bar{\mathbf{u}}_A^j \end{bmatrix} = \mathbf{0}$$

where

$$\begin{aligned}\mathbf{q} &= [\mathbf{q}^{i^{\mathrm{T}}} \quad \mathbf{q}^{j^{\mathrm{T}}}]^{\mathrm{T}} = [\mathbf{R}^{i^{\mathrm{T}}} \quad \theta^i \quad \mathbf{R}^{j^{\mathrm{T}}} \quad \theta^j]^{\mathrm{T}} \\ &= [R_x^i \quad R_y^i \quad \theta^i \quad R_x^j \quad R_y^j \quad \theta^j]^{\mathrm{T}}\end{aligned}$$

It follows that

$$\mathbf{C_q}\,\delta\mathbf{q} = \begin{bmatrix} \delta\mathbf{R}^i + \mathbf{A}_\theta^i\bar{\mathbf{u}}_O^i\,\delta\theta^i \\ \delta\mathbf{R}^i + \mathbf{A}_\theta^i\bar{\mathbf{u}}_A^i\,\delta\theta^i - \delta\mathbf{R}^j - \mathbf{A}_\theta^j\bar{\mathbf{u}}_A^j\,\delta\theta^j \end{bmatrix}$$

$$= \begin{bmatrix} \mathbf{I} & \mathbf{A}_\theta^i\bar{\mathbf{u}}_O^i & \mathbf{0} & \mathbf{0} \\ \mathbf{I} & \mathbf{A}_\theta^i\bar{\mathbf{u}}_A^i & -\mathbf{I} & -\mathbf{A}_\theta^j\bar{\mathbf{u}}_A^j \end{bmatrix} \begin{bmatrix} \delta\mathbf{R}^i \\ \delta\theta^i \\ \delta\mathbf{R}^j \\ \delta\theta^j \end{bmatrix} = \mathbf{0}$$

Since the system has two degrees of freedom, we may select θ^i and θ^j to be the independent coordinates, that is

$$\mathbf{q}_i = [\theta^i \quad \theta^j]^{\mathrm{T}}, \qquad \mathbf{q}_d = [R_x^i \quad R_y^i \quad R_x^j \quad R_y^j]^{\mathrm{T}}$$

According to this coordinate partitioning, the preceding equation can be rewritten as

$$\mathbf{C_q}\,\delta\mathbf{q} = \mathbf{C}_{\mathbf{q}_d}\,\delta\mathbf{q}_d + \mathbf{C}_{\mathbf{q}_i}\,\delta\mathbf{q}_i$$

$$= \begin{bmatrix} \mathbf{I} & \mathbf{0} \\ \mathbf{I} & -\mathbf{I} \end{bmatrix} \begin{bmatrix} \delta\mathbf{R}^i \\ \delta\mathbf{R}^j \end{bmatrix} + \begin{bmatrix} \mathbf{A}_\theta^i\bar{\mathbf{u}}_O^i & \mathbf{0} \\ \mathbf{A}_\theta^i\bar{\mathbf{u}}_A^i & -\mathbf{A}_\theta^j\bar{\mathbf{u}}_A^j \end{bmatrix} \begin{bmatrix} \delta\theta^i \\ \delta\theta^j \end{bmatrix} = \mathbf{0}$$

where the Jacobian matrices $\mathbf{C_q}$, $\mathbf{C}_{\mathbf{q}_d}$ and $\mathbf{C}_{\mathbf{q}_i}$ can be recognized as

$$\mathbf{C_q} = \begin{bmatrix} \mathbf{I} & \mathbf{A}_\theta^i\bar{\mathbf{u}}_O^i & \mathbf{0} & \mathbf{0} \\ \mathbf{I} & \mathbf{A}_\theta^i\bar{\mathbf{u}}_A^i & -\mathbf{I} & -\mathbf{A}_\theta^j\bar{\mathbf{u}}_A^j \end{bmatrix}$$

$$\mathbf{C}_{\mathbf{q}_d} = \begin{bmatrix} \mathbf{I} & \mathbf{0} \\ \mathbf{I} & -\mathbf{I} \end{bmatrix}, \qquad \mathbf{C}_{\mathbf{q}_i} = \begin{bmatrix} \mathbf{A}_\theta^i\bar{\mathbf{u}}_O^i & \mathbf{0} \\ \mathbf{A}_\theta^i\bar{\mathbf{u}}_A^i & -\mathbf{A}_\theta^j\bar{\mathbf{u}}_A^j \end{bmatrix}$$

The matrix $\mathbf{C}_{\mathbf{q}_d}$ is nonsingular and its inverse is

$$\mathbf{C}_{\mathbf{q}_d}^{-1} = \mathbf{C}_{\mathbf{q}_d} = \begin{bmatrix} \mathbf{I} & \mathbf{0} \\ \mathbf{I} & -\mathbf{I} \end{bmatrix}$$

The matrix $\mathbf{C}_{di}$ of Eq. 71 can then be defined as

$$\mathbf{C}_{di} = -\mathbf{C}_{\mathbf{q}_d}^{-1}\mathbf{C}_{\mathbf{q}_i} = -\begin{bmatrix} \mathbf{I} & \mathbf{0} \\ \mathbf{I} & -\mathbf{I} \end{bmatrix} \begin{bmatrix} \mathbf{A}_\theta^i\bar{\mathbf{u}}_O^i & \mathbf{0} \\ \mathbf{A}_\theta^i\bar{\mathbf{u}}_A^i & -\mathbf{A}_\theta^j\bar{\mathbf{u}}_A^j \end{bmatrix}$$

$$= \begin{bmatrix} -\mathbf{A}_\theta^i\bar{\mathbf{u}}_O^i & \mathbf{0} \\ \mathbf{A}_\theta^i(\bar{\mathbf{u}}_A^i - \bar{\mathbf{u}}_O^i) & -\mathbf{A}_\theta^j\bar{\mathbf{u}}_A^j \end{bmatrix}$$

The matrix $\mathbf{C}_{di}$ is a 4×2 matrix since the system has four dependent coordinates and two independent coordinates. The virtual change of the system coordinates can be expressed in terms of the virtual changes of the independent coordinates using Eq. 73, where in this example, the matrix $\mathbf{B}_i$ of Eq. 74 is given by

$$\mathbf{B}_i = \begin{bmatrix} -\mathbf{A}_\theta^i \bar{\mathbf{u}}_O^i & \mathbf{0} \\ \mathbf{A}_\theta^i (\bar{\mathbf{u}}_A^i - \bar{\mathbf{u}}_O^i) & -\mathbf{A}_\theta^j \bar{\mathbf{u}}_A^j \\ 1 & 0 \\ 0 & 1 \end{bmatrix}$$

This is a 6×2 matrix since the system has six coordinates and only two of them are independent. The matrix $\mathbf{B}_i$ can be written more explicitly as

$$\mathbf{B}_i = \begin{bmatrix} -l_O^i \sin\theta^i & 0 \\ l_O^i \cos\theta^i & 0 \\ -(l_O^i + l_A^i)\sin\theta^i & -l_A^j \sin\theta^j \\ (l_O^i + l_A^i)\cos\theta^i & l_A^j \cos\theta^j \\ 1 & 0 \\ 0 & 1 \end{bmatrix}$$

The mass matrix of the system, which was derived in Example 4, can be rearranged according to the partitioning of the coordinates as dependent and independent. This yields

$$\mathbf{M} = \begin{bmatrix} m^i & 0 & 0 & 0 & 0 & 0 \\ 0 & m^i & 0 & 0 & 0 & 0 \\ 0 & 0 & m^j & 0 & 0 & 0 \\ 0 & 0 & 0 & m^j & 0 & 0 \\ 0 & 0 & 0 & 0 & J^i & 0 \\ 0 & 0 & 0 & 0 & 0 & J^j \end{bmatrix}$$

It follows that

$$\mathbf{M}\mathbf{B}_i = \begin{bmatrix} -m^i l_O^i \sin\theta^i & 0 \\ m^i l_O^i \cos\theta^i & 0 \\ -m^j l^i \sin\theta^i & -m^j l_A^j \sin\theta^j \\ m^j l^i \cos\theta^i & m^j l_A^j \cos\theta^j \\ J^i & 0 \\ 0 & J^j \end{bmatrix}$$

where $l^i = l_O^i + l_A^i$ is the length of link i. The mass matrix associated with the independent coordinates can be evaluated using Eq. 85 as

$$\begin{aligned} \overline{\mathbf{M}}_i &= \mathbf{B}_i^{\mathrm{T}} \mathbf{M} \mathbf{B}_i \\ &= \begin{bmatrix} m(l_O^i)^2 + m^j (l^i)^2 + J^i & m^j l^i l_A^j \cos(\theta^i - \theta^j) \\ m^j l^i l_A^j \cos(\theta^i - \theta^j) & m^j (l_A^j)^2 + J^j \end{bmatrix} \end{aligned}$$

which is a 2×2 symmetric matrix, since the system has two degrees of freedom.

The constraint force vector $\mathbf{Q}_c$ of the system discussed in this example was defined in Example 4. If the components of this vector are rearranged according to the partitioning of the coordinates as dependent and independent, one obtains

$$\mathbf{Q}_c = \begin{bmatrix} F_x^{1i} - F_x^{ij} \\ F_y^{1i} - F_y^{ij} \\ F_x^{ij} \\ F_y^{ij} \\ (F_x^{1i} \sin\theta^i - F_y^{1i} \cos\theta^i) l_O^i + (F_x^{ij} \sin\theta^i - F_y^{ij} \cos\theta^i) l_A^i \\ (F_x^{ij} \sin\theta^j - F_y^{ij} \cos\theta^j) l_A^j \end{bmatrix}$$

It is easy to verify that

$$\mathbf{B}_i^{\mathrm{T}} \mathbf{Q}_c = \mathbf{0}$$

Identification of the System Degrees of Freedom In the computer analysis of large-scale mechanical systems, numerical methods are often used to identify the system-dependent and system-independent coordinates. Based on the numerical structure of the constraint Jacobian matrix, an optimum set of independent coordinates can be identified. Recall that for a *dynamically driven system*, the Jacobian matrix is an $n_c \times n$ nonsquare matrix, where n_c is the number of constraint equations and n is the total number of the system coordinates. If the constraints are linearly independent, the Jacobian matrix has a full row rank, and Gaussian elimination can be used to identify a nonsingular $n_c \times n_c$ sub-Jacobian. For instance, applying the Gaussian elimination method with complete or full pivoting on the constraint Jacobian matrix of Eq. 66 and assuming that no zero pivots are encountered, Eq. 65, after n_c steps, can be written in the following form:

$$\begin{bmatrix} 1 & C_{12} & \cdots & C_{1n_c} & C_{1(n_c+1)} & \cdots & C_{1n} \\ 0 & 1 & \cdots & C_{2n_c} & C_{2(n_c+1)} & \cdots & C_{2n} \\ \vdots & \vdots & \ddots & \vdots & \vdots & \ddots & \vdots \\ 0 & 0 & \cdots & 1 & C_{n_c(n_c+1)} & \cdots & C_{n_c n} \end{bmatrix} \begin{bmatrix} \delta u_1 \\ \delta u_2 \\ \vdots \\ \delta u_{n_c} \\ \delta v_1 \\ \delta v_2 \\ \vdots \\ \delta v_{n-n_c} \end{bmatrix} = \begin{bmatrix} 0 \\ 0 \\ \vdots \\ 0 \\ 0 \\ 0 \\ \vdots \\ 0 \end{bmatrix} \tag{6.91}$$

In the case of full pivoting, elementary column operations are used and the elements of the vector $\delta\mathbf{q}$ are reordered accordingly. Hence, the vectors $\delta\mathbf{u} = [\delta u_1 \quad \delta u_2 \cdots \delta u_{n_c}]^{\mathrm{T}}$ and $\delta\mathbf{v} = [\delta v_1 \quad \delta v_2 \cdots \delta v_{n-n_c}]^{\mathrm{T}}$ contain elements of the vector $\delta\mathbf{q}$, which are reordered as the result of the elementary column operations. The preceding equation can be written in a partitioned form as

$$\mathbf{U}\,\delta\mathbf{u} + \mathbf{V}\,\delta\mathbf{v} = \mathbf{0} \tag{6.92}$$

where

$$\mathbf{U} = \begin{bmatrix} 1 & C_{12} & \cdots & C_{1n_c} \\ 0 & 1 & \cdots & C_{2n_c} \\ \vdots & \vdots & \ddots & \vdots \\ 0 & 0 & \cdots & 1 \end{bmatrix}, \quad \mathbf{V} = \begin{bmatrix} C_{1(n_c+1)} & \cdots & C_{1n} \\ \vdots & \ddots & \vdots \\ C_{n_c(n_c+1)} & \cdots & C_{n_c n} \end{bmatrix} \tag{6.93}$$

The matrix $\mathbf{U}$ is an upper-triangular matrix with all the diagonal elements equal to one. Because this matrix is nonsingular, the vector $\delta\mathbf{u}$ can be expressed in terms of the components of the vector $\delta\mathbf{v}$. The elements of the vector $\delta\mathbf{v}$ can then be recognized as the independent coordinates and the elements of the vector $\delta\mathbf{u}$ are recognized as the dependent coordinates, that is,

$$\delta\mathbf{q}_d = \delta\mathbf{u}, \ \delta\mathbf{q}_i = \delta\mathbf{v} \tag{6.94}$$

The Gaussian elimination method can also be used to detect *redundant constraints*. If such redundant constraints exist, the constraint equations are no longer independent and the constraint Jacobian matrix does not have a full row rank. In this case, the Gaussian elimination procedure leads to zero rows, and the number of these zero rows is equal to the number of the dependent constraint equations which is the same as the row-rank deficiency of the constraint Jacobian matrix. The redundant constraints must be eliminated in order to determine a set of linearly independent constraint equations that yield a Jacobian matrix that has a full row rank.

6.6 LAGRANGE MULTIPLIERS

The embedding technique that leads to a minimum set of strongly coupled equations has several computational disadvantages. It requires finding the inverse of the sub-Jacobian matrix associated with the dependent coordinates and it also leads to a dense and highly nonlinear generalized mass matrix. In this section, some basic concepts in the force analysis of constrained systems of rigid bodies are discussed. These concepts, which are fundamental and are widely used in classical and computational mechanics, will allow us to obtain a system of loosely coupled dynamic equations in which the coefficient matrix is sparse and symmetric. The solution of this system of equations defines the accelerations and a set of multipliers that can be used to define the constraint forces. In order to introduce these multipliers, we first consider the simple system shown in Fig. 4. The system consists of two rigid bodies, body i and body j, which are *rigidly connected* at point P. The constraint equations for this system can be expressed in terms of the Cartesian coordinates as

$$\mathbf{R}^i + \mathbf{A}^i\bar{\mathbf{u}}_P^i - \mathbf{R}^j - \mathbf{A}^j\bar{\mathbf{u}}_P^j = \mathbf{0}, \ \theta^i - \theta^j = 0 \tag{6.95}$$

where $\mathbf{R}^i$ and $\mathbf{R}^j$ are, respectively, the global position vectors of the origins of the coordinate systems of bodies i and j, $\mathbf{A}^i$ and $\mathbf{A}^j$ are, respectively, the transformation matrices from the coordinate systems of body i and body j to the global coordinate system, $\bar{\mathbf{u}}_P^i$ and $\bar{\mathbf{u}}_P^j$ are, respectively, the local position vectors of point P with respect to the reference points

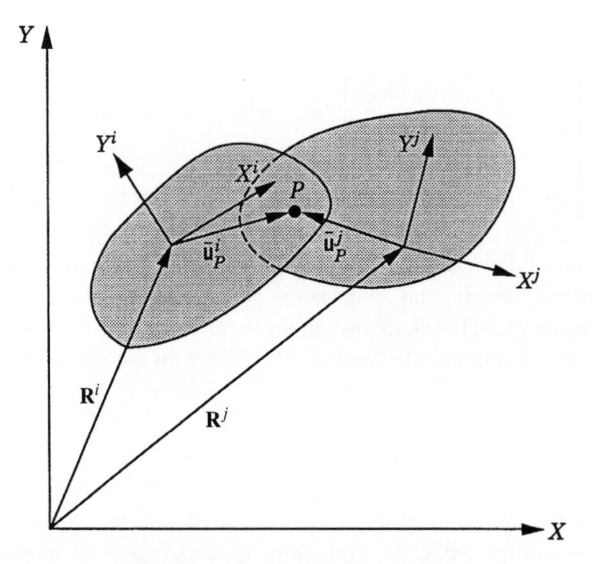

Figure 6.4 Two-body system

of body i and body j, and θ^i and θ^j are the angular orientations of bodies i and j. While Eq. 95 guarantees that there is no translational displacements between body i and body j, the second equation ensures that there is no relative rotations between the two bodies.

Equation 95 can be written as

$$\mathbf{C}(\mathbf{q}^i, \mathbf{q}^j) = \mathbf{0} \tag{6.96}$$

where $\mathbf{C}$ is the vector of constraint equations defined as

$$\mathbf{C} = \begin{bmatrix} \mathbf{R}^i + \mathbf{A}^i \bar{\mathbf{u}}_P^i - \mathbf{R}^j - \mathbf{A}^j \bar{\mathbf{u}}_P^j \\ \theta^i - \theta^j \end{bmatrix} = \mathbf{0} \tag{6.97}$$

The Jacobian matrix of these constraint equations can be written in a partitioned form as

$$\mathbf{C_q} = [\mathbf{C}_{\mathbf{q}^i} \quad \mathbf{C}_{\mathbf{q}^j}] \tag{6.98}$$

where

$$\mathbf{C}_{\mathbf{q}^i} = \begin{bmatrix} \mathbf{I} & \mathbf{A}_\theta^i \bar{\mathbf{u}}_P^i \\ 0 & 1 \end{bmatrix}, \ \mathbf{C}_{\mathbf{q}^j} = -\begin{bmatrix} \mathbf{I} & \mathbf{A}_\theta^j \bar{\mathbf{u}}_P^j \\ 0 & 1 \end{bmatrix} \tag{6.99}$$

where $\mathbf{I}$ is the 2×2 identity matrix and $\mathbf{A}_\theta^i$ and $\mathbf{A}_\theta^j$ are, respectively, the partial derivatives of the transformation matrices $\mathbf{A}^i$ and $\mathbf{A}^j$ with respect to the rotational coordinates θ^i and θ^j.

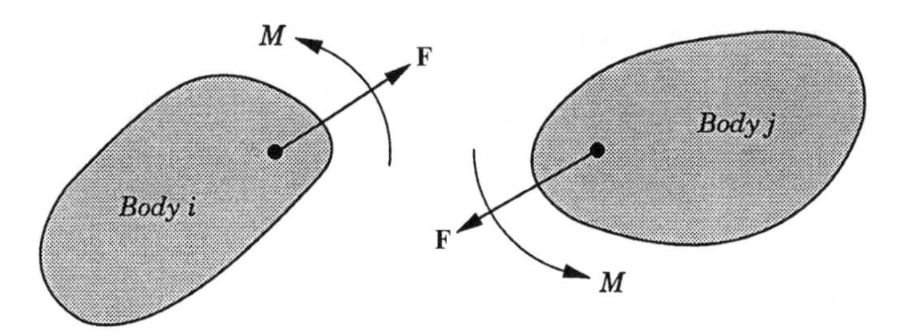

Figure 6.5 Constraint forces and moments

Equipollent Systems of Forces Figure 5 shows the actual reaction forces acting on bodies i and j of Fig. 4 as a result of the rigid connection between the two bodies. Let $\boldsymbol{\lambda}$ be the vector

$$\boldsymbol{\lambda} = -\begin{bmatrix} \mathbf{F} \\ M \end{bmatrix} \tag{6.100}$$

The reaction forces acting on body i and body j, which are equal in magnitude and opposite in direction, can be expressed, respectively, in a vector form as

$$\mathbf{F}^i = -\boldsymbol{\lambda} = \begin{bmatrix} \mathbf{F} \\ M \end{bmatrix} \quad \text{and} \quad \mathbf{F}^j = \boldsymbol{\lambda} = -\begin{bmatrix} \mathbf{F} \\ M \end{bmatrix} \tag{6.101}$$

The systems of the actual reaction forces of Eq. 101 are equipollent to other systems of generalized reaction forces defined at the origins of the coordinate systems of the two bodies and given by

$$\mathbf{Q}_c^i = \begin{bmatrix} \mathbf{F} \\ M + (\mathbf{A}^i \bar{\mathbf{u}}_P^i \times \mathbf{F}) \cdot \mathbf{k} \end{bmatrix}, \quad \mathbf{Q}_c^j = -\begin{bmatrix} \mathbf{F} \\ M + (\mathbf{A}^j \bar{\mathbf{u}}_P^j \times \mathbf{F}) \cdot \mathbf{k} \end{bmatrix} \tag{6.102}$$

Equation 102 is the result of the simple fact that a force $\mathbf{F}$ acting at a point P is equipollent to a system of forces that consists of the same force acting at another point O plus a moment defined as the cross product between the position vector of P with respect to O and the force vector $\mathbf{F}$.

Lagrange Multipliers Recall that

$$\left.\begin{aligned} [(\mathbf{A}^i \bar{\mathbf{u}}_P^i) \times \mathbf{F}] \cdot \mathbf{k} &= \bar{\mathbf{u}}_P^{i\mathrm{T}} \mathbf{A}_\theta^{i\mathrm{T}} \mathbf{F} \\ [(\mathbf{A}^j \bar{\mathbf{u}}_P^j) \times \mathbf{F}] \cdot \mathbf{k} &= \bar{\mathbf{u}}_P^{j\mathrm{T}} \mathbf{A}_\theta^{j\mathrm{T}} \mathbf{F} \end{aligned}\right\} \tag{6.103}$$

Using these identities and Eq. 102, one obtains

$$\mathbf{Q}_c^i = \begin{bmatrix} \mathbf{F} \\ M + \bar{\mathbf{u}}_P^{i\mathrm{T}} \mathbf{A}_\theta^{i\mathrm{T}} \mathbf{F} \end{bmatrix}, \quad \mathbf{Q}_c^j = -\begin{bmatrix} \mathbf{F} \\ M + \bar{\mathbf{u}}_P^{j\mathrm{T}} \mathbf{A}_\theta^{j\mathrm{T}} \mathbf{F} \end{bmatrix} \tag{6.104}$$

which can be written using matrix notation as

$$\mathbf{Q}_c^i = \begin{bmatrix} \mathbf{I} & 0 \\ \bar{\mathbf{u}}_P^{i\mathrm{T}} \mathbf{A}_\theta^{i\mathrm{T}} & 1 \end{bmatrix} \begin{bmatrix} \mathbf{F} \\ M \end{bmatrix}, \quad \mathbf{Q}_c^j = -\begin{bmatrix} \mathbf{I} & 0 \\ \bar{\mathbf{u}}_P^{j\mathrm{T}} \mathbf{A}_\theta^{j\mathrm{T}} & 1 \end{bmatrix} \begin{bmatrix} \mathbf{F} \\ M \end{bmatrix} \tag{6.105}$$

Comparing the matrices in these equations with the Jacobian matrices of Eq. 99, one concludes that the generalized reactions can be expressed in this example in terms of the actual reactions as

$$\mathbf{Q}_c^i = -\mathbf{C}_{\mathbf{q}^i}^{\mathrm{T}} \boldsymbol{\lambda}, \quad \mathbf{Q}_c^j = -\mathbf{C}_{\mathbf{q}^j}^{\mathrm{T}} \boldsymbol{\lambda} \tag{6.106}$$

Each equation of Eq. 106 contains three force components; two force components associated with the translation of the reference point and one component associated with the rotation of the body. Equation 106 implies that the generalized reaction forces as the result of the rigid connection between bodies i and j can be expressed in terms of the Jacobian matrices of the kinematic constraint equations. In this example, the vector $\boldsymbol{\lambda}$ was found to be the negative of the vector that contains the reaction force and moment. The vector $\boldsymbol{\lambda}$, whose dimension is equal to the number of constraint equations, is called the vector of *Lagrange multipliers*. While the form of Eq. 106 is valid for all types of constraints, the vector of Lagrange multipliers may not, in some applications, take the simple form of Eq. 100.

Example 6.6

Find the generalized reaction forces associated with the Cartesian coordinates of two bodies i and j connected by a revolute joint in terms of Lagrange multipliers.

Solution. The revolute joint between two bodies i and j allows only relative rotation. The constraint equations for this joint are

$$\mathbf{R}^i + \mathbf{A}^i \bar{\mathbf{u}}_P^i - \mathbf{R}^j - \mathbf{A}^j \bar{\mathbf{u}}_P^j = \mathbf{0}$$

where $\bar{\mathbf{u}}_P^i$ and $\bar{\mathbf{u}}_P^j$ are, respectively, the local position vectors of the joint definition points. The Jacobian matrix of the revolute joint constraints can be written as

$$\mathbf{C}_\mathbf{q} = [\mathbf{I} \quad \mathbf{A}_\theta^i \bar{\mathbf{u}}_P^i \quad -\mathbf{I} \quad -\mathbf{A}_\theta^j \bar{\mathbf{u}}_P^j]$$

which can be written as

$$\mathbf{C}_\mathbf{q} = [\mathbf{C}_{\mathbf{q}^i} \quad \mathbf{C}_{\mathbf{q}^j}]$$

in which

$$\mathbf{C}_{\mathbf{q}^i} = [\mathbf{I} \quad \mathbf{A}_\theta^i \bar{\mathbf{u}}_P^i], \quad \mathbf{C}_{\mathbf{q}^j} = [-\mathbf{I} \quad -\mathbf{A}_\theta^j \bar{\mathbf{u}}_P^j]$$

The generalized constraint reactions of the revolute joint associated with the translation of the reference points and the rotations of the bodies i and j are

$$\mathbf{Q}_c^i = -\mathbf{C}_{\mathbf{q}^i}^{\mathrm{T}} \boldsymbol{\lambda}, \quad \mathbf{Q}_c^j = -\mathbf{C}_{\mathbf{q}^j}^{\mathrm{T}} \boldsymbol{\lambda}$$

which can be written explicitly as

$$\mathbf{Q}_c^i = -\begin{bmatrix} 1 & 0 \\ 0 & 1 \\ -\bar{x}_P^i \sin\theta^i - \bar{y}_P^i \cos\theta^i & \bar{x}_P^i \cos\theta^i - \bar{y}_P^i \sin\theta^i \end{bmatrix} \begin{bmatrix} \lambda_1 \\ \lambda_2 \end{bmatrix}$$

$$= \begin{bmatrix} -\lambda_1 \\ -\lambda_2 \\ (\bar{x}_P^i \sin\theta^i + \bar{y}_P^i \cos\theta^i)\lambda_1 - (\bar{x}_P^i \cos\theta^i - \bar{y}_P^i \sin\theta^i)\lambda_2 \end{bmatrix}$$

$$\mathbf{Q}_c^j = -\begin{bmatrix} -1 & 0 \\ 0 & -1 \\ \bar{x}_P^j \sin\theta^j + \bar{y}_P^j \cos\theta^j & -\bar{x}_P^j \cos\theta^j + \bar{y}_P^j \sin\theta^j \end{bmatrix} \begin{bmatrix} \lambda_1 \\ \lambda_2 \end{bmatrix}$$

$$= \begin{bmatrix} \lambda_1 \\ \lambda_2 \\ -(\bar{x}_P^j \sin\theta^j + \bar{y}_P^j \cos\theta^j)\lambda_1 + (\bar{x}_P^j \cos\theta^j - \bar{y}_P^j \sin\theta^j)\lambda_2 \end{bmatrix}$$

where

$$\bar{\mathbf{u}}_P^i = [\bar{x}_P^i \quad \bar{y}_P^i]^{\mathrm{T}}, \quad \bar{\mathbf{u}}_P^j = [\bar{x}_P^j \quad \bar{y}_P^j]^{\mathrm{T}}$$

Example 6.7

A link in a mechanism is assumed to be fixed if it is subjected to the *ground constraints*, which do not allow the translational and rotational displacements of the link. These *ground constraints* for link i are given by

$$\mathbf{R}^i = \mathbf{C}_1, \quad \theta^i = C_2$$

where $\mathbf{C}_1$ and C_2 are, respectively, a constant vector and a constant scalar. The Jacobian matrix of the ground constraints is

$$\mathbf{C}_{\mathbf{q}^i} = \begin{bmatrix} \mathbf{I} & 0 \\ 0 & 1 \end{bmatrix} = \begin{bmatrix} 1 & 0 & 0 \\ 0 & 1 & 0 \\ 0 & 0 & 1 \end{bmatrix}$$

The generalized reactions as the result of imposing the ground constraints are

$$\mathbf{Q}_c^i = -\mathbf{C}_{\mathbf{q}^i}^{\mathrm{T}} \boldsymbol{\lambda} = -\begin{bmatrix} 1 & 0 & 0 \\ 0 & 1 & 0 \\ 0 & 0 & 1 \end{bmatrix} \begin{bmatrix} \lambda_1 \\ \lambda_2 \\ \lambda_3 \end{bmatrix} = -\begin{bmatrix} \lambda_1 \\ \lambda_2 \\ \lambda_3 \end{bmatrix}$$

Multiple Joints A body in a multibody system may be connected to other bodies by several joints. For example, in multibody vehicle systems the chassis of the vehicle is connected to the suspension elements by different types of joints. If a body in the mechanical system is connected with other bodies by more than one joint, one can use the procedure previously described in this section to define the contribution of each joint to the generalized reaction forces. For example, let body i be connected to other bodies in the system by n_i joints. The constraint equations that describe these joints can be expressed in vector forms as

$$\left.\begin{aligned} \mathbf{C}_1(\mathbf{q}, t) &= \mathbf{0} \\ \mathbf{C}_2(\mathbf{q}, t) &= \mathbf{0} \\ &\vdots \\ \mathbf{C}_{n_i}(\mathbf{q}, t) &= \mathbf{0} \end{aligned}\right\} \tag{6.107}$$

where $\mathbf{q} = [\mathbf{q}^{1^{\mathrm{T}}} \quad \mathbf{q}^{2^{\mathrm{T}}} \cdots \mathbf{q}^{n_b^{\mathrm{T}}}]^{\mathrm{T}}$ is the total vector of the system coordinates, t is time, and n_b is the total number of bodies in the system. Each of the vector equations in Eq. 107, which contains a number of scalar equations equal to the number of degrees of freedom eliminated by the corresponding joint, contributes to the vector of generalized forces of body i. By using Eq. 106, the generalized reactions of these constraint equations are

$$\left.\begin{aligned} \mathbf{Q}_1^i &= -(\mathbf{C}_1)_{\mathbf{q}^i}^{\mathrm{T}} \boldsymbol{\lambda}_1 \\ \mathbf{Q}_2^i &= -(\mathbf{C}_2)_{\mathbf{q}^i}^{\mathrm{T}} \boldsymbol{\lambda}_2 \\ &\vdots \\ \mathbf{Q}_{n_i}^i &= -(\mathbf{C}_{n_i})_{\mathbf{q}^i}^{\mathrm{T}} \boldsymbol{\lambda}_{n_i} \end{aligned}\right\} \tag{6.108}$$

where $\boldsymbol{\lambda}_k$ is the vector of Lagrange multipliers associated with the vector of constraints $\mathbf{C}_k$. The resultant generalized reaction forces due to all the constraints of Eq. 107 can be written as

$$\mathbf{Q}_c^i = \mathbf{Q}_1^i + \mathbf{Q}_2^i + \cdots + \mathbf{Q}_{n_i}^i = \sum_{k=1}^{n_i} \mathbf{Q}_k^i \tag{6.109}$$

which upon using Eq. 108, yields

$$\begin{aligned} \mathbf{Q}_c^i &= -(\mathbf{C}_1)_{\mathbf{q}^i}^{\mathrm{T}} \boldsymbol{\lambda}_1 - (\mathbf{C}_2)_{\mathbf{q}^i}^{\mathrm{T}} \boldsymbol{\lambda}_2 - \cdots - (\mathbf{C}_{n_i})_{\mathbf{q}^i}^{\mathrm{T}} \boldsymbol{\lambda}_{n_i} \\ &= -\sum_{k=1}^{n_i} (\mathbf{C}_k)_{\mathbf{q}^i}^{\mathrm{T}} \boldsymbol{\lambda}_k \end{aligned} \tag{6.110}$$

Equation 110 can also be written in a matrix form as

$$\mathbf{Q}_c^i = -[(\mathbf{C}_1)_{\mathbf{q}^i}^{\mathrm{T}} \quad (\mathbf{C}_2)_{\mathbf{q}^i}^{\mathrm{T}} \cdots (\mathbf{C}_{n_i})_{\mathbf{q}^i}^{\mathrm{T}}] \begin{bmatrix} \boldsymbol{\lambda}_1 \\ \boldsymbol{\lambda}_2 \\ \vdots \\ \boldsymbol{\lambda}_{n_i} \end{bmatrix} \tag{6.111}$$

It is clear that there will be no contribution from other joints that do not involve body i since the constraint equations that describe these joints are not explicit functions of the coordinates of body i. Equation 111 can then be written in a more general form as

$$\mathbf{Q}_c^i = -\mathbf{C}_{\mathbf{q}^i}^{\mathrm{T}} \boldsymbol{\lambda} \tag{6.112}$$

where $\mathbf{C}$ is the total vector of constraint equations of the system, and $\boldsymbol{\lambda}$ is the vector of the system Lagrange multipliers. Constraints which are not explicit functions of the coordinates of body i define zero rows in the Jacobian matrix $\mathbf{C}_{\mathbf{q}}^i$ of Eq. 112.

Equation 112 can be used to define the total vector of the system-generalized reactions. If the system consists of n_b bodies, one has

$$\mathbf{Q}_c = [\mathbf{Q}_c^{1^{\mathrm{T}}} \quad \mathbf{Q}_c^{2^{\mathrm{T}}} \quad \cdots \quad \mathbf{Q}_c^{n_b^{\mathrm{T}}}]^{\mathrm{T}} \tag{6.113}$$

which upon using Eq. 112, leads to

$$\mathbf{Q}_c = -\begin{bmatrix} \mathbf{C}_{\mathbf{q}^1}^{\mathrm{T}} \boldsymbol{\lambda} \\ \mathbf{C}_{\mathbf{q}^2}^{\mathrm{T}} \boldsymbol{\lambda} \\ \vdots \\ \mathbf{C}_{\mathbf{q}^{n_b}}^{\mathrm{T}} \boldsymbol{\lambda} \end{bmatrix} \tag{6.114}$$

By factoring out the vector of Lagrange multipliers $\boldsymbol{\lambda}$ and keeping in mind that

$$\mathbf{C}_{\mathbf{q}} = [\mathbf{C}_{\mathbf{q}^1} \quad \mathbf{C}_{\mathbf{q}^2} \quad \cdots \quad \mathbf{C}_{\mathbf{q}^{n_b}}] \tag{6.115}$$

where $\mathbf{C}_{\mathbf{q}}$ is the *constraint Jacobian matrix* of the system, Eq. 114 can be written as

$$\mathbf{Q}_c = -[\mathbf{C}_{\mathbf{q}^1} \quad \mathbf{C}_{\mathbf{q}^2} \quad \cdots \quad \mathbf{C}_{\mathbf{q}^{n_b}}]^{\mathrm{T}} \boldsymbol{\lambda} = -\mathbf{C}_{\mathbf{q}}^{\mathrm{T}} \boldsymbol{\lambda} \tag{6.116}$$

Since each joint is formulated in terms of the coordinates of the two bodies connected by this joint, the Jacobian matrix $\mathbf{C}_{\mathbf{q}}$ in a large-scale constrained mechanical system is a *sparse matrix*, which has a large number of zero entries.

Example 6.8

In order to demonstrate the use of Eq. 116, the slider crank mechanism shown in Fig. 6 is considered. The configuration of this mechanism can be identified using 12 Cartesian coordinates. There are, however, 11 constraint equations that describe the joints in the system. These constraint equations are the three ground constraints, the two revolute joint constraints at O, the two revolute joint constraints at A, the two revolute joint constraints at B, and the

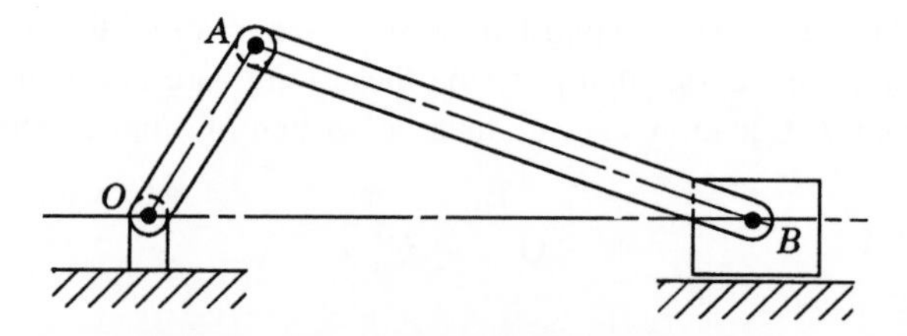

Figure 6.6 Slider crank mechanism

two constraints that allow only the translation of the slider block in the horizontal direction. The Jacobian matrix of the constraint equations of this mechanism was developed in Chapter 3 and is given by

$$\mathbf{C_q} = \begin{bmatrix} 1 & 0 & 0 & 0 & 0 & 0 & 0 & 0 & 0 & 0 & 0 & 0 \\ 0 & 1 & 0 & 0 & 0 & 0 & 0 & 0 & 0 & 0 & 0 & 0 \\ 0 & 0 & 1 & 0 & 0 & 0 & 0 & 0 & 0 & 0 & 0 & 0 \\ 0 & 0 & 0 & 1 & 0 & C_{4,6} & 0 & 0 & 0 & 0 & 0 & 0 \\ 0 & 0 & 0 & 0 & 1 & C_{5,6} & 0 & 0 & 0 & 0 & 0 & 0 \\ 0 & 0 & 0 & 1 & 0 & C_{6,6} & -1 & 0 & C_{6,9} & 0 & 0 & 0 \\ 0 & 0 & 0 & 0 & 1 & C_{7,6} & 0 & -1 & C_{7,9} & 0 & 0 & 0 \\ 0 & 0 & 0 & 0 & 0 & 0 & 1 & 0 & C_{8,9} & -1 & 0 & 0 \\ 0 & 0 & 0 & 0 & 0 & 0 & 0 & 1 & C_{9,9} & 0 & -1 & 0 \\ 0 & 0 & 0 & 0 & 0 & 0 & 0 & 0 & 0 & 0 & 1 & 0 \\ 0 & 0 & 0 & 0 & 0 & 0 & 0 & 0 & 0 & 0 & 0 & 1 \end{bmatrix}$$

where

$$C_{4,6} = \frac{l^2}{2} \sin\theta^2, \quad C_{5,6} = -\frac{l^2}{2} \cos\theta^2, \quad C_{6,6} = -\frac{l^2}{2} \sin\theta^2, \quad C_{7,6} = \frac{l^2}{2} \cos\theta^2$$

$$C_{6,9} = -\frac{l^3}{2} \sin\theta^3, \quad C_{7,9} = \frac{l^3}{2} \cos\theta^3, \quad C_{8,9} = -\frac{l^3}{2} \sin\theta^3, \quad C_{9,9} = \frac{l^3}{2} \cos\theta^3$$

The vector of system generalized reaction forces is

$$\mathbf{Q}_c = \begin{bmatrix} \mathbf{Q}_c^1 \\ \mathbf{Q}_c^2 \\ \mathbf{Q}_c^3 \\ \mathbf{Q}_c^4 \end{bmatrix} = -\mathbf{C}_\mathbf{q}^{\mathrm{T}} \boldsymbol{\lambda}$$

where

$$\boldsymbol{\lambda} = [\lambda_1 \quad \lambda_2 \quad \lambda_3 \quad \lambda_4 \quad \lambda_5 \quad \lambda_6 \quad \lambda_7 \quad \lambda_8 \quad \lambda_9 \quad \lambda_{10} \quad \lambda_{11}]^{\mathrm{T}}$$

The vector $\mathbf{Q}_c$ is given explicitly in terms of Lagrange multipliers by

$$\mathbf{Q}_c = \begin{bmatrix} -\lambda_1 \\ -\lambda_2 \\ -\lambda_3 \\ -(\lambda_4 + \lambda_6) \\ -(\lambda_5 + \lambda_7) \\ \dfrac{l^2}{2}[-(\lambda_4 - \lambda_6)\sin\theta^2 + (\lambda_5 - \lambda_7)\cos\theta^2] \\ \lambda_6 - \lambda_8 \\ \lambda_7 - \lambda_9 \\ \dfrac{l^3}{2}\,[(\lambda_6 + \lambda_8)\sin\theta^3 - (\lambda_7 + \lambda_9)\cos\theta^3] \\ \lambda_8 \\ \lambda_9 - \lambda_{10} \\ \lambda_{11} \end{bmatrix}$$

6.7 CONSTRAINED DYNAMIC EQUATIONS

The use of the generalized coordinate partitioning of the constraint Jacobian matrix to develop a minimum number of differential equations that govern the motion of the multibody system was demonstrated in Section 5. These equations are expressed in terms of the independent accelerations and, therefore, the constraint forces are automatically eliminated. An alternative approach for formulating the dynamic equations of the multibody systems is to use redundant coordinates which are related by the virtue of the kinematic constraints. This approach, in which the constraint forces appear in the final form of the equations of motion, leads to a larger system of loosely coupled equations which can be solved using sparse matrix techniques.

In this section, we discuss the *augmented formulation* in which the kinematic constraint equations are adjoined to the systems differential equations using the vector of Lagrange multipliers. This approach leads to a system of algebraic equations with a symmetric positive-definite coefficient matrix. This system can be solved for the accelerations and Lagrange multipliers. Lagrange multipliers can be used to determine the generalized reactions as discussed in the preceding section, while the accelerations can be integrated in order to determine the system coordinates and velocities.

In terms of the absolute Cartesian coordinates, the motion of a rigid body i in the multibody system is governed by Eq. 51, which is repeated here as

$$\mathbf{M}^i\ddot{\mathbf{q}}^i = \mathbf{Q}_e^i + \mathbf{Q}_c^i \tag{6.117}$$

where $\mathbf{M}^i$ is the mass matrix of the rigid body i, $\ddot{\mathbf{q}}^i = [\ddot{\mathbf{R}}^{i\mathrm{T}} \quad \ddot{\theta}^i]^{\mathrm{T}}$ is the acceleration vector, $\mathbf{Q}_e^i$ is the vector of generalized external forces, and $\mathbf{Q}_c^i$ is the vector of generalized constraint forces. If the reference point is taken to be the center of mass of the body, the mass matrix is diagonal and is given by

$$\mathbf{M}^i = \begin{bmatrix} m^i\mathbf{I} & \mathbf{0} \\ \mathbf{0} & J^i \end{bmatrix} \tag{6.118}$$

where m^i is the mass of the body, and J^i is the mass moment of inertia defined with respect to the body center of mass.

If the system consists of n_b interconnected bodies, a matrix equation similar to Eq. 117 can be developed for each body in the system. This leads to

$$\left.\begin{aligned} \mathbf{M}^1\ddot{\mathbf{q}}^1 &= \mathbf{Q}_e^1 + \mathbf{Q}_c^1 \\ \mathbf{M}^2\ddot{\mathbf{q}}^2 &= \mathbf{Q}_e^2 + \mathbf{Q}_c^2 \\ &\vdots \\ \mathbf{M}^i\ddot{\mathbf{q}}^i &= \mathbf{Q}_e^i + \mathbf{Q}_c^i \\ &\vdots \\ \mathbf{M}^{n_b}\ddot{\mathbf{q}}^{n_b} &= \mathbf{Q}_e^{n_b} + \mathbf{Q}_c^{n_b} \end{aligned}\right\} \tag{6.119}$$

These equations can be combined in one matrix equation as

$$\begin{bmatrix} \mathbf{M}^1 & & & & & \\ & \mathbf{M}^2 & & \mathbf{0} & & \\ & & \ddots & & & \\ & & & \mathbf{M}^i & & \\ & \mathbf{0} & & & \ddots & \\ & & & & & \mathbf{M}^{n_b} \end{bmatrix} \begin{bmatrix} \ddot{\mathbf{q}}^1 \\ \ddot{\mathbf{q}}^2 \\ \vdots \\ \ddot{\mathbf{q}}^i \\ \vdots \\ \ddot{\mathbf{q}}^{n_b} \end{bmatrix} = \begin{bmatrix} \mathbf{Q}_e^i \\ \mathbf{Q}_e^2 \\ \vdots \\ \mathbf{Q}_e^i \\ \vdots \\ \mathbf{Q}_e^{n_b} \end{bmatrix} + \begin{bmatrix} \mathbf{Q}_c^1 \\ \mathbf{Q}_c^2 \\ \vdots \\ \mathbf{Q}_c^i \\ \vdots \\ \mathbf{Q}_c^{n_b} \end{bmatrix} \tag{6.120}$$

which can be written as

$$\mathbf{M}\ddot{\mathbf{q}} = \mathbf{Q}_e + \mathbf{Q}_c \tag{6.121}$$

where

$$\left.\begin{aligned} \mathbf{M} &= \begin{bmatrix} \mathbf{M}^1 & & & & & \\ & \mathbf{M}^2 & & \mathbf{0} & & \\ & & \ddots & & & \\ & & & \mathbf{M}^i & & \\ & \mathbf{0} & & & \ddots & \\ & & & & & \mathbf{M}^{n_b} \end{bmatrix} \\ \mathbf{q} &= [\mathbf{q}^{1^{\mathrm{T}}} \quad \mathbf{q}^{2^{\mathrm{T}}} \cdots \mathbf{q}^{i^{\mathrm{T}}} \cdots \mathbf{q}^{n_b^{\mathrm{T}}}]^{\mathrm{T}} \\ \mathbf{Q}_e &= [\mathbf{Q}_e^{1^{\mathrm{T}}} \quad \mathbf{Q}_e^{2^{\mathrm{T}}} \cdots \mathbf{Q}_e^{i^{\mathrm{T}}} \cdots \mathbf{Q}_e^{n_b^{\mathrm{T}}}]^{\mathrm{T}} \\ \mathbf{Q}_c &= [\mathbf{Q}_c^{1^{\mathrm{T}}} \quad \mathbf{Q}_c^{2^{\mathrm{T}}} \cdots \mathbf{Q}_c^{i^{\mathrm{T}}} \cdots \mathbf{Q}_c^{n_b^{\mathrm{T}}}]^{\mathrm{T}} \end{aligned}\right\} \tag{6.122}$$

in which $\mathbf{M}$ is the system mass matrix, $\mathbf{q}$ is the total vector of system-generalized coordinates, $\mathbf{Q}_e$ is the vector of system-generalized external forces, and $\mathbf{Q}_c$ is the vector of the system-generalized constraint forces. It was shown in the preceding section that the vector $\mathbf{Q}_c$ of the system-generalized constraint forces can be written in terms of the system constraint Jacobian matrix and the vector of Lagrange multipliers as

$$\mathbf{Q}_c = -\mathbf{C}_{\mathbf{q}}^{\mathrm{T}}\boldsymbol{\lambda} \tag{6.123}$$

where $\mathbf{C}_{\mathbf{q}}$ is the constraint Jacobian matrix, $\boldsymbol{\lambda}$ is the n_c-dimensional vector of Lagrange multipliers, and n_c is the number of constraint equations. Substituting Eq. 123 into Eq. 121, one obtains $\mathbf{M}\ddot{\mathbf{q}} = \mathbf{Q}_e - \mathbf{C}_{\mathbf{q}}^{\mathrm{T}}\boldsymbol{\lambda}$ or equivalently,

$$\mathbf{M}\ddot{\mathbf{q}} + \mathbf{C}_{\mathbf{q}}^{\mathrm{T}}\boldsymbol{\lambda} = \mathbf{Q}_e \tag{6.124}$$

These are n second-order differential equations of motion, where n is total number of system coordinates.

Theoretical Proof In the preceding section, we used a simple example to introduce the technique of Lagrange multipliers. Before we start our discussion on the solution of the equations of motion, it may be helpful to provide the general theoretical derivation of Eq. 124 in order to demonstrate the generality of the technique of Lagrange multipliers. We make use of Eqs. 63 and 65, which are given, respectively, by

$$\delta\mathbf{q}^{\mathrm{T}}[\mathbf{M}\ddot{\mathbf{q}} - \mathbf{Q}_e] = 0 \tag{6.125}$$

and

$$\mathbf{C}_{\mathbf{q}}\,\delta\mathbf{q} = \mathbf{0} \tag{6.126}$$

Since the vector $\mathbf{C}_{\mathbf{q}}\,\delta\mathbf{q}$ is equal to zero from the second equation, the scalar product of this vector with any other vector is also equal to zero. Hence, the second equation yields

$$\boldsymbol{\lambda}^{\mathrm{T}}\mathbf{C}_{\mathbf{q}}\,\delta\mathbf{q} = 0 \tag{6.127}$$

for an arbitrary vector $\boldsymbol{\lambda}$, which has a dimension equal to the number of the constraint equations. When this equation is added to Eq. 125, one obtains

$$\delta\mathbf{q}^{\mathrm{T}}[\mathbf{M}\ddot{\mathbf{q}} - \mathbf{Q}_e + \mathbf{C}_{\mathbf{q}}^{\mathrm{T}}\boldsymbol{\lambda}] = 0 \tag{6.128}$$

The coefficients of the elements of the vector $\delta\mathbf{q}$ in this equation cannot be set equal to zero because the coordinates are not independent. Using the coordinate partitioning of Eq. 67, the preceding equation yields

$$[\delta\mathbf{q}_i^{\mathrm{T}} \quad \delta\mathbf{q}_d^{\mathrm{T}}]\left\{\begin{bmatrix}\mathbf{M}_{ii} & \mathbf{M}_{id}\\ \mathbf{M}_{di} & \mathbf{M}_{dd}\end{bmatrix}\begin{bmatrix}\ddot{\mathbf{q}}_i\\ \ddot{\mathbf{q}}_d\end{bmatrix} - \begin{bmatrix}\mathbf{Q}_{e_i}\\ \mathbf{Q}_{e_d}\end{bmatrix} + \begin{bmatrix}\mathbf{C}_{\mathbf{q}_i}^{\mathrm{T}}\\ \mathbf{C}_{\mathbf{q}_d}^{\mathrm{T}}\end{bmatrix}\boldsymbol{\lambda}\right\} = 0 \tag{6.129}$$

where the subscripts i and d refer, respectively, to independent and dependent coordinates. It follows that

$$\delta \mathbf{q}_i^{\mathrm{T}}[\mathbf{M}_{ii}\ddot{\mathbf{q}}_i + \mathbf{M}_{id}\ddot{\mathbf{q}}_d - \mathbf{Q}_{e_i} + \mathbf{C}_{\mathbf{q}_i}^{\mathrm{T}}\boldsymbol{\lambda}] = 0 \tag{6.130}$$

and

$$\delta \mathbf{q}_d^{\mathrm{T}}[\mathbf{M}_{di}\ddot{\mathbf{q}}_i + \mathbf{M}_{dd}\ddot{\mathbf{q}}_d - \mathbf{Q}_{e_d} + \mathbf{C}_{\mathbf{q}_d}^{\mathrm{T}}\boldsymbol{\lambda}] = 0 \tag{6.131}$$

As previously pointed out, the independent coordinates can be selected such that the matrix $\mathbf{C}_{\mathbf{q}_d}^{\mathrm{T}}$ is a square nonsingular matrix. The vector of Lagrange multipliers can then be selected to be the unique solution of the following system of algebraic equations:

$$\mathbf{C}_{\mathbf{q}_d}^{\mathrm{T}}\boldsymbol{\lambda} = \mathbf{Q}_{e_d} - \mathbf{M}_{di}\ddot{\mathbf{q}}_i - \mathbf{M}_{dd}\ddot{\mathbf{q}}_d \tag{6.132}$$

This choice of Lagrange multipliers guarantees that the coefficients of the elements of the vector $\delta\mathbf{q}_d$ in Eq. 131 are equal to zero. Furthermore, since in Eq. 130 the elements of the vector $\delta\mathbf{q}_i$ are independent, one has

$$\mathbf{M}_{ii}\ddot{\mathbf{q}}_i + \mathbf{M}_{id}\ddot{\mathbf{q}}_d + \mathbf{C}_{\mathbf{q}_i}^{\mathrm{T}}\boldsymbol{\lambda} = \mathbf{Q}_{e_i} \tag{6.133}$$

Combining the preceding two equations, one obtains

$$\begin{bmatrix} \mathbf{M}_{ii} & \mathbf{M}_{id} \\ \mathbf{M}_{di} & \mathbf{M}_{dd} \end{bmatrix}\begin{bmatrix} \ddot{\mathbf{q}}_i \\ \ddot{\mathbf{q}}_d \end{bmatrix} + \begin{bmatrix} \mathbf{C}_{\mathbf{q}_i}^{\mathrm{T}} \\ \mathbf{C}_{\mathbf{q}_d}^{\mathrm{T}} \end{bmatrix}\boldsymbol{\lambda} = \begin{bmatrix} \mathbf{Q}_{e_i} \\ \mathbf{Q}_{e_d} \end{bmatrix} \tag{6.134}$$

which leads to the general form of Eq. 124, $\mathbf{M}\ddot{\mathbf{q}} + \mathbf{C}_{\mathbf{q}}^{\mathrm{T}}\boldsymbol{\lambda} = \mathbf{Q}_e$.

Augmented Formulation If the vector of generalized external forces $\mathbf{Q}_e$ is known, the unknowns in Eq. 124 are the vector of accelerations $\ddot{\mathbf{q}}$ and the vector of Lagrange multipliers $\boldsymbol{\lambda}$. The number of unknowns in this case is $n + n_c$. Because Eq. 124 contains only n equations, in order to be able to solve this system one needs to have additional n_c equations. These n_c equations are the nonlinear algebraic constraint equations that represent the joints and the specified motion trajectories. These constraint equations, as previously shown, can be written as

$$\mathbf{C}(\mathbf{q}, t) = \mathbf{0} \tag{6.135}$$

which upon differentiation once and twice with respect to time, one obtains

$$\mathbf{C}_{\mathbf{q}}\dot{\mathbf{q}} = -\mathbf{C}_t, \quad \mathbf{C}_{\mathbf{q}}\ddot{\mathbf{q}} = \mathbf{Q}_d \tag{6.136}$$

where the subscript t denotes partial differentiation with respect to time and $\mathbf{Q}_d$ is a vector that absorbs first derivatives in the coordinates and is given by

$$\mathbf{Q}_d = -\mathbf{C}_{tt} - (\mathbf{C}_{\mathbf{q}}\dot{\mathbf{q}})_{\mathbf{q}}\dot{\mathbf{q}} - 2\mathbf{C}_{\mathbf{q}t}\dot{\mathbf{q}} \tag{6.137}$$

Equation 124 can be combined with the second equation of Eq. 136 in one matrix equation as

$$\begin{bmatrix} \mathbf{M} & \mathbf{C}_{\mathbf{q}}^{\mathrm{T}} \\ \mathbf{C}_{\mathbf{q}} & \mathbf{0} \end{bmatrix} \begin{bmatrix} \ddot{\mathbf{q}} \\ \boldsymbol{\lambda} \end{bmatrix} = \begin{bmatrix} \mathbf{Q}_e \\ \mathbf{Q}_d \end{bmatrix} \tag{6.138}$$

The vectors of accelerations and Lagrange multipliers can be obtained by solving Eq. 138 as

$$\begin{bmatrix} \ddot{\mathbf{q}} \\ \boldsymbol{\lambda} \end{bmatrix} = \begin{bmatrix} \mathbf{M} & \mathbf{C}_{\mathbf{q}}^{\mathrm{T}} \\ \mathbf{C}_{\mathbf{q}} & \mathbf{0} \end{bmatrix}^{-1} \begin{bmatrix} \mathbf{Q}_e \\ \mathbf{Q}_d \end{bmatrix} \tag{6.139}$$

By direct matrix multiplication, one can verify that

$$\begin{bmatrix} \mathbf{M} & \mathbf{C}_{\mathbf{q}}^{\mathrm{T}} \\ \mathbf{C}_{\mathbf{q}} & \mathbf{0} \end{bmatrix}^{-1} = \begin{bmatrix} \mathbf{H}_{qq} & \mathbf{H}_{q\lambda} \\ \mathbf{H}_{\lambda q} & \mathbf{H}_{\lambda\lambda} \end{bmatrix} \tag{6.140}$$

where

$$\left.\begin{aligned} \mathbf{H}_{\lambda\lambda} &= (\mathbf{C}_{\mathbf{q}}\mathbf{M}^{-1}\mathbf{C}_{\mathbf{q}}^{\mathrm{T}})^{-1} \\ \mathbf{H}_{qq} &= \mathbf{M}^{-1} + \mathbf{M}^{-1}\mathbf{C}_{\mathbf{q}}^{\mathrm{T}}\mathbf{H}_{\lambda\lambda}\mathbf{C}_{\mathbf{q}}\mathbf{M}^{-1} \\ \mathbf{H}_{q\lambda} &= \mathbf{H}_{\lambda q}^{\mathrm{T}} = -\mathbf{M}^{-1}\mathbf{C}_{\mathbf{q}}^{\mathrm{T}}\mathbf{H}_{\lambda\lambda} \end{aligned}\right\} \tag{6.141}$$

Substituting Eq. 140 into Eq. 139 and carrying out the matrix multiplication, one obtains

$$\left.\begin{aligned} \ddot{\mathbf{q}} &= \mathbf{H}_{qq}\mathbf{Q}_e + \mathbf{H}_{q\lambda}\mathbf{Q}_d \\ \boldsymbol{\lambda} &= \mathbf{H}_{\lambda q}\mathbf{Q}_e + \mathbf{H}_{\lambda\lambda}\mathbf{Q}_d \end{aligned}\right\} \tag{6.142}$$

For a given set of initial conditions, the vector $\ddot{\mathbf{q}}$ can be integrated in order to determine the coordinates and velocities. The vector $\boldsymbol{\lambda}$ of Eq. 142 can be used to determine the generalized constraint forces by using Eq. 123. These generalized constraint forces can be used to determine the actual reactions at the joints as described in the following section.

It is important to emphasize at this point that due to the approximations involved in the direct numerical integration, the resulting coordinates and velocities are not exact. One, therefore, expects that the constraints of Eq. 135 will be violated with a degree depending on the accuracy of the numerical integration method used. With the accumulation of the errors in some applications, the violation in the constraints of Eq. 135 may not be acceptable. In order to circumvent this difficulty, Wehage (1980) proposed a coordinate partitioning technique in which the independent accelerations are identified and integrated forward in time using a direct numerical integration method, thus defining the independent coordinates and velocities. By knowing the independent coordinates as a result of the direct numerical integration, Eq. 135 which can be considered as n_c nonlinear algebraic constraint equations in the n_c-dependent coordinates is then used to determine the dependent coordinates using a *Newton–Raphson algorithm*. Having also determined the independent velocities as the result of the numerical integration, the first equation in Eq. 136 can be used to determine the dependent velocities by partitioning the constraint Jacobian matrix, and rewriting this equation in the following form:

$$\mathbf{C}_{\mathbf{q}_d}\dot{\mathbf{q}}_d + \mathbf{C}_{\mathbf{q}_i}\dot{\mathbf{q}}_i = -\mathbf{C}_t \tag{6.143}$$

where $\mathbf{q}_d$ and $\mathbf{q}_i$ are, respectively, the vectors of dependent and independent coordinates that are selected in such a manner that $\mathbf{C}_{\mathbf{q}_d}$ is nonsingular. It follows that

$$\dot{\mathbf{q}}_d = -\mathbf{C}_{\mathbf{q}_d}^{-1}\mathbf{C}_{\mathbf{q}_i}\dot{\mathbf{q}}_i - \mathbf{C}_{\mathbf{q}_d}^{-1}\mathbf{C}_t \tag{6.144}$$

This equation defines the dependent velocities. Having determined both dependent and independent coordinates and velocities, Eq. 138 can be constructed and solved for the accelerations in order to advance the numerical integration.

Example 6.9

Figure 7 shows a two-body system that consists of body i and body j which are connected by a revolute joint at point P. In this case, the system mass matrix $\mathbf{M}$ is

$$\mathbf{M} = \begin{bmatrix} \mathbf{M}^i & \mathbf{0} \\ \mathbf{0} & \mathbf{M}^j \end{bmatrix}$$

where

$$\mathbf{M}^i = \begin{bmatrix} m^i & 0 & 0 \\ 0 & m^i & 0 \\ 0 & 0 & J^i \end{bmatrix}, \quad \mathbf{M}^j = \begin{bmatrix} m^j & 0 & 0 \\ 0 & m^j & 0 \\ 0 & 0 & J^j \end{bmatrix}$$

Therefore, the mass matrix $\mathbf{M}$ is

$$\mathbf{M} = \begin{bmatrix} m^i & 0 & 0 & 0 & 0 & 0 \\ 0 & m^i & 0 & 0 & 0 & 0 \\ 0 & 0 & J^i & 0 & 0 & 0 \\ 0 & 0 & 0 & m^j & 0 & 0 \\ 0 & 0 & 0 & 0 & m^j & 0 \\ 0 & 0 & 0 & 0 & 0 & J^j \end{bmatrix}$$

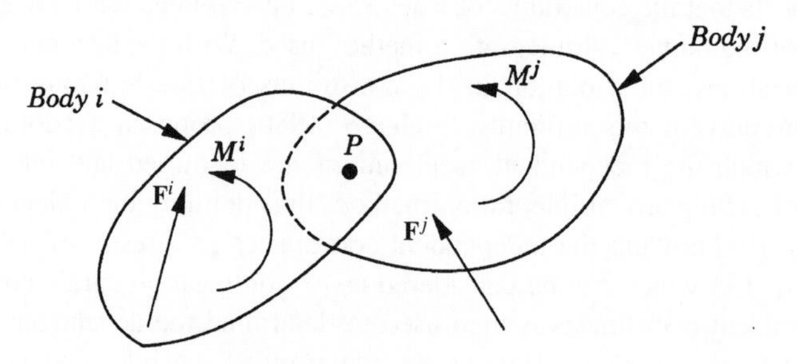

Figure 6.7 Two-body system

The Jacobian matrix of the revolute joint is

$$\mathbf{C_q} = [\mathbf{I} \quad \mathbf{A}^i_\theta \bar{\mathbf{u}}^i_P \quad -\mathbf{I} \quad -\mathbf{A}^j_\theta \bar{\mathbf{u}}^j_P] = \begin{bmatrix} 1 & 0 & C_{13} & -1 & 0 & C_{16} \\ 0 & 1 & C_{23} & 0 & -1 & C_{26} \end{bmatrix}$$

where

$$\begin{bmatrix} C_{13} \\ C_{23} \end{bmatrix} = \mathbf{A}^i_\theta \bar{\mathbf{u}}^i_P = \begin{bmatrix} -\sin\theta^i & -\cos\theta^i \\ \cos\theta^i & -\sin\theta^i \end{bmatrix} \begin{bmatrix} \bar{x}^i_P \\ \bar{y}^i_P \end{bmatrix} = \begin{bmatrix} -\bar{x}^i_P \sin\theta^i - \bar{y}^i_P \cos\theta^i \\ \bar{x}^i_P \cos\theta^i - \bar{y}^i_P \sin\theta^i \end{bmatrix}$$

$$\begin{bmatrix} C_{16} \\ C_{26} \end{bmatrix} = -\mathbf{A}^j_\theta \bar{\mathbf{u}}^j_P = -\begin{bmatrix} -\sin\theta^j & -\cos\theta^j \\ \cos\theta^j & -\sin\theta^j \end{bmatrix} \begin{bmatrix} \bar{x}^j_P \\ \bar{y}^j_P \end{bmatrix}$$

$$= \begin{bmatrix} \bar{x}^j_P \sin\theta^j + \bar{y}^j_P \cos\theta^j \\ -\bar{x}^j_P \cos\theta^j + \bar{y}^j_P \sin\theta^j \end{bmatrix}$$

It can also be shown that the vector $\mathbf{Q}_d$ of Eq. 137 is

$$\mathbf{Q}_d = \begin{bmatrix} (\mathbf{Q}_d)_1 \\ (\mathbf{Q}_d)_2 \end{bmatrix} = (\dot{\theta}^i)^2 \mathbf{A}^i \bar{\mathbf{u}}^i_P - (\dot{\theta}^j)^2 \mathbf{A}^j \bar{\mathbf{u}}^j_P$$

Equation 138 for this system can be written as

$$\begin{bmatrix} m^i & 0 & 0 & 0 & 0 & 0 & 1 & 0 \\ 0 & m^i & 0 & 0 & 0 & 0 & 0 & 1 \\ 0 & 0 & J^i & 0 & 0 & 0 & C_{13} & C_{23} \\ 0 & 0 & 0 & m^j & 0 & 0 & -1 & 0 \\ 0 & 0 & 0 & 0 & m^j & 0 & 0 & -1 \\ 0 & 0 & 0 & 0 & 0 & J^j & C_{16} & C_{26} \\ 1 & 0 & C_{13} & -1 & 0 & C_{16} & 0 & 0 \\ 0 & 1 & C_{23} & 0 & -1 & C_{26} & 0 & 0 \end{bmatrix} \begin{bmatrix} \ddot{R}^i_x \\ \ddot{R}^i_y \\ \ddot{\theta}^i \\ \ddot{R}^j_x \\ \ddot{R}^j_y \\ \ddot{\theta}^j \\ \lambda_1 \\ \lambda_2 \end{bmatrix} = \begin{bmatrix} F^i_x \\ F^i_y \\ M^i \\ F^j_x \\ F^j_y \\ M^j \\ (\mathbf{Q}_d)_1 \\ (\mathbf{Q}_d)_2 \end{bmatrix}$$

where $\mathbf{F}^i = [F^i_x \quad F^i_y]^{\mathrm{T}}$ and $\mathbf{F}^j = [F^j_x \quad F^j_y]^{\mathrm{T}}$ are, respectively, the vectors of forces acting at the center of masses of bodies i and j, and M^i and M^j are the moments acting, respectively, on body i and body j.

6.8 JOINT REACTION FORCES

The solution of Eq. 138 presented in the preceding section defines the vector of Lagrange multipliers which can be used in Eq. 123 to determine the vector of generalized constraint forces associated with the translation of the center of mass and the rotation of the body. While these generalized forces may not be the actual reaction forces of the joint, the generalized and actual constraint forces represent two equipollent systems of forces. This fact will be used to determine the actual joint forces in terms of the generalized constraint forces which are assumed, in the following discussion, to be known from the analytical or numerical solution of Eq. 138.

For a given joint k in the multibody system, the generalized constraint forces acting on body i, which is connected by this joint, are

$$(\mathbf{Q}_c^i)_k = -(\mathbf{C}_k)_{\mathbf{q}^i}^{\mathrm{T}} \boldsymbol{\lambda}_k = [F_{x_k}^i \quad F_{y_k}^i \quad M_k^i]^{\mathrm{T}} \tag{6.145}$$

where $F_{x_k}^i$, $F_{y_k}^i$, and M_k^i are the components of the vector $(\mathbf{Q}_c^i)_k$ which are assumed to be known, $\mathbf{C}_k$ is the vector of constraint equations of the joint k, and $\boldsymbol{\lambda}_k$ is the vector of Lagrange multipliers associated with these constraints.

The system of forces of Eq. 145 is equipollent to another system of forces defined on the joint surface. Let $\mathbf{u}_P^i$ be the position vector of the joint definition point with respect to the reference point. A system equipollent to the system of Eq. 145 can be obtained as

$$\mathbf{F}^i = \begin{bmatrix} F_{x_k}^i \\ F_{y_k}^i \\ M_k^i - (\mathbf{u}_P^i \times \mathbf{F}_k^i) \cdot \mathbf{k} \end{bmatrix} \tag{6.146}$$

where $\mathbf{F}_k^i = [F_{x_k}^i \quad F_{y_k}^i]^{\mathrm{T}}$, and $\mathbf{k}$ is a unit vector along the axis of rotation. Equation 146 defines the system of reaction forces at the joint.

Example 6.10

The constraint equations of the revolute joint of the pendulum shown in Fig. 8 are

$$\mathbf{R}^i + \mathbf{A}^i \bar{\mathbf{u}}_O^i = \mathbf{0}$$

and the Jacobian matrix of the constraints are

$$\mathbf{C}_{\mathbf{q}} = \begin{bmatrix} 1 & 0 & C_{13} \\ 0 & 1 & C_{23} \end{bmatrix}$$

where

$$\begin{bmatrix} C_{13} \\ C_{23} \end{bmatrix} = \mathbf{A}_\theta^i \bar{\mathbf{u}}_O^i = \begin{bmatrix} -\sin\theta^i & -\cos\theta^i \\ \cos\theta^i & -\sin\theta^i \end{bmatrix} \begin{bmatrix} \bar{x}_O^i \\ \bar{y}_O^i \end{bmatrix} = \begin{bmatrix} -\bar{x}_O^i \sin\theta^i - \bar{y}_O^i \cos\theta^i \\ \bar{x}_O^i \cos\theta^i - \bar{y}_O^i \sin\theta^i \end{bmatrix}$$

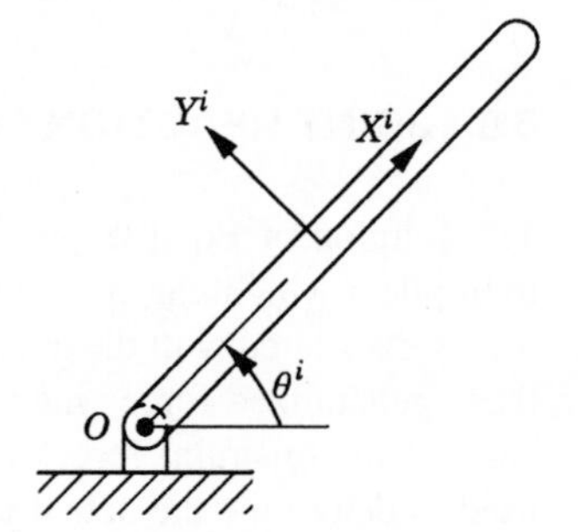

Figure 6.8 Pendulum motion

The generalized reaction forces can be expressed in terms of Lagrange multipliers as

$$\mathbf{Q}_c = -\mathbf{C}_{\mathbf{q}}^{\mathrm{T}}\boldsymbol{\lambda} = -\begin{bmatrix} 1 & 0 \\ 0 & 1 \\ C_{13} & C_{23} \end{bmatrix}\begin{bmatrix} \lambda_1 \\ \lambda_2 \end{bmatrix} = -\begin{bmatrix} \lambda_1 \\ \lambda_2 \\ C_{13}\lambda_1 + C_{23}\lambda_2 \end{bmatrix}$$

This system must be equipollent to a system of reaction forces acting at the joint. Using Eq. 146, the actual reaction forces can be written as

$$\mathbf{F}^i = \begin{bmatrix} -\lambda_1 \\ -\lambda_2 \\ -(C_{13}\lambda_1 + C_{23}\lambda_2) + M_O^i \end{bmatrix}$$

where M_O^i is the moment of the generalized reactions about point O and defined by the cross product

$$\begin{aligned} M_O^i &= -(\mathbf{u}_O^i \times \mathbf{F}_k^i) \cdot \mathbf{k} = -\{(\mathbf{A}^i\bar{\mathbf{u}}_O^i) \times [-\lambda_1 \quad -\lambda_2]^{\mathrm{T}}\} \cdot \mathbf{k} \\ &= -\left[\begin{vmatrix} \mathbf{i} & \mathbf{j} & \mathbf{k} \\ \bar{x}_O^i \cos\theta^i - \bar{y}_O^i \sin\theta^i & \bar{x}_O^i \sin\theta^i + \bar{y}_O^i \cos\theta^i & 0 \\ -\lambda_1 & -\lambda_2 & 0 \end{vmatrix}\right] \cdot \mathbf{k} \\ &= -\begin{bmatrix} 0 \\ 0 \\ -\lambda_2(\bar{x}_O^i \cos\theta^i - \bar{y}_O^i \sin\theta^i) + \lambda_1(\bar{x}_O^i \sin\theta^i + \bar{y}_O^i \cos\theta^i) \end{bmatrix} \cdot \mathbf{k} \\ &= \lambda_2(\bar{x}_O^i \cos\theta^i - \bar{y}_O^i \sin\theta^i) - \lambda_1(\bar{x}_O^i \sin\theta^i + \bar{y}_O^i \cos\theta^i) \end{aligned}$$

By substituting this expression into the definition of $\mathbf{F}^i$ and using the definition of C_{13} and C_{23}, one obtains

$$\mathbf{F}^i = \begin{bmatrix} -\lambda_1 \\ -\lambda_2 \\ 0 \end{bmatrix}$$

which implies that the actual reactions at the joints are the negative of Lagrange multipliers. Furthermore, the actual moment is equal to zero, which is an expected result for the revolute joint.

Virtual Work A more systematic, yet equivalent, procedure for determining the reaction forces at the joint definition points is to use the virtual work. Let $\mathbf{F}_k^i$ and M_k^i be, respectively, the generalized constraint force and moment associated with the reference coordinates as the result of a joint k. The virtual work of the constraint force and moment can be expressed as

$$\delta W_c^i = \mathbf{F}_k^{i\mathrm{T}}\,\delta\mathbf{R}^i + M_k^i\,\delta\theta^i \tag{6.147}$$

The global position vector of the joint definition point $\mathbf{r}_k^i$ can be expressed in terms of the reference coordinates as

$$\mathbf{r}_k^i = \mathbf{R}^i + \mathbf{A}^i\bar{\mathbf{u}}_k^i \tag{6.148}$$

where $\mathbf{A}^i$ is the planar transformation matrix of body i, and $\overline{\mathbf{u}}_k^i$ is the local position vector of the joint definition point with respect to the reference point. It follows that

$$\mathbf{R}^i = \mathbf{r}_k^i - \mathbf{A}^i \overline{\mathbf{u}}_k^i \tag{6.149}$$

and

$$\delta \mathbf{R}^i = \delta \mathbf{r}_k^i - \mathbf{A}_\theta^i \overline{\mathbf{u}}_k^i \, \delta\theta^i \tag{6.150}$$

The virtual work of the constraint forces can be expressed in terms of the virtual change in the coordinates of the joint definition point as

$$\delta W_c^i = \mathbf{F}_k^{i^{\mathrm{T}}} \delta \mathbf{r}_k^i + (M_k^i - \mathbf{F}_k^{i^{\mathrm{T}}} \mathbf{A}_\theta^i \overline{\mathbf{u}}_k^i) \, \delta\theta^i \tag{6.151}$$

The coefficients of $\delta \mathbf{r}_k^i$ and $\delta\theta^i$ in this equation define the reaction force and moment at the joint definition point. In the case of a revolute joint, the coefficient of $\delta\theta^i$ in the preceding equation is identically equal to zero as demonstrated by the preceding example.

6.9 ELIMINATION OF LAGRANGE MULTIPLIERS

While the use of the embedding technique is not recommended as a basis for developing general-purpose multibody computer programs because of the computational overhead, it is important to understand some of the basic concepts and techniques used in classical and computational dynamics to obtain a minimum set of independent differential equations. We have previously demonstrated that if $\mathbf{Q}_c$ is the vector of generalized constraint forces, then $\mathbf{B}_i^{\mathrm{T}} \mathbf{Q}_c = \mathbf{0}$ (Eq. 89), where $\mathbf{B}_i$ is the matrix defined by Eq. 74. In the method of Lagrange multipliers, the generalized constraint forces are expressed in terms of the Jacobian matrix of the kinematic constraints as $\mathbf{Q}_c = -\mathbf{C}_{\mathbf{q}}^{\mathrm{T}} \boldsymbol{\lambda}$. It follows that

$$\mathbf{B}_i^{\mathrm{T}} \mathbf{C}_{\mathbf{q}}^{\mathrm{T}} \boldsymbol{\lambda} = \mathbf{0} \tag{6.152}$$

This equation which is valid regardless of the values of Lagrange multipliers implies that the vector of generalized constraint forces $\mathbf{C}_{\mathbf{q}}^{\mathrm{T}} \boldsymbol{\lambda}$ is orthogonal to the columns of the matrix $\mathbf{B}_i$. To demonstrate that $\mathbf{B}_i^{\mathrm{T}} \mathbf{C}_{\mathbf{q}}^{\mathrm{T}} = \mathbf{0}$, we rewrite the constraint Jacobian matrix according to the coordinate partitioning as dependent and independent (see Eqs. 67 and 68) as

$$\mathbf{C}_{\mathbf{q}} = [\mathbf{C}_{\mathbf{q}_d} \quad \mathbf{C}_{\mathbf{q}_i}] \tag{6.153}$$

Recall that

$$\mathbf{B}_i = \begin{bmatrix} -\mathbf{C}_{\mathbf{q}_d}^{-1} \mathbf{C}_{\mathbf{q}_i} \\ \mathbf{I} \end{bmatrix} \tag{6.154}$$

Using this matrix, one gets

$$\mathbf{B}_i^{\mathrm{T}} \mathbf{C}_{\mathbf{q}}^{\mathrm{T}} = [-\mathbf{C}_{\mathbf{q}_i}^{\mathrm{T}} (\mathbf{C}_{\mathbf{q}_d}^{\mathrm{T}})^{-1} \quad \mathbf{I}] \begin{bmatrix} \mathbf{C}_{\mathbf{q}_d}^{\mathrm{T}} \\ \mathbf{C}_{\mathbf{q}_i}^{\mathrm{T}} \end{bmatrix} = \mathbf{0} \tag{6.155}$$

as previously stated.

The matrix $\mathbf{B}_i$ of Eq. 74 or 154 defines the relationship between the total vector of the system velocities and a smaller independent subset of the same velocities. There are other methods that can be used to express the system variables in terms of a more general set of independent coordinates, each of which can be a linear combination of the system coordinates. These methods can also be used to eliminate Lagrange multipliers and obtain a minimum set of independent differential equations. Among these methods are the **QR** *decomposition* and the *singular value decomposition*.

QR Decomposition Assuming that the constraint equations are independent such that the constraint Jacobian matrix has a full row rank, *Householder transformations* can be used, as described in Chapter 2, to write the transpose of the constraint Jacobian matrix as

$$\mathbf{C}_{\mathbf{q}}^{\mathrm{T}} = [\mathbf{Q}_1 \quad \mathbf{Q}_2] \begin{bmatrix} \mathbf{R}_1 \\ \mathbf{0} \end{bmatrix} \tag{6.156}$$

where $\mathbf{Q}_1$ and $\mathbf{Q}_2$ are $n \times n_c$ and $n \times (n - n_c)$ matrices, respectively, and $\mathbf{R}_1$ is an $n_c \times n_c$ upper-triangular matrix. It was demonstrated in Chapter 2 that the columns of the matrices $\mathbf{Q}_1$ and $\mathbf{Q}_2$ are orthogonal and

$$\mathbf{Q}_2^{\mathrm{T}} \mathbf{Q}_1 = \mathbf{0}, \quad \mathbf{Q}_2^{\mathrm{T}} \mathbf{C}_{\mathbf{q}}^{\mathrm{T}} = \mathbf{0} \tag{6.157}$$

In the **QR** decomposition, one can select the matrix $\mathbf{B}_i$, such that

$$\mathbf{B}_i = \mathbf{Q}_2 \tag{6.158}$$

With this choice of $\mathbf{B}_i$, we are guaranteed that $\mathbf{B}_i^{\mathrm{T}} \mathbf{C}_{\mathbf{q}}^{\mathrm{T}} \boldsymbol{\lambda} = \mathbf{0}$, and consequently, the velocity transformation $\dot{\mathbf{q}} = \mathbf{B}_i \dot{\mathbf{q}}_i$ can be used, as previously described, to obtain a minimum set of independent equations. The vector $\mathbf{q}_i$ in this case represents a new set of independent variables, each of which may be a combination of the system coordinates. The equations of motion can be expressed in terms of these new variables and their time derivatives and the solution can be obtained as described in later sections using the methods of numerical integration.

As pointed out in Chapter 2, while $\mathbf{Q}_1$ and $\mathbf{R}_1$ in the **QR** decomposition are unique, the matrix $\mathbf{Q}_2$ in this factorization is not unique. It is, therefore, numerically difficult to preserve the directional continuity of the bases represented by the columns of the transformation $\mathbf{Q}_2$. Kim and Vanderploeg (1986) defined constant orthogonal matrices $\mathbf{Q}_1$ and $\mathbf{Q}_2$ at the initial configuration and used the velocity constraint relationships to iteratively update these matrices in order to preserve the directional continuity of the null space of the constraint Jacobian matrix.

Singular Value Decomposition The singular value decomposition of the transpose of the Jacobian matrix can be written as

$$\mathbf{C}_{\mathbf{q}}^{\mathrm{T}} = \mathbf{Q}_1 \mathbf{B} \mathbf{Q}_2 \tag{6.159}$$

where $\mathbf{Q}_1$ and $\mathbf{Q}_2$ are two orthogonal matrices whose dimensions are $n \times n$ and $n_c \times n_c$, respectively, and $\mathbf{B}$ is an $n \times n_c$ matrix that contains the singular values along its diagonal.

It was shown in Chapter 2 that the preceding equation can be written in the following partitioned form:

$$\mathbf{C}_{\mathbf{q}}^{\mathrm{T}} = [\mathbf{Q}_{1d} \quad \mathbf{Q}_{1i}] \begin{bmatrix} \mathbf{B}_1 \\ \mathbf{0} \end{bmatrix} \mathbf{Q}_2 \tag{6.160}$$

where $\mathbf{B}_1$ is a diagonal matrix, and $\mathbf{Q}_{1d}$ and $\mathbf{Q}_{1i}$ are partitions of the matrix $\mathbf{Q}_1$. The columns of the matrices $\mathbf{Q}_{1d}$ and $\mathbf{Q}_{1i}$ are orthogonal vectors and

$$\mathbf{Q}_{1i}^{\mathrm{T}} \mathbf{Q}_{1d} = \mathbf{0} \tag{6.161}$$

It is also clear that

$$\mathbf{C}_{\mathbf{q}}^{\mathrm{T}} = \mathbf{Q}_{1d} \mathbf{B}_1 \mathbf{Q}_2 \tag{6.162}$$

The preceding two equations imply that

$$\mathbf{Q}_{1i}^{\mathrm{T}} \mathbf{C}_{\mathbf{q}}^{\mathrm{T}} = \mathbf{0} \tag{6.163}$$

If the matrix $\mathbf{B}_i$ is selected such that

$$\mathbf{B}_i = \mathbf{Q}_{1i} \tag{6.164}$$

then $\mathbf{B}_i^{\mathrm{T}} \mathbf{C}_{\mathbf{q}}^{\mathrm{T}} \boldsymbol{\lambda} = \mathbf{0}$, and the transformation $\dot{\mathbf{q}} = \mathbf{B}_i \dot{\mathbf{q}}_i$ can be used to eliminate Lagrange multipliers and obtain a minimum set of independent differential equations as described in Section 5.

Kim and Vanderploeg (1986) pointed out that the **QR** decomposition is about two times more computationally expensive than the **LU** decomposition, while the singular value decomposition is two to ten times more expensive than the **QR** decomposition, depending upon the size of the constraint Jacobian matrix. The **QR** and the singular value decompositions can also be used to obtain an identity-generalized mass matrix associated with the independent coordinates. For example, in the planar analysis using the absolute Cartesian coordinates, the mass matrix $\mathbf{M}$ of Eq. 122 is a diagonal matrix whose inverse can be easily defined. Multiplying Eq. 124 by the inverse of the mass matrix, one obtains

$$\ddot{\mathbf{q}} + \mathbf{M}^{-1} \mathbf{C}_{\mathbf{q}}^{\mathrm{T}} \boldsymbol{\lambda} = \mathbf{M}^{-1} \mathbf{Q}_e$$

Now, let $\mathbf{B}_i$ be the orthogonal velocity transformation obtained by the **QR** decomposition or the singular value decomposition of the matrix $\mathbf{M}^{-1} \mathbf{C}_{\mathbf{q}}^{\mathrm{T}}$. Substituting the transformation $\dot{\mathbf{q}} = \mathbf{B}_i \dot{\mathbf{q}}_i$ into the preceding equation and premultiplying by the transpose of the matrix $\mathbf{B}_i$ leads to an identity-generalized mass matrix associated with the independent coordinates since in this special case $\overline{\mathbf{M}}_i = \mathbf{B}_i^{\mathrm{T}} \mathbf{B}_i = \mathbf{I}$.

6.10 STATE SPACE REPRESENTATION

By now, it should be clear that there are two basic approaches for formulating the dynamic equations of multibody mechanical systems. These approaches are the embedding technique and the augmented formulation. In the embedding technique, the dynamic equations are formulated in terms of the system degrees of freedom, thereby eliminating the workless constraint forces. This approach leads to a system of linear algebraic equations in the *independent accelerations*. The coefficient matrix in this system of algebraic equations is the generalized mass matrix associated with the independent coordinates, and the right-hand side is the vector of externally applied and centrifugal forces which depend on the system coordinates, velocities, and possibly on time. By assuming that the mass matrix is positive definite, the inverse of this matrix can be obtained and used to express the independent accelerations in terms of the independent coordinates, velocities, and time. This procedure leads to Eq. 86, $\ddot{\mathbf{q}}_i = \overline{\mathbf{M}}_i^{-1}\overline{\mathbf{Q}}_i$, where $\overline{\mathbf{M}}_i$ is the system mass matrix associated with the independent coordinates, $\ddot{\mathbf{q}}_i$ is the vector of independent accelerations, and $\overline{\mathbf{Q}}_i$ is the vector of forces associated with the independent coordinates. The mass matrix can be a nonlinear function of the system coordinates, while the force vector, which contains externally applied forces and Coriolis and centrifugal forces is a function of the system coordinates, velocities, and possibly time. Since the dependent coordinates and velocities can always be expressed in terms of the independent variables, Eq. 86 can be written as

$$\ddot{\mathbf{q}}_i = \mathbf{G}_i(\mathbf{q}_i, \dot{\mathbf{q}}_i, t) \tag{6.165}$$

where $\mathbf{G}_i$ is the vector function $\mathbf{G}_i = \overline{\mathbf{M}}_i^{-1}\overline{\mathbf{Q}}_i$. In Eq. 165, the independent coordinates can be the joint variables or any set of independent coordinates identified based on the numerical structure of the constraint Jacobian matrix.

In the second approach, the augmented formulation is used, leading to the dynamic equations which are expressed in terms of the dependent and independent accelerations. The nonlinear algebraic constraint equations are adjoined to the dynamic equations using the vector of Lagrange multipliers. This formulation leads to a linear system of algebraic equations in the system accelerations and Lagrange multipliers. The coefficient matrix in this system of equations (Eq. 138) depends on the system mass matrix as well as the constraint Jacobian matrix. As shown in Section 7, the solution of this system of equations defines the vector of accelerations as well as the vector of Lagrange multipliers (Eqs. 142), which can be used to evaluate the generalized reaction forces, as discussed in the preceding section. Clearly, in this case the independent accelerations can be identified and expressed in the form of Eq. 165, and consequently, the form of Eq. 165 can be obtained by using the embedding technique, or by using the augmented formulation.

Having expressed the independent accelerations in terms of the independent coordinates, velocities, and time, one can proceed a step further in the direction of obtaining the numerical solution of the nonlinear equations of the multibody system. To this end, we define the *state vector* $\mathbf{y}$ as

$$\mathbf{y} = \begin{bmatrix} \mathbf{y}_1 \\ \mathbf{y}_2 \end{bmatrix} = \begin{bmatrix} \mathbf{q}_i \\ \dot{\mathbf{q}}_i \end{bmatrix} \tag{6.166}$$

The dimension of this vector is equal to twice the number of the system degrees of freedom. Differentiating Eq. 166 with respect to time and using the definition of Eq. 165, one obtains the following first-order *state equations*:

$$\dot{\mathbf{y}} = \begin{bmatrix} \dot{\mathbf{y}}_1 \\ \dot{\mathbf{y}}_2 \end{bmatrix} = \begin{bmatrix} \dot{\mathbf{q}}_i \\ \ddot{\mathbf{q}}_i \end{bmatrix} = \begin{bmatrix} \dot{\mathbf{y}}_1 \\ \mathbf{G}_i(\mathbf{q}_i, \dot{\mathbf{q}}_i, t) \end{bmatrix} \tag{6.167}$$

Since $\mathbf{y} = [\mathbf{q}_i^{\mathrm{T}} \quad \dot{\mathbf{q}}_i^{\mathrm{T}}]^{\mathrm{T}}$, the preceding equation can be written as

$$\dot{\mathbf{y}} = \begin{bmatrix} \dot{\mathbf{y}}_1 \\ \dot{\mathbf{y}}_2 \end{bmatrix} = \begin{bmatrix} \dot{\mathbf{y}}_1 \\ \mathbf{G}_i(\mathbf{y}, t) \end{bmatrix} \tag{6.168}$$

which can simply be written as

$$\dot{\mathbf{y}} = \mathbf{f}(\mathbf{y}, t) \tag{6.169}$$

where

$$\mathbf{f}(\mathbf{y}, t) = \begin{bmatrix} \dot{\mathbf{y}}_1 \\ \mathbf{G}_i(\mathbf{y}, t) \end{bmatrix} \tag{6.170}$$

Equation 169 represents the *state space equations* of the multibody system. These equations are first-order differential equations and their number is equal to twice the number of the system degrees of freedom. Therefore, in the *state space formulation*, the second-order differential equations associated with the independent coordinates are replaced by a system of first-order differential equations that has a number of equations equal to twice the number of the degrees of freedom of the system.

Example 6.11

In order to demonstrate the formulation of the state space equations in both cases of the augmented formulation and the formulation in terms of the degrees of freedom, we consider the simple system discussed in Example 10. If the dynamic relationships are formulated in terms of the degrees of freedom using the embedding technique, the system has one differential equation, which can be expressed in terms of this degree of freedom. If we select the angular orientation θ^i to be the degree of freedom, the differential equation of motion of the system is given by

$$J^i_O \ddot{\theta}^i = M^i - m^i g l^i_O \cos\theta^i$$

where $J^i_O = J^i + m^i (l^i_O)^2$ is the mass moment of inertia defined with respect to point O, J^i is the mass moment of inertia defined with respect to the center of mass, m^i is the total mass of the rod, M^i is the applied external moment acting on the rod, and l^i_O is the distance of point O from the center of mass of the rod. In this example, the vector $\mathbf{q}_i$ reduces to a scalar defined by the angle θ^i. The preceding equation then yields

$$\ddot{\mathbf{q}}_i = \ddot{\theta}^i = \frac{1}{J^i_O} [M^i - m^i g l^i_O \cos\theta^i]$$

From which the vector $\mathbf{G}_i$ of Eq. 165 reduces to a scalar and is recognized as

$$\mathbf{G}_i = \frac{1}{J_O^i}\left[M^i - m^i g l_O^i \cos\theta^i\right]$$

The state vector is

$$\mathbf{y} = \begin{bmatrix} \mathbf{y}_1 \\ \mathbf{y}_2 \end{bmatrix} = \begin{bmatrix} \theta^i \\ \dot{\theta}^i \end{bmatrix}$$

and the state equations can be defined using Eqs. 169 and 170 as

$$\dot{\mathbf{y}} = \begin{bmatrix} \dot{\mathbf{y}}_1 \\ \dot{\mathbf{y}}_2 \end{bmatrix} = \begin{bmatrix} \dot{\theta}^i \\ \ddot{\theta}^i \end{bmatrix} = \begin{bmatrix} \dot{\theta}^i \\ \dfrac{1}{J_O^i}\left[M^i - m^i g l_O^i \cos\theta^i\right] \end{bmatrix}$$

Let us now consider the second approach of the augmented formulation in which the equations are developed in terms of the absolute coordinates R_x^i, R_y^i, and θ^i. In this case, the mass matrix $\mathbf{M}^i$ is

$$\mathbf{M}^i = \begin{bmatrix} m^i & 0 & 0 \\ 0 & m^i & 0 \\ 0 & 0 & J^i \end{bmatrix}$$

The algebraic constraints of the revolute joint are

$$\mathbf{C}(\mathbf{q}, t) = \mathbf{R}^i + \mathbf{A}^i \bar{\mathbf{u}}_O^i = \mathbf{0}$$

and the constraint Jacobian matrix is

$$\mathbf{C}_{\mathbf{q}} = \begin{bmatrix} 1 & 0 & l_O^i \sin\theta^i \\ 0 & 1 & -l_O^i \cos\theta^i \end{bmatrix}$$

The vector $\mathbf{C}_t$ of Eq. 136 is the null vector while the vector $\mathbf{Q}_d$ of Eq. 137 is

$$\mathbf{Q}_d = (\dot{\theta}^i)^2 \mathbf{A}^i \bar{\mathbf{u}}_O^i = (\dot{\theta}^i)^2 \begin{bmatrix} l_O^i \cos\theta^i \\ l_O^i \sin\theta^i \end{bmatrix}$$

Equation 138 can be constructed for this system as

$$\begin{bmatrix} m^i & 0 & 0 & 1 & 0 \\ 0 & m^i & 0 & 0 & 1 \\ 0 & 0 & J^i & l_O^i \sin\theta^i & -l_O^i \cos\theta^i \\ 1 & 0 & l_O^i \sin\theta^i & 0 & 0 \\ 0 & 1 & -l_O^i \cos\theta^i & 0 & 0 \end{bmatrix} \begin{bmatrix} \ddot{R}_x^i \\ \ddot{R}_y^i \\ \ddot{\theta}^i \\ \lambda_1 \\ \lambda_2 \end{bmatrix} = \begin{bmatrix} 0 \\ -m^i g \\ M^i \\ (\dot{\theta}^i)^2 l_O^i \cos\theta^i \\ (\dot{\theta}^i)^2 l_O^i \sin\theta^i \end{bmatrix}$$

One can verify that the solution of this system of algebraic equations defines

$$\ddot{\mathbf{q}}_i = \ddot{\theta}^i = \frac{1}{J_O^i}\left[M^i - m^i g l_O^i \cos\theta^i\right]$$

which is the same equation obtained previously using the degree of freedom of the system as the generalized coordinate. Therefore, the state space equations are the same as obtained previously in this example.

6.11 NUMERICAL INTEGRATION

Many of the existing accurate numerical integration algorithms are developed for the solution of first-order differential equations. By putting the dynamic equations in the state space form, one can use many of the existing well-developed numerical integration methods to obtain the state of the multibody system over a specified period of simulation time. In this section, some of these numerical integration methods are discussed, and we shall start with a simple but a less accurate method called *Euler's method*.

Euler's Method Perhaps the simplest known numerical integration method is Euler's method. While this method is not accurate enough for practical applications, it can be used to demonstrate many of the features that are common in most numerical integration methods. The form of the state space equations given by Eq. 169 can be used to derive *Euler's formula* for the numerical integration. For this purpose, Eq. 169 is rewritten as

$$\frac{d\mathbf{y}}{dt} = \mathbf{f}(\mathbf{y}, t) \tag{6.171}$$

or $d\mathbf{y} = \mathbf{f}(\mathbf{y}, t)\, dt$ which upon integration leads to

$$\int_{\mathbf{y}_0}^{\mathbf{y}_1} d\mathbf{y} = \int_{t_0}^{t_1} \mathbf{f}(\mathbf{y}, t)\, dt \tag{6.172}$$

where

$$\mathbf{y}_0 = \mathbf{y}(t = t_0), \qquad \mathbf{y}_1 = \mathbf{y}(t = t_1) \tag{6.173}$$

in which $t_1 = t_0 + h$, where h is a selected time step, and $\mathbf{y}_0$ is the state vector that contains the initial conditions. Equation 172 can be written as

$$\mathbf{y}_1 = \mathbf{y}_0 + \int_{t_0}^{t_1} \mathbf{f}(\mathbf{y}, t)\, dt \tag{6.174}$$

If the time step h is selected to be very small, the integral on the right-hand side of Eq. 174 can be approximated as

$$\int_{t_0}^{t_1} \mathbf{f}(\mathbf{y}, t)\, dt = \mathbf{f}(\mathbf{y}_0, t_0)h \tag{6.175}$$

Substituting this equation into Eq. 174, one obtains

$$\mathbf{y}_1 = \mathbf{y}_0 + \mathbf{f}(\mathbf{y}_0, t_0)h \tag{6.176}$$

By using a similar procedure, one can also show that $\mathbf{y}_2 = \mathbf{y}_1 + h\mathbf{f}(\mathbf{y}_1, t_1)$. This procedure leads to Euler's method defined in its general form by the equation

$$\mathbf{y}_{n+1} = \mathbf{y}_n + h\mathbf{f}(\mathbf{y}_n, t_n) \tag{6.177}$$

where $\mathbf{y}_n = \mathbf{y}(t = t_n)$. Equation 177 implies that if the state vector of the system is known at time t_n and the step size h is assumed, the right-hand side of Eq. 177 can be evaluated and used to predict the state of the system at time t_{n+1}. Once $\mathbf{y}_{n+1}$ is determined, the procedure can be repeated to advance the numerical integration and evaluate the state of the system at time t_{n+2}. This procedure continues until the end of the simulation time is reached.

As pointed out earlier, Euler's method is not an accurate technique for the numerical integration because of its low order of integration. In order to demonstrate this fact, we write Taylor's expansion for the state vector $\mathbf{y}$ at time t_{n+1} as

$$\mathbf{y}(t_{n+1}) = \mathbf{y}(t_n) + h\dot{\mathbf{y}}(t_n) + \frac{(h)^2}{2!}\ddot{\mathbf{y}}(t_n) + \frac{(h)^3}{3!}\frac{d^3\mathbf{y}(t_n)}{dt^3} + \cdots \tag{6.178}$$

where

$$\left.\begin{aligned} \dot{\mathbf{y}}(t_n) &= \frac{d\mathbf{y}(t_n)}{dt} = \mathbf{f}(\mathbf{y}_n, t_n) \\ \ddot{\mathbf{y}}(t_n) &= \frac{d^2\mathbf{y}(t_n)}{dt^2} = \frac{d\mathbf{f}(\mathbf{y}_n, t_n)}{dt} \end{aligned}\right\} \tag{6.179}$$

By dropping terms of an order higher than the first order in h, Euler's method can be obtained from Taylor's series. In this case, the error in the numerical integration is defined by the equation

$$\mathbf{E}_n = \frac{(h)^2}{2!}\dot{\mathbf{f}}(\mathbf{y}_n, t_n) + \frac{(h)^3}{3!}\ddot{\mathbf{f}}(\mathbf{y}_n, t_n) + \cdots \tag{6.180}$$

Being a first-order method (straight-line approximation), Euler's method becomes inaccurate when the state vector is a rapidly varying function of time.

Example 6.12

In order to demonstrate the use of Euler's method for the numerical integration, we consider the simple system of Example 11. We assume that the mass of the body is 1 kg, the distance from the center of mass to point O is 0.5 m and its mass moment of inertia about point O is $0.3333\ \text{kg}\cdot\text{m}^2$. The joint torque M^i is assumed to be harmonic function that takes the form

$$M^i = 10 \sin 5t \quad \text{N}\cdot\text{m}$$

The initial conditions are assumed to be zeros, that is

$$\theta_0^i = \dot{\theta}_0^i = 0$$

The state space equations of this system were obtained in Example 11 as

$$\dot{\mathbf{y}} = \begin{bmatrix} \dot{y}_1 \\ \dot{y}_2 \end{bmatrix} = \begin{bmatrix} \dot{\theta}^i \\ \dfrac{1}{J_O^i}\left[M^i - m^i g l_O^i \cos\theta^i\right] \end{bmatrix}$$

Keeping in mind that $y_1 = \theta^i$ and $y_2 = \dot{\theta}^i = \dot{y}_1$, the state space equations can be written as

$$\dot{\mathbf{y}} = \begin{bmatrix} y_2 \\ \dfrac{1}{J_O^i}\left[M^i - m^i g l_O^i \cos y_1\right] \end{bmatrix} = \mathbf{f}(\mathbf{y}, t)$$

If $J_O^i = 0.3333\,\mathrm{kg}\cdot\mathrm{m}^2$, $m^i = 1\,\mathrm{kg}$, $g = 9.81\,\mathrm{m/s}^2$, and $l_O^i = 0.5\,\mathrm{m}$, one has

$$\dot{\mathbf{y}} = \mathbf{f}(\mathbf{y}, t) = \begin{bmatrix} y_2 \\ 30\sin 5t - 14.715\cos y_1 \end{bmatrix}$$

subject to the initial conditions

$$\mathbf{y}_0 = \begin{bmatrix} y_1(t=0) \\ y_2(t=0) \end{bmatrix} = \begin{bmatrix} \theta_0^i \\ \dot{\theta}_0^i \end{bmatrix} = \begin{bmatrix} 0 \\ 0 \end{bmatrix}$$

If we select the time step for the numerical integration to be $h = 0.01$ s, the solution at time $t_1 = t_0 + h = 0.01$ s is

$$\begin{aligned} \mathbf{y}_1 &= \mathbf{y}_0 + h\mathbf{f}(\mathbf{y}_0, t_0) \\ &= \begin{bmatrix} 0 \\ 0 \end{bmatrix} + 0.01\begin{bmatrix} 0 \\ 0 - 14.715 \end{bmatrix} = \begin{bmatrix} 0 \\ -0.14715 \end{bmatrix} \end{aligned}$$

The solution at $t_2 = t_1 + h = 0.02$ s can be obtained as

$$\begin{aligned} \mathbf{y}_2 &= \mathbf{y}_1 + h\mathbf{f}(\mathbf{y}_1, t_1) \\ &= \begin{bmatrix} 0 \\ -0.14715 \end{bmatrix} + 0.01\begin{bmatrix} -0.14715 \\ 30\sin(5)(0.01) - 14.715\cos 0 \end{bmatrix} \\ &= \begin{bmatrix} -0.0014715 \\ -0.279306 \end{bmatrix} \end{aligned}$$

Similarly, the solution at $t_3 = t_2 + h = 0.03$ s is

$$\begin{aligned} \mathbf{y}_3 &= \mathbf{y}_2 + h\mathbf{f}(\mathbf{y}_2, t_2) \\ &= \begin{bmatrix} -0.0014715 \\ -0.279306 \end{bmatrix} + 0.01\begin{bmatrix} -0.279306 \\ 30\sin(5)(0.02) - 14.715\cos(-0.0014715) \end{bmatrix} \\ &= \begin{bmatrix} -4.26456 \times 10^{-3} \\ -0.396506 \end{bmatrix}. \end{aligned}$$

This process continues until the desired end of the simulation time is reached.

Higher-Order Numerical Integration Methods Euler's method is a *single-step method* with an order of integration equal to 1. The error resulting from the use of this method is large, especially in the analysis of nonlinear systems. For rapidly varying functions, the use of low-order integration methods is not generally recommended since the error is a function of the frequency content and as the frequency increases the error also increases when these integration methods are used. In order to demonstrate this, we consider the simplest oscillatory system that consists of mass m, which is supported by a linear spring that has a stiffness coefficient k. The equation of motion that describes the free vibration of this system is given by

$$m\ddot{x} + kx = 0 \tag{6.181}$$

where x is the displacement of the mass. The exact solution of the preceding differential equation is

$$x = X\,\sin(\omega t + \phi) \tag{6.182}$$

where $\omega = \sqrt{k/m}$ is the natural frequency of the system, and X and ϕ are constants that can be determined using the initial conditions. For this system, the state vector is defined as

$$\mathbf{y} = \begin{bmatrix} y_1 \\ y_2 \end{bmatrix} = \begin{bmatrix} x \\ \dot{x} \end{bmatrix} \tag{6.183}$$

If we use the exact solution to substitute for x and $\dot{x}$, the state vector can be written in terms of the system natural frequency as

$$\mathbf{y} = \begin{bmatrix} y_1 \\ y_2 \end{bmatrix} = X \begin{bmatrix} \sin(\omega t + \phi) \\ \omega\cos(\omega t + \phi) \end{bmatrix} \tag{6.184}$$

and

$$\dot{\mathbf{y}} = \begin{bmatrix} \dot{y}_1 \\ \dot{y}_2 \end{bmatrix} = \mathbf{f}(\mathbf{y}, t) = X\omega \begin{bmatrix} \cos(\omega t + \phi) \\ -\omega\sin(\omega t + \phi) \end{bmatrix} \tag{6.185}$$

In Euler's method, first-order approximation is made. Second- and higher-order terms are neglected and the error is in the form defined by Eq. 180. It is clear from this error equation that the first term that appears in the error series is function of $\ddot{\mathbf{y}} = \dot{\mathbf{f}}(\mathbf{y}, t)$. For our simple oscillatory system, $\ddot{\mathbf{y}}$ is given by

$$\ddot{\mathbf{y}} = -X(\omega)^2 \begin{bmatrix} \sin(\omega t + \phi) \\ \omega\cos(\omega t + \phi) \end{bmatrix} = -(\omega)^2\mathbf{y} \tag{6.186}$$

As differentiation continues, the resulting terms become functions of higher power of the frequency ω and, therefore, as the frequency increases, the error resulting from the use of Euler's method increases. For that reason, very low-order numerical integration methods are hardly used in the analysis of nonlinear multibody systems. More accurate numerical integration methods such as the single-step *Runge–Kutta methods* and multistep *Adams–Bashforth* and *Adams–Moulton methods* are often used. While in single-step methods only knowledge of the numerical solution $\mathbf{y}_n$ is required in order to compute the next value $\mathbf{y}_{n+1}$, in multistep methods several previous values are required.

Runge–Kutta Methods Runge–Kutta methods are widely used in the numerical solutions of the nonlinear differential equations of mechanical systems. While the order of integration of Runge–Kutta methods is normally higher than one, only knowledge of the function $\mathbf{f}(\mathbf{y}, t)$ is required since the use of Runge–Kutta methods does not require determining the derivatives of the function $\mathbf{f}(\mathbf{y}, t)$. One of the most widely used Runge–Kutta formulas is

$$\mathbf{y}_{n+1} = \mathbf{y}_n + \tfrac{1}{6}\left(\mathbf{f}_1 + 2\mathbf{f}_2 + 2\mathbf{f}_3 + \mathbf{f}_4\right) \tag{6.187}$$

where

$$\left.\begin{aligned} &\mathbf{f}_1 = h\mathbf{f}(\mathbf{y}_n, t_n),\ \mathbf{f}_2 = h\mathbf{f}\left(\mathbf{y}_n + \tfrac{1}{2}\mathbf{f}_1, t_n + \tfrac{h}{2}\right), \\ &\mathbf{f}_3 = h\mathbf{f}\left(\mathbf{y}_n + \tfrac{1}{2}\mathbf{f}_2,\ t_n + \tfrac{h}{2}\right),\ \mathbf{f}_4 = h\mathbf{f}(\mathbf{y}_n + \mathbf{f}_3, t_n + h) \end{aligned}\right\} \tag{6.188}$$

where h is the time step size, and $\mathbf{y}_n$ and $\mathbf{y}_{n+1}$ are, respectively, the solutions at time t_n and t_{n+1}.

In order to demonstrate the use of Runge–Kutta methods, we consider the problem discussed in Example 11. For this example, the vector function $\mathbf{f}$ is

$$\mathbf{f}(\mathbf{y}, t) = \begin{bmatrix} y_2 \\ 30\sin 5t - 14.715\cos y_1 \end{bmatrix}$$

and $\mathbf{y}_0 = \mathbf{0}$. In order to advance the numerical integration using Runge–Kutta methods, we must evaluate the functions $\mathbf{f}_1$, $\mathbf{f}_2$, $\mathbf{f}_3$, and $\mathbf{f}_4$. If $h = 0.01$ s, one has

$$\mathbf{f}_1 = h\mathbf{f}(\mathbf{y}_0, 0) = 0.01\begin{bmatrix} 0 \\ -14.715 \end{bmatrix} = \begin{bmatrix} 0 \\ -0.14715 \end{bmatrix}$$

Using $\mathbf{f}_1$, one evaluates

$$\mathbf{y}_0 + \frac{1}{2}\mathbf{f}_1 = \begin{bmatrix} 0 \\ 0 \end{bmatrix} + \frac{1}{2}\begin{bmatrix} 0 \\ -0.14715 \end{bmatrix} = \begin{bmatrix} 0 \\ -0.073575 \end{bmatrix}$$

It follows that

$$\mathbf{f}_2 = h\mathbf{f}\left(\mathbf{y}_0 + \frac{1}{2}\mathbf{f}_1, t_0 + \frac{h}{2}\right) = 0.01\begin{bmatrix} -0.073575 \\ -13.96508 \end{bmatrix} = \begin{bmatrix} -0.73575 \times 10^{-3} \\ -0.1396508 \end{bmatrix}$$

The function $\mathbf{f}_2$ can be used to evaluate $\mathbf{y}_0 + \frac{1}{2}\mathbf{f}_2$ as

$$\mathbf{y}_0 + \frac{1}{2}\mathbf{f}_2 = \begin{bmatrix} 0 \\ 0 \end{bmatrix} + \frac{1}{2}\begin{bmatrix} -0.73575 \times 10^{-3} \\ -0.1396508 \end{bmatrix} = \begin{bmatrix} -0.367875 \times 10^{-3} \\ -0.069825 \end{bmatrix}$$

Therefore, $\mathbf{f}_3$ can be evaluated as

$$\mathbf{f}_3 = h\mathbf{f}\left(\mathbf{y}_0 + \frac{1}{2}\,\mathbf{f}_2, t_0 + \frac{h}{2}\right) = 0.01\begin{bmatrix} -0.069825 \\ -13.96508 \end{bmatrix} = \begin{bmatrix} -0.69825 \times 10^{-3} \\ -0.1396508 \end{bmatrix}$$

This vector can be used to evaluate $\mathbf{y}_0 + \mathbf{f}_3$ as

$$\mathbf{y}_0 + \mathbf{f}_3 = \begin{bmatrix} 0 \\ 0 \end{bmatrix} + \begin{bmatrix} -0.69825 \times 10^{-3} \\ -0.1396508 \end{bmatrix} = \begin{bmatrix} -0.69825 \times 10^{-3} \\ -0.1396508 \end{bmatrix}$$

and

$$\begin{aligned}\mathbf{f}_4 = h\mathbf{f}(\mathbf{y}_0 + \mathbf{f}_3, t_0 + h) &= 0.01\begin{bmatrix} -0.1396508 \\ -13.215621 \end{bmatrix} \\ &= \begin{bmatrix} -0.1396508 \times 10^{-2} \\ -0.13215621 \end{bmatrix}\end{aligned}$$

The solution at time $t_1 = t_0 + h = 0.01$ s can be obtained using the Runge–Kutta method as

$$\begin{aligned}\mathbf{y}_1 &= \mathbf{y}_0 + \frac{1}{6}(\mathbf{f}_1 + 2\mathbf{f}_2 + 2\mathbf{f}_3 + \mathbf{f}_4) \\ &= \frac{1}{6}\left\{\begin{bmatrix} 0 \\ -0.14715 \end{bmatrix} + 2\begin{bmatrix} -0.00073575 \\ -0.1396508 \end{bmatrix} + 2\begin{bmatrix} -0.00069825 \\ -0.1396508 \end{bmatrix}\right. \\ &\quad \left. + \begin{bmatrix} -0.001396508 \\ -0.13215621 \end{bmatrix}\right\} = \begin{bmatrix} -0.71075 \times 10^{-4} \\ -0.139652 \end{bmatrix}\end{aligned}$$

The vector $\mathbf{y}_1$ can then be used to predict the solution at time $t_2 = t_1 + h$. This process continues until the desired end of the simulation time is reached. Clearly, the use of the Runge–Kutta method requires many more calculations as compared to Euler's method, a price that must be paid for higher accuracy.

Multistep Methods In the case of a single function y, the general form of the multistep methods is given by (Atkinson, 1978)

$$y_{n+1} = \sum_{j=0}^{k} a_j y_{n-j} + h \sum_{j=-1}^{k} b_j f(t_{n-j}, y_{n-j}) \tag{6.189}$$

where h is the time step size, $t_n = t_0 + nh$, $a_0, \cdots, a_k, b_{-1}, b_0, \cdots$, and b_k are constants, and $k \geq 0$. The multistep method is called the $k+1$ step method if $k+1$ previous solution values are used to evaluate y_{n+1}. In this case, either $a_k \neq 0$ or $b_k \neq 0$. Note that Euler's method is an example of a one-step method with $k = 0$ and

$$a_0 = 1, \quad b_0 = 1, \quad b_{-1} = 0 \tag{6.190}$$

If the term y_{n+1} appears only on the left-hand side of Eq. 189, the method is said to be an *explicit method*. This is the case in which $b_{-1} = 0$. If $b_{-1} \neq 0$, y_{n+1} appears on both sides of Eq. 189 which, in this case, defines an *implicit method*. In general, iterative procedures are used to solve implicit methods.

An example of an explicit method is the two-step *midpoint method* defined by the formula

$$y_{n+1} = y_{n-1} + 2hf(t_n, y_n) \tag{6.191}$$

An example of an implicit method is the one-step *trapezoidal method* defined by the formula

$$y_{n+1} = y_n + \frac{h}{2}\left[f(t_n, y_n) + f(t_{n+1}, y_{n+1})\right] \tag{6.192}$$

The convergence of the approximate solution obtained by using multistep methods can be proved by defining an error function and requiring that this error function approaches zero as the step size approaches zero. This condition, which is called the *consistency condition*, can be used to obtain a set of algebraic equations that relate the constants a_j and b_j of Eq. 189. In general, there are two approaches for deriving higher order multistep methods. These are the *method of undetermined coefficients* and the *method of numerical integration*. In the method of undetermined coefficients, the algebraic equations obtained using the consistency condition are used to define the constants a_j and b_j, while in the method of numerical integration, polynomial approximations are employed. The methods based on the numerical integration are more popular and are used to derive the most widely used multistep methods such as the *predictor–corrector Adams methods*, where an explicit formula is used to predict the solution at t_{n+1} using the previous solution values. The predicted solution is then substituted into the implicit corrector formula to determine the corrected solution. Such a solution procedure can be used to control the size of the truncation error.

Adams Methods Adams methods are widely used in solving first-order ordinary differential equations. They are used to obtain predictor-corrector algorithms where the error is controlled by varying the step size and the order of the method. Adams methods can be derived using the method of numerical integration starting with the following equation:

$$y_{n+1} = y_n + \int_{t_n}^{t_{n+1}} f[t, y(t)]\, dt \tag{6.193}$$

Interpolation polynomials are used to approximate $f[t, y(t)]$, and by integrating these polynomials over the interval $[t_n, t_{n+1}]$, one obtains an approximation to y_{n+1}. Two formulas are often used; these are the predictor explicit *Adams–Bashforth methods* and the corrector implicit *Adams–Moulton methods*.

In the explicit or predictor *Adams–Bashforth methods*, interpolation polynomials $P_k(t)$ of a degree less than or equal to k are used to approximate $f(y, t)$ at $t_{n-k}, \ldots, t_n$. A convenient way of constructing the interpolation polynomials is to use the *Newton backward difference formula* expanded at t_n,

$$P_k(t) = f(t_n) + \frac{t - t_n}{h} \nabla f_n + \frac{(t - t_n)(t - t_{n-1})}{2!\,(h)^2} \nabla^2 f_n$$

$$+\cdots+\frac{(t-t_n)\cdots(t-t_{n-k+1})}{k!\,(h)^k}\nabla^k f_n \tag{6.194}$$

where $f_n = f(y_n, t_n)$ and the backward differences are defined as

$$\left.\begin{aligned}\nabla^1 f_n &= f_n - f_{n-1}\\ \nabla^2 f_n &= \nabla^1 f_n - \nabla^1 f_{n-1}\\ &\vdots\\ \nabla^k f_n &= \nabla^{k-1} f_n - \nabla^{k-1} f_{n-1}\end{aligned}\right\} \tag{6.195}$$

The integral of $P_k(t)$ is given by (Atkinson, 1978)

$$\int_{t_n}^{t_{n+1}} P_k(t) = h\sum_{j=0}^{k}\gamma_j \nabla^j f_n \tag{6.196}$$

where the coefficients γ_j are obtained using the formula

$$\gamma_j = \frac{1}{j!}\int_0^1 s(s+1)\cdots(s+j-1)\,ds \qquad j \geq 1 \tag{6.197}$$

in which $s = (t - t_n)/h$ with $\gamma_0 = 1$. It can also be shown that $\gamma_1 = 1/2$, $\gamma_2 = 5/12$, $\gamma_3 = 3/8$, $\gamma_4 = 251/720$, and $\gamma_5 = 95/288$.

The derivation of the *corrector implicit Adams–Moulton method* is exactly the same as the Adams–Bashforth method except we interpolate $f(y, t)$ at the $k + 1$ points $t_{n+1}, \ldots, t_{n-k+1}$. The formula for the implicit Adams–Moulton method is

$$y_{n+1} = y_n + h\sum_{j=0}^{k}\delta_j \nabla^j f_{n+1} \tag{6.198}$$

where

$$\delta_j = \frac{1}{j!}\int_0^1 (s-1)s(s+1)\cdots(s+j-2)ds \qquad j \geq 1 \tag{6.199}$$

with $\delta_0 = 1$. It can also be shown that $\delta_1 = -1/2$, $\delta_2 = -1/12$, $\delta_3 = -1/24$, $\delta_4 = -19/720$, and $\delta_5 = -3/160$. Note that the trapezoidal method is a special case of the Adams–Moulton formula with $k = 1$. Since the Adams–Moulton formula is an implicit method, the iterative procedure for solving it requires the use of a predictor formula such as the explicit Adams–Bashforth formula. One of the main reasons for using the implicit Adams–Moulton formulas is the fact that they have a much smaller truncation error as compared to the Adams–Bashforth formulas when comparable order is used. Computationally, it is desirable to make the order of the predictor formula less than the order of the corrector formula by one. Such a choice has the advantage that the predictor and corrector would both use derivative values at the same nodes. The second-order Adams–Moulton formula that uses the first-order Adams–Bashforth formula as a predictor is the same as the trapezoidal method that uses the Euler formula as a predictor.

Most of the predictor–corrector algorithms that are based on Adams method control the truncation error by varying both the order and the step size. By using a variable order, there is no difficulty in starting the numerical integration and obtain starting values for the higher order formulas. For instance, the numerical integration begins with the second-order trapezoidal method that uses Euler formula as a predictor, and as more starting values become available, the order of the method can be increased.

6.12 ALGORITHM AND SPARSE MATRIX IMPLEMENTATION

The formulation of the dynamic equations using the independent variables leads to the smallest system of strongly coupled equations. The numerical solution of this system that requires only the numerical integration of differential equations defines the independent velocities and coordinates which can be used to determine the dependent coordinates and velocities in a straightforward manner using the kinematic equations. In the augmented formulation, on the other hand, the kinematic constraint equations are adjoined to the system differential equations using the vector of Lagrange multipliers. This approach leads to a large system of loosely coupled equations that can be solved for the accelerations and Lagrange multipliers. The vector of Lagrange multipliers can be used to determine the generalized reaction forces while the independent accelerations are identified and integrated forward in time in order to determine the independent velocities and coordinates. In this approach, the solution for the dependent coordinates requires the solution of the algebraic system of nonlinear constraint equations:

$$\mathbf{C}(\mathbf{q}, t) = \mathbf{0} \tag{6.200}$$

where $\mathbf{C}$ is the vector of kinematic constraint equations, $\mathbf{q}$ is the total vector of system coordinates, and t is time. A Newton–Raphson algorithm can be used in order to solve Eq. 200 for the dependent coordinates. In this iterative Newton–Raphson algorithm, the independent coordinates are kept fixed to their values that are obtained from the numerical integration. That is, the Newton differences $\Delta\mathbf{q}_i$ associated with these independent coordinates are assumed to be zero. The dependent coordinates, on the other hand, are determined by solving the nonlinear algebraic constraint equations using the iterative Newton–Raphson procedure. In the sparse matrix implementation, one can use the following system of algebraic equations in the Newton iterations:

$$\begin{bmatrix} \mathbf{C}_{\mathbf{q}} \\ \mathbf{I}_d \end{bmatrix} \Delta\mathbf{q} = \begin{bmatrix} -\mathbf{C} \\ \mathbf{0} \end{bmatrix} \tag{6.201}$$

In this equation, $\mathbf{C}_{\mathbf{q}}$ is the constraint Jacobian matrix, $\Delta\mathbf{q}$ is the vector of the Newton differences of all coordinates, and $\mathbf{I}_d$ is a Boolean matrix that has zeros and ones only; with the ones in the locations that correspond to the independent coordinates in order to ensure that $\Delta\mathbf{q}_i = \mathbf{0}$. The square coefficient matrix in Eq. 201 is sparse, and therefore, sparse matrix techniques can be used to efficiently solve the preceding system of equations. In this sparse matrix implementation, one does not need to identify the sub-Jacobians $\mathbf{C}_{\mathbf{q}_i}$ and $\mathbf{C}_{\mathbf{q}_d}$ associated respectively, with the independent and dependent coordinates. Furthermore, there is no need to find the inverse of the matrix $\mathbf{C}_{\mathbf{q}_d}$ associated with the dependent coordinates.

Note also that, for a given constrained system, the locations of the nonzero elements of the sparse coefficient matrix in Eq. 201 remain the same throughout the simulation.

Once the dependent coordinates are determined, the dependent velocities can be determined using the first equation in Eq. 136, which can be written according to the partitioning of the coordinates as dependent and independent as

$$\mathbf{C}_{\mathbf{q}_d}\dot{\mathbf{q}}_d + \mathbf{C}_{\mathbf{q}_i}\dot{\mathbf{q}}_i = -\mathbf{C}_t \tag{6.202}$$

where the vector $\mathbf{C}_t$, which is the partial derivative of the constraint equations with respect to time, has dimension equal to the number of the kinematic constraints.

As pointed out in the preceding sections, if the kinematic constraints are linearly independent, the independent coordinates can be selected in such a manner that the matrix $\mathbf{C}_{\mathbf{q}_d}$ is a nonsingular matrix. Since the independent velocities are assumed to be known as a result of the numerical integration of the independent state equations, Eq. 202 can be considered as a linear system of algebraic equations in the dependent velocities. This equation can be used to define the dependent velocities as

$$\dot{\mathbf{q}}_d = -\mathbf{C}_{\mathbf{q}_d}^{-1}[\mathbf{C}_{\mathbf{q}_i}\dot{\mathbf{q}}_i + \mathbf{C}_t] \tag{6.203}$$

An alternative, yet equivalent approach for solving for the dependent velocities is to solve the following linear sparse system of algebraic equations in the total vector of system velocities:

$$\begin{bmatrix} \mathbf{C}_{\mathbf{q}} \\ \mathbf{I}_d \end{bmatrix} \dot{\mathbf{q}} = \begin{bmatrix} -\mathbf{C}_t \\ \dot{\mathbf{q}}_i \end{bmatrix} \tag{6.204}$$

where $\mathbf{I}_d$ is the same matrix used in Eq. 201. The right-hand side of Eq. 204 is assumed to be known since $\dot{\mathbf{q}}_i$ is determined from the numerical integration, and $\mathbf{C}_t$ depends on time and the coordinates that are assumed to be known from the position analysis. Note that Eq. 204 is simply the result of combining the two equations $\mathbf{C}_{\mathbf{q}}\dot{\mathbf{q}} = -\mathbf{C}_t, \dot{\mathbf{q}}_i = \dot{\mathbf{q}}_i$. While the use of Eq. 204 is equivalent to the use of Eq. 203, the advantage of using Eq. 204 is that one avoids the partitioning of the constraint Jacobian matrix and identifying the sub-Jacobians $\mathbf{C}_{\mathbf{q}_d}$ and $\mathbf{C}_{\mathbf{q}_i}$ as in the case of the position analysis. Furthermore, if the set of independent coordinates changes during the simulation time one has only to change the locations of the nonzero entries of the matrix $\mathbf{I}_d$, while the structure of the Jacobian matrix $\mathbf{C}_{\mathbf{q}}$ remains the same as previously mentioned.

Once the generalized coordinates and velocities are determined, the equations for the accelerations and Lagrange multipliers can be constructed as

$$\begin{bmatrix} \mathbf{M} & \mathbf{C}_{\mathbf{q}}^{\mathrm{T}} \\ \mathbf{C}_{\mathbf{q}} & \mathbf{0} \end{bmatrix} \begin{bmatrix} \ddot{\mathbf{q}} \\ \boldsymbol{\lambda} \end{bmatrix} = \begin{bmatrix} \mathbf{Q}_e \\ \mathbf{Q}_d \end{bmatrix} \tag{6.205}$$

where all the matrices and vectors that appear in this equation are as defined in Section 7. Equation 205 can be solved for the accelerations and Lagrange multipliers and the independent accelerations can be identified and used to define the independent state equations, which can be integrated forward in time to determine the independent coordinates and velocities.

The main steps for a numerical algorithm that can be used to solve the mixed system of differential and algebraic equations that appear in the analysis of multibody systems can then be summarized as follows:

1. An estimate of the initial conditions that define the initial configuration of the multibody system is made. The initial conditions that represent the initial coordinates and velocities must be a good approximation of the exact initial configuration.
2. Using the initial coordinates, the constraint Jacobian matrix can be constructed, and based on the numerical structure of this matrix an **LU** factorization algorithm can be used to identify a set of independent coordinates.
3. Using the values of the independent coordinates, the constraint equations can be considered as a nonlinear system of algebraic equations in the dependent coordinates. This system can be solved iteratively using Eq. 201 and a Newton–Raphson algorithm.
4. Using the total vector of system coordinates, which is assumed to be known from the previous step, one can construct Eq. 202 or equivalently, Eq. 204, which represents a linear system of algebraic equations in the velocities. The solution of this system of equations defines the dependent velocities.
5. Having determined the coordinates and velocities, Eq. 205 can be constructed and solved for the accelerations and Lagrange multipliers. The vector of Lagrange multipliers can be used to determine the generalized reaction forces.
6. The independent accelerations can be identified and used to define the state space equations which can be integrated forward in time using a direct numerical integration method. The numerical solution of the state equations defines the independent coordinates and velocities, which can be used to determine the dependent coordinates and velocities as discussed in steps 3 and 4.
7. This process continues until the desired end of the simulation time is reached.

6.13 DIFFERENTIAL AND ALGEBRAIC EQUATIONS

The selection of the set of independent coordinates is an important step in the computer solution of the constrained dynamic equations. This selection has a significant effect on the stability of the solution and also on reducing the accumulation of the numerical error when the algebraic kinematic constraint equations are solved for the dependent variables. In Section 5, a numerical procedure that utilizes the Gaussian elimination for identifying the set of independent coordinates was discussed. It is necessary, however, in many applications, to change the set of independent coordinates during the numerical integration of the equations of motion.

In order to demonstrate some of the difficulties encountered when the independent coordinates are not properly selected, we consider the closed kinematic chain shown in Fig. 9. Such a closed kinematic chain that consists of n_b links connected by revolute joints has n_b degrees of freedom. In order to define the chain configuration in the global coordinate system, at least two translational Cartesian coordinates must be selected as degrees of freedom, as shown in Fig. 9. The other remaining degrees of freedom can be selected as rotation angles, and hence there are two rotational coordinates for two links that must be treated as

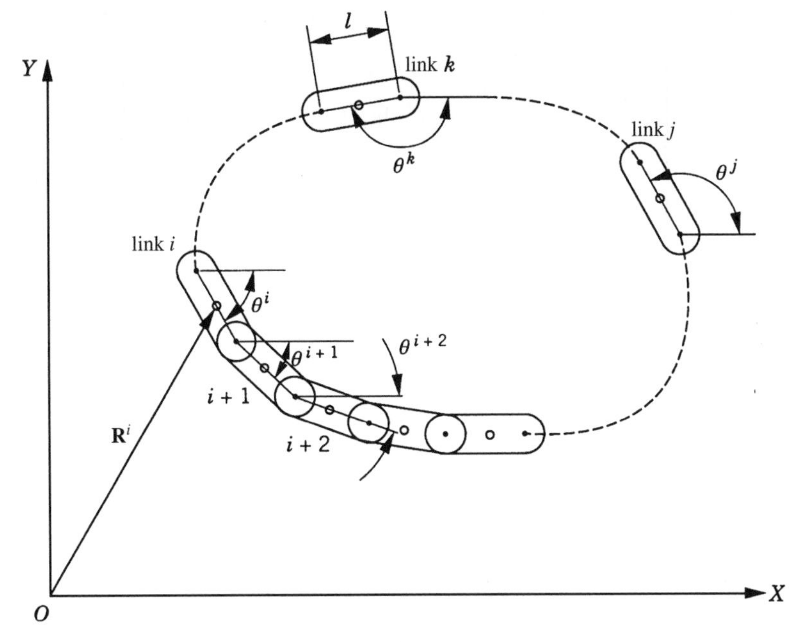

Figure 6.9 Singular configurations

dependent coordinates. The dependent rotational coordinates can be expressed in terms of the independent angles using the loop-closure equations.

$$\left.\begin{aligned} &\left\{\sum_{\substack{i=1\\(i\neq j,k)}}^{n_b} l\cos\theta^i\right\} + l\cos\theta^j + l\cos\theta^k = 0\\ &\left\{\sum_{\substack{i=1\\(i\neq j,k)}}^{n_b} l\sin\theta^i\right\} + l\sin\theta^j + l\sin\theta^k = 0 \end{aligned}\right\} \tag{6.206}$$

where n_b is the total number of the links, θ^j and θ^k are the dependent rotation angles of the links j and k, and l is the length of the link. For simplicity, we assumed here that all the links are of equal length. The preceding two equations can be written as

$$\left.\begin{aligned} l\cos\theta^j + l\cos\theta^k = A\\ l\sin\theta^j + l\sin\theta^k = B \end{aligned}\right\} \tag{6.207}$$

where

$$A = -\sum_{\substack{i=1\\(i\neq j,k)}}^{n_b} l\cos\theta^i, \quad B = -\sum_{\substack{i=1\\(i\neq j,k)}}^{n_b} l\sin\theta^i \tag{6.208}$$

By differentiating the resulting loop-closure equations with respect to time, one obtains

$$\left.\begin{aligned}\left\{\sum_{\substack{i=1\\(i\neq j,k)}}^{n_b} l\dot{\theta}^i \sin\theta^i\right\} + l\dot{\theta}^j \sin\theta^j + l\dot{\theta}^k \sin\theta^k = 0\\ \left\{\sum_{\substack{i=1\\(i\neq j,k)}}^{n_b} l\dot{\theta}^i \cos\theta^i\right\} + l\dot{\theta}^j \cos\theta^j + l\dot{\theta}^k \cos\theta^k = 0\end{aligned}\right\} \tag{6.209}$$

These two equations can be rewritten as

$$\left.\begin{aligned}l\dot{\theta}^j \sin\theta^j + l\dot{\theta}^k \sin\theta^k &= -A_d\\ l\dot{\theta}^j \cos\theta^j + l\dot{\theta}^k \cos\theta^k &= -B_d\end{aligned}\right\} \tag{6.210}$$

where

$$\left.\begin{aligned}A_d = \frac{dA}{dt} = \sum_{\substack{i=1\\(i\neq j,k)}}^{n_b} l\dot{\theta}^i \sin\theta^i\\ B_d = \frac{dB}{dt} = \sum_{\substack{i=1\\(i\neq j,k)}}^{n_b} l\dot{\theta}^i \cos\theta^i\end{aligned}\right\} \tag{6.211}$$

It follows that

$$\begin{bmatrix} l\sin\theta^j & l\sin\theta^k \\ l\cos\theta^j & l\cos\theta^k \end{bmatrix}\begin{bmatrix}\dot{\theta}^j\\ \dot{\theta}^k\end{bmatrix} = -\begin{bmatrix}A_d\\ B_d\end{bmatrix} \tag{6.212}$$

This system of equations can be solved for $\dot{\theta}^j$ and $\dot{\theta}^k$ as

$$\begin{bmatrix}\dot{\theta}^j\\ \dot{\theta}^k\end{bmatrix} = \frac{-1}{l\,\sin(\theta^j - \theta^k)}\begin{bmatrix} A_d\cos\theta^k - B_d\sin\theta^k \\ -A_d\cos\theta^j + B_d\sin\theta^j\end{bmatrix} \tag{6.213}$$

It is clear from this equation that singularities will be encountered when $\theta^j - \theta^k$ is close to or equal to 0 or π. In these situations, an alternate set of independent coordinates must be used; otherwise, a small error in the independent variables will lead to a very large error in the dependent variables.

It is clear from the closed-chain example that if the set of independent coordinates is defined only once at the beginning of the simulation, numerical difficulties may be encountered when the system configuration changes. If the error in the dependent coordinates becomes large, the number of the Newton–Raphson iterations required to solve the nonlinear kinematic constraint equations will significantly increase. Furthermore, the numerical errors in the dependent coordinates may lead to significant changes in the forces and system inertia, which, in turn, make the dynamic equations appear as being *stiff*, thereby forcing

Figure 6.10 Tracked vehicle (Photograph courtesy of Komatsu Ltd.)

the numerical integration method to select a smaller step size. For this reason, it is recommended that one redefines the set of independent coordinates every few time steps in order to avoid such numerical difficulties. An example of these difficulties was observed in the analysis of the tracked vehicle shown in Fig. 10. A planar model that consists of 54 rigid bodies was used in a numerical investigation (Nakanishi and Shabana, 1994). These bodies, as shown in Fig. 11, are the ground denoted as body 1, the chassis denoted as body 2, the sprocket denoted as body 3, the idler denoted as body 4, the seven lower rollers denoted as bodies 5–11, the upper roller denoted as body 12, and 42 track links denoted as bodies 13–54.

In the model shown in Fig. 11, the idler, sprocket, upper roller, and lower rollers are connected to the chassis using revolute joints. The track is modeled as a closed kinematic chain in which the track segments are connected by revolute joints. The equations of motion of the multibody tracked vehicle was obtained using the augmented approach where the kinematic constraints are adjoined to the dynamic equations using the technique of Lagrange multipliers. Contact force models which describe the interaction between the sprocket, the idler, the rollers, and the track segments were developed and were introduced to the dynamic formulation as a set of nonlinear generalized forces that depend on the system coordinates and velocities. Friction forces between the track segments and the ground were also considered (Nakanishi and Shabana, 1994). In the Lagrangian formulation, the dynamics of the vehicle is described using 162 absolute coordinates and 52 revolute joints that introduce 104 kinematic constraint equations in addition to three ground constraint equations. Because of the joint constraints, the vehicle has only 55 degrees of freedom. The vehicle, however, was driven by rotating the sprocket with a constant angular velocity, and the driving constraint of the sprocket when applied reduces the number of degrees of freedom by one. For the tracked vehicle model shown in Fig. 11, the coefficient matrix of Eq. 205 is a 270×270 symmetric matrix.

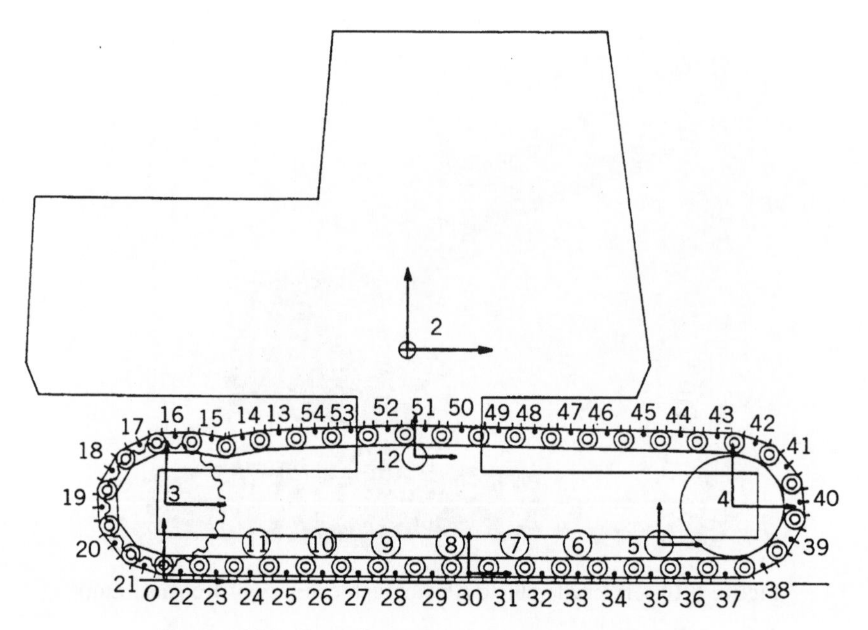

Figure 6.11 Two-dimensional tracked vehicle model

The numerical solution of the tracked vehicle equations was obtained first by using one set of independent coordinates throughout the dynamic simulation. Figure 12 shows the contact forces acting on some of the rollers, while Fig. 13 shows the acceleration of the chassis of the vehicle. From the results presented in these two figures, some unexpected dynamic behaviors can be observed. There was no reason to justify the sudden change in the contact forces and the accelerations shown in Figs. 12 and 13 after 4 s. The sudden increase in the contact forces was found to be the result of the motion of body 24, whose coordinates were all selected as dependent coordinates.

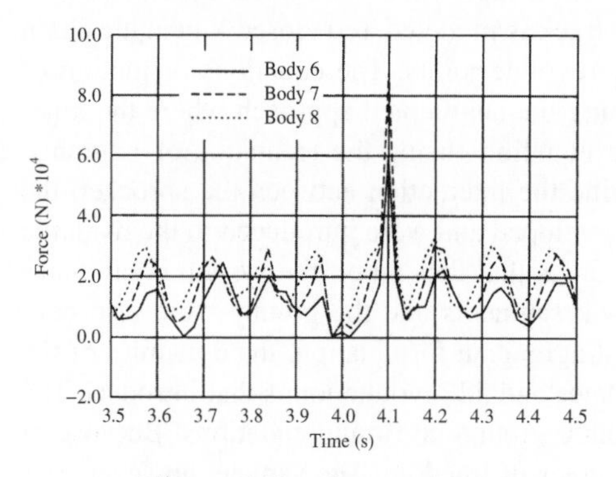

Figure 6.12 Contact forces

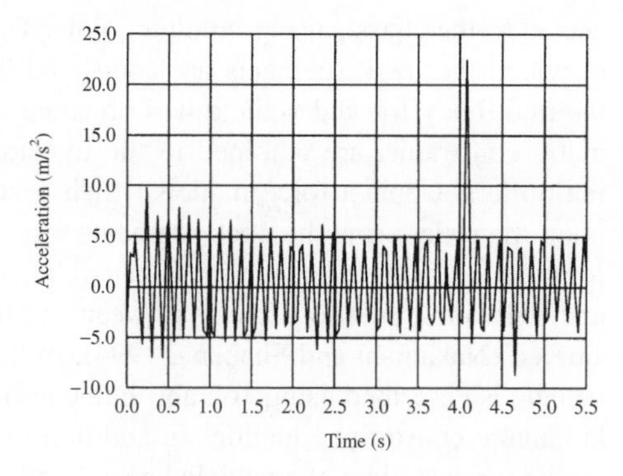

Figure 6.13 Chassis acceleration

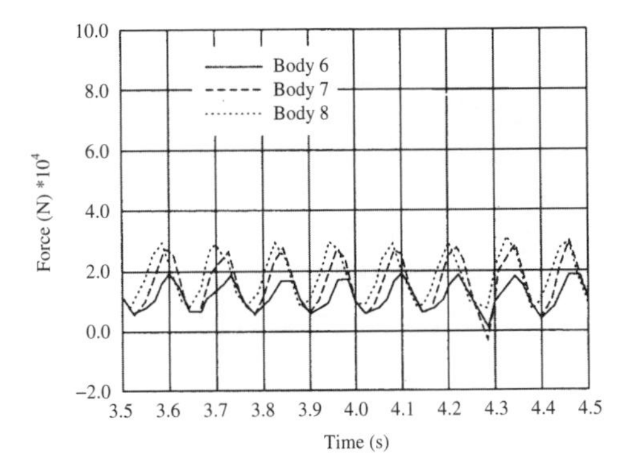

Figure 6.14 Contact forces

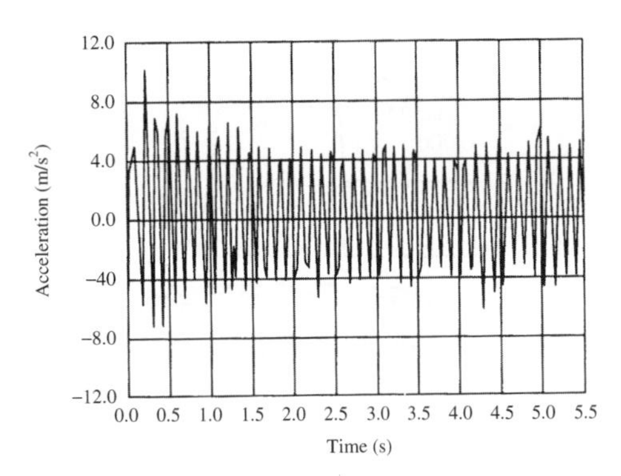

Figure 6.15 Chassis acceleration

In order to avoid the singular configuration, the dynamic simulation was carried out using different sets of independent coordinates. Before the time of the singular configuration was reached, the set of independent coordinates was changed to include the rotation of body 24. Significant improvements in the results were achieved as the result of changing the set of degrees of freedom. Figures 14 and 15 show, respectively, the contact forces acting on some of the rollers, and the acceleration of the center of the mass of the chassis when different sets of degrees of freedom were used to avoid the singular configurations. Figure 16 shows the results of a computer animation of the vehicle model.

Constraint Stabilization Methods The success of the numerical algorithm based on the coordinate partitioning method, which has been extensively used in multibody system dynamics literature, depends on the convergence of the iterative Newton–Raphson algorithm. A simple approach that can eliminate the need for using the Newton–Raphson method is *Baumgarte's constraint stabilization method*. Recall that the second time derivative of the constraints can be written as

$$\ddot{\mathbf{C}} = \mathbf{C_q}\ddot{\mathbf{q}} - \mathbf{Q}_d = \mathbf{0} \tag{6.214}$$

It is known, however, that the solution of the preceding equation can be an exponential growth, which is the case of an unstable solution. In Baumgarte's method, the equation above is modified and is written in the following form:

$$\ddot{\mathbf{C}} + 2\alpha\dot{\mathbf{C}} + (\beta)^2\mathbf{C} = \mathbf{0} \tag{6.215}$$

where $\alpha > 0$ and $\beta \neq 0$. The preceding two equations lead to

$$\mathbf{C_q}\ddot{\mathbf{q}} = \mathbf{Q}_d - 2\alpha(\mathbf{C_q}\dot{\mathbf{q}} + \mathbf{C}_t) - (\beta)^2\mathbf{C} \tag{6.216}$$

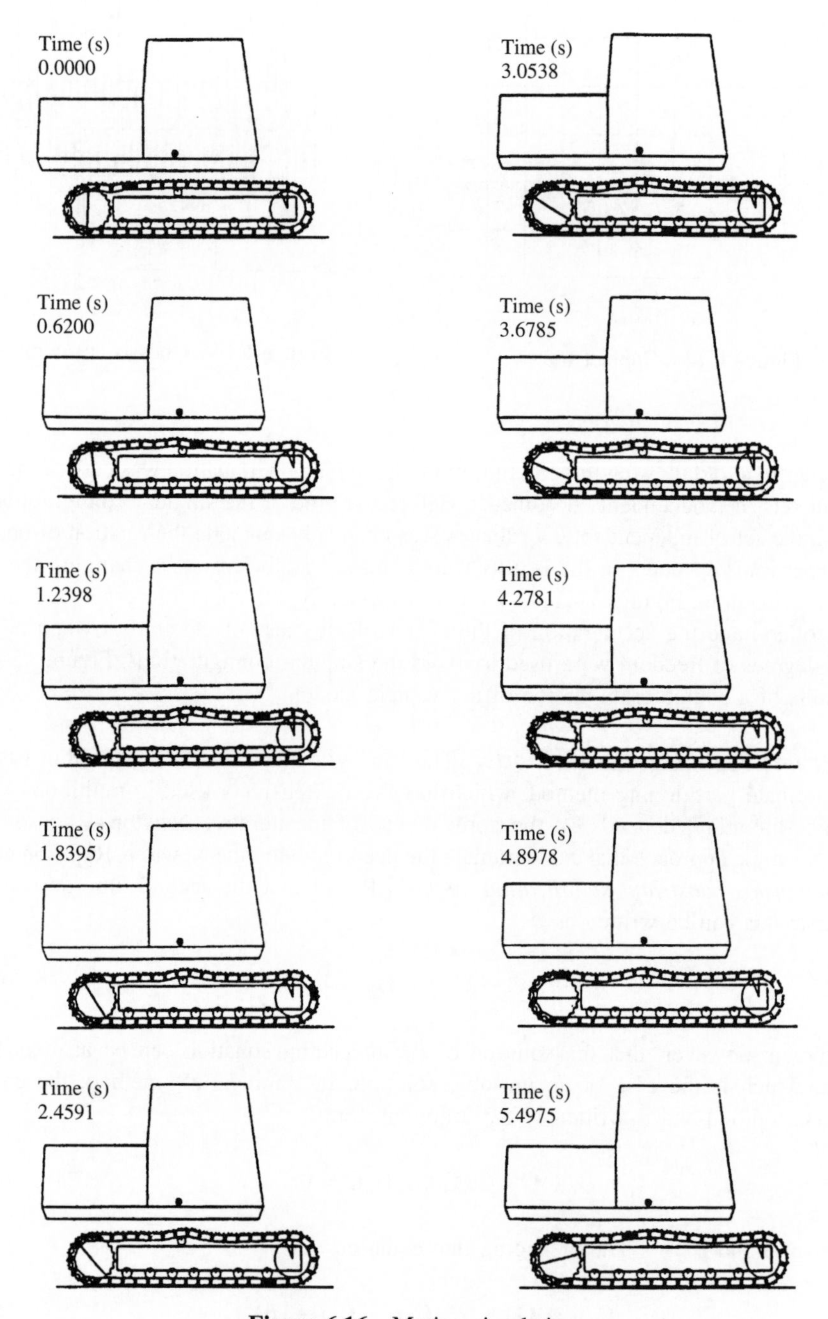

Figure 6.16 Motion simulation

The equations of motion of the system (Eq. 205) can be modified and written as

$$\begin{bmatrix} \mathbf{M} & \mathbf{C}_{\mathbf{q}}^{\mathrm{T}} \\ \mathbf{C}_{\mathbf{q}} & \mathbf{0} \end{bmatrix} \begin{bmatrix} \ddot{\mathbf{q}} \\ \boldsymbol{\lambda} \end{bmatrix} = \begin{bmatrix} \mathbf{Q}_e \\ \mathbf{Q}_d - 2\alpha(\mathbf{C}_{\mathbf{q}}\dot{\mathbf{q}} + \mathbf{C}_t) - (\beta)^2\mathbf{C} \end{bmatrix} \tag{6.217}$$

With this modification in the equations of motion, all the accelerations are numerically integrated without partitioning the coordinates into dependent and independent, thereby eliminating the need for the use of the iterative Newton–Raphson method.

While Baumgarte's constraint stabilization method gives accurate results in some applications, there is no known reliable method for selecting the coefficients α and β. Improper selection of these coefficients can lead to erroneous results.

6.14 INVERSE DYNAMICS

We discussed in Chapter 3 the case of kinematically driven systems in which the number of constraint equations is equal to the number of system-generalized coordinates. In this case, the Jacobian matrix is a square matrix, and if the constraint equations are linearly independent, the Jacobian matrix has a full row rank, and all the system coordinates can be considered as dependent. Equation 200 can then be solved for the system coordinates using a Newton–Raphson algorithm. Once the coordinates are determined, the first equation in Eq. 136 can be considered as a linear system of algebraic equations which can be solved for the total vector of system velocities. The accelerations, however, can be obtained by either using Eq. 136 as discussed in Chapter 3 or by using Eq. 205. In many multibody computer programs, Eq. 205 is used since its solution determines the vector of Lagrange multipliers, which can be used to evaluate the generalized constraint forces, including the generalized driving constraint forces.

The *inverse dynamics* is the problem of determining the driving joint forces that produce the desired motion trajectories. The procedure for solving the inverse dynamics problem is to define a *kinematically driven system* by introducing a set of driving constraints that define the prescribed motion. Hence, the position coordinates, velocities, and accelerations of the bodies that form the system can be determined using a standard kinematic analysis procedure as discussed in Chapter 3. Knowing the coordinates, velocities, and accelerations, the equations of motion of the system can be solved as a set of algebraic equations to determine the joint reaction forces as well as the driving constraint forces that are required to generate the prescribed motion. The obtained driving joint forces are often referred to as the *feed forward control law*. It is expected that when these forces are used to drive the system, the desired motion trajectories will be obtained.

The algorithms for the kinematic and dynamic analyses presented in Chapter 3 and in this chapter can be used to solve the inverse dynamics problem. In order to demonstrate the procedure for solving the inverse dynamics problem using the augmented formulation, we consider the slider crank mechanism shown in Fig. 17. One may be interested in determining the crankshaft torque that produces a specified desired motion of the slider block of the mechanism. Let us assume that the prescribed motion of the slider block is

$$R_x^4 = f(t) \tag{6.218}$$

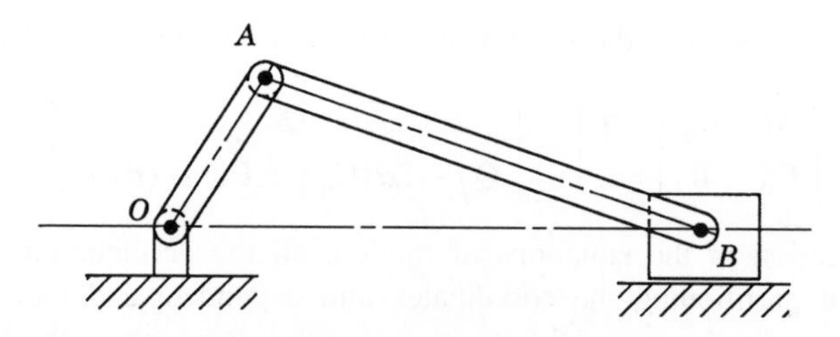

Figure 6.17 Inverse dynamics

where $f(t)$ is a given function of time. The objective then is to determine the crankshaft driving torque that produces the desired motion defined by the preceding equation. Since the slider crank mechanism has one degree of freedom, the use of the driving constraint defined by the preceding equation leads to the kinematically driven system discussed in Chapter 3. The coordinates and velocities of the bodies in the mechanism can be determined using a kinematic analysis procedure similar to the one discussed in Chapter 3. Once the coordinates and velocities are determined, the vectors of accelerations and Lagrange multipliers can be determined at every time step using Eq. 205. The solution of this equation defines Lagrange multipliers associated with the joint and driving constraints. Lagrange multipliers associated with the driving constraints can be identified and used to evaluate the driving constraint force.

For the slider crank mechanism example, the driving constraint can be written in the following form:

$$C_d = R_x^4 - f(t) = 0 \tag{6.219}$$

It is clear from this equation that the only nonzero component of the driving constraint force is the component associated with the motion of the slider block. This force component is defined as

$$F_x^4 = -\frac{\partial C_d}{\partial R_x^4}\lambda_d = -\lambda_d \tag{6.220}$$

where λ_d is the Lagrange multiplier associated with the driving constraint C_d. The virtual work of the driving constraint force is

$$\delta W_d = F_x^4\, \delta R_x^4 \tag{6.221}$$

In order to determine the *crankshaft torque* that is required to generate the desired motion of the slider block, we write δR_x^4 in terms of the virtual change in the crankshaft angle θ^2. The loop equations of the mechanism are

$$\left.\begin{aligned} l^2\cos\theta^2 + l^3\cos\theta^3 &= R_x^4 \\ l^2\sin\theta^2 + l^3\sin\theta^3 &= 0 \end{aligned}\right\} \tag{6.222}$$

where l^2 and l^3 are, respectively, the lengths of the crankshaft and the connecting rod, and θ^3 is the angle that defines the orientation of the connecting rod. Using the loop equations, one has

$$\delta R_x^4 = -l^2[\sin\theta^2 - \cos\theta^2 \tan\theta^3]\,\delta\theta^2 \tag{6.223}$$

Using this equation, the virtual work of the driving constraint force can be written as

$$\delta W_d = F_x^4\,\delta R_x^4 = -F_x^4 l^2[\sin\theta^2 - \cos\theta^2 \tan\theta^3]\,\delta\theta^2 \tag{6.224}$$

or

$$\delta W_d = M^2\,\delta\theta^2 \tag{6.225}$$

where M^2 is the crankshaft driving torque that is required to generate the desired motion of the slider block. This torque, which defines the *feedforward control law*, is given by

$$M^2 = -F_x^4 l^2[\sin\theta^2 - \cos\theta^2 \tan\theta^3] = \lambda_d l^2[\sin\theta^2 - \cos\theta^2 \tan\theta^3] \tag{6.226}$$

The simple one-degree-of-freedom slider crank mechanism example discussed in this section demonstrates the use of the augmented formulation and general-purpose multibody computer programs to solve the inverse dynamics problem. A similar procedure can be used in the case of multidegree of freedom systems. For such a multidegree of freedom system, one obtains the driving generalized joint forces associated with the independent joint coordinates.

6.15 STATIC ANALYSIS

Another case that can be considered as a special case of the general computational algorithm discussed earlier in this chapter is the case of the static analysis of constrained multibody systems. It is desirable in many applications to obtain the static equilibrium configuration prior to the dynamic simulation. Since in the case of static equilibrium, the velocities and accelerations are assumed to be equal to zero, Eq. 205 reduces to

$$\mathbf{C}_{\mathbf{q}}^{\mathrm{T}}\boldsymbol{\lambda} = \mathbf{Q}_e \tag{6.227}$$

This equation implies that if the multibody system is in static equilibrium, the generalized constraint forces must be equal to the generalized applied forces. Multiplying both sides of Eq. 227 by the vector $\delta\mathbf{q}$ yields

$$[\mathbf{Q}_e - \mathbf{C}_{\mathbf{q}}^{\mathrm{T}}\boldsymbol{\lambda}]^{\mathrm{T}}\,\delta\mathbf{q} = 0 \tag{6.228}$$

As described in Section 6 of this chapter, the vector $\delta\mathbf{q}$ of the virtual changes of the system-generalized coordinates can be expressed in terms of the virtual changes of the independent coordinates as

$$\delta\mathbf{q} = \mathbf{B}_i\,\delta\mathbf{q}_i \tag{6.229}$$

Substituting Eq. 229 into Eq. 228 leads to

$$[\mathbf{Q}_e - \mathbf{C}_{\mathbf{q}}^{\mathrm{T}}\boldsymbol{\lambda}]^{\mathrm{T}}\mathbf{B}_i\ \delta\mathbf{q}_i = 0 \tag{6.230}$$

Using the fact that $\mathbf{C}_{\mathbf{q}}\mathbf{B}_i = \mathbf{0}$, the preceding equation reduces to

$$\mathbf{Q}_e^{\mathrm{T}}\mathbf{B}_i\ \delta\mathbf{q}_i = 0 \tag{6.231}$$

Since the components of the vector $\delta\mathbf{q}_i$ are assumed to be independent, the preceding equation reduces to

$$\mathbf{B}_i^{\mathrm{T}}\mathbf{Q}_e = \mathbf{0} \tag{6.232}$$

This system, which has a number of equations equal to the number of independent coordinates, can be written as

$$\mathbf{R}_i(\mathbf{q}) = \mathbf{0} \tag{6.233}$$

where $\mathbf{R}_i = \mathbf{B}_i^{\mathrm{T}}\mathbf{Q}_e$. Equation 233 is a nonlinear system of equations in the system coordinates and can be solved using a Newton–Raphson approach.

Keeping in mind that the kinematic constraints can be used to express the dependent coordinates in terms of the independent ones, the numerical procedure for solving Eq. 233 starts by making a guess for the independent coordinates and use the constraint equations to determine the dependent coordinates. Being able to do this, allows us to write Eq. 233 in terms of the independent coordinates only as

$$\mathbf{R}_i(\mathbf{q}_i) = \mathbf{0} \tag{6.234}$$

This vector equation can be solved for the independent coordinates using an iterative Newton–Raphson algorithm.

Another alternative, yet equivalent, approach for the static analysis of a multibody system is to consider Eq. 227 with the constraint equations

$$\mathbf{C}(\mathbf{q}) = \mathbf{0} \tag{6.235}$$

Equations 227 and 235 represent $n + n_c$ nonlinear algebraic equations in the $n + n_c$ unknowns $\mathbf{q}$ and $\boldsymbol{\lambda}$. These two vector equations can be solved for the unknowns using a Newton–Raphson algorithm. The main difficulty in using this approach is the need for having an initial estimate of the vector of Lagrange multipliers.

PROBLEMS

1. Figure P1 shows a uniform slender rod that has mass 1 kg. At a given configuration, the velocity and acceleration of the center of mass are given by

$$\dot{\mathbf{R}} = [\dot{R}_x \quad \dot{R}_y]^{\mathrm{T}} = [5 \quad -15]^{\mathrm{T}}\ \mathrm{m/s}$$

$$\ddot{\mathbf{R}} = [\ddot{R}_x \quad \ddot{R}_y]^{\mathrm{T}} = [100 \quad 250]^{\mathrm{T}}\ \mathrm{m/s^2}$$

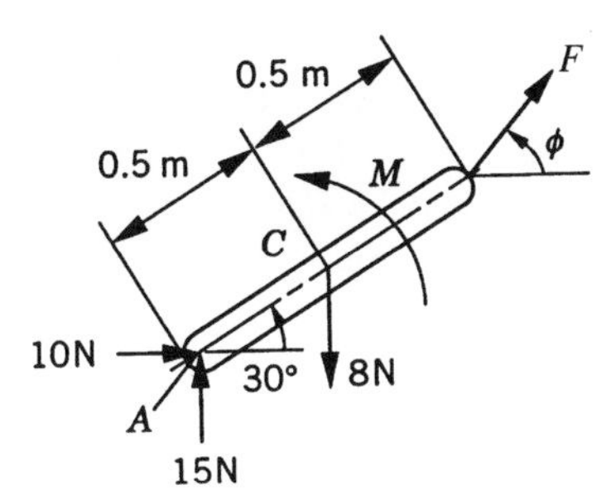

Figure P6.1

The angular velocity of the rod at this point in time is 150 rad/s, while the angular acceleration is zero. Determine the generalized inertia forces associated with the absolute coordinates of the center of mass as well as the rotation of the rod.

2. Repeat problem 1 assuming that the angular acceleration is $450\,\text{rad/s}^2$. What is the kinetic energy of the system and what is the mass matrix.

3. In problem 1 determine the generalized inertia forces associated with the absolute coordinates of point A as well as the rotation of the rod.

4. In problem 2 determine the generalized inertia forces associated with the absolute coordinates of point A as well as the rotation of the rod.

5. In problem 1, what will be the mass matrix if the generalized coordinates are selected to be the absolute coordinates of point A and the angular orientation of the rod? Also calculate the generalized centrifugal forces.

6. The slender rod shown in Fig. P1 has mass $m = 1\,\text{kg}$, $F = 5\,\text{N}$, $M = 10\,\text{N} \cdot \text{m}$, and $\phi = 45^\circ$. Determine the matrix equation of motion of this rod assuming that the generalized coordinates are the absolute Cartesian coordinates of the center of mass and the angular orientation of the rod. Using the obtained equations, determine the accelerations of the center of mass and the angular acceleration of the rod.

7. Repeat problem 6 assuming that the generalized coordinates are the absolute Cartesian coordinates of point A and the angular orientation of the rod.

8. For the system shown in Fig. P2, let $F = 10\,\text{N}$, $M = 15\,\text{N} \cdot \text{m}$, $\theta = 45^\circ$, $\phi = 80^\circ$, and assume that the rod is slender with mass $m = 1\,\text{kg}$, length $l = 1\,\text{m}$, and its angular velocity at the given configuration is 150 rad/s. If the generalized coordinates are selected to

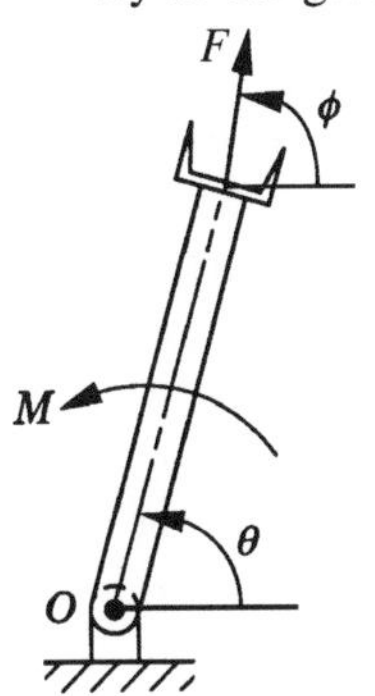

Figure P6.2

be the absolute coordinates of the center of mass and the angular orientation of the rod, obtain the matrix differential equations of the system in terms of these coordinates and the reaction forces. Express the absolute accelerations of the center of mass in terms of the angle θ. Use these kinematic relationships with the dynamic equations to determine the numerical values of the acceleration and the joint reaction forces.

9. In the preceding problem formulate the constraint equations of the revolute joint. Define the constraint Jacobian matrix and use the generalized coordinate partitioning to identify the Jacobian matrices associated with the dependent and independent coordinates. Use the coordinate partitioning to reduce the three differential equations obtained in the preceding problem to one that can be solved to determine the angular acceleration of the rod.

10. For the system shown in Fig. P3, let $M^2 = M^3 = 15\,\text{N}\cdot\text{m}$, $F = 10\,\text{N}$, $\theta^2 = 60°$, and $\theta^3 = 45°$. The angular velocities of the links are assumed to be $\dot{\theta}^2 = 10\,\text{rad/s}$ and $\dot{\theta}^3 = 5\,\text{rad/s}$. If the generalized coordinates are selected to be the absolute Cartesian coordinates of the center of mass of the slender rods and their angular orientations, obtain the matrix differential equations of the system in terms of these coordinates and the joint reaction forces. Express the absolute accelerations of the centers of mass in terms of the angular orientations of the two links. Use these kinematic equations with the obtained differential equations to solve for the accelerations and the joint reaction forces.

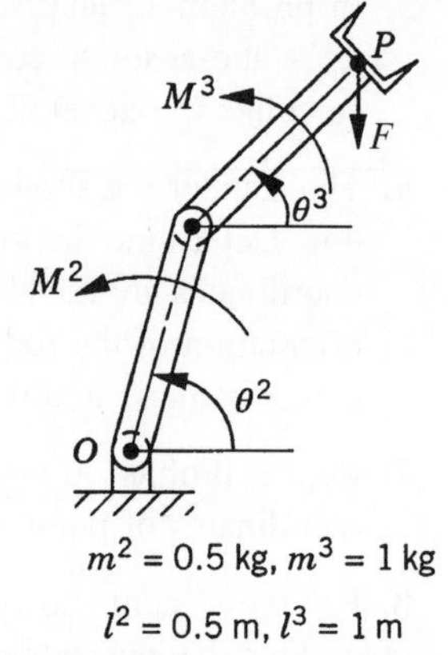

Figure P6.3

11. Use the generalized coordinate partitioning of the constraint Jacobian matrix to solve the preceding problem.

12. For the system shown in Fig. P4, let $M^2 = 10\,\text{N}\cdot\text{m}$, $M^3 = 8\,\text{N}\cdot\text{m}$, $M^4 = 5\,\text{N}\cdot\text{m}$, $F^4 = 5\,\text{N}$, $\phi = 30°$, $\theta^2 = 60°$, $\dot{\theta}^3 = 30°$, $\theta^4 = 45°$, $\dot{\theta}^2 = 4\,\text{rad/s}$, $\dot{\theta}^3 = \dot{\theta}^4 = 3\,\text{rad/s}$, and assume the links to be slender rods. If the generalized coordinates are selected to be the absolute coordinates of the center of mass and the angular orientations of the links, obtain the differential equations of motion of the system in terms of these coordinates and the reaction forces. Express the accelerations in terms of the angular orientations of the links. Use these kinematic relationships with the dynamic equations to solve for the accelerations and the joint reaction forces.

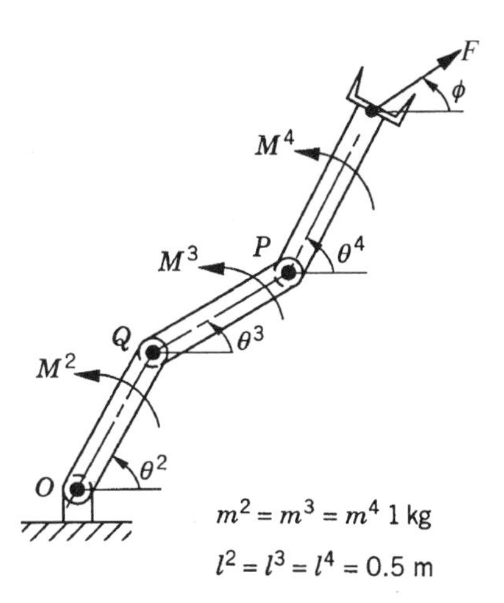

Figure P6.4

13. Use the generalized coordinate partitioning of the constraint Jacobian matrix to solve the preceding problem.

14. Develop the differential equations of motion of the slider crank mechanism shown in Fig. P5 in terms of the absolute coordinates. Assume that $M^2 = 10\,\mathrm{N \cdot m}$, $F^4 = 15\,\mathrm{N}$, $\theta^2 = 45^\circ$, and $\dot{\theta}^2 = 150\,\mathrm{rad/s}$. Use the generalized coordinate partitioning of the constraint Jacobian matrix to reduce the number of equations to one.

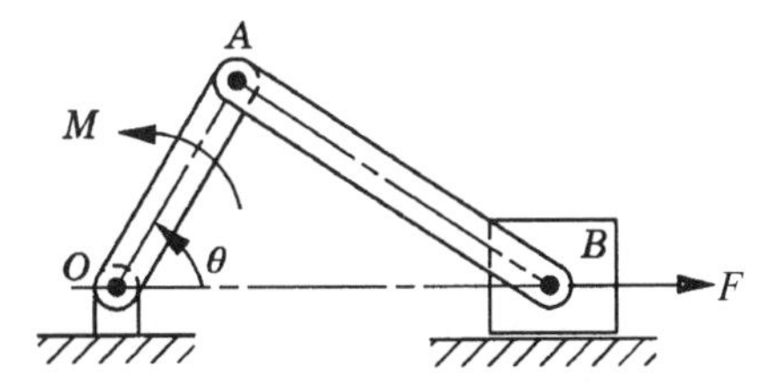

$m^2 = 0.5$ kg, $m^3 = 1$ kg, $m^4 = 0.2$ kg

$l^2 = 0.2$ m, $l^3 = 0.4$ m

Figure P6.5

15. Develop the differential equations of motion of the four-bar mechanism shown in Fig. P6 in terms of the absolute coordinates of each link. Assume $M^2 = M^4 = 5\,\mathrm{N \cdot m}$, $F^3 = 10\,\mathrm{N}$, $\theta^2 = 60^\circ$, and $\dot{\theta}^2 = 150\,\mathrm{rad/s}$. Use the generalized coordinate partitioning of the constraint Jacobian matrix to reduce the number of equations of motion to one.

16. Develop the equations of motion of the system of problem 12 in terms of Lagrange multipliers. Use the computer to solve the resulting system of equations for the accelerations and Lagrange multipliers, and use the obtained solution to determine the generalized and the actual reaction forces.

17. Develop the equations of motion of the slider crank mechanism of problem 14 in terms of Lagrange multipliers. Solve the resulting system of equations using computer methods in order to determine the accelerations and Lagrange multipliers, and determine the generalized and actual joint reaction forces.

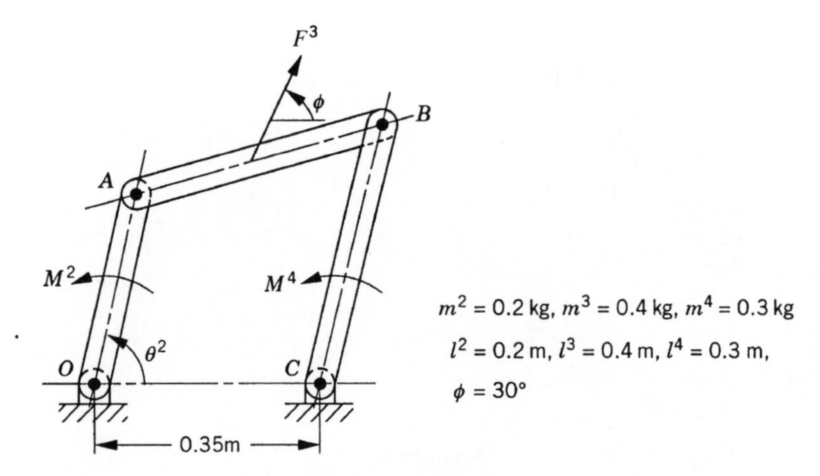

Figure P6.6

18. Obtain the differential equations of motion of the four-bar linkage of problem 15 in terms of Lagrange multipliers. Use the computer to solve the resulting system of equations for the accelerations and Lagrange multipliers, and determine the generalized and actual joint reaction forces.

19. Put the second-order differential equations of motion of the system of problem 6 in the state space form. Using the Runge–Kutta method, develop a computer program for the numerical integration of the state space equations of this system. Plot the angular displacement and velocity versus time for 1 s of simulation time. Use a step size $\Delta t = 0.01$ s, and assume that the initial conditions are $\dot{\mathbf{R}}_o = [5 \quad -15]^{\mathrm{T}}$ m/s and $\dot{\theta}_o = 150$ rad/s.

20. Develop the state space equations of the system of problem 8. Using the Runge–Kutta method, develop a computer program for the numerical integration of these equations. Plot the angular displacement and velocity of the rod for 1 s of simulation. Assume a step size of 0.005 s and use the configuration described in problem 8 as the initial configuration.

21. Find the state space equations of the system of problem 10. Develop a computer program based on the Runge–Kutta method for the numerical solution of this system. Plot the solution obtained using this program for 2 s of simulation with a step size $\Delta t = 0.01$ s. Use the configuration described in problem 10 as the initial configuration.

22. Obtain the state space equations of the slider crank mechanism of problem 14. Solve these equations numerically using the Runge–Kutta method and plot the position and orientation of the links for one revolution of the crankshaft. Use the configuration described in problem 14 as the initial configuration. Present the solution for two different time step sizes ($\Delta t = 0.001$ and 0.005 s) and compare the results.

23. Obtain the state space equations of the four-bar mechanism of problem 15. Solve these equations numerically using the Runge–Kutta method, and plot the angular displacements and velocities of the links for one revolution of the crankshaft. Use the configuration described in problem 15 as the initial configuration, and use a step size $\Delta t = 0.001$ s.

CHAPTER 7

SPATIAL DYNAMICS

In the planar analysis, the rotation of the rigid body can be described using one coordinate, such that the angular velocity of the body is defined as the time derivative of this orientation coordinate. Furthermore, the order of the finite rotation is commutative since the body rotation is about the same axis. Two consecutive rotations can be added and the sequence of performing these rotations is immaterial. One of the principal differences between the planar and the spatial kinematics is due to the complexity of defining the orientation of a body in a three-dimensional space. In the spatial analysis, the unconstrained motion of a rigid body is described using six coordinates; three coordinates describe the translation of a reference point on the body and three coordinates define the body orientation. The order of the finite rotation in the spatial analysis is not commutative and, consequently, the sequence of performing the rotations must be taken into consideration. Moreover, the angular velocities of a rigid body are not the time derivatives of a set of orientation coordinates. These angular velocities, however, can be expressed in terms of a selected set of orientation coordinates and their time derivatives.

In this chapter, methods for describing the orientation of rigid bodies in space are presented. The configuration of the rigid body in a multibody system is described using a set of generalized coordinates that define the global position vector of a reference point on the body as well as the body orientation. As in the planar analysis, coordinate transformations are defined in terms of the generalized orientation coordinates. The relationships between the angular velocity vectors and the time derivatives of the generalized orientation coordinates are developed and used to define the absolute velocity and acceleration vectors of an arbitrary point on the rigid body. These kinematic relationships are then used to develop the

Computational Dynamics, Third Edition Ahmed A. Shabana

dynamic equations of motion of multibody systems in terms of the system-generalized coordinates. As it is shown in the analysis presented in this chapter, the equations that describe the motion of a rigid body in three-dimensional space are quite complex as compared to the equations of the planar motion. Derivation of the dynamic equations of motion and definition of the mass matrix of spatial rigid body systems are presented and the simplifications in the dynamic relationships when the reference point is selected to be the body center of mass are discussed. This special case leads to the formulation of the *Newton–Euler equations*, in which there is no inertia coupling between the translation and rotation of the rigid body. The application of the augmented formulation and recursive methods to the dynamics of spatial multibody systems consisting of interconnected rigid bodies is also demonstrated in this chapter.

7.1 GENERAL DISPLACEMENT

In the three-dimensional analysis, the unconstrained motion of a rigid body is described using six independent coordinates; three independent coordinates describe the translation of the body and three independent coordinates define its orientation. The translational motion of the rigid body can be defined by the displacement of a selected *reference point* that is fixed in the rigid body. As shown in Fig. 1, which depicts a rigid body i in the three-dimensional space, the global position vector of an arbitrary point on the body can be written as

$$\mathbf{r}^i = \mathbf{R}^i + \mathbf{A}^i\bar{\mathbf{u}}^i \tag{7.1}$$

where $\mathbf{R}^i$ is the global position vector of the origin of the body reference $X^i Y^i Z^i$, $\mathbf{A}^i$ is the transformation matrix from the body coordinate system to the global XYZ coordinate system, and $\bar{\mathbf{u}}^i$ is the position vector of the arbitrary point with respect to the body coordinate

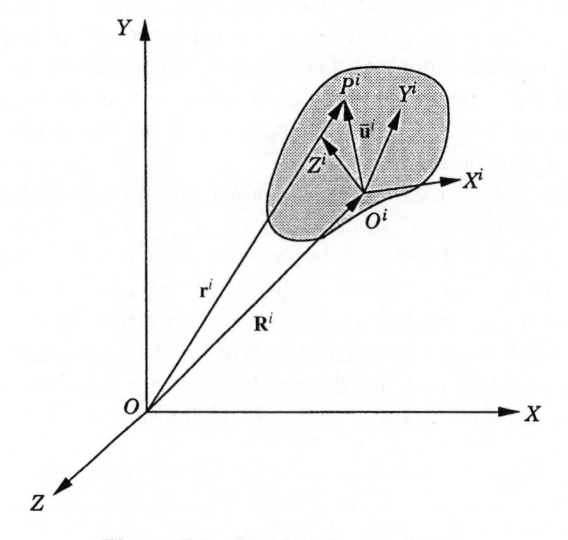

Figure 7.1 Rigid body coordinates

system. The transformation matrix $\mathbf{A}^i$ is a 3×3 matrix, and $\mathbf{R}^i$ and $\bar{\mathbf{u}}^i$ are three-dimensional vectors defined as

$$\mathbf{R}^i = [R_x^i \quad R_y^i \quad R_z^i]^{\mathrm{T}} \tag{7.2}$$

and

$$\bar{\mathbf{u}}^i = [\bar{u}_x^i \quad \bar{u}_y^i \quad \bar{u}_z^i]^{\mathrm{T}} = [x^i \quad y^i \quad z^i]^{\mathrm{T}} \tag{7.3}$$

In the case of pure translational motion, the orientation of the body does not change, and consequently, all the points on the rigid body move with equal velocities. On the other hand, if the reference point is fixed, the body does not have the freedom to translate and the remaining degrees of freedom are rotational.

7.2 FINITE ROTATIONS

In this section, a method for determining the transformation matrix $\mathbf{A}^i$ of Eq. 1 is presented. This matrix defines the orientation of the body coordinate system $X^iY^iZ^i$ with respect to the coordinate system XYZ. Figure 2 shows the coordinate system $X^iY^iZ^i$ of body i. The orientation of this coordinate system in the XYZ system can be described using the method of the *direction cosines*. Let $\mathbf{i}^i$, $\mathbf{j}^i$, and $\mathbf{k}^i$ be unit vectors along the X^i, Y^i, and Z^i axes, respectively, and let $\mathbf{i}$, $\mathbf{j}$, and $\mathbf{k}$ be unit vectors along the axes X, Y, and Z, respectively. Let β_1^i be the angle between the X^i and X axes, β_2^i be the angle between X^i and Y axes, and β_3^i be the angle between X^i and Z axes. The components of the unit vector $\mathbf{i}^i$ along the X, Y, and Z axes are given, respectively, by

$$\left.\begin{aligned} \alpha_{11} &= \cos\beta_1^i = \mathbf{i}^i \cdot \mathbf{i} \\ \alpha_{12} &= \cos\beta_2^i = \mathbf{i}^i \cdot \mathbf{j} \\ \alpha_{13} &= \cos\beta_3^i = \mathbf{i}^i \cdot \mathbf{k} \end{aligned}\right\} \tag{7.4}$$

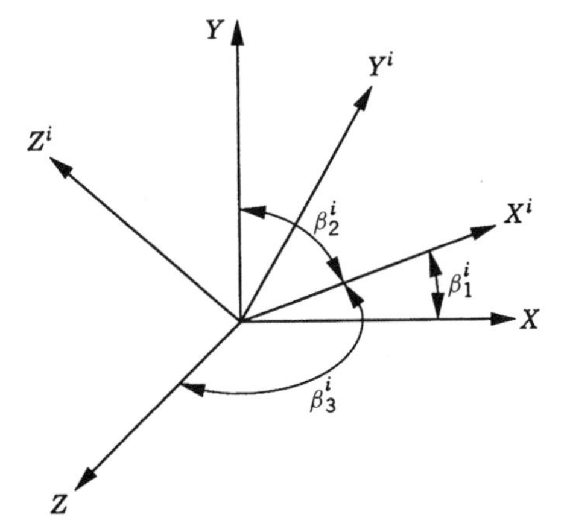

Figure 7.2 Direction cosines

where α_{11}, α_{12}, and α_{13} are the direction cosines of the X^i axis with respect to the X, Y, and Z axes, respectively. In a similar manner, the direction cosines α_{21}, α_{22}, and α_{23} of the axis Y^i, and the direction cosines α_{31}, α_{32}, and α_{33} of the axis Z^i, can be defined as

$$\left.\begin{aligned} \alpha_{21} &= \mathbf{j}^i \cdot \mathbf{i} \\ \alpha_{22} &= \mathbf{j}^i \cdot \mathbf{j} \\ \alpha_{23} &= \mathbf{j}^i \cdot \mathbf{k} \end{aligned}\right\} \tag{7.5}$$

and

$$\left.\begin{aligned} \alpha_{31} &= \mathbf{k}^i \cdot \mathbf{i} \\ \alpha_{32} &= \mathbf{k}^i \cdot \mathbf{j} \\ \alpha_{33} &= \mathbf{k}^i \cdot \mathbf{k} \end{aligned}\right\} \tag{7.6}$$

Since the direction cosines α_{ij} represent the components of the unit vectors $\mathbf{i}^i$, $\mathbf{j}^i$, and $\mathbf{k}^i$ along the axes X, Y, and Z, one has

$$\left.\begin{aligned} \mathbf{i}^i &= \alpha_{11}\mathbf{i} + \alpha_{12}\mathbf{j} + \alpha_{13}\mathbf{k} \\ \mathbf{j}^i &= \alpha_{21}\mathbf{i} + \alpha_{22}\mathbf{j} + \alpha_{23}\mathbf{k} \\ \mathbf{k}^i &= \alpha_{31}\mathbf{i} + \alpha_{32}\mathbf{j} + \alpha_{33}\mathbf{k} \end{aligned}\right\} \tag{7.7}$$

Let us now consider the vector $\mathbf{u}^i$ whose components in the body i coordinate system are denoted as $\overline{u}_x^i$, $\overline{u}_y^i$, and $\overline{u}_z^i$, and in the coordinate system XYZ, the components of the vector $\mathbf{u}^i$ are denoted as u_x^i, u_y^i, and u_z^i. The vector $\mathbf{u}^i$ can, therefore, have the following different representations:

$$\mathbf{u}^i = \overline{u}_x^i \mathbf{i}^i + \overline{u}_y^i \mathbf{j}^i + \overline{u}_z^i \mathbf{k}^i \tag{7.8}$$

or

$$\mathbf{u}^i = u_x^i \mathbf{i} + u_y^i \mathbf{j} + u_z^i \mathbf{k} \tag{7.9}$$

Substituting Eq. 7 into Eq. 8, one obtains

$$\begin{aligned} \mathbf{u}^i &= \overline{u}_x^i(\alpha_{11}\mathbf{i} + \alpha_{12}\mathbf{j} + \alpha_{13}\mathbf{k}) \\ &\quad + \overline{u}_y^i(\alpha_{21}\mathbf{i} + \alpha_{22}\mathbf{j} + \alpha_{23}\mathbf{k}) \\ &\quad + \overline{u}_z^i(\alpha_{31}\mathbf{i} + \alpha_{32}\mathbf{j} + \alpha_{33}\mathbf{k}) \end{aligned}$$

which leads to

$$\begin{aligned} \mathbf{u}^i &= (\alpha_{11}\overline{u}_x^i + \alpha_{21}\overline{u}_y^i + \alpha_{31}\overline{u}_z^i)\mathbf{i} \\ &\quad + (\alpha_{12}\overline{u}_x^i + \alpha_{22}\overline{u}_y^i + \alpha_{32}\overline{u}_z^i)\mathbf{j} \\ &\quad + (\alpha_{13}\overline{u}_x^i + \alpha_{23}\overline{u}_y^i + \alpha_{33}\overline{u}_z^i)\mathbf{k} \end{aligned} \tag{7.10}$$

By comparing Eqs. 9 and 10, one concludes

$$u_x^i = \alpha_{11}\overline{u}_x^i + \alpha_{21}\overline{u}_y^i + \alpha_{31}\overline{u}_z^i$$

$$u_y^i = \alpha_{12}\overline{u}_x^i + \alpha_{22}\overline{u}_y^i + \alpha_{32}\overline{u}_z^i$$
$$u_z^i = \alpha_{13}\overline{u}_x^i + \alpha_{23}\overline{u}_y^i + \alpha_{33}\overline{u}_z^i$$

That is, the relationship between the coordinates of the vector $\mathbf{u}^i$ in the XYZ coordinate system, and its coordinates in the $X^iY^iZ^i$ coordinate system, can be written in the following matrix form

$$\begin{bmatrix} u_x^i \\ u_y^i \\ u_z^i \end{bmatrix} = \begin{bmatrix} \alpha_{11} & \alpha_{21} & \alpha_{31} \\ \alpha_{12} & \alpha_{22} & \alpha_{32} \\ \alpha_{13} & \alpha_{23} & \alpha_{33} \end{bmatrix} \begin{bmatrix} \overline{u}_x^i \\ \overline{u}_y^i \\ \overline{u}_z^i \end{bmatrix} \tag{7.11}$$

In order to distinguish between the two different representations of the vector $\mathbf{u}^i$, this vector will be denoted as $\overline{\mathbf{u}}^i$ whenever its components are defined in the $X^iY^iZ^i$ coordinate system. That is

$$\mathbf{u}^i = [u_x^i \quad u_y^i \quad u_z^i]^{\mathrm{T}} \tag{7.12}$$

and

$$\overline{\mathbf{u}}^i = [\overline{u}_x^i \quad \overline{u}_y^i \quad \overline{u}_z^i]^{\mathrm{T}} \tag{7.13}$$

Using this notation, Eq. 11 can be rewritten as

$$\mathbf{u}^i = \mathbf{A}^i\overline{\mathbf{u}}^i \tag{7.14}$$

where $\mathbf{A}^i$ is recognized as the transformation matrix defined in terms of the direction cosines α_{ij}, $i, j = 1, 2, 3$ as

$$\mathbf{A}^i = \begin{bmatrix} \alpha_{11} & \alpha_{21} & \alpha_{31} \\ \alpha_{12} & \alpha_{22} & \alpha_{32} \\ \alpha_{13} & \alpha_{23} & \alpha_{33} \end{bmatrix} \tag{7.15}$$

Orthogonality of the Transformation Matrix The transformation matrix of Eq. 15 is expressed in terms of nine parameters α_{ij}, $i, j = 1, 2, 3$. The nine direction cosines α_{ij} are not totally independent because only three independent parameters are sufficient to describe the orientation of a rigid body in space. Since the direction cosines represent the components of three orthogonal unit vectors, they are related by the six algebraic equations

$$\alpha_{k1}\alpha_{l1} + \alpha_{k2}\alpha_{l2} + \alpha_{k3}\alpha_{l3} = \delta_{kl} \quad k, l = 1, 2, 3 \tag{7.16}$$

where δ_{kl} is the Kronecker delta defined as

$$\delta_{kl} = \begin{cases} 1 & \text{if } k = l \\ 0 & \text{if } k \neq l \end{cases} \tag{7.17}$$

Because of the six algebraic relationships of Eq. 16, there are only three independent components among the elements of the transformation matrix $\mathbf{A}^i$.

An important property of the transformation matrix is the *orthogonality*, that is

$$\mathbf{A}^{i^{\mathrm{T}}}\mathbf{A}^{i} = \mathbf{A}^{i}\mathbf{A}^{i^{\mathrm{T}}} = \mathbf{I} \tag{7.18}$$

where $\mathbf{I}$ is the 3×3 identity matrix. The orthogonality of the transformation matrix can be verified by direct matrix multiplication and the use of the identity of Eq. 16. The relationship of Eq. 18 remains valid regardless of the set of coordinates used to describe the orientation of the body in the three-dimensional space.

Example 7.1

If the axes X^i, Y^i, and Z^i of the coordinate system of body i are defined in the coordinate system XYZ by the vector $[1.0 \quad 0.0 \quad 1.0]^{\mathrm{T}}$, $[1.0 \quad 1.0 \quad -1.0]^{\mathrm{T}}$, and $[-1.0 \quad 2.0 \quad 1.0]^{\mathrm{T}}$, obtain the transformation matrix that defines the orientation of the coordinate system $X^iY^iZ^i$ with respect to the system XYZ.

Solution. The unit vectors $\mathbf{i}^i$, $\mathbf{j}^i$, and $\mathbf{k}^i$ along the axes X^i, Y^i, and Z^i are

$$\mathbf{i}^i = \left[\frac{1}{\sqrt{2}} \quad 0 \quad \frac{1}{\sqrt{2}}\right]^{\mathrm{T}}$$
$$\mathbf{j}^i = \left[\frac{1}{\sqrt{3}} \quad \frac{1}{\sqrt{3}} \quad \frac{-1}{\sqrt{3}}\right]^{\mathrm{T}}$$
$$\mathbf{k}^i = \left[\frac{-1}{\sqrt{6}} \quad \frac{2}{\sqrt{6}} \quad \frac{1}{\sqrt{6}}\right]^{\mathrm{T}}$$

The vectors $\mathbf{i}$, $\mathbf{j}$, and $\mathbf{k}$ are

$$\mathbf{i} = [1 \quad 0 \quad 0]^{\mathrm{T}}$$
$$\mathbf{j} = [0 \quad 1 \quad 0]^{\mathrm{T}}$$
$$\mathbf{k} = [0 \quad 0 \quad 1]^{\mathrm{T}}$$

It follows that

$$\alpha_{11} = \mathbf{i}^i \cdot \mathbf{i} = \frac{1}{\sqrt{2}}$$
$$\alpha_{12} = \mathbf{i}^i \cdot \mathbf{j} = 0$$
$$\alpha_{13} = \mathbf{i}^i \cdot \mathbf{k} = \frac{1}{\sqrt{2}}$$

Similarly,

$$\alpha_{21} = \frac{1}{\sqrt{3}}, \quad \alpha_{22} = \frac{1}{\sqrt{3}}, \quad \alpha_{23} = \frac{-1}{\sqrt{3}}$$

and

$$\alpha_{31} = \frac{-1}{\sqrt{6}}, \quad \alpha_{32} = \frac{2}{\sqrt{6}}, \quad \alpha_{33} = \frac{1}{\sqrt{6}}$$

The transformation matrix $\mathbf{A}^i$ that defines the orientation of the coordinate system $X^iY^iZ^i$ with respect to the coordinate system XYZ is

$$\mathbf{A}^i = \begin{bmatrix} \alpha_{11} & \alpha_{21} & \alpha_{31} \\ \alpha_{12} & \alpha_{22} & \alpha_{32} \\ \alpha_{13} & \alpha_{23} & \alpha_{33} \end{bmatrix} = \begin{bmatrix} \frac{1}{\sqrt{2}} & \frac{1}{\sqrt{3}} & \frac{-1}{\sqrt{6}} \\ 0 & \frac{1}{\sqrt{3}} & \frac{2}{\sqrt{6}} \\ \frac{1}{\sqrt{2}} & \frac{-1}{\sqrt{3}} & \frac{1}{\sqrt{6}} \end{bmatrix}$$

Note that

$$\mathbf{A}^i\mathbf{A}^{i^{\mathrm{T}}} = \begin{bmatrix} \frac{1}{\sqrt{2}} & \frac{1}{\sqrt{3}} & \frac{-1}{\sqrt{6}} \\ 0 & \frac{1}{\sqrt{3}} & \frac{2}{\sqrt{6}} \\ \frac{1}{\sqrt{2}} & \frac{-1}{\sqrt{3}} & \frac{1}{\sqrt{6}} \end{bmatrix} \begin{bmatrix} \frac{1}{\sqrt{2}} & 0 & \frac{1}{\sqrt{2}} \\ \frac{1}{\sqrt{3}} & \frac{1}{\sqrt{3}} & \frac{-1}{\sqrt{3}} \\ \frac{-1}{\sqrt{6}} & \frac{2}{\sqrt{6}} & \frac{1}{\sqrt{6}} \end{bmatrix}$$

$$= \begin{bmatrix} 1 & 0 & 0 \\ 0 & 1 & 0 \\ 0 & 0 & 1 \end{bmatrix} = \mathbf{I} = \mathbf{A}^{i^{\mathrm{T}}}\mathbf{A}^i$$

Simple Rotations We now consider the case of simple finite rotations of the coordinate system $X^iY^iZ^i$ about the axes of the coordinate system XYZ. Consider the case in which the axes of the two coordinate systems are initially parallel and the origins of the two systems coincide. First we consider the rotation of the coordinate system $X^iY^iZ^i$ about the Z axis by an angle ϕ^i. As shown in Fig. 3, as a result of this finite rotation

$$\begin{aligned} \alpha_{11} &= \mathbf{i}^i \cdot \mathbf{i} = \cos\phi^i \\ \alpha_{12} &= \mathbf{i}^i \cdot \mathbf{j} = \sin\phi^i \\ \alpha_{13} &= \mathbf{i}^i \cdot \mathbf{k} = 0 \end{aligned}$$

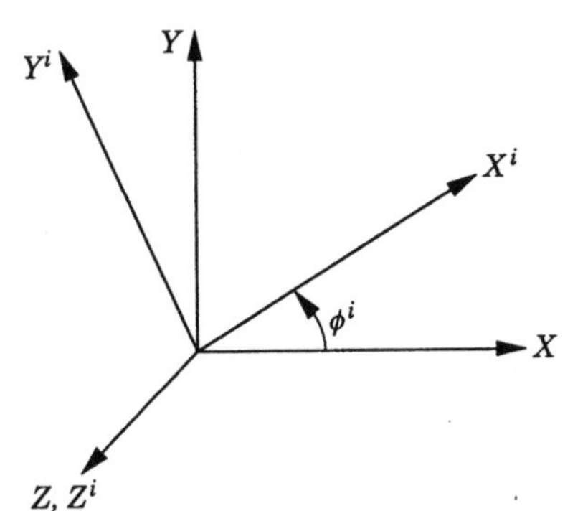

Figure 7.3 Simple rotation

Similarly, one can show that the other direction cosines are

$$\alpha_{21} = -\sin\phi^i, \quad \alpha_{22} = \cos\phi^i, \quad \alpha_{23} = 0$$
$$\alpha_{31} = 0, \quad \alpha_{32} = 0, \quad \alpha_{33} = 1$$

It follows that the transformation matrix that defines the orientation of the coordinate system $X^iY^iZ^i$ as the result of a rotation ϕ^i about the Z axis is given by

$$\mathbf{A}^i = \begin{bmatrix} \cos\phi^i & -\sin\phi^i & 0 \\ \sin\phi^i & \cos\phi^i & 0 \\ 0 & 0 & 1 \end{bmatrix} \tag{7.19}$$

The use of a similar procedure shows that if the coordinate system $X^iY^iZ^i$ rotates an angle θ^i about the Y axis, the transformation matrix is

$$\mathbf{A}^i = \begin{bmatrix} \cos\theta^i & 0 & \sin\theta^i \\ 0 & 1 & 0 \\ -\sin\theta^i & 0 & \cos\theta^i \end{bmatrix} \tag{7.20}$$

and in the case of a rotation ψ^i about the X axis, the resulting transformation matrix is

$$\mathbf{A}^i = \begin{bmatrix} 1 & 0 & 0 \\ 0 & \cos\psi^i & -\sin\psi^i \\ 0 & \sin\psi^i & \cos\psi^i \end{bmatrix} \tag{7.21}$$

Successive Rotations As demonstrated in Chapter 1, the order of the finite rotation is not commutative. An exception to this rule occurs only when the axes of rotation are parallel. Consider the case of three coordinate systems $X^1Y^1Z^1$, $X^2Y^2Z^2$, and $X^3Y^3Z^3$. As shown in Fig. 4, these three coordinate systems have different orientations. Let $\mathbf{A}^{32}$ be the transformation matrix that defines the orientation of the coordinate system $X^3Y^3Z^3$ with respect to the coordinate system $X^2Y^2Z^2$, and let $\mathbf{A}^{21}$ be the transformation matrix that defines the orientation of the coordinate system $X^2Y^2Z^2$ with respect to the coordinate system $X^1Y^1Z^1$. Let $\bar{\mathbf{u}}^3$ be a vector defined in the coordinate system $X^3Y^3Z^3$. The components of the vector $\bar{\mathbf{u}}^3$ can be defined in the coordinate system $X^2Y^2Z^2$ by the vector $\mathbf{u}^2$, where

$$\mathbf{u}^2 = \mathbf{A}^{32}\bar{\mathbf{u}}^3 \tag{7.22}$$

The components of the vector $\mathbf{u}^2$ can be defined in the coordinate system $X^1Y^1Z^1$ by the vector $\mathbf{u}^1$, where

$$\mathbf{u}^1 = \mathbf{A}^{21}\mathbf{u}^2 \tag{7.23}$$

Substituting Eq. 22 into Eq. 23, the vector $\mathbf{u}^1$ can be expressed in terms of the original vector $\bar{\mathbf{u}}^3$ as

$$\mathbf{u}^1 = \mathbf{A}^{21}\mathbf{A}^{32}\bar{\mathbf{u}}^3 \tag{7.24}$$

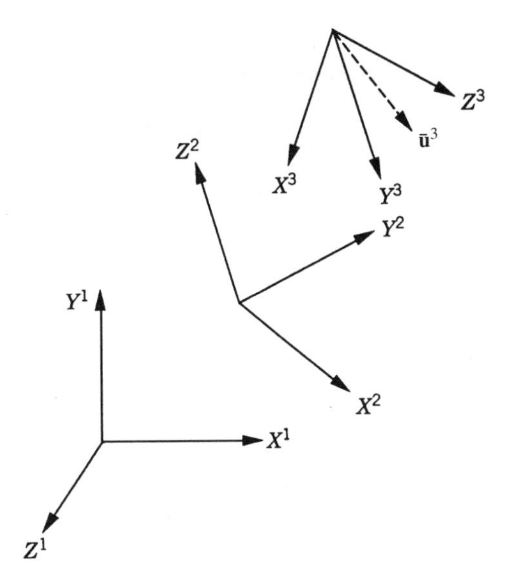

Figure 7.4 Successive rotations

which can be written as

$$\mathbf{u}^1 = \mathbf{A}^{31}\bar{\mathbf{u}}^3 \tag{7.25}$$

where, in this case, the transformation matrix $\mathbf{A}^{31}$ that defines the orientation of the coordinate system $X^3Y^3Z^3$ with respect to the coordinate system $X^1Y^1Z^1$ is given by

$$\mathbf{A}^{31} = \mathbf{A}^{21}\mathbf{A}^{32} \tag{7.26}$$

Similarly, in the case of n coordinate systems, one has

$$\mathbf{A}^{n1} = \mathbf{A}^{21}\mathbf{A}^{32} \ \cdots \ \mathbf{A}^{(n-1)(n-2)}\mathbf{A}^{n(n-1)} \tag{7.27}$$

where $\mathbf{A}^{ij}$ is the transformation matrix that defines the orientation of the ith coordinate system with respect to the jth coordinate system. Note that in general

$$\mathbf{A}^{ij}\mathbf{A}^{jk} \neq \mathbf{A}^{jk}\mathbf{A}^{ij}$$

as previously demonstrated.

Example 7.2

In the initial configuration, the axes X^i, Y^i, and Z^i of the coordinate system of body i are defined in the XYZ coordinate system by the vectors $[1.0 \quad 0.0 \quad 1.0]^{\mathrm{T}}$, $[1.0 \quad 1.0 \quad -1.0]^{\mathrm{T}}$, and $[-1.0 \quad 2.0 \quad 1.0]^{\mathrm{T}}$, respectively. The body then rotates an angle $\theta_1^i = 90^\circ$ about its Z^i axis followed by a rotation $\theta_2^i = 90^\circ$ about its X^i axis. Determine the transformation matrix that defines the orientation of the body i in the XYZ coordinate system as a result of the successive rotations. If $\bar{\mathbf{u}}^i = [0.0 \quad -1.0 \quad 0.0]^{\mathrm{T}}$ is a vector defined in the body coordinate system, define this vector in the XYZ coordinate system after the rotations θ_1^i and θ_2^i.

Solution. In the initial configuration, the method of the direction cosines can be used to determine the transformation $\mathbf{A}_0^i$ that defines the orientation of the body before the rotations θ_1^i and θ_2^i. It was shown in Example 1 that this transformation matrix is

$$\mathbf{A}_0^i = \begin{bmatrix} \frac{1}{\sqrt{2}} & \frac{1}{\sqrt{3}} & \frac{-1}{\sqrt{6}} \\ 0 & \frac{1}{\sqrt{3}} & \frac{2}{\sqrt{6}} \\ \frac{1}{\sqrt{2}} & \frac{-1}{\sqrt{3}} & \frac{1}{\sqrt{6}} \end{bmatrix}$$

As a result of the rotation θ_1^i, the body occupies a new position. The orientation of the body as a result of this rotation is defined with respect to the initial configuration by the matrix $\mathbf{A}_1^i$ defined as

$$\mathbf{A}_1^i = \begin{bmatrix} \cos\theta_1^i & -\sin\theta_1^i & 0 \\ \sin\theta_1^i & \cos\theta_1^i & 0 \\ 0 & 0 & 1 \end{bmatrix} = \begin{bmatrix} 0 & -1 & 0 \\ 1 & 0 & 0 \\ 0 & 0 & 1 \end{bmatrix}$$

Since the second rotation θ_2^i is about the X^i axis, the matrix $\mathbf{A}_2^i$ is given by

$$\mathbf{A}_2^i = \begin{bmatrix} 1 & 0 & 0 \\ 0 & \cos\theta_2^i & -\sin\theta_2^i \\ 0 & \sin\theta_2^i & \cos\theta_2^i \end{bmatrix} = \begin{bmatrix} 1 & 0 & 0 \\ 0 & 0 & -1 \\ 0 & 1 & 0 \end{bmatrix}$$

The final orientation of the body is defined by the matrix $\mathbf{A}^i$ given by

$$\mathbf{A}^i = \mathbf{A}_0^i\mathbf{A}_1^i\mathbf{A}_2^i = \begin{bmatrix} \frac{1}{\sqrt{2}} & \frac{1}{\sqrt{3}} & \frac{-1}{\sqrt{6}} \\ 0 & \frac{1}{\sqrt{3}} & \frac{2}{\sqrt{6}} \\ \frac{1}{\sqrt{2}} & \frac{-1}{\sqrt{3}} & \frac{1}{\sqrt{6}} \end{bmatrix} \begin{bmatrix} 0 & -1 & 0 \\ 1 & 0 & 0 \\ 0 & 0 & 1 \end{bmatrix} \begin{bmatrix} 0 & 0 & 0 \\ 0 & 0 & -1 \\ 0 & 1 & 0 \end{bmatrix}$$

$$= \begin{bmatrix} \frac{1}{\sqrt{3}} & \frac{-1}{\sqrt{6}} & \frac{1}{\sqrt{2}} \\ \frac{1}{\sqrt{3}} & \frac{2}{\sqrt{6}} & 0 \\ \frac{-1}{\sqrt{3}} & \frac{1}{\sqrt{6}} & \frac{1}{\sqrt{2}} \end{bmatrix} = \begin{bmatrix} 0.5773 & -0.4082 & 0.7071 \\ 0.5773 & 0.8165 & 0 \\ -0.5773 & 0.4082 & 0.7071 \end{bmatrix}$$

The components of the vector $\bar{\mathbf{u}}^i = [0.0 \quad -1.0 \quad 0.0]^{\mathrm{T}}$ in the coordinate system XYZ can then be obtained as

$$\mathbf{u}^i = \mathbf{A}^i\bar{\mathbf{u}}^i = \begin{bmatrix} \frac{1}{\sqrt{3}} & \frac{-1}{\sqrt{6}} & \frac{1}{\sqrt{2}} \\ \frac{1}{\sqrt{3}} & \frac{2}{\sqrt{6}} & 0 \\ \frac{-1}{\sqrt{3}} & \frac{1}{\sqrt{6}} & \frac{1}{\sqrt{2}} \end{bmatrix} \begin{bmatrix} 0 \\ -1 \\ 0 \end{bmatrix} = \begin{bmatrix} \frac{1}{\sqrt{6}} \\ \frac{-2}{\sqrt{6}} \\ \frac{-1}{\sqrt{6}} \end{bmatrix}$$

7.3 EULER ANGLES

The direction cosines are rarely used in describing the three-dimensional rotations of multibody systems. Among the most widely used parameters for describing the orientation are the three independent *Euler angles*. Several different sets of Euler angles are in common use in the analysis of mechanical and aerospace systems. In this section, the most widely used Euler angles are defined and the transformation matrix in terms of them is developed.

Euler angles involve three successive rotations about three axes that are not, in general, perpendicular. By performing these three successive rotations in the proper sequence, a coordinate system can reach any orientation. Consider the coordinate systems XYZ and $X^iY^iZ^i$, which initially coincide. The sequence starts as shown in Fig. 5, by rotating the system $X^iY^iZ^i$ an angle ϕ^i about the Z axis. Since ϕ^i is a rotation about the Z axis, the transformation matrix resulting from this rotation is given by

$$\mathbf{A}_1^i = \begin{bmatrix} \cos\phi^i & -\sin\phi^i & 0 \\ \sin\phi^i & \cos\phi^i & 0 \\ 0 & 0 & 1 \end{bmatrix} \tag{7.28}$$

The coordinate system $X^iY^iZ^i$ is then rotated an angle θ^i about the current X^i axis, which at the current position is called the *line of nodes*. The change in the orientation of the

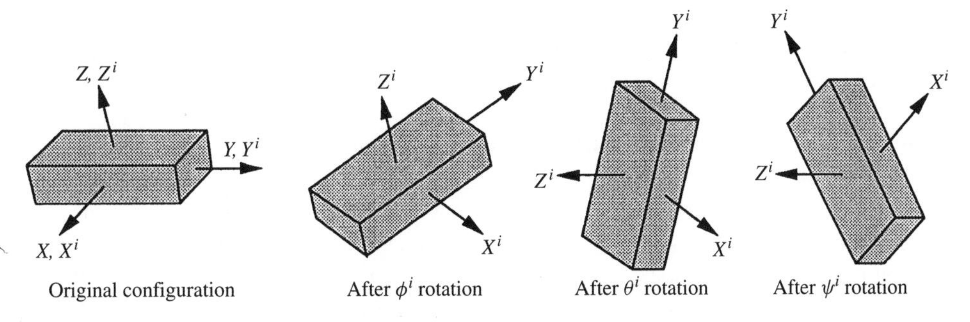

Figure 7.5 Euler angles

coordinate system $X^iY^iZ^i$ as a result of the rotation θ^i is described using the matrix

$$\mathbf{A}_2^i = \begin{bmatrix} 1 & 0 & 0 \\ 0 & \cos\theta^i & -\sin\theta^i \\ 0 & \sin\theta^i & \cos\theta^i \end{bmatrix} \tag{7.29}$$

Finally, the coordinate system $X^iY^iZ^i$ is rotated an angle ψ^i about the current Z^i axis. The change in the orientation of the system $X^iY^iZ^i$ as a result of this rotation is given by

$$\mathbf{A}_3^i = \begin{bmatrix} \cos\psi^i & -\sin\psi^i & 0 \\ \sin\psi^i & \cos\psi^i & 0 \\ 0 & 0 & 1 \end{bmatrix} \tag{7.30}$$

Using the transformation matrices $\mathbf{A}_1^i$, $\mathbf{A}_2^i$, and $\mathbf{A}_3^i$ given, respectively, by Eqs. 28, 29, and 30, the final orientation of the coordinate system $X^iY^iZ^i$ can be defined in the system XYZ by the transformation matrix $\mathbf{A}^i$ given by

$$\mathbf{A}^i = \mathbf{A}_1^i\mathbf{A}_2^i\mathbf{A}_3^i \tag{7.31}$$

This equation can be used to define the matrix $\mathbf{A}^i$ in terms of the angles ϕ^i, θ^i, and ψ^i as

$$\mathbf{A}^i = \begin{bmatrix} \cos\psi^i\cos\phi^i - \cos\theta^i\sin\phi^i\sin\psi^i & -\sin\psi^i\cos\phi^i - \cos\theta^i\sin\phi^i\cos\psi^i & \sin\theta^i\sin\phi^i \\ \cos\psi^i\sin\phi^i + \cos\theta^i\cos\phi^i\sin\psi^i & -\sin\psi^i\sin\phi^i + \cos\theta^i\cos\phi^i\cos\psi^i & -\sin\theta^i\cos\phi^i \\ \sin\theta^i\sin\psi^i & \sin\theta^i\cos\psi^i & \cos\theta^i \end{bmatrix} \tag{7.32}$$

The three angles ϕ^i, θ^i, and ψ^i are called the *Euler angles*. The orientation of any rigid body in space can be obtained by performing these three independent successive rotations.

In the discussion presented in this section, we considered the sequence of rotations about the Z^i, X^i, and Z^i axes. Other sequences that are also used to define Euler angles are rotations about Z^i, Y^i, and X^i, or rotations about Z^i, Y^i, and Z^i. The use of these sequences leads to transformation matrices that are different from the one presented in this section. The procedure used to define the transformation matrix using these sequences, however, is the same as described in this section and is left to the reader as an exercise. There are also other sets of rotational coordinates that are often used to describe the orientations of rigid bodies in space. Among these coordinates are the four *Euler parameters* and the three independent *Rodriguez parameters*. These sets will be introduced in the following chapter.

Relationship between Euler Angles and Direction Cosines Using the fact that the elements of the transformation matrix are the direction cosines of the axes X^i, Y^i, and Z^i, the nine direction cosines can be easily expressed in terms of Euler angles. For instance, the elements of the third column in the transformation matrix of Eq. 32 are the direction cosines of the Z^i axis. These three direction cosines are functions of the angle ϕ^i since the rotation θ^i is about an axis whose direction cosines are defined by the unit vector $[\cos\phi^i \quad -\sin\phi^i \quad 0]^{\mathrm{T}}$. Euler angles can also be expressed in terms of the direction cosines by equating the elements of the transformation matrices in the two cases. For example, using the last row and column, one has

$$
\begin{aligned}
\theta^i &= \cos^{-1}(\alpha_{33}) \\
\psi^i &= \cos^{-1}\left\{\frac{-\alpha_{23}}{\sin\theta^i}\right\} \\
\phi^i &= \cos^{-1}\left\{\frac{\alpha_{32}}{\sin\theta^i}\right\}
\end{aligned}
$$

The quadrants where the angles lie are selected so as to ensure that all the remaining elements in the Euler angle transformation matrix are the same as the elements of the transformation matrix evaluated using the direction cosines.

7.4 VELOCITY AND ACCELERATION

The general displacement of a rigid body in space is the result of translational and rotational displacements. In this case, the global position vector of an arbitrary point on the body is given by Eq. 1, which is reproduced here:

$$\mathbf{r}^i = \mathbf{R}^i + \mathbf{A}^i\bar{\mathbf{u}}^i \tag{7.33}$$

where $\mathbf{R}^i$ is the global position vector of the origin of the body fixed coordinate system, $\mathbf{A}^i$ is the transformation matrix that defines the orientation of the body in the global coordinate system, and $\bar{\mathbf{u}}^i$ is the position vector of the arbitrary point with respect to the origin of the body coordinate system. The vectors $\mathbf{R}^i$ and $\bar{\mathbf{u}}^i$ are defined by Eqs. 2 and 3, respectively. In the case of rigid body analysis, the components of the vector $\bar{\mathbf{u}}^i$ are constant.

Velocity The absolute velocity of an arbitrary point on the rigid body can be obtained by differentiating Eq. 33 with respect to time. This leads to

$$\dot{\mathbf{r}}^i = \dot{\mathbf{R}}^i + \dot{\mathbf{A}}^i\bar{\mathbf{u}}^i \tag{7.34}$$

Since the transformation matrix is orthogonal, one has

$$\mathbf{A}^i\mathbf{A}^{i\mathrm{T}} = \mathbf{I} \tag{7.35}$$

which upon differentiation leads to

$$\dot{\mathbf{A}}^i\mathbf{A}^{i\mathrm{T}} + \mathbf{A}^i\dot{\mathbf{A}}^{i\mathrm{T}} = \mathbf{0} \tag{7.36}$$

or

$$\dot{\mathbf{A}}^i\mathbf{A}^{i\mathrm{T}} = -\mathbf{A}^i\dot{\mathbf{A}}^{i\mathrm{T}} \tag{7.37}$$

This equation implies that

$$\dot{\mathbf{A}}^i \mathbf{A}^{i^{\mathrm{T}}} = -(\dot{\mathbf{A}}^i \mathbf{A}^{i^{\mathrm{T}}})^{\mathrm{T}} \tag{7.38}$$

A matrix that is equal to the negative of its transpose is a skew symmetric matrix. Therefore, Eq. 38 can be written as

$$\dot{\mathbf{A}}^i \mathbf{A}^{i^{\mathrm{T}}} = \tilde{\boldsymbol{\omega}}^i \tag{7.39}$$

where $\tilde{\boldsymbol{\omega}}^i$ is a skew symmetric matrix that can be written as

$$\tilde{\boldsymbol{\omega}}^i = \begin{bmatrix} 0 & -\omega_3^i & \omega_2^i \\ \omega_3^i & 0 & -\omega_1^i \\ -\omega_2^i & \omega_1^i & 0 \end{bmatrix} \tag{7.40}$$

and ω_1^i, ω_2^i, and ω_3^i are called the components of the *angular velocity vector* $\boldsymbol{\omega}^i$, that is

$$\boldsymbol{\omega}^i = [\omega_1^i \quad \omega_2^i \quad \omega_3^i]^{\mathrm{T}} \tag{7.41}$$

Postmultiplying both sides of Eq. 39 by the matrix $\mathbf{A}^i$ and using the orthogonality of the transformation matrix, one obtains

$$\dot{\mathbf{A}}^i = \tilde{\boldsymbol{\omega}}^i \mathbf{A}^i \tag{7.42}$$

Substituting this equation into Eq. 34 yields

$$\dot{\mathbf{r}}^i = \dot{\mathbf{R}}^i + \tilde{\boldsymbol{\omega}}^i \mathbf{A}^i \overline{\mathbf{u}}^i \tag{7.43}$$

which can also be written as

$$\dot{\mathbf{r}}^i = \dot{\mathbf{R}}^i + \tilde{\boldsymbol{\omega}}^i \mathbf{u}^i \tag{7.44}$$

where

$$\mathbf{u}^i = \mathbf{A}^i \overline{\mathbf{u}}^i \tag{7.45}$$

Using the cross product notation, Eq. 44 can be rewritten as

$$\dot{\mathbf{r}}^i = \dot{\mathbf{R}}^i + \boldsymbol{\omega}^i \times \mathbf{u}^i \tag{7.46}$$

Example 7.3

Consider the case in which the rigid body i rotates about the fixed Z axis. If the angle of rotation is denoted as θ^i, the transformation matrix is

$$\mathbf{A}^i = \begin{bmatrix} \cos\theta^i & -\sin\theta^i & 0 \\ \sin\theta^i & \cos\theta^i & 0 \\ 0 & 0 & 1 \end{bmatrix}$$

and

$$\dot{\mathbf{A}}^i = \dot{\theta}^i \begin{bmatrix} -\sin\theta^i & -\cos\theta^i & 0 \\ \cos\theta^i & -\sin\theta^i & 0 \\ 0 & 0 & 0 \end{bmatrix}$$

The skew-symmetric matrix $\tilde{\boldsymbol{\omega}}^i$ of Eq. 39 is

$$\tilde{\boldsymbol{\omega}}^i = \dot{\mathbf{A}}^i \mathbf{A}^{i^{\mathrm{T}}} = \dot{\theta}^i \begin{bmatrix} 0 & -1 & 0 \\ 1 & 0 & 0 \\ 0 & 0 & 0 \end{bmatrix}$$

That is,

$$\tilde{\boldsymbol{\omega}}^i = \begin{bmatrix} 0 & -\omega_3^i & \omega_2^i \\ \omega_3^i & 0 & -\omega_1^i \\ -\omega_2^i & \omega_1^i & 0 \end{bmatrix} = \dot{\theta}^i \begin{bmatrix} 0 & -1 & 0 \\ 1 & 0 & 0 \\ 0 & 0 & 0 \end{bmatrix}$$

which defines the angular velocity vector $\boldsymbol{\omega}^i$ as

$$\boldsymbol{\omega}^i = \begin{bmatrix} \omega_1^i \\ \omega_2^i \\ \omega_3^i \end{bmatrix} = \begin{bmatrix} 0 \\ 0 \\ \dot{\theta}^i \end{bmatrix} = \dot{\theta}^i \mathbf{k}$$

where $\mathbf{k}$ is a unit vector along the axis of rotation. The preceding equation, which is the familiar form of the angular velocity vector used in the planar analysis, is obtained here using the general development presented in this section.

An Alternative Representation Equation 39 defines the components of the angular velocity vector in the global coordinate system. The components of this vector can also be defined in the coordinate system of body i using the transformation

$$\overline{\boldsymbol{\omega}}^i = \mathbf{A}^{i^{\mathrm{T}}} \boldsymbol{\omega}^i \tag{7.47}$$

where $\overline{\boldsymbol{\omega}}^i$ is the absolute angular velocity vector defined in the coordinate system of body i.

An alternative form for the absolute velocity vector of an arbitrary point on the rigid body can be obtained by using the vector $\overline{\boldsymbol{\omega}}^i$. This can be achieved by directly using Eq. 47, or by utilizing the orthogonality condition of the transformation matrix. To demonstrate the use of the second route, the orthogonality condition is repeated here:

$$\mathbf{A}^{i^{\mathrm{T}}} \mathbf{A}^i = \mathbf{I} \tag{7.48}$$

which, upon differentiation, leads to

$$\mathbf{A}^{i^{\mathrm{T}}} \dot{\mathbf{A}}^i = -(\mathbf{A}^{i^{\mathrm{T}}} \dot{\mathbf{A}}^i)^{\mathrm{T}} \tag{7.49}$$

This implies that the matrix $\mathbf{A}^{i^{\mathrm{T}}}\dot{\mathbf{A}}^i$ is a skew-symmetric matrix that can be written as

$$\mathbf{A}^{i^{\mathrm{T}}}\dot{\mathbf{A}}^i = \tilde{\overline{\boldsymbol{\omega}}}^i \tag{7.50}$$

where

$$\tilde{\overline{\boldsymbol{\omega}}}^i = \begin{bmatrix} 0 & -\overline{\omega}_3^i & \overline{\omega}_2^i \\ \overline{\omega}_3^i & 0 & -\overline{\omega}_1^i \\ -\overline{\omega}_2^i & \overline{\omega}_1^i & 0 \end{bmatrix} \tag{7.51}$$

where $\overline{\omega}_1^i$, $\overline{\omega}_2^i$, and $\overline{\omega}_3^i$ are the components of the angular velocity vector defined in the coordinate system of body i, that is,

$$\overline{\boldsymbol{\omega}}^i = [\overline{\omega}_1^i \quad \overline{\omega}_2^i \quad \overline{\omega}_3^i]^{\mathrm{T}} \tag{7.52}$$

Premultiplying Eq. 50 with $\mathbf{A}^i$ and using the orthogonality of the transformation matrix, one obtains

$$\dot{\mathbf{A}}^i = \mathbf{A}^i\tilde{\overline{\boldsymbol{\omega}}}^i \tag{7.53}$$

Substituting this equation into Eq. 34, another form of the absolute velocity vector of an arbitrary point on the rigid body can be obtained as

$$\dot{\mathbf{r}}^i = \dot{\mathbf{R}}^i + \mathbf{A}^i\tilde{\overline{\boldsymbol{\omega}}}^i\overline{\mathbf{u}}^i = \dot{\mathbf{R}}^i + \mathbf{A}^i(\overline{\boldsymbol{\omega}}^i \times \overline{\mathbf{u}}^i) \tag{7.54}$$

The use of Eqs. 42 and 53 implies that $\tilde{\boldsymbol{\omega}}^i\mathbf{A}^i = \mathbf{A}^i\tilde{\overline{\boldsymbol{\omega}}}^i$, which leads to the following identities:

$$\tilde{\boldsymbol{\omega}}^i = \mathbf{A}^i\tilde{\overline{\boldsymbol{\omega}}}^i\mathbf{A}^{i^{\mathrm{T}}} \tag{7.55}$$

$$\tilde{\overline{\boldsymbol{\omega}}}^i = \mathbf{A}^{i^{\mathrm{T}}}\tilde{\boldsymbol{\omega}}^i\mathbf{A}^i \tag{7.56}$$

Example 7.4

We again consider the case of simple rotation of a rigid body i about the fixed Z axis by an angle θ^i. The transformation matrix $\mathbf{A}^i$ and its time derivative $\dot{\mathbf{A}}^i$ are

$$\mathbf{A}^i = \begin{bmatrix} \cos\theta^i & -\sin\theta^i & 0 \\ \sin\theta^i & \cos\theta^i & 0 \\ 0 & 0 & 1 \end{bmatrix}, \qquad \dot{\mathbf{A}}^i = \dot{\theta}^i \begin{bmatrix} -\sin\theta^i & -\cos\theta^i & 0 \\ \cos\theta^i & -\sin\theta^i & 0 \\ 0 & 0 & 0 \end{bmatrix}$$

Using Eq. 50, one has

$$\tilde{\overline{\boldsymbol{\omega}}}^i = \mathbf{A}^{i^{\mathrm{T}}}\dot{\mathbf{A}}^i = \dot{\theta}^i \begin{bmatrix} 0 & -1 & 0 \\ 1 & 0 & 0 \\ 0 & 0 & 0 \end{bmatrix} = \begin{bmatrix} 0 & -\overline{\omega}_3^i & \overline{\omega}_2^i \\ \overline{\omega}_3^i & 0 & -\overline{\omega}_1^i \\ -\overline{\omega}_2^i & \overline{\omega}_1^i & 0 \end{bmatrix}$$

which defines the components of the absolute angular velocity vector in the body coordinate system as

$$\bar{\boldsymbol{\omega}}^i = \begin{bmatrix} \bar{\omega}_1^i \\ \bar{\omega}_2^i \\ \bar{\omega}_3^i \end{bmatrix} = \dot{\theta}^i \begin{bmatrix} 0 \\ 0 \\ 1 \end{bmatrix}$$

Comparing these results with the results obtained in Example 3, we conclude in this special case of a simple rotation about a fixed axis that $\bar{\boldsymbol{\omega}}^i = \boldsymbol{\omega}^i$. This situation occurs only when the axis of rotation is fixed in space. In this special case, the relationships

$$\boldsymbol{\omega}^i = \mathbf{A}^i \bar{\boldsymbol{\omega}}^i, \quad \bar{\boldsymbol{\omega}}^i = \mathbf{A}^{i^{\mathrm{T}}} \boldsymbol{\omega}^i$$

are still in effect.

Relative Angular Velocities The transformation matrix that defines the orientation of an arbitrary body i can be expressed in terms of the transformation matrix that defines the orientation of another body j as

$$\mathbf{A}^i = \mathbf{A}^j \mathbf{A}^{ij}$$

It follows that

$$\begin{aligned} \tilde{\boldsymbol{\omega}}^i &= \dot{\mathbf{A}}^i \mathbf{A}^{i^{\mathrm{T}}} = (\dot{\mathbf{A}}^j \mathbf{A}^{ij} + \mathbf{A}^j \dot{\mathbf{A}}^{ij})(\mathbf{A}^j \mathbf{A}^{ij})^{\mathrm{T}} \\ &= (\tilde{\boldsymbol{\omega}}^j \mathbf{A}^j \mathbf{A}^{ij} + \mathbf{A}^j (\tilde{\boldsymbol{\omega}}^{ij})_j \mathbf{A}^{ij})(\mathbf{A}^j \mathbf{A}^{ij})^{\mathrm{T}} \end{aligned} \tag{7.57}$$

where $(\tilde{\boldsymbol{\omega}}^{ij})_j$ is the skew-symmetric matrix associated with the angular velocity of body i with respect to body j defined in the coordinate system of body j. Since the following identity:

$$\tilde{\boldsymbol{\omega}}^{ij} = \mathbf{A}^j (\tilde{\boldsymbol{\omega}}^{ij})_j \mathbf{A}^{j^{\mathrm{T}}}$$

holds (see Eqs. 55 and 56), where $\tilde{\boldsymbol{\omega}}^{ij}$ is the skew-symmetric matrix associated with the angular velocity of body i with respect to body j defined in the global coordinate system, Eq. 57 yields

$$\tilde{\boldsymbol{\omega}}^i = \tilde{\boldsymbol{\omega}}^j + \tilde{\boldsymbol{\omega}}^{ij}$$

which implies that

$$\boldsymbol{\omega}^i = \boldsymbol{\omega}^j + \boldsymbol{\omega}^{ij} \tag{7.58}$$

This equation states that the absolute angular velocity of body i is equal to the absolute angular velocity of body j plus the angular velocity of body i with respect to body j.

Acceleration The equation of the absolute acceleration can be obtained by differentiating Eq. 34 with respect to time. This leads to

$$\ddot{\mathbf{r}}^i = \ddot{\mathbf{R}}^i + \ddot{\mathbf{A}}^i\bar{\mathbf{u}}^i \tag{7.59}$$

Differentiating Eq. 42 with respect to time, one obtains

$$\ddot{\mathbf{A}}^i = \dot{\tilde{\omega}}^i\mathbf{A}^i + \tilde{\omega}^i\dot{\mathbf{A}}^i \tag{7.60}$$

Substituting Eq. 42 into Eq. 60, one obtains

$$\ddot{\mathbf{A}}^i = \dot{\tilde{\omega}}^i\mathbf{A}^i + \tilde{\omega}^i\tilde{\omega}^i\mathbf{A}^i = \tilde{\alpha}^i\mathbf{A}^i + (\tilde{\omega}^i)^2\mathbf{A}^i \tag{7.61}$$

where $\tilde{\alpha}^i$ is a skew symmetric matrix defined as

$$\tilde{\alpha}^i = \dot{\tilde{\omega}}^i \tag{7.62}$$

Substituting Eq. 61 into Eq. 59, the absolute acceleration of an arbitrary point on the rigid body i can be written as

$$\ddot{\mathbf{r}}^i = \ddot{\mathbf{R}}^i + \tilde{\alpha}^i\mathbf{A}^i\bar{\mathbf{u}}^i + (\tilde{\omega}^i)^2\mathbf{A}^i\bar{\mathbf{u}}^i \tag{7.63}$$

which, upon the use of Eq. 45 and the notation of the cross product, can be written as

$$\ddot{\mathbf{r}}^i = \ddot{\mathbf{R}}^i + \boldsymbol{\alpha}^i \times \mathbf{u}^i + \boldsymbol{\omega}^i \times (\boldsymbol{\omega}^i \times \mathbf{u}^i) \tag{7.64}$$

where $\boldsymbol{\alpha}^i = [\alpha_1^i \quad \alpha_2^i \quad \alpha_3^i]^{\mathrm{T}}$ is the *angular acceleration vector* of body i. The term $\boldsymbol{\alpha}^i \times \mathbf{u}^i$ on the right-hand side of Eq. 64 is called the *tangential component* of the acceleration, while the term $\boldsymbol{\omega}^i \times (\boldsymbol{\omega}^i \times \mathbf{u}^i)$ is called the *normal component*.

Equation 64 can also be written in an alternative form as

$$\ddot{\mathbf{r}}^i = \ddot{\mathbf{R}}^i + \mathbf{A}^i(\bar{\boldsymbol{\alpha}}^i \times \bar{\mathbf{u}}^i) + \mathbf{A}^i[\bar{\boldsymbol{\omega}}^i \times (\bar{\boldsymbol{\omega}}^i \times \bar{\mathbf{u}}^i)] \tag{7.65}$$

in which

$$\bar{\boldsymbol{\alpha}}^i = \mathbf{A}^{i^{\mathrm{T}}}\boldsymbol{\alpha}^i \tag{7.66}$$

7.5 GENERALIZED COORDINATES

The kinematic and dynamic relationships of multibody mechanical systems can be formulated in terms of the angular velocity and acceleration vectors. The angular velocities, however, are not the time derivatives of a set of orientation coordinates and, therefore, they cannot be integrated to obtain the orientation coordinates. For this reason it is desirable to formulate the dynamic equations using the rotational coordinates such as Euler angles.

In order to achieve this objective, Eq. 46, which defines the absolute velocity vector of an arbitrary point on the rigid body i, is written as

$$\dot{\mathbf{r}}^i = \dot{\mathbf{R}}^i - \mathbf{u}^i \times \boldsymbol{\omega}^i \tag{7.67}$$

which can be written using the skew-symmetric matrix notation as

$$\dot{\mathbf{r}}^i = \dot{\mathbf{R}}^i - \tilde{\mathbf{u}}^i \boldsymbol{\omega}^i \tag{7.68}$$

where $\tilde{\mathbf{u}}^i$ is the skew symmetric matrix

$$\tilde{\mathbf{u}}^i = \begin{bmatrix} 0 & -u_z^i & u_y^i \\ u_z^i & 0 & -u_x^i \\ -u_y^i & u_x^i & 0 \end{bmatrix} \tag{7.69}$$

and u_x^i, u_y^i, and u_z^i are the components of the vector $\mathbf{u}^i$.

In Section 3, the transformation matrix that defines the orientation of body i was developed in terms of Euler angles. By using this transformation matrix and the identity of Eq. 39, the angular velocity vector $\boldsymbol{\omega}^i$ defined in the global coordinate system can be expressed in terms of Euler angles and their time derivatives as

$$\boldsymbol{\omega}^i = \mathbf{G}^i \dot{\boldsymbol{\theta}}^i \tag{7.70}$$

where $\boldsymbol{\theta}^i$ is the set of Euler angles defined as

$$\boldsymbol{\theta}^i = [\phi^i \quad \theta^i \quad \psi^i]^{\mathrm{T}} \tag{7.71}$$

and

$$\mathbf{G}^i = \begin{bmatrix} 0 & \cos\phi^i & \sin\theta^i \sin\phi^i \\ 0 & \sin\phi^i & -\sin\theta^i \cos\phi^i \\ 1 & 0 & \cos\theta^i \end{bmatrix} \tag{7.72}$$

The columns of this matrix, which represent unit vectors along the axes about which the Euler angle rotations ϕ^i, θ^i, and ψ^i are performed, are vectors defined in the fixed coordinate system.

Substituting Eq. 70 into Eq. 68, the absolute velocity of an arbitrary point on the rigid body can be expressed in terms of Euler angles as

$$\dot{\mathbf{r}}^i = \dot{\mathbf{R}}^i - \tilde{\mathbf{u}}^i \mathbf{G}^i \dot{\boldsymbol{\theta}}^i \tag{7.73a}$$

which can be written using matrix partitioning as

$$\dot{\mathbf{r}}^i = [\mathbf{I} \quad -\tilde{\mathbf{u}}^i \mathbf{G}^i] \begin{bmatrix} \dot{\mathbf{R}}^i \\ \dot{\boldsymbol{\theta}}^i \end{bmatrix} \tag{7.73b}$$

Similarly, the absolute acceleration of the arbitrary point on the rigid body can also be expressed in terms of the generalized orientation coordinates using Eq. 64, which can be written using the skew symmetric matrix notation as

$$\ddot{\mathbf{r}}^i = \ddot{\mathbf{R}}^i - \tilde{\mathbf{u}}^i \boldsymbol{\alpha}^i + (\tilde{\boldsymbol{\omega}}^i)^2 \mathbf{u}^i \tag{7.74}$$

Differentiating Eq. 70 with respect to time, one obtains

$$\boldsymbol{\alpha}^i = \mathbf{G}^i \ddot{\boldsymbol{\theta}}^i + \dot{\mathbf{G}}^i \dot{\boldsymbol{\theta}}^i \tag{7.75}$$

Substituting this equation into Eq. 74, the absolute acceleration vector $\ddot{\mathbf{r}}^i$ can be written as

$$\ddot{\mathbf{r}}^i = \ddot{\mathbf{R}}^i - \tilde{\mathbf{u}}^i \mathbf{G}^i \ddot{\boldsymbol{\theta}}^i + \mathbf{a}_v^i \tag{7.76}$$

where $\mathbf{a}_v^i$ is a vector that absorbs terms which are quadratic in the velocities. This vector is defined as

$$\mathbf{a}_v^i = (\tilde{\boldsymbol{\omega}}^i)^2 \mathbf{u}^i - \tilde{\mathbf{u}}^i \dot{\mathbf{G}}^i \dot{\boldsymbol{\theta}}^i \tag{7.77}$$

The vector $\mathbf{a}_v^i$ absorbs the normal component of the acceleration as well as the portion of the tangential component that is quadratic in the velocities.

Another Representation In the development of the kinematic equations presented in this section, the angular velocity and acceleration vectors defined in the global coordinate system are used. Another alternate approach to the formulation of the kinematic equations is to use the expressions of the angular velocity and acceleration vectors as defined in the body coordinate system. By following this procedure, one can show that

$$\dot{\mathbf{r}}^i = \begin{bmatrix} \mathbf{I} & -\mathbf{A}^i \tilde{\bar{\mathbf{u}}}^i \overline{\mathbf{G}}^i \end{bmatrix} \begin{bmatrix} \dot{\mathbf{R}}^i \\ \dot{\boldsymbol{\theta}}^i \end{bmatrix} \tag{7.78}$$

and

$$\ddot{\mathbf{r}}^i = \begin{bmatrix} \mathbf{I} & -\mathbf{A}^i \tilde{\bar{\mathbf{u}}}^i \overline{\mathbf{G}}^i \end{bmatrix} \begin{bmatrix} \ddot{\mathbf{R}}^i \\ \ddot{\boldsymbol{\theta}}^i \end{bmatrix} + \mathbf{a}_v^i \tag{7.79}$$

where

$$\overline{\mathbf{G}}^i = \mathbf{A}^{i^{\mathrm{T}}} \mathbf{G}^i \tag{7.80}$$

The matrix $\overline{\mathbf{G}}^i$, in the case of Euler angles, is given by

$$\overline{\mathbf{G}}^i = \begin{bmatrix} \sin\theta^i \sin\psi^i & \cos\psi^i & 0 \\ \sin\theta^i \cos\psi^i & -\sin\psi^i & 0 \\ \cos\theta^i & 0 & 1 \end{bmatrix} \tag{7.81}$$

and

$$\tilde{\bar{\mathbf{u}}}^i = \begin{bmatrix} 0 & -\bar{u}_z^i & \bar{u}_y^i \\ \bar{u}_z^i & 0 & -\bar{u}_x^i \\ -\bar{u}_y^i & \bar{u}_x^i & 0 \end{bmatrix} = \begin{bmatrix} 0 & -\bar{z}^i & \bar{y}^i \\ \bar{z}^i & 0 & -\bar{x}^i \\ -\bar{y}^i & \bar{x}^i & 0 \end{bmatrix} \tag{7.82}$$

where $\bar{\mathbf{u}}^i = [\bar{u}_x^i \quad \bar{u}_y^i \quad \bar{u}_z^i]^{\mathrm{T}} = [\bar{x}^i \quad \bar{y}^i \quad \bar{z}^i]^{\mathrm{T}}$. The columns of the matrix $\overline{\mathbf{G}}^i$ define, in the body coordinate system, unit vectors along the axes about which the Euler angles ϕ^i, θ^i, and ψ^i are performed.

Using Eq. 47 or Eq. 50, it can be shown that the angular velocity vector $\bar{\boldsymbol{\omega}}^i$ defined in the body coordinate system can be written in terms of the matrix $\overline{\mathbf{G}}^i$ of Eq. 81 as

$$\bar{\boldsymbol{\omega}}^i = \overline{\mathbf{G}}^i \dot{\boldsymbol{\theta}}^i \tag{7.83}$$

Equations 78 and 79 can be obtained directly from Eqs. 73b and 76 by using the identity

$$\tilde{\mathbf{u}}^i = \mathbf{A}^i \tilde{\bar{\mathbf{u}}}^i \mathbf{A}^{i^{\mathrm{T}}} \tag{7.84}$$

Recall that $\boldsymbol{\omega}^i = \mathbf{A}^i \bar{\boldsymbol{\omega}}^i$, which upon differentiation and the use of the identity of Eq. 53 leads to

$$\boldsymbol{\alpha}^i = \mathbf{A}^i \dot{\bar{\boldsymbol{\omega}}}^i + \dot{\mathbf{A}}^i \bar{\boldsymbol{\omega}}^i = \mathbf{A}^i \dot{\bar{\boldsymbol{\omega}}}^i + \mathbf{A}^i \tilde{\bar{\boldsymbol{\omega}}}^i \bar{\boldsymbol{\omega}}^i$$

Since $\bar{\boldsymbol{\omega}}^i \times \bar{\boldsymbol{\omega}}^i = \mathbf{0}$, one has

$$\boldsymbol{\alpha}^i = \mathbf{A}^i \dot{\bar{\boldsymbol{\omega}}}^i = \mathbf{A}^i \overline{\mathbf{G}}^i \ddot{\boldsymbol{\theta}}^i + \mathbf{A}^i \dot{\overline{\mathbf{G}}}^i \dot{\boldsymbol{\theta}} \tag{7.85}$$

Comparing this equation with Eq. 75, one concludes that

$$\dot{\mathbf{G}}^i \dot{\boldsymbol{\theta}}^i = \mathbf{A}^i \dot{\overline{\mathbf{G}}}^i \dot{\boldsymbol{\theta}}^i \tag{7.86}$$

Therefore, the quadratic velocity vector of Eq. 77 that also appears in Eq. 79 can be written in another form using the identity of Eq. 84 as

$$\mathbf{a}_v^i = \mathbf{A}^i (\tilde{\bar{\boldsymbol{\omega}}}^i)^2 \bar{\mathbf{u}}^i - \mathbf{A}^i \tilde{\bar{\mathbf{u}}}^i \dot{\overline{\mathbf{G}}}^i \dot{\boldsymbol{\theta}}^i \tag{7.87}$$

The fact that $\dot{\mathbf{G}}^i \dot{\boldsymbol{\theta}}^i = \mathbf{A}^i \dot{\overline{\mathbf{G}}}^i \dot{\boldsymbol{\theta}}^i$ does not imply that $\dot{\mathbf{G}}^i = \mathbf{A}^i \dot{\overline{\mathbf{G}}}^i$ as will be demonstrated in Example 8.

Remarks It can be verified that the determinant of the matrix $\overline{\mathbf{G}}^i$ of Eq. 81 is equal to $-\sin\theta^i$, which is the same as the determinant of the matrix $\mathbf{G}^i$. Consequently, there is a *singularity* in the transformation using Euler angles when θ^i is equal to zero or π. In this case, the axes of rotation of the Euler angles ϕ^i and ψ^i are parallel and, therefore, these two angles are not distinct. In other words, in the singular configuration, the Euler angles rates cannot be represented in terms of three independent components of the angular velocity vector using Eq. 70 or Eq. 83. The transformation matrix $\mathbf{A}^i$ of Eq. 32 can be written in the case of the singular configuration as

$$\mathbf{A}^i = \begin{bmatrix} \cos(\psi^i + \phi^i) & -\sin(\psi^i + \phi^i) & 0 \\ \sin(\psi^i + \phi^i) & \cos(\psi^i + \phi^i) & 0 \\ 0 & 0 & 1 \end{bmatrix} \tag{7.88}$$

which is a nonsingular orthogonal matrix whose inverse remains equal to its transpose. All Euler angle representations suffer from the singularity problem, which is encountered when two rotations occur about two axes that have the same orientation in space. In this case, the two Euler angles are not independent. A similar singularity problem is encountered when any known method that employs three parameters to describe the orientation of the rigid bodies in space is used. For this reason, the four *Euler parameters* are often used in the computer-aided analysis of the spatial motion of rigid bodies.

Equations 81 and 83 can be used to demonstrate that the components of the angular velocities are not exact differentials. For example, the use of these two equations defines $\overline{\omega}_y^i$ in terms of Euler angles and their time derivatives as

$$\overline{\omega}_y^i = g_{21}\dot{\phi}^i + g_{22}\dot{\theta}^i + g_{23}\dot{\psi}^i \tag{7.89}$$

where

$$g_{21} = \sin\theta^i \cos\psi^i, \quad g_{22} = -\sin\psi^i, \quad g_{23} = 0 \tag{7.90}$$

It is clear from these definitions that

$$\frac{\partial g_{21}}{\partial \theta^i} \neq \frac{\partial g_{22}}{\partial \phi^i}, \quad \frac{\partial g_{21}}{\partial \psi^i} \neq \frac{\partial g_{23}}{\partial \phi^i} \tag{7.91}$$

which imply that $\overline{\omega}_y^i$ is not an exact differential. Similar comments apply to $\overline{\omega}_x^i$ and $\overline{\omega}_z^i$.

7.6 GENERALIZED INERTIA FORCES

There are several methods for developing the dynamic equations of motion of the rigid bodies. In this chapter, the *principle of virtual work in dynamics* will be used to obtain the differential equations that govern the spatial motion of the rigid bodies. First, we develop in this section an expression for the virtual work of the generalized inertia forces.

The virtual change in the position vector of an arbitrary point on the rigid body i is given by (see Eq. 78)

$$\delta\mathbf{r}^i = [\mathbf{I} \quad -\mathbf{A}^i\tilde{\overline{\mathbf{u}}}^i\overline{\mathbf{G}}^i] \begin{bmatrix} \delta\mathbf{R}^i \\ \delta\boldsymbol{\theta}^i \end{bmatrix} \tag{7.92}$$

The virtual work of the inertia forces of the rigid body is

$$\delta W_i^i = \int_{V^i} \rho^i \ddot{\mathbf{r}}^{i^{\mathrm{T}}} \delta \mathbf{r}^i \, dV^i \tag{7.93}$$

where ρ^i and V^i are, respectively, the mass density and volume of the rigid body i. Substituting Eqs. 79 and 92 into Eq. 93, one obtains

$$\delta W_i^i = [\ddot{\mathbf{R}}^{i^{\mathrm{T}}} \quad \ddot{\boldsymbol{\theta}}^{i^{\mathrm{T}}}] \left\{ \int_{V_i} \rho^i \left\{ \begin{bmatrix} \mathbf{I} \\ -\bar{\mathbf{G}}^{i^{\mathrm{T}}} \tilde{\bar{\mathbf{u}}}^{i^{\mathrm{T}}} \mathbf{A}^{i^{\mathrm{T}}} \end{bmatrix} [\mathbf{I} \quad -\mathbf{A}^i \tilde{\bar{\mathbf{u}}}^i \bar{\mathbf{G}}^i] + \mathbf{a}_v^{i^{\mathrm{T}}} [\mathbf{I} \quad -\mathbf{A}^i \tilde{\bar{\mathbf{u}}}^i \bar{\mathbf{G}}^i] \right\} dV^i \right\} \begin{bmatrix} \delta \mathbf{R}^i \\ \delta \boldsymbol{\theta}^i \end{bmatrix} \tag{7.94}$$

which can be written as

$$\delta W_i^i = [\ddot{\mathbf{q}}^{i^{\mathrm{T}}} \mathbf{M}^i - \mathbf{Q}_v^{i^{\mathrm{T}}}] \, \delta \mathbf{q}^i \tag{7.95}$$

where $\mathbf{q}^i$ is the vector of *generalized coordinates* of the rigid body i defined as

$$\mathbf{q}^i = [\mathbf{R}^{i^{\mathrm{T}}} \quad \boldsymbol{\theta}^{i^{\mathrm{T}}}]^{\mathrm{T}} \tag{7.96}$$

$\mathbf{M}^i$ is the symmetric mass matrix

$$\mathbf{M}^i = \int_{V^i} \rho^i \begin{bmatrix} \mathbf{I} & -\mathbf{A}^i \tilde{\bar{\mathbf{u}}}^i \bar{\mathbf{G}}^i \\ \text{symmetric} & \bar{\mathbf{G}}^{i^{\mathrm{T}}} \tilde{\bar{\mathbf{u}}}^{i^{\mathrm{T}}} \tilde{\bar{\mathbf{u}}}^i \bar{\mathbf{G}}^i \end{bmatrix} dV^i \tag{7.97}$$

and $\mathbf{Q}_v^i$ is the vector of the inertia forces that absorbs terms that are quadratic in the velocities. This vector is

$$\mathbf{Q}_v^i = -\int_{V^i} \rho^i \begin{bmatrix} \mathbf{I} \\ -\bar{\mathbf{G}}^{i^{\mathrm{T}}} \tilde{\bar{\mathbf{u}}}^{i^{\mathrm{T}}} \mathbf{A}^{i^{\mathrm{T}}} \end{bmatrix} \mathbf{a}_v^i \, dV^i \tag{7.98}$$

In developing Eq. 95, the origin of the body coordinate system (*reference point*) is assumed to be an arbitrary point on the rigid body, and therefore, the mass matrix of Eq. 97 and the inertia force vector of Eq. 98 are presented in their most general form. Equations 97 and 98 can be simplified if the reference point is chosen to be the center of mass of the body, which is the case of a *centroidal body coordinate system*. The use of a centroidal body coordinate system is one of the basic assumptions used in developing the well-known *Newton–Euler equations*, which are discussed in later sections.

Mass Matrix The symmetric mass matrix of the rigid body i defined by Eq. 97 can be written in the form

$$\mathbf{M}^i = \begin{bmatrix} \mathbf{m}_{RR}^i & \mathbf{m}_{R\theta}^i \\ \mathbf{m}_{\theta R}^i & \mathbf{m}_{\theta\theta}^i \end{bmatrix} \tag{7.99}$$

where

$$\left.\begin{aligned} \mathbf{m}^i_{RR} &= m^i\mathbf{I}, \quad \mathbf{m}^i_{\theta\theta} = \overline{\mathbf{G}}^{i\mathrm{T}}\overline{\mathbf{I}}^i_{\theta\theta}\overline{\mathbf{G}}^i \\ \mathbf{m}^i_{R\theta} &= \mathbf{m}^{i\mathrm{T}}_{\theta R} = -\mathbf{A}^i\left[\int_{V^i}\rho^i\tilde{\overline{\mathbf{u}}}^i\,dV^i\right]\overline{\mathbf{G}}^i \end{aligned}\right\} \tag{7.100}$$

where m^i is the total mass of the rigid body i, and $\overline{\mathbf{I}}^i_{\theta\theta}$ is a 3×3 symmetric matrix, called the *inertia tensor* of the rigid body, and is defined by the equation

$$\overline{\mathbf{I}}^i_{\theta\theta} = \int_{V^i}\rho^i\tilde{\overline{\mathbf{u}}}^{i\mathrm{T}}\tilde{\overline{\mathbf{u}}}^i\,dV^i \tag{7.101}$$

Using Eqs. 3, 82, and 101, the inertia tensor of the rigid body i can be written as

$$\overline{\mathbf{I}}^i_{\theta\theta} = \begin{bmatrix} i_{xx} & i_{xy} & i_{xz} \\ & i_{yy} & i_{yz} \\ \text{symmetric} & & i_{zz} \end{bmatrix} \tag{7.102}$$

where the elements i_{xx}, i_{yy}, i_{zz} are called the *moments of inertia* and i_{xy}, i_{xz}, i_{yz} are called the *products of inertia*. These elements are defined as

$$\left.\begin{aligned} i_{xx} &= \int_{V^i}\rho^i[(\overline{y}^i)^2 + (\overline{z}^i)^2]\,dV^i \\ i_{yy} &= \int_{V^i}\rho^i[(\overline{x}^i)^2 + (\overline{z}^i)^2]\,dV^i \\ i_{zz} &= \int_{V^i}\rho^i[(\overline{x}^i)^2 + (\overline{y}^i)^2]\,dV^i \\ i_{xy} &= -\int_{V^i}\rho^i\overline{x}^i\overline{y}^i\,dV^i \\ i_{xz} &= -\int_{V^i}\rho^i\overline{x}^i\overline{z}^i\,dV^i \\ i_{zy} &= -\int_{V^i}\rho^i\overline{y}^i\overline{z}^i\,dV^i \end{aligned}\right\} \tag{7.103}$$

It is clear that the moments of inertia satisfy the following identity:

$$\begin{aligned} i_{xx} + i_{yy} + i_{zz} &= 2\int_{V^i}\rho^i[(\overline{x}^i)^2 + (\overline{y}^i)^2 + (\overline{z}^i)^2]\,dV^i \\ &= 2\int_{V^i}\rho^i\overline{\mathbf{u}}^{i\mathrm{T}}\overline{\mathbf{u}}^i\,dV^i \end{aligned} \tag{7.104}$$

While the moments and products of inertia defined by Eq. 103 are constant since they are defined in the rigid body coordinate system, the matrix $\mathbf{m}^i_{\theta\theta}$ of Eq. 100 is a nonlinear matrix since it depends on the orientation coordinates of the rigid body. This matrix, upon the use of the identity of Eq. 80, can be written in an alternate form as

$$\mathbf{m}^i_{\theta\theta} = \mathbf{G}^{i^{\mathrm{T}}}\mathbf{A}^i\bar{\mathbf{I}}^i_{\theta\theta}\mathbf{A}^{i^{\mathrm{T}}}\mathbf{G}^i = \mathbf{G}^{i^{\mathrm{T}}}\mathbf{I}^i_{\theta\theta}\mathbf{G}^i \tag{7.105}$$

where

$$\mathbf{I}^i_{\theta\theta} = \mathbf{A}^i\bar{\mathbf{I}}^i_{\theta\theta}\mathbf{A}^{i^{\mathrm{T}}} \tag{7.106}$$

Note also that in the mass matrix of Eq. 99, there is a *dynamic* or *inertia coupling* between the translation and the rotation of the rigid body, and this coupling is represented by the off-diagonal nonlinear matrices, $\mathbf{m}^i_{R\theta}$ and $\mathbf{m}^i_{\theta R}$, which also depend on the orientation coordinates of the rigid body.

Example 7.5

Obtain the components of the inertia tensor of the *rectangular prism* shown in Fig. 6 with respect to a coordinate system whose origin is located at one of the corners, as shown in the figure.

Solution. The elements of the inertia tensor can be evaluated using Eq. 103. For the rectangular prism shown in the figure,

$$\begin{aligned} i_{xx} &= \int_{V^i} \rho^i[(\bar{y}^i)^2 + (\bar{z}^i)^2]\, dV^i \\ &= \int_0^c \int_0^b \int_0^a \rho^i[(\bar{y}^i)^2 + (\bar{z}^i)^2]\, d\bar{x}^i\, d\bar{y}^i\, d\bar{z}^i \\ &= \int_0^c \int_0^b \rho^i a[(\bar{y}^i)^2 + (\bar{z}^i)^2]\, d\bar{y}^i\, d\bar{z}^i \\ &= \frac{m^i}{3}[(b)^2 + (c)^2] \end{aligned}$$

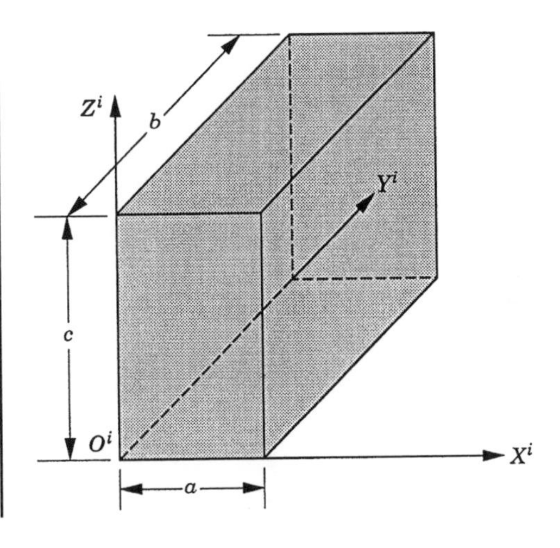

Figure 7.6 Solid prism

where m^i is the total mass of the rectangular prism. Similarly, one can show that

$$i_{yy} = \frac{m^i}{3}\,[(a)^2 + (c)^2], \quad i_{zz} = \frac{m^i}{3}\,[(a)^2 + (b)^2]$$

The product of inertia i_{xy} is

$$\begin{aligned} i_{xy} &= -\int_{V^i} \rho^i \bar{x}^i \bar{y}^i \, dV^i = -\int_0^c \int_0^b \int_0^a \rho^i \bar{x}^i \bar{y}^i \, d\bar{x}^i \, d\bar{y}^i \, d\bar{z}^i \\ &= -\int_0^c \int_0^b \rho^i \, \frac{(a)^2 \bar{y}^i}{2} \, d\bar{y}^i \, d\bar{z}^i = -\int_0^c \rho^i \, \frac{(a)^2 (b)^2}{4} \, d\bar{z}^i = -\frac{m^i ab}{4} \end{aligned}$$

Similarly,

$$\begin{aligned} i_{xz} &= -\int_{V^i} \rho^i \bar{x}^i \bar{z}^i \, dV^i = -\frac{m^i ac}{4} \\ i_{yz} &= -\int_{V^i} \rho^i \bar{y}^i \bar{z}^i \, dV^i = -\frac{m^i bc}{4} \end{aligned}$$

The inertia tensor of the rectangular prism defined in the coordinate system shown in the figure can be written as

$$\bar{\mathbf{I}}^i_{\theta\theta} = m^i \begin{bmatrix} \frac{1}{3}[(b)^2 + (c)^2] & -\frac{1}{4}(ab) & -\frac{1}{4}(ac) \\ -\frac{1}{4}(ab) & \frac{1}{3}[(a)^2 + (c)^2] & -\frac{1}{4}(bc) \\ -\frac{1}{4}(ac) & -\frac{1}{4}(bc) & \frac{1}{3}[(a)^2 + (b)^2] \end{bmatrix}$$

In the special case of a *homogeneous cube*, $a = b = c$, and the elements of the inertia tensor reduce to

$$\begin{aligned} i_{xx} = i_{yy} = i_{zz} &= \frac{2m^i (a)^2}{3} \\ i_{xy} = i_{xz} = i_{yz} &= -\frac{m^i (a)^2}{4} \end{aligned}$$

Parallel Axis Theorem The elements of the inertia tensor given by Eq. 103 are defined in a coordinate system whose origin is attached to an arbitrary point on the rigid body. In this coordinate system, let $\bar{\mathbf{u}}^i_c$ be the local position vector of the center of mass of the body. The vector $\bar{\mathbf{u}}^i$ that defines the location of an arbitrary point with respect to the reference point O^i can be written as shown in Fig. 7 as

$$\bar{\mathbf{u}}^i = \bar{\mathbf{u}}^i_c + \bar{\mathbf{u}}^i_r \tag{7.107}$$

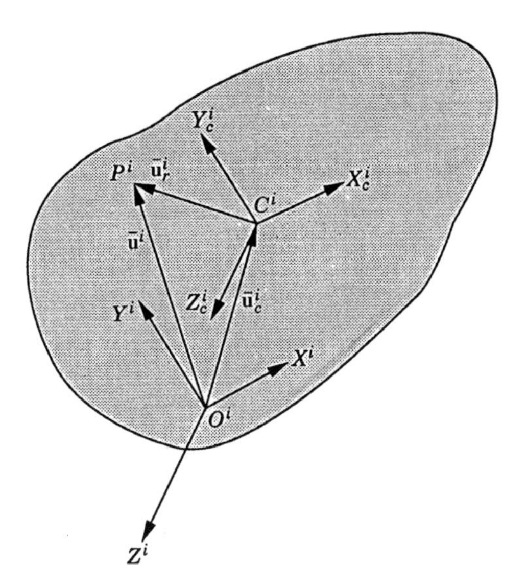

Figure 7.7 Centroidal coordinates

where $\bar{\mathbf{u}}_r^i$ is the position vector of the arbitrary point with respect to the center of mass. It follows from the use of the preceding equation and Eq. 101 that

$$\begin{aligned}\bar{\mathbf{I}}_{\theta\theta}^i &= \int_{V^i} \rho^i \tilde{\bar{\mathbf{u}}}^{i\mathrm{T}} \tilde{\bar{\mathbf{u}}}^i \, dV^i \\ &= \int_{V^i} \rho^i [(\tilde{\bar{\mathbf{u}}}_c^i + \tilde{\bar{\mathbf{u}}}_r^i)^{\mathrm{T}} (\tilde{\bar{\mathbf{u}}}_c^i + \tilde{\bar{\mathbf{u}}}_r^i)] \, dV^i \end{aligned} \tag{7.108}$$

or

$$\begin{aligned}\bar{\mathbf{I}}_{\theta\theta}^i = \tilde{\bar{\mathbf{u}}}_c^{i\mathrm{T}} \tilde{\bar{\mathbf{u}}}_c^i \left[\int_{V^i} \rho^i \, dV^i\right] + \tilde{\bar{\mathbf{u}}}_c^{i\mathrm{T}} \left[\int_{V^i} \rho^i \tilde{\bar{\mathbf{u}}}_r^i \, dV^i\right] \\ + \left[\int_{V^i} \rho^i \tilde{\bar{\mathbf{u}}}_r^{i\mathrm{T}} \, dV^i\right] \tilde{\bar{\mathbf{u}}}_c^i + \int_{V^i} \rho^i \tilde{\bar{\mathbf{u}}}_r^{i\mathrm{T}} \, \tilde{\bar{\mathbf{u}}}_r^i \, dV^i \end{aligned} \tag{7.109}$$

Since $\bar{\mathbf{u}}_r^i$ defines the position vector of an arbitrary point on the rigid body with respect to the center of mass, one has

$$\int_{V^i} \rho^i \bar{\mathbf{u}}_r^i \, dV^i = \mathbf{0} \tag{7.110}$$

and consequently,

$$\int_{V^i} \rho^i \tilde{\bar{\mathbf{u}}}_r^i \, dV^i = \mathbf{0} \tag{7.111}$$

The equation for the inertia tensor of the body becomes

$$\bar{\mathbf{I}}^i_{\theta\theta} = m^i \tilde{\bar{\mathbf{u}}}^{i\mathrm{T}}_c \tilde{\bar{\mathbf{u}}}^i_c + (\bar{\mathbf{I}}^i_{\theta\theta})_c \tag{7.112}$$

where the fact that

$$m^i = \int_{V^i} \rho^i \, dV^i \tag{7.113}$$

was utilized, and

$$(\bar{\mathbf{I}}^i_{\theta\theta})_c = \int_{V^i} \rho^i \tilde{\bar{\mathbf{u}}}^{i\mathrm{T}}_r \tilde{\bar{\mathbf{u}}}^i_r \, dV^i \tag{7.114}$$

is the inertia tensor of the body defined in a *centroidal* $X^i_c Y^i_c Z^i_c$ coordinate system whose axes are parallel to the axes of the coordinate system $X^i Y^i Z^i$ as shown in Fig. 7. Equation 112 is called the *parallel axis theorem*, which states that the mass moments of inertia defined with respect to a noncentroidal coordinate system $X^i Y^i Z^i$ are equal to the mass moments of inertia defined with respect to a centroidal coordinate system $X^i_c Y^i_c Z^i_c$ plus the mass of the body times the square of the distances between the respective axes. These moments of inertia as well as the products of inertia can be expressed in terms of those defined with respect to the centroidal coordinate system as

$$\left.\begin{aligned}
i_{xx} &= (i_{xx})_c + m^i[(\bar{y}^i_c)^2 + (\bar{z}^i_c)^2] \\
i_{yy} &= (i_{yy})_c + m^i[(\bar{x}^i_c)^2 + (\bar{z}^i_c)^2] \\
i_{zz} &= (i_{zz})_c + m^i[(\bar{x}^i_c)^2 + (\bar{y}^i_c)^2] \\
i_{xy} &= (i_{xy})_c - m^i \bar{x}^i_c \bar{y}^i_c \\
i_{xz} &= (i_{xz})_c - m^i \bar{x}^i_c \bar{z}^i_c \\
i_{yz} &= (i_{yz})_c - m^i \bar{y}^i_c \bar{z}^i_c
\end{aligned}\right\} \tag{7.115}$$

where $(i_{xx})_c$, $(i_{yy})_c$, $(i_{zz})_c$, $(i_{xy})_c$, $(i_{xz})_c$, and $(i_{yz})_c$ are the moments and products of inertia defined with respect to the center of mass, and $\bar{x}^i_c$, $\bar{y}^i_c$, and $\bar{z}^i_c$ are the components of the vector $\bar{\mathbf{u}}^i_c$ that defines the location of the center of mass with respect to the reference point O^i as shown in Fig. 7. Since the moments of inertia are always positive, it can be seen from the preceding equations that a translation of a coordinate system away from the center of mass always leads to an increase in the moments of inertia. The products of inertia, however, may increase or decrease depending on the direction of the translation. Table 1 shows the mass moments of inertia of some homogeneous solids. The moments of inertia presented in this table are defined with respect to a centroidal body coordinate system.

It is worth noting that since $\bar{\mathbf{u}}^i_r = \bar{\mathbf{u}}^i - \bar{\mathbf{u}}^i_c$, one can always determine the vector $\bar{\mathbf{u}}^i_c$ that defines the location of the center of mass with respect to the origin of the body coordinate system. To demonstrate this, we write

$$\int_{V^i} \rho^i \bar{\mathbf{u}}^i_r \, dV^i = \int_{V^i} \rho^i \bar{\mathbf{u}}^i \, dV^i - \int_{V^i} \rho^i \bar{\mathbf{u}}^i_c \, dV^i = \mathbf{0} \tag{7.116}$$

TABLE 7.1 Mass Moments of Inertia of Homogeneous Solids

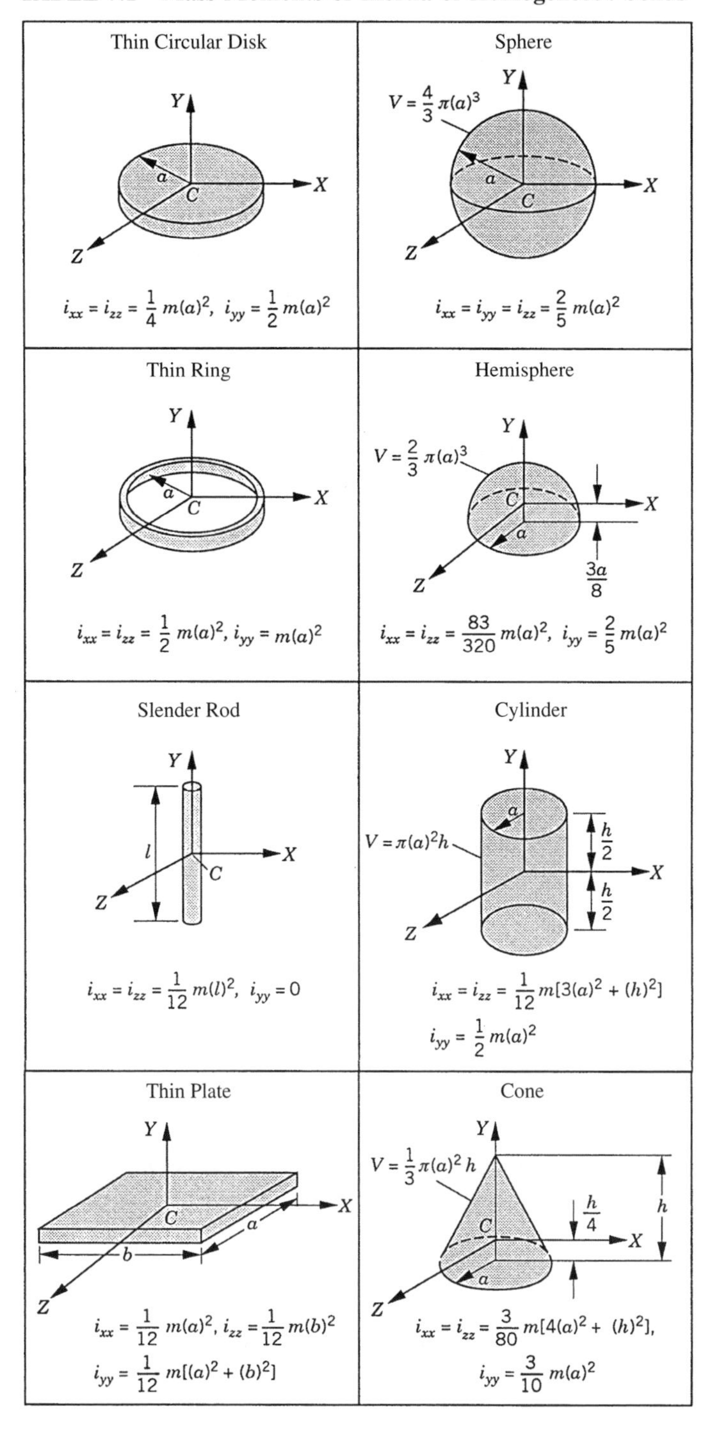

or

$$\bar{\mathbf{u}}_c^i \int_{V^i} \rho^i \, dV^i = \int_{V^i} \rho^i \bar{\mathbf{u}}^i \, dV^i \tag{7.117}$$

which leads to

$$\bar{\mathbf{u}}_c^i = \frac{1}{m^i} \int_{V^i} \rho^i \bar{\mathbf{u}}^i \, dV^i \tag{7.118}$$

This equation defines the position vector of the *center of mass* with respect to the origin of the body coordinate system.

Example 7.6

For the rectangular prism of Example 5, obtain the elements of the inertia tensor defined with respect to a centroidal coordinate system whose axes are parallel to the axes of the coordinate system shown in Fig. 6.

Solution. It is clear in this simple example that the center of mass of the rectangular prism in the $X^iY^iZ^i$ coordinate system is

$$\bar{\mathbf{u}}_c^i = \begin{bmatrix} \bar{x}_c^i \\ \bar{y}_c^i \\ \bar{z}_c^i \end{bmatrix} = \begin{bmatrix} a/2 \\ b/2 \\ c/2 \end{bmatrix}$$

This obvious result can also be obtained using the general equation

$$\bar{\mathbf{u}}_c^i = \frac{1}{m^i} \int_{V^i} \rho^i \bar{\mathbf{u}}^i \, dV^i$$

which, by assuming that the mass density ρ^i is constant, yields in this example

$$\bar{\mathbf{u}}_c^i = \begin{bmatrix} \bar{x}_c^i \\ \bar{y}_c^i \\ \bar{z}_c^i \end{bmatrix} = \frac{1}{m^i} \int_0^c \int_0^b \int_0^a \rho^i \begin{bmatrix} \bar{x}^i \\ \bar{y}^i \\ \bar{z}^i \end{bmatrix} dV^i = \frac{1}{2} \begin{bmatrix} a \\ b \\ c \end{bmatrix}$$

as expected.

The element $(i_{xx})_c$ of the inertia tensor defined with respect to the centroidal coordinate system can be evaluated using the results obtained in Example 5 as

$$\begin{aligned} (i_{xx})_c &= i_{xx} - m^i[(\bar{y}_c^i)^2 + (\bar{z}_c^i)^2] \\ &= \frac{m^i}{3}[(b)^2 + (c)^2] - m^i\left[\left(\frac{b}{2}\right)^2 + \left(\frac{c}{2}\right)^2\right] \\ &= \frac{m^i}{12}[(b)^2 + (c)^2] \end{aligned}$$

Similarly, one can show that

$$(i_{yy})_c = \frac{m^i}{12}\,[(a)^2 + (c)^2], \quad (i_{zz})_c = \frac{m^i}{12}\,[(a)^2 + (b)^2]$$

The product of inertia $(i_{xy})_c$ is

$$(i_{xy})_c = i_{xy} + m^i \bar{x}_c^i \bar{y}_c^i = i_{xy} + m^i\,\frac{ab}{4}$$

Using the results of the preceding example, one has

$$(i_{xy})_c = -m^i\left(\frac{ab}{4}\right) + m^i\left(\frac{ab}{4}\right) = 0$$

It can be also shown that

$$i_{xz} = i_{yz} = 0$$

Thus,

$$(\bar{\mathbf{I}}_{\theta\theta}^i)_c = \frac{m^i}{12}\begin{bmatrix} (b)^2 + (c)^2 & 0 & 0 \\ 0 & (a)^2 + (c)^2 & 0 \\ 0 & 0 & (a)^2 + (b)^2 \end{bmatrix}$$

The results obtained in this example using the parallel axis theorem can also be obtained by attaching the origin of the body coordinate system $X^iY^iZ^i$ to the center of mass and using Eq. 103. For example,

$$\begin{aligned} i_{xx} &= \int_{V^i} \rho^i[(\bar{y}^i)^2 + (\bar{z}^i)^2]\,dV^i \\ &= \int_{-c/2}^{c/2}\int_{-b/2}^{b/2}\int_{-a/2}^{a/2} \rho^i[(\bar{y}^i)^2 + (\bar{z}^i)^2]\,d\bar{x}^i\,d\bar{y}^i\,d\bar{z}^i \\ &= \frac{m^i}{12}\,[(b)^2 + (c)^2] \end{aligned}$$

which is the same result obtained previously by the application of the parallel axis theorem.

In the special case of a *homogeneous cube*, the moments and products of inertia reduce to

$$\begin{aligned} (i_{xx})_c &= (i_{yy})_c = (i_{zz})_c = \frac{m^i(a)^2}{6} \\ (i_{xy})_c &= (i_{xz})_c = (i_{yz})_c = 0 \end{aligned}$$

Principal Moments of Inertia The parallel axis theorem shows the effect of the translation of the coordinate system on the definition of the moments and products of inertia. In order to define the relationship between two inertia tensors defined with respect to two body-fixed coordinate systems that differ in their orientation, let $(\bar{\mathbf{I}}_{\theta\theta}^i)_1$ be the inertia tensor defined with respect to a body-fixed coordinate system $X_1^i Y_1^i Z_1^i$, and $(\bar{\mathbf{I}}_{\theta\theta}^i)_2$ be the inertia tensor defined with respect to another body-fixed coordinate system $X_2^i Y_2^i Z_2^i$. It is assumed that the origins of the two coordinate systems $X_1^i Y_1^i Z_1^i$ and $X_2^i Y_2^i Z_2^i$ coincide. The orientation of the coordinate system $X_2^i Y_2^i Z_2^i$ with respect to the coordinate system $X_1^i Y_1^i Z_1^i$ is defined by the constant transformation matrix $\mathbf{C}^i$. Let $\bar{\mathbf{u}}_1^i$ and $\bar{\mathbf{u}}_2^i$ be the position vectors of an arbitrary point on the rigid body i defined in the coordinate systems $X_1^i Y_1^i Z_1^i$ and $X_2^i Y_2^i Z_2^i$, respectively. Using an identity similar to the one of Eq. 84, one has

$$\tilde{\bar{\mathbf{u}}}_1^i = \mathbf{C}^i \tilde{\bar{\mathbf{u}}}_2^i \mathbf{C}^{i^{\mathrm{T}}} \tag{7.119}$$

Using this equation, it can be shown that the relationship between the inertia tensors defined in the two body-fixed coordinate systems can be written as

$$(\bar{\mathbf{I}}_{\theta\theta}^i)_1 = \mathbf{C}^i (\bar{\mathbf{I}}_{\theta\theta}^i)_2 \mathbf{C}^{i^{\mathrm{T}}} \tag{7.120}$$

Using the orthogonality of the transformation matrix, and postmultiplying both sides of this equation by $\mathbf{C}^i$, one obtains

$$(\bar{\mathbf{I}}_{\theta\theta}^i)_1 \mathbf{C}^i = \mathbf{C}^i (\bar{\mathbf{I}}_{\theta\theta}^i)_2 \tag{7.121}$$

It is possible to select the orientation of the body-fixed coordinate system $X_2^i Y_2^i Z_2^i$ such that all the products of inertia are equal to zeros. In this special case, the inertia tensor $(\bar{\mathbf{I}}_{\theta\theta}^i)_2$ is a diagonal matrix. In fact, this is the case that occurred in the preceding example. In this case, the axes of the body-fixed coordinate system $X_2^i Y_2^i Z_2^i$ are called the *principal axes* and the moments of inertia are referred to as the *principal moments of inertia*.

The principal axes and principal moments of inertia can be determined using Eq. 121. If X_2^i, Y_2^i, and Z_2^i are principal axes, the inertia tensor $(\bar{\mathbf{I}}_{\theta\theta}^i)_2$ can be written as

$$(\bar{\mathbf{I}}_{\theta\theta}^i)_2 = \begin{bmatrix} i_1 & 0 & 0 \\ 0 & i_2 & 0 \\ 0 & 0 & i_3 \end{bmatrix} \tag{7.122}$$

where i_1, i_2, and i_3 are the principal moments of inertia. The inertia tensor $(\bar{\mathbf{I}}_{\theta\theta}^i)_1$, on the other hand, takes the general form

$$(\bar{\mathbf{I}}_{\theta\theta}^i)_1 = \begin{bmatrix} i_{xx} & i_{xy} & i_{xz} \\ & i_{yy} & i_{yz} \\ \text{symmetric} & & i_{zz} \end{bmatrix} \tag{7.123}$$

Now let $\mathbf{C}_k^i$ be the kth column of the transformation matrix $\mathbf{C}^i$ in Eq. 121. Equation 121 can, therefore, be written as

$$[(\bar{\mathbf{I}}_{\theta\theta}^i)_1 - i_k \mathbf{I}]\mathbf{C}_k^i = \mathbf{0}, \qquad k = 1, 2, 3 \tag{7.124}$$

where $\mathbf{I}$ is the 3×3 identity matrix. Equation 124 is a system of homogeneous equations that can be solved for the vector $\mathbf{C}_k^i$. This system has a nontrivial solution if and only if the coefficient matrix is singular. That is,

$$|(\bar{\mathbf{I}}_{\theta\theta}^i)_1 - i_k \mathbf{I}| = 0 \tag{7.125}$$

or

$$\begin{vmatrix} i_{xx} - i_k & i_{xy} & i_{xz} \\ i_{xy} & i_{yy} - i_k & i_{yz} \\ i_{xz} & i_{yz} & i_{zz} - i_k \end{vmatrix} = 0 \tag{7.126}$$

This determinant defines a cubic polynomial in i_k. The roots of this polynomial define the principal moments of inertia i_k, $k = 1, 2, 3$, and the principal directions $\mathbf{C}_k^i$ can be defined using Eq. 124. Since $(\bar{\mathbf{I}}_{\theta\theta}^i)_1$ is real symmetric and positive definite matrix, the principal moments of inertia obtained by solving Eq. 125 are all real and nonnegative. Furthermore, one can show that the principal directions associated with distinctive mass moments of inertia are orthogonal, that is

$$\begin{aligned} \mathbf{C}_k^{i\mathrm{T}} \mathbf{C}_l^i &= 0 \quad \text{if } k \neq l \\ &\neq 0 \quad \text{if } k = l, \quad k, l = 1, 2, 3 \end{aligned}$$

Once $\mathbf{C}_k^i$ are determined, they can be used to determine unit vectors along the principal directions. These unit vectors define the vectors of the direction cosines, which form the columns of the transformation matrix $\mathbf{C}^i$ that defines the orientation of the coordinate system $X_2^i Y_2^i Z_2^i$ with respect to the coordinate system $X_1^i Y_1^i Z_1^i$. If two principal moments of inertia are equal, say $i_2 = i_3 \neq i_1$, the direction of the principal axis associated with i_1 is uniquely defined but any axis that lies in the plane whose normal is defined by $\mathbf{C}_1^i$ is a principal axis. In the case $i_1 = i_2 = i_3$, any three mutually perpendicular axes form the principal directions. An example of this special case is the case of a sphere.

We note, in general, that in the case of repeated roots, the substitution of the repeated root in the coefficient matrix of Eq. 124 reduces the rank of this matrix by the number of the equal roots. This makes the dimension of the *null space* of the resulting coefficient matrix equal to the number of the repeated roots, ensuring that Eq. 124 has a number of independent solutions equal to the number of the repeated roots. These independent solutions define the principal directions.

Centrifugal Forces Using Eq. 98, the vector $\mathbf{Q}_v^i$ that absorbs terms that are quadratic in the velocities can be written as

$$\mathbf{Q}_v^i = \begin{bmatrix} (\mathbf{Q}_v^i)_R \\ (\mathbf{Q}_v^i)_\theta \end{bmatrix} \tag{7.127}$$

which upon the use of the expression for the vector $\mathbf{a}_v^i$ presented at the end of the preceding section yields

$$\left.\begin{aligned}(\mathbf{Q}_v^i)_R &= -\mathbf{A}^i \int_{V^i} \rho^i [(\tilde{\bar{\boldsymbol{\omega}}}^i)^2 \bar{\mathbf{u}}^i - \tilde{\bar{\mathbf{u}}}^i \bar{\mathbf{G}}^i \dot{\boldsymbol{\theta}}^i]\, dV^i \\ (\mathbf{Q}_v^i)_\theta &= \bar{\mathbf{G}}^{i^{\mathrm{T}}} \int_{V^i} \rho^i [\tilde{\bar{\mathbf{u}}}^{i^{\mathrm{T}}} (\tilde{\bar{\boldsymbol{\omega}}}^i)^2 \bar{\mathbf{u}}^i - \tilde{\bar{\mathbf{u}}}^{i^{\mathrm{T}}} \tilde{\bar{\mathbf{u}}}^i \dot{\bar{\mathbf{G}}}^i \dot{\boldsymbol{\theta}}^i]\, dV^i\end{aligned}\right\} \tag{7.128}$$

The vectors $(\mathbf{Q}_v^i)_R$ and $(\mathbf{Q}_v^i)_\theta$ can be written as

$$(\mathbf{Q}_v^i)_R = -\mathbf{A}^i (\tilde{\bar{\boldsymbol{\omega}}}^i)^2 \left[\int_{V^i} \rho^i \bar{\mathbf{u}}^i \, dV^i\right] + \mathbf{A}^i \left[\int_{V^i} \rho^i \tilde{\bar{\mathbf{u}}}^i \, dV^i\right] \dot{\bar{\mathbf{G}}}^i \dot{\boldsymbol{\theta}}^i \tag{7.129}$$

and

$$(\mathbf{Q}_v^i)_\theta = \bar{\mathbf{G}}^{i^{\mathrm{T}}} \left[\int_{V^i} \rho^i \tilde{\bar{\mathbf{u}}}^{i^{\mathrm{T}}} (\tilde{\bar{\boldsymbol{\omega}}}^i)^2 \bar{\mathbf{u}}^i \, dV^i\right] - \bar{\mathbf{G}}^{i^{\mathrm{T}}} \bar{\mathbf{I}}_{\theta\theta}^i \dot{\bar{\mathbf{G}}}^i \dot{\boldsymbol{\theta}}^i \tag{7.130}$$

where the definition of $\bar{\mathbf{I}}_{\theta\theta}^i$ given by Eq. 101 or 102 is utilized.

The following vector and matrix identities can be verified

$$\left.\begin{aligned}(\tilde{\bar{\mathbf{u}}}^i \tilde{\bar{\boldsymbol{\omega}}}^i) &= \bar{\boldsymbol{\omega}}^i \bar{\mathbf{u}}^{i^{\mathrm{T}}} - \bar{\mathbf{u}}^{i^{\mathrm{T}}} \bar{\boldsymbol{\omega}}^i \mathbf{I} \\ (\widetilde{\tilde{\bar{\mathbf{u}}}^i \bar{\boldsymbol{\omega}}^i}) &= \bar{\boldsymbol{\omega}}^i \bar{\mathbf{u}}^{i^{\mathrm{T}}} - \bar{\mathbf{u}}^i \bar{\boldsymbol{\omega}}^{i^{\mathrm{T}}} = \tilde{\bar{\mathbf{u}}}^i \tilde{\bar{\boldsymbol{\omega}}}^i - \tilde{\bar{\boldsymbol{\omega}}}^i \tilde{\bar{\mathbf{u}}}^i \\ \tilde{\bar{\mathbf{u}}}^i \tilde{\bar{\boldsymbol{\omega}}}^i + \bar{\mathbf{u}}^i \bar{\boldsymbol{\omega}}^{i^{\mathrm{T}}} &= \tilde{\bar{\boldsymbol{\omega}}}^i \tilde{\bar{\mathbf{u}}}^i + \bar{\boldsymbol{\omega}}^i \bar{\mathbf{u}}^{i^{\mathrm{T}}}\end{aligned}\right\} \tag{7.131}$$

Using the last identity in the preceding equation, one has

$$\begin{aligned}\int_{V^i} \rho^i \tilde{\bar{\mathbf{u}}}^{i^{\mathrm{T}}} \tilde{\bar{\boldsymbol{\omega}}}^i \tilde{\bar{\boldsymbol{\omega}}}^i \bar{\mathbf{u}}^i \, dV^i &= \int_{V^i} \rho^i \tilde{\bar{\mathbf{u}}}^i \tilde{\bar{\boldsymbol{\omega}}}^i \tilde{\bar{\mathbf{u}}}^i \bar{\boldsymbol{\omega}}^i \, dV^i \\ &= \int_{V^i} \rho^i \{\tilde{\bar{\boldsymbol{\omega}}}^i \tilde{\bar{\mathbf{u}}}^i \tilde{\bar{\mathbf{u}}}^i \bar{\boldsymbol{\omega}}^i + \bar{\boldsymbol{\omega}}^i \bar{\mathbf{u}}^{i^{\mathrm{T}}} \tilde{\bar{\mathbf{u}}}^i \bar{\boldsymbol{\omega}}^i - \bar{\mathbf{u}}^i \bar{\boldsymbol{\omega}}^{i^{\mathrm{T}}} \tilde{\bar{\boldsymbol{\omega}}}^i \bar{\mathbf{u}}^i\}\, dV^i\end{aligned} \tag{7.132}$$

The last two terms in this equation are identically equal to zero. Thus

$$\begin{aligned}\int_{V^i} \rho^i \tilde{\bar{\mathbf{u}}}^{i^{\mathrm{T}}} \tilde{\bar{\boldsymbol{\omega}}}^i \tilde{\bar{\boldsymbol{\omega}}}^i \bar{\mathbf{u}}^i \, dV^i &= \int_{V^i} \rho^i \tilde{\bar{\boldsymbol{\omega}}}^i \tilde{\bar{\mathbf{u}}}^i \tilde{\bar{\mathbf{u}}}^i \bar{\boldsymbol{\omega}}^i \, dV^i \\ &= -\tilde{\bar{\boldsymbol{\omega}}}^i \bar{\mathbf{I}}_{\theta\theta}^i \bar{\boldsymbol{\omega}}^i = -\bar{\boldsymbol{\omega}}^i \times (\bar{\mathbf{I}}_{\theta\theta}^i \bar{\boldsymbol{\omega}}^i)\end{aligned} \tag{7.133}$$

Substituting this equation into Eq. 130, one obtains

$$(\mathbf{Q}_v^i)_\theta = -\bar{\mathbf{G}}^{i^{\mathrm{T}}} [\bar{\boldsymbol{\omega}}^i \times (\bar{\mathbf{I}}_{\theta\theta}^i \bar{\boldsymbol{\omega}}^i) + \bar{\mathbf{I}}_{\theta\theta}^i \dot{\bar{\mathbf{G}}}^i \dot{\boldsymbol{\theta}}^i] \tag{7.134}$$

7.7 GENERALIZED APPLIED FORCES

The generalized external forces of the rigid body i can be defined using the expression of the virtual work. Examples of these forces are the gravity, spring, damping, friction, actuator forces, and motor torques. Examples of some of these forces which can be nonlinear functions of the system variables are presented in this section and the formulation of the generalized applied forces associated with the generalized coordinates of the spatial rigid body systems are discussed.

Force Vector Let $\mathbf{F}^i$ be a force vector that acts at a point P^i on the rigid body i as shown in Fig. 8. This force vector is assumed to be defined in the global coordinate system. The virtual work of this force vector can be written as

$$\delta W_e^i = \mathbf{F}^{i^{\mathrm{T}}} \delta \mathbf{r}_P^i \tag{7.135}$$

where $\delta \mathbf{r}_P^i$ can be obtained using Eq. 92 as

$$\delta \mathbf{r}_P^i = \delta \mathbf{R}^i - \mathbf{A}^i \tilde{\bar{\mathbf{u}}}_P^i \bar{\mathbf{G}}^i \, \delta \boldsymbol{\theta}^i \tag{7.136}$$

where $\tilde{\bar{\mathbf{u}}}_P^i$ is the skew symmetric matrix associated with the vector $\bar{\mathbf{u}}_P^i$ that defines the local coordinates of the point P^i. Also note that Eq. 136 can be written as

$$\delta \mathbf{r}_P^i = \delta \mathbf{R}^i - \tilde{\mathbf{u}}_P^i \mathbf{G}^i \, \delta \boldsymbol{\theta}^i \tag{7.137}$$

This equation can be obtained from Eq. 136 by using the identity of Eq. 84. Substituting Eq. 137 into Eq. 135, one obtains

$$\delta W_e^i = \mathbf{F}^{i^{\mathrm{T}}} \delta \mathbf{R}^i - \mathbf{F}^{i^{\mathrm{T}}} \tilde{\mathbf{u}}_P^i \mathbf{G}^i \, \delta \boldsymbol{\theta}^i \tag{7.138}$$

or

$$\delta W_e^i = \mathbf{F}_R^{i^{\mathrm{T}}} \delta \mathbf{R}^i + \mathbf{F}_\theta^{i^{\mathrm{T}}} \delta \theta^i \tag{7.139}$$

in which

$$\left.\begin{aligned} \mathbf{F}_R^i &= \mathbf{F}^i \\ \mathbf{F}_\theta^i &= -\mathbf{G}^{i^{\mathrm{T}}} \tilde{\mathbf{u}}_P^{i^{\mathrm{T}}} \mathbf{F}^i \end{aligned}\right\} \tag{7.140}$$

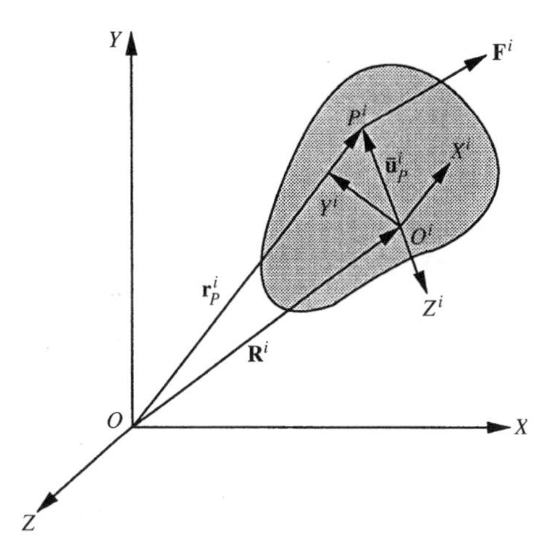

Figure 7.8 Force vector

Equation 139 implies that a force that acts at an arbitrary point on the rigid body i is equipollent to another system defined at the reference point that consists of the same force and a set of generalized forces, defined by the second equation of Eq. 140, associated with the orientation coordinates of the body.

Since $\tilde{\mathbf{u}}_P^i$ is a skew-symmetric matrix, it follows that $\tilde{\mathbf{u}}_P^i = -\tilde{\mathbf{u}}_P^{i^{\mathrm{T}}}$. Using this fact and the cross-product notation, the second equation of Eq. 140 leads to

$$\mathbf{F}_\theta^i = \mathbf{G}^{i^{\mathrm{T}}}(\mathbf{u}_P^i \times \mathbf{F}^i) \tag{7.141}$$

Recall that $\mathbf{u}_P^i \times \mathbf{F}^i$ is the Cartesian moment resulting from the application of the force $\mathbf{F}^i$, that is,

$$\mathbf{M}_a^i = \mathbf{u}_P^i \times \mathbf{F}^i \tag{7.142}$$

where $\mathbf{M}_a^i$ is the moment vector whose components are defined in the Cartesian coordinate system. Equation 141, therefore, defines the relationship between the generalized forces associated with the orientation coordinates and the Cartesian components of the moment as

$$\mathbf{F}_\theta^i = \mathbf{G}^{i^{\mathrm{T}}}\mathbf{M}_a^i \tag{7.143}$$

One can also show that if the components of the moment vector are defined in the body coordinate system, one has

$$\mathbf{F}_\theta^i = \overline{\mathbf{G}}^{i^{\mathrm{T}}}\overline{\mathbf{M}}_a^i \tag{7.144}$$

where $\overline{\mathbf{M}}_a^i = \mathbf{A}^{i^{\mathrm{T}}}\mathbf{M}_a^i$ is the moment vector whose components are defined in the coordinate system of body i.

A special case occurs when the reference point lies on the line of action of the force $\mathbf{F}^i$. In this special case, the vector $\mathbf{u}_P^i$ and the force $\mathbf{F}^i$ are parallel, and hence

$$\mathbf{u}_P^i \times \mathbf{F}^i = \mathbf{0} \tag{7.145}$$

It follows that $\mathbf{F}_\theta^i = \mathbf{0}$. This equation and Eq. 140 imply that the effect of the force does not change if its point of application is moved to an arbitrary position along its line of action. For this reason, the force is considered as a *sliding vector*.

Example 7.7

A force vector $\mathbf{F}^i = [5.0 \quad 0.0 \quad -3.0]^{\mathrm{T}}$ N is acting on body i whose orientation is defined by the Euler angles

$$\phi^i = \frac{\pi}{4}, \quad \theta^i = \frac{\pi}{2}, \quad \psi^i = 0$$

The local position vector of the point of application of the force is

$$\bar{\mathbf{u}}_P^i = [0.2 \quad 0 \quad -0.15]^{\mathrm{T}}\ \mathrm{m}$$

Using the absolute Cartesian coordinates and Euler angles as the generalized coordinates, define the generalized forces associated with the body-generalized coordinates.

Solution. The transformation matrix that defines the orientation of the body is

$$\mathbf{A}^i = \begin{bmatrix} 0.7071 & 0 & 0.7071 \\ 0.7071 & 0 & -0.7071 \\ 0 & 1 & 0 \end{bmatrix}$$

At the given configuration,

$$\overline{\mathbf{G}}^i = \begin{bmatrix} 0 & 1 & 0 \\ 1 & 0 & 0 \\ 0 & 0 & 1 \end{bmatrix}, \quad \mathbf{G}^i = \begin{bmatrix} 0 & 0.7071 & 0.7071 \\ 0 & 0.7071 & -0.7071 \\ 1 & 0 & 0 \end{bmatrix}$$

The virtual work of the force is

$$\mathbf{F}^{i^{\mathrm{T}}}\,\delta\mathbf{r}_P^i = \mathbf{F}^{i^{\mathrm{T}}}\,\delta\mathbf{R}^i - \mathbf{F}^{i^{\mathrm{T}}}\mathbf{A}^i\tilde{\bar{\mathbf{u}}}_P^i\overline{\mathbf{G}}^i\,\delta\boldsymbol{\theta}^i$$

where

$$\tilde{\bar{\mathbf{u}}}_P^i = \begin{bmatrix} 0 & 0.15 & 0 \\ -0.15 & 0 & -0.2 \\ 0 & 0.2 & 0 \end{bmatrix}$$

and

$$\mathbf{A}^i\tilde{\bar{\mathbf{u}}}_P^i\overline{\mathbf{G}}^i = \begin{bmatrix} 0.2475 & 0 & 0 \\ -0.0354 & 0 & 0 \\ 0 & -0.15 & -0.2 \end{bmatrix}$$

The generalized forces associated with the translation of the body reference are

$$\mathbf{Q}_R^i = \mathbf{F}^i = [5.0 \quad 0 \quad -3.0]^{\mathrm{T}}$$

and the generalized forces associated with the orientation coordinates are

$$\mathbf{Q}_\theta^i = -(\mathbf{A}^i\tilde{\bar{\mathbf{u}}}_P^i\overline{\mathbf{G}}^i)^{\mathrm{T}}\mathbf{F}^i$$

$$= -\begin{bmatrix} 0.2475 & -0.0354 & 0 \\ 0 & 0 & -0.15 \\ 0 & 0 & -0.2 \end{bmatrix}\begin{bmatrix} 5.0 \\ 0 \\ -3.0 \end{bmatrix} = \begin{bmatrix} -1.2375 \\ -0.45 \\ -0.6 \end{bmatrix}$$

The generalized forces associated with the orientation coordinates can be defined using an alternative approach by first defining the Cartesian moment $\mathbf{u}_P^i \times \mathbf{F}^i$, where

$$\mathbf{u}_P^i = \mathbf{A}^i \bar{\mathbf{u}}_P^i = \begin{bmatrix} 0.0354 \\ 0.2475 \\ 0 \end{bmatrix}$$

The Cartesian moment can then be defined as

$$\mathbf{u}_P^i \times \mathbf{F}^i = \begin{bmatrix} 0 & 0 & 0.2475 \\ 0 & 0 & -0.0354 \\ -0.2475 & 0.0354 & 0 \end{bmatrix} \begin{bmatrix} 5.0 \\ 0 \\ -3.0 \end{bmatrix} = \begin{bmatrix} -0.7425 \\ 0.1061 \\ -1.2375 \end{bmatrix}$$

The vector of generalized forces associated with the orientation coordinates is

$$\begin{aligned} \mathbf{Q}_\theta^i &= \mathbf{G}^{i^{\mathrm{T}}}(\mathbf{u}_P^i \times \mathbf{F}^i) \\ &= \begin{bmatrix} 0 & 0 & 1 \\ 0.7071 & 0.7071 & 0 \\ 0.7071 & -0.7071 & 0 \end{bmatrix} \begin{bmatrix} -0.7425 \\ 0.1061 \\ -1.2375 \end{bmatrix} = \begin{bmatrix} -1.2375 \\ -0.45 \\ -0.60 \end{bmatrix} \end{aligned}$$

which is the same vector obtained previously.

System of Forces and Moments If a rigid body i is subjected to a set of forces $\mathbf{F}_1^i$, $\mathbf{F}_2^i, \ldots, \mathbf{F}_{n_f}^i$ that act, respectively, at points whose position vectors are $\mathbf{r}_1^i, \mathbf{r}_2^i, \ldots, \mathbf{r}_{n_f}^i$, and a set of moments $\mathbf{M}_1^i, \mathbf{M}_2^i, \ldots, \mathbf{M}_{n_m}^i$, then the virtual work of these forces and moments can be written as

$$\begin{aligned} \delta W_e^i = {} & \mathbf{F}_1^{i^{\mathrm{T}}} \delta \mathbf{r}_1^i + \mathbf{F}_2^{i^{\mathrm{T}}} \delta \mathbf{r}_2^i + \cdots + \mathbf{F}_{n_f}^{i^{\mathrm{T}}} \delta \mathbf{r}_{n_f}^i \\ & + (\mathbf{M}_1^i + \mathbf{M}_2^i + \cdots + \mathbf{M}_{n_m}^i)^{\mathrm{T}} \mathbf{G}^i \, \delta \boldsymbol{\theta}^i \end{aligned} \tag{7.146}$$

which upon the use of the relationship presented previously in this section yields

$$\begin{aligned} \delta W_e^i = {} & (\mathbf{F}_1^i + \mathbf{F}_2^i + \cdots + \mathbf{F}_{n_f}^i)^{\mathrm{T}} \delta \mathbf{R}^i \\ & - (\mathbf{F}_1^{i^{\mathrm{T}}} \tilde{\mathbf{u}}_1^i + \mathbf{F}_2^{i^{\mathrm{T}}} \tilde{\mathbf{u}}_2^i + \cdots + \mathbf{F}_{n_f}^{i^{\mathrm{T}}} \tilde{\mathbf{u}}_{n_f}^i) \mathbf{G}^i \, \delta \boldsymbol{\theta}^i \\ & + (\mathbf{M}_1^i + \mathbf{M}_2^i + \cdots + \mathbf{M}_{n_m}^i)^{\mathrm{T}} \mathbf{G}^i \, \delta \boldsymbol{\theta}^i \end{aligned} \tag{7.147}$$

This equation can be written as

$$\delta W_e^i = (\mathbf{Q}_e^i)_R^{\mathrm{T}} \, \delta \mathbf{R}^i + (\mathbf{Q}_e^i)_\theta^{\mathrm{T}} \, \delta \boldsymbol{\theta}^i \tag{7.148}$$

where $(\mathbf{Q}_e^i)_R$ and $(\mathbf{Q}_e^i)_\theta$ are the vectors of generalized forces associated, respectively, with the generalized translation and orientation coordinates. These two vectors are defined as

$$\left.\begin{aligned}(\mathbf{Q}_e^i)_R &= \mathbf{F}_1^i + \mathbf{F}_2^i + \cdots + \mathbf{F}_{n_f}^i = \sum_{j=1}^{n_f} \mathbf{F}_j^i \\ (\mathbf{Q}_e^i)_\theta &= \mathbf{G}^{i^{\mathrm{T}}}[\mathbf{M}_1^i + \mathbf{M}_2^i + \cdots + \mathbf{M}_{n_m}^i + \mathbf{u}_1^i \times \mathbf{F}_1^i + \mathbf{u}_2^i \\ &\quad \times \mathbf{F}_2^i + \cdots + \mathbf{u}_{n_f}^i \times \mathbf{F}_{n_f}^i] \\ &= \mathbf{G}^{i^{\mathrm{T}}}\left[\sum_{j=1}^{n_m} \mathbf{M}_j^i + \sum_{k=1}^{n_f} (\mathbf{u}_k^i \times \mathbf{F}_k^i)\right]\end{aligned}\right\} \tag{7.149}$$

Spring–Damper–Actuator Element Figure 9 shows two bodies, body i and body j, connected by a spring–damper–actuator element. The attachment points of the spring–damper–actuator element on body i and body j are, respectively, P^i and P^j. The

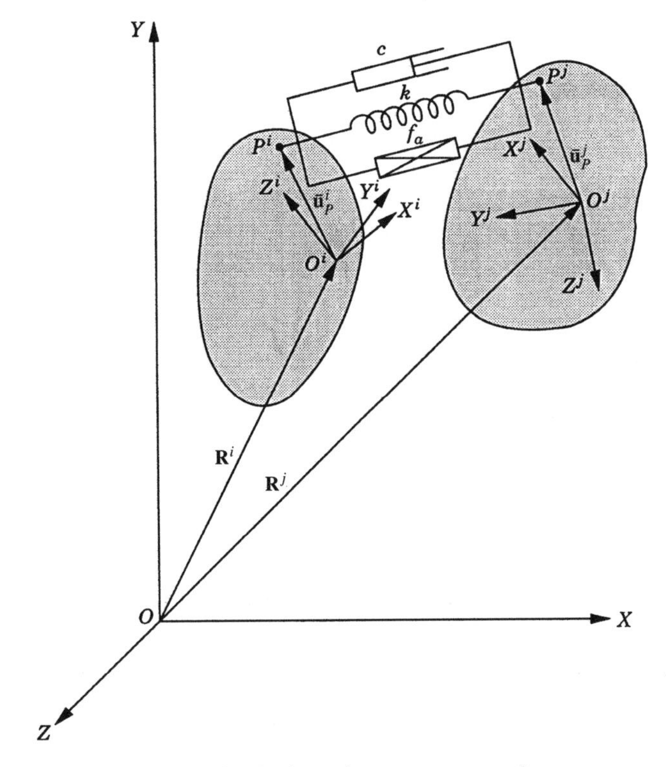

Figure 7.9 Spring–damper–actuator force

spring constant is k, the damping coefficient is c, and the actuator force acting along a line connecting points P^i and P^j is f_a. Here, f_a is a general actuator force that may depend on the system coordinates, on velocities, and possibly on time, and the coefficients k and c can also be nonlinear functions of the system variables. The underformed length of the spring is denoted as l_o. The component of the spring–damper–actuator force along a line connecting points P^i and P^j can then be written as

$$F_s = k(l - l_o) + c\dot{l} + f_a \tag{7.150}$$

where l is the current spring length. The first term on the right-hand side of Eq. 150 is the spring force, the second term is the damping force, and the third term is the actuator force. The virtual work of the force of Eq. 150 is

$$\delta W = -F_s \, \delta l \tag{7.151}$$

where

$$l = |\mathbf{r}_P^{ij}| = \sqrt{\mathbf{r}_P^{ij^{\mathrm{T}}} \mathbf{r}_P^{ij}} \tag{7.152}$$

and $\mathbf{r}_P^{ij}$ is the position vector of point P^i with respect to point P^j, that is,

$$\begin{aligned} \mathbf{r}_P^{ij} &= \mathbf{r}_P^i - \mathbf{r}_P^j \\ &= \mathbf{R}^i + \mathbf{A}^i \bar{\mathbf{u}}_P^i - \mathbf{R}^j - \mathbf{A}^j \bar{\mathbf{u}}_P^j \end{aligned} \tag{7.153}$$

where $\bar{\mathbf{u}}_P^i$ and $\bar{\mathbf{u}}_P^j$ are, respectively, the position vectors of points P^i and P^j defined in the coordinate system of the respective body, $\mathbf{A}^i$ and $\mathbf{A}^j$ are the transformation matrices of the two bodies, and $\mathbf{R}^i$ and $\mathbf{R}^j$ are the global position vectors of the origins of the coordinate systems of bodies i and j, respectively. Using Eq. 152, the virtual change in the spring length is

$$\delta l = (\mathbf{r}_P^{ij^{\mathrm{T}}} \mathbf{r}_P^{ij})^{-1/2} \mathbf{r}_P^{ij^{\mathrm{T}}} \, \delta \mathbf{r}_P^{ij} \tag{7.154}$$

which, upon using Eq. 152, yields

$$\begin{aligned} \delta l &= \frac{1}{l} \mathbf{r}_P^{ij^{\mathrm{T}}} \, \delta \mathbf{r}_P^{ij} \\ &= \frac{\mathbf{r}_P^{ij^{\mathrm{T}}}}{l} [\delta \mathbf{R}^i - \tilde{\mathbf{u}}_P^i \mathbf{G}^i \, \delta \boldsymbol{\theta}^i - \delta \mathbf{R}^j + \tilde{\mathbf{u}}_P^j \mathbf{G}^j \, \delta \boldsymbol{\theta}^j] \end{aligned} \tag{7.155}$$

where $\tilde{\mathbf{u}}_P^i$ and $\tilde{\mathbf{u}}_P^j$ are the skew symmetric matrices associated, respectively, with the vectors $\mathbf{A}^i \bar{\mathbf{u}}_P^i$ and $\mathbf{A}^j \bar{\mathbf{u}}_P^j$.

Let

$$\hat{\mathbf{r}}_P^{ij} = \frac{\mathbf{r}_P^{ij}}{l} \tag{7.156}$$

be a unit vector along the line of action of the force F_s. Using the preceding equation and Eqs. 150 and 155, the virtual work of Eq. 151 takes the form

$$\delta W = -F_s \hat{\mathbf{r}}_P^{ij^{\mathrm{T}}} [\delta \mathbf{R}^i - \tilde{\mathbf{u}}_P^i \mathbf{G}^i \, \delta \boldsymbol{\theta}^i - \delta \mathbf{R}^j + \tilde{\mathbf{u}}_P^j \mathbf{G}^j \, \delta \boldsymbol{\theta}^j] \tag{7.157}$$

which can be written as

$$\delta W = \mathbf{Q}_R^{i^{\mathrm{T}}} \, \delta \mathbf{R}^i + \mathbf{Q}_\theta^{i^{\mathrm{T}}} \, \delta \boldsymbol{\theta}^i + \mathbf{Q}_R^{j^{\mathrm{T}}} \, \delta \mathbf{R}^j + \mathbf{Q}_\theta^{j^{\mathrm{T}}} \, \delta \boldsymbol{\theta}^j \tag{7.158}$$

where the generalized forces $\mathbf{Q}_R^i$, $\mathbf{Q}_\theta^i$, $\mathbf{Q}_R^j$, and $\mathbf{Q}_\theta^j$ are

$$\left.\begin{aligned} \mathbf{Q}_R^i &= -F_s \hat{\mathbf{r}}_P^{ij} \\ \mathbf{Q}_\theta^i &= F_s \mathbf{G}^{i^{\mathrm{T}}} \tilde{\mathbf{u}}_P^{i^{\mathrm{T}}} \hat{\mathbf{r}}_P^{ij} \\ \mathbf{Q}_R^j &= F_s \hat{\mathbf{r}}_P^{ij} \\ \mathbf{Q}_\theta^j &= -F_s \mathbf{G}^{j^{\mathrm{T}}} \tilde{\mathbf{u}}_P^{j^{\mathrm{T}}} \hat{\mathbf{r}}_P^{ij} \end{aligned}\right\} \tag{7.159}$$

The virtual work of Eq. 158 can also be expressed in the following form:

$$\delta W = \mathbf{Q}^{i^{\mathrm{T}}} \, \delta \mathbf{q}^i + \mathbf{Q}^{j^{\mathrm{T}}} \, \delta \mathbf{q}^j \tag{7.160}$$

where $\mathbf{q}^i$ and $\mathbf{q}^j$ are the generalized coordinates of body i and body j, respectively, and

$$\mathbf{Q}^i = \begin{bmatrix} \mathbf{Q}_R^i \\ \mathbf{Q}_\theta^i \end{bmatrix} = \begin{bmatrix} -F_s \hat{\mathbf{r}}_P^{ij} \\ F_s \mathbf{G}^{i^{\mathrm{T}}} \tilde{\mathbf{u}}_P^{i^{\mathrm{T}}} \hat{\mathbf{r}}_P^{ij} \end{bmatrix} \tag{7.161}$$

$$\mathbf{Q}^j = \begin{bmatrix} \mathbf{Q}_R^j \\ \mathbf{Q}_\theta^j \end{bmatrix} = \begin{bmatrix} F_s \hat{\mathbf{r}}_P^{ij} \\ -F_s \mathbf{G}^{j^{\mathrm{T}}} \tilde{\mathbf{u}}_P^{j^{\mathrm{T}}} \hat{\mathbf{r}}_P^{ij} \end{bmatrix} \tag{7.162}$$

Rotational Spring–Damper–Actuator Element While the order of the finite rotation is not commutative and, consequently, such rotations cannot be treated as vector quantities, the order of the infinitesimal rotation is commutative. Thus, infinitesimal rotations can be treated as vectors. This can be demonstrated by considering two successive infinitesimal rotations that define the two transformation matrices $\mathbf{A}_1^i$ and $\mathbf{A}_2^i$. By using a first-order approximation, one can show that

$$\mathbf{A}_1^i \mathbf{A}_2^i = \mathbf{A}_2^i \mathbf{A}_1^i \tag{7.163}$$

We have shown previously that the angular velocity vector, defined in the global coordinate system, can be written in terms of the orientation coordinates and their time derivatives as

$$\boldsymbol{\omega}^i = \mathbf{G}^i \dot{\boldsymbol{\theta}}^i \tag{7.164}$$

where the matrix $\mathbf{G}^i$ in the case of *Euler angles* is defined by Eq. 72. We observe from the preceding equation that the angular velocity in the spatial analysis is not the time derivative of the orientation coordinates. The angular velocity vector, however, can be considered as the time rate of a set of infinitesimal rotations $\delta\pi^i$ about the axes of the global coordinate system. Therefore, the use of the preceding equation leads to

$$\boldsymbol{\omega}^i = \frac{\delta\boldsymbol{\pi}^i}{\delta t} = \mathbf{G}^i \frac{\delta\boldsymbol{\theta}^i}{\delta t} \tag{7.165}$$

which leads to the relationship between the infinitesimal virtual rotations about the axes of the global Cartesian coordinate system and the virtual change in the generalized orientation coordinates $\boldsymbol{\theta}^i$ as

$$\delta\boldsymbol{\pi}^i = \mathbf{G}^i\,\delta\boldsymbol{\theta}^i \tag{7.166}$$

We now consider the case of a *rotational spring–damper–actuator element* (Fig. 10) that connects two arbitrary bodies i and j in the multibody system. These two bodies may be connected by a revolute, screw, or cylindrical joint. Let θ^{ij} be the rotation of body i with respect to body j along the joint axis, and let $\mathbf{h}^{ij}$ be a unit vector along the joint axis. The virtual change $\delta\theta^{ij}$ can be expressed in terms of the generalized orientation coordinates of bodies i and j as

$$\delta\theta^{ij} = \mathbf{h}^{ij\mathrm{T}}(\delta\boldsymbol{\pi}^i - \delta\boldsymbol{\pi}^j) = \mathbf{h}^{ij\mathrm{T}}(\mathbf{G}^i\,\delta\boldsymbol{\theta}^i - \mathbf{G}^j\,\delta\boldsymbol{\theta}^j) \tag{7.167}$$

This equation implies that the virtual relative rotation $\delta\theta^{ij}$ is the projection of the relative Cartesian rotation $(\delta\boldsymbol{\pi}^i - \delta\boldsymbol{\pi}^j)$ on the joint axis.

The torque exerted on body i by the rotational spring–damper–actuator element as the result of the rotation θ^{ij} is

$$T^{ij} = (k_r\theta^{ij} + c_r\dot{\theta}^{ij} + T_a) \tag{7.168}$$

where k_r and c_r are, respectively, the rotational spring and damping coefficients, and T_a is the actuator torque. The coefficients k_r and c_r and the torque T_a can be nonlinear functions of the system coordinates, velocities, and time.

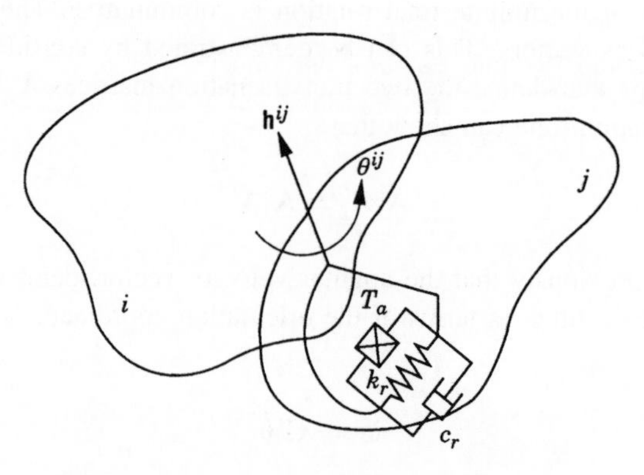

Figure 7.10 Torsional spring and damper

The virtual work of the torque T^{ij} is

$$\delta W = -T^{ij}\,\delta\theta^{ij} = -T^{ij}\mathbf{h}^{ij^{\mathrm{T}}}(\mathbf{G}^i\,\delta\boldsymbol{\theta}^i - \mathbf{G}^j\,\delta\boldsymbol{\theta}^j) \tag{7.169}$$

This equation can also be written as

$$\delta W = \mathbf{Q}^{i^{\mathrm{T}}}\,\delta\mathbf{q}^i + \mathbf{Q}^{j^{\mathrm{T}}}\,\delta\mathbf{q}^j \tag{7.170}$$

where the generalized forces $\mathbf{Q}^i$ and $\mathbf{Q}^j$ are given by

$$\mathbf{Q}^i = \begin{bmatrix} \mathbf{0} \\ -T^{ij}\mathbf{G}^{i^{\mathrm{T}}}\mathbf{h}^{ij} \end{bmatrix}, \quad \mathbf{Q}^j = \begin{bmatrix} \mathbf{0} \\ T^{ij}\mathbf{G}^{j^{\mathrm{T}}}\mathbf{h}^{ij} \end{bmatrix} \tag{7.171}$$

7.8 DYNAMIC EQUATIONS OF MOTION

In the preceding two sections, the virtual work of the inertia forces and the virtual work of the applied forces acting on the rigid body i were developed. It was shown that the virtual work of the inertia forces can be written as (Eq. 95)

$$\delta W_i^i = [\mathbf{M}^i\ddot{\mathbf{q}}^i - \mathbf{Q}_v^i]^{\mathrm{T}}\,\delta\mathbf{q}^i \tag{7.172}$$

where $\mathbf{M}^i$ is the symmetric mass matrix, $\mathbf{q}^i = [\mathbf{R}^{i^{\mathrm{T}}}\quad \boldsymbol{\theta}^{i^{\mathrm{T}}}]^{\mathrm{T}}$ is the vector of generalized coordinates, and $\mathbf{Q}_v^i$ is the vector of inertia forces that absorbs terms that are quadratic in the velocities.

The virtual work of the applied forces is

$$\delta W_e^i = \mathbf{Q}_e^{i^{\mathrm{T}}}\,\delta\mathbf{q}^i \tag{7.173}$$

where $\mathbf{Q}_e^i$ is the vector of generalized applied forces.

Using the *principle of virtual work in dynamics* for unconstrained motion, one has

$$\delta W_i^i = \delta W_e^i \tag{7.174}$$

Substituting Eqs. 172 and 173 into Eq. 174, one obtains

$$[\mathbf{M}^i\ddot{\mathbf{q}}^i - \mathbf{Q}_v^i]^{\mathrm{T}}\,\delta\mathbf{q}^i = \mathbf{Q}_e^{i^{\mathrm{T}}}\,\delta\mathbf{q}^i \tag{7.175}$$

or

$$[\mathbf{M}^i\ddot{\mathbf{q}}^i - \mathbf{Q}_v^i - \mathbf{Q}_e^i]^{\mathrm{T}}\,\delta\mathbf{q}^i = 0 \tag{7.176}$$

In the case of *unconstrained motion*, the elements of the vector $\delta\mathbf{q}^i$ are independent. In this case, Eq. 176 leads to

$$\mathbf{M}^i\ddot{\mathbf{q}}^i = \mathbf{Q}_e^i + \mathbf{Q}_v^i, \qquad i = 1, 2, \ldots, n_b \tag{7.177}$$

where n_b is the total number of rigid bodies in the system. The preceding equation can be written in a partitioned matrix form as

$$\begin{bmatrix} \mathbf{m}^i_{RR} & \mathbf{m}^i_{R\theta} \\ \mathbf{m}^i_{\theta R} & \mathbf{m}^i_{\theta\theta} \end{bmatrix} \begin{bmatrix} \ddot{\mathbf{R}}^i \\ \ddot{\boldsymbol{\theta}}^i \end{bmatrix} = \begin{bmatrix} (\mathbf{Q}^i_e)_R \\ (\mathbf{Q}^i_e)_\theta \end{bmatrix} + \begin{bmatrix} (\mathbf{Q}^i_v)_R \\ (\mathbf{Q}^i_v)_\theta \end{bmatrix}, \qquad i = 1, 2, \ldots, n_b \tag{7.178}$$

where $\mathbf{m}^i_{RR}$, $\mathbf{m}^i_{R\theta} = \mathbf{m}^{i\mathrm{T}}_{\theta R}$, and $\mathbf{m}^i_{\theta\theta}$ are defined by Eq. 100, and $(\mathbf{Q}^i_v)_R$ and $(\mathbf{Q}^i_v)_\theta$ are defined by Eqs. 129 and 134, respectively.

Centroidal Coordinate System Equation 178 is the matrix equation that governs the unconstrained motion of the rigid body. This equation can be simplified if the origin of the body coordinate system is rigidly attached to the center of mass of the body. In this case

$$\int_{V^i} \rho^i \bar{\mathbf{u}}^i \, dV^i = \int_{V^i} \rho^i [\bar{x}^i \quad \bar{y}^i \quad \bar{z}^i]^{\mathrm{T}} \, dV^i = \mathbf{0} \tag{7.179}$$

It follows also that

$$\int_{V^i} \rho^i \tilde{\bar{\mathbf{u}}}^i \, dV^i = \mathbf{0} \tag{7.180}$$

Substituting this equation into Eq. 100, one obtains

$$\mathbf{m}^i_{R\theta} = \mathbf{0}, \quad \mathbf{m}^i_{\theta R} = \mathbf{0} \tag{7.181}$$

which imply that in the case of a *centroidal body coordinate system*, there is no *inertia coupling* between the translation and the rotation of the rigid body. Furthermore, in this special case

$$(\mathbf{Q}^i_v)_R = \mathbf{0} \tag{7.182}$$

The use of Eqs. 181, 182, and 178 leads to the following dynamic equations:

$$\begin{bmatrix} \mathbf{m}^i_{RR} & \mathbf{0} \\ \mathbf{0} & \mathbf{m}^i_{\theta\theta} \end{bmatrix} \begin{bmatrix} \ddot{\mathbf{R}}^i \\ \ddot{\boldsymbol{\theta}}^i \end{bmatrix} = \begin{bmatrix} (\mathbf{Q}^i_e)_R \\ (\mathbf{Q}^i_e)_\theta \end{bmatrix} + \begin{bmatrix} \mathbf{0} \\ (\mathbf{Q}^i_v)_\theta \end{bmatrix}, \qquad i = 1, 2, \ldots, n_b \tag{7.183}$$

where, as previously defined by Eqs. 100 and 134,

$$\left.\begin{aligned} \mathbf{m}^i_{RR} &= m^i \mathbf{I}, \quad \mathbf{m}^i_{\theta\theta} = \bar{\mathbf{G}}^{i\mathrm{T}} \bar{\mathbf{I}}^i_{\theta\theta} \bar{\mathbf{G}}^i \\ (\mathbf{Q}^i_v)_\theta &= -\bar{\mathbf{G}}^{i\mathrm{T}} [\bar{\boldsymbol{\omega}}^i \times (\bar{\mathbf{I}}^i_{\theta\theta} \bar{\boldsymbol{\omega}}^i) + \bar{\mathbf{I}}^i_{\theta\theta} \dot{\bar{\mathbf{G}}}^i \dot{\boldsymbol{\theta}}^i] \end{aligned}\right\} \tag{7.184}$$

When Euler angles are used, the set of orientation coordinates $\boldsymbol{\theta}^i$ has three elements and the mass matrix in Eq. 183 is a 6×6 matrix.

Example 7.8

A force vector $\mathbf{F}^i = [5.0 \quad 0.0 \quad -3.0]^{\mathrm{T}}$ N is acting on unconstrained body i. The orientation of a centroidal body coordinate system is defined by the Euler angles

$$\phi^i = \frac{\pi}{4}, \quad \theta^i = \frac{\pi}{2}, \quad \psi^i = 0$$

The time rate of change of Euler angles at the given configuration is

$$\dot{\phi}^i = 20, \quad \dot{\theta}^i = 0, \quad \dot{\psi}^i = -35 \text{ rad/s}$$

The local position vector of the point of application of the force is

$$\bar{\mathbf{u}}_P^i = [0.2 \quad 0 \quad -0.15]^{\mathrm{T}} \text{ m}$$

The mass of the body is 2 kg and its inertia tensor is

$$\bar{\mathbf{I}}_{\theta\theta}^i = \begin{bmatrix} 0.3 & 0 & 0 \\ 0 & 0.3 & 0 \\ 0 & 0 & 0.3 \end{bmatrix} \text{kg} \cdot \text{m}^2$$

Write the equations of motion of this system at the given configuration and determine the accelerations.

Solution. The transformation matrix that defines the orientation of the body is

$$\mathbf{A}^i = \begin{bmatrix} 0.7071 & 0 & 0.7071 \\ 0.7071 & 0 & -0.7071 \\ 0 & 1 & 0 \end{bmatrix}$$

At the given configuration, one also has

$$\bar{\mathbf{G}}^i = \begin{bmatrix} 0 & 1 & 0 \\ 1 & 0 & 0 \\ 0 & 0 & 1 \end{bmatrix}, \quad \mathbf{G}^i = \begin{bmatrix} 0 & 0.7071 & 0.7071 \\ 0 & 0.7071 & -0.7071 \\ 1 & 0 & 0 \end{bmatrix}$$

It was shown in Example 7 that the generalized forces associated with the translation and the orientation coordinates of the body reference are

$$\mathbf{Q}_R^i = \mathbf{F}^i = [5.0 \quad 0 \quad -3.0]^{\mathrm{T}}$$

$$\mathbf{Q}_\theta^i = [-1.2375 \quad -0.45 \quad -0.6]^{\mathrm{T}}$$

Using the mass of the body, one has

$$\mathbf{m}_{RR}^i = m^i \mathbf{I} = \begin{bmatrix} 2 & 0 & 0 \\ 0 & 2 & 0 \\ 0 & 0 & 2 \end{bmatrix}$$

and

$$\mathbf{m}_{\theta\theta}^i = \bar{\mathbf{G}}^{i\mathrm{T}} \bar{\mathbf{I}}_{\theta\theta}^i \bar{\mathbf{G}}^i = \begin{bmatrix} 0.3 & 0 & 0 \\ 0 & 0.3 & 0 \\ 0 & 0 & 0.3 \end{bmatrix}$$

Since a centroidal body coordinate system is used

$$\mathbf{m}^i_{R\theta} = \mathbf{m}^{i^{\mathrm{T}}}_{\theta R} = \mathbf{0}$$

The angular velocity vector in the body coordinate system is

$$\bar{\boldsymbol{\omega}}^i = \bar{\mathbf{G}}^i \dot{\boldsymbol{\theta}}^i = [0 \quad 20 \quad -35]^{\mathrm{T}}$$

Using Euler angles and their time derivatives, one has

$$\dot{\bar{\mathbf{G}}}^i = \begin{bmatrix} \dot{\theta}^i \cos\theta^i \sin\psi^i + \dot{\psi}^i \sin\theta^i \cos\psi^i & -\dot{\psi}^i \sin\psi^i & 0 \\ \dot{\theta}^i \cos\theta^i \cos\psi^i - \dot{\psi}^i \sin\theta^i \sin\psi^i & -\dot{\psi}^i \cos\psi^i & 0 \\ -\dot{\theta}^i \sin\theta^i & 0 & 0 \end{bmatrix} = \begin{bmatrix} -35 & 0 & 0 \\ 0 & 35 & 0 \\ 0 & 0 & 0 \end{bmatrix}$$

It follows that

$$\dot{\bar{\mathbf{G}}}^i \dot{\boldsymbol{\theta}}^i = [-700 \quad 0 \quad 0]^{\mathrm{T}}$$

The quadratic velocity inertia force vector associated with the translation of the body reference is equal to zero since a centroidal body coordinate system is used, while the force vector associated with the orientation coordinates is

$$(\mathbf{Q}^i_v)_\theta = -\bar{\mathbf{G}}^{i^{\mathrm{T}}}[\bar{\boldsymbol{\omega}}^i \times (\bar{\mathbf{I}}^i_{\theta\theta}\bar{\boldsymbol{\omega}}^i) + \bar{\mathbf{I}}^i_{\theta\theta}\dot{\bar{\mathbf{G}}}^i \dot{\boldsymbol{\theta}}^i] = \begin{bmatrix} 0 \\ 210 \\ 0 \end{bmatrix}$$

Using the case of the centroidal body coordinate system, the matrix equation of motion of the body can be written as

$$\begin{bmatrix} 2 & 0 & 0 & 0 & 0 & 0 \\ 0 & 2 & 0 & 0 & 0 & 0 \\ 0 & 0 & 2 & 0 & 0 & 0 \\ 0 & 0 & 0 & 0.3 & 0 & 0 \\ 0 & 0 & 0 & 0 & 0.3 & 0 \\ 0 & 0 & 0 & 0 & 0 & 0.3 \end{bmatrix} \begin{bmatrix} \ddot{R}^i_x \\ \ddot{R}^i_y \\ \ddot{R}^i_z \\ \ddot{\phi}^i \\ \ddot{\theta}^i \\ \ddot{\psi}^i \end{bmatrix} = \begin{bmatrix} 5.0 \\ 0 \\ -3.0 \\ -1.2375 \\ -0.45 \\ -0.6 \end{bmatrix} + \begin{bmatrix} 0 \\ 0 \\ 0 \\ 0 \\ 210 \\ 0 \end{bmatrix}$$

or

$$\begin{bmatrix} 2 & 0 & 0 & 0 & 0 & 0 \\ 0 & 2 & 0 & 0 & 0 & 0 \\ 0 & 0 & 2 & 0 & 0 & 0 \\ 0 & 0 & 0 & 0.3 & 0 & 0 \\ 0 & 0 & 0 & 0 & 0.3 & 0 \\ 0 & 0 & 0 & 0 & 0 & 0.3 \end{bmatrix} \begin{bmatrix} \ddot{R}^i_x \\ \ddot{R}^i_y \\ \ddot{R}^i_z \\ \ddot{\phi}^i \\ \ddot{\theta}^i \\ \ddot{\psi}^i \end{bmatrix} = \begin{bmatrix} 5.0 \\ 0 \\ -3.0 \\ -1.2375 \\ 209.55 \\ -0.6 \end{bmatrix}$$

The solution of this system of equations defines the accelerations as

$$[\ddot{R}_x^i \quad \ddot{R}_y^i \quad \ddot{R}_z^i \quad \ddot{\phi}^i \quad \ddot{\theta}^i \quad \ddot{\psi}^i] = [2.5 \quad 0 \quad -1.5 \quad -4.125 \quad 698.5 \quad -2.0]$$

The results obtained in this example can also be used to show that

$$\mathbf{A}^i\dot{\overline{\mathbf{G}}}^i = \begin{bmatrix} -24.7485 & 0 & 0 \\ -24.7845 & 0 & 0 \\ 0 & 35 & 0 \end{bmatrix}$$

while

$$\dot{\mathbf{G}}^i = \begin{bmatrix} 0 & -\dot{\phi}^i \sin\phi^i & \dot{\theta}^i \cos\theta^i \sin\phi^i + \dot{\phi}^i \sin\theta^i \cos\phi^i \\ 0 & \dot{\phi}^i \cos\phi^i & -\dot{\theta}^i \cos\theta^i \cos\phi^i + \dot{\phi}^i \sin\theta^i \sin\phi^i \\ 0 & 0 & -\dot{\theta}^i \sin\theta^i \end{bmatrix}$$

$$= \begin{bmatrix} 0 & -14.142 & 14.142 \\ 0 & 14.142 & 14.142 \\ 0 & 0 & 0 \end{bmatrix}$$

Nonetheless,

$$\mathbf{A}^i\dot{\overline{\mathbf{G}}}^i\dot{\boldsymbol{\theta}}^i = \dot{\mathbf{G}}^i\dot{\boldsymbol{\theta}}^i = \begin{bmatrix} -494.97 \\ -494.97 \\ 0 \end{bmatrix}$$

remains in effect.

7.9 CONSTRAINED DYNAMICS

As in the case of planar analysis, there are different approaches for formulating the dynamic equations of constrained spatial multibody systems. The first approach, in which a set of independent coordinates are used in the formulation of the dynamic relationships, leads to the smallest set of differential equations expressed in terms of the system degrees of freedom. This method will be discussed in more detail in Section 14 of this chapter. An alternative approach that is discussed in this section is to use the *augmented formulation*, wherein the dynamic equations are formulated in terms of a set of dependent and independent coordinates. The kinematic relationships that describe mechanical joints and specified motion trajectories are adjoined to the system differential equations using the technique of *Lagrange multipliers*. This approach leads to a relatively large system of loosely coupled equations that can be solved using matrix, numerical, and computer methods.

Kinematic Equations Consider a multibody system that consists of n_b interconnected bodies. In the analysis presented in this section, the configuration of each body in the multibody system is described using the *absolute Cartesian coordinates* $\mathbf{R}^i$ and the *orientation* *coordinates* $\boldsymbol{\theta}^i$. The vector of system-generalized coordinates can be written as

$$\mathbf{q} = [\mathbf{R}^{1^{\mathrm{T}}} \quad \boldsymbol{\theta}^{2^{\mathrm{T}}} \quad \mathbf{R}^{2^{\mathrm{T}}} \quad \boldsymbol{\theta}^{2^{\mathrm{T}}} \quad \cdots \quad \mathbf{R}^{n_b^{\mathrm{T}}} \quad \boldsymbol{\theta}^{n_b^{\mathrm{T}}}]^{\mathrm{T}} \tag{7.185}$$

which can also be written as

$$\mathbf{q} = [\mathbf{q}^{1^{\mathrm{T}}} \quad \mathbf{q}^{2^{\mathrm{T}}} \quad \cdots \quad \mathbf{q}^{n_b^{\mathrm{T}}}]^{\mathrm{T}} \tag{7.186}$$

where

$$\mathbf{q}^i = [\mathbf{R}^{i^{\mathrm{T}}} \quad \boldsymbol{\theta}^{i^{\mathrm{T}}}]^{\mathrm{T}} \tag{7.187}$$

The kinematic relationships that describe mechanical joints and specified motion trajectories such as driving constraints can be written in the following vector form:

$$\mathbf{C}(\mathbf{q}, t) = \mathbf{0} \tag{7.188}$$

If the number of kinematic constraint equations is equal to the number of the generalized coordinates, the system is said to be *kinematically driven*, and in this case, Eq. 188 can be solved for the generalized coordinates using a *Newton–Raphson algorithm*.

The *velocity kinematic equations* can be obtained by differentiating Eq. 188 with respect to time, yielding

$$\mathbf{C}_{\mathbf{q}}\dot{\mathbf{q}} = -\mathbf{C}_t \tag{7.189}$$

where $\mathbf{C}_{\mathbf{q}}$ is the *constraint Jacobian matrix* and $\mathbf{C}_t$ is the vector of partial derivatives of the constraint equations with respect to time. The Jacobian matrix is obtained by differentiating the constraint equations with respect to the coordinates, while the vector $\mathbf{C}_t$ is defined as

$$\mathbf{C}_t = \frac{\partial \mathbf{C}}{\partial t} \tag{7.190}$$

The vector $\mathbf{C}_t$ is the zero vector if the constraint equations are not explicit functions of time. If the constraint equations are linearly independent, the constraint Jacobian matrix has a full row rank, and in the case of *kinematically driven systems*, the Jacobian matrix becomes a square matrix. In this special case, Eq. 189 can be considered as a linear system of algebraic equations in the velocities, and this system has a unique solution that can be determined assuming that the generalized coordinates are known from solving Eq. 188.

The kinematic *acceleration equations* can be obtained by differentiating Eq. 189 with respect to time. This leads to

$$\mathbf{C}_{\mathbf{q}}\ddot{\mathbf{q}} = \mathbf{Q}_d \tag{7.191}$$

where $\mathbf{Q}_d$ is a vector that absorbs terms that are quadratic in the velocities. This vector is defined as

$$\mathbf{Q}_d = -\mathbf{C}_{tt} - (\mathbf{C}_{\mathbf{q}}\dot{\mathbf{q}})_{\mathbf{q}}\dot{\mathbf{q}} - 2\mathbf{C}_{\mathbf{q}_t}\dot{\mathbf{q}} \tag{7.192}$$

Equation 191 can be considered as a linear system of algebraic equations in the accelerations, and in the case of kinematically driven systems, Eq. 191 has a unique solution that determines the vector of generalized accelerations.

It is clear that the kinematic equations of the spatial multibody systems are similar to the equations obtained in the case of planar systems. Consequently, the numerical algorithms used in the spatial kinematic analysis have the same steps as the algorithms used in the planar kinematic analysis. These algorithms were discussed in detail in Chapter 3.

Constrained Dynamic Equations In the formulation of the dynamic equations using the absolute coordinates, the technique of Lagrange multipliers is used to adjoin the kinematic constraint equations to the differential equations of motion. For body i in the system, the equations of motion can be written in a matrix form as

$$\mathbf{M}^i \ddot{\mathbf{q}}^i + \mathbf{C}_{\mathbf{q}^i}^{\mathrm{T}} \boldsymbol{\lambda} = \mathbf{Q}_e^i + \mathbf{Q}_v^i, \qquad i = 1, 2, \ldots, n_b \tag{7.193}$$

where $\mathbf{M}^i$ is the body mass matrix, $\mathbf{q}^i = [\mathbf{R}^{i^{\mathrm{T}}} \quad \boldsymbol{\theta}^{i^{\mathrm{T}}}]^{\mathrm{T}}$ is the vector of body-generalized coordinates, $\mathbf{C}_{\mathbf{q}^i}$ is the constraint Jacobian matrix, $\boldsymbol{\lambda}$ is the vector of Lagrange multipliers, $\mathbf{Q}_e^i$ is the vector of generalized applied forces, and $\mathbf{Q}_v^i$ is the vector of inertia forces that absorbs terms that are quadratic in the velocities. In the forward dynamics, the unknowns in Eq. 193 are the vectors of accelerations and Lagrange multipliers. The number of generalized coordinates is $6 \times n_b$, while the number of Lagrange multipliers is n_c, where n_c is the number of kinematic constraint equations. Therefore, the total number of unknowns is $6n_b + n_c$. Equation 193, when it is written for each body in the system, leads to $6n_b$ differential equations. The remaining n_c equations, which are required in order to be able to solve for the $6n_b + n_c$ unknowns, are defined by Eq. 191. This equation can be written in the following form:

$$[\mathbf{C}_{\mathbf{q}^1} \quad \mathbf{C}_{\mathbf{q}^2} \quad \cdots \quad \mathbf{C}_{\mathbf{q}^{n_b}}] \begin{bmatrix} \ddot{\mathbf{q}}^1 \\ \ddot{\mathbf{q}}^2 \\ \vdots \\ \ddot{\mathbf{q}}^{n_b} \end{bmatrix} = \mathbf{Q}_d \tag{7.194}$$

Equations 193 and 194 can be combined in order to obtain the following matrix equation:

$$\begin{bmatrix} \mathbf{M}^1 & \mathbf{0} & \cdots & \mathbf{0} & \mathbf{C}_{\mathbf{q}^1}^{\mathrm{T}} \\ \mathbf{0} & \mathbf{M}^2 & \cdots & \mathbf{0} & \mathbf{C}_{\mathbf{q}^2}^{\mathrm{T}} \\ \vdots & \vdots & \ddots & \vdots & \vdots \\ \mathbf{0} & \mathbf{0} & \cdots & \mathbf{M}^{n_b} & \mathbf{C}_{\mathbf{q}^{n_b}}^{\mathrm{T}} \\ \mathbf{C}_{\mathbf{q}^1} & \mathbf{C}_{\mathbf{q}^2} & \cdots & \mathbf{C}_{\mathbf{q}^{n_b}} & \mathbf{0} \end{bmatrix} \begin{bmatrix} \ddot{\mathbf{q}}^1 \\ \ddot{\mathbf{q}}^2 \\ \vdots \\ \ddot{\mathbf{q}}^{n_b} \\ \boldsymbol{\lambda} \end{bmatrix} = \begin{bmatrix} \mathbf{Q}_e^1 + \mathbf{Q}_v^1 \\ \mathbf{Q}_e^2 + \mathbf{Q}_v^2 \\ \vdots \\ \mathbf{Q}_e^{n_b} + \mathbf{Q}_v^{n_b} \\ \mathbf{Q}_d \end{bmatrix} \tag{7.195}$$

which can also be written as

$$\begin{bmatrix} \mathbf{M} & \mathbf{C}_{\mathbf{q}}^{\mathrm{T}} \\ \mathbf{C}_{\mathbf{q}} & \mathbf{0} \end{bmatrix} \begin{bmatrix} \ddot{\mathbf{q}} \\ \boldsymbol{\lambda} \end{bmatrix} = \begin{bmatrix} \mathbf{Q}_e + \mathbf{Q}_v \\ \mathbf{Q}_d \end{bmatrix} \tag{7.196}$$

This equation can be solved for the vectors of accelerations and Lagrange multipliers. The vector of Lagrange multipliers can be used to determine the generalized constraint forces

$\mathbf{C}_{\mathbf{q}}^{\mathrm{T}}\boldsymbol{\lambda}$, while the accelerations can be integrated forward in time in order to determine the generalized coordinates and velocities. The obtained numerical solution, however, has to satisfy the algebraic kinematic constraint relationships of Eq. 188. The numerical algorithm for solving this mixed system of differential and algebraic equations is the same as the one discussed in the preceding chapter.

7.10 FORMULATION OF THE JOINT CONSTRAINTS

In the analysis presented in this book, the kinematic constraints are classified as *joint* or *driving constraints*. Joint constraints define the connectivity between bodies in the system, while driving constraints describe the specified motion trajectories. The driving constraints may depend on time and may take any form depending on the particular application. On the other hand, in the case of using the absolute coordinates in the analysis of mechanical systems, the formulation of the kinematic constraints that describe a joint between two arbitrary bodies in the system can be made independent of the particular topological structure of that system since similar sets of coordinates are used to describe the motion of the bodies. A *computer library* that contains the formulations of a number of mechanical joints that are often encountered in the analysis of mechanical systems can be developed and used in the computer-aided analysis of a variety of applications. In this section, the formulations of some of the mechanical joints used in spatial multibody systems are discussed.

Spherical Joint Figure 11 shows two bodies, i and j, connected by a *spherical joint* which eliminates the freedom of relative translations between the two bodies, and it allows

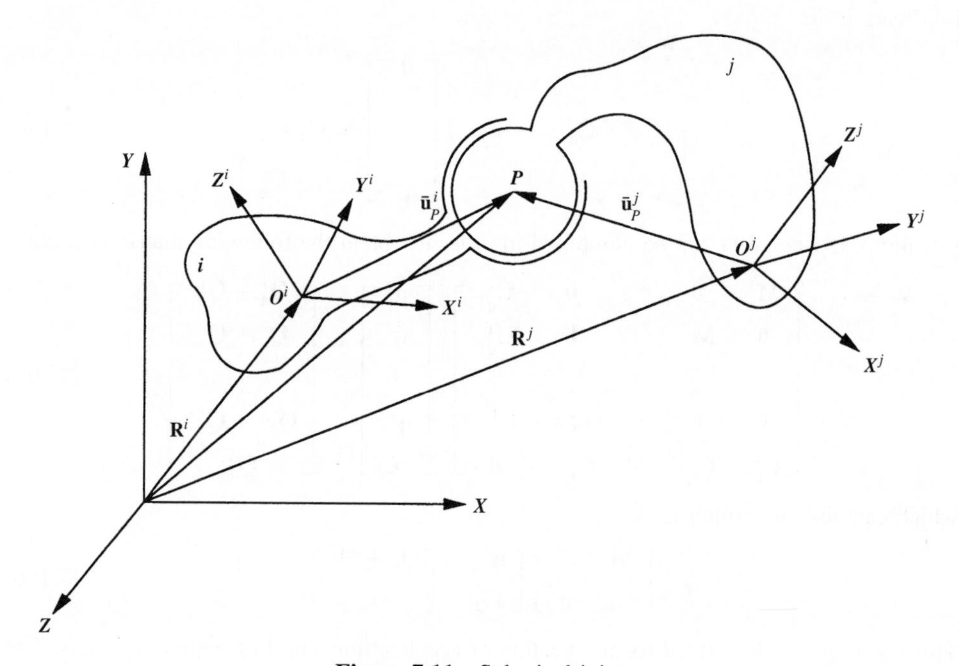

Figure 7.11 Spherical joint

only three degrees of freedom of relative rotations. The kinematic constraints of the spherical joint require that two points, P^i and P^j on bodies i and j, respectively, coincide throughout the motion. This condition can be written as

$$\mathbf{C}(\mathbf{q}^i, \mathbf{q}^j) = \mathbf{R}^i + \mathbf{A}^i\bar{\mathbf{u}}_P^i - \mathbf{R}^j - \mathbf{A}^j\bar{\mathbf{u}}_P^j = \mathbf{0} \tag{7.197}$$

where $\mathbf{R}^i$ and $\mathbf{R}^j$ are the global position vectors of the origins of the coordinate systems of bodies i and j, respectively; $\mathbf{A}^i$ and $\mathbf{A}^j$ are the transformation matrices of the two bodies; and $\bar{\mathbf{u}}_P^i$ and $\bar{\mathbf{u}}_P^j$ are the local position vectors of the joint definition points P^i and P^j, respectively. Recall that

$$\left.\begin{aligned} \frac{\partial(\mathbf{A}^i\bar{\mathbf{u}}_P^i)}{\partial\boldsymbol{\theta}^i} &= \tilde{\mathbf{u}}_P^{i\mathrm{T}}\mathbf{G}^i = \mathbf{A}^i(\tilde{\bar{\mathbf{u}}}_P^{i\mathrm{T}}\bar{\mathbf{G}}^i) \\ \frac{\partial(\mathbf{A}^j\bar{\mathbf{u}}_P^j)}{\partial\boldsymbol{\theta}^j} &= \tilde{\mathbf{u}}_P^{j\mathrm{T}}\mathbf{G}^j = \mathbf{A}^j(\tilde{\bar{\mathbf{u}}}_P^{j\mathrm{T}}\bar{\mathbf{G}}^j) \end{aligned}\right\} \tag{7.198}$$

where $\tilde{\mathbf{u}}_P^i$, $\tilde{\bar{\mathbf{u}}}_P^i$, $\tilde{\mathbf{u}}_P^j$, and $\tilde{\bar{\mathbf{u}}}_P^j$ are skew symmetric matrices associated with the vectors $\mathbf{u}_P^i$, $\bar{\mathbf{u}}_P^i$, $\mathbf{u}_P^j$, and $\bar{\mathbf{u}}_P^j$, respectively, and

$$\mathbf{u}_P^i = \mathbf{A}^i\bar{\mathbf{u}}_P^i, \quad \mathbf{u}_P^j = \mathbf{A}^j\bar{\mathbf{u}}_P^j \tag{7.199}$$

Therefore, a virtual change in the kinematic constraints of the spherical joint leads to

$$\delta\mathbf{C} = \delta\mathbf{R}^i + \tilde{\mathbf{u}}_P^{i\mathrm{T}}\mathbf{G}^i\,\delta\boldsymbol{\theta}^i - \delta\mathbf{R}^j - \tilde{\mathbf{u}}_P^{j\mathrm{T}}\mathbf{G}^j\,\delta\boldsymbol{\theta}^j = \mathbf{0} \tag{7.200}$$

This equation can be expressed in matrix form as

$$\delta\mathbf{C} = [\mathbf{I} \quad \tilde{\mathbf{u}}_P^{i\mathrm{T}}\mathbf{G}^i \quad -\mathbf{I} \quad -\tilde{\mathbf{u}}_P^{j\mathrm{T}}\mathbf{G}^j]\begin{bmatrix} \delta\mathbf{R}^i \\ \delta\boldsymbol{\theta}^i \\ \delta\mathbf{R}^j \\ \delta\boldsymbol{\theta}^j \end{bmatrix} = \mathbf{0} \tag{7.201}$$

which can be written as

$$\mathbf{C}_{\mathbf{q}}\,\delta\mathbf{q} = \mathbf{0} \tag{7.202}$$

where $\mathbf{C}_{\mathbf{q}}$ is the Jacobian matrix of the spherical joint constraints defined as

$$\mathbf{C}_{\mathbf{q}} = [\mathbf{I} \quad \tilde{\mathbf{u}}_P^{i\mathrm{T}}\mathbf{G}^i \quad -\mathbf{I} \quad -\tilde{\mathbf{u}}_P^{j\mathrm{T}}\mathbf{G}^j] \tag{7.203}$$

This Jacobian matrix can also be written as

$$\mathbf{C}_{\mathbf{q}} = [\mathbf{C}_{\mathbf{q}^i} \quad \mathbf{C}_{\mathbf{q}^j}] = [\mathbf{H}_P^i \quad -\mathbf{H}_P^j] \tag{7.204}$$

where

$$\left.\begin{aligned} \mathbf{H}_P^i &= [\mathbf{I} \quad \tilde{\mathbf{u}}_P^{i\mathrm{T}}\mathbf{G}^i] = [\mathbf{I} \quad \mathbf{A}^i\tilde{\bar{\mathbf{u}}}_P^{i\mathrm{T}}\overline{\mathbf{G}}^i] \\ \mathbf{H}_P^j &= [\mathbf{I} \quad \tilde{\mathbf{u}}_P^{j\mathrm{T}}\mathbf{G}^j] = [\mathbf{I} \quad \mathbf{A}^j\tilde{\bar{\mathbf{u}}}_P^{j\mathrm{T}}\overline{\mathbf{G}}^j] \end{aligned}\right\} \tag{7.205}$$

Example 7.9

Two bodies i and j are connected by a spherical joint. The orientation of the two bodies are defined by the Euler angles

$$\phi^i = \theta^i = \psi^i = 0,$$
$$\phi^j = \frac{\pi}{2}, \quad \theta^j = \psi^j = 0$$

The local position vectors of the joint definition point on bodies i and j are defined, respectively, by the vectors

$$\bar{\mathbf{u}}_P^i = [0 \quad 0 \quad 0.15]^\mathrm{T}, \quad \bar{\mathbf{u}}_P^j = [0.12 \quad 0 \quad 0]^\mathrm{T}$$

Using the absolute Cartesian coordinates and Euler angles as the generalized coordinates, obtain the Jacobian matrix of the kinematic constraints of this two-body system at the given configuration.

Solution. The transformation matrices that define the orientation of the two bodies are

$$\mathbf{A}^i = \begin{bmatrix} 1 & 0 & 0 \\ 0 & 1 & 0 \\ 0 & 0 & 1 \end{bmatrix}, \quad \mathbf{A}^j = \begin{bmatrix} 0 & -1 & 0 \\ 1 & 0 & 0 \\ 0 & 0 & 1 \end{bmatrix}$$

Using Eq. 81, it can be shown, at the given configuration, that

$$\overline{\mathbf{G}}^i = \overline{\mathbf{G}}^j = \begin{bmatrix} 0 & 1 & 0 \\ 0 & 0 & 0 \\ 1 & 0 & 1 \end{bmatrix}$$

The Jacobian matrix of the kinematic constraints is

$$\mathbf{C}_\mathbf{q} = [\mathbf{I} \quad \mathbf{A}^i\tilde{\bar{\mathbf{u}}}_P^{i\mathrm{T}}\overline{\mathbf{G}}^i \quad -\mathbf{I} \quad -\mathbf{A}^j\tilde{\bar{\mathbf{u}}}_P^{j\mathrm{T}}\overline{\mathbf{G}}^j]$$

where

$$\mathbf{q} = [R_x^i \quad R_y^i \quad R_z^i \quad \phi^i \quad \theta^i \quad \psi^i \quad R_x^j \quad R_y^j \quad R_z^j \quad \phi^j \quad \theta^j \quad \psi^j]^\mathrm{T}$$

$$\tilde{\bar{\mathbf{u}}}_P^i = \begin{bmatrix} 0 & -0.15 & 0 \\ 0.15 & 0 & 0 \\ 0 & 0 & 0 \end{bmatrix}, \quad \tilde{\bar{\mathbf{u}}}_P^j = \begin{bmatrix} 0 & 0 & 0 \\ 0 & 0 & -0.12 \\ 0 & 0.12 & 0 \end{bmatrix}$$

It can also be shown that

$$\mathbf{A}^i\tilde{\bar{\mathbf{u}}}_P^{i\mathrm{T}}\overline{\mathbf{G}}^i = -\begin{bmatrix} 0 & 0 & 0 \\ 0 & 0.15 & 0 \\ 0 & 0 & 0 \end{bmatrix}, \qquad \mathbf{A}^j\tilde{\bar{\mathbf{u}}}_P^{j\mathrm{T}}\overline{\mathbf{G}}^j = -\begin{bmatrix} 0.12 & 0 & 0.12 \\ 0 & 0 & 0 \\ 0 & 0 & 0 \end{bmatrix}$$

Thus,

$$\mathbf{C_q} = \begin{bmatrix} 1 & 0 & 0 & 0 & 0 & 0 & -1 & 0 & 0 & 0.12 & 0 & 0.12 \\ 0 & 1 & 0 & 0 & -0.15 & 0 & 0 & -1 & 0 & 0 & 0 & 0 \\ 0 & 0 & 1 & 0 & 0 & 0 & 0 & 0 & -1 & 0 & 0 & 0 \end{bmatrix}$$

where each row corresponds to one of the constraint equations of the spherical joint and each column corresponds to one of the generalized coordinates ordered as defined in the vector $\mathbf{q}$.

Cylindrical Joint Figure 12 shows a cylindrical joint that allows relative translation and rotation between body i and body j along the joint axis. The joint has two degrees of freedom because it eliminates the freedom of four possible independent relative displacements between the two bodies. The cylindrical joint constraints can, therefore, be described using four algebraic equations. Let $\mathbf{v}^i$ and $\mathbf{v}^j$ be two vectors defined along the joint axis on body

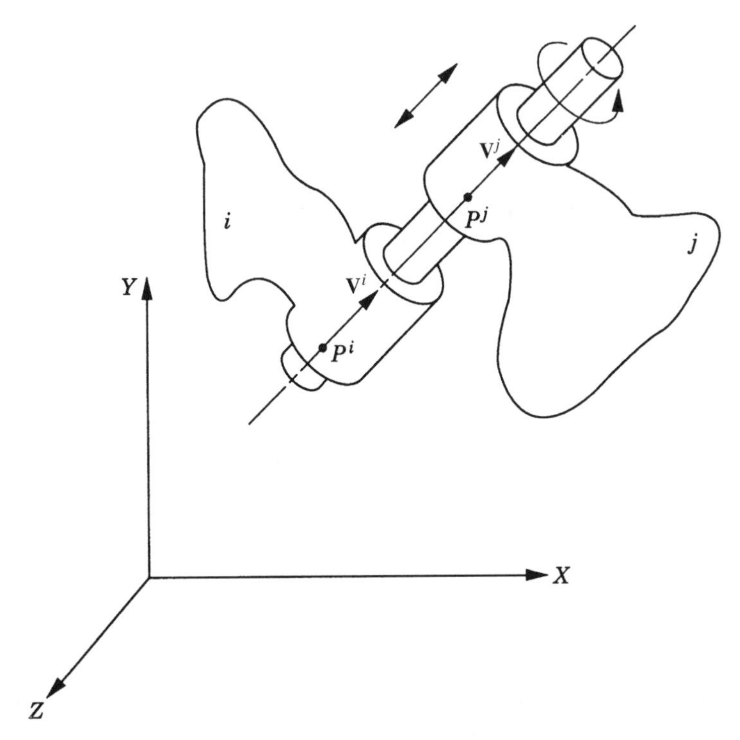

Figure 7.12 Cylindrical joint

i and body j, respectively, and let P^i and P^j be two points on body i and body j, defined along the joint axis. The constraint equations for the cylindrical joint can be defined as

$$\mathbf{C}(\mathbf{q}^i, \mathbf{q}^j) = \begin{bmatrix} \mathbf{v}^i \times \mathbf{v}^j \\ \mathbf{v}^i \times (\mathbf{r}_P^i - \mathbf{r}_P^j) \end{bmatrix} = \mathbf{0} \tag{7.206}$$

Since each of the cross products in the preceding equation defines two independent equations only, the preceding equation defines four independent kinematic relationships.

An alternative for the use of the cross product is to use two independent dot product equations as described in Chapter 2. In this case, two vectors $\mathbf{v}_1^i$ and $\mathbf{v}_2^i$ are defined on body i such that $\mathbf{v}^i$, $\mathbf{v}_1^i$, and $\mathbf{v}_2^i$ form an orthogonal triad. The constraint equations for the cylindrical joint can then be defined as

$$\mathbf{C}(\mathbf{q}^i, \mathbf{q}^j) = \begin{bmatrix} \mathbf{v}_1^{i\mathrm{T}} \mathbf{v}^j \\ \mathbf{v}_2^{i\mathrm{T}} \mathbf{v}^j \\ \mathbf{v}_1^{i\mathrm{T}} \mathbf{r}_P^{ij} \\ \mathbf{v}_2^{i\mathrm{T}} \mathbf{r}_P^{ij} \end{bmatrix} = \mathbf{0} \tag{7.207}$$

where $\mathbf{r}_P^{ij}$ is defined as

$$\mathbf{r}_P^{ij} = \mathbf{r}_P^i - \mathbf{r}_P^j \tag{7.208}$$

A simple computer procedure for defining the vectors $\mathbf{v}_1^i$ and $\mathbf{v}_2^i$ was described in Chapter 2.

The Jacobian matrix of the cylindrical joint constraint equations can be written as

$$\mathbf{C}_\mathbf{q} = [\mathbf{C}_{\mathbf{q}^i} \quad \mathbf{C}_{\mathbf{q}^j}] = \begin{bmatrix} \mathbf{v}^{j\mathrm{T}} \mathbf{H}_1^i & \mathbf{v}_1^{i\mathrm{T}} \mathbf{H}^j \\ \mathbf{v}^{j\mathrm{T}} \mathbf{H}_2^i & \mathbf{v}_2^{i\mathrm{T}} \mathbf{H}^j \\ \mathbf{r}_P^{ij\mathrm{T}} \mathbf{H}_1^i + \mathbf{v}_1^{i\mathrm{T}} \mathbf{H}_P^i & -\mathbf{v}_1^{i\mathrm{T}} \mathbf{H}_P^j \\ \mathbf{r}_P^{ij\mathrm{T}} \mathbf{H}_2^i + \mathbf{v}_2^{i\mathrm{T}} \mathbf{H}_P^i & -\mathbf{v}_2^{i\mathrm{T}} \mathbf{H}_P^j \end{bmatrix} \tag{7.209}$$

where $\mathbf{H}_P^i$ and $\mathbf{H}_P^j$ are as defined by Eq. 205, and

$$\left.\begin{aligned} \mathbf{H}_1^i &= \frac{\partial \mathbf{v}_1^i}{\partial \mathbf{q}^i} = \frac{\partial}{\partial \mathbf{q}^i}(\mathbf{A}^i \bar{\mathbf{v}}_1^i) \\ \mathbf{H}_2^i &= \frac{\partial \mathbf{v}_2^i}{\partial \mathbf{q}^i} = \frac{\partial}{\partial \mathbf{q}^i}(\mathbf{A}^i \bar{\mathbf{v}}_2^i) \\ \mathbf{H}^j &= \frac{\partial \mathbf{v}^j}{\partial \mathbf{q}^j} = \frac{\partial}{\partial \mathbf{q}^j}(\mathbf{A}^j \bar{\mathbf{v}}^j) \end{aligned}\right\} \tag{7.210}$$

in which $\bar{\mathbf{v}}_1^i$ and $\bar{\mathbf{v}}_2^i$ are the constant vectors defined in body i coordinate system, and $\bar{\mathbf{v}}^j$ is the constant vector defined in body j coordinate system.

Revolute Joint The revolute joint has one degree of freedom and can be considered as a special case of the cylindrical joint by eliminating the freedom of the relative translation between the two bodies. In order to preclude the relative translation between the two bodies, the distance between point P^i on body i and point P^j on body j (Fig. 12), both defined along the joint axis, must remain constant, that is,

$$\mathbf{r}_P^{ij\mathrm{T}} \mathbf{r}_P^{ij} - k_r = 0 \tag{7.211}$$

where k_r is a constant and

$$\mathbf{r}_P^{ij} = \mathbf{r}_P^i - \mathbf{r}_P^j \tag{7.212}$$

Equation 211 is a scalar equation that when added to the constraints of the cylindrical joint, as defined by Eq. 207, leads to

$$\mathbf{C}(\mathbf{q}^i, \mathbf{q}^j) = \begin{bmatrix} \mathbf{v}_1^{i\mathrm{T}} \mathbf{v}^j \\ \mathbf{v}_2^{i\mathrm{T}} \mathbf{v}^j \\ \mathbf{v}_1^{i\mathrm{T}} \mathbf{r}_P^{ij} \\ \mathbf{v}_2^{i\mathrm{T}} \mathbf{r}_P^{ij} \\ \mathbf{r}_P^{ij\mathrm{T}} \mathbf{r}_P^{ij} - k_r \end{bmatrix} = \mathbf{0} \tag{7.213}$$

This system of equations has five independent constraint equations.

The Jacobian matrix of the revolute joint constraints is

$$\mathbf{C}_\mathbf{q} = [\mathbf{C}_{\mathbf{q}^i} \quad \mathbf{C}_{\mathbf{q}^j}] = \begin{bmatrix} \mathbf{v}^{j\mathrm{T}} \mathbf{H}_1^i & \mathbf{v}_1^{i\mathrm{T}} \mathbf{H}^j \\ \mathbf{v}^{j\mathrm{T}} \mathbf{H}_2^i & \mathbf{v}_2^{i\mathrm{T}} \mathbf{H}^j \\ \mathbf{r}_P^{ij\mathrm{T}} \mathbf{H}_1^i + \mathbf{v}_1^{i\mathrm{T}} \mathbf{H}_P^i & -\mathbf{v}_1^{i\mathrm{T}} \mathbf{H}_P^j \\ \mathbf{r}_P^{ij\mathrm{T}} \mathbf{H}_2^i + \mathbf{v}_2^{i\mathrm{T}} \mathbf{H}_P^i & -\mathbf{v}_2^{i\mathrm{T}} \mathbf{H}_P^j \\ 2\mathbf{r}_P^{ij\mathrm{T}} \mathbf{H}_P^i & -2\mathbf{r}_P^{ij\mathrm{T}} \mathbf{H}_P^j \end{bmatrix} \tag{7.214}$$

where all the variables that appear in this equation are the same as the ones used in the case of the cylindrical joint.

Another alternate approach for formulating the revolute joint constraints is to consider it as a special case of the spherical joint in which the relative rotation between the two bodies is allowed only along the joint axis. If point P is the joint definition point as defined in the case of the spherical joint, and $\mathbf{v}^i$ and $\mathbf{v}^j$ are two vectors defined along the joint axis on bodies i and j, respectively, the constraint equations of the revolute joint can be written as

$$\mathbf{C}(\mathbf{q}^i, \mathbf{q}^j) = \begin{bmatrix} \mathbf{r}_P^i - \mathbf{r}_P^j \\ \mathbf{v}_1^{i\mathrm{T}} \mathbf{v}^j \\ \mathbf{v}_2^{i\mathrm{T}} \mathbf{v}^j \end{bmatrix} = \mathbf{0} \tag{7.215}$$

The last two equations in Eq. 215 guarantee that the two vectors $\mathbf{v}^i$ and $\mathbf{v}^j$ remain parallel, thereby eliminating the freedom of the relative rotation between the two bodies in two perpendicular directions.

The Jacobian matrix of the revolute joint constraints as defined by Eq. 215 is

$$\mathbf{C_q} = [\mathbf{C}_{\mathbf{q}^i} \quad \mathbf{C}_{\mathbf{q}^j}] = \begin{bmatrix} \mathbf{H}_P^i & -\mathbf{H}_P^j \\ \mathbf{v}^{j^{\mathrm{T}}}\mathbf{H}_1^i & \mathbf{v}_1^{i^{\mathrm{T}}}\mathbf{H}^j \\ \mathbf{v}^{j^{\mathrm{T}}}\mathbf{H}_2^i & \mathbf{v}_2^{i^{\mathrm{T}}}\mathbf{H}^j \end{bmatrix} \tag{7.216}$$

where the matrices $\mathbf{H}_P^i$, $\mathbf{H}_P^j$, $\mathbf{H}_1^i$, and $\mathbf{H}_2^i$ are as defined by Eqs. 205 and 210, respectively.

Prismatic Joint The single-degree-of-freedom prismatic joint can also be obtained as a special case of the cylindrical joint by eliminating the freedom of the relative rotation between the two bodies about the joint axis. The two orthogonal vectors $\mathbf{h}^i$ and $\mathbf{h}^j$ drawn perpendicular to the joint axis are defined on bodies i and j, respectively, as shown in Fig. 13. In order to preclude the relative rotation between the two bodies, one must have

$$\mathbf{h}^{i^{\mathrm{T}}}\mathbf{h}^j = 0 \tag{7.217}$$

This equation can be added to the constraint equations of the cylindrical joint, as defined by Eq. 207, in order to define the constraint equations of the prismatic joint as

$$\mathbf{C}(\mathbf{q}^i, \mathbf{q}^j) = \begin{bmatrix} \mathbf{v}_1^{i^{\mathrm{T}}}\mathbf{v}^j \\ \mathbf{v}_2^{i^{\mathrm{T}}}\mathbf{v}^j \\ \mathbf{v}_1^{i^{\mathrm{T}}}\mathbf{r}_P^{ij} \\ \mathbf{v}_2^{i^{\mathrm{T}}}\mathbf{r}_P^{ij} \\ \mathbf{h}^{i^{\mathrm{T}}}\mathbf{h}^j \end{bmatrix} = \mathbf{0} \tag{7.218}$$

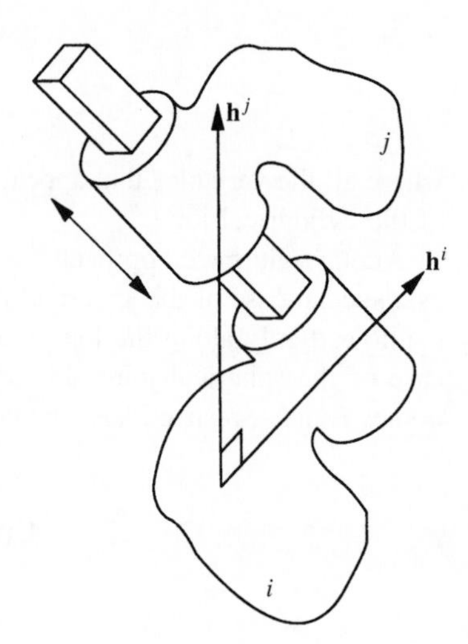

Figure 7.13 Translational joint

This vector equation has five independent constraint equations that depend only on the generalized coordinates of bodies i and j.

The Jacobian matrix of the prismatic joint constraints is

$$\mathbf{C_q} = [\mathbf{C_{q^i}} \quad \mathbf{C_{q^j}}] = \begin{bmatrix} \mathbf{v}^{j^{\mathrm{T}}}\mathbf{H}_1^i & \mathbf{v}_1^{i^{\mathrm{T}}}\mathbf{H}^j \\ \mathbf{v}^{j^{\mathrm{T}}}\mathbf{H}_2^i & \mathbf{v}_2^{i^{\mathrm{T}}}\mathbf{H}^j \\ \mathbf{r}_P^{ij^{\mathrm{T}}}\mathbf{H}_1^i + \mathbf{v}_1^{i^{\mathrm{T}}}\mathbf{H}_P^i & -\mathbf{v}_1^{i^{\mathrm{T}}}\mathbf{H}_P^j \\ \mathbf{r}_P^{ij^{\mathrm{T}}}\mathbf{H}_2^i + \mathbf{v}_2^{i^{\mathrm{T}}}\mathbf{H}_P^i & -\mathbf{v}_2^{i^{\mathrm{T}}}\mathbf{H}_P^j \\ \mathbf{h}^{j^{\mathrm{T}}}\mathbf{H}_h^i & \mathbf{h}^{i^{\mathrm{T}}}\mathbf{H}_h^j \end{bmatrix} \tag{7.219}$$

where

$$\left.\begin{aligned} \mathbf{H}_h^i &= [\mathbf{0} \quad \tilde{\mathbf{h}}^{i^{\mathrm{T}}}\mathbf{G}^i] = [\mathbf{0} \quad \mathbf{A}^i\tilde{\bar{\mathbf{h}}}^{i^{\mathrm{T}}}\bar{\mathbf{G}}^i] \\ \mathbf{H}_h^j &= [\mathbf{0} \quad \tilde{\mathbf{h}}^{j^{\mathrm{T}}}\mathbf{G}^j] = [\mathbf{0} \quad \mathbf{A}^j\tilde{\bar{\mathbf{h}}}^{j^{\mathrm{T}}}\bar{\mathbf{G}}^j] \end{aligned}\right\} \tag{7.220}$$

in which $\bar{\mathbf{h}}^i$ and $\bar{\mathbf{h}}^j$ contain the constant components of the vectors $\mathbf{h}^i$ and $\mathbf{h}^j$, respectively.

Universal Joint The universal joint shown in Fig. 14 is a two-degree-of-freedom joint. Let point P be the point of intersection of the two bars of the joint cross, as shown in the figure. The coordinates of this point in the coordinate systems of bodies i and j are constant. Let $\mathbf{h}^i$ and $\mathbf{h}^j$ be two orthogonal vectors defined along the intersecting axes of the joint on bodies i and j, respectively. The constraint equations of the universal joint can be written as

$$\mathbf{C}(\mathbf{q}^i, \mathbf{q}^j) = \begin{bmatrix} \mathbf{r}_P^i - \mathbf{r}_P^j \\ \mathbf{h}^{i^{\mathrm{T}}}\mathbf{h}^j \end{bmatrix} = \mathbf{0} \tag{7.221}$$

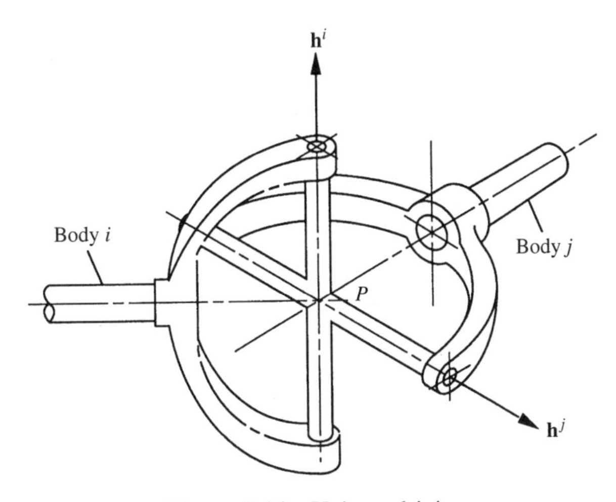

Figure 7.14 Universal joint

where $\mathbf{r}_P^i$ and $\mathbf{r}_P^j$ are the global position vectors of point P defined using the generalized coordinates of bodies i and j, respectively.

The Jacobian matrix of the universal joint constraints is

$$\mathbf{C_q} = [\mathbf{C}_{\mathbf{q}^i} \quad \mathbf{C}_{\mathbf{q}^j}] = \begin{bmatrix} \mathbf{H}_P^i & -\mathbf{H}_P^j \\ \mathbf{h}^{j^{\mathrm{T}}}\mathbf{H}_h^i & \mathbf{h}^{i^{\mathrm{T}}}\mathbf{H}_h^j \end{bmatrix} \tag{7.222}$$

where the matrices $\mathbf{H}_P^i$, $\mathbf{H}_P^j$, $\mathbf{H}_h^i$, and $\mathbf{H}_h^j$ are as previously defined.

Rigid Joint A *rigid joint* between two rigid bodies i and j does not allow any freedom of relative translation or relative rotation between the two bodies. The rigid joint is described using six constraint equations that can be written as

$$\mathbf{C}(\mathbf{q}^i, \mathbf{q}^j) = \begin{bmatrix} \mathbf{R}^i - \mathbf{R}^j - \mathbf{k}_r \\ \boldsymbol{\theta}^i - \boldsymbol{\theta}^j - \mathbf{k}_\theta \end{bmatrix} = \mathbf{0} \tag{7.223}$$

where $\mathbf{k}_r$ and $\mathbf{k}_\theta$ are two constant vectors.

The Jacobian matrix of the rigid joint can simply be written as

$$\mathbf{C_q} = [\mathbf{C}_{\mathbf{q}^i} \quad \mathbf{C}_{\mathbf{q}^j}] = [\mathbf{C}_{\mathbf{R}^i} \quad \mathbf{C}_{\boldsymbol{\theta}^i} \quad \mathbf{C}_{\mathbf{R}^j} \quad \mathbf{C}_{\boldsymbol{\theta}^j}] = \begin{bmatrix} \mathbf{I} & \mathbf{0} & -\mathbf{I} & \mathbf{0} \\ \mathbf{0} & \mathbf{I} & \mathbf{0} & -\mathbf{I} \end{bmatrix} \tag{7.224}$$

A special case of the rigid joint concerns the *ground constraints*. If, for example, body i is the *fixed link* (*ground*), one has

$$\mathbf{C}(\mathbf{q}^i) = \begin{bmatrix} \mathbf{R}^i - \mathbf{k}_r \\ \boldsymbol{\theta}^i - \mathbf{k}_\theta \end{bmatrix} = \mathbf{0} \tag{7.225}$$

where $\mathbf{k}_r$ and $\mathbf{k}_\theta$ are constant vectors. The preceding constraint equations eliminate the freedom of body i to translate or rotate.

Remarks It is clear from the discussion presented in this section that most of the joint kinematic constraint equations can be formulated in terms of simple vector addition and/or scalar product. For instance, the following basic vector operations were used in formulating the joint constraints between body i and body j:

$$\mathbf{r}_P^i - \mathbf{r}_P^j = \mathbf{c}, \quad \mathbf{v}^{i^{\mathrm{T}}}\mathbf{v}^j = 0, \quad \mathbf{v}^{i^{\mathrm{T}}}\mathbf{r}_P^{ij} = 0 \tag{7.226}$$

where $\mathbf{c}$ is a constant vector, $\mathbf{v}^i$ and $\mathbf{v}^j$ are vectors defined on body i and body j, respectively, and $\mathbf{r}_P^{ij}$ is as defined by Eq. 212. The preceding three basic equations and their partial derivatives with respect to the generalized coordinates can be used to develop a *computer library* that can be used in formulating many of the joint constraint equations described in this section.

7.11 NEWTON–EULER EQUATIONS

In the *Newton–Euler formulation*, the equations of motion of the rigid body are expressed in terms of the angular velocity and acceleration vectors. It is also assumed that the origin of the body coordinate system (reference point) is the center of mass of the body. Using this assumption and the relationships between the angular velocity and acceleration vectors, and the orientation coordinates, it can be shown that Eq. 183 leads to

$$m^i\ddot{\mathbf{R}}^i = (\mathbf{Q}_e^i)_R \tag{7.227}$$

$$\mathbf{I}_{\theta\theta}^i\boldsymbol{\alpha}^i = \mathbf{F}_\theta^i - \boldsymbol{\omega}^i \times (\mathbf{I}_{\theta\theta}^i\boldsymbol{\omega}^i) \tag{7.228}$$

where $\boldsymbol{\alpha}^i$ and $\boldsymbol{\omega}^i$ are, respectively, the angular acceleration and velocity vectors defined in the global coordinate system, m^i is the mass of the rigid body, $\mathbf{I}_{\theta\theta}^i = \mathbf{A}^i\bar{\mathbf{I}}_{\theta\theta}^i\mathbf{A}^{i\mathrm{T}}$ is the inertia tensor defined in the global coordinate system, and $\mathbf{F}_\theta^i$ is the vector of *Cartesian moments* that act on the rigid body i. Equations 227 and 228 can be combined in one matrix equation as

$$\begin{bmatrix} m^i\mathbf{I} & \mathbf{0} \\ \mathbf{0} & \mathbf{I}_{\theta\theta}^i \end{bmatrix}\begin{bmatrix} \ddot{\mathbf{R}}^i \\ \boldsymbol{\alpha}^i \end{bmatrix} = \begin{bmatrix} (\mathbf{Q}_e^i)_R \\ \mathbf{F}_\theta^i - \boldsymbol{\omega}^i \times (\mathbf{I}_{\theta\theta}^i\boldsymbol{\omega}^i) \end{bmatrix} \tag{7.229}$$

or

$$\mathbf{M}_d^i\ddot{\mathbf{P}}^i = \mathbf{F}_e^i + \mathbf{F}_v^i \tag{7.230}$$

where

$$\left.\begin{aligned} &\mathbf{M}_d^i = \begin{bmatrix} m^i\mathbf{I} & \mathbf{0} \\ \mathbf{0} & \mathbf{I}_{\theta\theta}^i \end{bmatrix}, \quad \ddot{\mathbf{P}}^i = [\ddot{\mathbf{R}}^{i\mathrm{T}} \quad \boldsymbol{\alpha}^{i\mathrm{T}}]^{\mathrm{T}} \\ &\mathbf{F}_e^i = [(\mathbf{Q}_e^i)_R^{\mathrm{T}} \quad \mathbf{F}_\theta^{i\mathrm{T}}]^{\mathrm{T}}, \quad \mathbf{F}_v^i = [\mathbf{0} \quad -[\boldsymbol{\omega}^i \times (\mathbf{I}_{\theta\theta}^i\boldsymbol{\omega}^i)]^{\mathrm{T}}]^{\mathrm{T}} \end{aligned}\right\} \tag{7.231}$$

As pointed out previously, the angular velocities in the spatial analysis are not, in general, the time derivatives of a set of orientation coordinates. For this reason, the angular accelerations obtained by solving Newton–Euler equations cannot be integrated directly in order to obtain the system coordinates. One must first determine the second derivatives of the generalized orientation coordinates as a function of the angular accelerations. The second derivatives of the generalized orientation coordinates can then be integrated to determine the generalized coordinates and velocities. In the recursive methods discussed in Section 14, however, the angular velocity and acceleration vectors are expressed in terms of the joint coordinates and their first and second time derivatives. The joint accelerations can be determined and can be integrated numerically in order to determine the joint angles and joint velocities.

The coefficient matrix in the Newton–Euler equations of Eq. 229 can be made constant if the angular acceleration is defined in the body coordinate system. Recall that $\boldsymbol{\alpha}^i = \mathbf{A}^i\overline{\boldsymbol{\alpha}}^i$. Substituting this equation into Euler equation (Eq. 228), premultiplying by the transpose of the transformation matrix, $\mathbf{A}^{i^{\mathrm{T}}}$, and utilizing the fact that $\overline{\mathbf{I}}^i_{\theta\theta} = \mathbf{A}^{i^{\mathrm{T}}}\mathbf{I}^i_{\theta\theta}\mathbf{A}^i$; one obtains the following alternate form of Euler equation:

$$\overline{\mathbf{I}}^i_{\theta\theta}\overline{\boldsymbol{\alpha}}^i = \overline{\mathbf{F}}^i_\theta - \overline{\boldsymbol{\omega}}^i \times (\overline{\mathbf{I}}^i_{\theta\theta}\overline{\boldsymbol{\omega}}^i) \tag{7.232}$$

In this equation, $\overline{\mathbf{F}}^i_\theta = \mathbf{A}^{i^{\mathrm{T}}}\mathbf{F}^i_\theta$ is the vector of Cartesian moments defined in the body coordinate system. Equations 227 and 232 can be combined to obtain the following alternate form to Eq. 229:

$$\begin{bmatrix} m^i\mathbf{I} & \mathbf{0} \\ \mathbf{0} & \overline{\mathbf{I}}^i_{\theta\theta} \end{bmatrix}\begin{bmatrix} \ddot{\mathbf{R}}^i \\ \overline{\boldsymbol{\alpha}}^i \end{bmatrix} = \begin{bmatrix} (\mathbf{Q}^i_e)_R \\ \overline{\mathbf{F}}^i_\theta - \overline{\boldsymbol{\omega}}^i \times (\overline{\mathbf{I}}^i_{\theta\theta}\overline{\boldsymbol{\omega}}^i) \end{bmatrix} \tag{7.233}$$

In this form of the Newton–Euler equations, the coefficient matrix is constant since the inertia tensor is defined in the body coordinate system.

7.12 D'ALEMBERT'S PRINCIPLE

In this chapter, the virtual work principle was used to obtain the Newton–Euler equations. As in the case of the planar analysis discussed in Chapter 4, the Newton–Euler equations can be derived in a straightforward manner by applying D'Alembert's principle. The procedure for doing that is outlined in this section.

Newton Equations If the rigid body i is assumed to consist of a large number of particles, each of which has mass $\rho^i dV^i$ where ρ^i is the mass density and dV^i the infinitesimal volume, the inertia force of a particle whose position vector $\mathbf{r}^i$ is equal to $(\rho^i dV^i)\ddot{\mathbf{r}}^i$. The inertia force of the body can be obtained by integrating the inertia forces of its particles over the volume of the body. By equating the inertia forces of the body to the applied forces, one obtains

$$\int_{V^i} \rho^i \ddot{\mathbf{r}}^i dV^i = (\mathbf{Q}^i_e)_R \tag{7.234}$$

In this equation, $(\mathbf{Q}^i_e)_R$ is the vector of resultant forces acting on the body. The preceding equation, as in the case of the planar analysis, allows using any point as the reference point. Recall that the absolute acceleration of an arbitrary point on the rigid body can be written as

$$\ddot{\mathbf{r}}^i = \ddot{\mathbf{R}}^i + \boldsymbol{\alpha}^i \times \mathbf{u}^i + \boldsymbol{\omega}^i \times (\boldsymbol{\omega}^i \times \mathbf{u}^i) \tag{7.235}$$

where $\mathbf{R}^i$ is the global position vector of the reference point on the body; $\boldsymbol{\omega}^i$ and $\boldsymbol{\alpha}^i$ are, respectively, the angular velocity and angular acceleration vectors of the body; $\mathbf{u}^i = \mathbf{A}^i\overline{\mathbf{u}}^i$ is the position vector of the arbitrary point with respect to the reference point; $\mathbf{A}^i$ is the transformation matrix that defines the orientation of the body; and $\overline{\mathbf{u}}^i$ is the position of the arbitrary point with respect to the reference point. If the reference point is selected to be the *center of mass* of the body, one has

$$\int_{V^i} \rho^i \mathbf{u}^i dV^i = \int_{V^i} \rho^i \mathbf{A}^i \bar{\mathbf{u}}^i dV^i = \mathbf{A}^i \int_{V^i} \rho^i \bar{\mathbf{u}}^i dV^i = \mathbf{0} \tag{7.236}$$

Substituting Eq. 235 into Eq. 234 and using the identity of Eq. 236 and the fact that $\boldsymbol{\omega}^i$ and $\boldsymbol{\alpha}^i$ do not depend on the location of the arbitrary point, one obtains

$$\int_{V^i} \rho^i \ddot{\mathbf{R}}^i dV^i = (\mathbf{Q}_e^i)_R \tag{7.237}$$

or

$$m^i \ddot{\mathbf{R}}^i = (\mathbf{Q}_e^i)_R \tag{7.238}$$

which is the Newton equation of motion.

Euler Equations Euler equation can be obtained from D'Alembert's principle by treating the inertia forces as the applied forces. To this end, it is assumed again that the rigid body i consists of a large number of particles. The moment of the inertia force of a particle about an arbitrary reference point is $\mathbf{u}^i \times \{(\rho^i dV^i)\ddot{\mathbf{r}}^i\}$. It follows from D'Alembert's principle that the moments of the inertia forces of the body are given by

$$\int_{V^i} \rho^i \mathbf{u}^i \times \ddot{\mathbf{r}}^i dV^i = \mathbf{F}_\theta^i \tag{7.239}$$

In this equation, $\mathbf{F}_\theta^i$ is as defined in the preceding section.

In order to obtain Euler equation, the reference point is chosen to be the body center of mass. Substituting Eq. 235 into Eq. 239, and using the identity of Eq. 236, one can show that

$$\left(\int_{V^i} \rho^i \tilde{\bar{\mathbf{u}}}^{i\mathrm{T}} \times \tilde{\bar{\mathbf{u}}}^i dV^i\right) \boldsymbol{\alpha}^i = \mathbf{F}_\theta^i - \boldsymbol{\omega}^i \times \left(\left(\int_{V^i} \rho^i \tilde{\bar{\mathbf{u}}}^{i\mathrm{T}} \times \tilde{\bar{\mathbf{u}}}^i dV^i\right) \boldsymbol{\omega}^i\right) \tag{7.240}$$

This equation can be written as

$$\mathbf{I}_{\theta\theta}^i \boldsymbol{\alpha}^i = \mathbf{F}_\theta^i - \boldsymbol{\omega}^i \times (\mathbf{I}_{\theta\theta}^i \boldsymbol{\omega}^i) \tag{7.241}$$

which is the Euler equation obtained in the preceding section.

7.13 LINEAR AND ANGULAR MOMENTUM

Newton–Euler equations can also be derived using the *principle of linear and angular momentum*. The linear momentum of a rigid body is defined as

$$\mathbf{p}^i = \int_{V^i} \rho^i \dot{\mathbf{r}}^i \, dV^i \tag{7.242}$$

where $\dot{\mathbf{r}}^i$ is the velocity vector of an arbitrary point on the rigid body as defined by Eq. 68 in terms of the angular velocity vector and in Eq. 73a in terms of the generalized orientation coordinates. Upon the use of the identities of Eqs. 84 and 180, the linear momentum reduces in the case of a centroidal body coordinate system to

$$\mathbf{p}^i = \int_{V^i} \rho^i \dot{\mathbf{R}}^i \, dV^i = m^i \dot{\mathbf{R}}^i \tag{7.243}$$

where $\mathbf{R}^i$, in this case, is the global position vector of the center of mass of the rigid body.

Newton's law of motion states that the rate of change of the linear momentum is equal to the vector of the resultant force acting on the body, that is

$$\frac{d\mathbf{p}^i}{dt} = \dot{\mathbf{p}}^i = (\mathbf{Q}_e^i)_R \tag{7.244}$$

which leads to *Newton's equations*

$$m^i \ddot{\mathbf{R}}^i = (\mathbf{Q}_e^i)_R \tag{7.245}$$

The angular momentum of the rigid body i is defined as

$$\begin{aligned} \mathbf{\Omega}^i &= \int_{V^i} \rho^i \mathbf{r}^i \times \dot{\mathbf{r}}^i \, dV^i \\ &= \int_{V^i} \rho^i \{(\mathbf{R}^i + \mathbf{u}^i) \times (\dot{\mathbf{R}}^i + \boldsymbol{\omega}^i \times \mathbf{u}^i)\} \, dV^i \end{aligned} \tag{7.246}$$

which upon the use of Eqs. 84 and 180 and the definition of the linear momentum reduces to

$$\mathbf{\Omega}^i = \mathbf{R}^i \times \mathbf{p}^i + \mathbf{\Omega}_r^i \tag{7.247}$$

where $\mathbf{\Omega}_r^i$ is the angular momentum defined with respect to the center of mass of the body, and is given by

$$\begin{aligned} \mathbf{\Omega}_r^i &= \int_{V^i} \rho^i \mathbf{u}^i \times (\boldsymbol{\omega}^i \times \mathbf{u}^i) \, dV^i = \int_{V^i} \rho^i \mathbf{A}^i \tilde{\bar{\mathbf{u}}}^{i\mathrm{T}} \tilde{\bar{\mathbf{u}}}^i \bar{\boldsymbol{\omega}}^i \, dV^i \\ &= \mathbf{A}^i \bar{\mathbf{I}}_{\theta\theta}^i \bar{\boldsymbol{\omega}}^i \end{aligned} \tag{7.248}$$

where $\bar{\mathbf{I}}_{\theta\theta}^i$ is defined by Eq. 101. It follows that

$$\begin{aligned} \frac{d\mathbf{\Omega}_r^i}{dt} &= \dot{\mathbf{A}}^i \bar{\mathbf{I}}_{\theta\theta}^i \bar{\boldsymbol{\omega}}^i + \mathbf{A}^i \bar{\mathbf{I}}_{\theta\theta}^i \dot{\bar{\boldsymbol{\omega}}}^i \\ &= \tilde{\boldsymbol{\omega}}^i \mathbf{A}^i \bar{\mathbf{I}}_{\theta\theta}^i \mathbf{A}^{i\mathrm{T}} \boldsymbol{\omega}^i + \mathbf{A}^i \bar{\mathbf{I}}_{\theta\theta}^i \mathbf{A}^{i\mathrm{T}} \boldsymbol{\alpha}^i \end{aligned} \tag{7.249}$$

where the fact that $\bar{\boldsymbol{\omega}}^i = \mathbf{A}^{i^{\mathrm{T}}}\boldsymbol{\omega}^i$ and $\bar{\boldsymbol{\alpha}}^i = \mathbf{A}^{i^{\mathrm{T}}}\boldsymbol{\alpha}^i$ is utilized. Recall that $\mathbf{I}^i_{\theta\theta} = \mathbf{A}^i\bar{\mathbf{I}}^i_{\theta\theta}\mathbf{A}^{i^{\mathrm{T}}}$. Using this equation, the time rate of change of the angular momentum can be written as

$$\frac{d\boldsymbol{\Omega}^i_r}{dt} = \tilde{\boldsymbol{\omega}}^i\mathbf{I}^i_{\theta\theta}\boldsymbol{\omega}^i + \mathbf{I}^i_{\theta\theta}\boldsymbol{\alpha}^i = \boldsymbol{\omega}^i \times (\mathbf{I}^i_{\theta\theta}\boldsymbol{\omega}^i) + \mathbf{I}^i_{\theta\theta}\boldsymbol{\alpha}^i \tag{7.250}$$

The rate of change of the angular momentum $\boldsymbol{\Omega}^i_r$ is equal to the applied torques. Therefore,

$$\frac{d\boldsymbol{\Omega}^i_r}{dt} = \mathbf{F}^i_\theta \tag{7.251}$$

which leads to *Euler equations*

$$\mathbf{I}^i_{\theta\theta}\boldsymbol{\alpha}^i = \mathbf{F}^i_\theta - \boldsymbol{\omega}^i \times (\mathbf{I}^i_{\theta\theta}\boldsymbol{\omega}^i) \tag{7.252}$$

This equation can also be expressed in terms of vectors defined in the centroidal body coordinate system as

$$\mathbf{A}^i\bar{\mathbf{I}}^i_{\theta\theta}\mathbf{A}^{i^{\mathrm{T}}}\mathbf{A}^i\bar{\boldsymbol{\alpha}}^i = \mathbf{F}^i_\theta - \mathbf{A}^i[\bar{\boldsymbol{\omega}}^i \times (\bar{\mathbf{I}}^i_{\theta\theta}\bar{\boldsymbol{\omega}}^i)] \tag{7.253}$$

which upon premultiplying by $\mathbf{A}^{i^{\mathrm{T}}}$ yields

$$\bar{\mathbf{I}}^i_{\theta\theta}\bar{\boldsymbol{\alpha}}^i = \bar{\mathbf{F}}^i_\theta - \bar{\boldsymbol{\omega}}^i \times (\bar{\mathbf{I}}^i_{\theta\theta}\bar{\boldsymbol{\omega}}^i) \tag{7.254}$$

where $\bar{\mathbf{F}}^i_\theta = \mathbf{A}^{i^{\mathrm{T}}}\mathbf{F}^i_\theta$ is the vector of moments defined in the coordinate system of body i.

It is clear from the preceding discussion that if there are no forces or moments acting on the rigid body, one gets

$$\frac{d\mathbf{p}^i}{dt} = \mathbf{0}, \quad \frac{d\boldsymbol{\Omega}^i_r}{dt} = \mathbf{0} \tag{7.255}$$

which imply that the linear and angular momentum are *constants of motion*.

Example 7.10

Consider the rigid body whose inertia tensor defined in a centroidal body coordinate system is

$$\bar{\mathbf{I}}^i_{\theta\theta} = \begin{bmatrix} 1.5 & 0.0 & -1.0 \\ 0.0 & 2.0 & 0.0 \\ -1.0 & 0.0 & 2.5 \end{bmatrix} \mathrm{kg}\cdot\mathrm{m}^2$$

The body rotates with a constant angular velocity such that

$$\bar{\boldsymbol{\omega}}^i = \begin{bmatrix} 15.0 \\ 0.0 \\ 0.0 \end{bmatrix} \text{ rad/s}$$

Determine the components of the external moments applied to this body.

Solution. Euler's equation defined in the body coordinate system is

$$\bar{\mathbf{I}}^i_{\theta\theta}\bar{\boldsymbol{\alpha}}^i = \bar{\mathbf{F}}^i_\theta - \bar{\boldsymbol{\omega}}^i \times (\bar{\mathbf{I}}^i_{\theta\theta}\bar{\boldsymbol{\omega}}^i)$$

Since the angular velocity vector is constant, it follows that

$$\bar{\boldsymbol{\alpha}}^i = \mathbf{0}$$

Euler's equation reduces in this case to

$$\begin{aligned} \bar{\mathbf{F}}^i_\theta &= \tilde{\bar{\boldsymbol{\omega}}}^i(\bar{\mathbf{I}}^i_{\theta\theta}\bar{\boldsymbol{\omega}}^i) \\ &= \begin{bmatrix} 0.0 & 0.0 & 0.0 \\ 0.0 & 0.0 & -15.0 \\ 0.0 & 15.0 & 0.0 \end{bmatrix} \begin{bmatrix} 1.5 & 0.0 & -1.0 \\ 0.0 & 2.0 & 0.0 \\ -1.0 & 0.0 & 2.5 \end{bmatrix} \begin{bmatrix} 15.0 \\ 0.0 \\ 0.0 \end{bmatrix} \\ &= \begin{bmatrix} 0.0 \\ 225.0 \\ 0.0 \end{bmatrix} \text{ N}\cdot\text{m} \end{aligned}$$

It can be seen from these results that the applied moment is constant in the body coordinate system. This moment is equal in magnitude and opposite in direction to the inertial moment due to the centrifugal force.

7.14 RECURSIVE METHODS

As pointed out previously, one of the major advantages of using the absolute coordinates is that the motion of each body in the multibody system is described using similar sets of generalized coordinates that do not depend on the topological structure of the system. The mass matrices of the bodies in the system have similar form and dimensions, such that a computer program with simple structure can be developed based on the augmented formulation. A library of standard joint constraints can also be developed and used as a module in this computer program. One disadvantage, however, of using the augmented formulation is the complexity of the numerical algorithm that must be used to solve the resulting mixed system of differential and algebraic equations.

Other alternate approaches for formulating the equations of motion of constrained mechanical systems are the *recursive methods*, wherein the equations of motion are formulated in terms of the *joint degrees of freedom*. This formulation leads to a minimum set of differential equations from which the workless constraint forces are automatically eliminated. The numerical procedure used in solving these differential equations is much simpler than the procedure used in the solution of the mixed system of differential and algebraic equations resulting from the use of the augmented formulation.

There are several techniques for formulating the recursive dynamic equations of multibody systems. These techniques eventually lead to the same equations if the same set of joint variables is used. In fact, the equivalence of these techniques can be demonstrated using simple coordinate transformations. One of the approaches used for formulating the dynamic recursive equations is based on *Newton–Euler equations* that are expressed in terms of the *angular accelerations*. We discuss this approach in this section in order to have a better understanding of the basic joint–force relationships in the analysis of interconnected bodies. By so doing, we will have an appreciation of the principle of virtual work in dynamics, which can also be used to obtain the same recursive dynamic equations presented in this section.

Recursive Kinematic Equations In order to illustrate the development of the recursive kinematic equations, we consider the two bodies $i-1$ and i, which are connected by a *cylindrical joint* as shown in Fig. 15. The two-degree-of-freedom cylindrical joint allows

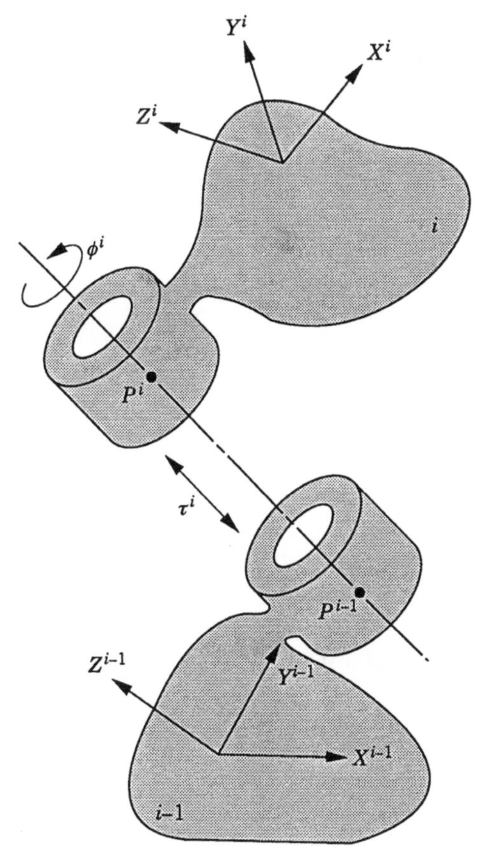

Figure 7.15 Relative motion

relative translation along, and relative rotation about the joint axis. If τ^i and ϕ^i denote, respectively, the relative translation and rotation between the two bodies, the following kinematic relationships between the absolute and relative coordinates hold:

$$\left.\begin{aligned} \mathbf{R}^i + \mathbf{A}^i\bar{\mathbf{u}}_P^i - \mathbf{R}^{i-1} - \mathbf{A}^{i-1}\bar{\mathbf{u}}_P^{i-1} &= \mathbf{v}^{i-1}\tau^i \\ \boldsymbol{\omega}^i &= \boldsymbol{\omega}^{i-1} + \boldsymbol{\omega}^{i,i-1} \end{aligned}\right\} \tag{7.256}$$

where $\bar{\mathbf{u}}_P^i$ and $\bar{\mathbf{u}}_P^{i-1}$ are the local position vectors of the joint definition points on bodies i and $i-1$, respectively, $\mathbf{v}^{i-1}$ is a unit vector defined along the axis of rotation, and $\boldsymbol{\omega}^{i,i-1}$ is the angular velocity vector of body i with respect to body $i-1$. The vector $\boldsymbol{\omega}^{i,i-1}$ can be written as

$$\boldsymbol{\omega}^{i,i-1} = \mathbf{v}^{i-1}\dot{\phi}^i \tag{7.257}$$

and the vector $\mathbf{v}^{i-1}$ can be written as

$$\mathbf{v}^{i-1} = \mathbf{A}^{i-1}\bar{\mathbf{v}}^{i-1} \tag{7.258}$$

where $\bar{\mathbf{v}}^{i-1}$ is a unit vector along the axis of rotation and is defined in the coordinate system of body $i-1$. The components of the unit vector $\bar{\mathbf{v}}^{i-1}$ are constant. It follows that

$$\left.\begin{aligned} \dot{\mathbf{v}}^{i-1} &= \tilde{\boldsymbol{\omega}}^{i-1}\mathbf{A}^{i-1}\bar{\mathbf{v}}^{i-1} = \boldsymbol{\omega}^{i-1} \times \mathbf{v}^{i-1} \\ \ddot{\mathbf{v}}^{i-1} &= \boldsymbol{\alpha}^{i-1} \times \mathbf{v}^{i-1} + \boldsymbol{\omega}^{i-1} \times (\boldsymbol{\omega}^{i-1} \times \mathbf{v}^{i-1}) \end{aligned}\right\} \tag{7.259}$$

Similarly, differentiating Eq. 257 with respect to time, one obtains

$$\dot{\boldsymbol{\omega}}^{i,i-1} = \mathbf{v}^{i-1}\ddot{\phi}^i + (\boldsymbol{\omega}^{i-1} \times \mathbf{v}^{i-1})\dot{\phi}^i \tag{7.260}$$

Differentiating the first equation in Eq. 256 twice with respect to time and the second equation once with respect to time and using Eqs. 259 and 260, one obtains

$$\begin{aligned} &\ddot{\mathbf{R}}^i + \boldsymbol{\alpha}^i \times \mathbf{u}_P^i + \boldsymbol{\omega}^i \times (\boldsymbol{\omega}^i \times \mathbf{u}_P^i) \\ &\quad -\ddot{\mathbf{R}}^{i-1} - \boldsymbol{\alpha}^{i-1} \times \mathbf{u}_P^{i-1} - \boldsymbol{\omega}^{i-1} \times (\boldsymbol{\omega}^{i-1} \times \mathbf{u}_P^{i-1}) \\ &\quad = (\boldsymbol{\alpha}^{i-1} \times \mathbf{v}^{i-1})\tau^i + \boldsymbol{\omega}^{i-1} \times (\boldsymbol{\omega}^{i-1} \times \mathbf{v}^{i-1})\tau^i + \mathbf{v}^{i-1}\ddot{\tau}^i + 2\dot{\mathbf{v}}^{i-1}\dot{\tau}^i \end{aligned} \tag{7.261}$$

and

$$\boldsymbol{\alpha}^i = \boldsymbol{\alpha}^{i-1} + \mathbf{v}^{i-1}\ddot{\phi}^i + (\boldsymbol{\omega}^{i-1} \times \mathbf{v}^{i-1})\dot{\phi}^i \tag{7.262}$$

Using the skew-symmetric matrix notation, the preceding two equations can be written as

$$\ddot{\mathbf{R}}^i - \tilde{\mathbf{u}}_P^i\boldsymbol{\alpha}^i = \ddot{\mathbf{R}}^{i-1} - (\tilde{\mathbf{u}}_P^{i-1} + \tau^i\tilde{\mathbf{v}}^{i-1})\boldsymbol{\alpha}^{i-1} + \mathbf{v}^{i-1}\ddot{\tau}^i + \boldsymbol{\gamma}_R^i \tag{7.263}$$

$$\boldsymbol{\alpha}^i = \boldsymbol{\alpha}^{i-1} + \mathbf{v}^{i-1}\ddot{\phi}^i + \boldsymbol{\gamma}_\theta^i \tag{7.264}$$

where $\boldsymbol{\gamma}_R^i$ and $\boldsymbol{\gamma}_\theta^i$ are vectors that absorb terms that are quadratic in the velocities. Those vectors are defined as

$$\left.\begin{aligned}\boldsymbol{\gamma}_R^i &= -\boldsymbol{\omega}^i \times (\boldsymbol{\omega}^i \times \mathbf{u}_P^i) + \boldsymbol{\omega}^{i-1} \times (\boldsymbol{\omega}^{i-1} \times \mathbf{u}_P^{i-1}) \\ &\quad + \boldsymbol{\omega}^{i-1} \times (\boldsymbol{\omega}^{i-1} \times \mathbf{v}^{i-1})\tau^i + 2\dot{\mathbf{v}}^{i-1}\dot{\tau}^i \\ \boldsymbol{\gamma}_\theta^i &= (\boldsymbol{\omega}^{i-1} \times \mathbf{v}^{i-1})\dot{\phi}^i\end{aligned}\right\} \tag{7.265}$$

Equations 263 and 264 can be combined into one matrix equation as

$$\begin{aligned}\begin{bmatrix}\mathbf{I} & -\tilde{\mathbf{u}}_P^i \\ \mathbf{0} & \mathbf{I}\end{bmatrix}\begin{bmatrix}\ddot{\mathbf{R}}^i \\ \boldsymbol{\alpha}^i\end{bmatrix} &= \begin{bmatrix}\mathbf{I} & -(\tilde{\mathbf{u}}_P^{i-1} + \tau^i\tilde{\mathbf{v}}^{i-1}) \\ \mathbf{0} & \mathbf{I}\end{bmatrix}\begin{bmatrix}\ddot{\mathbf{R}}^{i-1} \\ \boldsymbol{\alpha}^{i-1}\end{bmatrix} \\ &\quad + \begin{bmatrix}\mathbf{v}^{i-1} & \mathbf{0} \\ \mathbf{0} & \mathbf{v}^{i-1}\end{bmatrix}\begin{bmatrix}\ddot{\tau}^i \\ \ddot{\phi}^i\end{bmatrix} + \begin{bmatrix}\boldsymbol{\gamma}_R^i \\ \boldsymbol{\gamma}_\theta^i\end{bmatrix}\end{aligned} \tag{7.266}$$

Note that

$$\begin{bmatrix}\mathbf{I} & -\tilde{\mathbf{u}}_P^i \\ \mathbf{0} & \mathbf{I}\end{bmatrix}^{-1} = \begin{bmatrix}\mathbf{I} & \tilde{\mathbf{u}}_P^i \\ \mathbf{0} & \mathbf{I}\end{bmatrix} \tag{7.267}$$

Using this equation, Eq. 266 leads to

$$\begin{aligned}\begin{bmatrix}\ddot{\mathbf{R}}^i \\ \boldsymbol{\alpha}^i\end{bmatrix} &= \begin{bmatrix}\mathbf{I} & \tilde{\mathbf{u}}_P^i - (\tilde{\mathbf{u}}_P^{i-1} + \tau^i\tilde{\mathbf{v}}^{i-1}) \\ \mathbf{0} & \mathbf{I}\end{bmatrix}\begin{bmatrix}\ddot{\mathbf{R}}^{i-1} \\ \boldsymbol{\alpha}^{i-1}\end{bmatrix} \\ &\quad + \begin{bmatrix}\mathbf{v}^{i-1} & \tilde{\mathbf{u}}_P^i\mathbf{v}^{i-1} \\ \mathbf{0} & \mathbf{v}^{i-1}\end{bmatrix}\begin{bmatrix}\ddot{\tau}^i \\ \ddot{\phi}^i\end{bmatrix} + \begin{bmatrix}\boldsymbol{\gamma}_R^i + \tilde{\mathbf{u}}_P^i\boldsymbol{\gamma}_\theta^i \\ \boldsymbol{\gamma}_\theta^i\end{bmatrix}\end{aligned} \tag{7.268}$$

which can also be written as

$$\ddot{\mathbf{P}}^i = \mathbf{D}^i\ddot{\mathbf{P}}^{i-1} + \mathbf{H}^i\ddot{\mathbf{P}}_r^i + \boldsymbol{\gamma}^i \tag{7.269}$$

where

$$\left.\begin{aligned}\ddot{\mathbf{P}}^{i-1} &= [\ddot{\mathbf{R}}^{i-1^{\mathrm{T}}} \quad \boldsymbol{\alpha}^{i-1^{\mathrm{T}}}]^{\mathrm{T}}, \quad \ddot{\mathbf{P}}^i = [\ddot{\mathbf{R}}^{i^{\mathrm{T}}} \quad \boldsymbol{\alpha}^{i^{\mathrm{T}}}]^{\mathrm{T}} \\ \mathbf{D}^i(P^i, P^{i-1}) &= \begin{bmatrix}\mathbf{I} & \tilde{\mathbf{u}}_P^i - (\tilde{\mathbf{u}}_P^{i-1} + \tau^i\tilde{\mathbf{v}}^{i-1}) \\ \mathbf{0} & \mathbf{I}\end{bmatrix} \\ \mathbf{H}^i &= \begin{bmatrix}\mathbf{v}^{i-1} & \tilde{\mathbf{u}}_P^i\mathbf{v}^{i-1} \\ \mathbf{0} & \mathbf{v}^{i-1}\end{bmatrix} \\ \mathbf{P}_r^i &= [\tau^i \quad \phi^i]^{\mathrm{T}}, \quad \boldsymbol{\gamma}^i = [(\boldsymbol{\gamma}_R^i + \tilde{\mathbf{u}}_P^i\boldsymbol{\gamma}_\theta^i)^{\mathrm{T}} \quad \boldsymbol{\gamma}_\theta^{i^{\mathrm{T}}}]^{\mathrm{T}}\end{aligned}\right\} \tag{7.270}$$

The matrix $\mathbf{D}^i$ can be written as

$$\mathbf{D}^i(P^i, P^j) = \begin{bmatrix} \mathbf{I} & \tilde{\mathbf{r}}_P^{ij} \\ \mathbf{0} & \mathbf{I} \end{bmatrix} \tag{7.271}$$

and $\tilde{\mathbf{r}}_P^{ij}$ is the skew-symmetric matrix associated with the position vector of the origin of body i with respect to the origin of body j with $j = i - 1$, that is,

$$\mathbf{r}_P^{ij} = \mathbf{u}_P^i - \mathbf{u}_P^j - \tau^i \mathbf{v}^j \tag{7.272}$$

The matrix $\mathbf{D}^i$ satisfies the following identities:

$$\left.\begin{aligned} [\mathbf{D}^i(P^i, P^j)]^{-1} &= \mathbf{D}^j(P^j, P^i) \\ \mathbf{D}^i(P^i, P^k) &= \mathbf{D}^i(P^i, P^j)\mathbf{D}^j(P^j, P^k) \\ \mathbf{D}^i(P^i, P^i) &= \mathbf{I} \end{aligned}\right\} \tag{7.273}$$

In Eq. 269, the vector of accelerations of body i is expressed in terms of the accelerations of body $i - 1$ and the vector of joint accelerations $\ddot{\mathbf{P}}_r^i$. The dimension of the vector $\mathbf{P}_r^i$ is equal to the number of the joint degrees of freedom. For a multibody system consisting of a set of interconnected rigid bodies, a matrix equation similar to Eq. 269 can be obtained for any pair of bodies connected by a joint. The form of the matrices $\mathbf{D}^i$ and $\mathbf{H}^i$ depends on the joint type, and consequently, equations similar to Eq. 269 can be developed in the cases of spherical, universal, prismatic, and revolute joints. The case of revolute joint can be considered as a special case of the cylindrical joint in which the translation τ^i is constant, and the case of the prismatic joint can also be considered as a special case of the cylindrical joint in which the rotation ϕ^i is assumed to be constant.

Since Eq. 269 is developed for two arbitrary bodies, one also has

$$\left.\begin{aligned} \ddot{\mathbf{P}}^{i-1} &= \mathbf{D}^{i-1}\ddot{\mathbf{P}}^{i-2} + \mathbf{H}^{i-1}\ddot{\mathbf{P}}_r^{i-1} + \boldsymbol{\gamma}^{i-1} \\ \ddot{\mathbf{P}}^{i-2} &= \mathbf{D}^{i-2}\ddot{\mathbf{P}}^{i-3} + \mathbf{H}^{i-2}\ddot{\mathbf{P}}_r^{i-2} + \boldsymbol{\gamma}^{i-2} \\ &\vdots \\ \ddot{\mathbf{P}}^2 &= \mathbf{D}^2\ddot{\mathbf{P}}^1 + \mathbf{H}^2\ddot{\mathbf{P}}_r^2 + \boldsymbol{\gamma}^2 \end{aligned}\right\} \tag{7.274}$$

where body 1 is considered as the base body.

Substituting Eq. 274 into Eq. 269, the accelerations of body i can be expressed in terms of the accelerations of the base body and the joint accelerations as

$$\ddot{\mathbf{P}}^i = \mathbf{D}_t^i\ddot{\mathbf{P}}^1 + \mathbf{H}_t^i\ddot{\mathbf{P}}_r + \boldsymbol{\gamma}_t^i \tag{7.275}$$

where $\mathbf{P}_r$ is the vector of the system joint degrees of freedom, $\mathbf{D}_t^i$ and $\mathbf{H}_t^i$ are *velocity influence coefficient matrices*, and $\boldsymbol{\gamma}_t^i$ is a vector that absorbs terms that are quadratic in the velocities. The vector $\mathbf{P}_r$ is given by

$$\mathbf{P}_r = [\mathbf{P}_r^{2^{\mathrm{T}}} \quad \mathbf{P}_r^{3^{\mathrm{T}}} \quad \cdots \quad \mathbf{P}_r^{i^{\mathrm{T}}}]^{\mathrm{T}} \tag{7.276}$$

where $\mathbf{P}_r^k$ is the vector of the degrees of freedom of the joint connecting bodies k and $k-1$, and

$$\left.\begin{aligned} \mathbf{D}_t^i &= \mathbf{D}^i\mathbf{D}^{i-1}\mathbf{D}^{i-2}\cdots\mathbf{D}^2 = \prod_{j=2}^{i}\mathbf{D}^{i+2-j} \\ \mathbf{H}_t^i &= [\mathbf{H}_2^i \quad \mathbf{H}_3^i \quad \cdots \quad \mathbf{H}_i^i] \\ \boldsymbol{\gamma}_t^i &= \mathbf{D}^i\mathbf{D}^{i-1}\cdots\mathbf{D}^3\boldsymbol{\gamma}^2 + \mathbf{D}^i\mathbf{D}^{i-1}\cdots\mathbf{D}^4\boldsymbol{\gamma}^3 \\ &\quad + \cdots + \mathbf{D}^i\boldsymbol{\gamma}^{i-1} + \boldsymbol{\gamma}^i \end{aligned}\right\} \tag{7.277}$$

in which

$$\mathbf{H}_k^i = \mathbf{D}^i\mathbf{D}^{i-1}\cdots\mathbf{D}^{k+1}\mathbf{H}^k \tag{7.278}$$

If the motion of the base body is specified, Eq. 275 can be written as

$$\ddot{\mathbf{P}}^i = \mathbf{H}_t^i\ddot{\mathbf{P}}_r + \bar{\boldsymbol{\gamma}}_t^i \tag{7.279}$$

where

$$\bar{\boldsymbol{\gamma}}_t^i = \boldsymbol{\gamma}_t^i + \mathbf{D}_t^i\ddot{\mathbf{P}}^1 \tag{7.280}$$

Example 7.11

Figure 16 shows a two-degree-of-freedom *robotic manipulator* which consists of three bodies including the ground (body 1). Body 2 is connected to body 1 by a revolute joint at point Q, and the axis of this joint is along the Y^1 axis. Body 3 is connected to body 2 by a revolute joint at point P, and the axis of this revolute joint is assumed to be parallel to the Z^2 axis. Obtain the recursive kinematic relationships of this system.

Solution. Since body 2 is rotating about the Y^1 axis, the transformation matrix that defines the orientation of this body is given by

$$\mathbf{A}^2 = \begin{bmatrix} \cos\phi^2 & 0 & \sin\phi^2 \\ 0 & 1 & 0 \\ -\sin\phi^2 & 0 & \cos\phi^2 \end{bmatrix}$$

where ϕ^2 is the degree of freedom of the joint at point Q. Similarly, since body 3 rotates about an axis parallel to the Z^2 axis, the orientation of this body with respect to body 2 is defined by the matrix

$$\mathbf{A}^{32} = \begin{bmatrix} \cos\phi^3 & -\sin\phi^3 & 0 \\ \sin\phi^3 & \cos\phi^3 & 0 \\ 0 & 0 & 1 \end{bmatrix}$$

where ϕ^3 is the degree of freedom of the second revolute joint.

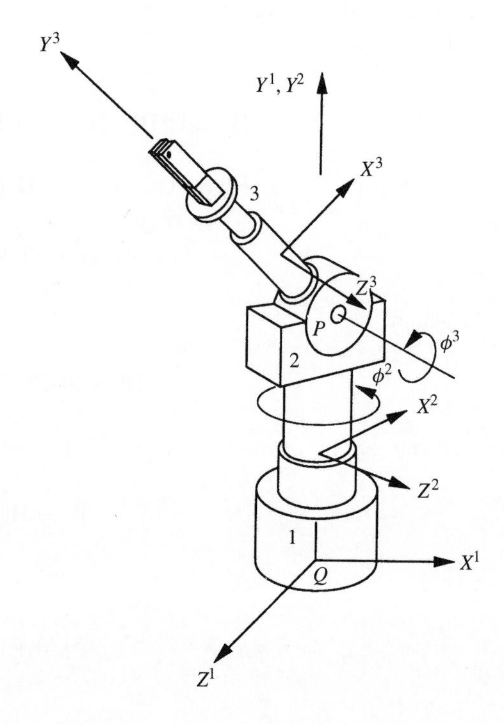

Figure 7.16 Manipulator example

The transformation matrix that defines the orientation of body 3 with respect to body 1 is then given by

$$\mathbf{A}^3 = \mathbf{A}^2\mathbf{A}^{32} = \begin{bmatrix} \cos\phi^2 & 0 & \sin\phi^2 \\ 0 & 1 & 0 \\ -\sin\phi^2 & 0 & \cos\phi^2 \end{bmatrix} \begin{bmatrix} \cos\phi^3 & -\sin\phi^3 & 0 \\ \sin\phi^3 & \cos\phi^3 & 0 \\ 0 & 0 & 1 \end{bmatrix}$$

$$= \begin{bmatrix} \cos\phi^2\cos\phi^3 & -\cos\phi^2\sin\phi^3 & \sin\phi^2 \\ \sin\phi^3 & \cos\phi^3 & 0 \\ -\sin\phi^2\cos\phi^3 & \sin\phi^2\sin\phi^3 & \cos\phi^2 \end{bmatrix}$$

Equation 256 can be used to describe the connectivity between bodies 2 and 3 as

$$\mathbf{R}^3 + \mathbf{A}^3\bar{\mathbf{u}}_P^3 = \mathbf{R}^2 + \mathbf{A}^2\bar{\mathbf{u}}_P^2$$

$$\boldsymbol{\omega}^3 = \boldsymbol{\omega}^2 + \boldsymbol{\omega}^{32}$$

where $\bar{\mathbf{u}}_P^2$ and $\bar{\mathbf{u}}_P^3$ are the local position vectors of point P defined in the coordinate systems of body 2 and 3, respectively, and $\boldsymbol{\omega}^{32}$ is the angular velocity vector of body 3 with respect to body 2. The vector $\boldsymbol{\omega}^{32}$ is given by

$$\boldsymbol{\omega}^{32} = \dot{\phi}^3\mathbf{v}^2$$

in which

$$\mathbf{v}^2 = \begin{bmatrix} \cos\phi^2 & 0 & \sin\phi^2 \\ 0 & 1 & 0 \\ -\sin\phi^2 & 0 & \cos\phi^2 \end{bmatrix} \begin{bmatrix} 0 \\ 0 \\ 1 \end{bmatrix} = \begin{bmatrix} \sin\phi^2 \\ 0 \\ \cos\phi^2 \end{bmatrix}$$

This defines $\boldsymbol{\omega}^{32}$ as

$$\boldsymbol{\omega}^{32} = \dot{\phi}^3 \begin{bmatrix} \sin\phi^2 \\ 0 \\ \cos\phi^2 \end{bmatrix}$$

which upon differentiation yields

$$\dot{\boldsymbol{\omega}}^{32} = \ddot{\phi}^3 \begin{bmatrix} \sin\phi^2 \\ 0 \\ \cos\phi^2 \end{bmatrix} + \dot{\phi}^2\dot{\phi}^3 \begin{bmatrix} \cos\phi^2 \\ 0 \\ -\sin\phi^2 \end{bmatrix} = \ddot{\phi}^3\mathbf{v}^2 + \boldsymbol{\gamma}_\theta^3$$

where

$$\boldsymbol{\gamma}_\theta^3 = \dot{\phi}^2\dot{\phi}^3 \begin{bmatrix} \cos\phi^2 \\ 0 \\ -\sin\phi^2 \end{bmatrix}$$

By differentiating the kinematic relationships of the revolute joint at P, one obtains

$$\ddot{\mathbf{R}}^3 - \tilde{\mathbf{u}}_P^3\boldsymbol{\alpha}^3 = \ddot{\mathbf{R}}^2 - \tilde{\mathbf{u}}_P^2\boldsymbol{\alpha}^2 - (\tilde{\boldsymbol{\omega}}^3)^2\mathbf{u}_P^3 + (\tilde{\boldsymbol{\omega}}^2)^2\mathbf{u}_P^2$$
$$\boldsymbol{\alpha}^3 = \boldsymbol{\alpha}^2 + \dot{\boldsymbol{\omega}}^{32}$$

or

$$\begin{bmatrix} \mathbf{I} & -\tilde{\mathbf{u}}_P^3 \\ \mathbf{0} & \mathbf{I} \end{bmatrix} \begin{bmatrix} \ddot{\mathbf{R}}^3 \\ \boldsymbol{\alpha}^3 \end{bmatrix} = \begin{bmatrix} \mathbf{I} & -\tilde{\mathbf{u}}_P^2 \\ \mathbf{0} & \mathbf{I} \end{bmatrix} \begin{bmatrix} \ddot{\mathbf{R}}^2 \\ \boldsymbol{\alpha}^2 \end{bmatrix} + \begin{bmatrix} \mathbf{0} \\ \mathbf{v}^2 \end{bmatrix} \ddot{\phi}^3 + \begin{bmatrix} \boldsymbol{\gamma}_R^3 \\ \boldsymbol{\gamma}_\theta^3 \end{bmatrix}$$

where $\tilde{\mathbf{u}}_P^2$ and $\tilde{\mathbf{u}}_P^3$ are the skew-symmetric matrices associated with the vectors

$$\mathbf{u}_P^2 = \mathbf{A}^2\bar{\mathbf{u}}_P^2, \quad \mathbf{u}_P^3 = \mathbf{A}^3\bar{\mathbf{u}}_P^3$$

and

$$\boldsymbol{\gamma}_R^3 = (\tilde{\boldsymbol{\omega}}^2)^2\mathbf{u}_P^2 - (\tilde{\boldsymbol{\omega}}^3)^2\mathbf{u}_P^3$$

It follows that

$$\begin{bmatrix} \ddot{\mathbf{R}}^3 \\ \boldsymbol{\alpha}^3 \end{bmatrix} = \begin{bmatrix} \mathbf{I} & \tilde{\mathbf{u}}_P^3 - \tilde{\mathbf{u}}_P^2 \\ \mathbf{0} & \mathbf{I} \end{bmatrix} \begin{bmatrix} \ddot{\mathbf{R}}^2 \\ \boldsymbol{\alpha}^2 \end{bmatrix} + \begin{bmatrix} \tilde{\mathbf{u}}_P^3\mathbf{v}^2 \\ \mathbf{v}^2 \end{bmatrix} \ddot{\phi}^3 + \begin{bmatrix} \boldsymbol{\gamma}_R^3 + \tilde{\mathbf{u}}_P^3\boldsymbol{\gamma}_\theta^3 \\ \boldsymbol{\gamma}_\theta^3 \end{bmatrix}$$

This equation can also be obtained as a special case of Eq. 268 or equivalently, Eq. 269, that describes the more general *two-parameter screw motion*. It is clear from the preceding equation that the matrix $\mathbf{H}^3$ reduces in the case of revolute joint to a six-dimensional vector since the revolute joint has only one degree of freedom.

For the revolute joint between body 2 and body 1, similar kinematic relationships can be obtained. Nonetheless, the resulting equations can be simplified since body 1 is fixed in space. In this case, one has

$$\begin{bmatrix} \ddot{\mathbf{R}}^2 \\ \boldsymbol{\alpha}^2 \end{bmatrix} = \begin{bmatrix} \tilde{\mathbf{u}}_Q^2 \mathbf{v}^1 \\ \mathbf{v}^1 \end{bmatrix} \ddot{\phi}^2 + \begin{bmatrix} \boldsymbol{\gamma}_R^2 \\ \mathbf{0} \end{bmatrix}$$

where $\tilde{\mathbf{u}}_Q^2$ is the skew-symmetric matrix associated with the vector

$$\mathbf{u}_Q^2 = \mathbf{A}^2 \bar{\mathbf{u}}_Q^2$$

and

$$\mathbf{v}^1 = [0 \quad 1 \quad 0]^{\mathrm{T}}, \qquad \boldsymbol{\gamma}_R^2 = -(\tilde{\boldsymbol{\omega}}^2)^2 \mathbf{u}_Q^2$$

The absolute acceleration of body 3 can then be expressed in terms of the joint variables as

$$\begin{aligned} \begin{bmatrix} \ddot{\mathbf{R}}^3 \\ \boldsymbol{\alpha}^3 \end{bmatrix} &= \begin{bmatrix} \mathbf{I} & \tilde{\mathbf{u}}_P^3 - \tilde{\mathbf{u}}_P^2 \\ \mathbf{0} & \mathbf{I} \end{bmatrix} \left\{ \begin{bmatrix} \tilde{\mathbf{u}}_Q^2 \mathbf{v}^1 \\ \mathbf{v}^1 \end{bmatrix} \ddot{\phi}^2 + \begin{bmatrix} \boldsymbol{\gamma}_R^2 \\ \mathbf{0} \end{bmatrix} \right\} \\ &\quad + \begin{bmatrix} \tilde{\mathbf{u}}_P^3 \mathbf{v}^2 \\ \mathbf{v}^2 \end{bmatrix} \ddot{\phi}^3 + \begin{bmatrix} \boldsymbol{\gamma}_R^3 + \tilde{\mathbf{u}}_P^3 \boldsymbol{\gamma}_\theta^3 \\ \boldsymbol{\gamma}_\theta^3 \end{bmatrix} \end{aligned}$$

which can be written in the form of Eq. 279 as

$$\ddot{\mathbf{P}}^3 = \mathbf{H}_t^3 \ddot{\mathbf{P}}_r + \bar{\boldsymbol{\gamma}}_t^3$$

where

$$\begin{aligned} &\ddot{\mathbf{P}}^3 = [\ddot{\mathbf{R}}^{3^{\mathrm{T}}} \quad \boldsymbol{\alpha}^{3^{\mathrm{T}}}]^{\mathrm{T}}, \quad \mathbf{P}_r = [\phi^2 \quad \phi^3]^{\mathrm{T}} \\ &\mathbf{H}_t^3 = \begin{bmatrix} (\tilde{\mathbf{u}}_Q^2 + \tilde{\mathbf{u}}_P^3 - \tilde{\mathbf{u}}_P^2)\mathbf{v}^1 & \tilde{\mathbf{u}}_P^3 \mathbf{v}^2 \\ \mathbf{v}^1 & \mathbf{v}^2 \end{bmatrix}, \quad \bar{\boldsymbol{\gamma}}_t^3 = \begin{bmatrix} \boldsymbol{\gamma}_R^2 + \boldsymbol{\gamma}_R^3 + \tilde{\mathbf{u}}_P^3 \boldsymbol{\gamma}_\theta^3 \\ \boldsymbol{\gamma}_\theta^3 \end{bmatrix} \end{aligned}$$

We note also from the preceding equations that the angular velocity of body 3 can be expressed in terms of the joint rates as

$$\begin{aligned} \boldsymbol{\omega}^3 &= \boldsymbol{\omega}^2 + \boldsymbol{\omega}^{32} = \dot{\phi}^2 \mathbf{v}^1 + \dot{\phi}^3 \mathbf{v}^2 \\ &= \dot{\phi}^2 \begin{bmatrix} 0 \\ 1 \\ 0 \end{bmatrix} + \dot{\phi}^3 \begin{bmatrix} \sin\phi^2 \\ 0 \\ \cos\phi^2 \end{bmatrix} = \begin{bmatrix} \dot{\phi}^3 \sin\phi^2 \\ \dot{\phi}^2 \\ \dot{\phi}^3 \cos\phi^2 \end{bmatrix} \end{aligned}$$

Using the results obtained in this example, one can also verify that

$$\tilde{\boldsymbol{\omega}}^3 \mathbf{A}^3 = \dot{\mathbf{A}}^3$$

and

$$\boldsymbol{\omega}^{32^{\mathrm{T}}} \mathbf{v}^2 = \dot{\phi}^3$$

Dynamic Equations Equation 279, in which the accelerations of body i are expressed in terms of the joint accelerations, can be used with Newton–Euler equations to obtain a minimum set of differential equations expressed in terms of the joint degrees of freedom. If Eq. 279 is substituted into Newton–Euler equations of body i as defined by Eq. 230, one gets

$$\mathbf{M}_d^i(\mathbf{H}_t^i\ddot{\mathbf{P}}_r + \overline{\boldsymbol{\gamma}}_t^i) = \mathbf{F}_e^i + \mathbf{F}_v^i + \mathbf{F}_c^i \tag{7.281}$$

where $\mathbf{F}_c^i$ is the vector of joint reaction forces acting on body i. Premultiplying Eq. 281 by $\mathbf{H}_t^{i\mathrm{T}}$, and rearranging terms, one obtains

$$\mathbf{H}_t^{i\mathrm{T}}\mathbf{M}_d^i\mathbf{H}_t^i\ddot{\mathbf{P}}_r = \mathbf{H}_t^{i\mathrm{T}}(\mathbf{F}_e^i + \mathbf{F}_v^i + \mathbf{F}_c^i - \mathbf{M}_d^i\overline{\boldsymbol{\gamma}}_t^i) \tag{7.282}$$

which can be written as

$$\mathbf{M}_r^i\ddot{\mathbf{P}}_r = \mathbf{Q}_r^i + \mathbf{H}_t^{i\mathrm{T}}\mathbf{F}_c^i, \qquad i = 1, 2, \ldots, n_b \tag{7.283}$$

where n_b is the total number of bodies, and

$$\mathbf{M}_r^i = \mathbf{H}_t^{i\mathrm{T}}\mathbf{M}_d^i\mathbf{H}_t^i, \quad \mathbf{Q}_r^i = \mathbf{H}_t^{i\mathrm{T}}(\mathbf{F}_e^i + \mathbf{F}_v^i - \mathbf{M}_d^i\overline{\boldsymbol{\gamma}}_t^i) \tag{7.284}$$

Note that $\mathbf{M}_r^i$ is a square matrix whose dimension is the same as the number of the system joint degrees of freedom. Since Eq. 283 is developed for an arbitrary body in the system, one has

$$\sum_{i=1}^{n_b}\mathbf{M}_r^i\ddot{\mathbf{P}}_r = \sum_{i=1}^{n_b}\mathbf{Q}_r^i + \sum_{i=1}^{n_b}\mathbf{H}_t^{i\mathrm{T}}\mathbf{F}_c^i \tag{7.285}$$

Utilizing the fact that

$$\sum_{i=1}^{n_b}\mathbf{H}_t^{i\mathrm{T}}\mathbf{F}_c^i = \mathbf{0}, \tag{7.286}$$

one obtains

$$\mathbf{M}_r\ddot{\mathbf{P}}_r = \mathbf{Q}_r \tag{7.287}$$

where $\mathbf{M}_r$ is the generalized system mass matrix associated with the joint degrees of freedom, and $\mathbf{Q}_r$ is the vector of generalized forces. The matrix $\mathbf{M}_r$ and the vector $\mathbf{Q}_r$ are given by

$$\mathbf{M}_r = \sum_{i=1}^{n_b}\mathbf{M}_r^i, \quad \mathbf{Q}_r = \sum_{i=1}^{n_b}\mathbf{Q}_r^i \tag{7.288}$$

Since the kinetic energy is a positive definite quadratic form, the system mass matrix $\mathbf{M}_r$ is nonsingular, and Eq. 287 can be solved for the joint accelerations as

$$\ddot{\mathbf{P}}_r = \mathbf{M}_r^{-1}\mathbf{Q}_r \tag{7.289}$$

These accelerations can be integrated numerically forward in time using a direct numerical integration method. The numerical solution defines the joint coordinates and velocities. Once the joint variables are determined, the absolute variables can be determined by using the kinematic relationships.

Example 7.12

Use the recursive kinematic relationships obtained in Example 11 to derive the independent differential equations of motion of the two degree of freedom robotic manipulator shown in Fig. 16.

Solution. Newton–Euler equations of body 3 can be written in the following matrix form:

$$\mathbf{M}_d^3 \ddot{\mathbf{P}}^3 = \mathbf{F}_e^3 + \mathbf{F}_v^3 + \mathbf{F}_c^3$$

or

$$\begin{bmatrix} m^3\mathbf{I} & \mathbf{0} \\ \mathbf{0} & \mathbf{I}_{\theta\theta}^3 \end{bmatrix} \begin{bmatrix} \ddot{\mathbf{R}}^3 \\ \boldsymbol{\alpha}^3 \end{bmatrix} = \begin{bmatrix} (\mathbf{F}_e^3)_R + (\mathbf{F}_c^3)_R \\ (\mathbf{F}_e^3)_\theta - \boldsymbol{\omega}^3 \times (\mathbf{I}_{\theta\theta}^3 \boldsymbol{\omega}^3) + (\mathbf{F}_c^3)_\theta \end{bmatrix}$$

This equation can be expressed in terms of the joint degrees of freedom using the kinematic equations obtained in Example 11. It was shown that the acceleration kinematic relationships are

$$\ddot{\mathbf{P}}^3 = \mathbf{H}_t^3 \ddot{\mathbf{P}}_r + \bar{\boldsymbol{\gamma}}_t^3$$

where

$$\mathbf{H}_t^3 = \begin{bmatrix} (\tilde{\mathbf{u}}_Q^2 + \tilde{\mathbf{u}}_P^3 - \tilde{\mathbf{u}}_P^2)\mathbf{v}^1 & \tilde{\mathbf{u}}_P^3 \mathbf{v}^2 \\ \mathbf{v}^1 & \mathbf{v}^2 \end{bmatrix}$$

and

$$\bar{\boldsymbol{\gamma}}_t^3 = \begin{bmatrix} \boldsymbol{\gamma}_R^2 + \boldsymbol{\gamma}_R^3 + \tilde{\mathbf{u}}_P^3 \boldsymbol{\gamma}_\theta^3 \\ \boldsymbol{\gamma}_\theta^3 \end{bmatrix}$$

Using Eq. 283, one has

$$\mathbf{M}_r^3 \ddot{\mathbf{P}}_r = \mathbf{Q}_r^3 + \mathbf{H}_t^{3^\mathrm{T}} \mathbf{F}_c^3$$

where

$$\begin{aligned} \mathbf{M}_r^3 = \mathbf{H}_t^{3^\mathrm{T}} \mathbf{M}_d^3 \mathbf{H}_t^3 &= \begin{bmatrix} \mathbf{v}^{1^\mathrm{T}} (\tilde{\mathbf{u}}_Q^2 + \tilde{\mathbf{u}}_P^3 - \tilde{\mathbf{u}}_P^2)^\mathrm{T} & \mathbf{v}^{1^\mathrm{T}} \\ \mathbf{v}^{2^\mathrm{T}} \tilde{\mathbf{u}}_P^{3^\mathrm{T}} & \mathbf{v}^{2^\mathrm{T}} \end{bmatrix} \begin{bmatrix} m^3\mathbf{I} & \mathbf{0} \\ \mathbf{0} & \mathbf{I}_{\theta\theta}^3 \end{bmatrix} \\ &\quad \cdot \begin{bmatrix} (\tilde{\mathbf{u}}_Q^2 + \tilde{\mathbf{u}}_P^3 - \tilde{\mathbf{u}}_P^2)\mathbf{v}^1 & \tilde{\mathbf{u}}_P^3 \mathbf{v}^2 \\ \mathbf{v}^1 & \mathbf{v}^2 \end{bmatrix} \\ &= \begin{bmatrix} m_{11}^3 & m_{12}^3 \\ m_{21}^3 & m_{22}^3 \end{bmatrix} \end{aligned}$$

in which

$$m_{11}^3 = m^3 \mathbf{v}^{1^{\mathrm{T}}}(\tilde{\mathbf{u}}_Q^2 + \tilde{\mathbf{u}}_P^3 - \tilde{\mathbf{u}}_P^2)^{\mathrm{T}}(\tilde{\mathbf{u}}_Q^2 + \tilde{\mathbf{u}}_P^3 - \tilde{\mathbf{u}}_P^2)\mathbf{v}^1 + \mathbf{v}^{1^{\mathrm{T}}}\mathbf{I}_{\theta\theta}^3\mathbf{v}^1$$

$$m_{12}^3 = m_{21}^3 = \mathbf{v}^{1^{\mathrm{T}}}\{m^3(\tilde{\mathbf{u}}_Q^2 + \tilde{\mathbf{u}}_P^3 - \tilde{\mathbf{u}}_P^2)^{\mathrm{T}}\tilde{\mathbf{u}}_P^3 + \mathbf{I}_{\theta\theta}^3\}\mathbf{v}^2$$

$$m_{22}^3 = \mathbf{v}^{2^{\mathrm{T}}}[m^3\tilde{\mathbf{u}}_P^{3^{\mathrm{T}}}\tilde{\mathbf{u}}_P^3 + \mathbf{I}_{\theta\theta}^3]\mathbf{v}^2$$

The vector $\mathbf{Q}_r^3$ is given by

$$\begin{aligned}
\mathbf{Q}_r^3 &= \mathbf{H}_t^{3^{\mathrm{T}}}[\mathbf{F}_e^3 + \mathbf{F}_v^3 - \mathbf{M}_d^3\bar{\boldsymbol{\gamma}}_t^3] \\
&= \begin{bmatrix} \mathbf{v}^{1^{\mathrm{T}}}(\tilde{\mathbf{u}}_Q^2 + \tilde{\mathbf{u}}_P^3 - \tilde{\mathbf{u}}_P^2)^{\mathrm{T}} & \mathbf{v}^{1^{\mathrm{T}}} \\ \mathbf{v}^{2^{\mathrm{T}}}\tilde{\mathbf{u}}_P^{3^{\mathrm{T}}} & \mathbf{v}^{2^{\mathrm{T}}} \end{bmatrix} \begin{bmatrix} (\mathbf{F}_e^3)_R - m^3(\boldsymbol{\gamma}_R^2 + \boldsymbol{\gamma}_R^3 + \tilde{\mathbf{u}}_P^3\boldsymbol{\gamma}_\theta^3) \\ (\mathbf{F}_e^3)_\theta - \boldsymbol{\omega}^3 \times (\mathbf{I}_{\theta\theta}^3\boldsymbol{\omega}^3) - \mathbf{I}_{\theta\theta}^3\boldsymbol{\gamma}_\theta^3 \end{bmatrix} \\
&= \begin{bmatrix} (\mathbf{Q}_r^3)_1 \\ (\mathbf{Q}_r^3)_2 \end{bmatrix}
\end{aligned}$$

where

$$\begin{aligned}
(\mathbf{Q}_r^3)_1 &= \mathbf{v}^{1^{\mathrm{T}}}(\tilde{\mathbf{u}}_Q^2 + \tilde{\mathbf{u}}_P^3 - \tilde{\mathbf{u}}_P^2)^{\mathrm{T}}\{(\mathbf{F}_e^3)_R - m^3(\boldsymbol{\gamma}_R^2 + \boldsymbol{\gamma}_R^3 + \tilde{\mathbf{u}}_P^3\boldsymbol{\gamma}_\theta^3)\} \\
&\quad + \mathbf{v}^{1^{\mathrm{T}}}\{(\mathbf{F}_e^3)_\theta - \boldsymbol{\omega}^3 \times (\mathbf{I}_{\theta\theta}^3\boldsymbol{\omega}^3) - \mathbf{I}_{\theta\theta}^3\boldsymbol{\gamma}_\theta^3\} \\
(\mathbf{Q}_r^3)_2 &= \mathbf{v}^{2^{\mathrm{T}}}\tilde{\mathbf{u}}_P^3\{(\mathbf{F}_e^3)_R - m^3(\boldsymbol{\gamma}_R^2 + \boldsymbol{\gamma}_R^3 + \tilde{\mathbf{u}}_P^3\boldsymbol{\gamma}_\theta^3)\} \\
&\quad + \mathbf{v}^{2^{\mathrm{T}}}\{(\mathbf{F}_e^3)_\theta - \boldsymbol{\omega}^3 \times (\mathbf{I}_{\theta\theta}^3\boldsymbol{\omega}^3) - \mathbf{I}_{\theta\theta}^3\boldsymbol{\gamma}_\theta^3\}
\end{aligned}$$

Similarly, for body 2, one has

$$\ddot{\mathbf{P}}^2 = \mathbf{H}_t^2\ddot{\mathbf{P}}_r + \bar{\boldsymbol{\gamma}}_t^2$$

where, as shown in Example 11,

$$\mathbf{H}_t^2 = \begin{bmatrix} \tilde{\mathbf{u}}_Q^2\mathbf{v}^1 & \mathbf{0} \\ \mathbf{v}^1 & \mathbf{0} \end{bmatrix}, \quad \bar{\boldsymbol{\gamma}}_t^2 = \begin{bmatrix} \boldsymbol{\gamma}_R^2 \\ \mathbf{0} \end{bmatrix}$$

It follows that

$$\begin{aligned}
\mathbf{M}_r^2 &= \mathbf{H}_t^{2^{\mathrm{T}}}\mathbf{M}_d^2\mathbf{H}_t^2 \\
&= \begin{bmatrix} \mathbf{v}^{1^{\mathrm{T}}}\tilde{\mathbf{u}}_Q^{2^{\mathrm{T}}} & \mathbf{v}^{1^{\mathrm{T}}} \\ \mathbf{0} & \mathbf{0} \end{bmatrix} \begin{bmatrix} m^2\mathbf{I} & \mathbf{0} \\ \mathbf{0} & \mathbf{I}_{\theta\theta}^2 \end{bmatrix} \begin{bmatrix} \tilde{\mathbf{u}}_Q^2\mathbf{v}^1 & \mathbf{0} \\ \mathbf{v}^1 & \mathbf{0} \end{bmatrix} = \begin{bmatrix} m_{11}^2 & 0 \\ 0 & 0 \end{bmatrix}
\end{aligned}$$

where

$$m_{11}^2 = \mathbf{v}^{1^{\mathrm{T}}}[m^2\tilde{\mathbf{u}}_Q^{2^{\mathrm{T}}}\tilde{\mathbf{u}}_Q^2 + \mathbf{I}_{\theta\theta}^2]\mathbf{v}^1$$

and

$$\mathbf{Q}_r^2 = \mathbf{H}_t^{2^{\mathrm{T}}}(\mathbf{F}_e^2 + \mathbf{F}_v^2 - \mathbf{M}_d^2\overline{\boldsymbol{\gamma}}_t^2)$$

$$= \begin{bmatrix} \mathbf{v}^{1^{\mathrm{T}}}\tilde{\mathbf{u}}_Q^{2^{\mathrm{T}}} & \mathbf{v}^{1^{\mathrm{T}}} \\ \mathbf{0} & \mathbf{0} \end{bmatrix} \begin{bmatrix} (\mathbf{F}_e^2)_R - m^2\boldsymbol{\gamma}_R^2 \\ (\mathbf{F}_e^2)_\theta - \boldsymbol{\omega}^2 \times (\mathbf{I}_{\theta\theta}^2\boldsymbol{\omega}^2) \end{bmatrix} = \begin{bmatrix} (\mathbf{Q}_r^2)_1 \\ 0 \end{bmatrix}$$

where

$$(\mathbf{Q}_r^2)_1 = \mathbf{v}^{1^{\mathrm{T}}}\tilde{\mathbf{u}}_Q^{2^{\mathrm{T}}}\{(\mathbf{F}_e^2)_R - m^2\boldsymbol{\gamma}_R^2\} + \mathbf{v}^{1^{\mathrm{T}}}\{(\mathbf{F}_e^2)_\theta - \boldsymbol{\omega}^2 \times (\mathbf{I}_{\theta\theta}^2\boldsymbol{\omega}^2)\}$$

The system equations of motion can then be written in terms of the joint degrees of freedom using Eq. 287 as

$$\begin{bmatrix} m_{11}^2 + m_{11}^3 & m_{12}^3 \\ m_{21}^3 & m_{22}^3 \end{bmatrix} \begin{bmatrix} \ddot{\phi}^2 \\ \ddot{\phi}^3 \end{bmatrix} = \begin{bmatrix} (\mathbf{Q}_r^2)_1 + (\mathbf{Q}_r^3)_1 \\ (\mathbf{Q}_r^3)_2 \end{bmatrix}$$

The recursive method presented in this section can also be extended to the analysis of open-chain multibody systems with *multiple branches*. The use of this approach in the dynamic analysis of such systems has two major computational advantages over the augmented formulations that employ Lagrange multipliers. The first advantage is that a minimum set of differential equations is obtained. As a result of that, the constraint forces are automatically eliminated since the dynamic equations are expressed in terms of the joint degrees of freedom. The second advantage is the simplicity of the numerical scheme used for the solution of the dynamic equations developed using the recursive methods. There is no need for using a Newton–Raphson algorithm since the recursive formulation of the equations of open kinematic chains leads only to a set of differential equations. One disadvantage of using the recursive method, however, is that the dynamic equations are expressed in terms of a set of joint variables that depend on the topological structure of the multibody system. For that reason, it is more difficult to develop general-purpose multibody computer programs based on the recursive methods. These methods also become less attractive in the analysis of *closed-chain mechanical systems*. One approach for dealing with closed-chain systems using the recursive methods is to make *cuts* at selected *secondary joints* to form spanning tree structures. The method of analysis presented in this section can then be used to develop the equations of motion of the resulting open-chain branches. Connectivity conditions between the branches at the secondary joints can be handled by either eliminating the dependent variables or by using the method of Lagrange multipliers. In the case of Lagrange multipliers, the recursive formulation leads to a mixed system of differential and algebraic equations that must be solved using the same numerical procedure used in the case of the augmented formulation.

An Alternative Matrix Approach Another elegant approach similar, in principle, to the methods discussed in Chapter 6 can be used for deriving the recursive kinematic and dynamic equations of spatial mechanical systems. In this approach, Eq. 274 can be written for n_b bodies as

$$\left.\begin{aligned} \ddot{\mathbf{P}}^{n_b} - \mathbf{D}^{n_b}\ddot{\mathbf{P}}^{n_b-1} &= \mathbf{H}^{n_b}\ddot{\mathbf{P}}_r^{n_b} + \boldsymbol{\gamma}^{n_b} \\ \ddot{\mathbf{P}}^{n_b-1} - \mathbf{D}^{n_b-1}\ddot{\mathbf{P}}^{n_b-2} &= \mathbf{H}^{n_b-1}\ddot{\mathbf{P}}_r^{n_b-1} + \boldsymbol{\gamma}^{n_b-1} \\ &\vdots \\ \ddot{\mathbf{P}}^{i} - \mathbf{D}^{i}\ddot{\mathbf{P}}^{i-1} &= \mathbf{H}^{i}\ddot{\mathbf{P}}_r^{i} + \boldsymbol{\gamma}^{i} \\ &\vdots \\ \ddot{\mathbf{P}}^{2} - \mathbf{D}^{2}\ddot{\mathbf{P}}^{1} &= \mathbf{H}^{2}\ddot{\mathbf{P}}_r^{2} + \boldsymbol{\gamma}^{2} \\ \ddot{\mathbf{P}}^{1} &= \ddot{\mathbf{P}}^{1} \end{aligned}\right\} \tag{7.290}$$

which leads to

$$\begin{bmatrix} \mathbf{I} & \mathbf{0} & \mathbf{0} & \mathbf{0} & \cdots & \mathbf{0} & \mathbf{0} \\ -\mathbf{D}^2 & \mathbf{I} & \mathbf{0} & \mathbf{0} & \cdots & \mathbf{0} & \mathbf{0} \\ \mathbf{0} & -\mathbf{D}^3 & \mathbf{I} & \mathbf{0} & \cdots & \mathbf{0} & \mathbf{0} \\ \vdots & \vdots & \vdots & \vdots & \vdots & \ddots & \vdots \\ \mathbf{0} & \mathbf{0} & \mathbf{0} & \mathbf{0} & \cdots & -\mathbf{D}^{n_b} & \mathbf{I} \end{bmatrix} \begin{bmatrix} \ddot{\mathbf{P}}^1 \\ \ddot{\mathbf{P}}^2 \\ \ddot{\mathbf{P}}^3 \\ \vdots \\ \ddot{\mathbf{P}}^{n_b} \end{bmatrix}$$

$$= \begin{bmatrix} \mathbf{I} & & & & \\ & \mathbf{H}^2 & & \mathbf{0} & \\ & & \mathbf{H}^3 & & \\ & \mathbf{0} & & \ddots & \\ & & & & \mathbf{H}^{n_b} \end{bmatrix} \begin{bmatrix} \ddot{\mathbf{P}}^1 \\ \ddot{\mathbf{P}}_r^2 \\ \ddot{\mathbf{P}}_r^3 \\ \vdots \\ \ddot{\mathbf{P}}_r^{n_b} \end{bmatrix} + \begin{bmatrix} \mathbf{0} \\ \boldsymbol{\gamma}^2 \\ \boldsymbol{\gamma}^3 \\ \vdots \\ \boldsymbol{\gamma}^{n_b} \end{bmatrix} \tag{7.291}$$

This equation can be written as

$$\mathbf{D}\ddot{\mathbf{P}} = \mathbf{H}\ddot{\mathbf{q}}_i + \boldsymbol{\gamma} \tag{7.292}$$

where

$$\left.\begin{aligned} \mathbf{D} &= \begin{bmatrix} \mathbf{I} & \mathbf{0} & \mathbf{0} & \mathbf{0} & \cdots & \mathbf{0} & \mathbf{0} \\ -\mathbf{D}^2 & \mathbf{I} & \mathbf{0} & \mathbf{0} & \cdots & \mathbf{0} & \mathbf{0} \\ \mathbf{0} & -\mathbf{D}^3 & \mathbf{I} & \mathbf{0} & \cdots & \mathbf{0} & \mathbf{0} \\ \vdots & \vdots & \vdots & \vdots & \vdots & \ddots & \vdots \\ \mathbf{0} & \mathbf{0} & \mathbf{0} & \mathbf{0} & \cdots & -\mathbf{D}^{n_b} & \mathbf{I} \end{bmatrix} \\ \mathbf{H} &= \begin{bmatrix} \mathbf{I} & & & & \\ & \mathbf{H}^2 & & \mathbf{0} & \\ & & \mathbf{H}^3 & & \\ & \mathbf{0} & & \ddots & \\ & & & & \mathbf{H}^{n_b} \end{bmatrix} \\ \ddot{\mathbf{P}} &= [\ddot{\mathbf{P}}^{1^{\mathrm{T}}} \quad \ddot{\mathbf{P}}^{2^{\mathrm{T}}} \quad \ddot{\mathbf{P}}^{3^{\mathrm{T}}} \quad \cdots \quad \ddot{\mathbf{P}}^{n_b^{\mathrm{T}}}]^{\mathrm{T}} \\ \ddot{\mathbf{q}}_i &= [\ddot{\mathbf{P}}^{1^{\mathrm{T}}} \quad \ddot{\mathbf{P}}_r^{2^{\mathrm{T}}} \quad \ddot{\mathbf{P}}_r^{3^{\mathrm{T}}} \quad \cdots \quad \ddot{\mathbf{P}}_r^{n_b^{\mathrm{T}}}]^{\mathrm{T}} \\ \boldsymbol{\gamma} &= [\mathbf{0} \quad \boldsymbol{\gamma}^{2^{\mathrm{T}}} \quad \boldsymbol{\gamma}^{3^{\mathrm{T}}} \quad \cdots \quad \boldsymbol{\gamma}^{n_b^{\mathrm{T}}}]^{\mathrm{T}} \end{aligned}\right\} \tag{7.293}$$

The matrix $\mathbf{D}$ can be written as the product of $n_b - 1$ matrices as follows:

$$\mathbf{D} = \begin{bmatrix} \mathbf{I} & & & & & \\ -\mathbf{D}^2 & \mathbf{I} & & & & \\ & \mathbf{0} & \mathbf{I} & & & \\ & & \mathbf{0} & & & \\ & & & \ddots & \ddots & \\ & & & & \mathbf{0} & \mathbf{I} \end{bmatrix} \begin{bmatrix} \mathbf{I} & & & & & \\ \mathbf{0} & \mathbf{I} & & & & \\ & -\mathbf{D}^3 & \mathbf{I} & & & \\ & & \mathbf{0} & & & \\ & & & \ddots & \ddots & \\ & & & & \mathbf{0} & \mathbf{I} \end{bmatrix}$$

$$\cdots \begin{bmatrix} \mathbf{I} & & & & & \\ \mathbf{0} & \mathbf{I} & & & & \\ & \mathbf{0} & \mathbf{I} & & & \\ & & \mathbf{0} & & & \\ & & & \ddots & \ddots & \\ & & & & -\mathbf{D}^{n_b} & \mathbf{I} \end{bmatrix} \tag{7.294}$$

from which

$$\mathbf{D}^{-1} = \begin{bmatrix} \mathbf{I} & & & & & \\ \mathbf{0} & \mathbf{I} & & & & \\ & \mathbf{0} & \mathbf{I} & & & \\ & & \mathbf{0} & & & \\ & & & \ddots & \ddots & \\ & & & & \mathbf{D}^{n_b} & \mathbf{I} \end{bmatrix} \begin{bmatrix} \mathbf{I} & & & & & \\ \mathbf{0} & \mathbf{I} & & & & \\ & \mathbf{0} & & & & \\ & & \ddots & \ddots & & \\ & & & \mathbf{D}^{n_b-1} & \mathbf{I} & \\ & & & & \mathbf{0} & \mathbf{I} \end{bmatrix}$$

$$\cdots \begin{bmatrix} \mathbf{I} & & & & & \\ \mathbf{D}^2 & \mathbf{I} & & & & \\ & \mathbf{0} & \mathbf{I} & & & \\ & & \mathbf{0} & & & \\ & & & \ddots & \ddots & \\ & & & & \mathbf{0} & \mathbf{I} \end{bmatrix} \tag{7.295}$$

or

$$\mathbf{D}^{-1} = \begin{bmatrix} \mathbf{I} & \mathbf{0} & \mathbf{0} & \cdots & \mathbf{0} \\ \mathbf{D}_1^2 & \mathbf{I} & \mathbf{0} & \cdots & \mathbf{0} \\ \mathbf{D}_2^3 & \mathbf{D}_1^3 & \mathbf{I} & \cdots & \mathbf{0} \\ \mathbf{D}_3^4 & \mathbf{D}_2^4 & \mathbf{D}_1^4 & \cdots & \mathbf{0} \\ \vdots & \vdots & \vdots & \ddots & \vdots \\ \mathbf{D}_{n_b-1}^{n_b} & \mathbf{D}_{n_b-2}^{n_b} & \mathbf{D}_{n_b-3}^{n_b} & \cdots & \mathbf{I} \end{bmatrix} \tag{7.296}$$

where

$$\mathbf{D}_r^k = \mathbf{D}^k\mathbf{D}^{k-1} \cdots \mathbf{D}^{k-r+1} \tag{7.297}$$

The absolute accelerations can then be expressed in terms of the joint accelerations as

$$\ddot{\mathbf{P}} = \mathbf{B}_i\ddot{\mathbf{q}}_i + \boldsymbol{\gamma}_i \tag{7.298}$$

where $\mathbf{B}_i = \mathbf{D}^{-1}\mathbf{H}$ and $\boldsymbol{\gamma}_i = \mathbf{D}^{-1}\boldsymbol{\gamma}$. In the case of constrained motion, the constraint forces can be added to Eq. 230 and the system equations of motion can be written in terms of the absolute variables as

$$\mathbf{M}_d\ddot{\mathbf{P}} = \mathbf{F}_e + \mathbf{F}_v + \mathbf{F}_c \tag{7.299}$$

where $\mathbf{M}_d$ is the block diagonal mass matrix, $\mathbf{F}_e$ is the vector of applied forces, $\mathbf{F}_v$ is the vector of centrifugal forces, and $\mathbf{F}_c$ is the vector of constraint forces. The matrix $\mathbf{M}_d$ is defined as

$$\mathbf{M}_d = \begin{bmatrix} \mathbf{M}_d^1 & & & \\ & \mathbf{M}_d^2 & \mathbf{0} & \\ & \mathbf{0} & \ddots & \\ & & & \mathbf{M}_d^{nb} \end{bmatrix}, \tag{7.300}$$

and the vectors $\mathbf{F}_e$, $\mathbf{F}_v$, and $\mathbf{F}_c$ are defined as

$$\mathbf{F}_e = \begin{bmatrix} \mathbf{F}_e^1 \\ \mathbf{F}_e^2 \\ \vdots \\ \mathbf{F}_e^{nb} \end{bmatrix}, \quad \mathbf{F}_v = \begin{bmatrix} \mathbf{F}_v^1 \\ \mathbf{F}_v^2 \\ \vdots \\ \mathbf{F}_v^{nb} \end{bmatrix}, \quad \mathbf{F}_c = \begin{bmatrix} \mathbf{F}_c^1 \\ \mathbf{F}_c^2 \\ \vdots \\ \mathbf{F}_c^{nb} \end{bmatrix} \tag{7.301}$$

Substituting the kinematic acceleration equations into the dynamic equations of motion and premultiplying by $\mathbf{B}_i^{\mathrm{T}}$, one obtains

$$\mathbf{B}_i^{\mathrm{T}}[\mathbf{M}_d(\mathbf{B}_i\ddot{\mathbf{q}}_i + \boldsymbol{\gamma}_i)] = \mathbf{B}_i^{\mathrm{T}}(\mathbf{F}_e + \mathbf{F}_v + \mathbf{F}_c) \tag{7.302}$$

Since these equations are expressed in terms of the independent joint accelerations one must have

$$\mathbf{B}_i^{\mathrm{T}}\mathbf{F}_c = \mathbf{0} \tag{7.303}$$

Using the preceding two equations, the independent differential equations of motion of the system reduce to

$$\overline{\mathbf{M}}_i\ddot{\mathbf{q}}_i = \overline{\mathbf{Q}}_i \tag{7.304}$$

where

$$\overline{\mathbf{M}}_i = \mathbf{B}_i^{\mathrm{T}}\mathbf{M}_d\mathbf{B}_i, \qquad \overline{\mathbf{Q}}_i = \mathbf{B}_i^{\mathrm{T}}(\mathbf{F}_e + \mathbf{F}_v - \mathbf{M}_d\boldsymbol{\gamma}_i) \tag{7.305}$$

PROBLEMS

1. If the axes X^i, Y^i, and Z^i of the coordinate system of body i are defined in the coordinate system XYZ by the vectors $[0.5 \quad 0.0 \quad 0.5]^T$, $[0.25 \quad 0.25 \quad -0.25]^T$, and $[-2.0 \quad 4.0 \quad 2.0]^T$, respectively, use the method of direction cosines to determine the transformation matrix that defines the orientation of body i in the coordinate system XYZ.

2. The axes X^i, Y^i, and Z^i of the coordinate system of body i are defined in the coordinate system XYZ by the vectors $[0.0 \quad 1.0 \quad 1.0]^T$, $[-1.0 \quad 1.0 \quad -1.0]^T$, and $[-2.0 \quad -1.0 \quad 1.0]^T$, respectively. Use the method of the direction cosines to determine the transformation matrix that defines the orientation of the rigid body i in the coordinate system XYZ.

3. The axes X^i, Y^i, and Z^i of the coordinate system of the rigid body i are defined in the coordinate system XYZ by the vectors $[0.0 \quad -1.0 \quad 1.0]^T$, $[-1.0 \quad 1.0 \quad 1.0]^T$, and $[-2.0 \quad -1.0 \quad -1.0]^T$, respectively. Determine the transformation matrix of body i using the method of the direction cosines.

4. The axes X^i, Y^i, and Z^i of the coordinate system of the rigid body i are defined in the coordinate system XYZ by the vectors $[0.0 \quad 1.0 \quad 1.0]^T$, $[-1.0 \quad 1.0 \quad -1.0]^T$, and $[-2.0 \quad -1.0 \quad 1.0]^T$, respectively. The axes X^j, Y^j, and Z^j of the coordinate system of body j are defined in the coordinate system XYZ by the vectors $[0.0 \quad -1.0 \quad 1.0]^T$, $[-1.0 \quad 1.0 \quad 1.0]^T$, and $[-2.0 \quad -1.0 \quad -1.0]^T$, respectively. Use the method of the direction cosines to define the orientation of body i with respect to body j. Define also the orientation of body j with respect to body i.

5. The axes of the coordinate system of body i are defined in the XYZ coordinate system by the vectors $[1.0 \quad 0.0 \quad 1.0]^T$, $[1.0 \quad 1.0 \quad -1.0]^T$, and $[-1.0 \quad 2.0 \quad 1.0]^T$. Use the method of the direction cosines to define the orientation of body i in the coordinate system of body j whose axes are defined by the vectors $[0.0 \quad -1.0 \quad 1.0]^T$, $[-1.0 \quad 1.0 \quad 1.0]^T$, and $[-2.0 \quad -1.0 \quad -1.0]^T$.

6. In problem 1, if body i rotates an angle $\theta_1^i = 45^\circ$ about its Z^i axis followed by another rotation $\theta_2^i = 60^\circ$ about its X^i axis, determine the transformation matrix that defines the orientation of the body in the XYZ coordinate system as the result of these two consecutive rotations.

7. In problem 2, if body i rotates an angle $\theta_1^i = 90^\circ$ about its Y^i axis followed by a rotation $\theta_2^i = 30^\circ$ about its X^i axis, determine the transformation matrix that defines the orientation of the rigid body as the result of these two consecutive rotations.

8. If the rigid body i in problem 3 rotates an angle $\theta_1^i = 60^\circ$ about its X^i axis, and an angle $\theta_2^i = 45^\circ$ about its Z^i axis, determine the transformation matrix that defines the final orientation of the body in the XYZ coordinate system.

9. Obtain the transformation matrix in terms of Euler angles if the sequence of rotations is defined as follows: a rotation ϕ^i about Z^i axis, a rotation θ^i about Y^i axis, and a rotation ψ^i about X^i axis.

10. Obtain the transformation matrix in terms of Euler angles if the sequence of rotation is defined as follows: a rotation ϕ^i about Z^i axis, a rotation θ^i about Y^i axis, and a rotation ψ^i about Z^i axis.

11. Use the transformation obtained in problem 1 to extract the three Euler angles by considering the sequence of rotation described in Section 3. If the origin of the body coordinate system is defined by the vector $\mathbf{R}^i = [-1.5 \quad 0.4 \quad 3.2]^{\mathrm{T}}$, determine the global position vector of point P^j whose local coordinates are defined by the vector $\bar{\mathbf{u}}_P^i = [0.3 \quad 0.1 \quad -0.5]^{\mathrm{T}}$.

12. Use the transformation matrix obtained by solving problem 2 to extract the three Euler angles using the sequence of rotation described in Section 3. If the origin of the body coordinate system is defined by the vector $\mathbf{R}^i = [2.1 \quad 3.4 \quad -11.0]^{\mathrm{T}}$, determine the global position vector of point P^i whose position vector in the body coordinate system is defined by the vector $\bar{\mathbf{u}}_P^i = [0.1 \quad -0.2 \quad 0.35]^{\mathrm{T}}$.

13. Use the general development presented in Section 4 to define the angular velocity vector $\boldsymbol{\omega}^i$ in the following two cases: (a) a simple rotation about the global X axis, and (b) a simple rotation about the global Y axis.

14. Use the general development presented in Section 4 to define the angular velocity vector $\bar{\boldsymbol{\omega}}^i$ in the following two cases: (a) a simple rotation about the global X axis, and (b) a simple rotation about the global Y axis.

15. The angular velocity in the body coordinate system is defined by Eq. 83 as $\bar{\boldsymbol{\omega}}^i = \bar{\mathbf{G}}^i\dot{\boldsymbol{\theta}}^i$. Using this equation, show that $\bar{\boldsymbol{\alpha}}^i = \dot{\bar{\boldsymbol{\omega}}}^i = \mathbf{A}^{i^{\mathrm{T}}}\boldsymbol{\alpha}^i$.

16. Use Eq. 118 to determine the location of the center of mass of the *hemisphere* and *right circular cone* shown in Fig. P1.

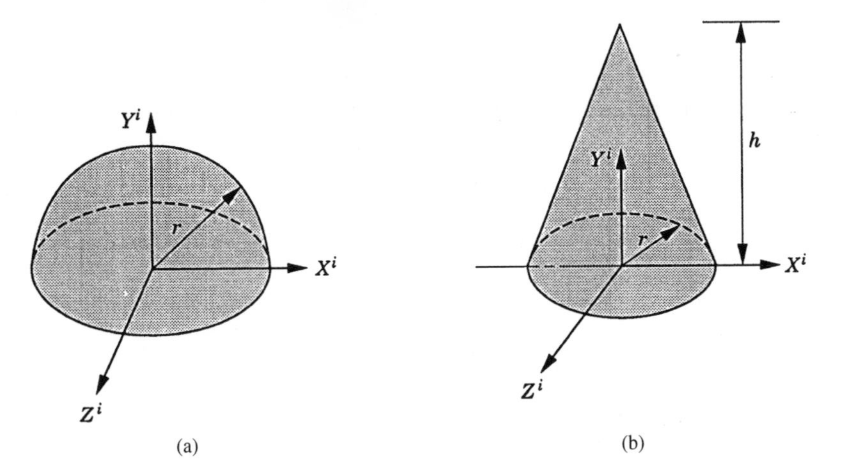

Figure P7.1

17. Determine the elements of the inertia tensor of a *right circular cylinder* with radius r and length h with respect to a centroidal coordinate system.

18. Determine the elements of the inertia tensor of *hollow cylinder* with inner and outer radii r_i and r_o, respectively, and length h. Use a centroidal coordinate system.

19. Using a centroidal coordinate system, determine the elements of the inertia tensor of the *hemisphere* and *right circular cone* shown in Fig. P1. Use the parallel axes theorem to determine the moments of inertia in the coordinate system $X^iY^iZ^i$ shown in the figure.

20. Determine the elements of the inertia tensor of the composite bodies shown in Fig. P2 in the coordinate systems shown in the figure.

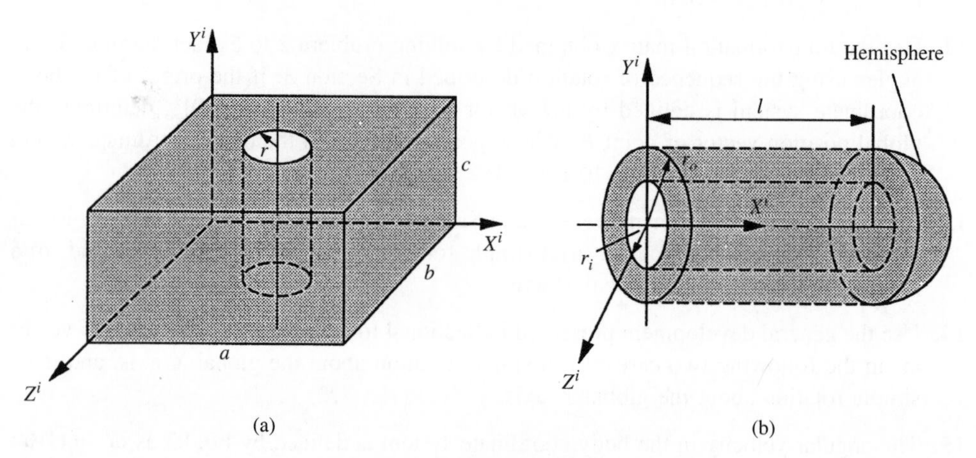

Figure P7.2

21. In problem 11, let the rigid body be subjected to a moment whose components are defined in the global coordinate system by

$$\mathbf{M}_a^i = [3.0 \quad -11.0 \quad 0.0]^{\mathrm{T}}\ \mathrm{N \cdot m}$$

Obtain the generalized forces associated with Euler angles as the result of the application of the moment $\mathbf{M}_a^i$.

22. In problem 11, let the rigid body be subjected to the forces

$$\mathbf{F}_1^i = [10.0 \quad 0.0 \quad 35.0)]^{\mathrm{T}}\ \mathrm{N}, \quad \mathbf{F}_2^i = [5.0 \quad 12.0 \quad -25.0]^{\mathrm{T}}\ \mathrm{N}$$

and the moment

$$\mathbf{M}_a^i = [3.0 \quad -11.0 \quad 0.0]^{\mathrm{T}}\ \mathrm{N \cdot m}$$

The coordinates of the point of application of the forces $\mathbf{F}_1^i$ and $\mathbf{F}_2^i$ are given, respectively, by

$$\bar{\mathbf{u}}_1^i = [0.0 \quad -0.25 \quad 0.8]^{\mathrm{T}}\ \mathrm{m}, \quad \bar{\mathbf{u}}_2^i = [0.15 \quad -0.3 \quad 0.65]^{\mathrm{T}}\ \mathrm{m}$$

Assuming that the forces and the moment are defined in the global coordinate system, obtain the generalized forces associated with the translation and orientation coordinates of the rigid body. Use Euler angles as the orientation coordinates.

23. Repeat the preceding problem assuming that the forces and the moment are defined in the body coordinate system.

24. A force vector $\mathbf{F}^i = [3.0 \quad 0.0 \quad -8.0]^\mathrm{T}$ N is acting on unconstrained body i. The orientation of a centroidal body coordinate system is defined by the Euler angles

$$\phi^i = \frac{\pi}{4}, \quad \theta^i = \frac{\pi}{2}, \quad \psi^i = \pi$$

The time rate of change of Euler angles at the given configuration is

$$\dot{\phi}^i = 15, \quad \dot{\theta}^i = 80, \quad \dot{\psi}^i = -35 \text{ rad/s}$$

The local position vector of the point of the application of the force is

$$\bar{\mathbf{u}}^i = [0.1 \quad 0 \quad -0.15]^\mathrm{T} \text{ m}$$

The mass of the body is 5 kg and its inertia tensor is

$$\bar{\mathbf{I}}^i_{\theta\theta} = \begin{bmatrix} 4.5 & -1.5 & 0 \\ -1.5 & 7.5 & 0 \\ 0 & 0 & 9.0 \end{bmatrix} \text{kg} \cdot \text{m}^2$$

Write the equations of motion of this system at the given configuration and determine the accelerations.

25. A force vector $\mathbf{F}^i = [9.0 \quad 7.0 \quad -12.0]^\mathrm{T}$ N is acting on unconstrained body i. The orientation of a centroidal body coordinate system is defined by the Euler angles

$$\phi^i = 0, \quad \theta^i = \frac{\pi}{2}, \quad \psi^i = \frac{\pi}{4}$$

The time rate of change of Euler angles at the given configuration is

$$\dot{\phi}^i = \dot{\theta}^i = \dot{\psi}^i = 0$$

The local position vector of the point of the application of the force is

$$\bar{\mathbf{u}}^i = [0.2 \quad 0 \quad -0.15]^\mathrm{T} \text{ m}$$

The mass of the body is 3 kg and its inertia tensor is

$$\bar{\mathbf{I}}^i_{\theta\theta} = \begin{bmatrix} 15 & 0 & -10 \\ 0 & 25 & 0 \\ -10 & 0 & 30 \end{bmatrix} \text{kg} \cdot \text{m}^2$$

Write the equations of motion of this system at the given configuration and determine the accelerations.

26. The orientation of a centroidal body coordinate system of unconstrained body i is defined by the Euler angles

$$\phi^i = \frac{\pi}{4}, \quad \theta^i = \frac{\pi}{2}, \quad \psi^i = 0$$

The time rate of change of Euler angles at the given configuration is

$$\dot{\phi}^i = 20, \quad \dot{\theta}^i = 0, \quad \dot{\psi}^i = -35 \text{ rad/s}$$

and the second time derivatives of the coordinates are

$$[\ddot{R}_x^i \quad \ddot{R}_y^i \quad \ddot{R}_z^i] = [0 \quad 150 \quad -28] \text{ m/s}^2$$
$$[\ddot{\phi}^i \quad \ddot{\theta}^i \quad \ddot{\psi}^i] = [300 \quad 10 \quad -15] \text{ rad/s}^2$$

The mass of the body is 4 kg and its inertia tensor is

$$\bar{\mathbf{I}}_{\theta\theta}^i = \begin{bmatrix} 15 & 0 & -10 \\ 0 & 25 & 0 \\ -10 & 0 & 20 \end{bmatrix} \text{kg} \cdot \text{m}^2$$

Determine the resultant external force vector acting at the center of mass as well as the external moment vector acting on the body.

27. Determine the vector $\mathbf{Q}_d$ of Eq. 191 in the case of the *spherical joint*.

28. Determine the vector $\mathbf{Q}_d$ of Eq. 191 in the case of the *cylindrical joint*.

29. Determine the vector $\mathbf{Q}_d$ of Eq. 191 in the case of the *revolute joint*.

30. Two bodies i and j are connected by a spherical joint. Let

$$\mathbf{R}^i = [R_x^i \quad R_y^i \quad R_z^i]^{\mathrm{T}} = [0.95 \quad 3.0 \quad -1.5]^{\mathrm{T}} \text{ m}$$
$$\boldsymbol{\theta}^i = [\phi^i \quad \theta^i \quad \psi^i]^{\mathrm{T}} = [30^\circ \quad 60^\circ \quad 150^\circ]^{\mathrm{T}}$$
$$\boldsymbol{\theta}^j = [\phi^j \quad \theta^j \quad \psi^j]^{\mathrm{T}} = [60^\circ \quad 30^\circ \quad 90^\circ]^{\mathrm{T}}$$

The position vectors of the joint definition points on body i and body j defined in the respective body coordinate systems are given by

$$\bar{\mathbf{u}}_P^i = [0.1 \quad 0.1 \quad 0.25]^{\mathrm{T}} \text{ m}$$
$$\bar{\mathbf{u}}_P^j = [-0.3 \quad -0.15 \quad 0.0]^{\mathrm{T}} \text{ m}$$

Determine the vector $\mathbf{R}^j$ and the Jacobian matrix of the spherical joint constraints.

31. Using Eq. 183 as the starting point, derive Newton–Euler equations.

32. The inertia tensor of a rigid body defined in a centroidal body coordinate system is given by

$$\bar{\mathbf{I}}^i_{\theta\theta} = \begin{bmatrix} 2.0 & 1.5 & -1.0 \\ 1.5 & 1.8 & 0.5 \\ -1.0 & 0.5 & 3.0 \end{bmatrix} \text{kg} \cdot \text{m}^2$$

The body rotates with a constant angular velocity equal to 20 rad/s about its Y^i axis. Determine using Newton–Euler equations the external moments applied to this body.

33. Determine the velocity influence coefficient matrices $\mathbf{D}^i$ and $\mathbf{H}^i$ and the vector $\boldsymbol{\gamma}^i$ of Eq. 269 in the case of the revolute joint.

34. Determine the velocity influence coefficient matrices $\mathbf{D}^i$ and $\mathbf{H}^i$ and the vector $\boldsymbol{\gamma}^i$ of Eq. 269 in the case of prismatic joint.

35. Determine the velocity influence coefficient matrices $\mathbf{D}^i$ and $\mathbf{H}^i$ and the vector $\boldsymbol{\gamma}^i$ of Eq. 269 in the case of spherical joint.

36. Use the principle of virtual work in dynamics to obtain the recursive formulation of Eq. 287.

37. Figure P3 shows a disk, denoted as body 3, which rotates about its X^3 axis with an angle θ^3. The supporting arm, denoted as body 2, rotates with an angle θ^2 about its Z^2 axis. Determine the absolute angular velocity vector of the disk in the fixed $X^1Y^1Z^1$ coordinate system. Determine also the components of this angular velocity vector in the disk coordinate system.

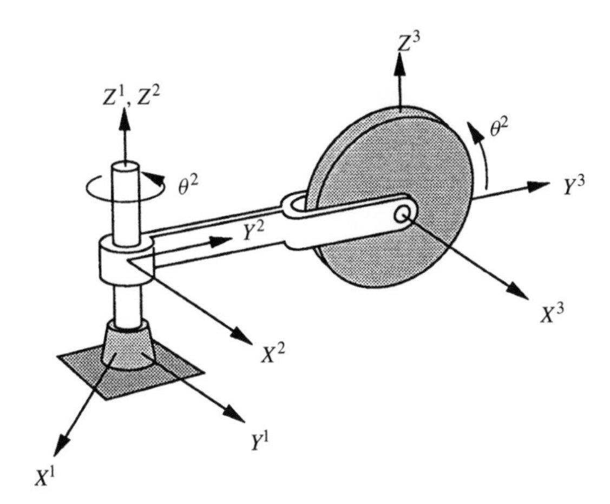

Figure P7.3

38. Obtain the recursive kinematic relationships of the system of Example 11, if body 1 rotates about its X^1 axis with an angle ϕ^1. Assume that the X^1 axis is fixed in space. Determine the absolute angular velocity and acceleration vectors of bodies 2 and 3 in terms of the joint variables ϕ^1, ϕ^2, and ϕ^3 and their time derivatives.

39. In the preceding problem, determine the recursive dynamic equations using the Newton–Euler formulation. Obtain the system of independent differential equations and identify the mass matrix associated with the joint degrees of freedom.

CHAPTER 8

SPECIAL TOPICS IN DYNAMICS

In this chapter, several topics in dynamics are presented. In the first section, the use of Euler angles to study the gyroscopic motion is discussed. In the following sections, several alternative methods for defining the orientations of the rigid bodies in space are presented. In Section 2, the *Rodriguez formula*, which is expressed in terms of the angle of rotation and a unit vector along the axis of rotation, is presented. *Euler parameters*, which are widely used in general-purpose multibody computer programs to avoid the singularities associated with Euler angles, are introduced in Section 3. *Rodriguez parameters* are discussed in Section 4 for the sake of completeness. Euler parameters can be considered as an example of the *quaternions*, which are introduced in Section 5. In Section 6, the problem of nonimpulsive contact between rigid bodies is discussed. A method for the stability and eigenvalue analysis of constrained multibody system is presented in Section 7.

8.1 GYROSCOPES AND EULER ANGLES

The study of the gyroscopic motion is one of the most interesting problems in spatial dynamics. This problem occurs when the orientation of the axis of rotation of a rigid body changes. The gyroscope shown in Fig. 1 consists of a rotor that spins about its axis of rotational symmetry Z^3 which is mounted on a ring called the *inner gimbal*. As shown in the figure, the rotor is free to rotate about its axis of symmetry relative to the inner gimbal, and the inner gimbal rotates freely about the axis X^2, which is perpendicular to the axis of the rotor. The axis X^2 is mounted on a second gimbal, called the *outer gimbal*, which is

Computational Dynamics, Third Edition Ahmed A. Shabana

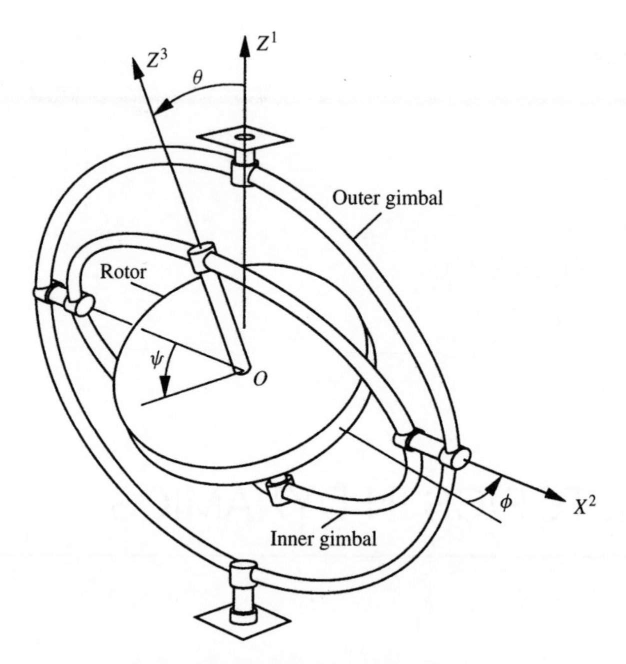

Figure 8.1 Gyroscope

free to rotate about the axis Z^1. The rotor whose center of gravity remains fixed may attain any arbitrary position as shown in Fig. 1 by the following three successive rotations.

1. A rotation ϕ of the outer gimbal about the axis Z^1.
2. A rotation θ of the inner gimbal about the axis X^2.
3. A rotation ψ of the rotor about its own axis Z^3.

These three Euler angles are called the *precession*, the *nutation*, and the *spin*, and the type of mounting used in the gyroscope is called a *cardan suspension*.

The angular velocity of the rotor can be written as

$$\boldsymbol{\omega} = \dot{\phi}\mathbf{k}^1 + \dot{\theta}\mathbf{i}^2 + \dot{\psi}\mathbf{k}^3 \tag{8.1}$$

where $\mathbf{k}^1$ is a unit vector along the Z^1 axis, $\mathbf{i}^2$ is a unit vector along the X^2 axis, and $\mathbf{k}^3$ is a unit vector along the Z^3 axis. The unit vector $\mathbf{k}^1$ is

$$\mathbf{k}^1 = [0 \quad 0 \quad 1]^{\mathrm{T}} \tag{8.2}$$

Since the rotation ϕ is about the Z^1 axis, the unit vector $\mathbf{i}^2$ is defined as

$$\mathbf{i}^2 = \begin{bmatrix} \cos\phi & -\sin\phi & 0 \\ \sin\phi & \cos\phi & 0 \\ 0 & 0 & 1 \end{bmatrix} \begin{bmatrix} 1 \\ 0 \\ 0 \end{bmatrix} = \begin{bmatrix} \cos\phi \\ \sin\phi \\ 0 \end{bmatrix} \tag{8.3}$$

Since the rotation θ is about the X^2 axis, the unit vector $\mathbf{k}^3$ is defined as

$$\mathbf{k}^3 = \begin{bmatrix} \cos\phi & -\sin\phi & 0 \\ \sin\phi & \cos\phi & 0 \\ 0 & 0 & 1 \end{bmatrix} \begin{bmatrix} 1 & 0 & 0 \\ 0 & \cos\theta & -\sin\theta \\ 0 & \sin\theta & \cos\theta \end{bmatrix} \begin{bmatrix} 0 \\ 0 \\ 1 \end{bmatrix}$$
$$= \begin{bmatrix} \sin\phi\sin\theta \\ -\cos\phi\sin\theta \\ \cos\theta \end{bmatrix} \tag{8.4}$$

The angular velocity of the rotor can then be written as

$$\boldsymbol{\omega} = \dot{\phi}\begin{bmatrix} 0 \\ 0 \\ 1 \end{bmatrix} + \dot{\theta}\begin{bmatrix} \cos\phi \\ \sin\phi \\ 0 \end{bmatrix} + \dot{\psi}\begin{bmatrix} \sin\phi\sin\theta \\ -\cos\phi\sin\theta \\ \cos\theta \end{bmatrix} \tag{8.5}$$

which can be written in a matrix form as

$$\boldsymbol{\omega} = \begin{bmatrix} 0 & \cos\phi & \sin\phi\sin\theta \\ 0 & \sin\phi & -\cos\phi\sin\theta \\ 1 & 0 & \cos\theta \end{bmatrix} \begin{bmatrix} \dot{\phi} \\ \dot{\theta} \\ \dot{\psi} \end{bmatrix} \tag{8.6}$$

This equation can be written as

$$\boldsymbol{\omega} = \mathbf{G}\dot{\boldsymbol{\gamma}} \tag{8.7}$$

where $\mathbf{G}$ is the matrix whose columns are the unit vectors $\mathbf{k}^1$, $\mathbf{i}^2$, and $\mathbf{k}^3$. This matrix was previously defined in terms of Euler angles in the preceding chapter as

$$\mathbf{G} = \begin{bmatrix} 0 & \cos\phi & \sin\phi\sin\theta \\ 0 & \sin\phi & -\cos\phi\sin\theta \\ 1 & 0 & \cos\theta \end{bmatrix} \tag{8.8}$$

and

$$\boldsymbol{\gamma} = [\phi \quad \theta \quad \psi]^{\mathrm{T}} \tag{8.9}$$

Differentiating the angular velocity vector with respect to time, one obtains the absolute angular acceleration vector $\boldsymbol{\alpha}$ of the rotor as

$$\boldsymbol{\alpha} = \mathbf{G}\ddot{\boldsymbol{\gamma}} + \dot{\mathbf{G}}\dot{\boldsymbol{\gamma}} \tag{8.10}$$

where $\dot{\mathbf{G}}\dot{\boldsymbol{\gamma}}$ can be written explicitly as

$$\dot{\mathbf{G}}\dot{\boldsymbol{\gamma}} = \begin{bmatrix} -\dot{\phi}\dot{\theta}\sin\phi + \dot{\psi}(\dot{\phi}\cos\phi\sin\theta + \dot{\theta}\sin\phi\cos\theta) \\ \dot{\phi}\dot{\theta}\cos\phi + \dot{\psi}(\dot{\phi}\sin\phi\sin\theta - \dot{\theta}\cos\phi\cos\theta) \\ -\dot{\theta}\dot{\psi}\sin\theta \end{bmatrix} \tag{8.11}$$

The equations of motion of the rotor of the gyroscope can be conveniently derived using *Lagrange's equation*. To this end, we first define the angular velocity vector in the rotor coordinate system as

$$\overline{\boldsymbol{\omega}} = \overline{\mathbf{G}}\dot{\boldsymbol{\gamma}} = \begin{bmatrix} \sin\theta\sin\psi & \cos\psi & 0 \\ \sin\theta\cos\psi & -\sin\psi & 0 \\ \cos\theta & 0 & 1 \end{bmatrix} \begin{bmatrix} \dot{\phi} \\ \dot{\theta} \\ \dot{\psi} \end{bmatrix}$$

$$= \begin{bmatrix} \dot{\phi}\sin\theta\sin\psi + \dot{\theta}\cos\psi \\ \dot{\phi}\sin\theta\cos\psi - \dot{\theta}\sin\psi \\ \dot{\psi} + \dot{\phi}\cos\theta \end{bmatrix} \tag{8.12}$$

Because of the symmetry of the rotor about its X^3 axis, its products of inertia are equal to zero, and $i_{xx} = i_{yy}$. The inertia tensor of the rotor defined in the rotor coordinate system is

$$\overline{\mathbf{I}}_{\theta\theta} = \begin{bmatrix} i_{xx} & 0 & 0 \\ 0 & i_{xx} & 0 \\ 0 & 0 & i_{zz} \end{bmatrix} \tag{8.13}$$

Since the center of mass of the rotor is fixed, the kinetic energy of the rotor is given by

$$T = \tfrac{1}{2}\overline{\boldsymbol{\omega}}^{\mathrm{T}}\overline{\mathbf{I}}_{\theta\theta}\overline{\boldsymbol{\omega}} = \tfrac{1}{2}\{i_{xx}[(\dot{\phi})^2\sin^2\theta + (\dot{\theta})^2] + i_{zz}(\dot{\psi} + \dot{\phi}\cos\theta)^2\} \tag{8.14}$$

Using the Eulerian angles ϕ, θ, and ψ as the generalized coordinates of the rotor, the equations of motion of the rotor are given by

$$\left.\begin{aligned} \frac{d}{dt}\left(\frac{\partial T}{\partial \dot{\phi}}\right) - \frac{\partial T}{\partial \phi} &= M_\phi \\ \frac{d}{dt}\left(\frac{\partial T}{\partial \dot{\theta}}\right) - \frac{\partial T}{\partial \theta} &= M_\theta \\ \frac{d}{dt}\left(\frac{\partial T}{\partial \dot{\psi}}\right) - \frac{\partial T}{\partial \psi} &= M_\psi \end{aligned}\right\} \tag{8.15}$$

where M_ϕ, M_θ and M_ψ are the components of the generalized applied torque associated with the angles ϕ, θ, and ψ, respectively. Using the expression previously obtained for the kinetic energy, one can show that the equations of motion of the rotor are

$$\left.\begin{aligned} \frac{d}{dt}\,[i_{xx}\dot{\phi}\sin^2\theta + i_{zz}(\dot{\psi} + \dot{\phi}\cos\theta)\cos\theta] &= M_\phi \\ i_{xx}\ddot{\theta} - i_{xx}(\dot{\phi})^2\sin\theta\cos\theta + i_{zz}(\dot{\psi} + \dot{\phi}\cos\theta)\dot{\phi}\sin\theta &= M_\theta \\ \frac{d}{dt}\,[i_{zz}(\dot{\psi} + \dot{\phi}\cos\theta)] &= M_\psi \end{aligned}\right\} \tag{8.16}$$

In what follows, several important special cases are discussed.

Ignorable Coordinates It can be seen that the kinetic energy of the rotor is not an explicit function of the precession and spin angles ϕ and ψ, that is,

$$\frac{\partial T}{\partial \phi} = \frac{\partial T}{\partial \psi} = 0 \tag{8.17}$$

If the torques M_ϕ and M_ψ are equal to zero, Lagrange's equation yields

$$\frac{d}{dt}\left(\frac{\partial T}{\partial \dot{\phi}}\right) = 0, \qquad \frac{d}{dt}\left(\frac{\partial T}{\partial \dot{\psi}}\right) = 0 \tag{8.18}$$

or

$$\left(\frac{\partial T}{\partial \dot{\phi}}\right) = d_1, \qquad \left(\frac{\partial T}{\partial \dot{\psi}}\right) = d_2 \tag{8.19}$$

where d_1 and d_2 are constants. In this special case, the precession and spin angles ϕ and ψ are called *ignorable coordinates* since the generalized momentum associated with these coordinates is conserved. The use of the preceding equations yields the following integrals of motion:

$$\left.\begin{aligned} i_{xx}\dot{\phi}\sin^2\theta + i_{zz}(\dot{\psi} + \dot{\phi}\cos\theta)\cos\theta = d_1 \\ i_{zz}(\dot{\psi} + \dot{\phi}\cos\theta) = d_2 \end{aligned}\right\} \tag{8.20}$$

where the constants d_1 and d_2 can be determined using the initial conditions. In this special case, the equation for the nutation angle θ can be written as

$$i_{xx}[\ddot{\theta} - (\dot{\phi})^2\sin\theta\cos\theta] + d_2\dot{\phi}\sin\theta = M_\theta \tag{8.21}$$

Precession at a Steady Rate Consider the special case in which the rotor precesses at a steady rate $\dot{\phi}$ at a constant angle θ and with a constant spin velocity $\dot{\psi}$. These conditions can be expressed as

$$\left.\begin{aligned} \dot{\phi} &= \text{constant}, & \ddot{\phi} &= 0 \\ \theta &= \text{constant}, & \dot{\theta} &= \ddot{\theta} = 0 \\ \dot{\psi} &= \text{constant}, & \ddot{\psi} &= 0 \end{aligned}\right\} \tag{8.22}$$

In this case, the equations of the rotor reduce to

$$\left.\begin{aligned} &0 &&= M_\phi \\ &(\dot{\phi})^2(i_{zz} - i_{xx})\sin\theta\cos\theta + i_{zz}\dot{\phi}\dot{\psi}\sin\theta &&= M_\theta \\ &0 &&= M_\psi \end{aligned}\right\} \tag{8.23}$$

It is clear that in this case $M_\phi = M_\psi = 0$, and the only nonzero torque acting on the rotor is the constant torque associated with the nutation angle θ. The axis of this torque is perpendicular to the axis of precession. If we further assume that $\theta = \pi/2$, the torque M_θ takes the simple form $M_\theta = \dot{\phi}\dot{\psi}i_{zz}$.

8.2 RODRIGUEZ FORMULA

Euler's theorem states that the most general displacement of a body with one point fixed is a rotation about an axis called the instantaneous axis of rotation. According to this theorem, the coordinate transformation can be defined by a single rotation about the instantaneous axis of rotation. The components of a unit vector along the instantaneous axis of rotation as well as the angle of rotation of the rigid body about this axis can be used to develop the transformation matrix that defines the body orientation. The obtained transformation matrix, in this case, is expressed in terms of four parameters; the three components of the unit vector along the instantaneous axis of rotation and the angle of rotation.

Figure 2 shows the initial position of a vector $\bar{\mathbf{r}}^i$ on the rigid body i. The final position of the vector $\bar{\mathbf{r}}^i$ as the result of a rotation θ^i about an axis of rotation defined by the unit vector $\mathbf{v}^i = [v_1^i \quad v_2^i \quad v_3^i]^{\mathrm{T}}$ is defined by the vector $\mathbf{r}^i$. It is clear from Fig. 2 that the vector $\mathbf{r}^i$ can be expressed as

$$\mathbf{r}^i = \bar{\mathbf{r}}^i + \Delta\bar{\mathbf{r}}^i \tag{8.24}$$

It can be shown that (Shabana, 1998)

$$\Delta\bar{\mathbf{r}}^i = (\mathbf{v}^i \times \bar{\mathbf{r}}^i)\sin\theta^i + 2[\mathbf{v}^i \times (\mathbf{v}^i \times \bar{\mathbf{r}}^i)]\sin^2\frac{\theta^i}{2} \tag{8.25}$$

It follows that

$$\mathbf{r}^i = \bar{\mathbf{r}}^i + (\mathbf{v}^i \times \bar{\mathbf{r}}^i)\sin\theta^i + 2[\mathbf{v}^i \times (\mathbf{v}^i \times \bar{\mathbf{r}}^i)]\sin^2\frac{\theta^i}{2} \tag{8.26}$$

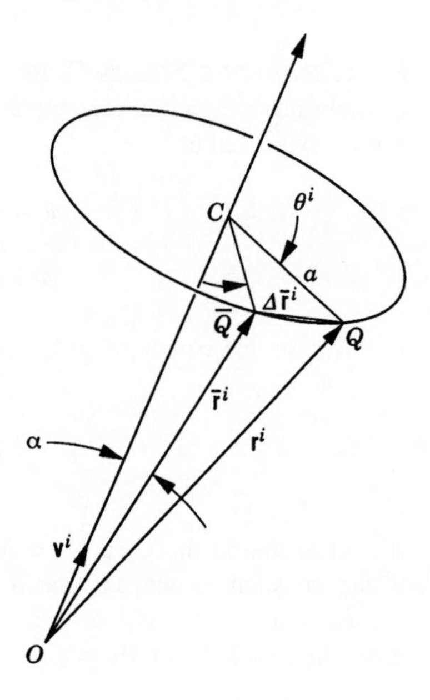

Figure 8.2 Rodriguez formula

Using the skew-symmetric matrix notation, the vector $\mathbf{r}^i$ can be written as

$$\mathbf{r}^i = \bar{\mathbf{r}}^i + \tilde{\mathbf{v}}^i \bar{\mathbf{r}}^i \sin\theta^i + 2(\tilde{\mathbf{v}}^i)^2 \bar{\mathbf{r}}^i \sin^2(\theta^i/2) \tag{8.27}$$

which can also be written as

$$\mathbf{r}^i = \mathbf{A}^i \bar{\mathbf{r}}^i \tag{8.28}$$

where $\mathbf{A}^i$ is the transformation matrix defined in terms of the unit vector $\mathbf{v}^i$ and the rotation angle θ^i as

$$\mathbf{A}^i = \mathbf{I} + \tilde{\mathbf{v}}^i \sin\theta^i + 2(\tilde{\mathbf{v}}^i)^2 \sin^2(\theta^i/2) \tag{8.29}$$

This equation, which is called *Rodriguez formula*, is expressed in terms of the four parameters v_1^i, v_2^i, v_3^i, and θ^i which are not totally independent since

$$\mathbf{v}^{i^{\mathrm{T}}} \mathbf{v}^i = (v_1^i)^2 + (v_2^i)^2 + (v_3^i)^2 = 1 \tag{8.30}$$

Angular Velocity The components of the angular velocity vector $\boldsymbol{\omega}^i$ of the rigid body i can be defined using the relationship,

$$\tilde{\boldsymbol{\omega}}^i = \dot{\mathbf{A}}^i \mathbf{A}^{i^{\mathrm{T}}} \tag{8.31}$$

from which

$$\boldsymbol{\omega}^i = 2\mathbf{v}^i \times \dot{\mathbf{v}}^i \sin^2(\theta^i/2) + \dot{\mathbf{v}}^i \sin\theta^i + \dot{\theta}^i \mathbf{v}^i \tag{8.32}$$

which can be expressed in a matrix form as

$$\boldsymbol{\omega}^i = \mathbf{G}^i \dot{\boldsymbol{\theta}}^i \tag{8.33}$$

where

$$\left.\begin{aligned} \boldsymbol{\theta}^i &= [\mathbf{v}^{i^{\mathrm{T}}} \quad \theta^i]^{\mathrm{T}} = [v_1^i \quad v_2^i \quad v_3^i \quad \theta^i]^{\mathrm{T}} \\ \mathbf{G}^i &= \left[\left(\mathbf{I}\sin\theta^i + 2\tilde{\mathbf{v}}^i \sin^2\frac{\theta^i}{2}\right) \quad \mathbf{v}^i\right] \end{aligned}\right\} \tag{8.34}$$

Similarly, the components of the angular velocity vector of the rigid body i defined in the body coordinate system can be obtained using the relationship

$$\tilde{\bar{\boldsymbol{\omega}}}^i = \mathbf{A}^{i^{\mathrm{T}}} \dot{\mathbf{A}}^i, \tag{8.35}$$

from which

$$\bar{\boldsymbol{\omega}}^i = -2\mathbf{v}^i \times \dot{\mathbf{v}}^i \sin^2(\theta^i/2) + \dot{\mathbf{v}}^i \sin\theta^i + \dot{\theta}^i \mathbf{v}^i \tag{8.36}$$

which can also be written in a matrix form as

$$\overline{\boldsymbol{\omega}}^i = \overline{\mathbf{G}}^i \dot{\boldsymbol{\theta}}^i \tag{8.37}$$

where

$$\overline{\mathbf{G}}^i = \left[\left(\mathbf{I} \sin\theta^i - 2\tilde{\mathbf{v}}^i \sin^2 \frac{\theta^i}{2} \right) \quad \mathbf{v}^i \right] \tag{8.38}$$

Equations 32 and 36 clearly demonstrate that the magnitude of the angular velocity is $\dot{\theta}^i$ only if the axis of rotation is fixed in space. If the axis of rotation is not fixed, the angular velocity vector has two additional components along the two perpendicular vectors $\dot{\mathbf{v}}^i$ and $\mathbf{v}^i \times \dot{\mathbf{v}}^i$.

Application of Rodriguez Formula If the axis of rotation is parallel to one of the axes of the coordinate system, the use of Rodriguez formula leads to the definition of the simple rotation matrices defined in the preceding chapter. Interestingly, Rodriguez formula suggests that there are two different procedures for solving the problem of successive rotations. In order to illustrate the use of these two different procedures, we consider the robotic manipulator shown in Fig. 3. It is clear from this figure that the orientation of object 4 with

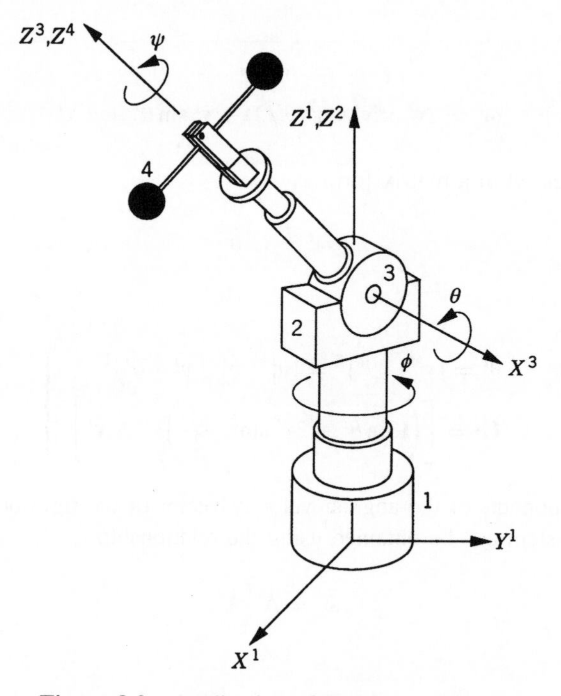

Figure 8.3 Application of Rodriguez formula

respect to link 3 is defined by the transformation matrix

$$\mathbf{A}^{43} = \begin{bmatrix} \cos\psi & -\sin\psi & 0 \\ \sin\psi & \cos\psi & 0 \\ 0 & 0 & 1 \end{bmatrix} \tag{8.39}$$

The orientation of link 3 with respect to link 2 is defined by the transformation matrix

$$\mathbf{A}^{32} = \begin{bmatrix} 1 & 0 & 0 \\ 0 & \cos\theta & -\sin\theta \\ 0 & \sin\theta & \cos\theta \end{bmatrix} \tag{8.40}$$

and the orientation of link 2 with respect to link 1 is

$$\mathbf{A}^{21} = \begin{bmatrix} \cos\phi & -\sin\phi & 0 \\ \sin\phi & \cos\phi & 0 \\ 0 & 0 & 1 \end{bmatrix} \tag{8.41}$$

The transformation matrix that defines the orientation of object 4 with respect to the fixed link (link 1) can then be written as

$$\mathbf{A}^{41} = \mathbf{A}^{21}\mathbf{A}^{32}\mathbf{A}^{43} \tag{8.42}$$

Matrix multiplications show that the matrix $\mathbf{A}^{41}$ takes the same form as the Euler angle transformation matrix given in the preceding chapter.

Using the Rodriguez formula, a different procedure can be used to define the orientation of object 4. In this procedure, the joint axes of rotations are defined in the fixed coordinate system as

$$\left.\begin{aligned} \mathbf{v}^1 &= [0 \quad 0 \quad 1]^{\mathrm{T}} \\ \mathbf{v}^2 &= [\cos\phi \quad \sin\phi \quad 0]^{\mathrm{T}} \\ \mathbf{v}^3 &= [\sin\phi\sin\theta \quad -\cos\phi\sin\theta \quad \cos\theta]^{\mathrm{T}} \end{aligned}\right\} \tag{8.43}$$

Using these three unit vectors, the following three transformation matrices are defined:

$$\left.\begin{aligned} \mathbf{A}^1 &= \mathbf{I} + \tilde{\mathbf{v}}^1 \sin\phi + 2(\tilde{\mathbf{v}}^1)^2 \sin^2\frac{\phi}{2} \\ \mathbf{A}^2 &= \mathbf{I} + \tilde{\mathbf{v}}^2 \sin\theta + 2(\tilde{\mathbf{v}}^2)^2 \sin^2\frac{\theta}{2} \\ \mathbf{A}^3 &= \mathbf{I} + \tilde{\mathbf{v}}^3 \sin\psi + 2(\tilde{\mathbf{v}}^3)^2 \sin^2\frac{\psi}{2} \end{aligned}\right\} \tag{8.44}$$

It is left to the reader as an exercise to show that the transformation matrix $\mathbf{A}^{41}$ can also be defined as

$$\mathbf{A}^{41} = \mathbf{A}^3\mathbf{A}^2\mathbf{A}^1 \tag{8.45}$$

This sequence of transformations differs from the sequence discussed previously in the sense that after each rotation the orientation of the body is redefined in the fixed coordinate system using Rodriguez formula (Shabana, 1998).

8.3 EULER PARAMETERS

An alternative set of four parameters that can be used to describe the orientation of the rigid bodies is the set of *Euler parameters*. The four Euler parameters can be expressed in terms of the components of the unit vector along the instantaneous axis of rotation as well as the angle of rotation as

$$\left.\begin{aligned} \beta_0^i &= \cos(\theta^i/2) \\ \beta_1^i &= v_1^i \sin(\theta^i/2) \\ \beta_2^i &= v_2^i \sin(\theta^i/2) \\ \beta_3^i &= v_3^i \sin(\theta^i/2) \end{aligned}\right\} \tag{8.46}$$

The four Euler parameters are not totally independent since

$$\sum_{k=0}^{3} (\beta_k^i)^2 = 1 \tag{8.47}$$

Using Rodriguez formula, the transformation matrix $\mathbf{A}^i$ can be expressed in terms of Euler parameters as

$$\mathbf{A}^i = \mathbf{I} + 2\tilde{\boldsymbol{\beta}}_s^i(\beta_0^i \mathbf{I} + \tilde{\boldsymbol{\beta}}_s^i) \tag{8.48}$$

where $\tilde{\boldsymbol{\beta}}_s^i$ is the skew-symmetric matrix associated with the vector

$$\boldsymbol{\beta}_s^i = [\beta_1^i \quad \beta_2^i \quad \beta_3^i]^{\mathrm{T}} \tag{8.49}$$

The transformation matrix can be written explicitly in terms of Euler parameters as

$$\mathbf{A}^i = \begin{bmatrix} 1 - 2(\beta_2^i)^2 - 2(\beta_3^i)^2 & 2(\beta_1^i\beta_2^i - \beta_0^i\beta_3^i) & 2(\beta_1^i\beta_3^i + \beta_0^i\beta_2^i) \\ 2(\beta_1^i\beta_2^i + \beta_0^i\beta_3^i) & 1 - 2(\beta_1^i)^2 - 2(\beta_3^i)^2 & 2(\beta_2^i\beta_3^i - \beta_0^i\beta_1^i) \\ 2(\beta_1^i\beta_3^i - \beta_0^i\beta_2^i) & 2(\beta_2^i\beta_3^i + \beta_0^i\beta_1^i) & 1 - 2(\beta_1^i)^2 - 2(\beta_2^i)^2 \end{bmatrix} \tag{8.50}$$

In terms of Euler parameters, the angular velocity vector of the rigid body i is defined as

$$\boldsymbol{\omega}^i = \mathbf{G}^i \dot{\boldsymbol{\beta}}^i \tag{8.51}$$

where

$$\left.\begin{aligned} \boldsymbol{\beta}^i &= [\beta_0^i \quad \beta_1^i \quad \beta_2^i \quad \beta_3^i]^{\mathrm{T}} \\ \mathbf{G}^i &= 2\begin{bmatrix} -\beta_1^i & \beta_0^i & -\beta_3^i & \beta_2^i \\ -\beta_2^i & \beta_3^i & \beta_0^i & -\beta_1^i \\ -\beta_3^i & -\beta_2^i & \beta_1^i & \beta_0^i \end{bmatrix} \end{aligned}\right\} \tag{8.52}$$

The angular velocity vector defined in the body coordinate system is

$$\bar{\boldsymbol{\omega}}^i = \bar{\mathbf{G}}^i \dot{\boldsymbol{\beta}}^i \tag{8.53}$$

where

$$\bar{\mathbf{G}}^i = 2\begin{bmatrix} -\beta_1^i & \beta_0^i & \beta_3^i & -\beta_2^i \\ -\beta_2^i & -\beta_3^i & \beta_0^i & \beta_1^i \\ -\beta_3^i & \beta_2^i & -\beta_1^i & \beta_0^i \end{bmatrix} \tag{8.54}$$

The transformation matrix $\mathbf{A}^i$ can be expressed in terms of the matrices $\mathbf{G}^i$ and $\bar{\mathbf{G}}^i$ as

$$\mathbf{A}^i = \tfrac{1}{4}\mathbf{G}^i\bar{\mathbf{G}}^{i\mathrm{T}} \tag{8.55}$$

Note also that

$$\left.\begin{aligned} \mathbf{G}^i\mathbf{G}^{i\mathrm{T}} &= \bar{\mathbf{G}}^i\bar{\mathbf{G}}^{i\mathrm{T}} = 4\mathbf{I} \\ \mathbf{G}^{i\mathrm{T}}\mathbf{G}^i &= \bar{\mathbf{G}}^{i\mathrm{T}}\bar{\mathbf{G}}^i = 4(\mathbf{I}_4 - \boldsymbol{\beta}^i\boldsymbol{\beta}^{i\mathrm{T}}) \\ \mathbf{G}^i\dot{\boldsymbol{\beta}}^i &= \bar{\mathbf{G}}^i\dot{\boldsymbol{\beta}}^i = \mathbf{0} \\ \boldsymbol{\beta}^{i\mathrm{T}}\dot{\boldsymbol{\beta}}^i &= 0 \end{aligned}\right\} \tag{8.56}$$

where $\mathbf{I}_4$ is the 4×4 identity matrix. Differentiating Eqs. 51 and 53 and using the identities of Eq. 56, it can be shown that the angular acceleration vectors can be expressed in terms of Euler parameters as

$$\boldsymbol{\alpha}^i = \mathbf{G}^i\ddot{\boldsymbol{\beta}}^i, \quad \bar{\boldsymbol{\alpha}}^i = \bar{\mathbf{G}}^i\ddot{\boldsymbol{\beta}}^i \tag{8.57}$$

Clearly, Euler parameters have one redundant variable since they are related by Eq. 47. Nonetheless, Euler parameters are used in several general-purpose multibody computer programs because they do not suffer from the singularity problem associated with the three-parameter representation. Euler parameters are also bounded because they are defined in terms of sine and cosine functions. Moreover, the time derivatives of Euler parameters can be determined at any configuration of the rigid body, if the angular velocity vector is given. This is demonstrated by the following example.

Example 8.1

The orientation of a rigid body is defined by the four Euler parameters

$$\beta_0^i = 0.9239, \qquad \beta_1^i = \beta_2^i = \beta_3^i = 0.2209$$

At the given configuration, the body has an instantaneous angular velocity defined in the global coordinate system by the vector

$$\boldsymbol{\omega}^i = [120.72 \quad 75.87 \quad -46.59]^{\mathrm{T}}\ \mathrm{rad/s}$$

Find the time derivatives of Euler parameters.

Solution. Using Eq. 51, one has

$$\boldsymbol{\omega}^i = \mathbf{G}^i \dot{\boldsymbol{\beta}}^i$$

Multiplying both sides of this equation by $\mathbf{G}^{i^{\mathrm{T}}}$ and using the identities of Eq. 56, one obtains

$$\mathbf{G}^{i^{\mathrm{T}}} \boldsymbol{\omega}^i = \mathbf{G}^{i^{\mathrm{T}}} \mathbf{G}^i \dot{\boldsymbol{\beta}}^i = 4(\mathbf{I}_4 - \boldsymbol{\beta}^i \boldsymbol{\beta}^{i^{\mathrm{T}}}) \dot{\boldsymbol{\beta}}^i = 4\dot{\boldsymbol{\beta}}^i$$

which yields

$$\dot{\boldsymbol{\beta}}^i = \frac{1}{4} \mathbf{G}^{i^{\mathrm{T}}} \boldsymbol{\omega}^i = \frac{1}{4} \begin{bmatrix} -0.4418 & -0.4418 & -0.4418 \\ 1.8478 & 0.4418 & -0.4418 \\ -0.4418 & 1.8478 & 0.4418 \\ 0.4418 & -0.4418 & 1.8478 \end{bmatrix} \cdot \begin{bmatrix} 120.72 \\ 75.87 \\ -46.59 \end{bmatrix} = \begin{bmatrix} -16.5675 \\ 69.2923 \\ 16.5688 \\ -16.5686 \end{bmatrix}$$

This solution must satisfy the identity

$$\boldsymbol{\beta}^{i^{\mathrm{T}}} \dot{\boldsymbol{\beta}}^i = 0$$

Using Euler parameters given in this example, one can show that the transformation matrix $\mathbf{A}^i$ is

$$\mathbf{A}^i = \begin{bmatrix} 0.8048 & -0.3106 & 0.5058 \\ 0.5058 & 0.8048 & -0.3106 \\ -0.3106 & 0.5058 & 0.8048 \end{bmatrix}$$

Using this transformation matrix, it can be shown that the angular velocity vector defined in the body coordinate system is

$$\bar{\boldsymbol{\omega}}^i = \mathbf{A}^{i^{\mathrm{T}}} \boldsymbol{\omega}^i = [150 \quad 0.0 \quad 0.0]^{\mathrm{T}}$$

The time derivatives of Euler parameters can also be evaluated using the relationship

$$\dot{\boldsymbol{\beta}}^i = \frac{1}{4} \bar{\mathbf{G}}^{i^{\mathrm{T}}} \bar{\boldsymbol{\omega}}^i$$

8.4 RODRIGUEZ PARAMETERS

In Rodriguez formula and in the case of Euler parameters, the transformation matrix is expressed in terms of four parameters. It was demonstrated in the preceding chapter that the transformation matrix can be expressed in terms of the three independent Euler angles. In what follows, an alternate representation that uses three parameters called *Rodriguez parameters*, is developed. The three Rodriguez parameters are defined in terms of the components of a unit vector along the axis of rotation and the angle of rotation as

$$\boldsymbol{\gamma}^i = [\gamma_1^i \quad \gamma_2^i \quad \gamma_3^i]^{\mathrm{T}} = \mathbf{v}^i \tan(\theta^i/2) \tag{8.58}$$

That is

$$\gamma_1^i = v_1^i \tan(\theta^i/2), \quad \gamma_2^i = v_2^i \tan(\theta^i/2), \quad \gamma_3^i = v_3^i \tan(\theta^i/2) \tag{8.59}$$

Using the definitions of the preceding equation and the Rodriguez formula of Eq. 29, one obtains the transformation matrix expressed in terms of Rodriguez parameters as

$$\mathbf{A}^i = \mathbf{I} + \frac{2}{1 + (\gamma^i)^2} [\tilde{\boldsymbol{\gamma}}^i + (\tilde{\boldsymbol{\gamma}}^i)^2] \tag{8.60}$$

where $\tilde{\boldsymbol{\gamma}}^i$ is the skew symmetric matrix associated with the vector $\boldsymbol{\gamma}^i$ and

$$\gamma^i = \sqrt{\boldsymbol{\gamma}^{i\mathrm{T}} \boldsymbol{\gamma}^i} \tag{8.61}$$

The angular velocity vectors can be expressed in terms of Rodriguez parameters as

$$\boldsymbol{\omega}^i = \mathbf{G}^i \dot{\boldsymbol{\gamma}}^i, \quad \overline{\boldsymbol{\omega}}^i = \overline{\mathbf{G}}^i \dot{\boldsymbol{\gamma}}^i \tag{8.62}$$

where

$$\mathbf{G}^i = \frac{2}{1 + (\gamma^i)^2} \begin{bmatrix} 1 & -\gamma_3^i & \gamma_2^i \\ \gamma_3^i & 1 & -\gamma_1^i \\ -\gamma_2^i & \gamma_1^i & 1 \end{bmatrix} \tag{8.63}$$

and

$$\overline{\mathbf{G}}^i = \frac{2}{1 + (\gamma^i)^2} \begin{bmatrix} 1 & \gamma_3^i & -\gamma_2^i \\ -\gamma_3^i & 1 & \gamma_1^i \\ \gamma_2^i & -\gamma_1^i & 1 \end{bmatrix} \tag{8.64}$$

Example 8.2

The orientation of a rigid body is defined by the four Euler parameters

$$\beta_0^i = 0.9239, \qquad \beta_1^i = \beta_2^i = \beta_3^i = 0.2209$$

At the given configuration, the body has an instantaneous absolute angular velocity defined by the vector

$$\boldsymbol{\omega}^i = [120.72 \quad 75.87 \quad -46.59]^{\mathrm{T}} \text{ rad/s}$$

Find the time derivatives of Rodriguez parameters.

Solution. Rodriguez parameters at the given orientation are

$$\gamma_1^i = \frac{\beta_1^i}{\beta_0^i}, \qquad \gamma_2^i = \frac{\beta_2^i}{\beta_0^i}, \qquad \gamma_3^i = \frac{\beta_3^i}{\beta_0^i}$$

Using the values of Euler parameters given in this example, one has

$$\gamma_1^i = \gamma_2^i = \gamma_3^i = \frac{0.2209}{0.9239} = 0.2391$$

Recall that

$$\boldsymbol{\omega}^i = \mathbf{G}^i \dot{\boldsymbol{\gamma}}^i$$

where

$$\mathbf{G}^i = \frac{2}{1 + (\gamma^i)^2} \begin{bmatrix} 1 & -\gamma_3^i & \gamma_2^i \\ \gamma_3^i & 1 & -\gamma_1^i \\ -\gamma_2^i & \gamma_1^i & 1 \end{bmatrix}$$

The time derivatives of Rodriguez parameters can then be expressed in terms of the angular velocity as

$$\dot{\boldsymbol{\gamma}}^i = (\mathbf{G}^i)^{-1} \boldsymbol{\omega}^i$$

where

$$(\mathbf{G}^i)^{-1} = \tfrac{1}{2} \begin{bmatrix} 1 + (\gamma_1^i)^2 & \gamma_1^i\gamma_2^i + \gamma_3^i & \gamma_1^i\gamma_3^i - \gamma_2^i \\ \gamma_1^i\gamma_2^i - \gamma_3^i & 1 + (\gamma_2^i)^2 & \gamma_2^i\gamma_3^i + \gamma_1^i \\ \gamma_1^i\gamma_3^i + \gamma_2^i & \gamma_2^i\gamma_3^i - \gamma_1^i & 1 + (\gamma_3^i)^2 \end{bmatrix}$$

$$= \begin{bmatrix} 0.5286 & 0.1481 & -0.0910 \\ -0.0910 & 0.5286 & 0.1481 \\ 0.1481 & -0.0910 & 0.5286 \end{bmatrix}$$

The time derivatives of Rodriguez parameters are

$$\dot{\boldsymbol{\gamma}}^i = \begin{bmatrix} \dot{\gamma}_1^i \\ \dot{\gamma}_2^i \\ \dot{\gamma}_3^i \end{bmatrix} = (\mathbf{G}^i)^{-1}\boldsymbol{\omega}^i = \begin{bmatrix} 0.5286 & 0.1481 & -0.0910 \\ -0.0910 & 0.5286 & 0.1481 \\ 0.1481 & -0.0910 & 0.5286 \end{bmatrix} \begin{bmatrix} 120.72 \\ 75.87 \\ -46.59 \end{bmatrix}$$

$$= \begin{bmatrix} 79.2886 \\ 22.2194 \\ -13.6530 \end{bmatrix}$$

8.5 QUATERNIONS

Quaternion algebra can be used to study the rotations of the rigid bodies in space. This algebra can serve as a convenient way for describing the relative rotations between different bodies. For instance, Euler parameters discussed previously in this chapter can be considered as an example of the quaternions. Before demonstrating the use of the quaternions to describe the three-dimensional rotations, a brief introduction to quaternion algebra is presented (Megahed, 1993).

Quaternion Algebra A quaternion q_t is defined using a scalar and a vector as follows:

$$q_t = s + v_1\mathbf{i} + v_2\mathbf{j} + v_3\mathbf{k} = s + \mathbf{v} \tag{8.65}$$

where s is a scalar, $\mathbf{i}$, $\mathbf{j}$, and $\mathbf{k}$ are unit vectors along three perpendicular Cartesian axes, and $\mathbf{v}$ is the vector which is defined as

$$\mathbf{v} = [v_1 \quad v_2 \quad v_3]^{\mathrm{T}} \tag{8.66}$$

Given two quaternion q_{t1} and q_{t2} defined as

$$q_{t1} = s_1 + \mathbf{v}_1, \qquad q_{t2} = s_2 + \mathbf{v}_2 \tag{8.67}$$

quaternion addition and subtraction follows the rule

$$q_t = q_{t1} \pm q_{t2} = (s_1 \pm s_2) + (\mathbf{v}_1 \pm \mathbf{v}_2) \tag{8.68}$$

The multiplication of the quaternions is noncommutative and follows the rule

$$\begin{aligned} q_t &= q_{t1}q_{t2} = (s_1 + \mathbf{v}_1)(s_2 + \mathbf{v}_2) \\ &= s_1s_2 + s_1\mathbf{v}_2 + s_2\mathbf{v}_1 + \mathbf{v}_1 \times \mathbf{v}_2 - \mathbf{v}_1 \cdot \mathbf{v}_2 \\ &= (s_1s_2 - \mathbf{v}_1 \cdot \mathbf{v}_2) + (s_1\mathbf{v}_2 + s_2\mathbf{v}_1 + \mathbf{v}_1 \times \mathbf{v}_2) \\ &= s + \mathbf{v} \end{aligned} \tag{8.69}$$

where $(\cdot)$ indicates a dot product, $\times$ indicates a cross product, and the scalar s and the vector $\mathbf{v}$ are defined as

$$\left.\begin{aligned} s &= s_1s_2 - \mathbf{v}_1 \cdot \mathbf{v}_2 \\ \mathbf{v} &= s_1\mathbf{v}_2 + s_2\mathbf{v}_1 + \mathbf{v}_1 \times \mathbf{v}_2 \end{aligned}\right\} \tag{8.70}$$

The *conjugate* of the quaternion q_t is defined as

$$q_t^* = s - \mathbf{v} \tag{8.71}$$

The *norm* of the quaternion q_t is defined as

$$|q_t| = \sqrt{q_t q_t^*} \tag{8.72}$$

Using the rule of multiplication of the quaternions, it can be shown that

$$|q_t| = \sqrt{q_t q_t^*} = \sqrt{(s)^2 + \mathbf{v} \cdot \mathbf{v}} \tag{8.73}$$

If the scalar part of the quaternion is equal to zero, the quaternion norm reduces to the definition of the length of vectors used in conventional vector algebra.

Three-Dimensional Rotations The set of Euler parameters can be considered as a quaternion with a unit norm. This fact can easily be demonstrated by writing the set of Euler parameters of a rigid body or a frame of reference i in the following quaternion form:

$$q_t^i = \beta_0^i + \boldsymbol{\beta}_s^i \tag{8.74}$$

where the scalar and vector components of this quaternion are defined in Section 3. The conjugate of the quaternion defined in the preceding equation is

$$q_t^{i*} = \beta_0^i - \boldsymbol{\beta}_s^i \tag{8.75}$$

and its norm is

$$|q_t^i| = \sqrt{q_t^i q_t^{i*}} = \sqrt{(\beta_0^i)^2 + \boldsymbol{\beta}_s^i \cdot \boldsymbol{\beta}_s^i} = 1 \tag{8.76}$$

If $\mathbf{A}^1$, $\mathbf{A}^2$, and $\mathbf{A}^3$ are three orthogonal transformation matrices defined in terms of the sets of Euler parameters $(\beta_0^1, \boldsymbol{\beta}_s^1)$, $(\beta_0^2, \boldsymbol{\beta}_s^2)$, and $(\beta_0^3, \boldsymbol{\beta}_s^3)$, respectively, matrix multiplication and quaternion multiplication are in one-to-one correspondence; that is, if

$$\mathbf{A}^1 = \mathbf{A}^2\mathbf{A}^3 \tag{8.77}$$

then

$$\boldsymbol{\beta}^1 = \boldsymbol{\beta}^2\boldsymbol{\beta}^3 \tag{8.78}$$

or

$$\left.\begin{aligned} \beta_0^1 &= \beta_0^2\beta_0^3 - \boldsymbol{\beta}_s^2 \cdot \boldsymbol{\beta}_s^3 \\ \boldsymbol{\beta}_s^1 &= \beta_0^2\boldsymbol{\beta}_s^3 + \beta_0^3\boldsymbol{\beta}_s^2 + \boldsymbol{\beta}_s^2 \times \boldsymbol{\beta}_s^3 \end{aligned}\right\} \tag{8.79}$$

This important result from quaternion algebra can be used to study the relative motion between rigid bodies or coordinate systems. For instance, let $\mathbf{A}^i$ and $\mathbf{A}^j$ be the transformation matrices that define the orientation of two coordinate systems i and j, respectively, and let

$\mathbf{A}^{ji}$ be the matrix that defines the orientation of the coordinate system j with respect to the coordinate system i. It follows that

$$\mathbf{A}^j = \mathbf{A}^i \mathbf{A}^{ji} \tag{8.80}$$

or

$$\mathbf{A}^{ji} = \mathbf{A}^{i^{\mathrm{T}}} \mathbf{A}^j \tag{8.81}$$

Recall that if $(\beta_0^i, \boldsymbol{\beta}_s^i)$ is the set of Euler parameters of $\mathbf{A}^i$, the set of Euler parameters associated with the transpose of $\mathbf{A}^i$ is $(\beta_0^i, -\boldsymbol{\beta}_s^i)$ (Shabana, 1998). It follows that the set of Euler parameters associated with the relative transformation matrix $\mathbf{A}^{ji}$ can easily be obtained using the quaternion algebra as

$$\left.\begin{aligned} \beta_0^{ji} &= \beta_0^i \beta_0^j + \boldsymbol{\beta}_s^i \cdot \boldsymbol{\beta}_s^j \\ \boldsymbol{\beta}_s^{ji} &= \beta_0^i \boldsymbol{\beta}_s^j - \beta_0^j \boldsymbol{\beta}_s^i - \boldsymbol{\beta}_s^i \times \boldsymbol{\beta}_s^j \end{aligned}\right\} \tag{8.82}$$

As an example that illustrates the use of the preceding equations, we assume that body j rotates with respect to body i with a constant angular velocity ω^{ji}. It follows from the first equation of the preceding set of equations that

$$\cos\left(\frac{\omega^{ji} t}{2}\right) = \beta_0^i \beta_0^j + \beta_1^i \beta_1^j + \beta_2^i \beta_2^j + \beta_3^i \beta_3^j \tag{8.83}$$

Clearly, this constraint equation is of the holonomic type.

Example 8.3

As an example of the use of the quaternion algebra, we consider a rigid body whose orientation is defined at the initial configuration before displacement by the transformation matrix $\mathbf{B}$ expressed in terms of the four Euler parameters β_0, β_1, β_2, and β_3. In this example, we consider three cases of the body rotation with a constant angular velocity. The first is the rotation of the body about the global X axis, the second is the rotation of the body about the global Y axis, and the third is the rotation about the global Z axis. In the three cases, we assume that the axes of rotations are defined in the global system. For simplicity of notation, we drop the superscript that indicates the body number.

Rotation about the Global X Axis: If the body rotates with a constant angular velocity ω about the X axis, the final orientation of the body is defined by the transformation matrix $\mathbf{C}$ given by

$$\mathbf{C} = \mathbf{AB}$$

where $\mathbf{A}$ is the transformation matrix resulting from the rotation of the body about the global X axis with a constant angular velocity. The order of the matrix multiplication used in the preceding equation is due to the fact that the axis of rotation is defined in the global system. It can be shown that the four Euler parameters that define the matrix $\mathbf{A}$ are as follows:

$$\alpha_0 = \cos\frac{\omega t}{2}, \qquad \alpha_1 = \sin\frac{\omega t}{2}, \qquad \alpha_2 = 0, \qquad \alpha_3 = 0$$

Using the quaternion multiplication role, it can be shown that the four Euler parameters γ_0, γ_1, γ_2, and γ_3 that define the transformation matrix **C** are given by

$$\gamma_0 = \alpha_0\beta_0 - \alpha_1\beta_1, \quad \gamma_1 = \alpha_0\beta_1 + \alpha_1\beta_0$$

$$\gamma_2 = \alpha_0\beta_2 - \alpha_1\beta_3, \quad \gamma_3 = \alpha_0\beta_3 + \alpha_1\beta_2$$

Rotation about the Global Y Axis: In the case of the rotation with a constant angular velocity about the Y axis, the four Euler parameters that define the matrix **A** are given by

$$\alpha_0 = \cos\frac{\omega t}{2}, \qquad \alpha_1 = 0, \qquad \alpha_2 = \sin\frac{\omega t}{2}, \qquad \alpha_3 = 0$$

The four Euler parameters that define the matrix **C** are then given by

$$\gamma_0 = \alpha_0\beta_0 - \alpha_2\beta_2, \quad \gamma_1 = \alpha_0\beta_1 + \alpha_2\beta_3$$

$$\gamma_2 = \alpha_0\beta_2 + \alpha_2\beta_0, \quad \gamma_3 = \alpha_0\beta_3 + \alpha_2\beta_1$$

Rotation about the Global Z Axis: If the rotation is about the global Z axis, the four Euler parameters that define the matrix **A** are given by

$$\alpha_0 = \cos\frac{\omega t}{2}, \qquad \alpha_1 = 0, \qquad \alpha_2 = 0, \qquad \alpha_3 = \sin\frac{\omega t}{2}$$

In this case, the four Euler parameters of the matrix **C** that defines the final orientation of the body are as follows:

$$\gamma_0 = \alpha_0\beta_0 - \alpha_3\beta_3, \quad \gamma_1 = \alpha_0\beta_1 - \alpha_3\beta_2$$

$$\gamma_2 = \alpha_0\beta_2 + \alpha_3\beta_1, \quad \gamma_3 = \alpha_0\beta_3 + \alpha_3\beta_0$$

8.6 RIGID BODY CONTACT

The nonimpulsive contact between rigid bodies does not result in instantaneous change in the system velocities and momentum. Examples of this type of nonimpulsive contact are cam and follower contact, wheel and rail contact, disk rolling on a flat surface, and nonimpulsive contact between the end effector of a robot and a surface, among others. In this section, the kinematic equations that describe the nonimpulsive contact between two surfaces of two rigid bodies in the multibody system are formulated in terms of the system generalized coordinates and the surface parameters. Each contact surface is defined using two independent parameters that completely define the tangent and normal vectors at an arbitrary point on the body surface. The surface parameters can be considered as a set of *nongeneralized coordinates* since there are no inertia or external forces associated with them (Shabana and Sany, 2001). In the contact model discussed in this section, the location of the contact points on the two surfaces is determined by solving the nonlinear differential and algebraic equations of the constrained multibody system. The equations of motion of the bodies in contact are developed using the principle of mechanics, and the contact constraint equations are adjoined to the system dynamic equations using the technique of Lagrange multipliers. Lagrange multipliers associated with the contact constraints are used to determine the generalized contact constraint forces.

Parameterization of the Contact Surfaces The two bodies in contact are denoted as bodies i and j. Contact between the two surfaces of the bodies is assumed to occur at point P. Two coordinate systems $X^iY^iZ^i$ and $X^jY^jZ^j$ are introduced for bodies i and j, respectively, as shown in Fig. 4. At a given instant of time, the location of the contact point P with respect to the coordinate systems of the two bodies is defined by the following two vectors:

$$\bar{\mathbf{u}}_P^i = \begin{bmatrix} \bar{x}^i \\ \bar{y}^i \\ \bar{z}^i \end{bmatrix}, \qquad \bar{\mathbf{u}}_P^j = \begin{bmatrix} \bar{x}^j \\ \bar{y}^j \\ \bar{z}^j \end{bmatrix} \tag{8.84}$$

The contact surface of body i is assumed to be defined by the two surface parameters s_1^i and s_2^i, while the contact surface of body j is defined in terms of the surface parameters s_1^j and s_2^j. Therefore, the local position vectors of the contact point, defined in Eq. 84, can be expressed in terms of the surface parameters as follows:

$$\bar{\mathbf{u}}_P^i(s_1^i, s_2^i) = \begin{bmatrix} \bar{x}^i(s_1^i, s_2^i) \\ \bar{y}^i(s_1^i, s_2^i) \\ \bar{z}^i(s_1^i, s_2^i) \end{bmatrix}, \qquad \bar{\mathbf{u}}_P^j(s_1^j, s_2^j) = \begin{bmatrix} \bar{x}^j(s_1^j, s_2^j) \\ \bar{y}^j(s_1^j, s_2^j) \\ \bar{z}^j(s_1^j, s_2^j) \end{bmatrix} \tag{8.85}$$

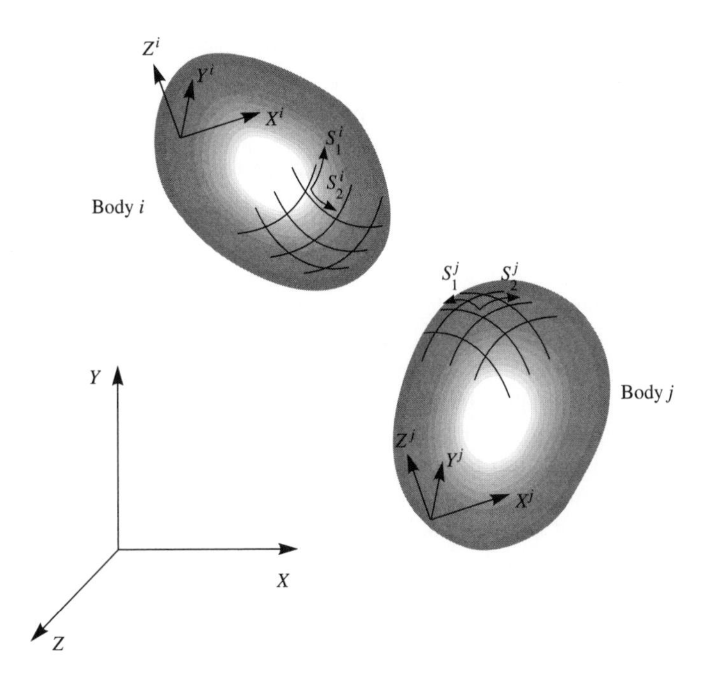

Figure 8.4 Contact surfaces

The tangents to the surface at the contact point on body i are defined in the body coordinate system by the following two independent, not necessarily orthogonal vectors:

$$\bar{\mathbf{t}}_1^i = \frac{\partial \bar{\mathbf{u}}_P^i}{\partial s_1^i}, \qquad \bar{\mathbf{t}}_2^i = \frac{\partial \bar{\mathbf{u}}_P^i}{\partial s_2^i} \tag{8.86}$$

The normal to the surface at the contact point on body i is defined in the body coordinate system as

$$\bar{\mathbf{n}}^i = \bar{\mathbf{t}}_1^i \times \bar{\mathbf{t}}_2^i \tag{8.87}$$

Similarly, the tangents and normal to the surface of body j at the contact point, defined in the body coordinate system, are given by

$$\bar{\mathbf{t}}_1^j = \frac{\partial \bar{\mathbf{u}}_P^j}{\partial s_1^j}, \qquad \bar{\mathbf{t}}_2^j = \frac{\partial \bar{\mathbf{u}}_P^j}{\partial s_2^j}, \qquad \bar{\mathbf{n}}^j = \bar{\mathbf{t}}_1^j \times \bar{\mathbf{t}}_2^j \tag{8.88}$$

The tangent and normal vectors and their first and second derivatives with respect to the surface parameters are required to enforce the contact kinematic constraints discussed below.

Contact Constraints In the multibody contact formulation presented in this section, it is assumed that the body motion is described using absolute Cartesian and orientation coordinates. For an arbitrary body k in the multibody system, the three-dimensional vector of absolute Cartesian coordinates $\mathbf{R}^k$ $(k = i \text{ or } j)$ is used to define the global location of the origin of the kth body coordinate system, while the orientation coordinates $\boldsymbol{\theta}^k$ are used to define the orientation of the body coordinate system. Using this motion description, the global position vectors of the contact point on bodies i and j can be defined, respectively, as follows:

$$\left.\begin{aligned} \mathbf{r}_P^i &= \mathbf{R}^i + \mathbf{A}^i \bar{\mathbf{u}}_P^i \\ \mathbf{r}_P^j &= \mathbf{R}^j + \mathbf{A}^j \bar{\mathbf{u}}_P^j \end{aligned}\right\} \tag{8.89}$$

where $\mathbf{A}^i$ and $\mathbf{A}^j$ are the spatial transformation matrices that define the orientation of bodies i and j, respectively, in a global inertial frame of reference. These matrices are functions of the orientation coordinates $\boldsymbol{\theta}^i$ and $\boldsymbol{\theta}^j$. Using this motion description, each unconstrained body has six independent coordinates. In the general case of contact where slipping is allowed, the contact conditions eliminate only one degree of freedom. Therefore, body i has five degrees of freedom with respect to body j.

To formulate the contact conditions, five constraint equations expressed in terms of the generalized coordinates of the two bodies and the four surface parameters are introduced. These constraint equations impose the following two conditions (Roberson and Schwertassek, 1988; Litvin, 1994):

1. Two points on the two contact surfaces coincide. This implies that the global position coordinates of the contact point P evaluated using the generalized coordinates of

body i are the same as the coordinates of the same point evaluated using the generalized coordinates of body j. This condition can be expressed mathematically as follows:

$$\mathbf{r}_P^i = \mathbf{r}_P^j \tag{8.90}$$

where the two vectors in this equation are defined explicitly in Eq. 89.

2. The normals to the two surfaces at the point of contact are parallel. This condition can be stated mathematically as follows:

$$\mathbf{n}^i = \alpha \mathbf{n}^j \tag{8.91}$$

where α is a constant and $\mathbf{n}^i$ and $\mathbf{n}^j$ are the normals at the contact point defined in the global coordinate system, that is,

$$\mathbf{n}^i = \mathbf{A}^i \bar{\mathbf{n}}^i, \qquad \mathbf{n}^j = \mathbf{A}^j \bar{\mathbf{n}}^j \tag{8.92}$$

It is important to point out that Eq. 91 leads to two independent equations only. An alternative to Eq. 91 is to use the unit normals instead of introducing the constant α. Another alternative to Eq. 91 is to use the tangent vectors to write two independent constraint equations which guarantee that the normals to the two surfaces remain parallel. These equations can be written as follows:

$$\mathbf{n}^{j^{\mathrm{T}}} \mathbf{t}_1^i = 0, \qquad \mathbf{n}^{j^{\mathrm{T}}} \mathbf{t}_2^i = 0 \tag{8.93}$$

where

$$\mathbf{t}_1^i = \mathbf{A}^i \bar{\mathbf{t}}_1^i, \qquad \mathbf{t}_2^i = \mathbf{A}^i \bar{\mathbf{t}}_2^i \tag{8.94}$$

Equations 90 and 91 or, alternatively, Eqs. 90 and 93, represent five independent nonlinear equations which can be used to impose the contact conditions. Note that the normal and tangent vectors in Eq. 94 are expressed in terms of the first derivatives of the local coordinates of the contact point with respect to the surface parameters. Imposing the constraints of Eq. 93 at the acceleration level requires the evaluation of the third partial derivatives of the contact point coordinates with respect to the surface parameters. Note also that the five independent contact constraint equations can be used to identify five dependent variables (including the four surface parameters s_1^i, s_2^i, s_1^j, and s_2^j). Therefore, the surface parameters and one generalized coordinate can be eliminated using a coordinate partitioning scheme and the embedding technique as discussed previously in this book. In the remainder of this section, however, an alternative technique based on the augmented formulation that employs Lagrange multipliers is discussed.

Multibody System Formulation The contact constraint formulation presented in the preceding section can be implemented in general-purpose multibody system algorithms. Recall that these contact constraints are formulated in terms of a mixed set of generalized and nongeneralized surface parameters. Using the principle of virtual work, the following

Lagrange–D'Alembert form can be obtained:

$$\{\mathbf{M}\ddot{\mathbf{q}} - \mathbf{Q}\}^{\mathrm{T}}\delta\mathbf{q} = 0 \tag{8.95}$$

where $\mathbf{M}$ is the system mass matrix, $\mathbf{q}$ is the vector of the system generalized coordinates, and $\mathbf{Q}$ is the vector of all forces acting on the system, excluding the constraint forces, which are eliminated automatically using the virtual work principle. The generalized force vector $\mathbf{Q}$ includes externally applied forces, gravity forces, and spring, damper, and actuator forces as well as friction forces. The constraint equations that describe mechanical joints and specified motion trajectories as well as the contact constraints can be written in the following form:

$$\mathbf{C}(\mathbf{q}, \mathbf{s}, t) = \mathbf{0} \tag{8.96}$$

where $\mathbf{s}$ is the vector of the parameters that describe the geometry of the surfaces in contact. Virtual changes in the system coordinates and the surface parameters, which are consistent with the kinematic constraints, lead to

$$\mathbf{C_q}\delta\mathbf{q} + \mathbf{C_s}\delta\mathbf{s} = \mathbf{0} \tag{8.97}$$

which for an arbitrary nonzero vector $\boldsymbol{\lambda}$ leads to

$$\boldsymbol{\lambda}^{\mathrm{T}}\{\mathbf{C_q}\delta\mathbf{q} + \mathbf{C_s}\delta\mathbf{s}\} = 0 \tag{8.98}$$

Adding Eqs. 95 and 98, one obtains

$$\delta\mathbf{q}^{\mathrm{T}}\{\mathbf{M}\ddot{\mathbf{q}} + \mathbf{C_q^{\mathrm{T}}}\boldsymbol{\lambda} - \mathbf{Q}\} + \delta\mathbf{s}^{\mathrm{T}}\mathbf{C_s^{\mathrm{T}}}\boldsymbol{\lambda} = 0 \tag{8.99}$$

The procedure used to obtain the augmented formulation when only generalized coordinates are used (see Chapter 6) can be generalized for the case of system with nongeneralized coordinates. To demonstrate this, the preceding equation can be written as follows:

$$\delta\mathbf{p}^{\mathrm{T}}(\mathbf{H} + \mathbf{C_p^{\mathrm{T}}}\boldsymbol{\lambda}) = 0 \tag{8.100}$$

where

$$\mathbf{H} = \begin{bmatrix} \mathbf{M}\ddot{\mathbf{q}} - \mathbf{Q} \\ \mathbf{0} \end{bmatrix}, \quad \mathbf{p} = \begin{bmatrix} \mathbf{q} \\ \mathbf{s} \end{bmatrix} \tag{8.101}$$

The vector $\mathbf{p}$, which includes generalized coordinates and nongeneralized surface parameters, can be partitioned into two sets: the set of independent coordinates $\mathbf{p}_i$ and the set of dependent coordinates $\mathbf{p}_d$, that is,

$$\mathbf{p} = \begin{bmatrix} \mathbf{p}_i \\ \mathbf{p}_d \end{bmatrix} \tag{8.102}$$

Each set, independent or dependent, may include nongeneralized surface parameters. According to this coordinate partitioning, Eq. 100 can be written as follows:

$$[\delta \mathbf{p}_i^{\mathrm{T}} \quad \delta \mathbf{p}_d^{\mathrm{T}}] \begin{bmatrix} \mathbf{H}_i + \mathbf{C}_{\mathbf{p}_i}^{\mathrm{T}} \boldsymbol{\lambda} \\ \mathbf{H}_d + \mathbf{C}_{\mathbf{p}_d}^{\mathrm{T}} \boldsymbol{\lambda} \end{bmatrix} = 0 \tag{8.103}$$

Assuming that the constraint sub-Jacobian $\mathbf{C}_{\mathbf{p}_d}$ associated with the dependent generalized and nongeneralized coordinates has a full row rank, the vector of Lagrange multipliers $\boldsymbol{\lambda}$ can be selected to be the solution of the following system of algebraic equations:

$$\mathbf{C}_{\mathbf{p}_d}^{\mathrm{T}} \boldsymbol{\lambda} = -\mathbf{H}_d \tag{8.104}$$

It follows from Eqs. 103 and 104 that

$$\delta \mathbf{p}_i^{\mathrm{T}} (\mathbf{C}_{\mathbf{p}_i}^{\mathrm{T}} \boldsymbol{\lambda} + \mathbf{H}_i) = 0 \tag{8.105}$$

Since the element of the vector $\mathbf{p}_i$ are assumed to be independent, Eq. 105 leads to

$$\mathbf{C}_{\mathbf{p}_i}^{\mathrm{T}} \boldsymbol{\lambda} + \mathbf{H}_i = \mathbf{0} \tag{8.106}$$

Combining Eqs. 104 and 106 and using the definition of $\mathbf{H}$ given by Eq. 101, one obtains

$$\left.\begin{aligned} \mathbf{M}\ddot{\mathbf{q}} + \mathbf{C}_{\mathbf{q}}^{\mathrm{T}} \boldsymbol{\lambda} &= \mathbf{Q} \\ \mathbf{C}_{\mathbf{s}}^{\mathrm{T}} \boldsymbol{\lambda} &= \mathbf{0} \end{aligned}\right\} \tag{8.107}$$

Differentiating the constraints of Eq. 96 twice with respect to time yields

$$\mathbf{C}_{\mathbf{q}} \ddot{\mathbf{q}} + \mathbf{C}_{\mathbf{s}} \ddot{\mathbf{s}} = \mathbf{Q}_d \tag{8.108}$$

where $\mathbf{Q}_d$ is a vector that absorbs terms which are quadratic in the first time derivatives of the coordinates and the surface parameters. Combining Eqs. 107 and 108 yields (Shabana and Sany, 2001)

$$\begin{bmatrix} \mathbf{M} & \mathbf{0} & \mathbf{C}_{\mathbf{q}}^{\mathrm{T}} \\ \mathbf{0} & \mathbf{0} & \mathbf{C}_{\mathbf{s}}^{\mathrm{T}} \\ \mathbf{C}_{\mathbf{q}} & \mathbf{C}_{\mathbf{s}} & \mathbf{0} \end{bmatrix} \begin{bmatrix} \ddot{\mathbf{q}} \\ \ddot{\mathbf{s}} \\ \boldsymbol{\lambda} \end{bmatrix} = \begin{bmatrix} \mathbf{Q} \\ \mathbf{0} \\ \mathbf{Q}_d \end{bmatrix} \tag{8.109}$$

It is important to point out that the sub-Jacobian $\mathbf{C}_{\mathbf{s}}$ associated with the surface parameters must have rank equal to the number of these surface parameters to be able to solve for the second derivatives of the coordinates and surface parameters as well as the vector of Lagrange multipliers. Therefore, the surface profiles must be chosen to guarantee a point contact. In the case of a line contact, such as in the case of a contact between a cylinder and a flat surface, there are infinite number of solutions, and the matrix $\mathbf{C}_{\mathbf{s}}$ has a rank that is not equal to the number of the surface parameters. This special case can also be handled using the algorithm proposed in this chapter by changing the number of equations and variables required to describe the contact.

The case of a point contact is described using five independent constraint equations and four independent surface parameters. Nonetheless, the second equation of Eq. 107 implies that for each contact, there is only one nontrivial solution since the rank of the matrix $\mathbf{C}_s^T$ is four times the number of contacts. Therefore, there is only one independent Lagrange multiplier associated with each contact. That is, there is only one independent generalized contact force, which can easily be visualized in simple contact configurations to be the normal force at the point of contact. The *nonconformal kinematic conditions* defined by Eqs. 90 and 93 do not allow separation or penetration of the two contact surfaces. These conditions when used in multibody system algorithms must be satisfied at the position, velocity, and acceleration levels. Another alternative for using the constraint contact formulation is to use an elastic contact method that employs a compliant force model to describe the interaction between the two contact surfaces. This elastic force model, which can be defined in terms of the penetration and stiffness and damping coefficients, can be introduced to the dynamic equations of motion using generalized applied forces instead of kinematic constraints. When elastic contact methods are used, separation and penetration of the two surfaces in the small contact region are allowed. Therefore, the elastic contact method does not lead to elimination of degrees of freedom as in the case of the constraint contact formulation (Khulief and Shabana, 1987; Shabana et al., 2008).

8.7 STABILITY AND EIGENVALUE ANALYSIS

The equations of motion that govern the dynamics of multibody systems are highly nonlinear as the results of the finite rotations and the nonlinear kinematic constraints imposed on the motion of the bodies. One method that can be used to study the stability of constrained multibody systems is to perform the time domain dynamic simulations and examine the stability of the system using numerical results obtained from the simulation. In addition to the time domain simulations that do not involve linearization of the dynamic equations, another method that can also be used is to linearize the nonlinear dynamic equations at certain configurations (points in time) and obtain the eigen solution that can be used to shed light on the system stability. For a multi-degree-of-freedom system with velocity dependent forces, as it is the case in many multibody system applications; the eigenvalues can be real and/or complex conjugates. As it is known from the theory of vibration, any eigenvalue with a positive real part leads to unstable solution, zero real parts lead to what is known as *critically stable solutions*, whereas negative real parts lead to stable solutions. Because the complete solution of a linear system can be written as a linear combination of mode shapes associated with the eigenvalues, the system becomes unstable if any of the eigenvalues has a positive real part. In order to obtain the eigenvalue solution, the constraint forces and dependent coordinates are eliminated using the embedding technique as discussed in Chapter 6 of this book. In this case, one obtains the following system of equations of motion expressed in terms of the degrees of freedom:

$$\overline{\mathbf{M}}_i(\mathbf{q}_i)\ddot{\mathbf{q}}_i(t) = \overline{\mathbf{Q}}_i(\mathbf{q}_i, \dot{\mathbf{q}}_i, t) \tag{8.110}$$

In this equation $\ddot{\mathbf{q}}_i$ is the vector of system independent coordinates or degrees of freedom, $\overline{\mathbf{M}}_i$ and $\overline{\mathbf{Q}}_i$ are, respectively, the mass matrix and generalized forces associated with these degrees of freedom. Using linearization techniques, one can define the following damping and stiffness matrices at a certain selected point of time during the dynamic simulation:

$$\overline{\mathbf{D}}_i = -\frac{\partial \overline{\mathbf{Q}}_i}{\partial \dot{\mathbf{q}}_i}, \quad \overline{\mathbf{K}}_i = -\frac{\partial \overline{\mathbf{Q}}_i}{\partial \mathbf{q}_i} \tag{8.111}$$

Using these two symmetric matrices and assuming that the mass matrix $\overline{\mathbf{M}}_i$ does not significantly change in the neighborhood of the selected points of time, one can write the following matrix equation that governs the linear vibration of the multibody system at the selected time points:

$$\overline{\mathbf{M}}_i \ddot{\mathbf{q}}_i + \overline{\mathbf{D}}_i \dot{\mathbf{q}}_i + \overline{\mathbf{K}}_i \mathbf{q}_i = \mathbf{0} \tag{8.112}$$

This linearized system of equations has a number of equations n_d equal to the number of the system degrees of freedom. All the matrices that appear in Eq. 112 are real and symmetric matrices.

The general purpose computer code SAMS/2000 described in Chapter 9 allows for solving two eigenvalue problems based on the system given by Eq. 112. In the first eigenvalue problem, the effect of the damping is neglected, that is, $\overline{\mathbf{D}}_i$ is assumed to be zero. The resulting system defines the natural frequencies of the system at the selected configurations. In the second eigenvalue problem, the case of general damping is considered, and the state space formulation is used to determine all the eigenvalues and eigenvectors, which can be real and/or complex conjugates. These two different eigenvalue problems are discussed below.

System Natural Frequencies In order to determine the system natural frequencies at selected points in time, the effect of the damping forces is neglected. In this case, Eq. 112 leads to

$$\overline{\mathbf{M}}_i \ddot{\mathbf{q}}_i + \overline{\mathbf{K}}_i \mathbf{q}_i = \mathbf{0} \tag{8.113}$$

In order to determine the eigenvalues and eigenvectors, one assumes a solution in the form $\mathbf{q}_i = \mathbf{S}e^{st}$, where $\mathbf{S}$ is the vector of amplitudes and s is the frequency. Substituting this assumed solution into Eq. 113, one obtains

$$(\overline{\mathbf{K}}_i - (s)^2 \overline{\mathbf{M}}_i)\mathbf{S} = \mathbf{0} \tag{8.114}$$

This equation defines a *generalized eigenvalue problem*. In order to have a nontrivial solution, one must have the following condition:

$$|\overline{\mathbf{K}}_i - (s)^2 \overline{\mathbf{M}}_i| = 0 \tag{8.115}$$

This determinant defines the characteristic equation which has n_d roots, $(s_1)^2, (s_2)^2, \ldots, (s_{n_d})^2$, called the *eigenvalues*. Associated with each root $(s_k)^2$, there is an eigenvector

$\mathbf{S}_k$, which can be determined to within an arbitrary constant using the following system of homogeneous equations:

$$(\overline{\mathbf{K}}_i - (s_k)^2\overline{\mathbf{M}}_i)\mathbf{S}_k = \mathbf{0} \tag{8.116}$$

Since the mass and stiffness matrices $\overline{\mathbf{M}}_i$ and $\overline{\mathbf{K}}_i$ are real and symmetric, all the eigenvalues and eigenvectors must be real. Furthermore, the eigenvectors are orthogonal. Note that the mass matrix $\overline{\mathbf{M}}_i$ is always positive definite. If the stiffness matrix $\overline{\mathbf{K}}_i$ is positive semidefinite then all the eigenvalues $(s_1)^2, (s_2)^2, \ldots, (s_{n_d})^2$ must be nonnegative.

State Space Formulation In the case of general damping matrix, another procedure is used to determine the eigenvalues and eigenvectors of the multibody system. In this more general case, some of the eigenvalues and eigenvectors can be complex conjugates. Since the coordinates are real, a procedure for defining real mode shapes is described in this section. In the case of general damping forces, the linearized system of Eq. 112 is also used as the starting point. In the state space formulation, one can define the state vector

$$\mathbf{z} = [\,\mathbf{q}_i^{\mathrm{T}} \quad \dot{\mathbf{q}}_i^{\mathrm{T}}\,]^{\mathrm{T}} \tag{8.117}$$

Note that the dimension of the vector $\mathbf{z}$ is $2 \times n_d$. Using Eqs. 112 and 117, one can write the following system of equations:

$$\dot{\mathbf{z}} = \begin{bmatrix} \dot{\mathbf{q}}_i \\ \ddot{\mathbf{q}}_i \end{bmatrix} = \begin{bmatrix} \mathbf{0} & \mathbf{I} \\ -\overline{\mathbf{M}}_i^{-1}\overline{\mathbf{K}}_i & -\overline{\mathbf{M}}_i^{-1}\overline{\mathbf{D}}_i \end{bmatrix} \begin{bmatrix} \mathbf{q}_i \\ \dot{\mathbf{q}}_i \end{bmatrix} \tag{8.118}$$

This equation can be written as

$$\dot{\mathbf{z}} = \mathbf{B}\mathbf{z} \tag{8.119}$$

where

$$\mathbf{B} = \begin{bmatrix} \mathbf{0} & \mathbf{I} \\ -\overline{\mathbf{M}}_i^{-1}\overline{\mathbf{K}}_i & -\overline{\mathbf{M}}_i^{-1}\overline{\mathbf{D}}_i \end{bmatrix} \tag{8.120}$$

In order to determine the eigenvalues and eigenvectors of the matrix $\mathbf{B}$ of Eq. 119, a solution of the following form is assumed:

$$\mathbf{z} = \mathbf{S}e^{st} \tag{8.121}$$

In this equation, the vector $\mathbf{S}$ has dimension $2 \times n_d$, and s is a constant to be determined. Substituting Eq. 121 into Eq. 119, one obtains the following system of homogeneous equations:

$$(\mathbf{B} - s\mathbf{I})\mathbf{S} = \mathbf{0} \tag{8.122}$$

This system of homogeneous equations has a nontrivial solution if and only if

$$|\mathbf{B} - s\mathbf{I}| = 0 \tag{8.123}$$

This is the characteristic equation that has roots $s_k, k = 1, 2, \ldots, 2 \times n_d$. These roots define the eigenvalues of the system of Eq. 122. Associated with each eigenvalue s_k, there is an eigenvector $\mathbf{S}_k$ that can be determined to within an arbitrary constant using the following system of homogeneous equations:

$$(\mathbf{B} - s_k\mathbf{I})\mathbf{S}_k = \mathbf{0} \tag{8.124}$$

In the case of multibody systems the roots s_k of the characteristic equation can be real and/or complex. Complex roots will appear as pairs of complex conjugates, which can have negative, zero, or positive real parts. The imaginary parts define the frequency of oscillations, while the real parts define the system stability. Any eigenvalue with a positive real part will render the system unstable. For linear problems, the solution can be written as a linear combination of the eigenvectors (mode shapes). The part of the solution due to eigenvectors associated with eigenvalues, which have negative real parts converges to zero, and therefore, such eigenvectors do not render the system unstable. The linearized system of Eq. 112, therefore, can shed light on the behavior of the multibody system in the neighborhood of the time-point at which the linearization is made.

As previously mentioned, complex eigenvalues appear as complex conjugates. For example, if s_k and s_{k+1} are two complex eigenvalues, they will take the following form:

$$s_k = p_k + i\omega_k, \quad s_{k+1} = p_k - i\omega_k \tag{8.125}$$

where $i = \sqrt{-1}$ is the imaginary operator. Since the coordinates and the velocities are real numbers, one must be able to write these coordinates and velocities in terms of real mode shapes regardless of whether or not the eigenvectors are complex. In order to demonstrate this fact, we write the part of the solution $\mathbf{z}_k$, which is a combination of the two modes $\mathbf{S}_k$ and $\mathbf{S}_{k+1}$ as

$$\mathbf{z}_k = \mathbf{S}_k e^{s_k t} + \mathbf{S}_{k+1} e^{s_{k+1} t} \tag{8.126}$$

Substituting Eq. 125 into Eq. 126 and rearranging the terms, one obtains

$$\mathbf{z}_k = (\mathbf{S}_k e^{i\omega_k t} + \mathbf{S}_{k+1} e^{-i\omega_k t})e^{p_k t} \tag{8.127}$$

Euler's formula for complex variables can be used to write

$$e^{i\omega_k t} = \cos\omega_k t + i\sin\omega_k t, \quad e^{-i\omega_k t} = \cos\omega_k t - i\sin\omega_k t \tag{8.128}$$

Substituting this equation into Eq. 127, one obtains

$$\mathbf{z}_k = ((\mathbf{S}_k + \mathbf{S}_{k+1})\cos\omega_k t + i(\mathbf{S}_k - \mathbf{S}_{k+1})\sin\omega_k t)e^{p_k t} \tag{8.129}$$

Since $\mathbf{z}$ must be real, $\mathbf{S}_k$ and $\mathbf{S}_{k+1}$ must appear as complex conjugate vectors. That is, $i(\mathbf{S}_k - \mathbf{S}_{k+1})$ in the preceding equation is a vector of real numbers. Using this fact, the real mode shapes associated with complex eigenvalues can be extracted using the eigenvectors obtained from the linearized multibody system equations of Eq. 122. It is important, however, to point out that such a linearized solution may not give in some applications accurate prediction of the stability of the multibody systems whose dynamics are, in general, governed by highly nonlinear differential equations.

Lightly Damped Modes In the case of lightly damped (under damped) modes, the eigenvalues are complex and take the form given in Eq. 125. Using the elementary theory of vibrations, one can write p_k and ω_k of Eq. 125 as

$$p_k = -\xi_k \omega_{nk}, \quad \omega_k = \omega_{nk}\sqrt{1-(\xi_k)^2} \tag{8.130}$$

In this equation, ξ_k is recognized as the modal *damping ratio*, ω_{nk} is the modal *natural frequency* of oscillation, and ω_k as the modal *damped frequency*; all associated with mode k. The preceding two equations can be solved for ω_{nk} and ξ_k as

$$\omega_{nk} = \sqrt{(p_k)^2 + (\omega_k)^2}, \quad \xi_k = -\frac{p_k}{\omega_{nk}} \tag{8.131}$$

Heavily damped modes are, in general, real and do not contribute to the oscillatory motion of the system.

PROBLEMS

1. Use the Rodriguez formula to prove the orthogonality of the rotation matrix.
2. Using the Rodriguez formula, show that the axis of rotation is an eigenvector of the spatial rotation matrix. Determine the associated eigenvalue.
3. Use the Rodriguez formula to determine the form of the spatial rotation matrix in the case of infinitesimal rotations.
4. Use the Rodriguez formula to show that two general consecutive three dimensional rotations are not commutative.
5. Discuss the singularity associated with Euler angles and show that this singularity problem does not arise when Euler parameters are used.
6. Use Newton–Euler equations to define the equations of motion of a rigid body in space in terms of Euler parameters.
7. Discuss the singularity problem associated with Rodriguez parameters.
8. Use Newton–Euler equations to define the equations of motion of a rigid body in space in terms of Rodriguez parameters.
9. Find the relationships between Euler and Rodriguez parameters.
10. The orientation of a rigid body is defined by the four Euler parameters

$$\beta_0^i = 0.8660, \quad \beta_1^i = \beta_2^i = \beta_3^i = 0.35355$$

At the given orientation, the body has an instantaneous absolute angular velocity defined by the vector

$$\boldsymbol{\omega}^i = [60.36 \quad 37.935 \quad -23.295]^{\mathrm{T}} \text{ rad/s}$$

Find the time derivatives of Euler parameters. Find also the time derivatives of Rodriguez parameters.

11. In the preceding problem, find the time derivatives of Euler angles.

12. Use the Rodriguez formula and unit vectors along the joint axes of rotation, defined in a fixed coordinate system, to obtain the transformation matrix that defines the orientation of body 4 shown in Fig. P1. Show that the form of this transformation matrix is the same as the Euler angle transformation matrix.

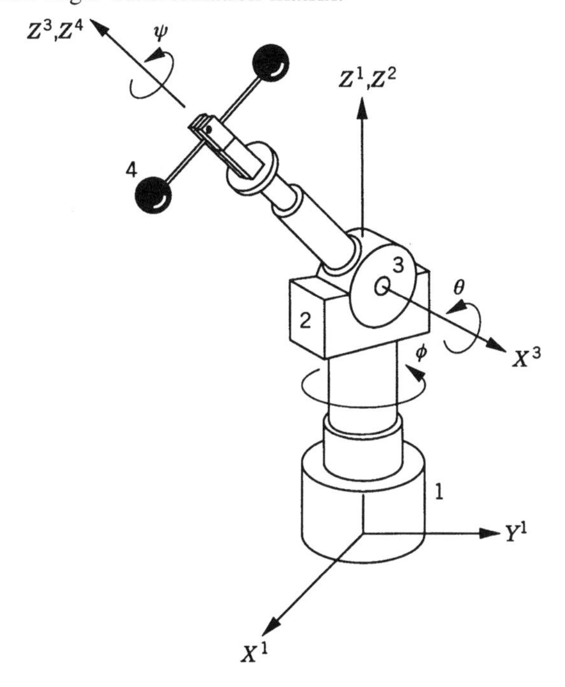

Figure P8.1

13. Discuss the use of the quaternions in describing the relative rotations between rigid bodies.

CHAPTER 9

MULTIBODY SYSTEM COMPUTER CODES

General purpose multibody system computer programs are widely used in the analysis of many industrial, technological, and biological applications. These computer programs can be used in virtual prototyping, design, performance evaluation, and analysis of complex systems. In this chapter, the general purpose multibody system computer code **SAMS/2000** (**S**ystematic **A**nalysis of **M**ultibody **S**ystems) is used as an example to introduce the reader to the capabilities and main features of these computer programs. Many of the formulations and algorithms discussed in various chapters of this book are implemented in SAMS/2000. This computer program allows the user to build complex multibody system models for planar and spatial multibody systems. The code has also advanced multibody system capabilities based on formulations and techniques that are beyond what is covered in this introductory text. If the user is interested in using the more advanced capabilities such as flexible body and rail dynamics, he/she can consult with other texts that provide complete documentations of the formulations and algorithms implemented in SAMS/2000 (Shabana, 2005, 2008; Shabana et al., 2008). This chapter provides an introduction to SAMS/2000 capabilities and features that allow the user to systematically build multibody system models. More details can be found in the online Help Manual of the code; and most panels of the code interface have an *Information Window* that provides detailed description of the data that can be provided by the user. The users of the code are encouraged to carefully read the description provided in these Information Windows before entering the data of their model. The panels used to input the data for body properties, constraints, and force elements have similar structure, thereby allowing the user to become quickly familiar with different data structures required to build a complex model.

Computational Dynamics, Third Edition Ahmed A. Shabana

The *educational version* of SAMS/2000 is limited to only four rigid bodies and does not include flexible body and rail simulation capabilities. It does not also include some other simulation options that are discussed in this chapter.

9.1 INTRODUCTION TO SAMS/2000

The objective of this chapter is to describe the use of the computer program SAMS/2000, which can be used in the computer simulation of nonlinear multibody system applications including mechanical, aerospace, and biomechanical systems. SAMS/2000 that is designed to systematically model complex systems of interconnected rigid and deformable bodies, can be used in the virtual prototyping, design, performance evaluation, and stability analysis. This code has many features that distinguish it from other existing general purpose multibody system computer programs. In particular, SAMS/2000 has capabilities based on formulations and algorithms that are more advanced than what is covered in this book. Nonetheless, most of the formulations and algorithms covered in various chapters of this book are implemented in SAMS/2000, allowing the user to solve many of the examples and exercise problems using this code. In order to be able to accurately model flexible components in multibody systems, small and large deformation finite element and multibody system algorithms are integrated, allowing the user to build models that include significant details that cannot be captured using rigid body analysis only (Shabana, 2005, 2008). The theory used to develop SAMS/2000 is documented in several books that explain clearly the formulations and computer algorithms implemented. Using these books, the user can have a clear understanding of the structure of the equations of motion used as well as the formulations of various force and constraint elements included in the code library. Some features of SAMS/2000 are briefly summarized below.

Planar and Spatial Systems SAMS/2000 can be used in the simulation of planar two-dimensional (2D) and spatial (3D) multibody systems. Nonlinear equations of motion of the multibody systems are automatically generated and numerically solved for both planar and spatial systems. The program includes a separate standard force and constraint libraries for each type of analysis in order to improve the computational efficiency in the case of the simpler planar analysis.

Type of Analysis The program has the capability of performing static, dynamic, and kinematic analyses of multibody systems. Determining the static configuration can be important in some applications before the dynamic simulation is performed. When all the system degrees of freedom are prescribed, the nonlinear constraint algebraic equations are solved using the kinematic analysis module of SAMS/2000. SAMS/2000 provides the user with several options for performing the nonlinear dynamic analysis of multibody systems as will be described in this chapter.

Joint Constraints The program can be used in the simulation of multibody system applications that can have different topological structures. Joints such as spherical, revolute, prismatic, cylindrical, etc. are standard elements of the library of the program. These different types of joints, which define the connectivity between the multibody system components, can be systematically modeled using SAMS/2000. Most of the joint formulations implemented in the code can be used to connect rigid, flexible, and very flexible bodies. These joint formulations take into account the effect of small and large deformations of the flexible bodies.

Force Elements Standard force elements such as gravity, bearing, bushing, spring, damper, and actuator forces, which are included in the library of SAMS/2000, can be systematically modeled. Several of these forces can be applied to rigid, flexible, and very flexible bodies. The force formulations implemented in SAMS/2000 automatically account for the small and large deformations of the bodies in the system.

User Constraints and Forces The program has a set of user subroutines that allow the user to introduce arbitrary nonlinear constraints and forces that are not standard elements of the SAMS/2000 constraint and force libraries. The user forces and constraints can depend on the system coordinates and velocities as well as time. The user has also access to the elastic coordinates and velocities of the bodies in order to allow for the formulation of forces and constraints that are function of the body deformations.

Solution Procedures SAMS/2000 offers several procedures that can be used by the user to solve the nonlinear dynamic equations of motion of constrained multibody systems. In one procedure, only the independent accelerations are integrated and the kinematic constraint equations are satisfied at the position, velocity, and acceleration levels. In a second procedure, all the accelerations (independent and dependent) are integrated and the constraint equations are satisfied at the position, velocity, and acceleration levels. In a third procedure, all the accelerations are integrated, while the constraint equations are satisfied only at the acceleration level. While the user can use any of these options among others offered by SAMS/2000, the first procedure, which is the default is recommended. In most solution procedures implemented in SAMS/2000, sparse matrix techniques are used in order to obtain efficient solution of the position, velocity, and acceleration equations.

Flexible Body Modeling SAMS/2000 has advanced flexible body modeling capabilities. The program allows modeling deformable bodies using the finite element method or experimental identification techniques. It also allows the use of the nodal or modal coordinates and lumped or consistent mass formulations. The program automatically generates the equations of motion that include all the nonlinear terms that represent the inertia coupling between the rigid body and elastic displacements. SAMS/2000 can be used in the analysis of small and large deformations in multibody system applications. Both the small deformation *floating frame of reference* (FFR) formulation and the large deformation *absolute nodal coordinate formulation* (ANCF) are implemented in SAMS/2000.

PRESAMS Preprocessor SAMS/2000 has a preprocessor, **Pre**processor for the **S**ystematic **A**nalysis of **M**ultibody **S**ystems (PRESAMS) that can be used to link SAMS/2000 with existing commercial finite element programs. It is clearly explained in the *Help Files* of the code what are the data required from the finite element programs at this preprocessing stage. Detailed derivations of the matrices and vectors required from the finite element programs are presented in the books that document the theory used to develop SAMS/2000. The flexible body formulations and algorithms are not discussed in this book.

Impact Dynamics SAMS/2000 allows modeling impact between the components of the multibody system. The generalized impulse momentum equations that predict the jump discontinuity in the joint forces as a result of the impact are automatically generated by the code. The code is capable of treating problems with multiple impacts. The user can also develop a continuous contact force model using spring-damper force elements (Khulief and Shabana, 1987). In this case the user must provide the stiffness and damping coefficients that enter into the formulation of the compliant forces.

User-Differential Equations SAMS/2000 allows the user to provide first-order differential equations. These equations that may depend on the system coordinates and velocities are integrated simultaneously with the differential equations of motion of the system. This feature of the code can be used to include differential equations that describe other models such as control elements or models in multiphysics problems. Some of these user-differential equations can be different from the standard differential equations that are already implemented in SAMS/2000.

9.2 CODE STRUCTURE

SAMS/2000 is divided into four main modules: the *Mass Module* (MASMOD), the *Constraint Module* (CONMOD), the *Force Module* (FRCMOD), and the *Numerical Module* (NUMMOD). The functions of these four modules are explained below.

Mass Module (MASMOD) The MASMOD is used to evaluate the mass matrix and the quadratic velocity centrifugal and Coriolis inertial forces of the rigid and flexible bodies in the multibody systems. Sparse matrix techniques are used to store the mass matrix in order to obtain an efficient solution of the acceleration equations. In the case of a flexible body modeled using the FFR formulation, the modal mass matrix and the constant terms that enter into the formulation of the inertia coupling between the rigid body motion and the elastic deformation are obtained from the finite element preprocessor. In the case of the large deformation, in which bodies are modeled using the ANCF, a diagonal mass matrix is obtained using the Cholesky coordinates, as described in other texts that document the theory of the code.

Constraint Module (CONMOD) The CONMOD constructs the nonlinear algebraic constraint equations, which describe mechanical joints. In this module, the Jacobian matrix of the kinematic constraints and the time derivatives of the constraint equations are computer generated. A set of standard joint constraints is available in the program library that can be utilized by the user. Nonstandard constraints can also be provided by the user using a set of user subroutines. The constraints equations are formulated to account for the effect of the deformation in the case of flexible or very flexible bodies.

Force Module (FRCMOD) The FRCMOD evaluates the generalized forces associated with the system-generalized coordinates. A set of standard force elements that can be utilized by the user are available in the library. These force elements include the gravity, bearing, bushing; and spring, damper, and actuator forces. Other nonlinear and nonstandard forces that may depend on the system coordinates, velocities, and time can also be introduced to the program using user subroutines. Most of the force element formulations implemented in SAMS/2000 account also for the body deformations in the case of flexible bodies.

Numerical Module (NUMMOD) The NUMMOD contains subroutines for solving system of algebraic equations, subroutines for solving differential equations, and subroutines for identifying the independent variables associated with a given set of algebraic equations. The code also allows for using sparse matrix techniques for the efficient solution of the governing equations at the position, velocity, and acceleration levels as discussed in Chapter 6 of this book. Different options of direct numerical integrators are provided by the code. SAMS/2000 also allows for the use of Spline function representations to describe coefficients and variables that are provided as a function of time or function of other system coordinates or variables.

As previously mentioned, the multibody system components can be treated as rigid or flexible. While this book is concerned only with the rigid body dynamic formulations, a multibody system modeled using SAMS/2000 may contain both types of components. For flexible bodies, the preprocessor PRESAMS must be used to generate the inertia shape integral matrices and the constant terms that appear in the system equations of motion based on lumped or consistent mass formulations. The use of the preprocessor PRESAMS is explained in the Help Manual of the code.

9.3 SYSTEM IDENTIFICATION AND DATA STRUCTURE

In general, multibody systems can be represented by an abstract drawing similar to the one shown in Fig. 1. The user of the code must have a good understanding of the topological structure of the system as well as the structure of the data that the code requires to model the system.

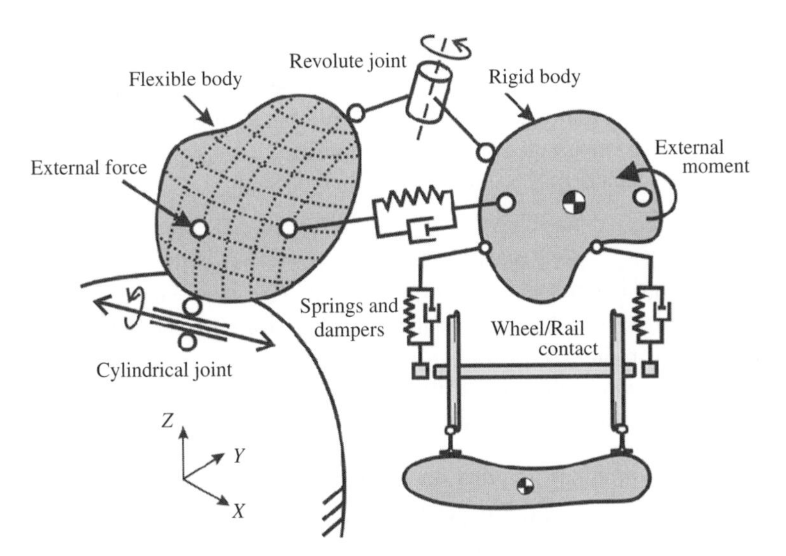

Figure 9.1 Multibody systems

System Identification The interface of SAMS/2000 allows the user to systematically build multibody system models. In order to develop such models, the user must provide specific information about the model including the following:

1. System Type: The user must choose the type of analysis to indicate whether the model is 2D or 3D (planar or spatial). The interface of the code allows the user to select the type of analysis.
2. Body Numbering: Each body, flexible and rigid, must be given a distinct number between 1 and n_b, where n_b is the total number of bodies in the multibody system.
3. Body Coordinate Systems: Each rigid body and each deformable body modeled using the FFR formulation or the ANCF must be assigned a local reference coordinate system. For rigid bodies, the origin of the reference coordinate system must be placed at the center of mass of the body. For flexible bodies, the origin of the reference coordinate system can be arbitrarily selected. The positions and orientations of the body coordinate systems at the initial configuration must be provided by the user.
4. Mechanical Joints and Constraints: All joints and constraints of the same kind must be grouped together. All elements of the group must be numbered consecutively from 1 to n_j, where n_j is the total number of joints or constraints of that type.
5. Force Elements: Forces are grouped according to the force type (spring-damper-actuator, bushing, bearing, etc.). All elements of one group must be numbered consecutively from 1 to n_f, where n_f is the total number of forces in this group or force type.
6. Specified Motion Trajectories: The prescribed motion trajectories can be introduced to SAMS/2000 using a set of user subroutines described in a later section of this chapter.
7. Externally Applied Forces and Moments: SAMS/2000 allows the user to introduce nonstandard forces that depend on system-generalized coordinates, velocities, and time. These forces can be introduced to the model using user subroutines.
8. Impact Pairs: Each impact pair must be given a number $l, l = 1, 2, \ldots, n_i$, where n_i is the total number of impact pairs in the multibody system.
9. Wheel/Rail Contact Models: SAMS/2000 can be used to model complex railroad vehicle systems. In the code several wheel/rail contact formulations are implemented. Each contact must be given a number $M, M = 1, 2, \ldots, n_i$, where n_i is the total number of wheel/rail contacts in the multibody vehicle system.

Data Structure The input data of SAMS/2000 is generated by the code interface. The interface produces the ASCII data file *SamsData.dat* that contains a description of the model. The user is encouraged to examine this file before performing the simulation. The input data of SAMS/2000 in *SamsData.dat* is divided into independent segments. A multibody application may not require the use of all these segments. Examples of the segments that are used in the input data are as follows:

1. A segment that provides a verbal description of the problem under investigation.
2. A segment that contains the control parameters, which define the type of analysis, the numerical integrator, the method of dealing with the redundant coordinates, and so on.

3. A segment that defines the inertia properties, constant forces, and initial coordinates and velocities of the bodies.
4. A segment that defines the degrees of freedom to be constrained for the static analysis.
5. A segment for each joint type that defines the bodies connected by this type of joint and the location of the joint definition points on the bodies.
6. A segment for each force element that describes the parameters that will be used in the formulation of this force element.
7. A segment that describes the impact pairs.
8. A segment that describes the inertia and stiffness characteristics of deformable bodies in the system. The data provided in this segment are obtained using the preprocessor PRESAMS, which is described in the online Help Manual of the code.
9. A segment that defines the parameters for the numerical routines such as the time of the start of the simulation, the time of the end of the simulation, and the reporting interval.
10. A segment that defines the wheel/rail contact parameters in the case of railroad vehicle systems.
11. Segments related to the use of the user subroutines. The use of these user subroutines requires that the user has a Fortran compiler that allows for editing these subroutines and linking them with the code library. The user can consult with the author of the book regarding the access to these user subroutines.

These segments and others can be easily identified in the ASCII data file *SamsData.dat*, which is the main input data file of SAMS/2000. Depending on the model, other data files may be required as described in the online Help Manual of the code.

9.4 INSTALLING THE CODE AND THEORETICAL BACKGROUND

After properly installing the code in the directory *C:\SAMS2000*, one can start using the code for the computer simulation of multibody systems that consist of interconnected rigid and flexible bodies. Note that SAMS/2000 must be installed to the directory *C:\SAMS2000.* If it is installed in a different directory, the code will not function properly. The code must be used according to the instructions provided in this chapter and the online Help Manual. An educational version of the code with limited capabilities can still be used to effectively demonstrate the use of multibody system formulations in the virtual prototyping of engineering and physics systems.

When the user starts SAMS/2000, the panel shown in Fig. 2 will appear. Using this panel, the user can select one of the following SAMS/2000 modes: Multibody Simulation, SAMS Utilities, Preprocessor PRESAMS, or Special SAMS/2000 Modules. In the case of multibody system simulation, the user must select *Multibody Simulation*. In the educational version of SAMS/2000, some other modes may be disabled. The preprocessor PRESAMS can be used to generate the inertia and stiffness characteristics of the deformable bodies in the system. If the user is interested only in rigid body simulation, there is no need for using PRESAMS.

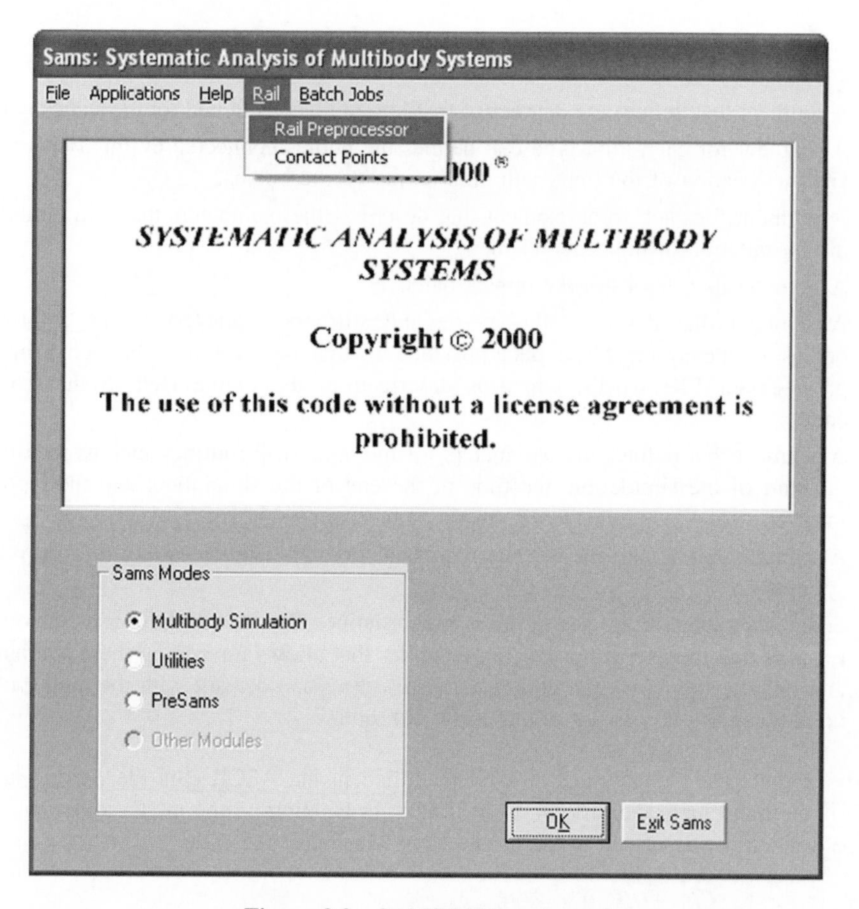

Figure 9.2 SAMS/2000 start panel

Theoretical Foundation The theory used in developing SAMS/2000 is documented in several texts and a large number of archival publications. These texts and publications provide detailed information on the formulations implemented in the code. This detailed documentation of the methods distinguishes SAMS/2000 from other commercial codes and allows the user to have a good understanding of the models developed. In addition to this book which provides an introduction to the rigid body formulations and the numerical algorithms implemented in the code, the user can consult with other two texts when more advanced capabilities of the code are used (Shabana, 2005, 2008). These two texts are more advanced and cover the theory used in the analysis of flexible multibody systems. The large deformation theory used in SAMS/2000 can be also found in these two texts. SAMS/2000 successfully integrates small and large deformation finite element and multibody system algorithms.

SAMS/2000 has advanced capabilities for modeling railroad vehicle systems. Several wheel/rail contact formulations are implemented in the code, as previously mentioned. Detailed documentation of the formulations used in SAMS/Rail module can be also found in the literature (Shabana et al., 2008).

In addition to the texts that document the formulations and algorithms implemented in SAMS/2000, the *Help* menu of SAMS/2000 includes notes for short courses to introduce the user to rigid and flexible multibody dynamics. The user is encouraged to read these notes in order to become familiar with the formulations used. The Help menu of SAMS/2000 has also several user files that further explain the structure of the input data in the case of the use of specific features of the code.

9.5 SAMS/2000 SETUP

SAMS/2000 is designed to work in a Window environment. The user must have a directory C:\WINDOWS on his/her computer. In order to install the code, the user can double-click on *SETUP.EXE* file in SAMS/2000 CD. This will start the process of SAMS/2000 Setup by displaying the panel shown in Fig. 3. The user can then click on the ***OK*** button to proceed. SAMS/2000 Setup will display ***C:\SAMS2000*** as the default Directory. The user must accept this directory and click the button shown in Fig. 4 in order to continue with the program setup. When the setup process is completed successfully, SAMS/2000 Setup will give a message that the setup is finished successfully.

After the installation of SAMS/2000 is completed, the user will need to manually adjust two configuration files: *SAMSWORD.CFG* and *SAMSFORT.CFG*. Both are located in *C:\SAMS2000* directory.

SAMSWORD.CFG In this file the user must provide the full path of the word processor program that SAMS/2000 will use as the default word processor. For example, the user can write C:\Program Files\Microsoft Office\Office\winword1.exe.

SAMSFORT.CFG In this file the user must provide the full path of the executable file that launches the Fortran code editor or the word processor that will be used to edit the code of the user subroutines of SAMS/2000. The order of these two directories is important.

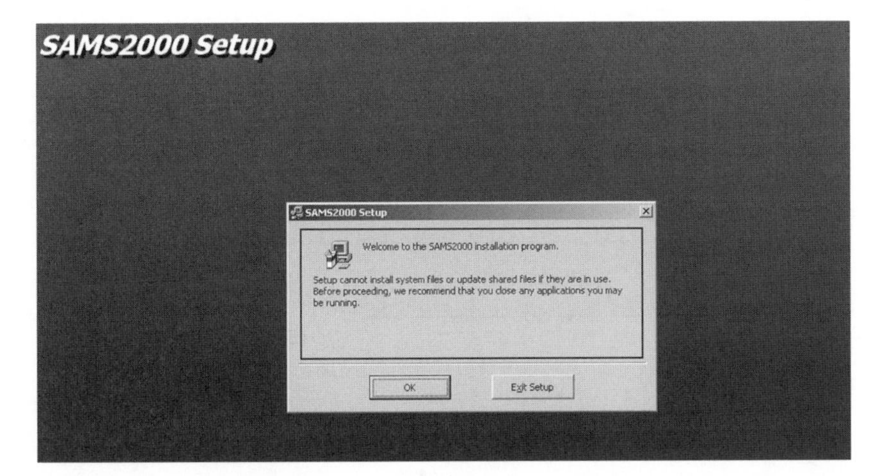

Figure 9.3 SAMS/2000 installation

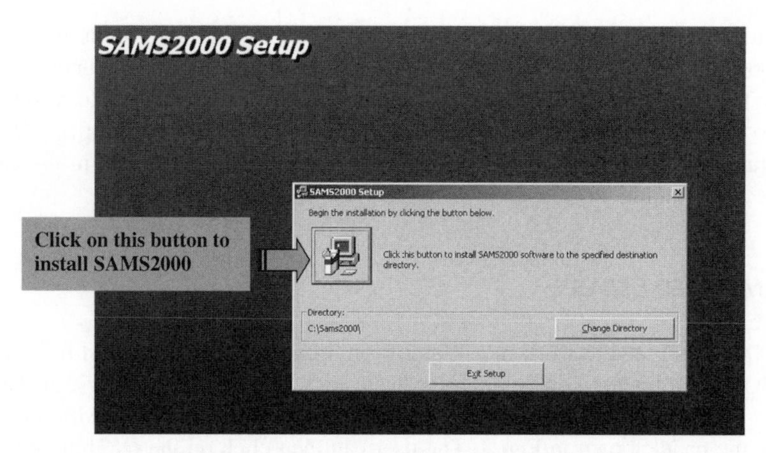

Figure 9.4 SAMS/2000 default directory

For example, the user can write as follows:

C:\msdev\bin\msdev.exe
C:\WINDOWS\

9.6 USE OF THE CODE

The use of the SAMS/2000 can be described using a simple example that is familiar to the reader. To this end, a planar slider crank mechanism is used. This simple slider crank mechanism example can be used to introduce the user to the input and output data of SAMS/2000 as well as its interface panels of the code.

Problem Description The slider crank mechanism shown in Fig. 5 has the following specifications:

Body Name	Mass (kg)	Mass Moment of Inertia (kg·m^2)
Ground	0.0	0.0
Crankshaft	1.0	0.1
Connecting rod	2.0	0.1
Slider block	3.0	0.1

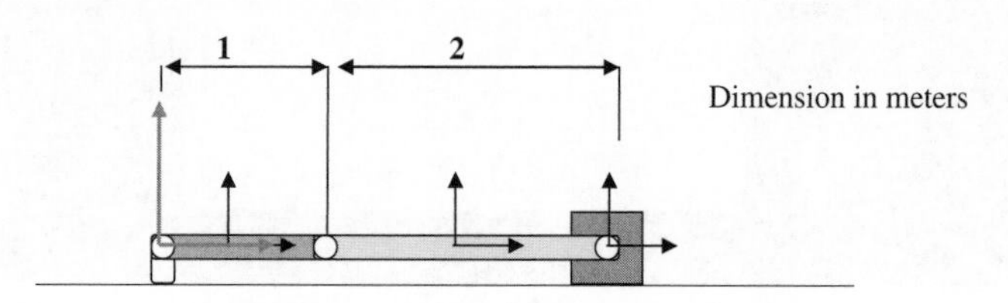

Figure 9.5 Slider crank mechanism

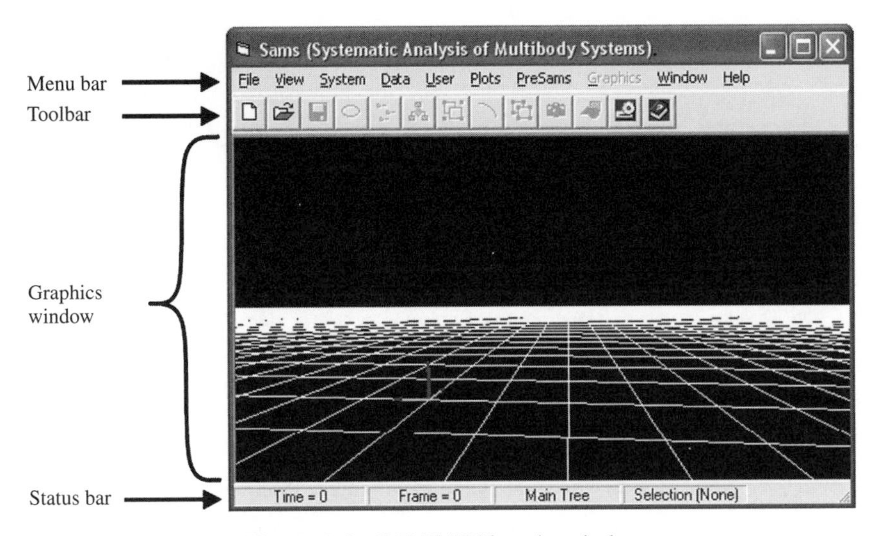

Figure 9.6 SAMS/2000 main window

In order to build the SAMS/2000 model for the slider crank mechanism assuming that all the bodies are rigid, a centroidal body coordinate system is assigned to each body as shown in Fig. 5. The gray arrows represent the global fixed frame, which is also assumed to represent the coordinate system of the ground body. The mechanism has four bodies, which have restricted motion because of the joint constraints. The mechanism has one ground (bracket) joint, three revolute joints, and one prismatic joint. In order to build the computer model for this mechanism, the user can start SAMS/2000 by clicking on Program/SAMS2000/SAMS2000 located in the Windows Start menu. SAMS/2000 will start and the screen shown in Fig. 2 will appear. Choose SAMS Mode to be *Multibody Simulation* and click the ***OK*** button. SAMS/2000 will display the main screen shown in Fig. 6. This screen has the following main items:

1. Menu bar
2. Toolbar
3. Graphic window
4. Status bar

In this chapter, menu items will be indicated with italic letters using the following notation: *Menu/Item* where "*Menu*" is the name of the menu and "*Item*" is the name of the element in the menu. Some menu items have toolbar icons associated with them. In such cases, an image of the toolbar button associated with the menu item may also be displayed in the text. In case a text appears in the status bar at the bottom of the SAMS/2000 interface window that is associated with a menu item, this text is presented in this chapter inside the curly brackets, {}.

Building the Model The user can start a new model by clicking on menu *File/New*\ . A new screen will appear to allow the user to start providing the input data of the mechanism

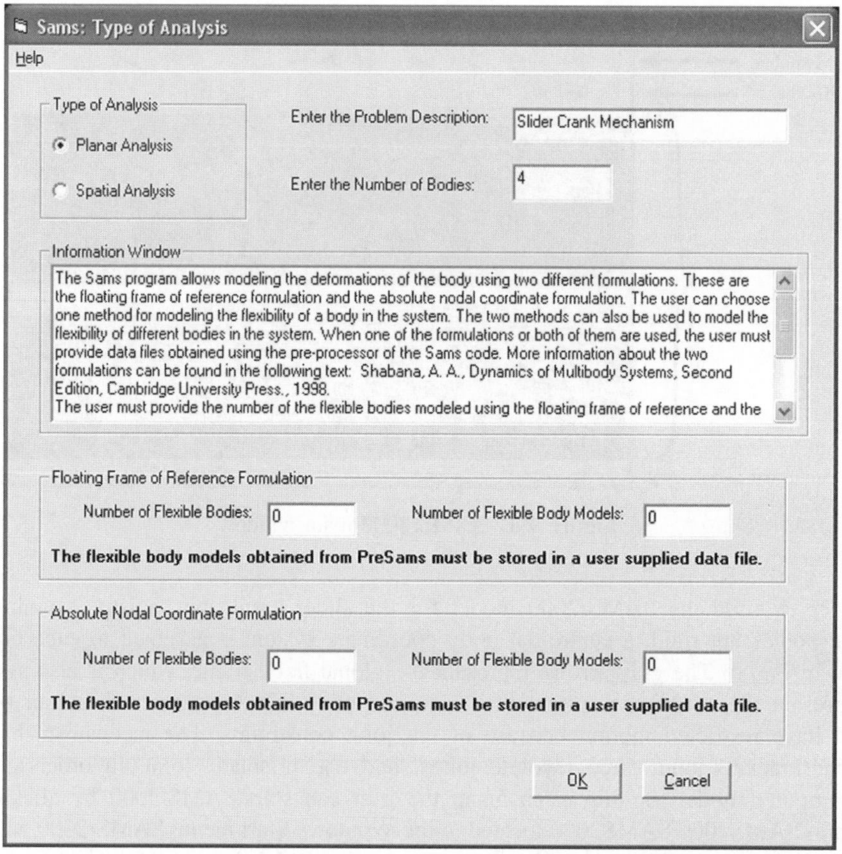

Figure 9.7 Analysis type

shown in Fig. 5. In this screen, the user can input the problem description "Slider Crank Mechanism" in the text box as shown in Fig. 7, and choose the *Planar* option for the *Type of Analysis* field since the mechanism is planar. The user can input 4 for the number of bodies, and click the ***OK*** button. The main SAMS/2000 window will display a tree at the left as shown in Fig. 8. This tree can be used to build the model.

The user can click on the (+) sign associated with the bodies to expand this part of the tree as shown in Fig. 9, and then double-click on each body to start inputting the body data. As an example, Body # 2 (Crankshaft) is chosen. Similar procedures can be used for other bodies.

9.7 BODY DATA

In order to input the body data, the user can double-click on *Body # 2* in the *Main Tree*, as shown in Fig. 9. This allows the user to access the forms in Figs. 10–15. These forms

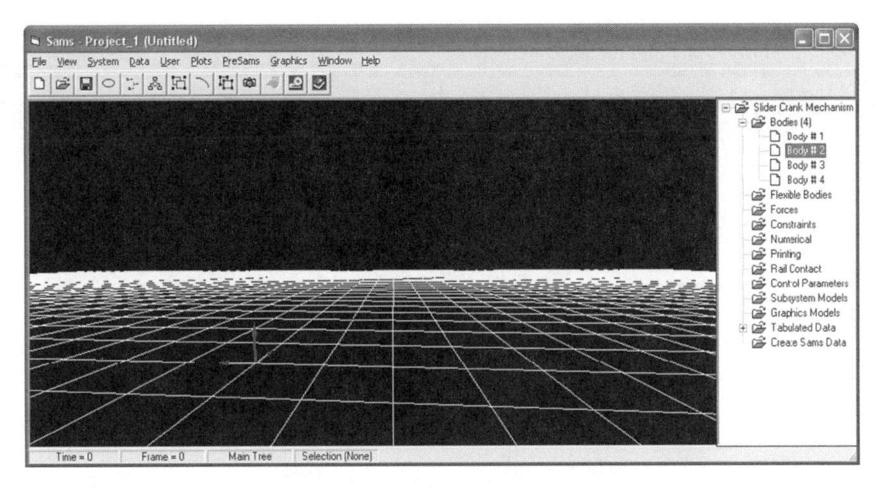

Figure 9.8 SAMS/2000 main tree

can be used to input the body data. Using these forms, the user can access and change the data for a given body. After the data for each tab has been entered, the user must make sure to click the ***Apply*** button before switching from one tab to another on the form. If the ***Apply*** button is not clicked before switching tabs, the changes of the data that correspond to a tab will not be saved. Note also that SAMS/2000 allows the use of different system of units and does not have a predefined unit system. The units are selected by the user and must be consistent when providing the input data. In this example, the unit of length is the

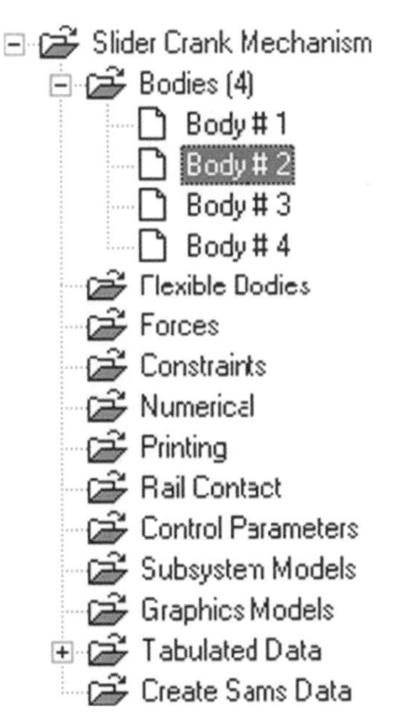

Figure 9.9 Main tree

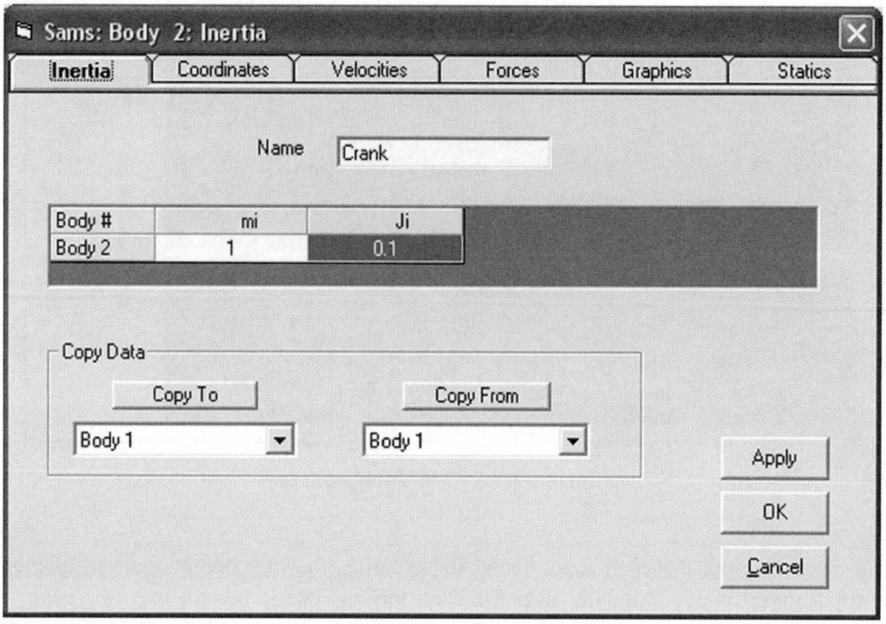

Sams: Body 2: Inertia
Inertia
Coordinates
Velocities
Forces
Graphics
Statics
Name
Crank
Body #
mi
Ji
Body 2
1
0.1
Copy Data
Copy To
Copy From
Body 1
Body 1
Apply
OK
Cancel

Figure 9.10 Inertia tab

Sams: Body 2: Initial Coordinates
Inertia
Coordinates
Velocities
Forces
Graphics
Statics
Name
Crank
Body #
Rxi
Ryi
Thetai
Body 2
0.5
0
0
Copy Data
Copy To
Copy From
Body 1
Body 1
Apply
OK
Cancel

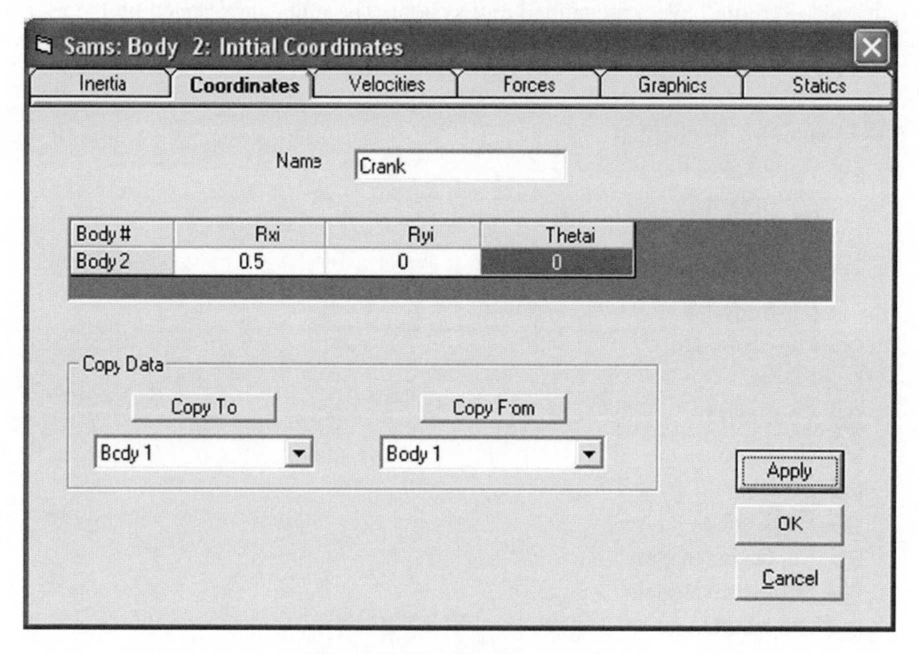

Figure 9.11 Coordinates tab

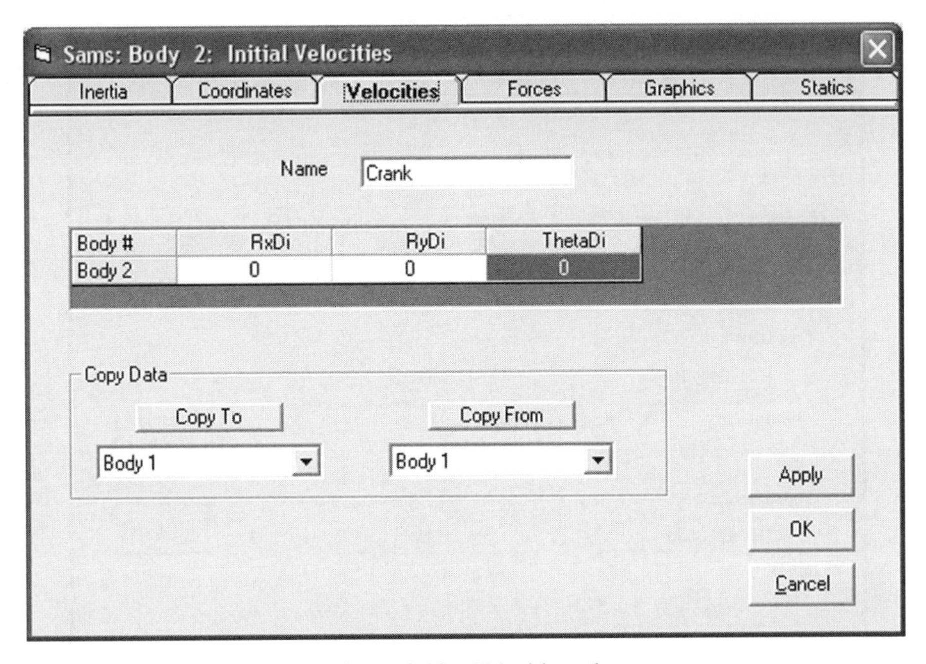

Figure 9.12 Velocities tab

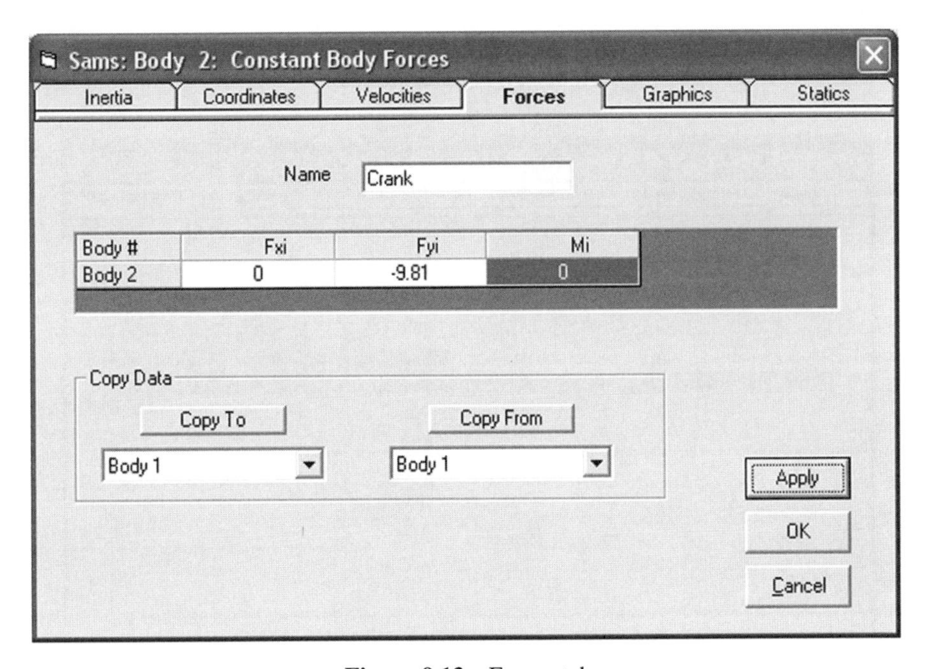

Figure 9.13 Forces tab

Sams: Body 2: Body Graphics
Inertia | Coordinates | Velocities | Forces | Graphics | Statics
Name Crank

Body #	Length	Width	Thickness	Plan
Body 2	1	0.2	0.1	X-Y Pl

Copy Data
Copy To
Body 1
Copy From
Body 1
Apply
OK
Cancel
Body Graphics
Model Shape
Rod
Model Color
Material
Jade

Figure 9.14 Graphics tab

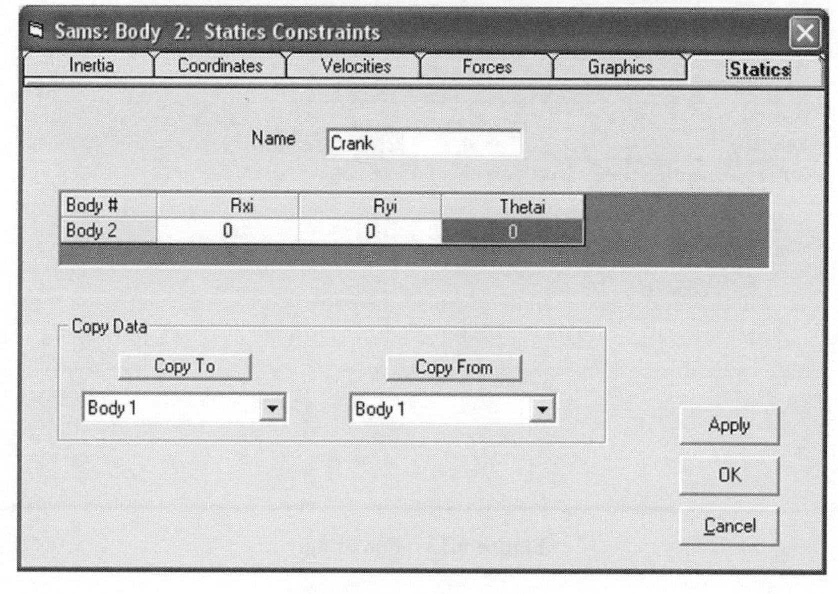

Figure 9.15 Statics tab

meter, the unit of time is the second, and the unit of mass is the kilogram. All other units are derived from these three basic units.

Body 2 of the mechanism is the crankshaft in this example. Therefore, in the *Name* text box of the body, the user can enter *Crank*. Note that the *Name* text box is displayed regardless of what tab is selected. The user can assign arbitrary names for the bodies. Below, a description of the tabs used for the body data is provided.

Inertia Tab This tab is used to input the mass and mass moment of inertia of the body. Mass of the body *mi* is assumed to be 1 kg in this example. Mass moment of inertia *Ji* is assumed to be 0.1 kg.m^2 in this example.

Coordinates Tab This tab is used to input the initial coordinates of the body that define the initial conditions at the beginning of the simulation. *Rxi* defines the initial global X coordinate of the center of mass, which is assumed in this example to be 0.5 m. *Ryi* defines the initial global Y coordinate of the center of mass, which is assumed to be 0 m. *Thetai* defines the initial rotational angle, which is assumed to be 0 rad.

Velocities Tab This tab is used to input the initial velocities of the body at the beginning of the simulation. *RxDi* is the initial global X velocity of the center of mass, which is assumed to be 0. *RyDi* is the initial global Y velocity of the center of mass, which is assumed to be 0. *ThetaDi* is the initial angular velocity, which is assumed to be 0.

Forces Tab This tab is used to input the constant forces acting on the body. These forces must be defined in the global coordinate system, and can include constant forces such as gravity and load forces. *Fxi* is the constant global force acting on the body in the X direction, which is assumed to be 0 in this example. *Fyi* is the constant global force acting on the body in the Y direction, which is assumed to be -9.81 N. *Mi* is the constant torque acting on the body, which is assumed to be 0 in this example.

Graphics Tab This tab can be used to define the graphics and shape of the body. The shape and dimensions selected in this form have no effect on the simulation results; they are used only for visualization. It is not necessary that the user assigns graphics data for a body in the system. Below is a description of some of the parameters that can be defined by the user.

1. Model Shape: The shape used to represent the body. Rod is selected for the crankshaft.
2. Material: The color of the body. Jade color is selected.
3. Length: The length of the rod. 1 unit is given.
4. Width: The width of the rod. 0.2 units is given.
5. Height: The height/thickness of the rod. 0.1 units is given.

Note that the *Model Shape* must be chosen before the *Length, Width*, and *Height* parameters become available to the user because the parameters are specific to the *Model Shape*. Different shapes are defined using different parameters.

Statics Tab This tab can be used to input the data for the static equilibrium analysis prior to running a simulation. In this example, static equilibrium will not be used, and the default zero values are used.

A procedure similar to the one used for inputing the data of the crankshaft can be used for other bodies of the mechanism. In order to complete the body data, the user can double-click on *Body # 1* in the *Main Tree* to input the data for the ground body. The user can set its name to "Ground." No other default values need to be changed because all the degrees of freedom of the ground body will be constrained in this example. Therefore, its mass, rotational inertia, initial velocities, constant forces, and statics can assume their zero default values. The initial coordinates $Rxi = 0$, $Ryi = 0$, $Thetai = 0$ are used for the ground body.

For *Body # 3*, the connecting rod of the mechanism, the user can select its name to be "Connecting Rod." For the inertia, initial conditions and forces, the user can input $mi = 2\,\text{kg}$, $Ji = 0.1\,\text{kg.m}^2$, $Rxi = 2\,\text{m}$, and $Fyi = -19.62\,\text{N}$. All other initial conditions and forces assume the zero default values. For the graphics, the user can select Rod for the *Model*, Turquoise for the *Material*, 2 for the *Length*, 0.2 for the *Width*, and 0.1 for the *Thickness*.

The user can then double-click on *Body # 4* in the *Main Tree* to enter the data for the slider block. The user can select the body name to be "Block." For the inertia, initial conditions and forces, the user can input $mi = 3\,\text{kg}$, $Ji = 0.1\,\text{kg}\cdot\text{m}^2$, $Rxi = 3\,\text{m}$, and $Fyi = -29.43\,\text{N}$. All other initial conditions and forces assume the zero default values. For the graphics, the user can select Solid for the *Model Shape*, Emerald for the *Material*, 0.5 for the *Length*, 0.5 for the *Width*, and 0.5 for the *Height*.

9.8 CONSTRAINT DATA

To input the data for the constraints and the joints of the model, the total number of elements of each constraint type must be identified. The user can double-click on *Constraints* in the tree list or click on the Menu *SYSTEM/Constraints*\. The constraints screen will appear as shown in Fig. 16. For the slider crank mechanism example, the user can input 1 for the simple constraint, 3 for the revolute joint, and 1 for the prismatic joint; and then click ***OK***. In the tree list, click on the (+) sign to expand the constraint list as shown in Fig. 17. The following abbreviations are used: **GENC** for the Generalized Coordinate Constraint (Simple Constraint), **REVJ** for the Revolute (Pin) Joint, and **TRNJ** for the Translational (Prismatic) Joint.

The user can then double-click on each constraint in order to obtain the associated input screen. For example, when the user double clicks on *GENC*, the screen shown in Fig. 18 appears. In this screen, the user can define the ground constraints by inputting 1 for all coordinates; a coordinate is fixed to assume its initial value when the user inputs 1 in the cell that corresponds to this coordinate. If the coordinate is not fixed, the zero default value is used.

Because the ground body was selected to be Body # 1, the user can input 1 for *IB*. Also, the user can constrain all the coordinates of the ground body by inputting 1 in the cells that correspond to all the coordinates as shown Fig. 18. The user can then click ***Apply***.

For the revolute joint, the user can double-click on *REVJ #1* to obtain the screen shown in Fig. 19. In this screen, the user can define the data of REVJ joints that eliminate the

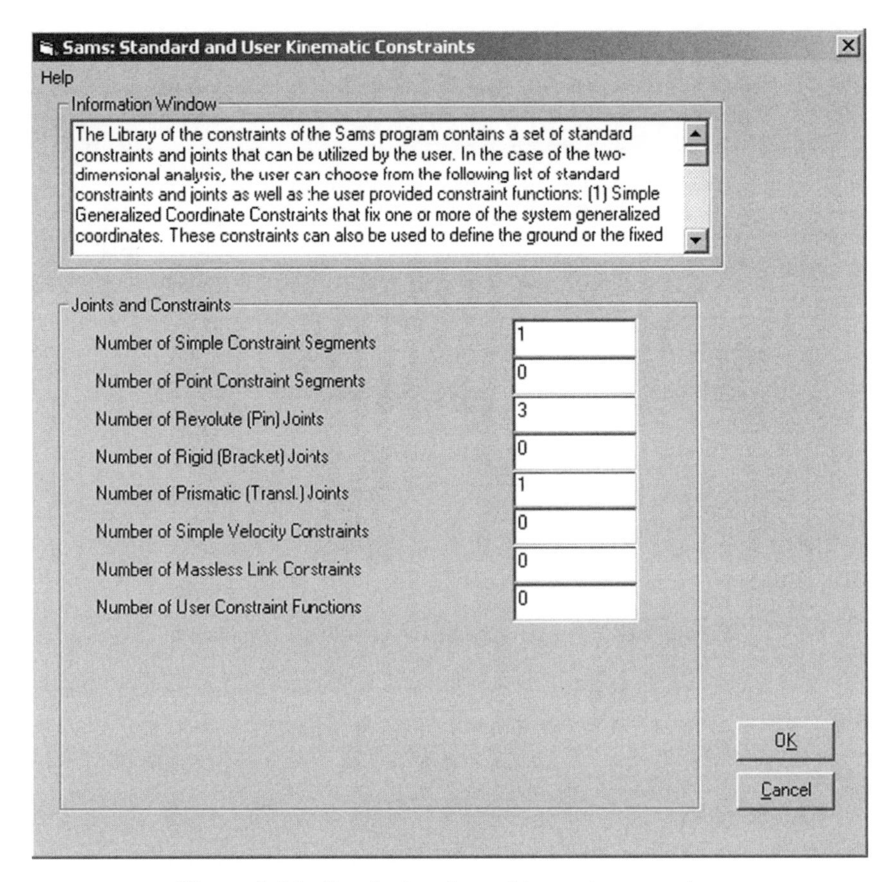

Figure 9.16 Standard and user kinematic constraints

relative translations between the two bodies connected by this type of joint. The revolute joint allows only one relative rotational degree of freedom between the two bodies. The user must define the two bodies *IB* and *JB* connected by this joint, the nodes *Ni* and *Nj* on the two bodies at which this joint is defined (enter zero in the case of a rigid body), and the local coordinates (*xi*, *yi*) and (*xj*, *yj*) of the joint definition points on the two bodies defined in the *IB* and *JB* body coordinate systems, respectively. For the revolute joint between the ground and the crankshaft, since the ground is Body # 1 and the crankshaft is Body # 2, the user can enter 1 for *IB* and 2 for *JB*. *xi* and *yi* are the coordinates of the position of

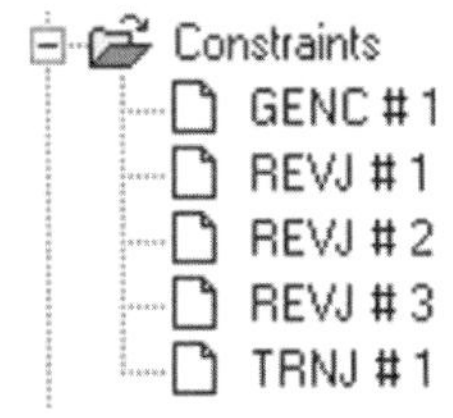

Figure 9.17 Constraints in the main tree

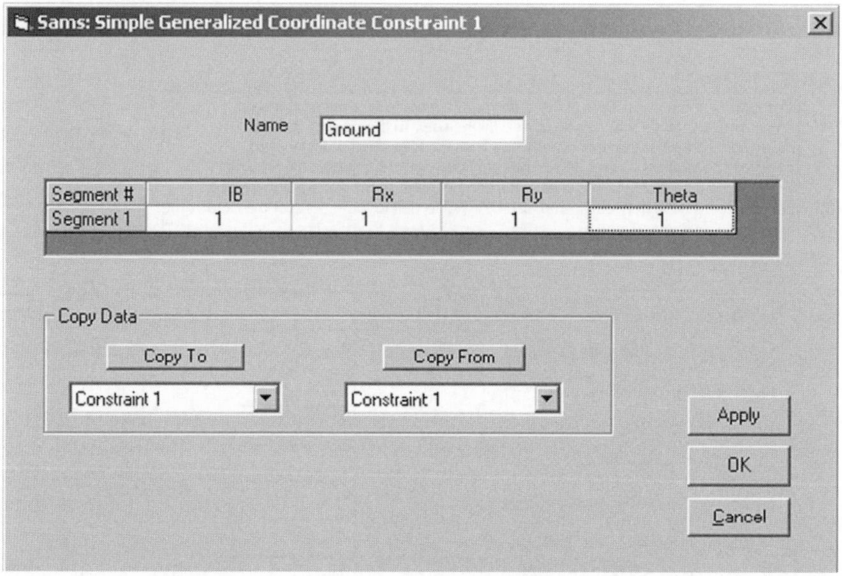

Sams: Simple Generalized Coordinate Constraint 1

Name Ground

Segment #	IB	Rx	Ry	Theta
Segment 1	1	1	1	1

Copy Data
Copy To
Copy From
Constraint 1
Constraint 1
Apply
OK
Cancel

Figure 9.18 Simple generalized coordinate constraints

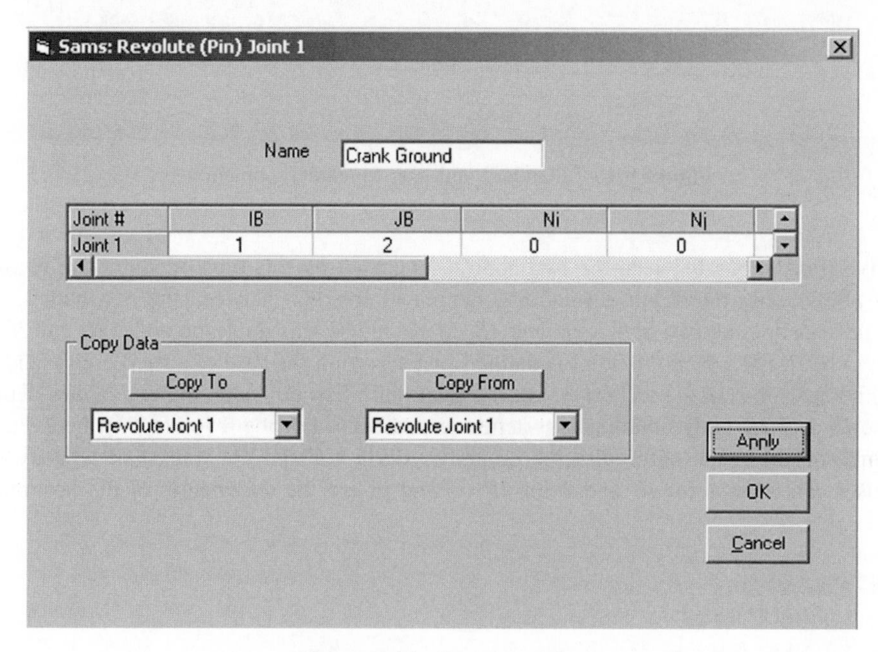

Figure 9.19 Revolute joint

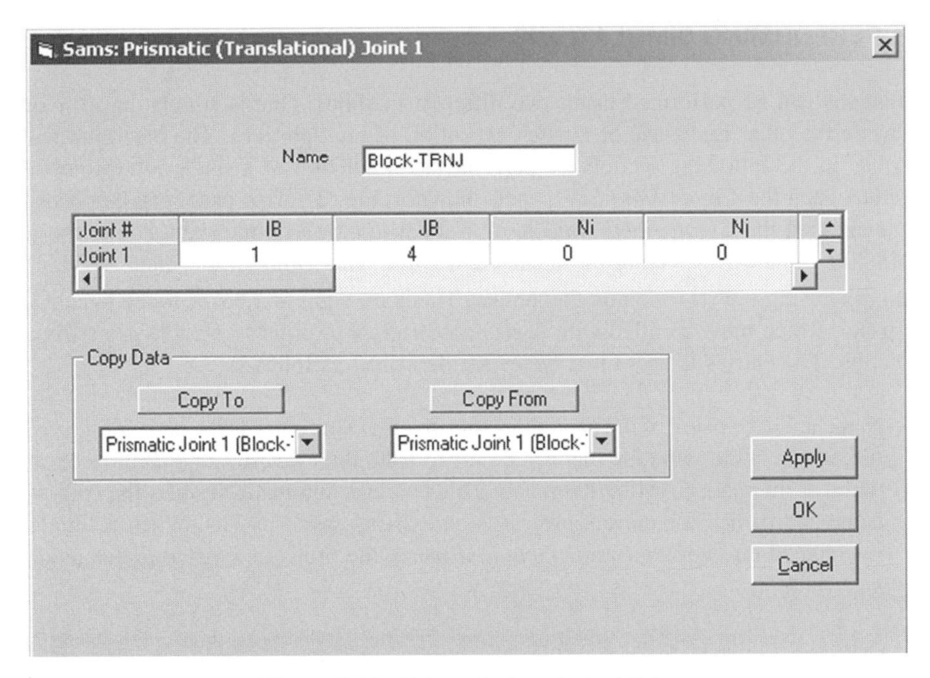

Figure 9.20 Prismatic (translation) joint

the revolute joint with respect to body *IB*. Similarly, *xj* and *yj* are the coordinates of the revolute joint with respect to body *JB*. For *xi* and *yi*, the user can input 0 and 0 respectively. For *xj* and *yj*, the user can input −0.5 and 0 respectively. The user can then click ***Apply***.

A similar procedure can be used for defining the data of the revolute joint between the crankshaft and connecting rod. The user can input 2 for *IB*, 3 for *JB*, input 0.5 and 0 for *xi* and *yi*, respectively; and −1 and 0 for *xj* and *yj*, respectively. Similarly, for the revolute joint between the connecting rod and the slider block, the user can enter 3 for *IB* and 4 for *JB*, 1 and 0 for *xi* and *yi* respectively, and 0 and 0 for *xj* and *yj*, respectively.

For the prismatic joint, the user can double-click on *TRNJ #1* to obtain the screen shown in Fig. 20. Using this screen, the user can define the data of the TRNJ joints that eliminate the relative rotation between the two bodies connected by this type of joint. This joint allows for one relative translation degree of freedom between the two bodies connected by this type of joint. The user must define the two bodies *IB* and *JB* connected by this joint, the nodes *Ni* and *Nj* to which the translational joint axis is attached on the two bodies (enter zero in the case of a rigid body), the local coordinates (*x1i*, *y1i*) and (*x2i*, *y2i*) of two points on the joint axis defined in body *IB* coordinate system, and the local coordinates of two other points (*x1j*, *y1j*) and (*x2j*, *y2j*) that lie on the joint axis and defined in body *JB* coordinate system. For the joint between the ground and the slider block, the user can enter 1 for *IB* and 4 for *JB*. For *x1i* and *y1i*, the user can enter 0 and 0, respectively. For *x2i* and *y2i*, the user can enter 1 and 0, respectively. For *x1j* and *y1j*, the user can enter 0 and 0, respectively. For *x2j* and *y2j*, the user can enter 1 and 0, respectively.

9.9 PERFORMING SIMULATIONS

Simulations can be performed using two different methods. One is simulation of a single job, while the other is the use of a batch execution of multiple jobs. The batch job will be described in the following section. To perform the simulation of a single job (project), the user must open the *Create Data File* panel, shown in Fig. 21. This panel can be opened by double-clicking the *Create Sams Data* item in the *Main Tree* on the right side of the main SAMS/2000 window, selecting the menu item *Data/Sams Data* in the main SAMS/2000 menu, or clicking the \ icon on the toolbar. There are eight text boxes in the *Create Data File* panel, which must be filled with appropriate entries in order to be able to perform the simulation. The entries to these text boxes are described as follows:

1. Name and Extension of the Data File (**Required**): This entry is the name of the project file, which is the same file that can be saved from the *File/Save* or *File/Save As* menu items of the main SAMS/2000 menu. This text box automatically lists the correct file name and path if the current project is an existing saved project. When a simulation is executed through the *Create Data File* panel, the project file listed in this field will

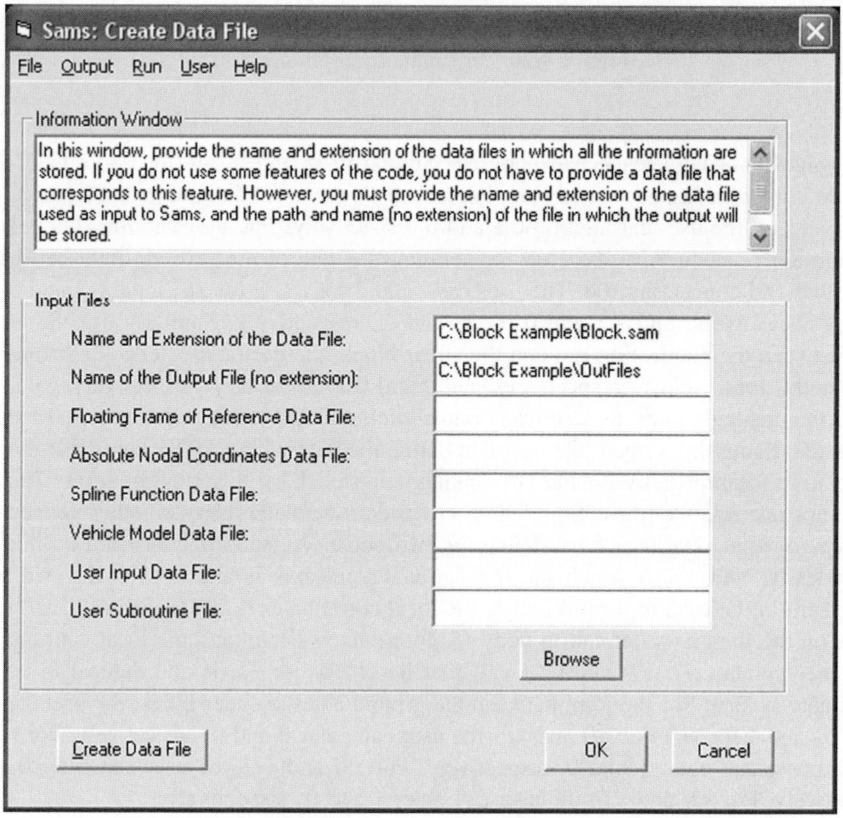

Figure 9.21 Create Data File screen

be re-saved with the current data created by the SAMS/2000 interface when the new SamsData.dat file is created.

2. Name of Output File (no extension) (**Required**): This field is used to identify the name and path of the two output files that SAMS/2000 creates for the user after the simulation is completed. The files that are saved using this path and file name are copies of the two files *C:\SAMS2000\RES.OUT* and *C:\SAMS2000\RECORD.DAT* which are also created in the C:\SAMS2000 directory. If the user wishes to have other copies of these files saved using other path and file name, the user must select the menu item *Output/Copy Results Files* in the menu of the *Create Data File* panel. The files saved will be *<path\filename.out>* and *<path\filename.rec>*. If the user does not wish to save these files, then the user can input any text in this field, though there must be text in this field in order to perform a simulation.
3. *FFR Data File*: This field is required only if the project includes flexible bodies modeled using the small deformation FFR formulation. The user must provide in this text box the path, name, and extension of the file obtained from PRESAMS that includes the flexible body data.
4. Absolute Nodal Coordinate Data File: This field is required only if the project includes flexible bodies modeled using the large deformation ANCF. The user must provide in this text box the path, name, and extension of the file obtained from PRESAMS that includes the flexible body data.
5. Spline Function Data File: This entry is required only if the project includes a Spline data file. The path, name, and extension of this file must be provided in this text box.
6. Vehicle Model Data File: In this field, the path, name, and extension of a file that includes additional data about vehicle components such as belt drives and rubber tracks can be specified. In order to understand the structure of these data, the user can consult with the *User Files* in the *Help menu* of the program.
7. User Input Data File: The user of SAMS/2000 can store user data in a file and have access to these data from the user subroutines. These data that can be prepared in the user-selected format, can be stored in the arrays of the code or in user-defined arrays. For more information about this feature of the code, the user can examine Subroutine USRINP, which is included in the User Subroutines of SAMS/2000. Using this field, the user can provide the path, name, and extension of this data file if the project requires the use of such a file, otherwise this field is left blank.
8. User Subroutine File: If the user makes changes in the user subroutines of SAMS/2000, the user must provide the path, name, and extension of the new Fortran file in this field, otherwise this field is left blank. The file specified in this field is copied to the file *C:\SAMS2000\SamsUser.FOR* just before the project execution, overwriting the copy of *C:\SAMS2000\SamsUser.FOR* in the C:\SAMS2000 directory. If the user does not provide a file name in this field, SAMS/2000 creates a copy of the original user subroutine file and overwrites *C:\SAMS2000\SamsUser.FOR* that exists in the C:\SAMS2000 directory.

The numerical data including the start and end time of the simulation can be provided by the user using the *System Parameters* panel which can be accessed by double-clicking *Numerical* of the main tree. After providing the system data and the necessary file names, the user can perform a simulation by clicking the ***Create Data File*** button. If there are no errors, SAMS/2000 interface displays a message that the data file has been successfully created. The user can click the ***OK*** button on the message box that appears. Another dialog box, in which the user is asked whether or not to perform the simulation, appears. If the user clicks the ***Yes*** button, SAMS/2000 performs the simulation. If the user clicks ***No***, then a new button labeled ***Run Sams*** appears. The user can click this button at any time to perform the simulation.

9.10 BATCH JOBS

SAMS/2000 allows the user to submit a list of several jobs that can be executed sequentially by the code using the *Batch Job* execution capabilities. The simulation result files of all the submitted jobs are stored allowing the user to have access to these files. In order to submit a batch job, the user must first create data files for all the jobs using SAMS/2000 interface as described in the online manual of the code. After creating the project data files, the user must restart SAMS/2000, by using the *File/Restart* menu in the main SAMS/2000 window, or by closing and restarting SAMS/2000. The *Batch Job* menu can be found on the opening panel that appears when starting SAMS/2000, as shown in Fig. 2. By clicking the *Batch Jobs* menu, the *Batch Jobs* screen shown in Fig. 22 is displayed.

In order to submit a batch job, the user must create for each simulation (job) a data row in the table shown in Fig. 22. To create a row in this table, the user must click first in the dark gray area of the *Batch Jobs* panel; this is the area, which has the titles of the columns. By clicking next the ***Add Element*** button, a new row will appear as shown in Fig. 23. The user can continue to click the ***Add Element*** button in order to have a number of rows equal to the number of the jobs in the batch.

The cells of each row are used to provide information that defines the files, which will be used in each simulation. The data that must be entered in the table are as follows:

1. Job Name: A name given by the user to identify the simulation (job).
2. Input Data File: The main SAMS/2000 project data file that will be used in this simulation. Note that the output files at the end of each simulation will all be copied to the directory that contains the project data file used for that simulation.
3. FFR Data File: The small deformation FFR data file obtained for flexible bodies using PRESAMS.
4. ANCF Data File: The large deformation ANCF data file obtained using PRESAMS for the very flexible bodies.
5. Spline Data File: The Spline data file used in this simulation.
6. Vehicle Data File: The vehicle data file used in this simulation.
7. User Input Data File: The file of the user-defined input data used in this simulation.
8. Track Data File: For the simulations of railroad vehicle systems, this is the file obtained using the SAMS/2000 track preprocessor. This file defines the track geometry.
9. User Subroutine File: The Fortran file that contains the user subroutines used in this simulation.

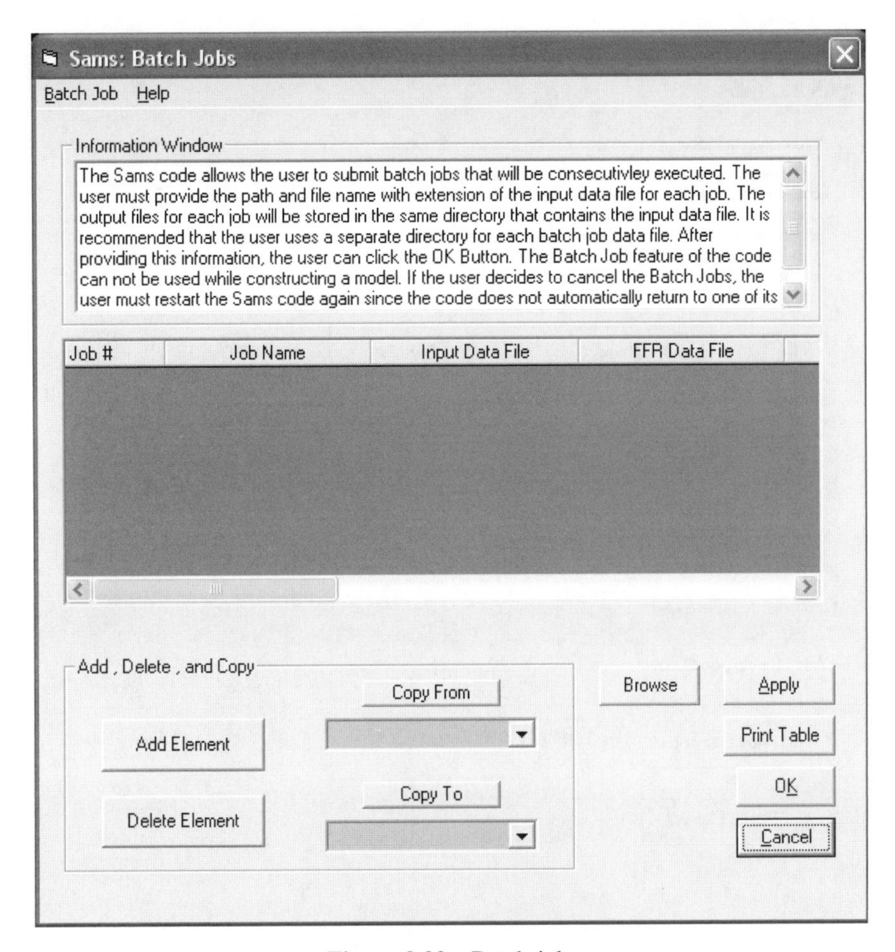

Figure 9.22 Batch jobs

Only the *Input Data File* is required to run a simulation, as previously mentioned. Other files may or may not be required depending on the application. If a file is not required, the corresponding cell must be left blank.

After entering the information for all the batch jobs, the user can save this information in a file that can be opened and used at a later time for executing this batch job. Selecting the menu *Batch Job/Save Batch List* opens a *Save File* dialog box, allowing the user to save the batch job in a text file. The user can also open a saved batch job file by selecting the menu *Batch Job/Open Existing Batch List*. The information stored in this file will be displayed in the cells of the batch job table. In order to execute the batch job, the user must select the menu *Batch Job/Execute Batch Job*. If the user wishes to prematurely terminate the simulations in this batch job, the user must close the SAMS/2000 interface.

The output files of each simulation are copied to the directory of the *Input Data File*. While the output files are all named with the same file name as the project data file, each file is given a different extension for the purpose of identification. For this reason, it is

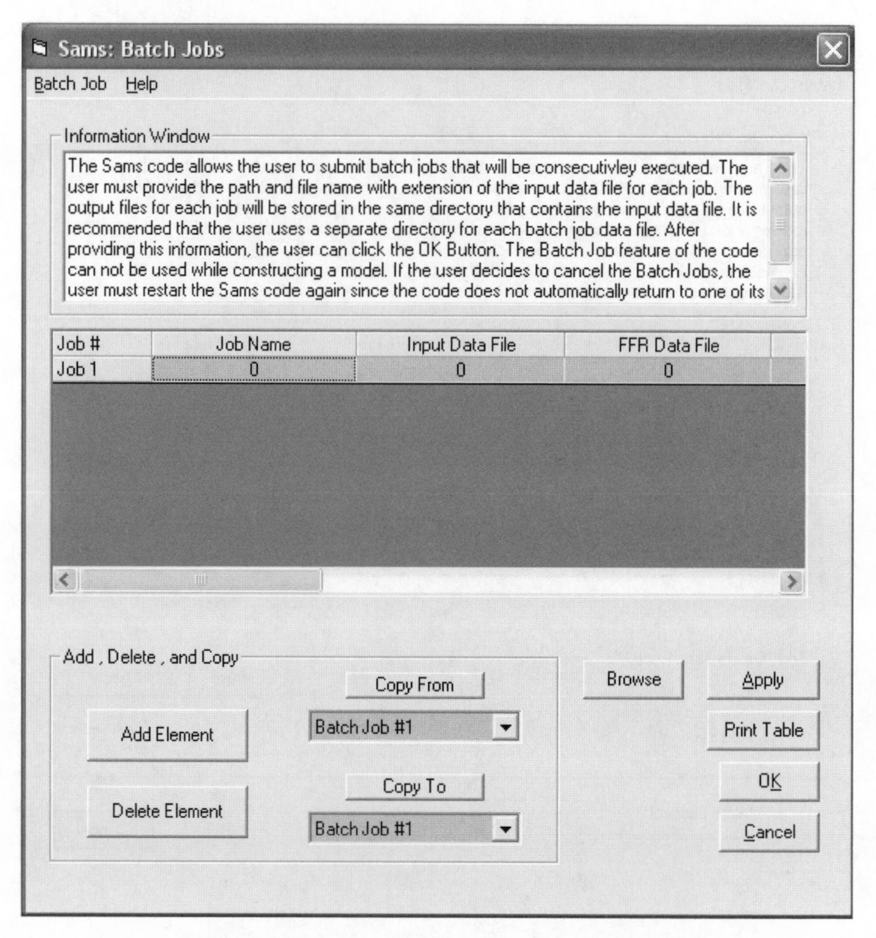

Figure 9.23 Adding rows to the batch job table

recommended that each *Input Data File* has its own directory in order to be able to easily identify the result files and avoid that some of the result files of one job to be overwritten by the files of the following job in the batch. When the batch job ends, SAMS/2000 displays a message box indicating the end of the simulations.

9.11 GRAPHICS CONTROL

SAMS/2000 allows the user to control the 3D graphics of a project using the *Graphics* panel. Three of the tabs of the graphics panel are shown in Figs. 24–26. To display the *Graphics* panel, the user can select the menu *Graphics/Properties* from the main SAMS/2000 menu bar, or right-click anywhere in the 3D graphical screen. Both actions open a similar pop-up menu, from which select *Properties*. The *Graphics* panel appears with the *Appearance* tab selected, as shown in Fig. 24.

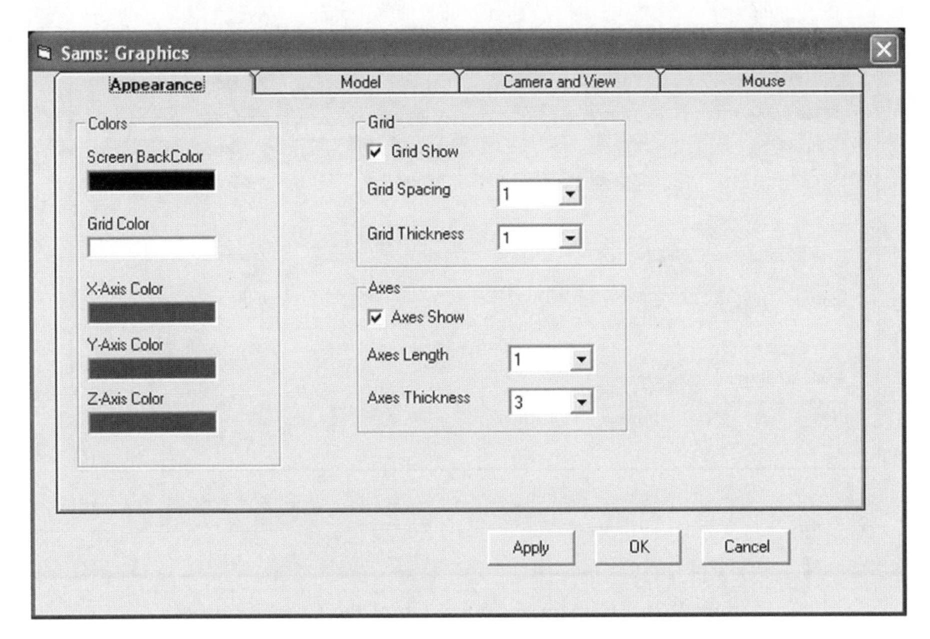

Figure 9.24 Appearance tab of the Graphics panel

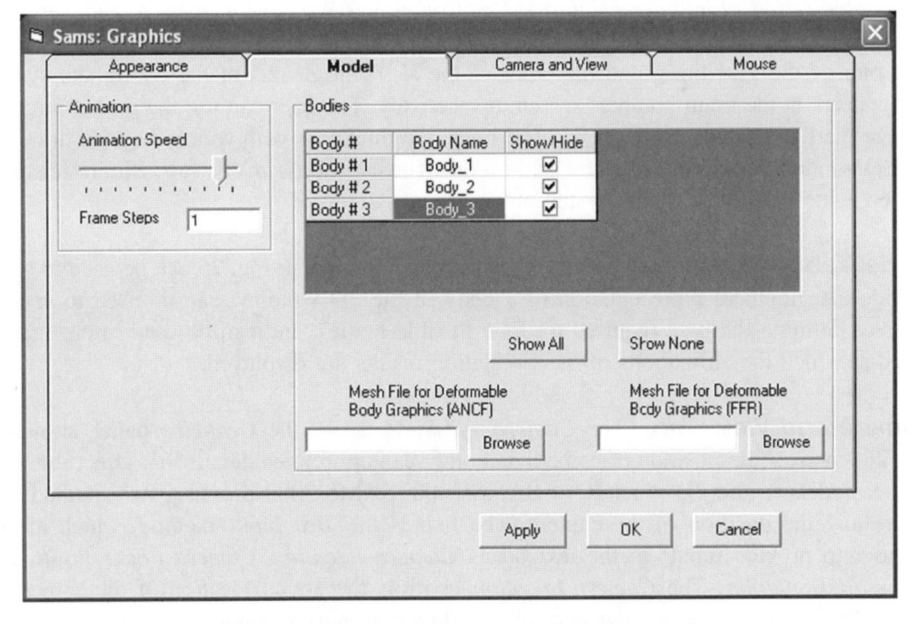

Figure 9.25 Model tab of the Graphics panel

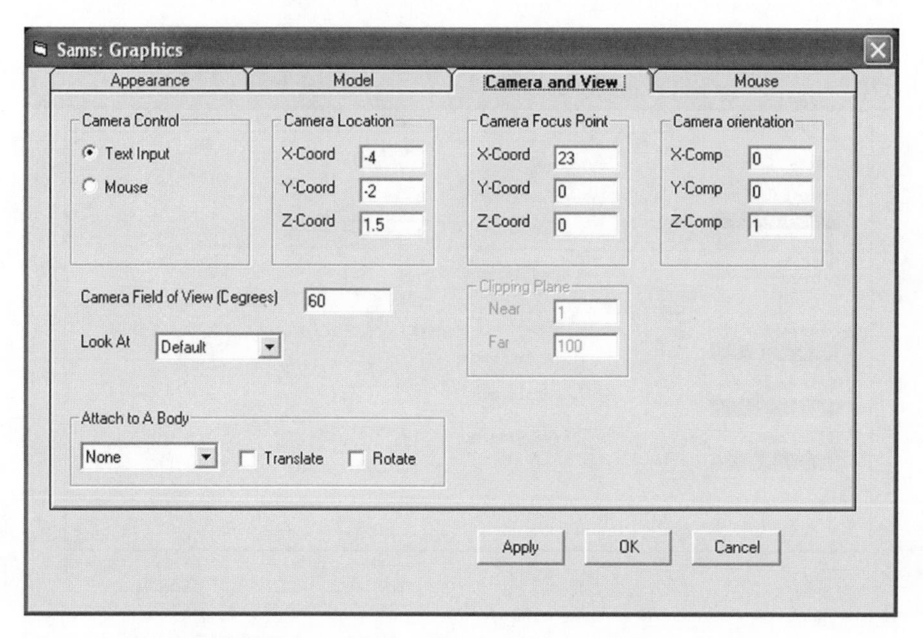

Figure 9.26 Camera and View tab of the Graphics panel

Appearance Tab The *Appearance* tab allows the user to change the color and size of the grid and the coordinate system axes. Double-clicking the colored box below *Screen BackColor, Grid Color, X-Axis Color, Y-Axis Color*, or *Z-Axis Color* opens a color selection dialog box, which can be used to change the color of the background of the 3D window, the color of the grid lines, and the color of the X, Y, or Z axis of the coordinate system that appears in the main graphics screen, respectively. The user can use the grid controls to change the line spacing between the grid lines, the thickness with which they are drawn in the 3D window, or simply turn the grid on or off using the *Grid* controls. Similar changes can be made in the coordinate system using the *Axes* control.

Model Tab The *Model* tab of the *Graphics* panel, shown in Fig. 25 can be used to show or hide the graphical representation of a body in the 3D window, can be used to specify the files defining the data required to draw flexible bodies, and can be used to change the speed at which the animations of the simulation results are displayed.

Camera and View Tab The *Camera and View* tab of the *Graphics* panel, shown in Fig. 26, controls the position of camera and other camera-related details that affect the view of the models in the 3D window of SAMS/2000. SAMS/2000 provides two methods for controlling the position of the camera. The first is the *Text Input* method, which allows the user to provide inputs in the text boxes *Camera Location, Camera Focus Point*, and *Camera Orientation*. The *Camera Location* controls the actual location of the camera in space. The *Camera Focus Point* controls the direction that the camera is pointed at in this space as well as the maximum distance that can be seen in the 3D window. The *Camera Orientation* controls the vertical direction in the perspective of the camera.

The second method for controlling the camera is the *Mouse* method, which allows the user by left-clicking the mouse to drag objects. Left-clicking and dragging while holding no buttons down rotates the camera. Left-clicking and dragging while holding down the *Ctrl* key on the keyboard will move the camera in a direction perpendicular to the screen. Left-clicking and dragging while holding the *Shift* key change the distance from the camera to the camera target. Left-clicking and dragging while simultaneously holding the *Shift* and the *Ctrl* keys will push or pull the camera in the direction the user is currently directing the camera. Note that when using the *Mouse* camera control option, a target that has the shape of a coordinate system will be displayed. The camera rotates around this target. If the user wants the camera to rotate around a particular body, then the target should be placed inside that body. Left-clicking and dragging with the *Ctrl* key pressed or the *Ctrl* key and the *Shift* key pressed simultaneously moves the target. Note that left-clicking and dragging with only the *Shift* key does not move the target. It only moves the camera closer to or farther from the target. The minimum and maximum distance that the user can see from the camera must be controlled manually when using the *Mouse* camera control. The *Near* and *Far* text boxes in the *Clipping Plane* group allow the user to specify how close or how far, respectively, that objects should be visible to the camera.

The user can also attach the camera to a body so that it moves in the global space with that body when SAMS/2000 is displaying an animation of the simulation results. To attach the camera to a body, place a check in the *translate* box if it is desired that the camera translates with a body, and place a check in the *rotate* box if it is desired that the camera rotates with the body. Note that if both the *translate* and the *rotate* check boxes are checked, then the camera will appear to have a constant position relative to the body. If using the *Mouse* camera control option, the user can still move the camera relative to the body it has been attached to by left-clicking and dragging. The user selects the body to attach the camera by using the *Attach to a Body* drop-down.

If the user desires that the camera always be pointed at a certain body, then the user must select a body to point the camera toward from the *Look At* drop-down. If the camera has been commanded to *Look At* a body while using *Mouse* camera control, the rotation of the camera using the mouse will be disabled.

9.12 ANIMATION CAPABILITIES

Once a simulation has been successfully completed, SAMS/2000 can animate the results to allow the user to visualize the motion of the bodies. In order to animate the results of a simulation, the *Play, Pause*, and *Stop* controls in the *Graphics/Animation* submenu can be used. The user can create an Audio Video Interleave (AVI) movie file of an animation by selecting the menu *File/Create AVI File*. A dialog box that allows the user to select a codec and set AVI file creation parameters will appear. The AVI creation process uses the current camera configuration specified in the *Graphics* panel.

9.13 GENERAL USE OF THE INPUT DATA PANELS

For conformity, the data of most standard SAMS/2000 elements are entered using data input panels similar to the one shown in Fig. 27 that was also used in the planar slider crank

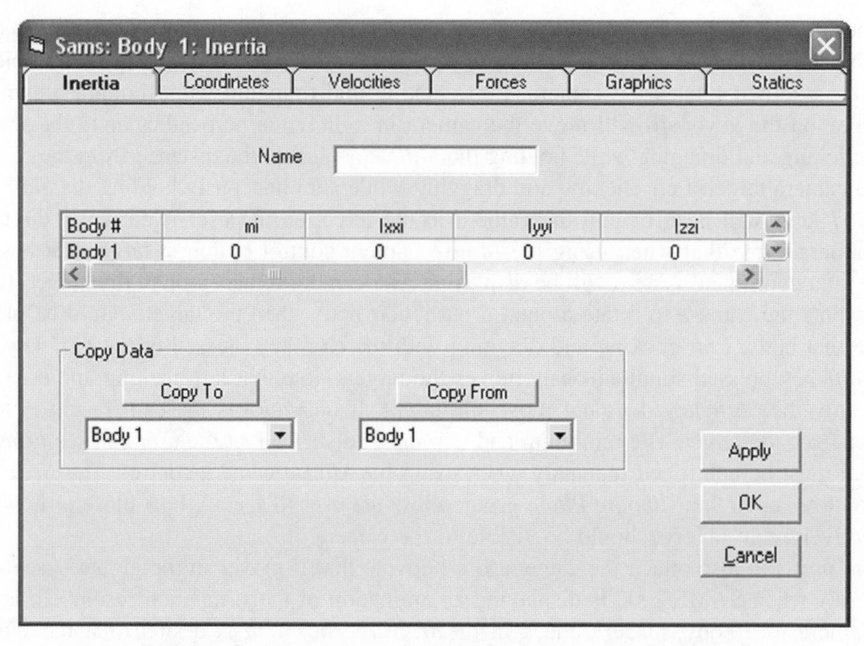

Figure 9.27 Example of individual element input data panel

mechanism example discussed previously in this chapter. There are two panel types that are used by the user to enter the input data of various elements. These are the *Individual Element Panel*, as the one shown in Fig. 27, and the *Tabulated Data Panel*, as the one shown in Fig. 28. The general use of these two panel types will be discussed in this section.

Almost all panels that accept input from the user have the buttons ***Apply, OK***, and ***Cancel***. If the user changes any data in the panel, the ***Apply*** button must be clicked before closing the panel or switching tabs in the panel. Changes made to the data in a panel will not be applied unless the ***Apply*** button is clicked. In most panels, the function of the ***OK*** and the ***Cancel*** buttons are identical.

Many panels include the *Copy Data* controls. These controls have a ***Copy To*** button with a corresponding list of similar elements in a drop-down list, and similarly, they have a ***Copy From*** button with a corresponding drop-down list of similar elements. These controls allow the data on the current tab of the current panel to be copied to or from another similar element. For example, the tab of the panel showed in Fig. 27 displays the body mass and the components of the rotational inertia of a 3D body. If the user wishes to fill this tab, and this tab only, with data from another body, the user can select that body from the drop-down list below the ***Copy From*** button, and then click the ***Copy From*** button. This will fill the body mass and inertia component fields with data from another body. Note that the user must still click the ***Apply*** button before changing tabs or closing this panel or else the changes will be lost. Similarly, if the user wishes to copy these mass and inertia data to another body, overwriting the current mass and inertia data of that body, the user can select the target body in the drop-down list below the ***Copy To*** button, and then click the ***Copy To*** button. Note, although, that if the data in this panel has been changed but the ***Apply*** button has not been clicked, the user must click the ***Apply*** button before clicking the ***Copy To*** button, or

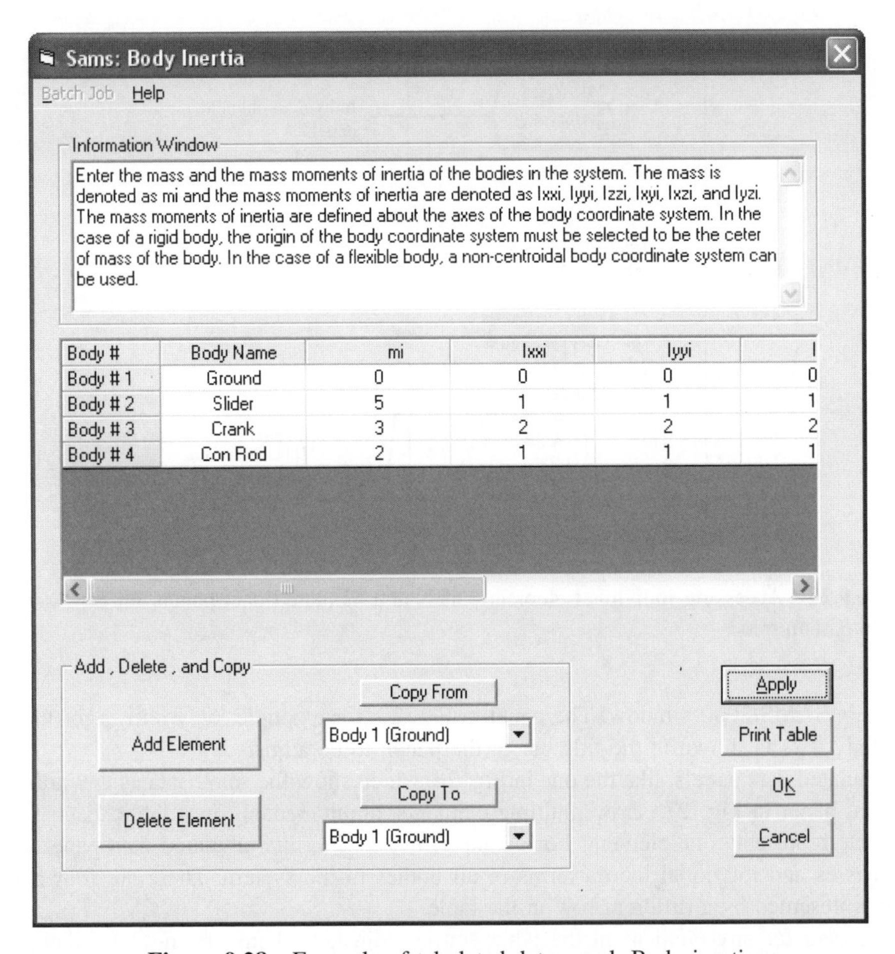

Figure 9.28 Example of tabulated data panel: Body inertia

else the data copied to the target body will be the data before the changes in the tab were made.

Individual Element Panels and Tabulated Data Panels As previously mentioned, there are two common panel types that are used to display the majority of the data of a SAMS/2000 project. One kind is the individual element panel, which is shown in Fig. 27. The other kind is the tabulated data panel, which is shown in Fig. 28. In some cases, the same data can be accessed through both kinds of panels, as is the case with the body data being shown in Figs. 27 and 28. Figure 29 shows where in the control tree to find each kind of panel for body data.

Individual element panels, like the one in Fig. 27, only display data for a single element such as a single body or a single force element. These panels frequently have more than one tab. Note, as stated in the previous section, that if the data on one tab is altered before moving to another tab, the user must click the ***Apply*** button, or the changes will be lost. The user typically accesses individual element panels from the control tree on the right of

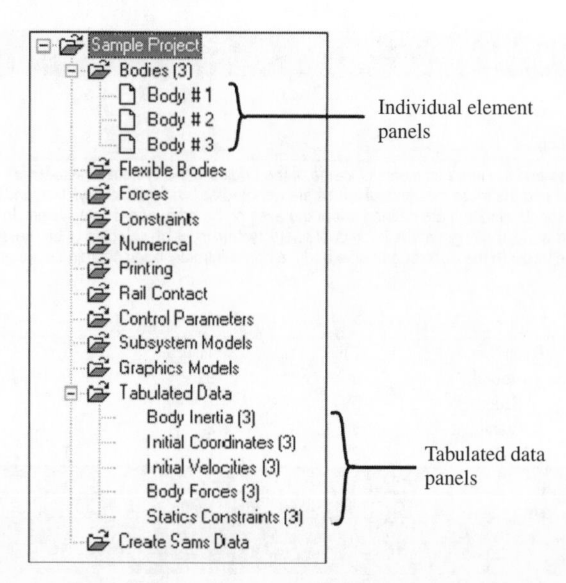

Figure 9.29 Accessing individual element panels and tabulated data panels for the body data from the main tree

the main SAMS/2000 window. The panel in Fig. 27, for example, only allows for the data for Body # 2, as shown in the title bar of the panel, to be adjusted.

Tabulated data panels, like the one in Fig. 28, which show the same data as the individual element panel in Fig. 27, show multiple elements simultaneously in a table. Each row in the table represents one element. For example, in Fig. 28, this tabulated data panel shows the masses and rotational inertia terms of all bodies in the system. There are four bodies, each represented by a different row in the table.

The data for any element in the table can be adjusted when viewing a tabulated data panel. The data can be adjusted by selecting a cell in the table and by directly modifying it. New elements can be added to the project by clicking the ***Add Element*** button. First, the user selects the row after which the new element will be added. Then, clicking the ***Add Element*** button will add a new row below the row that was selected. If no rows currently exist in the table, click in the dark gray area before clicking the ***Add Element*** button. Similarly, elements can be deleted from the project when viewing the elements in a tabulated data panel by selecting the row corresponding to the element that is to be deleted and then clicking the ***Delete Element*** button. Note that if changes to the data are made before adding a row or deleting a row, the user must click the ***Apply*** button before clicking the ***Add Element*** or ***Delete Element*** buttons.

9.14 SPATIAL ANALYSIS

Using the individual element and tabulated data panels discussed in the preceding section, the input data for the spatial systems can be introduced in a manner similar to the planar systems. For example, in the case of the body data, the user must provide the following

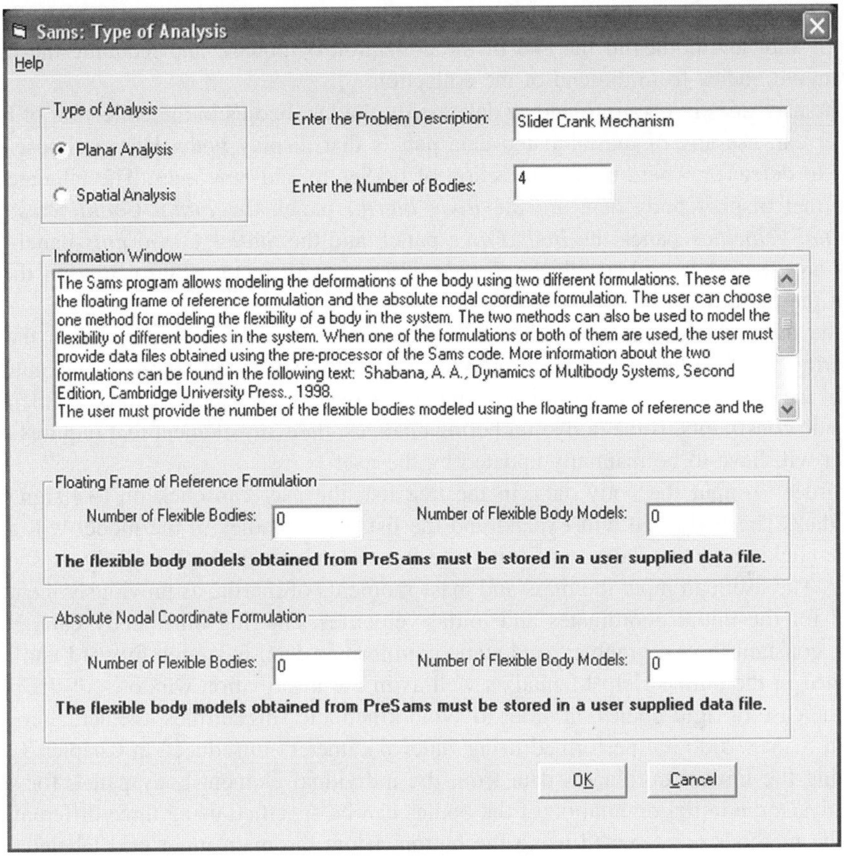

Sams: Type of Analysis

Help

Type of Analysis

Planar Analysis

Spatial Analysis

Enter the Problem Description: Slider Crank Mechanism

Enter the Number of Bodies: 4

Information Window

The Sams program allows modeling the deformations of the body using two different formulations. These are the floating frame of reference formulation and the absolute nodal coordinate formulation. The user can choose one method for modeling the flexibility of a body in the system. The two methods can also be used to model the flexibility of different bodies in the system. When one of the formulations or both of them are used, the user must provide data files obtained using the pre-processor of the Sams code. More information about the two formulations can be found in the following text: Shabana, A. A., Dynamics of Multibody Systems, Second Edition, Cambridge University Press., 1998.
The user must provide the number of the flexible bodies modeled using the floating frame of reference and the

Floating Frame of Reference Formulation

Number of Flexible Bodies: 0 Number of Flexible Body Models: 0

The flexible body models obtained from PreSams must be stored in a user supplied data file.

Absolute Nodal Coordinate Formulation

Number of Flexible Bodies: 0 Number of Flexible Body Models: 0

The flexible body models obtained from PreSams must be stored in a user supplied data file.

OK Cancel

Figure 9.30 Type of Analysis panel

information: mass and mass moments of inertia, initial coordinates, initial velocities, constant forces, graphics data, and static equilibrium data. These data can be provided or accessed through either individual element panel or from various tabulated data panels. The *graphics data* can be provided or accessed only from the individual element panels. As in the case of the planar systems, the individual element panels for each body are found under the *Bodies* element of the control tree on the right of the SAMS/2000 window. The tabulated data panels for this body data are found beneath the *Tabulated Data* item in the control tree.

For both planar and spatial systems, the user can add or delete bodies through two methods. One method is through the *Type of Analysis* panel, as shown in Fig. 30. This panel can be opened by double-clicking the *Bodies* element of the control tree, which is shown in Fig. 29, and is found at the right of the main SAMS/2000 window. This panel can also be opened by clicking the toolbar item \ or the menu item *System/Bodies*. Note that this panel is also displayed immediately after the user clicks the *New* toolbar button or the menu item *File/New* in the main SAMS/2000 window, which creates a new project. The text box labeled *Enter the Number of Bodies* controls the total number of bodies in the project. Changing this number changes the number of bodies in the project. If a project already exists, then increasing the number of bodies in the *Enter the Number of Bodies* text box will add new bodies to the end of the collection of bodies, and reducing the number will remove bodies from the end of the collection.

If the user does not want to add or delete only the last bodies in the collection of bodies, the user can use any of the tabulated data panels that display body data to choose which bodies to delete or where in the collection of bodies to add new ones. The tabulated data panels that display body data are the *Body Inertia* panel, the *Initial Coordinates* panel, the *Initial Velocities* panel, the *Body Force* panel, and the *Statics Constraints* panel. These panels are opened from beneath the *Tabulated Data* item in the control tree on the right side of the main SAMS/2000 window.

If the user adds or deletes bodies from other than the end of the list, then the body numbering will be altered, and each body is assigned a new body number. This number is used in the data defining many other elements of the project, such as force elements and kinematic constraints. If the body numbering changes, then any element that requires a body number will have to be manually updated by the user.

In order to input the body data, in the tree list, the user can click on (+) sign beside the bodies. The body tree will expand and the list of the bodies in the model will appear. Double-clicking on any body will bring the input data screen for this body as shown in Fig. 27. The user can input the mass and mass moments of inertia, as previously mentioned. Except for the initial coordinates and initial velocities, entering other body data, such as inertia, constant forces, graphics, and static equilibrium data, is straightforward and is well explained in the online Help Manual as well as in the information windows of the panels.

In the case of rigid bodies, in order to avoid kinematic singularities, the actual computations in SAMS/2000 are performed using Euler parameters introduced in Chapter 8. When accessing the initial coordinates data from the individual element body panel, for spatial, 3D analysis, the initial orientation of the bodies can be specified using three different methods. The methods are selected using the button below the main input grid labeled, either, ***Euler Angles***, ***Two Axes***, or ***Three Axes***. The button can be seen in Fig. 31. Clicking the input selection method button changes the input method. When inputting the orientation parameters using Euler angles, there will be three input fields for the following three Euler Angles: rotation angle ϕ (*Phi*) around the Z axis, rotation angle θ (*Theta*) around the X axis, and rotation angle ψ (*Psi*) around the new Z axis that will be obtained after the first two rotations. Note that when using the *Initial Coordinates* tabulated data panel to access the initial orientations of bodies, the data is displayed only in the *Euler Angles* format. The *Two Axes* and *Three Axes* options are not available. If using the *Two Axes* or *Three Axes* method of input, then the *Coordinates* tab of the individual element body panel will display entry fields that will allow the user to define the axes of the body coordinate system in the initial configuration.

The *Velocities* tab of the body individual element panel or the *Initial Velocities* tabulated data panel displays the velocities of the bodies at the beginning of the simulation. In the case of 3D analysis, three components of the absolute velocity vector of the reference point of the body in the initial configuration must be defined. The components of the absolute angular velocity vector defined in the global coordinate system must also be defined at the initial configuration.

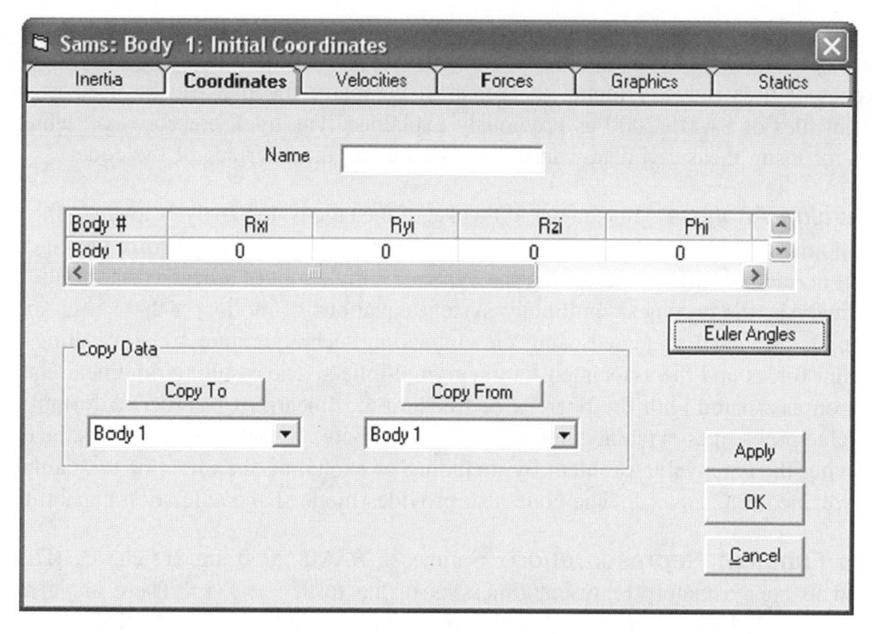

Figure 9.31 Coordinates tab of the body individual element panel

9.15 SPECIAL MODULES AND FEATURES OF THE CODE

As previously mentioned, SAMS/2000 has constraint and force libraries that include a large number of standard elements all of which cannot be covered in this chapter. The formulations of some of the standard constraint and force elements implemented in SAMS/2000 are discussed in previous chapters of this book. In addition to the *Main Tree* that provides access to some of these elements, there are three other control trees that allow the user to have access to other standard constraint and force elements that are part of the SAMS/2000 libraries. These trees are the *Vehicle, Flexibility*, and *Numerical Trees*. These trees are accessible from the main menu item *View* or by clicking on the corresponding icon on the status bar. The online Help Manual provides a detailed list of the standard constraint and force elements that can be accessed using the control trees of the code. SAMS/2000 has special modules and advanced features also; some of which will be discussed in this section. More detailed information on the special modules and advanced features of SAMS/2000 can be found in the online Help Manual.

Rail Module SAMS/2000 has advanced capabilities for modeling complex railroad vehicle systems. Several 3D wheel/rail contact formulations (Shabana et al., 2008) that allow for predicting the locations of the wheel/rail contact points online are implemented in SAMS/2000. The wheel/rail contact forces, including the tangential creep forces and spin moments, are calculated and introduced to the nonlinear dynamic equations as generalized and/or constraint forces. The wheel/rail contact algorithms implemented in SAMS/2000 allow the use of the Spline functions to describe the wheel and rail profiles. SAMS/2000 has a preprocessor that can be used to construct tracks with arbitrary geometry. The track preprocessor uses industry inputs to prepare a geometry file that can be used as one of the input files of SAMS/2000 as previously explained. The track preprocessor, which also allows for using measured data, can be accessed using the interface of the code.

Eigenvalue Analysis In Chapter 8, the use of the eigenvalue analysis to study the stability of multibody systems was discussed. SAMS/2000 can be used to perform the eigenvalue analysis at selected time-points during the dynamic simulation. In order to perform the eigenvalue analysis, the nonlinear multibody system equations of motion at these user-specified time-points are first formulated using the embedding technique, thereby eliminating all the constraint forces and the associated Lagrange multipliers. The resulting nonlinear equations of motion associated with the degrees of freedom are linearized in order to formulate the eigenvalue problem as explained in Chapter 8 of this book. SAMS/2000 provides the options for solving the eigenvalue problem by including or excluding the effect of the damping as also explained in Chapter 8. The code also provides mode shape animation capabilities.

Spline Function Representation Splines in SAMS/2000 are 2D curves. They can be used to approximately represent functions in the form $y = f(x)$. There are a number of standard SAMS/2000 constraint and force elements that already use Spline representations. Splines, however, can also be used through the user subroutines to allow the user to define any function relationship, including nonlinear spring, damper, and actuator force/displacement relationships. A single Spline is represented in SAMS/2000 as a series of piecewise continuous cubic polynomials. In general, a single Spline is defined as a series of input points (x, y) and the slope dy/dx at the start and end points. These points are referred to as *nodes* on the Spline. Between any two nodes, a single cubic polynomial is found that represents the Spline between the nodes. The cubic polynomials will start at the first node and end at the last node, thereby creating a single curve starting at the first node, passing through all intermediate nodes, and ending at the last node (Press et al., 1992).

Two kinds of Splines can be represented in SAMS/2000. They are the *single Spline* and the *double Spline*. A single Spline defines the relationship $y = f(x)$, where x is the independent variable that is specified by the user, and y is the value returned by the Spline to the user. A double Spline defines the two relationships $x = f_x(s)$ and $y = f_y(s)$, where s is the independent variable that represents an arc length and both x and y are calculated from s. Note that the user can create the same functionality as a double Spline through the use of two single Splines. As discussed in the online Help Manual of the code, the use of the arc length s instead of x can help in avoiding singularities in some applications. For example, in the case of a vertical straight line, for a given x there is no corresponding unique y value. Such a problem can be avoided by using the double spline. It is also important to point out that the interface of SAMS/2000 allows the user to manipulate the Spline data by inserting and/or deleting points. The code interface can also be used to plot the existing and new Spline data and their derivatives in order to provide enough information that allows the user to understand the degree of the smoothness of the Spline curves used in his/her application.

Integration Methods There are several numerical integration methods available for solving the differential equations of motion of the constrained multibody systems. The methods include *Adams-Bashforth-Moulton method* (Shampine and Gordon, 1975), *Runge Kutta method* (Atkinson, 1978), and *HHT method* (Hilber et al., 1977; Hussein et al., 2008). The integration method to be used is selected from the *System Parameters* dialog box. In the current version of the code, it is recommended to use *Adams Method* since the other methods are still under development. The *Adams Method* is the preferred method in many other multibody system codes designed to solving the differential equations of motion. This method, which is a multistep, predictor/corrector method, automatically attempts to generate a solution that meets user-specified error tolerances by automatically adjusting the order and time step size. The *Runge Kutta Method 1* should only be used if the *Adams Method* fails; it does not adjust its time step size and it does not estimate the solution error.

Change of the Degrees of Freedom SAMS/2000 allows the user to keep one set of degrees of freedom during the entire dynamic simulation or change this set at equally spaced points in time. In many simulation scenarios, it might be necessary to change the degrees of freedom in order to avoid singular configurations as previously discussed in this book. The constraint Jacobian matrix is used by the code to identify the set of independent coordinates (degrees of freedom). In order for the user to allow for the change of the set of the system degrees of freedom, the text box labeled *# of Degree of Freedom Change* of the *System Parameters* panel can be used. The value in this text box indicates to SAMS/2000 how many times during the simulation that the system should be reevaluated to determine the coordinates that are used as the degrees of freedom. The *System Parameters* panel can be accessed by double-clicking on the *Numerical* item in the main tree. The identification of the coordinates to use as the degrees of freedom is done automatically by the code, as previously mentioned. Therefore, the user does not need to supply which coordinates to use. Nonetheless, SAMS/2000 allows also the user to select and change the degrees of freedom manually using the user subroutines. The use of this option, however, is not recommended because in the case of complex multibody systems, the use of the numerical structure of the constraint Jacobian matrix will always lead to an optimum set of degrees of freedom.

Solution Procedures The user can also select different solution procedures for the constrained dynamic equations. The first three procedures employ the augmented form of the equations of motion expressed in terms of Lagrange multipliers. The first procedure is based on identifying a minimum set of independent coordinates and ensures that the constraints are satisfied at the position, velocity, and acceleration levels. When this method is used, only the independent accelerations are integrated. In the second procedure, all the accelerations are integrated, and a check on the constraint violation is made to ensure that the constraints are satisfied at the position, velocity, and acceleration levels. In the third procedure, all the accelerations are integrated with no check made on the constraint violation at the position and velocity levels, that is, there is no guarantee that the constraints are satisfied at the position and velocity levels while they are satisfied at the acceleration level. The fourth solution procedure is based on the penalty method (still under development) and does not require the use of algebraic equations and Lagrange multipliers. The fifth procedure is based on a recursive algorithm that utilizes the joint coordinates (under development). In the last three solution procedures, the iterative Newton–Raphson method is not used. The recursive method in the current version, which is still under development, can only be used in the case of rigid body analysis of multibody systems that include only ground constraints, spherical, revolute, prismatic, and cylindrical joints. This version of the recursive formulation does not support other joint types or flexible bodies.

Subsystem Models SAMS/2000 gives the user the capability to duplicate any model. This tool enables the user to create large models with minimum effort if this model can be divided into a set of repeated systems that can also be altered after the assembly. To start adding a subsystem, the user should first create the model that he/she will use by using SAMS/2000. The user can double-click on the *Subsystem Model* to obtain the *Subsystem Models* screen and start adding subsystem models. Using this screen, which is similar to the tabulated data screen, the user can add models as previously described. A row of cells will be created to start adding the subsystem model. The user can use the ***Browse*** button to choose the data file of the subsystem. The coordinate systems and units used in the data file of the subsystem must be consistent with the coordinate systems and units used in the main model. After selecting the subsystem data file, the user can define the location of the subsystem in the multibody model by using the translation and rotation variables provided in the table. The rotation can be performed about any axis defined by a unit vector that the user can define. All the bodies in the subsystem will be translated and rotated by the translations and rotations specified by the user. The user can then click on ***Read*** button to start reading the subsystem, and can click on ***Manipulate*** to adjust the subsystem position and orientation by the translational and rotational values. The user can also delete subsystems by choosing the subsystem and clicks on the ***Delete Element*** button. If this option is used, all the bodies included in the deleted subsystem will be removed.

User Subroutines The user can obtain standard output from SAMS/2000 by providing standard input data, as previously explained in this chapter and in the online Help Manual of the code. The library of the code includes several standard force elements and constraint functions that can be used by the user. In addition to these standard features that can be used in developing complex multibody system models, SAMS/2000 has a large list of user subroutines that allow the user to provide arbitrary force and constraint functions that are not part of the code library. Because all the user subroutines cannot be explained in this chapter, few features that can be used by the user through the user subroutines are discussed in this section. Some of the capabilities provided by the user subroutines are the following:

1. Provide *User's Writing Statements*
2. Provide *User's Nonlinear Forcing Functions* that depend on the system-generalized coordinates, velocities, and time
3. Provide *User's Constraint Functions* that describe specified motion trajectories or nonstandard joints that are not included in the code library
4. Provide *User's nonlinear characteristics* for rectilinear and torsional spring, damping, actuator, bushing, and bearing force coefficients
5. Provide *User's input data* that can be used in the formulation of user-provided force and CONMODs
6. Provide *User's Differential Equations* in the first-order form
7. Select the *System Degrees of Freedom* manually

An experienced user of SAMS/2000 can make use of many of these features by using a set of *User Subroutines* provided in the file *SAMSUSER.for* in the directory *C:\SAMS2000*. The user can edit the file SAMSUSER.for using the SAMS/2000 interface and an appropriate Fortran editor. The user's Fortran editor can be provided in the file *SAMSFORT.cfg* as described in the online Help Manual of the code where examples are given.

Flexible Body Modeling SAMS/2000 has advanced flexible body modeling capabilities that are based on a successful integration of finite element and multibody system algorithms. Both small and large deformation multibody system applications can be handled using SAMS/2000. The integration of *small deformation* finite element and multibody system algorithms is accomplished by using the finite element *FFR formulation*. The integration of *large deformation* finite element and multibody system algorithms in SAMS/2000 is based on the *ANCF*. While the subject of flexible body dynamics is not covered in this book, it is important that the reader becomes aware of the potential use of general purpose multibody computer codes and their advanced capabilities.

The simulation of multibody systems that consist of interconnected deformable bodies modeled using the FFR formulation requires the use of structural dynamics preprocessor to evaluate the inertia shape integrals that appear in the nonlinear dynamic equations of motion (Shabana, 2005). For this reason, an interface between general purpose finite element and multibody system computer programs needs to be established. The interface or the preprocessor used with SAMS/2000 is called *PRESAMS*. The preprocessor PRESAMS is used to generate a set of data required for the dynamic analysis of the flexible components in the multibody systems. The output of this preprocessor is stored in files that are read by SAMS/2000 which is considered as the *main processor* for the dynamic analysis of constrained multibody systems. No modifications in the output of the preprocessor PRESAMS need to be made since the output is in the correct format that is directly acceptable by SAMS/2000. In PRESAMS, all the shape integral matrices such as the mass and stiffness matrices associated with the elastic coordinates as well as the constant matrices required to evaluate the nonlinear inertia coupling between the rigid body motion and the elastic deformation are generated. These matrices are evaluated once in advance before the dynamic analysis in order to reduce the number of operations performed during the dynamic simulation. With the information provided by the preprocessor PRESAMS, the multibody system identification can be completed by defining the initial configuration of each flexible and rigid component; the joint types and the locations of the joint definition points with respect to a chosen body coordinate system; and the attachment points of the springs, dampers, and actuators and other force elements as described in the online Help Manual of the code. Since flexible body dynamics is beyond the scope of this book, the interested reader can consult with the online Help Manual of the code as well as more advanced books and technical papers on this important subject.

REFERENCES

1. Ambrosio, J. A. C., and Gonclaves, J. P. C., "Complex Flexible Multibody Systems with Application to Vehicle Dynamics," *Multibody System Dynamics*, vol. 6, 2001, pp. 163–182.

2. Anderson, K. S., and Duan, S. Z., "Highly Parallel Algorithm for Motion Simulation of Complex Multi-Rigid-Body Mechanical Systems," *AIAA Journal of Guidance, Control and Dynamics*, vol. 23, 2000, pp. 355–364.

3. Andriacchi, T. P., and Alexander, E. J., "Studies of Human Locomotion: Past, Present and Future," *Journal of Biomechanics*, vol. 33, 2000, pp. 1217–1224.

4. Arczewski, K., and Blajer, W., "A Unified Approach to the Modelling of Holonomic and Nonholonomic Mechanical Systems," *Mathematical and Computer Modelling of Dynamical Systems: Methods, Tools and Applications in Engineering and Related Sciences*, vol. 2, 1996, pp. 157–174.

5. Arnold, M., and Bruls, O., "Convergence of the Generalized-α Scheme for Constrained Mechanical Systems," *Multibody System Dynamics*, vol. 18, 2007, pp. 185–202.

6. Atkinson, K. E., *An Introduction to Numerical Analysis*, John Wiley & Sons, 1978.

7. Bae, D. S., Han, J. M., Choi, J. H., and Yang, S. M., "A generalized Recursive Formulation for Constrained Flexible Multibody Dynamics," *International Journal for Numerical Methods in Engineering*, vol. 50, 2001, pp. 1841–1859.

Computational Dynamics, Third Edition Ahmed A. Shabana

8. Barhorst, A. A., and Everett, L. J., "Modeling Hybrid Parameter Multiple Body Systems: A Different Approach," *International Journal of Nonlinear Mechanics*, vol. 30, 1995, pp. 1–21.

9. Bauchau, O. A., and Rodriguez, J., "Modeling of Joints with Clearance in Flexible Multibody Systems," *International Journal of Solids and Structures*, vol. 39, 2002, pp. 41–63.

10. Bae, D. S., Lee, J. K., Cho, H. J., and Yae, H., "An Explicit Integration Method for Realtime Simulation of Multibody Vehicle Models," *Computer Methods in Applied Mechanics and Engineering*, vol. 187, 2000, pp. 337–350.

11. Bayo, E., García de Jalón, J., and Serna, M. A., "A Modified Lagrangian Formulation for the Dynamic Analysis of Constrained Mechanical Systems," *Computer Methods in Applied Mechanics and Engineering*, vol. 71, 1988, pp. 183–195.

12. Berbyuk, V., "Towards Dynamics of Controlled Multibody Systems with Magnetostrictive Transducers," *Multibody System Dynamics*, vol. 18, 2007, pp. 203–216.

13. Blajer, W., "A Projection Method Approach to Constrained Dynamic Analysis," *Journal of Applied Mechanics*, vol. 59, 1991, pp. 643–649.

14. Blajer, W., "A Geometric Unification of Constrained System Dynamics," *Multibody System Dynamics*, vol. 1, 1997, pp. 3–21.

15. Borri, M., Bottasso, C. L., and Trainelli, L., "Integration of Elastic Multibody Systems by Invariant Conserving/Dissipating Algorithms, I. Formulation," *Computer Methods in Applied Mechanics and Engineering*, vol. 190, 2001, pp. 3669–3699.

16. Bottasso, C. L., Borri, M., and Trainelli, L., "Integration of Elastic Multibody Systems by Invariant Conserving/Dissipating Algorithms, II. Numerical Schemes and Applications," *Computer Methods in Applied Mechanics and Engineering*, vol. 190, 2001, pp. 3701–3733.

17. Brul, O., and Eberhard, P., "Sensitivity Analysis for Dynamic Mechanical Systems with Finite Rotations," *International Journal for Numerical Methods in Engineering*, vol. 74, 2007, pp. 1897–1927.

18. Chen, K., and Beale, D., "A New Method to Determine the Base Inertial Parameters of Planar Mechanisms," *Mechanism and Machine Theory*, vol. 39, no. 9, 2002, pp. 971–984.

19. Choi, J. H., Shabana, A. A., and Wehage, R. A., "Propagation of Nonlinearities in the Inertia Matrix of Tracked vehicles", Technical Report prepared for the U.S. Army Tank Automotive Command, TCN 93097, August 1993.

20. Cuadrado, J., Gutierrez, R., Naya, M. A., and Gonzalez, M., "Experimental Validation of a Flexible MBS Dynamic Formulation through Comparison between Measured and Calculated Stresses on a Prototype Car," *Multibody System Dynamics*, vol. 11, 2004, pp. 147–166.

21. Eberhard, P., and Schiehlen, W., "Hierarchical Modeling in Multibody Dynamics," *Archive of Applied Mechanics*, vol. 68, 1998, pp. 237–246.

22. Garcia de Jalon, J., and Bayo, E., *Kinematic and Dynamic Simulation of Multibody Systems*, Springer-Verlag, New York, 1993.

23. Ginsberg, J., *Engineering Dynamics*, Cambridge University Press, New York, 2008.

24. Goldstein, H., *Classical Mechanics*, Addison-Wesley, 1950.

25. Greenwood, D. T., *Principles of Dynamics*, 2nd ed., Prentice Hall, Englewood Cliffs, NJ, 1988.

26. Haug, E. J., *Computer Aided Kinematics and Dynamics of Mechanical Systems*, Allyn and Bacon, Boston, MA, 1989.

27. Hilber, H. M., Hughes, T. J. R., and Taylor, R. L., "Improved Numerical Dissipation for Time Integration Algorithms in Structural Dynamics," *Earthquake Engineering and Structural Dynamics*, vol. 5, 1977, pp. 283–292.

28. Hilber, M., and Kecskemethy, A, "A Computer-oriented Approach for the Automatic Generation and Solution of the Equations of Motion for Complex Mechanisms", Proceedings of the 7th IFFTOMM World Congress on the Theory of Machines and Mechanisms, Selvia, 1987.

29. Hussein, B., Negrut, D., and Shabana, A. A., "Implicit and Explicit Integration in the Solution of the Absolute Nodal Coordinate Differential/Algebraic Equations," *Nonlinear Dynamics*, vol. 54, no. 4, 2008, pp. 283–296.

30. Huston, R. L., *Multibody Dynamics*, Butterworth-Heinemann, Stoneham, MA, 1990.

31. Kaplan, W., *Advanced Calculus*, 4th ed., Addison-Wesley, Reading, MA, 1991.

32. Khulief, Y. A., and Shabana, A. A., "A Continuous Force Model for the Impact Analysis of Flexible Multi-Body Systems," *Mechanism and Machine Theory*, vol. 22, no. 3, 1987, pp. 213–224.

33. Kim, S. S., and Vanderploeg, M. J., "QR Decomposition for State Space Representation of Constrained Mechanical Dynamic Systems," *ASME Journal of Mechanisms, Transmissions, and Automation in Design*, vol. 108, 1986, pp. 183–188.

34. Kovecses, J., and Angeles, J., "The Stiffness Matrix in Elastically Articulated Rigid-Body Systems," *Multibody System Dynamics*, vol. 18, 2007, pp. 169–184.

35. Lankrani, H. M., "A Poisson-Based Formulation for Frictional Impact Analysis of Multibody Mechanical Systems With Open or Closed Kinematic Chains," *Journal of Mechanical Design*, vol. 122, 2000, pp. 489–497.

36. Leamy, M. J., and Wasfy, T. M., "Analysis of Belt-Driven Mechanics Using a Creep-Rate-Dependent Friction Law," *Journal of Applied Mechanics*, vol. 69, 2002, pp. 763–771.

37. Litvin, F. L., *Gear Geometry and Applied Theory*, Prentice Hall, Englewood Cliffs, NJ, 1994.

38. Mani, N. K., Haug, E. J., and Atkinson, K. E., "Application of Singular Value Decomposition for Analysis of Mechanical System Dynamics," *ASME Journal of Mechanisms, Transmissions, and Automation in Design*, vol. 107, 1985, pp. 82–87.

39. McPhee, J., "Virtual Prototyping of Multibody Systems with Linear Graph Theory and Symbolic Computing," in *Virtual Nonlinear Multibody Systems*, ed. W. Schiehlen, and M. Valasek, Kluwer Academic, Dordrecht, 2003.

40. Megaged, S. M., *Principles of Robot Modeling and Simulations*, Wiley & Sons, New York, 1993.

41. Meriam, J. L., and Kraige, L. G., *Engineering Mechanics, Vol. II: Dynamics*, John Wiley & Sons, New York, 1987.

42. Nakanishi, T., and Shabana, A. A., "The Use of Recursive Methods in the Dynamic Analysis of Tracked Vehicles," Technical Report KMTR-92-002, Department of Mechanical Engineering, University of Illinois at Chicago, 1994.

43. Negrut, D., Jay, L. O., and Khude, N., "A Discussion of Low-Order Numerical Integration Formulas for Rigid and Flexible Multibody Dynamics," *ASME Journal of Computational and Nonlinear Dynamics*, vol. 4, 2009, pp. 021008-1–021008-11.

44. Nikravesh, P. E., *Computer Aided Analysis of Mechanical Systems*, Prentice Hall, Englewood Cliffs, NJ, 1988.

45. O'Reilly, O. M., "The Dynamics of Rolling Diskd and Sliding Disks," *Nonlinear Dynamics*, vol. 10, 1996, pp. 287–305.

46. Orlandea, N. V., "From Newtonian Dynamics to Sparse Tableaux Formulation and Multibody Dynamics," *IMechE Journal of Multibody Dynamics*, vol. 222, 2008, pp. 301–314.

47. Orlandea, N., Chace, M. A., and Calahan, D. A., "A Sparsity Oriented Approach to Dynamic Analysis and Design of Mechanical Systems," *ASME Journal of Engineering for Industry*, vol. 99, 1977, pp. 773–784.

48. Paul, B., *Kinematics and Dynamics of Planar Machinery*, Prentice Hall, Englewood Cliffs, NJ, 1979.

49. Pennestri, E., and Freudenstein, F., "The Mechanical Efficiency of Epicyclic Gear Trains," *ASME Journal of Mechanical Design*, vol. 115, 1993, pp. 645–651.

50. Pereira, M. F. O. S., and Ambrosio, J. A. C. (Eds.), *Computer Aided Analysis of Rigid and Flexible Mechanical Systems*, Kluwer Academic Publishers, Norwell, MA, 1994.

51. Pfeiffer, F., *Mechanical System Dynamics*, Springer-Verlag, Berlin, 2005.

52. Pfeiffer, F., and Glocker, C., *Multibody Dynamics with Unilateral Contacts*, John Wiley & Sons, New York, 1996.

53. Pogorelov, D., "Differential Algebraic Equations in Multibody System Modeling," *Numerical Algorithms*, vol. 19, 2004, pp. 183–194.

54. Press, W. H., Teukolsky, S. A., Vetterling, W. T., and Flannery, B. P., *Numerical Recipes*, 3rd ed., Cambridge University Press, New York, 1992.

55. Rahnejat, H., "Physics of Causality and Continuum: Questioning Nature," *IMechE Journal of Multibody Dynamics*, vol. 222, 2008, pp. 255–263.

56. Rismantab-Sany, J., and Shabana, A. A., "On the Numerical Solution of Differential and Algebraic Equations of Deformable Mechanical Systems with Nonholonmic Constraints," *Computers & Structures*, vol. 33, 1989, pp. 1017–1029.

57. Roberson, R. E., and Schwertassek, R., *Dynamics of Multibody Systems*, Springer-Verlag, Berlin, 1988.

58. Schiehlen, W. O., "Dynamics of Complex Multibody Systems," *SM Archives*, vol. 9, 1982, pp. 297–308.

59. Schiehlen, W. O., *Technische Dynamik*, B.G. Teubner, Stuttgart, 1986.

60. Schiehlen, W. O. (Ed.), *Multibody Systems Handbook*, Springer-Verlag, Berlin, 1990.

61. Schiehlen, W. O., "Multibody System Dynamics: Roots and Perspectives," *Multibody System Dynamics*, vol. 1, 1997, pp. 149–188.

62. Shabana, A. A., *Theory of Vibration: An Introduction*, 2nd ed., Springer-Verlag, New York, 1996.

63. Shabana, A. A., *Vibration of Discrete and Continuous Systems*, 2nd ed., Springer-Verlag, New York, 1997.

64. Shabana, A. A., "Flexible Multibody Dynamics: Review of Past and Recent Developments," *Multibody System Dynamics*, vol. 1, 1997, pp. 189–222.

65. Shabana, A. A., *Dynamics of Multibody Systems*, 3rd ed., Cambridge University Press, 2005.

66. Shabana, A. A., and Sany, J. R., "An Augmented Formulation for Mechanical Systems with Non-Generalized Coordinates: Application to Rigid Body Contact problems," *Nonlinear Dynamics*, vol. 24, no. 2, 2001, pp. 183–204.

67. Shabana, A. A., Zaazaa, K. E., and Sugiyama, H., *Railroad Vehicle Dynamics: A Computational Approach*, Taylor & Francis/CRC, 2008.

68. Shampine, L., and Gordon, M., *Computer Solution of ODE: The Initial Value Problem*, W.H. Freeman, San Francisco, CA, 1975.

69. Sheth, P. N., and Uicker, J. J., "IMP (Integrated Mechanism Program), A Computer Aided Design Analysis System for Mechanism Linkages," *ASME Journal of Engineering for Industry*, vol. 94, 1972, pp. 454–464.

70. Singh, R. P., and Likins, P. W., "Singular Value Decomposition for Constrained Dynamical Systems," *ASME Journal of Applied Mechanics*, vol. 52, 1985, pp. 943–948.

71. Strang, G., *Linear Algebra and its Applications*, 3rd ed., Saunders College Publishing, New York, 1988.

72. Tasora, A., Negrut, D., and Anitescu, "Large Scale Parallel Multibody Dynamics with Frictional Contact on the Graphical Processing Unit," *IMechE Journal of Multibody Dynamics*, vol. 222, 2008, pp. 315–326.

73. Teodorescu, M., and Rahnejat, H., "Newtonian Mechanics in Scale of Minutia," *IMechE Journal of Multibody Dynamics*, vol. 222, 2008, pp. 393–405.

74. Terze, Z., Lefeber, D., and Mufic, O., "Null Space Integration Method for Constrained Multibody Systems with No Constraint Violation," *Multibody System Dynamics*, vol. 6, 2001, pp. 229–243.

75. Udwadia, F. E., "Equations of Motion for Constrained Multibody Systems and their Control," *Journal of Optimization Theory and Applications*, vol. 127, no. 3, 2005, pp. 627–638.

76. Udwadia, F. E., and Phohomsiri, P., "Recursive Formulas for the Generalized LM-Inverse of a Matrix," *Journal of Optimization Theory and Applications*, vol. 131, no. 1, 2006, pp. 1–16.

77. Uicker, J. J., Pennock, G. R., and Shigley, J. E., *Theory of Machines and Mechanisms*, 3rd ed., Oxford University Press, New York, 2003.

78. Wehage, R. A., "Generalized Coordinate Partitioning in Dynamic Analysis of Mechanical Systems," Ph.D. thesis, University of Iowa, Iowa City, IA, 1980.

INDEX